RF and Microwave Measurements

device characterization, signal integrity
and spectrum analysis

RF and Microwave Measurements

device characterization, signal integrity and spectrum analysis

First Edition

Andrea Mariscotti

ASTM

ASTM Analysis, Simulation, Test and Measurement Sagl
Via Comacini, 7 - 6830 Chiasso (Switzerland)

Printed by CreateSpace

Typeset in 10 pt. Charter. Cover pictures: reproduced Spin Electronics large inventory and with permission from Smiths Connectors (top-right image).

ISBN 978-88-941091-0-8

A designer knows that he has achieved perfection not when there is nothing left to add, but when there is nothing left to take away.

Antoine de Saint-Exupery

Preface

This is a book of applied theory. Basic theory and techniques are concentrated mostly in the first four chapters, where definitions, formulas and references are collected aiming at giving a thorough overview of the most relevant topics: circuit theory, material properties, transmission lines, signal analysis and spectral analysis, including random processes, probability and statistics. The central chapters 5, 6 and 7 deals with three important elements of setups and experiments: cables, printed circuit boards and connectors. The influence on the overall measurement, their modeling and characterization are discussed, keeping an eye on applicable standards. The last four chapters cover advanced aspects of scattering parameters, differential lines and mixed modes, and the use and performance of spectrum analyzer and vector network analyzer.

The subdivision of topics in an ordered and structured scheme was not trivial, trying to have the talk flowing naturally avoiding the rigidity of textbooks, that are however an example of clearness and organization: it's like when you step into your colleague's office asking for help and he/she knows where to start from with explanations and suggestions, that is not from the basic theory. On the other hand, the support we are looking for is partly theoretical and partly practical: it shall work, it shall be accessible and we need the possibility of demonstrating its correctness with a bit of theory. The first edition contains the result of a thorough literature search for reliable and accurate information with examples spread over the chapters.

A major challenge is that of achieving a balanced and comprehensive presentation of complex and articulated subjects such as signal integrity, high frequency design, interconnects, device characterization, metrology and electrical measurements, where hardware and instrument performance, calibration, statistics and signal processing have equal dignity and relevance. At some point a decision was taken, to go or not to go for a topic, to extend more or less a chapter, selecting references and examples, using cross-references and indexing to guide the reader through the chapters.

The experimental side is the true reason of this book. When it comes to do something, that shall work and that shall be the best solution to our problem, we face with the problem of spotting out and obtain the right materials and components, making tradeoffs for what is desirable, what is available, and what is cheap enough, and then using them correctly.

During production personnel is well trained and procedures are optimized and executed routinely. During design and prototyping or when unusual requests come from other departments or customers, there may be concepts, methods, setups that we do not master for several reasons. When considering modern technologies employed in complex products, the skills and background required to designers and technicians are rarely covered by a single university curriculum. Moreover, it is often necessary to possess previous knowledge and experience to select the most promising approaches and to decide what is needed in terms of instrumentation, methods, setups and support. For this reason the focus of this book is on applied theory, using e.g. statistics, signal and spectral analysis to evaluate the consistency and the uncertainty of measurements, to improve test setups, to use at best the available instruments, to judge on the applicability of a specific technique or if it is preferable to others. Simple rules of thumb are attractive, but may have short legs.

Often we have collected so many books, papers, application notes and then stuck on the implementation, trying to understand which formula is really applicable to our case and under which circumstances, which degree of approximation is tolerable, writing code and using experimental data for confirmation. We need to become soon experts to understand and discriminate the many sources of information, spotting out the original, the most reliable, those balancing theory and practice, where good reusable results are available. The process is time consuming and requires many trial-and-error attempts. In these moments we would like to ask our skilled and experienced colleague exactly the right question, straight to the problem, and not listening to a whole course back from the basics; for this reason the introductory material is kept to a minimum, conveniently pointing at good references.

This book is for engineers, technicians and students approaching problems of signal integrity, device characterization and product testing in the field of weak RF signals, mixing experimental techniques and data processing, relying on rigorous methods with only a fraction of the complexity of the theory behind, but with the backup of good references for further reading, possibly down to the page or section number, maybe niggling, but saving reader's time from generic citations. I hope to have hit the target, giving readers clear usable material.

Feedback on errors, questions, as well as suggestions for new problems and additional references are warmly welcome.

Chiasso, August 2015

Do, or do not. There is no try.

–Star Wars, The Empire strikes again

Contents

"Begin at the beginning," the King said, gravely, "and go on till you come to the end: then, stop."

Lewis Carroll, *Alice in Wonderland*

1

Circuits and Basic Relationships

This chapter, together with the successive other three, covers topics and concepts that are the basis for the on-going discussion. The review of the basic elements is not all comprehensive and exhaustive, as it should be for yet another textbook (and there are already very good around). Inevitably, the exposure of some topics might turn out fragmentary, in order to give only the most relevant information for selected topics.

1.1 Basic relationships

Definitions of electrical quantities (voltage, current, power) are skipped as well as the most relevant theorems and relationships of electrotechnics, such as Ohm's and Kirchhoff's laws, circuit theory theorems and solution methods, single and three phase systems, etc.. The same fate has befallen to constitutive equations of materials and components, demonstration of capacitance and inductance and basic electromagnetic theory[1].

---

[1] Electrotechnics and network theory are well presented and explained in several textbooks, such as: M.E. van Valkenburg, *Network Analysis*, 3rd ed., Prentice Hall, 1974; C.A. Desoer and E S. Kuh, *Basic Circuit Theory*, McGraw Hill, 1987.

1.1.1 Materials properties: conductivity, dielectric permittivity and magnetic permeability

The focus is on some material properties that might turn out useful when considering performance and behavior of setup elements: cables, planar transmission lines on PCBs and connectors. Basic material properties are briefly reviewed, together with the typical behavior of resistance, inductance and capacitance with frequency, with emphasis on the experimental determination.

1.1.1.1 Conductivity

For the frequencies of interest the electrons inside a metal can be approximated as a classical gas of charged particles floating bounded within the surfaces of the metal with a chaotic movement. When electric field is applied, electrons will drift in the direction of the field. Because of the high density of electrons, no electron is free enough to travel far before a collision with another electron or with an atom occurs. Atoms are arranged in a quite regular pattern that is the lattice of the metal. Due to collisions, some of the applied energy is lost to the random electron motion and atomic vibrations, which is simply the microscopic manifestation of heat. The energy loss into heat is proportional to the resistivity of the metal (some values of conductivity are reported in Table 1.1.1) multiplied by the current squared. For dc or ac signals and the resulting electric field the effect is basically the same, except that for ac fields the electrons are moving back and forth synchronized with the changing field.

If the frequency is large enough, as at optical wavelengths starting from the infrared region, the sloshing of electrons is so quick that the cloud of electrons can wave in the opposite direction before some of the electrons undergo a collision. The frequency at which this phenomenon begins is called the *relaxation frequency* (or *damping frequency*) of the metal. For copper this occurs at about 4 THz. At higher frequencies, such as in the visible range, electrons wave following the applied field without having any more collisions than they would without the external field, and there is no heat produced by the electric field. A metal at these frequencies is said to behave as a collision-less plasma. Metals are reflective both below and above relaxation frequency.

However, metals exhibit a second change in properties that occurs at a higher frequency, called *plasma frequency*, above the visible range. Beyond this frequency a metal becomes transparent, as it happens for most metals in the ultraviolet range. This phenomenon occurs also in the ionosphere, that consists of a gas of ions and free electrons, forming a plasma: at low radio frequencies (e.g. radio-amateur band and radio and TV broadcast frequencies) the ionosphere reflects radiation, allowing skip-mode propagation and global waveguide effects; on the contrary, the ionosphere is transparent to microwaves, allowing for communication between the Earth and orbiting satellites, since its plasma frequency occurs in the VHF band.

1.1.1.2 Dielectric permittivity

Dielectrics are materials that do not allow dc conduction because, simplistically saying, there are no free charges, but are known to conduct displacement current at ac, thus

Metal	Conductivity [10^7 S/m]
Copper	5.8
Aluminum	3.475
Brass, annealed, Cu 90%	2.52
Brass, annealed, Cu 80%	1.88
Brass, annealed, Cu 70%	1.51
Brass, hard drawn, Cu 70%	1.22
Bronze-Phosphor	0.8 - 2.5
Copper - Beryllium	1.0 - 1.4
Chromium	2.23
Gold	4.10
Lead	0.454
Magnesium	2.175
Nickel	1.28
Platinum	0.99
Silver	6.275
Silver - Copper 10%	4.96
Tin	0.654
Tungsten	1.78
Zinc	1.73
Graphite	0.013
Water (pure)	$2\,10^{-11}$
Sea water	$3-5\,10^{-7}$

Table 1.1.1 – Materials properties: conductivity of metals and alloys (choosing a scale suitable for common metals).

responding to the time-varying applied electric field. At the molecular or atomic scale, charge is displaced via rotation or separation. Another type of reaction is ionic polarization; the bond between atoms that make up a molecule is stretched and relaxed when placed in an electric field. The third other major type of polarization is electronic polarization: the electron cloud in the orbitals around the nucleus is stretched by the force exerted by the applied electric field. In all of these cases polarization cannot occur instantaneously: when the frequency of the applied field is low compared to the characteristic frequency of polarization (resonance frequency), the polarizing charge can perfectly track the changing electric field; when the frequency goes above the resonant frequency, polarization becomes less and less effective and the resulting dielectric constant drops in magnitude.

Dielectric losses are associated to all these mechanisms, because of friction at microscopic scale between molecules and because of vibrations and hysteresis of the polarization vector. Impurities and imperfections contribute with additional scattering and absorption. All real materials combine so many different effects.

Example 1.1: Dielectric properties of water

Water is an example of dielectric: a simple molecule with a varied frequency response. At low radio frequencies water is a good insulator, it is transparent to radiation and is low loss; as frequency increases, so does loss. Water is quite lossy at microwaves and is extremely lossy in the infrared, making it opaque to radiation. The loss drops dramatically in the near infrared, leading to the well-known transparent nature of water at visible frequencies, and it becomes opaque again in the ultraviolet. Any dissolved chemicals, such as salt, dramatically alter the properties of water, making it conductive at dc and low frequency, but much lossier at radio frequencies.

In the figure below[a] the dielectric complex permittivity of water is plotted over an extended frequency range, including RF and millimeter waves as well as infrared, visible and ultraviolet light. Since the behavior at optical frequencies is described by the complex index of refraction $(n+j\kappa)$, data regarding the latter are in general to be converted back to complex permittivity $(\varepsilon' - j\varepsilon'')$ values assuming non-magnetic behavior $(\mu_r = 1)$ and using the following:

$$\varepsilon' = n^2 - \kappa^2 \qquad \varepsilon'' = 2n\kappa$$

In reality, a bimodal expression was suggested to cover the entire frequency range and include water in different states: the two time constants are about 18 ps (reducing with temperature due to the hydrogen bonds becoming weaker) and 1 ps or less (0.2 ps), depending on the source of experimental data, and possibly broken in two relaxation constants of 1 and 0.1 ps. [b]

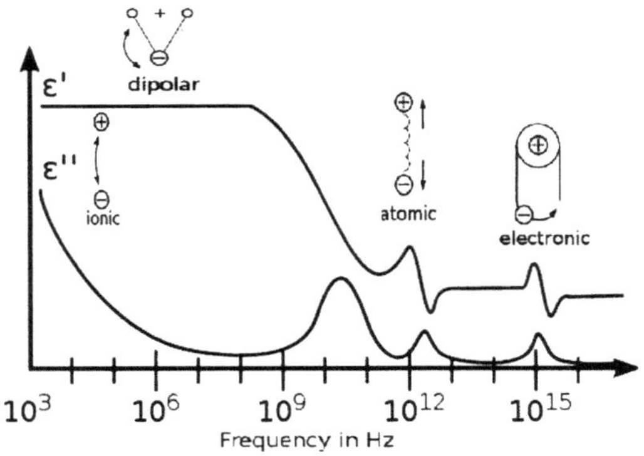

[a] Source Zulkifli Ahmad in Dielectric material, M.A. Silaghi (ed.), InTech Open, 2012 `http://www.intechopen.com/books/dielectric-material`

[b] Source `http://www1.lsbu.ac.uk/water/complex_dielectric.html`.

Since the number of molecules and free electrons depends on temperature, the resulting contribution to the dielectric constant of the material $\varepsilon_{r,1}$ is also in principle temperature-dependent. Its change with temperature may be related to the thermal expansion coefficient α of the material by

$$\frac{\Delta\varepsilon_{r,1}}{\varepsilon_{r,1}} = (\varepsilon_{r,1} - 1)(\varepsilon_{r,1} + 2)\,\alpha\,\Delta T \qquad (1.1.1)$$

where ΔT is the temperature change. The resulting fractional change is often less than a tenth of % per degree.

In addition to the induced dipoles created by the applied field, there exist permanent dipoles inside many materials, and also these dipoles tend to follow the applied field, proportionally to their mobility: depending on the molecule weight and the status of the dielectric material (gas, liquid, solid) the resonance frequency of such phenomenon may be located in the microwave or sub-visible range. When resonance is approached, friction between molecules becomes relevant. In general, when the molecules are following the externally applied and changing field, they will always lag behind the field, more and more as the frequency increase; the phenomenon is stigmatized by the relaxation constant appearing when the applied field is suddenly removed, the dipoles lose slowly their orientation and the polarization decays exponentially. This increased lagging with frequency not only reduces the contribution of dipoles to the dielectric constant of the material, but also increases the amount of losses and the loss tangent. Since a real material with a complex enough molecule may contain different types of dipoles of different size and weight, there is a range of relaxation time values and resonance frequencies, such that the reduction of the dielectric permittivity and the increase of the loss tangent does not follow a single exponential decay, but is smoother and covers a wider frequency range.

Last, for some dielectrics that contain electrolytes or featuring a particularly large conductivity, ionic or electronic conduction may influence the dielectric properties: at lower frequency the loss tangent is larger. It is the case when measuring batteries and electrolytic capacitors, and even soil (including the electrolytes in it and the amount of water as moisture).

If conduction is not completely free and is influenced by the inclusion of small conductive particles, electric field distortion occurs and this may result in a larger equivalent dielectric permittivity as seen by the observer.

The dielectric properties of a material are described and quantified by its complex *permittivity* (or *dielectric constant*)

$$\varepsilon = \varepsilon' - j\varepsilon'' \tag{1.1.2}$$

where the first term ε' is the one retained as the dielectric permittivity (often expressed relative to that of vacuum as $\varepsilon' = \varepsilon_r \varepsilon_0$) and the second one ε'' indicates the losses in the material (often expressed as a fraction of the first terms by the *loss tangent* $\tan\delta = \varepsilon''/\varepsilon'$).

Many materials are anisotropic and the permittivity (or dielectric constant) changes depending on the spatial orientation, so that it is better described by a three-by-three matrix rather than a scalar. Anisotropy may come from fabrication, when the material is a compound, such as fiber + resin and the fibers have a preferred orientation, or from the crystal and molecular properties of the pure material. Anisotropy is considered in sec. 6.1.2 for dielectric materials used in printed circuit boards fabrication: for transmission lines and circuits developing horizontally along the copper planes, distinction is made between the dielectric properties parallel to the plane and orthogonal to it. So, depending on the adopted test method and the foreseen use of the dielectric material, the dielectric properties may change slightly: roughly speaking variations in the order of 10-30% may be expected and this may relevant for some kinds of planar transmission line, such as microstrips (see sec. 6.3.1).

The behavior of dielectrics (for dielectric permittivity and in particular the loss tangent) is also dependent on the frequency of the applied electric field (some examples are reported in Table 1.1.2). Small changes to the fabrication process and small amounts of additives may largely influence dielectric properties, so that in many cases the same dielectric material may be reported with different values of permittivity and $\tan\delta$. In sec. 6.1.2 the characteristics of PCB materials are considered in more detail.

1.1.1.3 Magnetic permeability

For practical reasons materials used for cables, connectors and printed circuit boards considered in the rest of the book are non-ferromagnetic and all have magnetic permeability close to that of free space. The reason is mainly that ferromagnetic materials have a more pronounced skin effect (see sec. 1.1.2 and 1.1.3) and may cause intermodulation distortion (see sec. 5.7.2 and 7.4.3), both undesirable characteristics. This problem will be introduced in sec. 7.1.1.1 and 7.1.1.2 and the consequences reviewed in the rest of Chapter 7 for connectors.

There are, or were, applications using magnetic materials, such as windings with magnet wire to create delay lines and other special transmission lines[2], and phase shifters and isolators, as described in [108].

For these reasons the magnetic properties of materials are not further considered.

1.1.2 Resistance

Conduction, or "ohmic", resistance of conductors is well known: a good starting point for typical resistivity values is Table 1.1.1. Conduction represents a relevant factor especially for small conductor cross-sections, as it is for planar transmission lines (see Chapter 6): in the various sections where typical planar transmission line geometries are analyzed, there is a subsection where attenuation is estimated, one of the two relevant components being the attenuation due to conduction. For cables conduction resistance of the inner conductor determines the low frequency attenuation as well, but including the cable shield is also the low frequency part of the transfer impedance (see sec. 5.1.7), bound to increase for skin effect on the resistive part and most of all because of inductive effects.

The mechanism that determines the overall resistance of finished parts in contact is however more complex: this is generally termed "contact resistance" and includes aspects such as the finishing of surfaces, the creation of potentials and potential barriers, migration of materials, etc.. Contact resistance for connectors is briefly reviewed in sec. 7.2.1, being of marginal interest for radiofrequency and microwave signal applications, becoming relevant when voltage drop, self heating and current rating are concerned (covered by the standards in sec. 7.2, and in particular sec. 7.2.4). Switches

[2] P. Lefferson, "Twisted Magnet Wire Transmission Line," *IEEE Transactions on Parts, Hybrids, and Packaging*, Vol. 7, no. 4, Dec. 1971, pp. 148-154.

Material	@ 1 GHz		3 GHz		10 GHz	
	ε_r	$\tan\delta$	ε_r	$\tan\delta$	ε_r	$\tan\delta$
Mica	5.4	0.00003				
Polystyrene	2.5-2.55	0.25-0.45	2.5-2.55	0.25-0.5	2.5-2.55	0.3-0.7
Polystyrene (Laolin)	2.5	0.1	2.49	0.22	2.49	0.3
Polystyrene (foam)	1.05	0.00003				
Polyethylene (solid)	2.23-2.26	0.22-0.37	2.21-2.26	0.2-0.4	2.15-2.25	0.2-0.5
Polyethylene (cellular)	1.3-1.7	0.0004-0.0006				
Quartz (fused)					3.78	0.0001
Polytetrafluoroethylene (PTFE)	2.1-2.2	0.0004			2.1-2.2	0.0002
Fluorinated Ethylene Propylene (FEP)	2.1	0.001	2.1	0.001	2.1	0.0007
Polyvinyl chloride (PVC)	4-9	0.05				
Steatite	6.25	0.55	6.25	0.55		
Titanium dioxide	99	1			90	0.002
Boron nitride	4.0-4.6	0.0003-0.0005				
(hot pressed / pyrolitic)	3.0-3.6	0.0001-0.0003				
Glass (Corning Glass)	4.7-5.05	0.0048-0.0088	4.7-5.0	0.0052-0.0089	4.7-4.9	0.0061-0.0089

Table 1.1.2 – Properties of materials: dielectric permittivity and dielectric losses.

and relays, more than connectors, are subject to variability of their contact performance if heavy duty employed, such as in power and industrial applications, much less for signal applications, where they share the same problems of surface wearing and oxidation of connectors.

We have seen that besides the dc or low frequency conduction, the most relevant aspect for increasing frequency is skin effect: the quantity δ indicates the skin depth

$$\delta = \sqrt{\frac{2}{\omega\mu\sigma}} \tag{1.1.3}$$

that for the form of the solution nearly exponential takes the meaning of a "space constant": the current that at a given frequency resides in the peripheral layer of thickness δ is nearly the 63% of the total current.

The profile of $J_z(r)$ as per eq. (1.1.10) goes indeed to zero nearly exponentially, going from the periphery to the center (of course for a high enough frequency to observe a pronounced skin effect).

Example 1.2: Reference values of skin depth

Some values of skin depth are exemplified to give the reader the order of magnitude of the phenomenon: for copper with a conductivity of $5.8\,10^7$ S/m and relative magnetic permeability of 1, the skin depth at 10^0, 10^3, 10^6 and 10^9 Hz is 6.6 cm, 2.1 mm, 66 µm and 2.1 µm. For magnetic steel with a conductivity of 1/7 to 1/8 that of copper and a magnetic permeability 20 to 50 times larger, the skin depth reduces by a factor of 3 to 6.
Thus ohmic losses versus frequency are higher in steel than in copper or aluminum. Conversely, the penetration of electromagnetic waves is smaller, being smaller the skin depth.

1.1.3 Inductance

In this and the next section there are discussed a few concepts used in other parts of the book. For a more complete discussion, especially for theory and fundamentals, there are many renown texts of electric circuits and electromagnetic theory[3].

1.1.3.1 Self and mutual inductance formulations

A few formulations for simple geometries are reported, but a more complete discussion may be found in [255], sec. 7.5.5, and in Grover [141], especially to address large setups and connections; for compact geometries and arrangements, that lend themselves to meshing with finite elements, several tools exist as part of electromagnetic solver packages and programs. Additional references, quite useful to review available methods and to establish their accuracy, are Aebischer [1] and Hoer and Love [162].

For a cylindrical straight wire of non-magnetic material with length l and radius R

[3] C.A. Balanis, *Advanced Engineering Electromagnetics*, John Wiley & Sons, 1989 (chap. 2);
H.P. Neff, Jr., *Introductory Electromagnetics*, John Wiley & Sons, 1991 (chap. 1,2,3);
A.E. Fitzgerald, A. Kusko and C. Kingsley, *Electric Machinery*, McGraw Hill, 2002.

$$L = 2l\left[\ln\left(\frac{2l}{R}\right) - \frac{3}{4}\right] \tag{1.1.4}$$

If the conductor is complemented by a return circuit made of a similar parallel conductor at distance d, by neglecting end effects, the self inductance L is

$$L = 4l\left[\ln\left(\frac{d}{r}\right) - \frac{d}{l} + \frac{1}{4}\right] \tag{1.1.5}$$

If a ground plane is used replacing the return conductor, the self inductance is

$$L = l\,\frac{\mu_0}{2\pi}\ln\left(\frac{2h}{r}\right) \tag{1.1.6}$$

and the mutual inductance M with another conductor at distance d and same height h above the ground plane

$$M = l\,\frac{\mu_0}{4\pi}\ln\left[1 + \left(\frac{2h}{d}\right)^2\right] \tag{1.1.7}$$

1.1.3.2 Internal and external inductance

The distributed inductance of a conductor is always made of an internal part (called internal inductance, due to the magnetic field lines that link inside the conductor itself, due to the distribution of the flowing current in the not infinitesimally small cross-section of the conductor) and an external part (normally referred to as "inductance", due to the larger part of magnetic flux lines that link outside the conductor).

Let's take a straight circular conductor of radius R: the conductor is modeled as a series of concentric annular rings of thickness δr and located at a distance r from the center going from 0 to R; by observing that the magnetic flux for the i-th cylindrical layer at the radial position $r_i = i\,\delta r$ is the total flux due to the inner cylinders from 0 to the last one before it, after integration over the entire conductor from 0 to R the total internal inductance for a straight isolated circular conductor with uniform current distribution (e.g. at dc) may be derived:

$$L_{i,dc} = \frac{\mu_0}{8\pi} \tag{1.1.8}$$

By using the same formulation, it is possible to demonstrate that the inductance of internal cylinders is higher than those located at the periphery of the conductor; this is in support of the fact that increasing the frequency the current will distribute on the conductor periphery rather than in the center.

An accurate formulation in electromagnetic terms brings to the exact equation of the current density component along the longitudinal axis $J_z(r)$

$$\frac{\partial^2 J_z(r)}{\partial r^2} + \frac{1}{r}\frac{\partial J_z(r)}{\partial r} - j\omega\mu\sigma J_z(r) = 0 \tag{1.1.9}$$

that is a known equation called "modified Bessel equation of order zero": the order is zero because the term $-\nu^2 J_z(r)/r^2$ in a generic Bessel equation of order ν is zero; it is "modified" because the coefficient of the term $J_z(r)$ is a negative imaginary number, rather than a positive real number. The general solution is

$$J_z(r) = A_1 \left[\mathrm{ber}\left(\frac{r\sqrt{2}}{\delta}\right) + j\,\mathrm{bei}\left(\frac{r\sqrt{2}}{\delta}\right)\right] + A_2 \left[\mathrm{ker}\left(\frac{r\sqrt{2}}{\delta}\right) + j\,\mathrm{kei}\left(\frac{r\sqrt{2}}{\delta}\right)\right] \tag{1.1.10}$$

having indicated with $\mathrm{ber}(x)$ the real part of $J_0\left(x\sqrt{-j}\right)$, $\mathrm{bei}(x)$ the imaginary part of $J_0\left(x\sqrt{-j}\right)$, $\mathrm{ker}(x)$ the real part of $Y_0\left(x\sqrt{-j}\right)$, $\mathrm{kei}(x)$ the imaginary part of $Y_0\left(x\sqrt{-j}\right)$ and with δ the skin depth. The functions $J_0(\cdot)$[4] and $Y_0(\cdot)$ are the Bessel functions of order 0 of the first and second kind, respectively.

It is interesting to observe that the term on the right does not appear in the solution for a solid cylindrical conductor, since the function $\mathrm{ker}(x)$ goes to infinite for $x \to 0$ and thus the coefficient A_2 is 0, being a null argument compatible with the fact that the position $r=0$ is admissible because inside the conductor. This does not occur for a tubular conductor, that thus has a more complex expression.

Referring to Grover, the high frequency expressions for the inductance of cylindrical and tubular conductors show that the latter has a much smaller variation; both have the same limit inductance for infinite frequency:

$$L_\infty = 2l\left(\ln\frac{2l}{R} - 1\right) \tag{1.1.11}$$

where R is again the radius of the conductor and l its length, and the resulting inductance is in nH if the dimensions are measured in cm.

1.1.3.3 Proximity effect

When a go and a return conductor are considered, depending on the reciprocal distance there may be the need for a relevant correction due to the proximity effect. Proximity effect is due to the eddy current induced in nearby conductors due to the ac magnetic field of a given conductor. Proximity effect may be relevant for high frequency and/or close separation of conductors, since we will see it is proportional to the ratio of distance over conductor radius; in particular, in transformers proximity effect may increase losses and equivalent resistance, especially in high-voltage windings with a large number of turns.

---

4 The use of the letter J for both the current density and the Bessel function is misleading, but it is the normally adopted notation.

Grover [141] gives in his chap. 24 a complicated way to estimate proximity effect as a worst case for infinite frequency and including the already considered skin effect.

For the extreme case of the three insulated conductors inside a three-phase power cable minimizing the d/r ratio (nearly 4), proximity effect was estimated by comparing the results of a FEM model with the complete expression including skin effect reported in the previous section[5]: differences are 4.8% at 5 kHz, 11.6% at 10 kHz, 20% at 100 kHz and 22% at 1 MHz.

Nan and Sullivan[6] report thoroughly on several approaches (in particular Dowell[7] and Ferreira[8]) and present a useful analytical formulation that is commented and compared to the previous ones with a few examples. They indicate a general agreement between the other two methods, in particular with Dowell's and Ferreira's methods underestimating and overestimating proximity effects at the highest frequencies, respectively; the former is more accurate for closely packed conductors, while the latter is better applied to farther apart conductors. In reality already at low frequency there is a discrepancy between the two reference formulations of 4.7% due to a difference in the definition of the equivalent conductor, that for Dowell matches dc resistance only: the agreement with FEM results in the low frequency range is better than 1% for Ferreira and about 5% for Dowell, confirming thus a steady difference due to this discrepancy.

1.1.3.4 Experimental determination

The aim of this section is to briefly review some methods useful to determine the inductance of a component or a circuit.

Use of RCL bridge This is the simplest and most straightforward method, provided that the RCL bridge with adequate performance is available: sensitivity may be high enough to be able to measure parasitic values; to appreciate skin effect and internal inductance variation the frequency range shall be extended enough, usually up to some MHz or higher. This method is quite successful for components in standard packages, for which high performance bridges have specific test fixtures that hold the components in place. In all cases the parasitics of the connection (e.g. short wires and crocodiles) shall be zeroed out by making a short-circuited measurement for the correction of stray resistance and inductance and an open-circuited measurement for the stray capacitance. For already installed components and parts of circuits, a direct measurement of circuit voltages is the easiest and most convenient way.

---

[5] A. Pagnetti, A. Xemard, F. Paladian and C.A. Nucci, "Evaluation of the impact of proximity effect in the calculation of the internal impedance of cylindrical conductors," XXXth URSI General Assembly and Scientific Symposium, 2011, pp. 1-4. doi: 10.1109/URSIGASS.2011.6050734.

[6] Xi Nan and C.R. Sullivan, "An Improved Calculation of Proximity-Effect Loss in High-Frequency Windings of Round Conductors," IEEE Power Electronics Specialists Conference, June 2003, pp. 853-860. doi: 10.1109/PESC.2003.1218168

[7] P.L. Dowell, "Effects of eddy currents in transformer windings," *Proceedings of the IEEE*, vol. 113, no. 8, pp. 1387–1394, Aug. 1966. doi: 10.1049/piee.1966.0236

[8] J.A. Ferreira, "Improved analytical modeling of conductive losses in magnetic components," *IEEE Transactions on Power Electronics*, vol. 9, no. 1, pp. 127–131, Jan. 1994. doi: 10.1109/63.285503

Resonance with a known capacitance (frequency domain, narrow-band) This method works well for many inductors that have non-standard package (e.g. home made inductors), including circuits and cables, and when the frequency range is extended, but not too much. The problem is the availability of the resonating capacitor with the necessary value: not too small to avoid the influence of a non-negligible parasitic capacitance (and in case of a wound inductor, there is some) and not too large, limiting the choice to high-Q capacitors with good dielectric materials, such as ceramic, polypropylene, polyester, thus staying below some tens of µF.

This method requires that the frequency of the sinusoidal signal fed to the LC circuit is adjusted until the maximum of the current is observed, that corresponds to a pure resistive behavior of the LC circuit under resonance conditions.

Resonance and self-relaxation (time domain) For the high end of the frequency range and when the stray influence due to connections and external elements is relevant, it may be desirable to have the circuit under test resonating only with a feeding capacitance connected as close as possible. The feeding capacitor C is charged by short pulses through the diode D and the source circuit; the source circuit may be a square wave generator with a CR high-pass pulse-forming circuit or directly a pulse generator. During half-cycles when the pulse is negative, the diode is isolated and the charged capacitor may resonate with the circuit under test: diode D and capacitor C shall be soldered as close as possible to the circuit terminals minimizing stray inductance; the junction capacitance of the diode D shall be conveniently low, e.g. selecting a switching diode with a few pF of transition capacitance. The amplitude accuracy is not relevant because what is needed is the resonance frequency f_r, that is the period of the damped oscillation: as usual the unknown inductance is calculated as $L = 1/(2\pi f_r)^2 C$, neglecting diode and inductor parasitic capacitance.

Sloped current under dc voltage feeding This method exploits directly the inductor constitutive equation $v_L = L\dfrac{di_L}{dt}$: if the inductor is fed with a constant positive voltage E, the current will rise up linearly as $i_L(t) = E/L$. If the value of E is too large, the rise of current will be too fast to be easily seen on the oscilloscope and will be significantly disturbed by the closing of the switch that starts the experiment: the switch may be mechanical or a relay and we need thus some ms to stabilize it; otherwise, for faster operation a FET switch may be used, paying attention to a small enough "on" channel resistance (there are some with sub-Ωresistance values for telecommunication applications). Moreover, voltage sources with a significant internal resistance will cause a time constant to appear, probably in the range of time values where the slope evaluation is to be carried out.

This method is useful because it is suitable to track inductor saturation, as soon as the current waveform deviates from the linear slope, increasing because of inductance reduction; it is also fast enough not to have significant heating even in case of large current, and with adequate protection of the source (e.g. a battery), a large current may be fed for a short time driving the source in a controlled short-circuit operation. The protection may be a resettable fuse or a circuit breaker.

The drawback is that only the low frequency value of the inductance is measured. If a measurement of the high frequency behavior under different dc biasing points is desired, then the method is again of the "resonance, frequency domain" type above, and the high frequency signal shall be coupled to the power circuit in some way: direct injection is possible with a high-pass/low-pass network [126], or magnetic coupling through a reversed current transformer.

Step voltage excitation: time decay This method does not modify the circuit, by adding extraneous elements: a step voltage source $E(t)$ is used, as the one for oscilloscope calibration, that is standardized to a 1 ms period, and a voltage measurement is taken across one of the elements of the RL circuit under measurement. Fast rise times may be reached for modern oscilloscopes, so that the steep rising edge is as ideal as we need for our calculations. A known series resistance R may be added to the internal resistance R_g of the generator, if necessary.

The experiment is performed by triggering on the rising edge of the square wave, that we assume as time $t=0$, and after a short transient due to the combined effect of parasitic elements, the circuit will behave like a RL circuit. The measured voltage may be the one across the inductor v_L or the one across the burden resistor v_R. The former will be maximum at the beginning and then will decay to zero with time constant $\tau=L/(R+R_g)$; keeping $R \gg R_g$ the estimation is simplified.

Of course methods of this kind are exposed to noise and the difficulty in visually positioning the tangent at $t=0$, hitting the horizontal axis in the sought $t=\tau$ value. Some mathematical workout may be useful, first downloading the waveform to a computer and then finding the exponential curve that fits the measured values e.g. in the least mean square (LMS) sense, having the foresight of discarding the first samples too close to the initial transients and the last ones after about three τ, when the curve becomes too flat (and again showing undesirable influence of offsets and noise).

Step voltage excitation: area Once one recurs to downloading the waveform and doing some maths on it, a better method is that based on the evaluation of the area of the transient, that is far more robust to noise [199]. What is needed is the same circuit test setup as the "step voltage excitation" method above and the following consideration. Invoking the constitutive equation of the inductor, the area of the pulsed voltage appearing across it may be obtained by integration.

$$A=\int_{0}^{+\infty} v_L(t)\,dt=L\int_{0}^{+\infty}\frac{di_L}{dt}\,dt=L\,[i_L(+\infty)-i_L(0)] \qquad (1.1.12)$$

We observe that the current flowing in the circuit is initially zero (we know that theoretically this is the correct solution) and then it goes to the dc short circuit current value $I=E/(R+R_g)$. Thus the inductance is

$$L=\frac{A}{E}(R+R_g) \qquad (1.1.13)$$

1.1.4 Capacitance

Similarly, capacitance can be defined and considered in many ways, that are exhaustively considered in textbooks and electromagnetics references: energy based definitions, electrostatic problem formulation, dynamic and high frequency aspects, details related to element geometry, behavior of materials, etc. The relevance of the latter for the performance of interconnecting elements is considered in Chapter 5, 6 and 7.

As for inductance, we may talk of nominal capacitance value or stray capacitance of components (e.g. turn-to-turn capacitance for wound inductors), or of mutual capacitance between components, traces and in general different parts of a circuit. Thus, capacitance may be associated to air or a dielectric material with higher permittivity, or a combination of the two. Capacitance tends to be quite constant with frequency, as long as the dielectric permittivity is not changing, that for many materials occur at very high frequency; on the contrary, when very high permittivity dielectrics are used (e.g. those employed in polarized capacitors, often also strongly anisotropic), frequency dependency is quite evident even at low frequency.

1.1.4.1 Simple theoretical expressions

The calculation of self and mutual capacitance terms for conductors that are cylindrical (or approximately so) is done passing through the potential functions, that is an electrostatic problem is setup and solved with the Charge Image Method (CIM) [279]. Pairs of conductors i and j are considered above a ground plane and self and mutual potential coefficients p_{ii} and p_{ij} (that form the matrix $\mathbf{P}$ of the same size as the number of conductors) are defined:

$$p_{ii} = \frac{1}{2\pi\varepsilon}\ln\frac{2h_i}{r_i} \qquad\qquad p_{ij} = p_{ji} = \frac{1}{2\pi\varepsilon}\ln\frac{D_{ij}}{d_{ij}} \tag{1.1.14}$$

where d_{ij} is the distance between conductors i and j, D_{ij} is the distance between conductor i and image of conductor j beneath the ground plane, and vice-versa, h_i is the height above ground of conductor i, r_i is the radius of conductor i.

$$\mathbf{P} = \begin{bmatrix} \ddots & \vdots & & \vdots & \\ \cdots & p_{ii} & & p_{ij} & \cdots \\ & & \ddots & & \\ \cdots & p_{ji} & & p_{jj} & \cdots \\ & \vdots & & \vdots & \ddots \end{bmatrix} \tag{1.1.15}$$

The capacitance matrix $\mathbf{C}$ is obtained by inversion of the potential matrix $\mathbf{P}$: $\mathbf{C} = \mathbf{P}^{-1}$

1.1.4.2 Experimental determination

Some methods are briefly reviewed that turn out useful to estimate capacitance of a component or circuit. They may be applied, for example, for the effective permittivity of PCBs and to integrate the identification of planar transmission line response and equivalent circuit representation, as seen in sec. 6.1.2.1 and 6.5.1.3.

Use of RCL bridge As for inductance, this is a simple and straightforward method, provided that a RCL bridge with adequate performance is available; capacitance may be also measured with modern multimeters, with a fairly broad dynamic range (from a hundred pF to several hundreds μF or even a few mF. While multimeters give only the capacitance value, RCL bridges measure losses as well, represented as a series or a parallel equivalent resistance; the latter is the preferred equivalent circuit representation and can be directly related to $\tan\delta$. For many capacitors featuring high performance dielectrics, losses can be appreciated only at a sufficiently high frequency, usually well in the MHz range; the typical reference frequency value used to express capacitor performance is 1 MHz. Polarized capacitors, on the contrary, have significant $\tan\delta$ values already at tens of kHz; for these capacitors the preferred representation is that of the equivalent series resistance (ESR), which a maximum tolerable current is associated to (often called maximum ripple current and normally given at the second harmonic of the mains frequency, to relate it to the most common application, that of leveling the output voltage of a single-phase diode rectifier).

The use of RCL bridge is quite straightforward: high performance bridges have test fixtures for components in standard packages to hold them in place. In all cases connection parasitics (e.g., in particular, short wires and crocodiles) shall be zeroed out by making a short-circuited measurement for the correction of stray resistance and inductance and an open-circuited measurement for the stray capacitance.

A cheap, yet very accurate, method was proposed in [234], where the balance of voltage drop across the unknown capacitance and a reference resistance is sought by adjusting the frequency of the voltage source: this method is based on relative amplitude reading (see sec. 9.2.12.3) and the adjustable frequency may be determined with much more accuracy than amplitude, using accurate frequency meters. The uncertainty analysis is reported in the paper.

When stray capacitance is measured, values are usually so low that any calibration inconsistency and drift may affect significantly the overall accuracy: few or fractions of pF are commonplace in low capacitance windings and circuits. In these cases, if the problem and the test setup allow this technique, a two-step procedure may be followed: the circuit is made resonating (or interacts in other way) with a set of known capacitance values, larger than instrumentation stray capacitance and externally added; then extrapolation to zero external capacitance gives the desired result regarding stray capacitance effects only. This method was applied in [232] to measure stray capacitance of radiofrequency coils: the paper reports the uncertainty analysis and the propagation through the least-mean-square regression algorithm.

Step voltage excitation: time decay This method applies directly a step voltage $E(t)$ to the capacitor under test, using for example as voltage source the one for oscilloscope calibration, standardized to a 1 ms period. A voltage measurement is taken across one of the elements of the so formed RC circuit under measurement. Fast rise times may be reached for modern oscilloscopes, so that the steep rising edge is as ideal as we need for our calculations. A known series resistance R may be added to the internal resistance R_g of the generator, if necessary. Attention shall be given not to drive the generator into saturation and as a consequence a non-linear distorting behavior: the initial current, as known, is given by $E/(R+R_g)$.

The experiment is performed by triggering on the rising edge of the square wave, that we assume as time $t=0$, and after a short transient due to the combined effect of parasitic elements, the circuit will behave like a RC circuit. The measured voltage may be the one across the capacitor v_C or the one across the burden resistor v_R. The former will be initially zero and then will increases exponentially to the final value E with time constant $\tau=(R+R_g)C$; keeping $R \gg R_g$ the estimation is simplified.

Step voltage excitation: area Similarly to the inductive circuit, calculating the area of a suitable transient waveform inside the circuit is a more robust method with respect to noise. A suitable waveform is the current flowing in the circuit, that can be measured as the voltage drop across the series resistor R. Invoking the constitutive equation of the capacitor, the area of the pulsed current flowing through it may be obtained by integration.

$$A=\int_0^{+\infty} i_C(t)\,dt=C\int_0^{+\infty}\frac{dv_C}{dt}\,dt=C\,[v_C(+\infty)-v_C(0)] \tag{1.1.16}$$

We observe that the voltage across the capacitor is initially zero and then it reaches the final steady value of the applied voltage step E. Thus the capacitance is

$$C=\frac{A}{E} \tag{1.1.17}$$

as it is clear by observing that the total current flowing through the capacitor is the accumulated charge, and it is related to the capacitance by the voltage across it.

1.1.5 Decibel and distribution of measured quantities

1.1.5.1 Definition

The decibel (abbreviated dB) is a logarithmic representation of ratios, so that values, ranging several orders of magnitude, may be kept on a handy scale. The use of dB is so widespread since the application of gain or attenuation is translated directly into sum and difference operations.

The dB is defined as $10\log_{10}(X_2/X_1)$, where X_1 and X_2 represent two homogeneous variables with a unit of power. It is evident that if X_2 is ten times X_1, then their ratio equals 10 dB. The variables X_1 and X_2 may represent two variables of a circuit or system (for example the input and the output power of a module, so that their ratio represents gain or attenuation) or, if X_1 is taken equal to a reference value that expresses the reference unit of measure, then X_2 becomes expressed in dB of this unit of measure: so, if $X_1 = 1\,\mathrm{W}$, then X_2 is expressed in dBW; if $X_1 = 1\,\mathrm{mW}$, then X_2 is expressed in dBmW or more commonly dBm; it is immediate to relate watts to milliwatts by observing that $1\,\mathrm{W} = 30\,\mathrm{mW}$.

By recalling the power law that for electric variables puts power in relationship with either voltage or current, the dB representation of the latter is obtained by a coefficient of 20 in front of the log operation: $20\log_{10}(Y_2/Y_1)$. Again, the expression may represent a true ratio between two variables in a circuit or system, or the ratio of a variable with respect to a reference variable, thus obtaining dBV if $X = 1\,\mathrm{V}$, dBmV if $X = 1\,\mathrm{mV}$, dBA if $X = 1\,\mathrm{A}$ and so on; the relationship between different powers of the same measuring unit is obtained with the criterion of 20 dB/decade, so that $1\,\mathrm{dBV} = 60\,\mathrm{mV}$.

In this transformation we have momentarily neglected the value of the resistance R, which the voltage or current is applied to; power may be related to voltage or current when the reference resistance value is known.

Example 1.3: Relationship between power and voltage in dB

The very common transformation between $X\,[\mathrm{dBm}]$ and $Y\,[\mathrm{dBV}]$ commonly used in radio frequency measurements and calculations is developed for $R = 50\,\Omega$, the reference resistance for radio frequency equipment, such as antennas, cables, spectrum analyzers, amplifiers.

$$
\begin{aligned}
P &= 1\,\mathrm{mW}; \qquad P = V^2/R \\
X\,[\mathrm{dBm}] &= 10\log(P) = 10\log(V^2) - 10\log(R) = \\
&= 20\log(V) - 10\log(50\,\Omega) = Y\,[\mathrm{dBV}] - 17
\end{aligned}
$$

As known, $Y\,[\mathrm{dBV}]$ may be transformed to $Y_2\,[\mathrm{dB\mu V}]$ by adding 120 dB.

If the reference resistance value had been different, let's say $R_3 = 150\,\Omega$, then the new voltage $Y_2\,[\mathrm{dB\mu V}]$ would be larger by $K = 10\log(150\,\Omega) - 10\log(50\,\Omega) = 21.8 - 17\,\mathrm{dB} = 4.8\,\mathrm{dB}$. Pay attention that the so calculated K is applied to a 20 dB/decade variable, $Y_2\,[\mathrm{dB\mu V}]$, to get back another 20 dB/decade variable, $Y_3\,[\mathrm{dB\mu V}]$.

1.1.5.2 Effect on linearity and probability distributions

The dB scale is used by most RF equipment, such as Spectrum Analyzers and Vector Network Analyzers, and the most common attitude is to use it as a linear scale, forgetting that the value expressed in dB is obtained by means of a log operation. In particular, there are two questions that deserve an answer:

1. a confidence interval for a symmetric distribution is symmetric on the original measurement unit: is it still uniform if expressed in dB scale? if not, what is the expected skewness?

2. how the assumption on the original PDF of a variable (e.g. white Gaussian noise as an input to a Spectrum Analyzer) preserves while going through the measurement and getting out finally represented on a dB scale?

It is necessary to go into the details of the log operation to answer both questions. Reference is made to the content of Chapter 4 and in particular sec. 4.1.3. Let's assume an input random variable x with a given probability distribution function PDF $p(x)$, that might represent the input noise signal samples or a set of lumped readings to be transformed in dB. The resulting dB value is $y = 10\ \log_{10}(x)$ if x is a power, or $y = 20\ \log_{10}(x)$ if x is a voltage or current. Let's proceed assuming that x is a voltage without losing generality. It is evident that if the span of x values is not large (i.e. its dispersion is small with respect to its mean value, or in other words the normalized dispersion σ/μ is small), a linear approximation may be valid: the first order Taylor's approximation of $\log_{10}(x)$ holds and y distribution does not deviate much from that of x.

Let's consider an intermediate r.v. $y' = \ln(x)$; using the expressions of sec. 4.1.3.5 the mean and standard deviation of y' can be determined starting from those of x.

Then, y is related to y' by a linear transformation, that is a multiplying factor of 20 (or 10), together with $\log_{10}(e) \simeq 0.434$: $y = y'\,[20\log_{10}(e)]$.

When considering the intensity of a Gaussian random process passing through a Fourier transform, with normally distributed in-phase and in-quadrature components, the resulting intensity (i.e. the modulus) is Rayleigh distributed (please see sec. 4.3.2.3 for a graphical interpretation and quantitative evaluation). The Rayleigh PDF is not symmetric, has its maximum in a point near zero and has non-negative values. Thus the amplitude on a linear scale looks asymmetric, with a distribution far from the original normally distributed noise. In sec. 4.1.3 it is shown that taking the log of a Rayleigh causes a long left tail in the negative dB value, thus having the resulting distribution much different from a Gaussian. It is of course not really an issue nor noticeable, when the attention is focused on the largest values around and above the mean, that should look satisfactorily Gaussian.

1.2 Attenuators

Attenuators are largely used, as their name says, to attenuate signal amplitude to better feed it to an instrument or circuit; attenuation, of course, may be expressed in terms of voltage attenuation, or more rarely as power attenuation. Normally they are designed for a given reference impedance value that is $50\,\Omega$, but we will see that they are useful also when there is an impedance difference (or mismatch) between two circuits, or parts of a network, and they are thus used to ease impedance matching at the expense of some attenuation.

Attenuator behavior and design with respect to its main characteristics (that is attenuation and impedance matching at input and output) may be considered and evaluated from different viewpoints. Different degrees of simplification are possible, if symmetry or a known reference impedance are assumed.

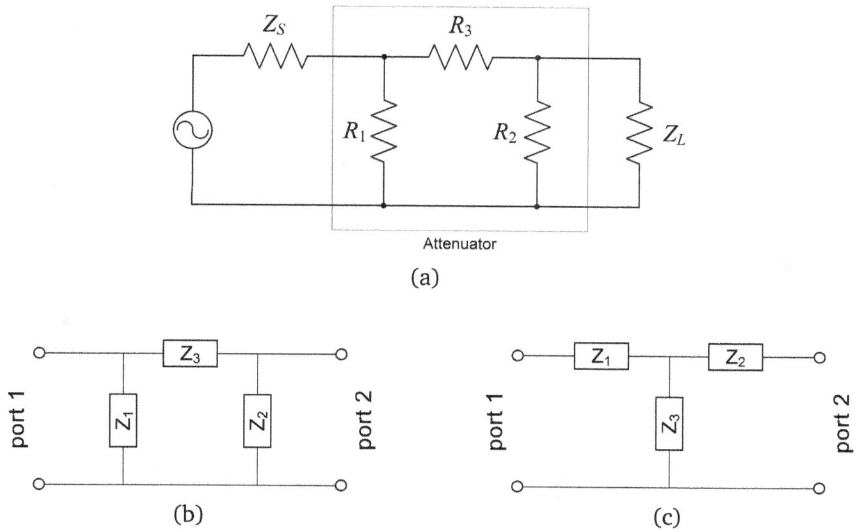

Figure 1.2.1 – (a) Attenuator π-cell and complete circuit including source and load; (b) general equivalent circuit for π-cell; (c) general equivalent circuit for T-cell.

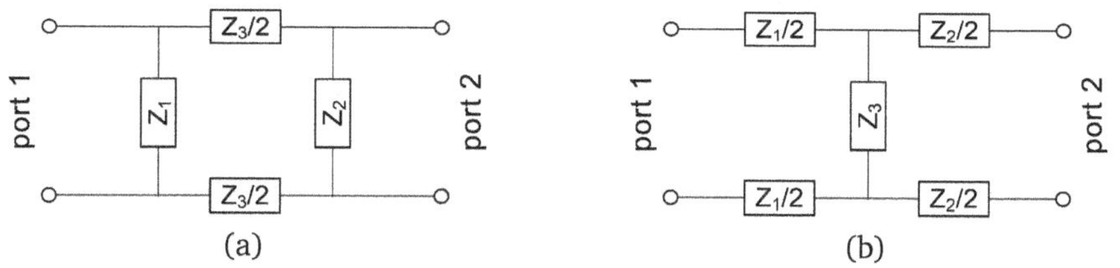

Figure 1.2.2 – Balanced versions of (a) π-cell attenuator and (b) T-cell attenuator.

1.2.1 Circuit approach and calculation

The most common circuits used to build an attenuator are the "Pi" (or π) and "T" cell, as shown in Figure 1.2.1 (only for generality the elements of the cells are indicated by "Z" rather than by "R", but attenuators are normally built around resistive elements).

Since they are used in circuits where the nominal impedance of devices and equipment is the network reference impedance (i.e. $50\,\Omega$), the same termination at the source and at the load is assumed, even if for generality source and load impedances are kept separate; for this reason attenuators of this kind are symmetric, so that $R_1 = R_2$, but again for generality we will keep them distinct in the calculations.

For the two attenuators in π and T configuration, there exist balanced counterparts with symmetric structure, where both terminals of the input and output ports are loaded in the same way (balanced). These configurations are obtained by splitting the longitudinal resistors in two halves and moving one of them to the lower part of the circuit, as shown in Figure 1.2.2: this brings benefits for common mode signals, but has no effect for differential signals.

The relevant quantities that characterize the attenuator when inserted in a larger circuit are the input impedance Z_{in} seen looking into either port 1 or port 2, and the attenuation A. The input impedance calculation is initially done for matched load conditions (where the opposite port is terminated on the correct reference impedance of $50\,\Omega$), but will be extended also to mismatched conditions, i.e. terminating the opposite port e.g. on a short circuit or open circuit. Similarly, attenuation shall be always determined in matched load conditions, and then we will check what happens when the opposite port is terminated in extreme conditions of short circuit and open circuit. Calculations are developed for the generic T and π networks, distinguishing the general case of arbitrary impedance system and that of $50\,\Omega$ reference impedance[9].

Equations for the π attenuator

$$\begin{aligned} Z_{in} &= Z_1//(Z_3+(Z_2//Z_L)) = \\ &= \frac{Z_1Z_2Z_L+Z_1Z_2Z_3+Z_1Z_3Z_L}{Z_1Z_2+Z_1Z_L+Z_2Z_L+Z_2Z_3+Z_3Z_L} \end{aligned} \tag{1.2.1}$$

$$V_{in} = E\frac{Z_{in}}{Z_S+Z_{in}}$$

The Thevenin quantities are:

$$V_{1,th} = E\frac{Z_1}{Z_S+Z_1} \qquad Z_{1,th} = \frac{Z_S+Z_1}{Z_S+Z_1}$$

$$\begin{aligned} V_{out} &= (Z_2//Z_L)\frac{V_{1,th}}{Z_{1,th}+Z_3+Z_2//Z_L} = \\ &= E\frac{Z_1Z_2Z_L}{Z_1Z_2Z_S+Z_1Z_SZ_L+Z_3(Z_1+Z_S)(Z_2+Z_L)+Z_1Z_2Z_L+Z_2Z_SZ_L} \end{aligned}$$

$$A = \frac{V_{out}}{V_{in}} = \frac{Z_2Z_L}{Z_2Z_3+Z_2Z_L+Z_3Z_L} \tag{1.2.2}$$

Equations for the T attenuator

$$Z_{in} = Z_1+\frac{Z_3(Z_2+Z_L)}{Z_3+Z_2+Z_L} \tag{1.2.3}$$

$$V_{in} = E\frac{Z_{in}}{Z_S+Z_{in}}$$

[9] One may wonder why repeating these calculations here, when they are surely available in other textbooks and the Web. The reason is simply that the result of a web search indicated that there is some confusion and disagreement, where attenuation is not clearly indicated whether expressed in voltage or power, if the effects of the source and load terminations are taken into account, and there is rarely a verification of the attenuator performance in extreme conditions, such as short circuit and open circuit, or mismatched load (and this is one of the attractive features of attenuators, that of mitigating mismatched conditions, especially with long cables).

$$V_{3,th} = E\frac{Z_3}{Z_S + Z_1 + Z_3} \qquad Z_{3,th} = \frac{Z_3(Z_S + Z_1)}{Z_S + Z_1 + Z_3}$$

$$V_{out} = Z_L \frac{V_{3,th}}{Z_{3,th} + Z_2 + Z_L}$$

$$\begin{aligned} A &= \frac{V_{out}}{V_{in}} = \frac{\dfrac{Z_L Z_3}{Z_S + Z_1 + Z_3}}{\dfrac{Z_3(Z_S + Z_1)}{Z_S + Z_1 + Z_3} + Z_2 + Z_L} \frac{Z_S + Z_{in}}{Z_{in}} = \\ &= \frac{Z_3 Z_L}{Z_2 Z_3 + Z_3 Z_L + Z_1(Z_2 + Z_3 + Z_L)} \end{aligned} \tag{1.2.4}$$

The results for the π and T attenuator are reported in Figure 1.2.3 and 1.2.4: the scale and the size of the figures were chosen to fit the wide range of reference impedance and attenuation values; for more precise estimations the reader is invited to write down and solve the equations above or to have a look at the next section 1.2.2.

Validation of attenuator equations Validation of the above expressions was done with two approaches: first, after the first writing of the basic equations, simplifications were checked with Matlab® Symbolic Toolbox; second, for the T attenuator the example appearing in Pozar's book [273], page 198, was considered, with $R_1^T = 8.56\,\Omega$ and $R_3^T = 141.8\,\Omega$ giving 3 dB attenuation in a 50 Ω system. Conversely, for the π attenuator, the wye-delta transformation (see sec. 1.3.3.3) of Pozar's example gives $R_1^\pi = 292.2\,\Omega$ and $R_3^\pi = 17.6\,\Omega$ for the same attenuation of 3 dB. Numeric results match within the numeric accuracy of round-offs, better than 1%.

1.2.2 Design formulas [328]

Given the required attenuation A (expressed in linear units starting from dB voltage) and the reference impedance Z_{ref} (with which input and output impedance shall be matched, so assuming that $Z_S = Z_L = Z_{\text{ref}}$), the following expressions are available for the determination of the internal resistor values[10]. For the Pi and T attenuator respectively, we have:

$$R_1^\pi = R_2^\pi = Z_{\text{ref}} \frac{1+A}{1-A} \qquad R_3^\pi = \frac{2R_1^\pi}{\left(\dfrac{R_1^\pi}{Z_{\text{ref}}}\right)^2 - 1} \tag{1.2.5}$$

for which it is clear that R_1 (and R_2) is larger than the reference source impedance Z_S and that R_3 is always smaller than R_1, and

[10] See also Wikepedia: `http://en.wikipedia.org/wiki/Attenuator_(electronics)`.

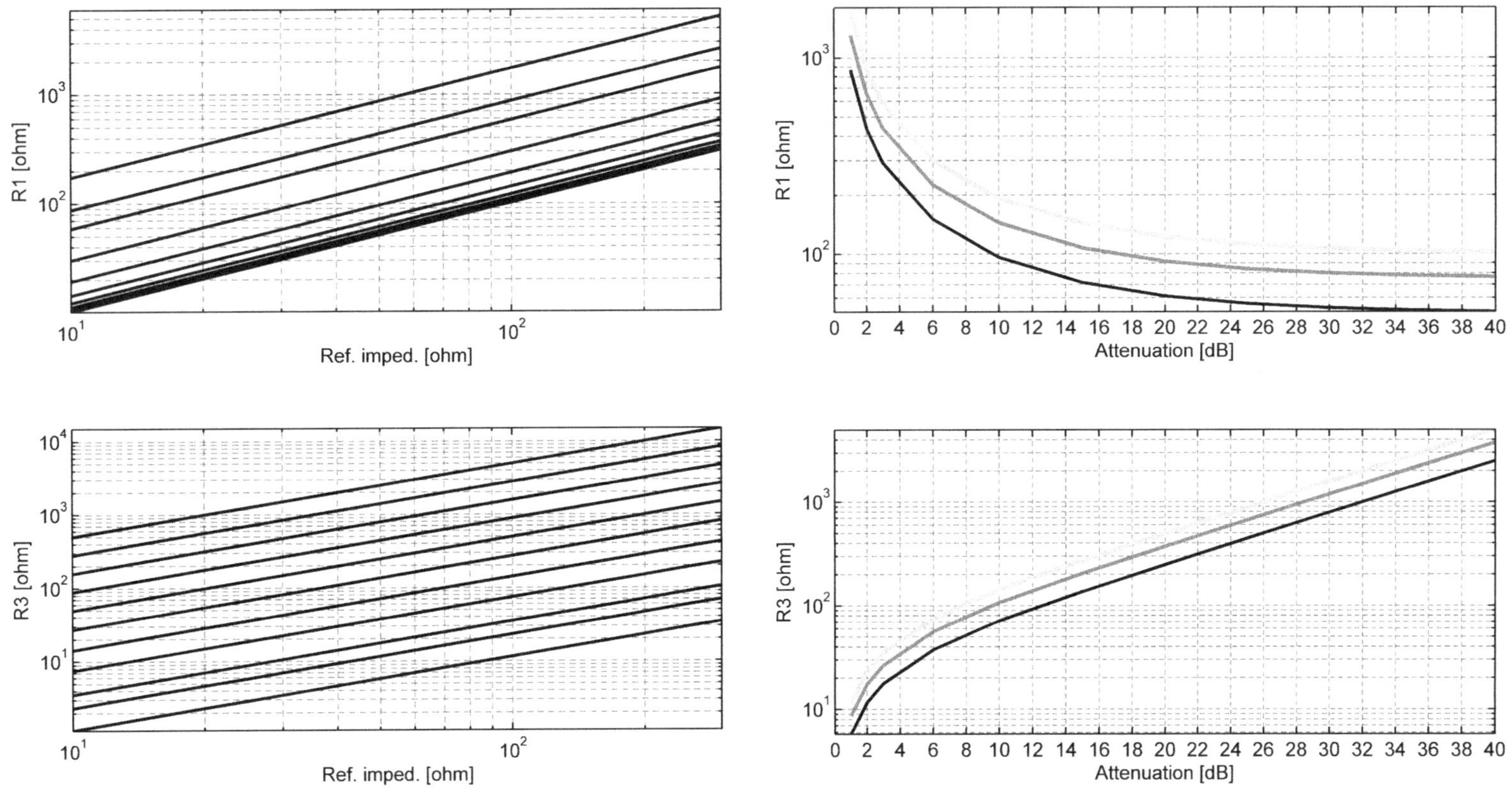

(a) (b)

Figure 1.2.3 – π attenuator charts for (a) variable reference impedance in abscissa and attenuation (1, 2, 3, 6, 10, 15, 20, 25, 30, 35, 40 dB from bottom to top), (b) variable attenuation and $Z_S = Z_L = 50\,\Omega$ (black line), $75\,\Omega$ (dark gray line), $100\,\Omega$ (light gray line).

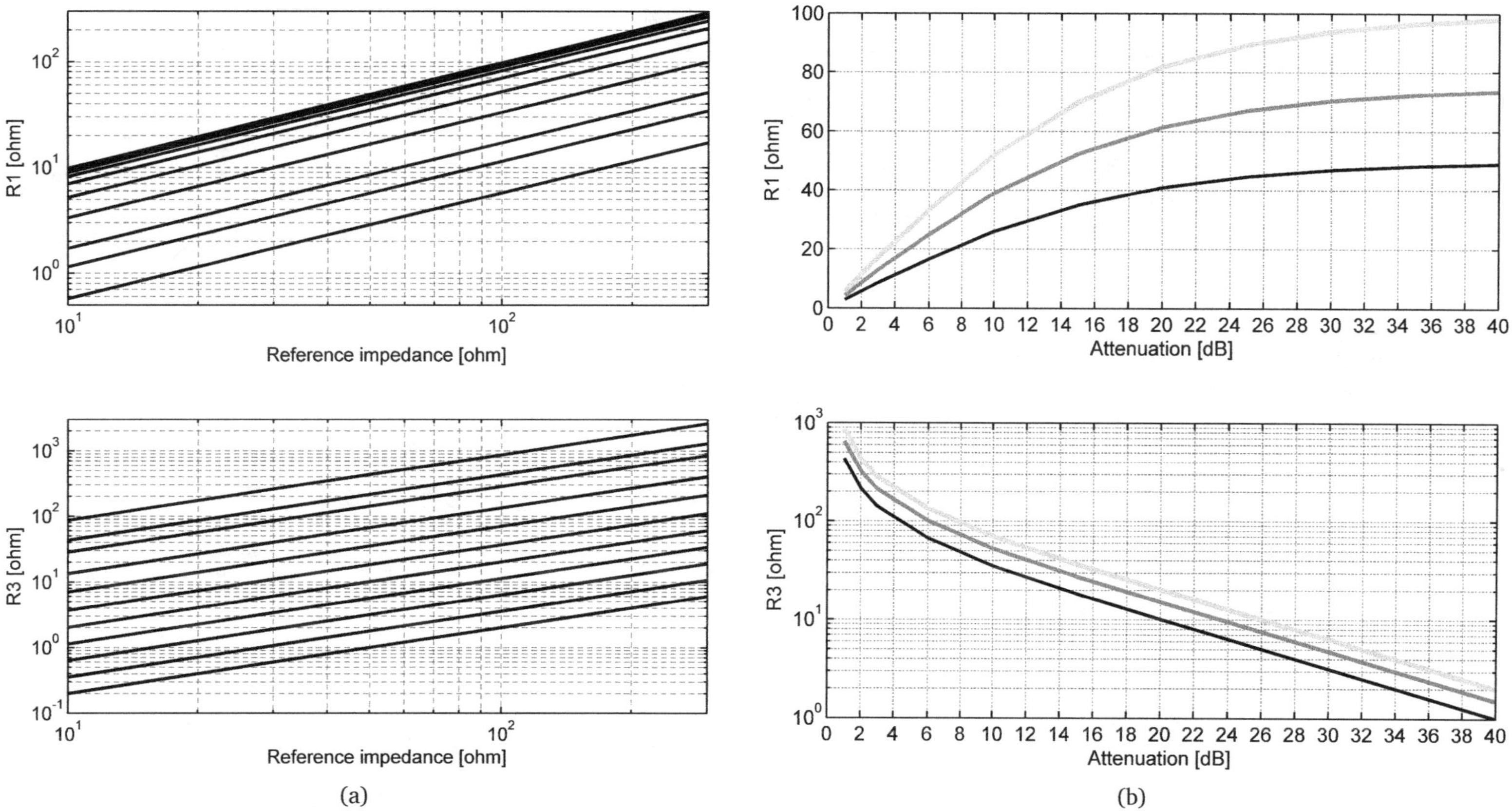

Figure 1.2.4 – T attenuator charts for (a) variable reference impedance in abscissa and attenuation (1, 2, 3, 6, 10, 15, 20, 25, 30, 35, 40 dB from bottom to top), (b) variable attenuation and $Z_S = Z_L = 50\,\Omega$ (black line), $75\,\Omega$ (dark gray line), $100\,\Omega$ (light gray line).

$$R_1^T = R_2^T = Z_{\mathrm{ref}}\,\frac{1-A}{1+A} \qquad R_3^T = \frac{Z_{\mathrm{ref}}^2 - \left(R_1^T\right)^2}{2R_1^T} \tag{1.2.6}$$

where, conversely, R_1 (and R_2) is smaller than the source impedance Z_S and R_3 is larger than Z_S. The results were checked against equations in sec. 1.2.1.

1.2.2.1 Improving matching with attenuators

It is a known practice that of placing attenuators at the two ends of a cable that is going to drive a mismatched load (such as e.g. an antenna, featuring a highly varying Voltage Standing Wave Ratio (VSWR)): the reason is to reduce mismatching at the interfaces and reduce thus cable reflections and resonances. The used attenuation values are a compromise between good matching and limited attenuation, so that usually 3 dB or 6 dB attenuators are chosen. In this section we verify attenuator performance in this sense, by terminating it with a varying Z_L, below and above the reference impedance ($50\,\Omega$).

Anticipating thus the subject of transmission lines, impedance matching and VSWR that the reader may find in Chapter 2, and in particular in sec. 2.3 and 2.5.3, the matching effect ensured by attenuators is considered, when the load impedance Z_L is free to vary over a 5:1 range below and above the reference impedance (in this case $50\,\Omega$). Such a Z_L variation will cause an impedance mismatch characterized by a $(Z_L-50)/(Z_L+50)$ reflection coefficient ranging between -0.667 and $+0.667$, and correspondingly a VSWR going up to 5:1. Results are shown in Figure 1.2.5: it may be seen that already a 6 dB attenuator limits the resulting VSWR to about 1.4 when Z_L gets to 10 or 250 Ω; a tighter control of the VSWR may be achieved then with a 10 dB attenuator, or even a 20 dB one; going above these values it is absolutely unnecessary in any case.

1.3 Network matrix representation

It is quite common that circuits and networks are represented in a compact form for input/output signals at ports by means of matrix representations. There are many more or less equivalent representations that may be suitable for some types of networks or for a specific use of them. For example, when working with cascadable blocks for sub-networks and circuits, transmission matrix is chosen, indicated by the letter "T" (or sometimes "A"). When the network is going to be solved by standard circuit techniques, impedance (Z) and admittance (Y) representations are used. Transistors (most often consisting of a three terminals network for which a two-port representation is identified, by sharing one terminal in common between input and output) are normally represented by hybrid parameters, collected in the matrix H. When finally incident and reflected waves are of interest, scattering parameters are used, indicated by the letter S (see also Chapter 10). All these matrix representations of the same network are equivalent (and shall be!) and there exist relationships, so that it is easy to pass from one form to the other and combine available information.

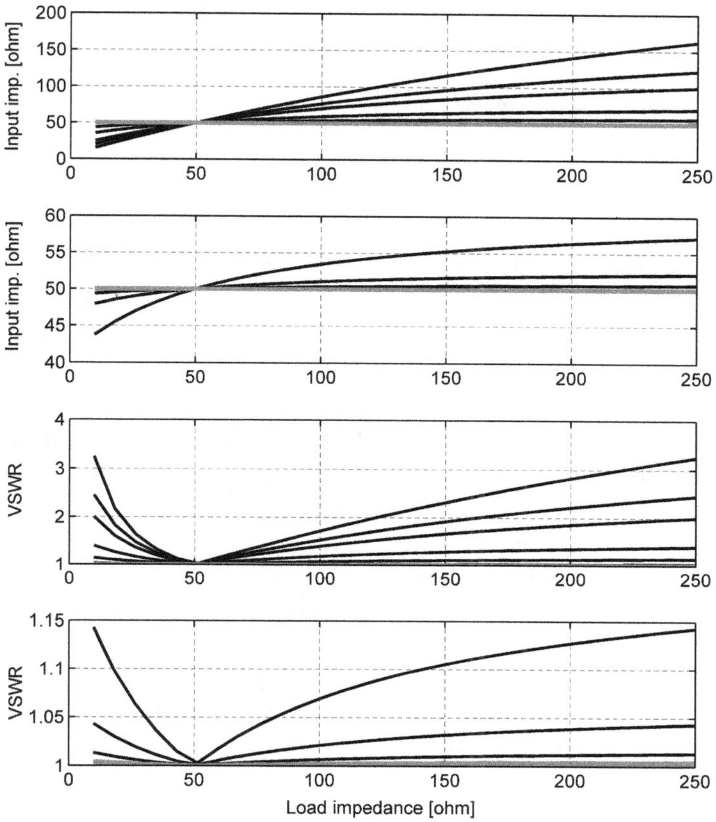

Figure 1.2.5 – Matching of input impedance enforced by T attenuators connected to a varying load: Att. $= 1, 2, 3, 6, 10, 15, 20, 25, 30, 35, 40\,\text{dB}$, $Z_L = 10 \div 250\,\Omega$, $Z_{\text{ref}} = 50\,\Omega$. First two subplots show the input impedance: (top) the five heavy black curves refer to Att. $= 1, 2, 3, 6, 10\,\text{dB}$, (bottom) the other three heavy black curves refer to Att. $= 10, 15, 20\,\text{dB}$; other two subplots show the resulting VSWR at the attenuator input: (top) the five heavy black curves refer to Att. $= 1, 2, 3, 6, 10\,\text{dB}$, (bottom) the other three heavy black curves refer to Att. $= 10, 15, 20\,\text{dB}$. The gray thick curve indicates the area occupied by the other values not explicitly taken into account.

1.3.1 Ports and matrix representation

In several disciplines it is useful and comfortable to use the concept of port, made of two terminal wires, where the input currents are bounded to sum to zero (or, in other words, are required to be equal in magnitude and of opposite polarity). Several electric circuits and equipment are suitable to be treated on a port basis (such as amplifiers, cables, transformers, etc.), which several matrix representations are related to. In the following a two-port network like in Figure 1.3.1 is considered, but n-port networks may be treated using the examples developed herein.

Depending on the choice of the input and output variables, starting from the four available variables, that is the voltages and currents of the two ports, V_1, I_1, V_2 and I_2, some matrix representations are possible.

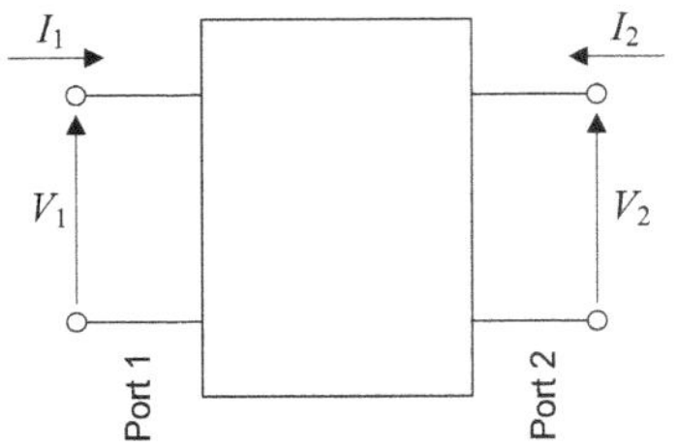

Figure 1.3.1 – Two-port network and related variables.

1.3.1.1 Impedance matrix

This is the more "natural" representation of a two-port. We see immediately that there are many equivalent representations, and that impedance representation is not the most handy for transistors (see sec. 1.3.1.6). The input variables are the two port currents I_1 and I_2; the output variables are the two port voltages V_1 and V_2.

$$\begin{bmatrix} V_1 \\ V_2 \end{bmatrix} = \begin{bmatrix} Z_{11} & Z_{12} \\ Z_{21} & Z_{22} \end{bmatrix} \begin{bmatrix} I_1 \\ I_2 \end{bmatrix} = [Z] \begin{bmatrix} I_1 \\ I_2 \end{bmatrix} \tag{1.3.1}$$

The relationships that define the terms inside the $[Z]$ matrix are very straightforward and identify a volt-amperometric test on the two-port network, while setting one quantity to a known condition.

$$Z_{11} = \left. \frac{V_1}{I_1} \right|_{I_2=0} \qquad Z_{12} = \left. \frac{V_1}{I_2} \right|_{I_1=0} \qquad Z_{21} = \left. \frac{V_2}{I_1} \right|_{I_2=0} \qquad Z_{22} = \left. \frac{V_2}{I_2} \right|_{I_1=0} \tag{1.3.2}$$

This representation is suitable for electric supply networks and to express the concept of input or loading impedance in various circuits (probes, transformers, etc.).

1.3.1.2 Admittance matrix

The input variables are the two voltages V_1 and V_2. The output variables are the two currents I_1 and I_2.

$$\begin{bmatrix} I_1 \\ I_2 \end{bmatrix} = \begin{bmatrix} Y_{11} & Y_{12} \\ Y_{21} & Y_{22} \end{bmatrix} \begin{bmatrix} V_1 \\ V_2 \end{bmatrix} = [Y] \begin{bmatrix} V_1 \\ V_2 \end{bmatrix} \tag{1.3.3}$$

The relationships that define the terms inside the $[Y]$ matrix are directly related to the one above for the impedance terms. The admittance matrix $[Y]$ is the inverse of the impedance matrix $[Z] = [Y]^{-1}$ and this can be easily demonstrated by substituting (1.3.1) into (1.3.3), or vice-versa.

The use of this representation is the same as that of the impedance matrix above; the choice between the two is based on the availability of either the two sets of variables when defining the problem, but for transistors admittance representation is particularly attractive, especially when including parasitic capacitance.

1.3.1.3 Transmission matrix

The input variables are the current and voltage on one port I_1 and V_1. The output variables are the current and voltage on the other port $-I_2$ and V_2.

$$\begin{bmatrix} V_2 \\ I_2 \end{bmatrix} = \begin{bmatrix} T_{11} & T_{12} \\ T_{21} & T_{22} \end{bmatrix} \begin{bmatrix} V_1 \\ I_1 \end{bmatrix} = [T] \begin{bmatrix} V_1 \\ I_1 \end{bmatrix} \tag{1.3.4}$$

The $[T]$ matrix contains non homogeneous terms, that are either the ratio of two voltages or two currents (so dimensionless), or the ratio of one current and one voltage, thus taking the dimension of an impedance or admittance. In analogy with the impedance terms they are defined as

$$T_{11} = \left.\frac{V_2}{V_1}\right|_{I_1=0} \qquad T_{12} = \left.\frac{V_2}{I_1}\right|_{V_1=0} \qquad T_{21} = \left.\frac{I_2}{V_1}\right|_{I_1=0} \qquad T_{22} = \left.\frac{I_2}{I_1}\right|_{V_1=0} \tag{1.3.5}$$

The opposite relationships between port 2 variables on the right-hand side and port 1 variables on the left-hand side are not used since they would lead to expressions that couldn't be realized, when trying to adopt the procedure for the measurement of the transmission parameters: for example, $T_{11} = V_1/V_2|_{I_2=0}$ requires the application of a driving signal V_2 to a port where also the condition $I_2 = 0$ applies.

This representation is suitable to assemble a series of building blocks to model more complex electrical networks, such as electric supply networks, complex circuits, etc. The choice of the input and output variables allows the direct cascade connections of different blocks, resulting simply in the chain multiplication of the respective $[T]$ matrices; to this aim one of the two currents is reversed in polarity and marked positive while exiting the port.

1.3.1.4 Hybrid parameters matrix

The input variables are hybrid, one current and one voltage, and precisely the port-1 current I_1 and the port-2 voltage V_2. The output variables are the remaining voltage V_1 and current I_2. For this reason - as it is known - the elements of the matrix H are hybrid and do not share the same physical dimensions

$$\begin{bmatrix} V_1 \\ I_2 \end{bmatrix} = \begin{bmatrix} h_{11} & h_{12} \\ h_{21} & h_{22} \end{bmatrix} \begin{bmatrix} I_1 \\ V_2 \end{bmatrix} = [H] \begin{bmatrix} I_1 \\ V_2 \end{bmatrix} \tag{1.3.6}$$

The relationships that define the terms inside the $[H]$ matrix are very straightforward and identify a volt-amperometric test on the two-port network, while holding a variable to a known condition.

$$h_{11} = \left.\frac{V_1}{I_1}\right|_{V_2=0} \qquad h_{12} = \left.\frac{V_1}{V_2}\right|_{I_1=0} \qquad h_{21} = \left.\frac{I_2}{I_1}\right|_{V_2=0} \qquad h_{22} = \left.\frac{I_2}{V_2}\right|_{I_1=0} \tag{1.3.7}$$

This representation is suitable for semiconductor devices, such as bipolar transistors. Field-effect transistors are better described by the complementary hybrid representation, identified by the letter "G", sometimes called improperly "conductance matrix", even if, as before, it is made of parameters with different physical units.

$$\begin{bmatrix} I_1 \\ V_2 \end{bmatrix} = \begin{bmatrix} g_{11} & g_{12} \\ g_{21} & g_{22} \end{bmatrix} \begin{bmatrix} V_1 \\ I_2 \end{bmatrix} = [G] \begin{bmatrix} V_1 \\ I_2 \end{bmatrix} \tag{1.3.8}$$

The relationships that define the terms inside the $[G]$ matrix are very straightforward and identify a volt-amperometric test on the two-port network, while holding a variable to a known condition.

$$g_{11} = \left.\frac{I_1}{V_1}\right|_{I_2=0} \qquad g_{12} = \left.\frac{I_1}{I_2}\right|_{V_1=0} \qquad g_{21} = \left.\frac{V_2}{V_1}\right|_{I_2=0} \qquad g_{22} = \left.\frac{V_2}{I_2}\right|_{V_1=0} \tag{1.3.9}$$

1.3.1.5 Wave parameters matrix

We anticipate a little what will be considered in detail in Chapter 10, regarding wave representations and scattering parameters, to complete the overview on matrix representations and input-output relationships. The focus here is on the relationship with the other representations and the underlying assumptions, rather then on their physical meaning and interpretation.

With a port perspective, the two sides of the matrix relationship are characterized by the pair of voltage and current at port 1 and at port 2 respectively:

$$\begin{bmatrix} V_1 \\ I_1 \end{bmatrix} = \begin{bmatrix} a_{11} & a_{12} \\ a_{21} & a_{22} \end{bmatrix} \begin{bmatrix} V_2 \\ -I_2 \end{bmatrix} = [A] \begin{bmatrix} V_2 \\ -I_2 \end{bmatrix} \tag{1.3.10}$$

$$\begin{bmatrix} V_2 \\ -I_2 \end{bmatrix} = \begin{bmatrix} b_{11} & b_{12} \\ b_{21} & b_{22} \end{bmatrix} \begin{bmatrix} V_1 \\ I_1 \end{bmatrix} = [B] \begin{bmatrix} V_1 \\ I_1 \end{bmatrix} \tag{1.3.11}$$

These two representations are the so-called *chain parameter* matrices.

For each port, once the reference impedance is established (corresponding usually to the characteristic impedance of the lines), called Z_{01} and Z_{02}, possibly different, we may write a set of expressions, relating current and voltage at each port, as governed by the reference impedance property. This is described in more detail in sec. 10.1.1.

$$a_1 = \frac{V_1 + Z_{01} I_1}{2\sqrt{Z_{01}}} \qquad b_1 = \frac{V_1 - Z_{01} I_1}{2\sqrt{Z_{01}}} \tag{1.3.12}$$

$$a_2 = \frac{V_2 + Z_{02} I_2}{2\sqrt{Z_{02}}} \qquad b_2 = \frac{V_2 - Z_{02} I_2}{2\sqrt{Z_{02}}} \tag{1.3.13}$$

These wave quantities, identified by letters a and b (maybe looking as parameters of matrices $[A]$ and $[B]$) are used to define the scattering parameters:

$$\begin{bmatrix} b_1 \\ b_2 \end{bmatrix} = \begin{bmatrix} s_{11} & s_{12} \\ s_{21} & s_{22} \end{bmatrix} \begin{bmatrix} a_1 \\ a_2 \end{bmatrix} = [S] \begin{bmatrix} a_1 \\ a_2 \end{bmatrix} \tag{1.3.14}$$

and going back to a port perspective, we obtain a scattering transfer matrix (or chain matrix) representations by means of the two matrices $[T]$ and $[U]$, its inverse:

$$\begin{bmatrix} b_1 \\ a_1 \end{bmatrix} = \begin{bmatrix} t_{11} & t_{12} \\ t_{21} & t_{22} \end{bmatrix} \begin{bmatrix} b_2 \\ a_2 \end{bmatrix} = [T] \begin{bmatrix} b_2 \\ a_2 \end{bmatrix} \tag{1.3.15}$$

$$\begin{bmatrix} b_2 \\ a_2 \end{bmatrix} = \begin{bmatrix} u_{11} & u_{12} \\ u_{21} & u_{22} \end{bmatrix} \begin{bmatrix} b_1 \\ a_1 \end{bmatrix} = [U] \begin{bmatrix} b_1 \\ a_1 \end{bmatrix} \tag{1.3.16}$$

In reality the parameters appearing in eq. (1.3.10) are indicated in [131], and confirmed in [115], as ABCD parameters, where $A = a_{11}$, $B = a_{12}$, $C = a_{21}$ and $D = a_{22}$.

1.3.1.6 Equivalence between matrix representations

It is possible finally to derive a set of relationships between the various matrix representations of two-port networks reviewed so far, that are extremely useful when combining circuits and networks, when comparing to datasheets and measurement results, and when including the effect of test setup in radiofrequency measurements (see e.g. sec. 11.7 on embedding/de-embedding external circuits when using VNA). The results appearing here below are mainly derived from [99], where the relationships are given clearly in tabular form, and crosschecked either by hand or by comparison with [137], Fig. 1.8.1, and [243], chap. 2.

Regarding the usefulness or appropriateness of the so many representations, Prof. Deane [99] expresses himself clearly and concisely: "The Z, Y and S parameters are widely used for their relationships to the familiar quantities of impedance, admittance and forward/backward wave ratios. The H and G parameters are widely used in the context of active devices, where the parameters are readily identifiable with familiar concepts. However, none of these matrices can be meaningfully multiplied together to give the overall matrix for n two-ports cascaded. For this purpose, the A, B, T and U matrices are required."

Many of these representations may be realized by Matlab functions of RF toolbox, such as `sparameters()`, `zparameters()`, `abcdparameters()`, etc., but they all pass through an internal `rftbxobj` object not allowing a direct conversion and the code implementation is not accessible.

Wave parameters in terms of wave parameters

$$S = \frac{1}{t_{22}} \begin{bmatrix} t_{12} & \Delta_T \\ 1 & -t_{21} \end{bmatrix} \qquad S = \frac{1}{u_{11}} \begin{bmatrix} -u_{12} & 1 \\ \Delta_U & u_{21} \end{bmatrix} \tag{1.3.17}$$

$$T = \frac{1}{s_{21}} \begin{bmatrix} -\Delta_S & s_{11} \\ -s_{22} & 1 \end{bmatrix} \qquad T = \frac{1}{\Delta_U} \begin{bmatrix} u_{22} & -u_{12} \\ -u_{21} & u_{11} \end{bmatrix} \tag{1.3.18}$$

$$U = \frac{1}{s_{12}} \begin{bmatrix} 1 & -s_{11} \\ s_{22} & -\Delta_S \end{bmatrix} \qquad U = \frac{1}{\Delta_T} \begin{bmatrix} t_{22} & -t_{12} \\ -t_{21} & t_{11} \end{bmatrix} \tag{1.3.19}$$

Circuit parameters in terms of circuit parameters

$$\begin{gathered} Z = \frac{1}{\Delta_Y} \begin{bmatrix} y_{22} & -y_{12} \\ -y_{21} & y_{11} \end{bmatrix} \; Z = \frac{1}{h_{22}} \begin{bmatrix} \Delta_H & h_{12} \\ -h_{21} & 1 \end{bmatrix} \; Z = \frac{1}{g_{11}} \begin{bmatrix} 1 & -g_{12} \\ g_{21} & \Delta_G \end{bmatrix} \\ Z = \frac{1}{a_{21}} \begin{bmatrix} a_{11} & \Delta_A \\ 1 & a_{22} \end{bmatrix} \qquad Z = \frac{-1}{b_{21}} \begin{bmatrix} b_{22} & 1 \\ \Delta_B & b_{11} \end{bmatrix} \end{gathered} \tag{1.3.20}$$

$$\begin{gathered} Y = \frac{1}{\Delta_Z} \begin{bmatrix} z_{22} & -z_{12} \\ -z_{21} & z_{11} \end{bmatrix} \; Y = \frac{1}{h_{11}} \begin{bmatrix} 1 & -h_{12} \\ h_{21} & \Delta_H \end{bmatrix} \; Y = \frac{1}{g_{22}} \begin{bmatrix} \Delta_G & g_{12} \\ -g_{21} & 1 \end{bmatrix} \\ Y = \frac{1}{a_{12}} \begin{bmatrix} a_{22} & -\Delta_A \\ -1 & a_{11} \end{bmatrix} \quad Y = \frac{1}{b_{12}} \begin{bmatrix} -b_{11} & 1 \\ \Delta_B & -b_{22} \end{bmatrix} \end{gathered} \tag{1.3.21}$$

$$\begin{gathered} H = \frac{1}{z_{22}} \begin{bmatrix} \Delta_Z & z_{12} \\ -z_{21} & 1 \end{bmatrix} \; H = \frac{1}{y_{11}} \begin{bmatrix} 1 & -y_{12} \\ y_{21} & \Delta_Y \end{bmatrix} \; H = \frac{1}{\Delta_G} \begin{bmatrix} g_{22} & -g_{12} \\ -g_{21} & g_{11} \end{bmatrix} \\ H = \frac{1}{a_{22}} \begin{bmatrix} a_{12} & \Delta_A \\ -1 & a_{21} \end{bmatrix} \quad H = \frac{-1}{b_{11}} \begin{bmatrix} b_{12} & -1 \\ \Delta_B & b_{21} \end{bmatrix} \end{gathered} \tag{1.3.22}$$

$$\begin{gathered} G = \frac{1}{z_{11}} \begin{bmatrix} 1 & -z_{12} \\ z_{21} & \Delta_Z \end{bmatrix} \; G = \frac{1}{y_{22}} \begin{bmatrix} \Delta_Y & y_{12} \\ -y_{21} & 1 \end{bmatrix} \; G = \frac{1}{\Delta_H} \begin{bmatrix} h_{22} & -h_{12} \\ -h_{21} & h_{11} \end{bmatrix} \\ G = \frac{1}{a_{11}} \begin{bmatrix} a_{21} & -\Delta_A \\ 1 & a_{12} \end{bmatrix} \; G = \frac{-1}{b_{22}} \begin{bmatrix} b_{21} & 1 \\ -\Delta_B & b_{12} \end{bmatrix} \end{gathered} \tag{1.3.23}$$

$$\begin{gathered} A = \frac{1}{z_{21}} \begin{bmatrix} z_{11} & \Delta_Z \\ 1 & z_{22} \end{bmatrix} \; A = \frac{-1}{y_{21}} \begin{bmatrix} y_{22} & 1 \\ \Delta_Y & y_{11} \end{bmatrix} \quad A = \frac{-1}{h_{21}} \begin{bmatrix} \Delta_H & h_{11} \\ h_{22} & 1 \end{bmatrix} \\ A = \frac{1}{g_{21}} \begin{bmatrix} 1 & g_{22} \\ g_{11} & \Delta_G \end{bmatrix} \; A = \frac{1}{\Delta_B} \begin{bmatrix} b_{22} & -b_{12} \\ -b_{21} & -b_{11} \end{bmatrix} \end{gathered} \tag{1.3.24}$$

$$\begin{gathered} B = \frac{1}{z_{12}} \begin{bmatrix} z_{22} & -\Delta_Z \\ -1 & z_{11} \end{bmatrix} \; Y = \frac{1}{y_{12}} \begin{bmatrix} -y_{11} & 1 \\ \Delta_Y & -y_{22} \end{bmatrix} \; Y = \frac{1}{h_{12}} \begin{bmatrix} 1 & -h_{11} \\ -h_{22} & \Delta_H \end{bmatrix} \\ B = \frac{1}{g_{12}} \begin{bmatrix} -\Delta_G & g_{22} \\ g_{11} & -1 \end{bmatrix} \; A = \frac{1}{\Delta_A} \begin{bmatrix} a_{22} & -a_{12} \\ -a_{21} & a_{11} \end{bmatrix} \end{gathered} \tag{1.3.25}$$

Wave parameters in terms of circuit parameters

$$
\begin{aligned}
S &= \frac{1}{(z_{11}+Z_{01})(z_{22}+Z_{02})-z_{12}z_{21}}\begin{bmatrix} (z_{11}-Z_{01})(z_{22}+Z_{02})-z_{12}z_{21} & 2z_{12}\sqrt{Z_{01}Z_{02}} \\ 2z_{21}\sqrt{Z_{01}Z_{02}} & (z_{11}+Z_{01})(z_{22}-Z_{02})-z_{12}z_{21} \end{bmatrix} \\
S &= \frac{1}{Z_{01}Z_{02}\Delta_Y+Z_{01}y_{11}+Z_{02}y_{22}+1}\begin{bmatrix} -Z_{01}Z_{02}\Delta_Y-Z_{01}y_{11}+Z_{02}y_{22}+1 & -2y_{12}\sqrt{Z_{01}Z_{02}} \\ -2y_{21}\sqrt{Z_{01}Z_{02}} & -Z_{01}Z_{02}\Delta_Y+Z_{01}y_{11}-Z_{02}y_{22}+1 \end{bmatrix} \\
S &= \frac{1}{Z_{02}\Delta_H+Z_{01}(Z_{02}h_{22}+1)+h_{11}}\begin{bmatrix} Z_{02}\Delta_H-Z_{01}(Z_{02}h_{22}+1)+h_{11} & 2h_{12}\sqrt{Z_{01}Z_{02}} \\ -2h_{21}\sqrt{Z_{01}Z_{02}} & -Z_{02}\Delta_H-Z_{01}(Z_{02}h_{22}-1)+h_{11} \end{bmatrix} \\
S &= \frac{1}{Z_{01}\Delta_G+Z_{02}(Z_{01}g_{11}+1)+g_{22}}\begin{bmatrix} -Z_{01}\Delta_G-Z_{02}(Z_{01}g_{11}-1)+g_{22} & -2g_{12}\sqrt{Z_{01}Z_{02}} \\ 2g_{21}\sqrt{Z_{01}Z_{02}} & Z_{01}\Delta_G-Z_{02}(Z_{01}g_{11}+1)+g_{22} \end{bmatrix} \\
S &= \frac{1}{Z_{02}(Z_{01}a_{21}+a_{11})+Z_{01}a_{22}+a_{12}}\begin{bmatrix} -Z_{02}(Z_{01}a_{21}-a_{11})-Z_{01}a_{22}+a_{12} & 2\Delta_A\sqrt{Z_{01}Z_{02}} \\ 2\sqrt{Z_{01}Z_{02}} & -Z_{02}(Z_{01}a_{21}+a_{11})+Z_{01}a_{22}+a_{12} \end{bmatrix} \\
S &= \frac{1}{Z_{02}(Z_{01}b_{21}-b_{22})-Z_{01}b_{11}+b_{12}}\begin{bmatrix} -Z_{02}(Z_{01}b_{21}+b_{22})+Z_{01}b_{11}+b_{12} & -2\sqrt{Z_{01}Z_{02}} \\ -2\Delta_B\sqrt{Z_{01}Z_{02}} & -Z_{02}(Z_{01}b_{21}-b_{22})-Z_{01}b_{11}+b_{12} \end{bmatrix}
\end{aligned}
\tag{1.3.26}
$$

Wave parameters in terms of circuit parameters (continued)

$$
\begin{aligned}
T &= \frac{1}{2z_{21}\sqrt{Z_{01}Z_{02}}}\begin{bmatrix} -\Delta_Z + Z_{01}z_{22} + Z_{02}z_{11} - Z_{01}Z_{02} & \Delta_Z - Z_{01}z_{22} + Z_{02}z_{11} - Z_{01}Z_{02} \\ -\Delta_Z - Z_{01}z_{22} + Z_{02}z_{11} + Z_{01}Z_{02} & \Delta_Z + Z_{01}z_{22} + Z_{02}z_{11} + Z_{01}Z_{02} \end{bmatrix} \\
T &= \frac{1}{2y_{21}\sqrt{Z_{01}Z_{02}}}\begin{bmatrix} Z_{01}Z_{02}\Delta_Y - Z_{01}y_{11} - Z_{02}y_{22} + 1 & Z_{01}Z_{02}\Delta_Y + Z_{01}y_{11} - Z_{02}y_{22} - 1 \\ -Z_{01}Z_{02}\Delta_Y + Z_{01}y_{11} - Z_{02}y_{22} + 1 & -Z_{01}Z_{02}\Delta_Y - Z_{01}y_{11} - Z_{02}y_{22} - 1 \end{bmatrix} \\
T &= \frac{1}{2h_{21}\sqrt{Z_{01}Z_{02}}}\begin{bmatrix} -Z_{02}\Delta_H + Z_{01}Z_{02}h_{22} - Z_{01} + h_{11} & -Z_{02}\Delta_H + Z_{01}Z_{02}h_{22} + Z_{01} - h_{11} \\ -Z_{02}\Delta_H - Z_{01}Z_{02}h_{22} + Z_{01} + h_{11} & -Z_{02}\Delta_H - Z_{01}Z_{02}h_{22} - Z_{01} - h_{11} \end{bmatrix} \\
T &= \frac{1}{2g_{21}\sqrt{Z_{01}Z_{02}}}\begin{bmatrix} Z_{01}\Delta_G - Z_{01}Z_{02}g_{11} + Z_{02} - g_{22} & -Z_{01}\Delta_G - Z_{01}Z_{02}g_{11} + Z_{02} + g_{22} \\ -Z_{01}\Delta_G + Z_{01}Z_{02}g_{11} + Z_{02} - g_{22} & Z_{01}\Delta_G + Z_{01}Z_{02}g_{11} + Z_{02} + g_{22} \end{bmatrix} \\
T &= \frac{1}{2\sqrt{Z_{01}Z_{02}}}\begin{bmatrix} -Z_{01}Z_{02}a_{21} + Z_{02}a_{11} + Z_{01}a_{22} - a_{12} & -Z_{01}Z_{02}a_{21} + Z_{02}a_{11} - Z_{01}a_{22} - a_{12} \\ Z_{01}Z_{02}a_{21} + Z_{02}a_{11} - Z_{01}a_{22} - a_{12} & Z_{01}Z_{02}a_{21} + Z_{02}a_{11} + Z_{01}a_{22} + a_{12} \end{bmatrix} \\
T &= \frac{1}{2\Delta_B\sqrt{Z_{01}Z_{02}}}\begin{bmatrix} Z_{01}Z_{02}b_{21} + Z_{02}b_{22} + Z_{01}b_{11} + b_{12} & Z_{01}Z_{02}b_{21} + Z_{02}b_{22} - Z_{01}b_{11} - b_{12} \\ -Z_{01}Z_{02}b_{21} + Z_{02}b_{22} - Z_{01}b_{11} + b_{12} & -Z_{01}Z_{02}b_{21} + Z_{02}b_{22} + Z_{01}b_{11} - b_{12} \end{bmatrix}
\end{aligned}
\tag{1.3.27}
$$

Wave parameters in terms of circuit parameters (continued)

$$
\begin{aligned}
U &= \frac{1}{2z_{12}\sqrt{Z_{01}Z_{02}}}\begin{bmatrix} \Delta_Z + Z_{01}z_{22} + Z_{02}z_{11} + Z_{01}Z_{02} & -\Delta_Z + Z_{01}z_{22} - Z_{02}z_{11} + Z_{01}Z_{02} \\ \Delta_Z + Z_{01}z_{22} - Z_{02}z_{11} - Z_{01}Z_{02} & -\Delta_Z + Z_{01}z_{22} + Z_{02}z_{11} - Z_{01}Z_{02} \end{bmatrix} \\
U &= \frac{1}{2y_{12}\sqrt{Z_{01}Z_{02}}}\begin{bmatrix} -Z_{01}Z_{02}\Delta_Y - Z_{01}y_{11} - Z_{02}y_{22} - 1 & -Z_{01}Z_{02}\Delta_Y - Z_{01}y_{11} + Z_{02}y_{22} + 1 \\ Z_{01}Z_{02}\Delta_Y - Z_{01}y_{11} + Z_{02}y_{22} - 1 & Z_{01}Z_{02}\Delta_Y - Z_{01}y_{11} - Z_{02}y_{22} + 1 \end{bmatrix} \\
U &= \frac{1}{2h_{12}\sqrt{Z_{01}Z_{02}}}\begin{bmatrix} Z_{02}\Delta_H + Z_{01}Z_{02}h_{22} + Z_{01} + h_{11} & -Z_{02}\Delta_H + Z_{01}Z_{02}h_{22} + Z_{01} - h_{11} \\ -Z_{02}\Delta_H - Z_{01}Z_{02}h_{22} + Z_{01} + h_{11} & Z_{02}\Delta_H - Z_{01}Z_{02}h_{22} + Z_{01} - h_{11} \end{bmatrix} \\
U &= \frac{1}{2g_{12}\sqrt{Z_{01}Z_{02}}}\begin{bmatrix} -Z_{01}\Delta_G - Z_{01}Z_{02}g_{11} - Z_{02} - g_{22} & -Z_{01}\Delta_G - Z_{01}Z_{02}g_{11} + Z_{02} + g_{22} \\ -Z_{01}\Delta_G + Z_{01}Z_{02}g_{11} + Z_{02} - g_{22} & -Z_{01}\Delta_G + Z_{01}Z_{02}g_{11} - Z_{02} + g_{22} \end{bmatrix} \\
U &= \frac{1}{2\Delta_A\sqrt{Z_{01}Z_{02}}}\begin{bmatrix} Z_{01}Z_{02}a_{21} + Z_{02}a_{11} + Z_{01}a_{22} + a_{12} & Z_{01}Z_{02}a_{21} - Z_{02}a_{11} + Z_{01}a_{22} - a_{12} \\ -Z_{01}Z_{02}a_{21} - Z_{02}a_{11} + Z_{01}a_{22} + a_{12} & -Z_{01}Z_{02}a_{21} + Z_{02}a_{11} + Z_{01}a_{22} - a_{12} \end{bmatrix} \\
U &= \frac{1}{2\sqrt{Z_{01}Z_{02}}}\begin{bmatrix} -Z_{01}Z_{02}b_{21} + Z_{02}b_{22} + Z_{01}b_{11} - b_{12} & -Z_{01}Z_{02}b_{21} - Z_{02}b_{22} + Z_{01}b_{11} + b_{12} \\ Z_{01}Z_{02}b_{21} - Z_{02}b_{22} + Z_{01}b_{11} - b_{12} & Z_{01}Z_{02}b_{21} + Z_{02}b_{22} + Z_{01}b_{11} + b_{12} \end{bmatrix}
\end{aligned}
\tag{1.3.28}
$$

Circuit parameters in terms of wave parameters

$$Z = \frac{1}{\Delta_S - s_{11} - s_{22} + 1} \begin{bmatrix} -Z_{01}(\Delta_S - s_{11} + s_{22} - 1) & 2s_{12}\sqrt{Z_{01}Z_{02}} \\ 2s_{21}\sqrt{Z_{01}Z_{02}} & -Z_{02}(\Delta_S + s_{11} - s_{22} - 1) \end{bmatrix}$$

$$Z = \frac{1}{-t_{11} - t_{12} + t_{21} + t_{22}} \begin{bmatrix} Z_{01}(t_{11} + t_{12} + t_{21} + t_{22}) & 2\Delta_T\sqrt{Z_{01}Z_{02}} \\ 2\sqrt{Z_{01}Z_{02}} & Z_{02}(t_{11} - t_{12} - t_{21} + t_{22}) \end{bmatrix} \tag{1.3.29}$$

$$Z = \frac{1}{-u_{11} - u_{12} + u_{21} + u_{22}} \begin{bmatrix} -Z_{01}(u_{11} - u_{12} - u_{21} + u_{22}) & -2\sqrt{Z_{01}Z_{02}} \\ -2\Delta_U\sqrt{Z_{01}Z_{02}} & -Z_{02}(u_{11} + u_{12} + u_{21} + u_{22}) \end{bmatrix}$$

$$Y = \frac{1}{Z_{01}Z_{02}(\Delta_S + s_{11} + s_{22} + 1)} \begin{bmatrix} Z_{02}(-\Delta_S - s_{11} + s_{22} + 1) & -2s_{12}\sqrt{Z_{01}Z_{02}} \\ -2s_{21}\sqrt{Z_{01}Z_{02}} & Z_{01}(-\Delta_S + s_{11} - s_{22} + 1) \end{bmatrix}$$

$$Y = \frac{1}{Z_{01}Z_{02}(-t_{11} + t_{12} - t_{21} + t_{22})} \begin{bmatrix} Z_{02}(t_{11} - t_{12} - t_{21} + t_{22}) & -2\Delta_T\sqrt{Z_{01}Z_{02}} \\ -2\sqrt{Z_{01}Z_{02}} & Z_{01}(t_{11} + t_{12} + t_{21} + t_{22}) \end{bmatrix} \tag{1.3.30}$$

$$Y = \frac{1}{Z_{01}Z_{02}(-u_{11} + u_{12} - u_{21} + u_{22})} \begin{bmatrix} -Z_{02}(u_{11} + u_{12} + u_{21} + u_{22}) & 2\sqrt{Z_{01}Z_{02}} \\ 2\Delta_U\sqrt{Z_{01}Z_{02}} & -Z_{01}(u_{11} - u_{12} - u_{21} + u_{22}) \end{bmatrix}$$

Circuit parameters in terms of wave parameters (continued)

$$H = \frac{1}{Z_{02}(\Delta_S + s_{11} - s_{22} - 1)} \begin{bmatrix} -Z_{01}Z_{02}(\Delta_S + s_{11} + s_{22} + 1) & -2s_{12}\sqrt{Z_{01}Z_{02}} \\ 2s_{21}\sqrt{Z_{01}Z_{02}} & -\Delta_S + s_{11} + s_{22} - 1 \end{bmatrix}$$

$$H = \frac{1}{Z_{02}(t_{11} - t_{12} - t_{21} + t_{22})} \begin{bmatrix} Z_{01}Z_{02}(-t_{11} + t_{12} - t_{21} + t_{22}) & 2\Delta_T\sqrt{Z_{01}Z_{02}} \\ -2\sqrt{Z_{01}Z_{02}} & -t_{11} - t_{12} + t_{21} + t_{22} \end{bmatrix} \tag{1.3.31}$$

$$H = \frac{1}{Z_{02}(u_{11} + u_{12} + u_{21} + u_{22})} \begin{bmatrix} Z_{01}Z_{02}(u_{11} - u_{12} + u_{21} - u_{22}) & 2\sqrt{Z_{01}Z_{02}} \\ -2\Delta_U\sqrt{Z_{01}Z_{02}} & u_{11} + u_{12} - u_{21} - u_{22} \end{bmatrix}$$

$$G = \frac{1}{Z_{01}(\Delta_S - s_{11} + s_{22} - 1)} \begin{bmatrix} -\Delta_S + s_{11} + s_{22} - 1 & 2s_{12}\sqrt{Z_{01}Z_{02}} \\ -2s_{21}\sqrt{Z_{01}Z_{02}} & -Z_{01}Z_{02}(\Delta_S + s_{11} + s_{22} + 1) \end{bmatrix}$$

$$G = \frac{1}{Z_{01}(t_{11} + t_{12} + t_{21} + t_{22})} \begin{bmatrix} -t_{11} - t_{12} + t_{21} + t_{22} & -2\Delta_T\sqrt{Z_{01}Z_{02}} \\ 2\sqrt{Z_{01}Z_{02}} & Z_{01}Z_{02}(-t_{11} + t_{12} - t_{21} + t_{22}) \end{bmatrix} \tag{1.3.32}$$

$$G = \frac{1}{Z_{01}(u_{11} - u_{12} - u_{21} + u_{22})} \begin{bmatrix} u_{11} + u_{12} - u_{21} - u_{22} & -2\sqrt{Z_{01}Z_{02}} \\ 2\Delta_U\sqrt{Z_{01}Z_{02}} & Z_{01}Z_{02}(u_{11} - u_{12} + u_{21} - u_{22}) \end{bmatrix}$$

Circuit parameters in terms of wave parameters (continued)

$$A = \frac{1}{2s_{21}\sqrt{Z_{01}Z_{02}}} \begin{bmatrix} -Z_{01}(\Delta_S - s_{11} + s_{22} - 1) & Z_{01}Z_{02}(\Delta_S + s_{11} + s_{22} + 1) \\ \Delta_S - s_{11} - s_{22} + 1 & Z_{02}(-\Delta_S - s_{11} + s_{22} + 1) \end{bmatrix}$$

$$A = \frac{1}{2\sqrt{Z_{01}Z_{02}}} \begin{bmatrix} Z_{01}(t_{11} + t_{12} + t_{21} + t_{22}) & Z_{01}Z_{02}(-t_{11} + t_{12} - t_{21} + t_{22}) \\ -t_{11} - t_{12} + t_{21} + t_{22} & Z_{02}(t_{11} - t_{12} - t_{21} + t_{22}) \end{bmatrix} \tag{1.3.33}$$

$$A = \frac{1}{2\Delta_U\sqrt{Z_{01}Z_{02}}} \begin{bmatrix} Z_{01}(u_{11} - u_{12} - u_{21} + u_{22}) & Z_{01}Z_{02}(u_{11} - u_{12} + u_{21} - u_{22}) \\ u_{11} + u_{12} - u_{21} - u_{22} & Z_{02}(u_{11} + u_{12} + u_{21} + u_{22}) \end{bmatrix}$$

$$B = \frac{1}{2s_{12}\sqrt{Z_{01}Z_{02}}} \begin{bmatrix} -Z_{02}(\Delta_S + s_{11} - s_{22} - 1) & -Z_{01}Z_{02}(\Delta_S + s_{11} + s_{22} + 1) \\ -\Delta_S + s_{11} + s_{22} - 1 & Z_{01}(-\Delta_S + s_{11} - s_{22} + 1) \end{bmatrix}$$

$$B = \frac{1}{2\Delta_T\sqrt{Z_{01}Z_{02}}} \begin{bmatrix} Z_{02}(t_{11} - t_{12} - t_{21} + t_{22}) & Z_{01}Z_{02}(t_{11} - t_{12} + t_{21} - t_{22}) \\ t_{11} + t_{12} - t_{21} - t_{22} & Z_{01}(t_{11} + t_{12} + t_{21} + t_{22}) \end{bmatrix} \tag{1.3.34}$$

$$B = \frac{1}{2\sqrt{Z_{01}Z_{02}}} \begin{bmatrix} Z_{02}(u_{11} + u_{12} + u_{21} + u_{22}) & Z_{01}Z_{02}(-u_{11} + u_{12} - u_{21} + u_{22}) \\ -u_{11} - u_{12} + u_{21} + u_{22} & Z_{02}(u_{11} - u_{12} - u_{21} + u_{22}) \end{bmatrix}$$

1.3.2 Propagation of uncertainty between network matrix transformations

The subject of uncertainty, that will be considered in more detail at the end of the book in sec. 10.5 and 11.11, is anticipated for what is relevant to the use of matrix representation and transformations between them, to automate a little the process of evaluating the propagation of uncertainty during measurements. In [309] Stenarson and Yhland propose a matrix formulation of conversions between network representations and they substantiate the demonstration of the correctness with an example. They show that once the uncertainty in one of the two representations is determined (they indicate this initial representation as $\mathbf{Q}$ and the uncertainty as $\delta\mathbf{Q}$), the uncertainty in the target matrix representation (that they indicate as $\delta\mathbf{Q}'$ and $\mathbf{Q}'$, respectively) is straightforwardly determined by the relationship

$$\delta\mathbf{Q}' = \mathbf{A}^{-1}\delta\mathbf{Q}\mathbf{A}' \tag{1.3.35}$$

The matrices $\mathbf{A}$ and $\mathbf{A}'$ are determined taking sub-matrices of the transformation between $\mathbf{Q}$ and $\mathbf{Q}'$:

$$\mathbf{Q}' = \mathbf{P}\mathbf{Q} = \begin{bmatrix} P_{11} & P_{12} \\ P_{21} & P_{22} \end{bmatrix} \mathbf{Q} \qquad \mathbf{A} = \mathbf{P}_{11} - \mathbf{Q}\mathbf{P}_{21} \quad \mathbf{A}' = \mathbf{P}_{21}\mathbf{Q}' + \mathbf{P}_{22} \tag{1.3.36}$$

The authors underline that the method can be automated, once the relationships between matrix representations are put in matrix form; the reader may use the matrix transformations given in the previous section and verify they are correctly input and framed into the work.

The authors also identify a weakness in the adopted simplification: correlation between quantities was assumed negligible (put to zero) and the method cannot be used twice consecutively chaining conversions because the method does not preserve the correlation between the converted parameters. They observe that when H and Z representations are used as the intermediate ones the final uncertainty is the maximum one, even slightly above the worst-case for the former hybrid parameter representation.

1.3.3 Choice of the two-port matrix representation

All two-port representations are equivalent for linear networks; the choice is thus for a simplification of calculations or better interpretation of results. If using series feedback, then Z parameters are preferred; if shunt feedback is applied, then Y parameters are preferable; with a combination of series and shunt connection, H or G parameters may be used instead. We will see that de-embedding of DUT parasitics (see sec. 11.7.1) proceeds with Z and Y matrices for series inductance and shunt capacitance terms.

Going back to the definitions of each representation, we may see that some of the matrix terms express very simple and straightforward concepts, such as y_{11} is the short-circuit input admittance, while z_{11} is the open-circuit input impedance; conversely g_{11} and h_{11} represent the same input admittance and impedance with opposite

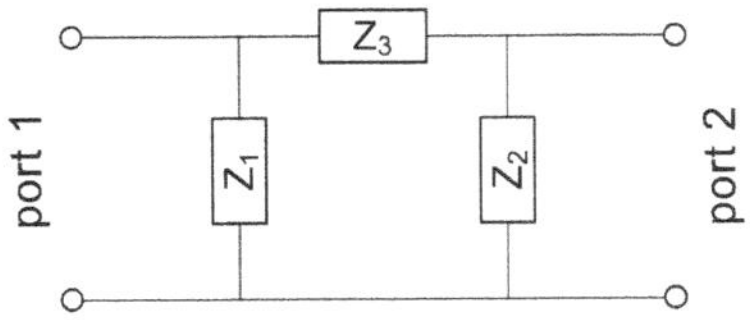

Figure 1.3.2 – π circuit equivalent of a two-port network

conditions at the other port, that is open-circuited for admittance and short-circuited for impedance terms; the forward current gain is directly expressed by h_{21} and the forward voltage gain by g_{21}.

1.3.3.1 Calculations in a π circuit

A π model is mostly useful for example to represent with a lumped approach transmission line sections or to model transistor devices. Since for transmission lines the followed approach is that of distributed parameter modeling, equivalent π cells are of scarce interest. It is possible instead to include all the significant characteristics of a transistor model (named here using the most common terms, even if slightly inappropriate sometimes): input impedance, output conductance (or impedance), transconductance (or current gain), output-to-input internal feedback through the base-collector capacitance. The equivalent π circuit is shown in Figure 1.3.2.

The input impedance Z_{11} with the output port 2 left open ($I_2 = 0$) is readily

$$Z_{11} = \frac{Z_1(Z_2 + Z_3)}{Z_1 + Z_2 + Z_3} \tag{1.3.37}$$

The output impedance Z_{22} with the input port 1 left open is perfectly specular.

The element Z_3 determines the longitudinal voltage drop. If the forward current gain with the output in short-circuit conditions is desired, Z_2 is shunted and ruled out by the short-circuit and we obtain

$$A_I = \frac{I_2}{I_1} = \frac{\dfrac{1}{Z_3}}{\dfrac{1}{Z_1} + \dfrac{1}{Z_3}} = \frac{Z_1}{Z_1 + Z_3} \tag{1.3.38}$$

The voltage gain A_V may be determined as well, even if it is not so straightforward: the output voltage with the port 2 left open is the voltage drop across Z_2, for which the current flow is determined by current sharing between Z_1 and $Z_2 + Z_3$, as just done above for the current gain; the input voltage is applied at port 1 directly on the parallel connection of the two branches Z_1 and $Z_2 + Z_3$.

$$A_V = Z_2 \frac{Z_1}{Z_1 + Z_2 + Z_3} \frac{1}{Z_{11}} = \frac{Z_1 Z_2}{Z_1(Z_2 + Z_3)} \tag{1.3.39}$$

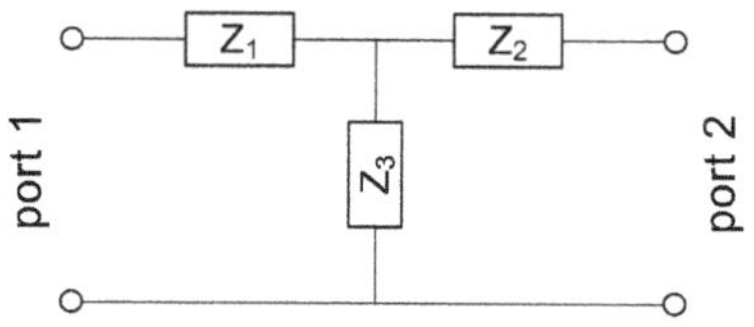

Figure 1.3.3 – T circuit equivalent of a two-port network

1.3.3.2 Calculations in a T circuit

T cells are complementary to the just seen π cells and it is possible to easily pass from one representation to the other.

The input impedance at port 1 with the opposite port left open is

$$Z_{11} = Z_1 + Z_3 \tag{1.3.40}$$

and symmetrically the output impedance involves Z_2.

When solving configurations where one port is left open with zero current, this rules out either of Z_1 or Z_2: e.g. calculating again the voltage gain, the voltage at port 2 with no current flowing is the voltage across Z_3, so

$$A_V = \frac{Z_3}{Z_1 + Z_3} \tag{1.3.41}$$

that is the symmetric equivalent of the current gain in the former π cell representation. For this reason, sometimes, the elements of the T cell are indicated as admittances Y, even if the keep their units and values as impedances.

The equivalence between π and T circuits is considered in the next section using the known wye-delta transformations.

1.3.3.3 Transform between π and T circuit representation

The π and T circuit representations are often used to model and characterize many devices and pieces of equipment (attenuators, filters, semiconductors, etc.) and also to design simple circuits (e.g. attenuators and matching pads); it is extremely useful to know how to pass from one representation to the other. The transformation between the two representations is nothing more than a wye-delta (or star-delta) transform, often used in three-phase systems and electrotechnics.

For the π-to-T transformation the criterion taking an element of the π circuit is to have at the numerator the impedance of the two T elements that insist on the same node and at the denominator the sum of all the three elements of the circuit:

$$Z_1^T = \frac{Z_1^\pi Z_3^\pi}{Z_1^\pi + Z_2^\pi + Z_3^\pi} \tag{1.3.42}$$

$$Z_2^T = \frac{Z_2^\pi Z_3^\pi}{Z_1^\pi + Z_2^\pi + Z_3^\pi} \tag{1.3.43}$$

$$Z_3^T = \frac{Z_1^\pi Z_2^\pi}{Z_1^\pi + Z_2^\pi + Z_3^\pi} \tag{1.3.44}$$

For the T-to-π transformation the criterion for taking an element of the T circuit is to have at the numerator the sum of all binary products (pair of terms) of the impedances of the circuit and at the denominator the opposite element, or in other words, the element of the π circuit that insists on the other two nodes:

$$Z_1^\pi = \frac{Z_1^T Z_2^T + Z_2^T Z_3^T + Z_3^T Z_1^T}{Z_2^T} \tag{1.3.45}$$

$$Z_2^\pi = \frac{Z_1^T Z_2^T + Z_2^T Z_3^T + Z_3^T Z_1^T}{Z_1^T} \tag{1.3.46}$$

$$Z_3^\pi = \frac{Z_1^T Z_2^T + Z_2^T Z_3^T + Z_3^T Z_1^T}{Z_3^T} \tag{1.3.47}$$

The behavior of the indexes is due to the particular naming scheme of the elements of the two circuit representations.

1.3.3.4 Common emitter amplifier example

We take a closer look to the common emitter (or common source, quite similarly) amplifier example, already considered when introducing π circuit in sec. 1.3.3.1 above: the circuit representation requires that to the π circuit an internal current generator is added.

Y parameters The calculation is performed with the Y parameters [253], because they are far handier than Z parameters when elements are connected in parallel.

The total output current i_2 is given by the sum of the two admittance terms y_{22} and y_L multiplied by the output voltage v_2. In this case Y parameters offer themselves to the rapid and simple connection of elements in parallel.

The output current in the circuit is drawn by the generator through the forward transconductance, driven by the input voltage v_1. So, finally, $(y_{22}+Y_L)v_2 = -y_{21}v_1$, and the forward voltage gain is immediately $A_v = \frac{v_2}{v_1} = \frac{-y_{21}}{y_{22}+Y_L}$.

The total input admittance is by definition

$$Y_{in} = \frac{i_1}{v_1} = y_{11} + y_{12}\frac{v_2}{v_1} = y_{11} - \frac{y_{12}y_{21}}{y_{22}+Y_L} \tag{1.3.48}$$

Symmetrically, the total output admittance is

$$Y_{out} = \frac{i_2}{v_2} = y_{22} + y_{12}\frac{v_1}{v_2} = y_{22} - \frac{y_{12}y_{21}}{y_{11}+Y_S} \tag{1.3.49}$$

An amplifier is said "unilateral" if $y_{12}=0$, that is no effect is reflected back to the input side from the output side, or in other words its operation is "forward". In this case the above expressions simplify to $Y_{in}=y_{11}$ and $Y_{out}=y_{22}$.

If we want to include the effect of the source and the load, calculating the total voltage gain between e_S and $v_L=v_2$, we write

$$\begin{aligned} A'_v &= \frac{v_2}{v_1}\frac{v_1}{e_S} = A_v \frac{Y_S}{Y_{in}+Y_S} = \\ &= \frac{-Y_S\, y_{21}}{(Y_{in}+Y_S)(y_{22}+Y_L)} = \frac{-Y_S\, y_{21}}{(y_{11}+Y_S)(y_{22}+Y_L)-y_{12}y_{21}} \end{aligned} \tag{1.3.50}$$

1.3.3.5 Determination of power-related terms

The scattering parameters are best introduced if we consider calculating power terms, such as power gain in its various forms. The demonstration of their usefulness is developed by first introducing the most significant power gain terms quantified in terms of external and internal Y parameters, taking into account different properties of the amplifier, and then showing an equivalent formalization in terms of scattering parameters.

Gain terms and Y parameters The most straightforward definition is the ratio of the power delivered to the load and the power input to the amplifier:

$$G_P = \frac{P_L}{P_{in}} = f(Y_L, y_{ij}) \tag{1.3.51}$$

This gain depends on both the load admittance and the two-port admittance terms, but not on the source admittance. What happens if there is source mismatch and the power transferred to the amplifier is only a fraction (maybe small) of the available source power? This cannot be detected by inspecting the G_P gain. Rather, the *transducer gain* term

$$G_T = \frac{P_L}{P_{av,S}} = f(Y_L, Y_S, y_{ij}) \tag{1.3.52}$$

measures the power delivered to the load divided by the power available at the source, including the source-input match.

If the interest is for the maximum performance of the amplifier irrespective of the load matching conditions, then the *available power gain* term is used:

$$G_A = \frac{P_{av,L}}{P_{av,S}} = f(Y_S, y_{ij}) \tag{1.3.53}$$

The three terms may be calculated as a function of the circuit parameters at terminals, and of course as a function of the internal two-port parameters:

$$P_{in} = \frac{|V_1|^2}{2} Re(Y_{in}) \qquad P_L = \frac{|V_2|^2}{2} \mathrm{Re}(Y_L) \tag{1.3.54}$$

Gain terms and S parameters If we consider S parameters (see Chapter 10), the above expressions may be rewritten and simplified, while limiting our representation to the terminal variables and parameters of the circuit, that is what is normally available for measurements and experimental characterization. This part is treated extensively in Gonzalez ([137], chap. 3) and in the Hewlett-Packard Application Note 95 [156], that represent probably the best references for amplifier design principles in microwave applications.

2

Transmission Lines

This subject is covered in textbooks, but the approaches are not homogeneous as some parts might not receive the due attention or be considered at all. Telegrapher's equations are the starting point, with the intuitive approach that may be found in [82, 332] or the more formal one for example appearing in [264]. The continuation of this chapter is Chapter 8 with multi-port transmission lines and then Chapter 10, where scattering parameters are discussed and applied, including multi-port and mixed-mode representations. Transmission lines may be in general modeled and analyzed in two ways:

- a circuit approach with the cascade connection of infinitesimal cells, including resistance, inductance, capacitance and conductance terms, each describing the physical properties of the transmission line components, i.e. conductors and dielectric;
- a waveguide approach, where coupled voltage and current waves propagate along the transmission line, governed by the same rules of the electromagnetic field propagation in a guided medium.

The second approach is more elegant, while the first one is more intuitive. Telegrapher's differential equations that solve the propagation problem are here derived using the first approach, going from the discrete to the continuous representation, letting the size of each cell go to zero.

2.1 Telegrapher's equations

Let's consider an infinitesimal section of line of length $\mathrm{d}z$ at the longitudinal position z. Starting from the p.u.l. parameters r, l, c and g, assumed constant over the entire transmission line, we may say that this section has a series resistance $r\,\mathrm{d}z$, series inductance $l\,\mathrm{d}z$, shunt capacitance $c\,\mathrm{d}z$ and shunt conductance $g\,\mathrm{d}z$.

For the most general approach, transmission line equations must be in time domain, to account for transients and arbitrary waveforms. Using instantaneous quantities for voltage and current at the left and right ports of the section, Kirchhoff equations give

$$v(z+\mathrm{d}z,t)-v(z,t)=-r\,\mathrm{d}z\,i(z,t)-l\,\mathrm{d}z\,\frac{\partial i(z,t)}{\partial t} \tag{2.1.1}$$

$$i(z+\mathrm{d}z,t)-i(z,t)=-g\,\mathrm{d}z\,v(z,t)-c\,\mathrm{d}z\,\frac{\partial v(z,t)}{\partial t} \tag{2.1.2}$$

Since the infinitesimal section has the shunt elements on the right, one may argue that the voltage and its derivative that produce the current in the second equation is in reality $v(z+\mathrm{d}z,\,t)$. The objection is, first, that we couldn't do otherwise to derive a symmetric set of equations, and, second, that the term $v(z+\mathrm{d}z,\,t)$ may be developed in Taylor's series and all the higher order terms would be anyway negligible.

Dividing by $\mathrm{d}z$ and letting $\mathrm{d}z$ approach zero gives the well known set of linear partial differential equations

$$\frac{\partial v(z,t)}{\partial t}=-r\,i(z,t)-l\,\frac{\partial i(z,t)}{\partial t} \tag{2.1.3}$$

$$\frac{\partial i(z,t)}{\partial t}=-g\,v(z,t)-c\,\frac{\partial v(z,t)}{\partial t} \tag{2.1.4}$$

whose solution will explicit v and i in terms of z and t, once the boundary conditions are known, that is the characteristics of source at $z=0$ and load at $z=L$.

If the line section is analyzed in the frequency domain, then phasors $V(z)$ and $I(z)$ take the place of the instantaneous quantities $v(z,t)$ and $i(z,t)$, and similarly

$$\frac{\mathrm{d}V(z)}{\mathrm{d}z}=-(r+j\omega l)\,I(z) \qquad \frac{\mathrm{d}I(z)}{\mathrm{d}z}=-(g+j\omega c)\,V(z) \tag{2.1.5}$$

The solution of equations ((2.1.5)) leads to second order differential equations that admit a time-harmonic integral of the form

$$V(z)=V_1\exp(-\gamma z)+V_2\exp(+\gamma z) \tag{2.1.6}$$

$$I(z)=I_1\exp(-\gamma z)+I_2\exp(+\gamma z) \tag{2.1.7}$$

The solution indicates that there exist two voltage waves (and two current waves) traveling along the transmission line, the leftmost one in the direction of increasing z and the rightmost one in that of decreasing z: they are usually named *progressive wave* and *regressive wave*, or *forward wave* and *backward wave*.

2.1.1 Propagation, attenuation and phase constants

The quantity γ, named *propagation constant*, is derived from the p.u.l. parameters as

$$\gamma = \sqrt{(r + j\omega l)(g + j\omega c)} \tag{2.1.8}$$

and is in general a complex number. Its real and imaginary parts

$$\gamma = \alpha + j\beta \tag{2.1.9}$$

are named *attenuation constant* and *phase constant* respectively (or better, especially for α, attenuation coefficient and phase propagation coefficient, since the former is in reality far from being constant because of skin effect and other power loss effects), and their function is clarified if eq. ((2.1.9)) is substituted into eq. ((2.1.6)):

$$V(z) = V_1 \exp(-\alpha z)\exp(-j\beta z) + V_2 \exp(+\alpha z)\exp(+j\beta z) \tag{2.1.10}$$

Going to the instantaneous voltage by application of the $\mathrm{Re}\{\cdot\}$ operator

$$\begin{aligned} v(z,t) &= |\mathrm{V}_1| \exp(-\alpha z)\,\mathrm{Re}\{\exp(j(\omega t - \beta z + \varphi_1))\} + \\ &\quad |\mathrm{V}_2| \exp(+\alpha z)\,\mathrm{Re}\{\exp(j(\omega t + \beta z + \varphi_2))\} \end{aligned} \tag{2.1.11}$$

the amplitude (given by the peak values $|\mathrm{V}_1|$ and $|\mathrm{V}_1|$ multiplied by the attenuation terms $\exp(-\alpha z)$ and $\exp(+\alpha z)$) and the orientation (given by the combination of angular pulsation ωt, initial phase φ_1 and φ_2, and phase rotation $\pm\beta z$) are separated.

The explicit expression of α and β with respect to line per-unit-length parameters is not simple and is not given in textbooks. A low frequency approximation exists

$$\alpha = \frac{1}{2}\sqrt{lc}\left(\frac{r}{l} + \frac{g}{c}\right) = \frac{1}{2}\left(rY_c + gZ_c\right) \tag{2.1.12}$$

$$\beta = \omega\sqrt{lc} \tag{2.1.13}$$

From the expression above for α derives the usual distinction of conduction loss (the first term that depends on the conductor series resistance r) and dielectric loss (the second term that contains the shunt conductance through the dielectric g).

While losses have already been introduced using α, for β a more complete form exists that includes terms up to the second order in ω and that is valid for almost all cases:

$$\beta = \omega\sqrt{lc}\left[1+\frac{r^2}{8\omega^2 l^2}-\frac{rg}{4\omega^2 lc}+\frac{g^2}{8\omega^2 C^2}\right] = \omega\sqrt{lc}\left[1+\frac{1}{2}\left(\frac{r}{2\omega l}-\frac{g}{2\omega c}\right)^2\right] \tag{2.1.14}$$

In this expression the terms r and g appear and make β slightly dependent on losses: by grouping and putting in evidence line losses, a compact expression is obtained where the term $\beta^0 = \omega\sqrt{lc}$ stand for the value valid for a lossless line,

$$\beta = \beta^0\left[1+\frac{1}{2}\left(\frac{\alpha_c}{\beta^0}-\frac{\alpha_d}{\beta^0}\right)^2\right] \tag{2.1.15}$$

and analogously the wavelength

$$\lambda = \lambda^0\left[1-\frac{1}{2}\left(\frac{\alpha_c}{\lambda^0}-\frac{\alpha_d}{\lambda^0}\right)^2\right] \tag{2.1.16}$$

The line is slightly "slower" in general if losses are considered, but there exists one case in which losses have no impact on the propagation, that is when

$$\frac{r}{\omega l} = \frac{g}{\omega c} \tag{2.1.17}$$

also known as Heaviside's relationship.

The propagation constant may also be expressed as amplitude $|\gamma|$ and phase $\angle\gamma$, that have an obvious relationship with α and β, and may also be related back to the per-unit-length parameters of the line:

$$|\gamma| = \sqrt{\sqrt{(r^2+\omega^2 l^2)}+\sqrt{(g^2+\omega^2 c^2)}} \tag{2.1.18}$$

$$\angle\gamma = \frac{1}{2}\left[\arctan\left(\frac{\omega l}{r}\right)+\arctan\left(\frac{\omega c}{g}\right)\right] \tag{2.1.19}$$

The measuring unit of α and β is 1/m, since both appears multiplied by z in the exponential argument, that must be dimensionless. Yet, some distinction is needed. It was felt to be convenient to find a unit for the dimensionless argument αz, that expresses the attenuation, and this unit for α is the neper/m: so, a voltage or current is said to undergo an attenuation of X nepers[1] when its magnitude changes by a factor $\exp(-X)$. The phase constant β analogously is measured in radians/m; by definition of wavelength λ, as the length in space traveled by a sinusoidal wave after a phase rotation of 2π, the known relationship $\beta = 2\pi/\lambda$ derives.

Both nepers and dB are used to express attenuation with respect to two voltages (or currents) V_a and V_b taken at two positions z_a and z_b. By definition of neper the atten-

[1] The name "neper" derives from the latinization of Napier, the Scottish mathematician who invented the logarithms.

uation between V_a and V_b is X neper, if $V_a/V_b = \exp(-X)$; then, by definition of dB $A = 20\log(V_a/V_b) = 20\log(\exp(-X)) = 8.686X$.

2.1.2 Line velocity and propagation time

Let's consider a "point" at the longitudinal position z_0 and time t_0 and the voltage of the first forward wave; at a slightly later instant of time $t_0 + \mathrm{d}t$ the point has moved from z_0 to $z_0 + \mathrm{d}z$ and the voltage value in z_0 is no longer the same. By neglecting attenuation of amplitude, and thus equating the angular positions to have (z_0, t_0) and $(z_0 + \mathrm{d}z, t_0 + \mathrm{d}t)$ points correspond with a simple horizontal shift, we may write $\omega t_0 - \beta z_0 + \varphi_1 = \omega(t_0 + \mathrm{d}t) - \beta(z_0 + \mathrm{d}z) + \varphi_1$, that leads to $\omega dt - \beta dz = 0$: the velocity, represented by the ratio $\mathrm{d}z/\mathrm{d}t$, with $\mathrm{d}z$ and $\mathrm{d}t$ approaching zero, is called *phase velocity*.

$$v_{ph} = \frac{\omega}{\beta} \tag{2.1.20}$$

The same result is obtained for the backward wave, except for the sign of the direction of propagation. So, the wave looks still to an observer moving at its velocity; if the observer is fixed in space observing a longitudinal position, the time variation of the instantaneous voltage is sinusoidal.

There is another type of velocity, named *group velocity*, that refers to several frequency components (belonging to a radio transmission, such as the modulated signal transmitted over a physical radio channel) that have not the same phase velocity. This case is however not of interest here (see later sec. 2.7.3.3 and 3.3.1) and for non-dispersive lines the two velocity quantities are equal; we will drop the subscript "ph." from the phase velocity whenever we can.

For a transmission line of length L the time for the wave to travel is called propagation delay, propagation time, or traveling time, and is given by length divided by velocity:

$$\tau = L/v_{ph} \tag{2.1.21}$$

2.2 The Characteristic Impedance

There is a beautiful demonstration of what the characteristic impedance is in [199], p. 142. Starting from the definition of the transmission line delay (or propagation delay, given above in eq. (2.1.21)), they imagine to charge the line capacitance with a step voltage of amplitude V and evaluate the amount of current flowing in the line. The step voltage is applied and propagates along the line; let's consider 1 m of line, so that the capacitance C is exactly the p.u.l. line capacitance c, the charge across it equals the capacitance C multiplied by the voltage difference, that is equal to V (since the step hasn't "arrived" yet ahead of 1 m). The time during which that line section is charged is the propagation delay multiplied by the length, in our case 1 m; the total charge divided by that time gives the current

$$I = \frac{cV}{\sqrt{lc}} = V/\sqrt{\frac{l}{c}} \tag{2.2.1}$$

and the last term is right the characteristic impedance!

Thus, the characteristic impedance Z_c is ideally calculated from a uniform line with no reflected waves ($V_2 = 0$ and $I_2 = 0$), by taking the ratio of the voltage and current waves, and it may be demonstrated that this ratio is constant at any point along the transmission line.

By taking eq. ((2.1.10)) with only the forward wave for simplicity and by differentiating with respect to z

$$\frac{\mathrm{d}V}{\mathrm{d}z} = -(\alpha + j\beta)V_1 \exp(-\alpha z)\exp(-j\beta z) \tag{2.2.2}$$

The direct comparison with eq. ((2.1.5)) gives

$$Z_c = \frac{V}{I} = \frac{V_1}{I_1} = \frac{r + j\omega l}{\alpha + j\beta} = \sqrt{\frac{r + j\omega l}{g + j\omega c}} = \sqrt{\frac{Z_{\text{series}}}{Y_{\text{shunt}}}} \tag{2.2.3}$$

where the last part has been added to recall quite a frequent way of expressing Z_c, as the square root of the ratio of the series impedance and the shunt admittance terms, and that will be used soon in sec. 2.4.

The quantity Z_c has the dimension of an impedance and it is a characteristic of the line behavior (neglecting reflections), so that it was named *characteristic impedance*. Its dependency on frequency is twofold: direct, since the numerator and denominator show the reactive terms multiplied by $j\omega$, and indirect, since the p.u.l. parameters may depend on frequency. If the resistance r and the conductance g are considered negligible (as it is, for example, for high quality coaxial cables and for the lowest frequencies), then the characteristic impedance becomes a pure resistance $Z_c = \sqrt{l/c}$. At low frequency, when the resistive part at the numerator is relevant if compared with the inductive reactance term on the right, we say that the line operates with a RC behavior; at higher frequency the characteristic impedance becomes constant. The "critical" frequency that distinguishes these two intervals is given by $f_c = r/2\pi l$.

There are practical examples of lines used below the critical frequency f_c in their RC region, such as long lines for relatively low data-rate transmissions, including telephone lines, serial lines used in automation, cable connecting parts and equipment, etc.. In this case the attention is often on the capacitive loading on the driving generator: it may be worth recalling the known remedy of putting a series resistor when driving a coax to avoid both excessive capacitive load and instability of the operational amplifier used as the driving source and going beyond the operational amplifier output current capability, affecting slew rate and distorting wave-fronts.

If one is interested in keeping line losses low, of course, the longitudinal resistance r shall be low (including the effect of the skin effect) and the conductance term g shall be low (that is high quality dielectrics shall be used, like for transmission lines in air).

For the vast majority of cables the attenuation will follow a curve that is dictated by the skin effect, so with a square-root dependency on frequency.

At low frequency the characteristic impedance is thus not real valued, as we may expect, but has a non negligible imaginary component, that depends on the loss parameters, resistance and conductance. This occurs for lines featuring small and large cross-sections and different type of materials[2]. The large value of the characteristic impedance at low frequency due to the preponderant imaginary part is of no consequence, being no exigency of impedance matching at low frequency. The characteristic impedance has an asymptotic behavior with the imaginary part vanishing above the critical frequency f_c. This is the macroscopic variation of the characteristic impedance.

Additionally, there is frequency dependency also for the inductance (due to skin effect and to the variation of internal inductance[3]) and capacitance (slight reduction of the dielectric permittivity and increase of dielectric losses, see sec. 6.1.2). The reduction of the inductance with frequency makes the characteristic impedance slightly increase; for high permeability materials magnetic losses also increase with frequency and this adds an imaginary component to the already stabilized characteristic impedance; analogously and more commonly the same effect occurs for the increase of dielectric losses, that never ceases nor saturates with frequency.

It is finally commented that the characteristic impedance is a somewhat intangible entity, in that it cannot be measured directly with a RCL bridge and its resistive part is not dissipative. It may be determined indirectly by some impedance measurements (at least two), as it will be seen shortly (see sec. 2.4).

The line velocity v_0 (that corresponds to the *phase velocity* $v_{ph.}$) also can be expressed in p.u.l. parameters, and, again under the assumption of negligible r and g parameters, the known formula is derived:

$$v_{ph.} = 1/\sqrt{lc} \tag{2.2.4}$$

It is quite natural to ask why $50\,\Omega$ was chosen as the reference impedance for RF applications. The answer is actually quite simple, and it derives from the characteristics of air-filled coaxial cables at that time. The geometry of coaxial cables is such that an air-filled cable designed for maximum power handling (i.e., designed optimizing the curvature and distance of conductors such that the breakdown voltage of the air is highest) had a characteristic impedance of $30\,\Omega$. An air-filled coaxial cable designed for maximum power efficiency (least loss) had an impedance of $77\,\Omega$. Therefore $50\,\Omega$ was a good compromise between the two optimal values for air-filled lines. Furthermore, for typical solid dielectrics such as Teflon, the impedance for least loss is about $50\,\Omega$; it then follows that typical $50\,\Omega$ cables are generally optimized for least loss. A

[2] The variability of the characteristic impedance was investigated for traction lines in A. Mariscotti and P. Pozzobon, "Synthesis of line impedance expressions for railway traction systems", *IEEE Transactions on Vehicular Technology*, vol. 52 n. 2, March 2003, pp. 420-430. A traction line has several non-idealities, related to the heavy skin effect in the running rails and shunt conductance and capacitance influenced by the rails supports.

[3] See A. Mariscotti and P. Pozzobon, "Resistance and Internal Inductance of traction rails: a survey", *IEEE Transactions on Vehicular Technology*, vol. 53 n. 4, July 2004, pp. 1069-1075, for a discussion of skin effect and internal parameters for heavy ferromagnetic conductors, i.e. running rails.

further reason for choosing $50\,\Omega$ is that an ideal monopole antenna[4] goes down to $37.5\,\Omega$ of input impedance at resonance and is slightly larger around it for not too a high Q factor, making such coaxial cables a good match.

2.3 Input impedance and reflection coefficient

The *input impedance* of a transmission line is the ratio of voltage and current at the input port, that is $Z_{in} = V_1/I_1$, so that for a non reflective transmission line the input impedance and the characteristic impedance coincide. Since it was said that the ratio V/I is constant over the entire line, at the far end $z = L$, the same relationship holds for the voltage and current phasors across the terminating load Z_t, so that $Z_t = Z_0$ becomes the condition for a non-reflecting transmission line termination (this condition is often referred to as *matched load condition*).

In general, the input impedance and the matching conditions at either end (source and/or load side) may be evaluated in the following way.

By differentiation of eq. ((2.1.10)) with respect to z, and writing the equality with eq. ((2.1.5)), the following expression for the current I is derived:

$$I(z) = \frac{\gamma}{r + j\omega l} V_1 \exp(-\gamma z) - \frac{\gamma}{r + j\omega l} V_2 \exp(+\gamma z) \tag{2.3.1}$$

where the term $\gamma/(r + j\omega l)$ is equal to $1/Z_0$. The phasor current coefficients for the forward and backward waves are thus related to the corresponding coefficients of the phasor voltage by

$$I_1 = \frac{V_1}{Z_0} \qquad I_2 = -\frac{V_2}{Z_0} \tag{2.3.2}$$

In the general case of possibly mismatched source and load impedances, the impedance at any point along the line is given by

$$Z(z) = Z_0 \frac{V_1 \exp(-\gamma z) + V_2 \exp(+\gamma z)}{V_1 \exp(-\gamma z) - V_2 \exp(+\gamma z)} \tag{2.3.3}$$

Let's assume now that at the end of the line a terminating impedance Z_t is connected; when the traveling wave arrives at $z = L$ where the terminating impedance is connected, $V(L)$ and $I(L)$ shall be such that their ratio is equal to Z_t and eq. (2.3.3) shall be satisfied for all positions and in particular for $z = L$. The same holds for the wave traveling in the opposite direction that meets the terminating impedance at $z = 0$: in this case calculations are much easier because the exponents are all equal to one and we adopt this configuration to conclude the reasoning.

---

[4] Monopole antennas are basic robust antennas used in many applications, especially for omni-directional radio transmission, with various types of ground plane: solid, radial stubs or the ground itself.

$$Z_t = \frac{V(0)}{I(0)} = Z_0 \frac{V_1 + V_2}{V_1 - V_2} \tag{2.3.4}$$

By solving this equation for V_1 and V_2 it is possible to obtain a similar relationship that gives the ratio of the forward and backward waves as a function of Z_t and Z_0:

$$\frac{V_2}{V_1} = \frac{Z_t - Z_0}{Z_t + Z_0} \tag{2.3.5}$$

This quickly demonstrates where the reflection comes from: a traveling wave, called *incident wave*, with its amplitude set by the characteristic impedance of the line approaching a longitudinal position where a terminating impedance is connected, give rise to a forward and a backward wave, whose amplitude is determined by the expression above. Now the attention is focused on the reflection coefficient and its relationship with the characteristic impedance and the terminating impedance is derived.

The *reflection coefficient*, either indicated by ρ or Γ, is defined as the ratio of the values of the reflected wave, going in the direction of decreasing z and associated to the term $\exp(+\gamma z)$, and of the incident wave, going in the direction of increasing z and associated to the term $\exp(-\gamma z)$. At the load port $z = L$, the reflection coefficient Γ_t is

$$\Gamma_t = (V_2/V_1)\exp(2\gamma L) \tag{2.3.6}$$

and is complex. Taking eq. ((2.3.3)) at $z = L$ and dividing all the terms by the same quantity $(V_1 \exp(-\gamma z))$, it is possible to isolate the reflection coefficient at the load as

$$\frac{Z_t}{Z_0} = \frac{1 + \Gamma_t}{1 - \Gamma_t} \tag{2.3.7}$$

that solved for Γ_t gives

$$\Gamma_t = \frac{Z_t - Z_0}{Z_t + Z_0} \tag{2.3.8}$$

For some extreme terminating conditions physical reasoning also helps in finding the value of the reflection coefficient. If $Z_t = 0$, that is a short circuit, then the incident and reflected voltage wave must be identically equal at $z = L$, since their sum must be zero; the reflection coefficient is thus $\Gamma_t = -1$. On the contrary, if the line is terminated on an open circuit, $Z_t = \infty$, the same reasoning should be applied to the phasor current waves; the equation above confirms that the reflection coefficient is $\Gamma_t = +1$. If Z_t is a generic reactive impedance the modulus of the reflection coefficient is still unity, but the phase is variable; this is reasonable observing that a reactive impedance is not dissipative, cannot absorb power from the incident wave and is thus totally reflecting.

Correspondingly, it is sometimes defined a *transmission coefficient* T_t that complements the reflection coefficient: $T_t = 1 + \Gamma_t$. The transmission coefficient gives the amount of the incident wave transmitted beyond an interface (e.g. between two line sections). Just to summarize, the reflection coefficient is always calculated as difference divided by sum of the two characteristic impedances, taking as first the one for the line section ahead where the wave is directed; the same applies if one of the line sections is replaced by a lumped circuit. Reversing the direction changes the sign of the reflection coefficient. Following the example above with a terminating Z_t impedance

$$T_t = 1 + \Gamma_t = \frac{2Z_t}{Z_t + Z_0} \tag{2.3.9}$$

Dividing eq. ((2.3.3)) by V_1 and substituting the ratio V_2/V_1 rewriting eq. ((2.3.6)), the line impedance $Z(z)$ is

$$Z(z) = Z_0 \frac{\exp(-\gamma z) + \Gamma_t \exp(-2\gamma L)\exp(+\gamma z)}{\exp(-\gamma z) - \Gamma_t \exp(-2\gamma L)\exp(+\gamma z)} \tag{2.3.10}$$

Multiplying all the terms at numerator and denominator by $\exp(\gamma L)$, the expression is put in a form where the characteristic impedance, the reflection coefficient at the terminating load and the distance $x = L - z$ between the point z and the load at $z = L$ are the sole parameters used.

$$Z(x) = Z_0 \frac{\frac{Z_t}{Z_0}[\exp(+\gamma x) + \exp(-\gamma x)] + [\exp(+\gamma x) - \exp(-\gamma x)]}{[\exp(+\gamma x) + \exp(-\gamma x)] + \frac{Z_t}{Z_0}[\exp(+\gamma x) - \exp(-\gamma x)]} \tag{2.3.11}$$

The input impedance at $z = 0$, that is $x = L$, may be readily obtained, where the complex exponential is replaced by the hyperbolic tangent:

$$Z_{in} = Z_0 \frac{\frac{Z_t}{Z_0} + \tanh(\gamma L)}{1 + \frac{Z_t}{Z_0}\tanh(\gamma L)} = Z_0 \frac{Z_t + Z_0 \tanh(\gamma L)}{Z_0 + Z_t \tanh(\gamma L)} \tag{2.3.12}$$

When the line is terminated with two extreme cases of short circuit ($\Gamma_t = -1$) and open circuit ($\Gamma_t = +1$), the input impedance expressions are as follows:

$$Z_{in,sc} = jZ_0 \tan(\gamma L) \tag{2.3.13}$$

$$Z_{in,oc} = -jZ_0 \cot(\gamma L) \tag{2.3.14}$$

When an integer multiple n of half the wavelength $\lambda/2$ fits line length $L = n\lambda/2$, then the terminating impedance Z_t is seen directly at the line input port, as if there is no line in between, because the term $\tanh(\cdot) = 0$ and vanishes.

When an odd integer multiple $(2n+1)$ of a quarter wavelength $\lambda/4$ fits line length $L = (2n+1)\lambda/4$, then it is said that the line is resonating in quarter-wave conditions and the input impedance becomes

$$Z_{in(\lambda/4)} = \frac{Z_0^2}{Z_t} \tag{2.3.15}$$

The line behaves like a transformer (quarter-wave transformer), so that a large (e.g. larger than Z_0) Z_t value at the far end is transformed into a small impedance value, proportionally smaller than Z_0. This of course works for a narrow frequency interval around the $\lambda/4$ condition. Quarter-wave matching is achieved when the line characteristic impedance is equal to geometric mean (square root of product) of the terminating impedance Z_t at the far end and the desired impedance value at the input Z_{in}^* (e.g. equal to that of the connected source, or to the complex conjugate of it, depending on the desired match).

It is now possible to consider the problem of the determination of the line characteristics on the basis of impedance measurements.

2.4 Line parameters from measured input impedance

Under the two extreme conditions of short circuit and open-circuit termination, the measured input impedance is expressed as

$$Z_{in,sc} = Z_c \tanh(\gamma L) \qquad Z_{in,oc} = \frac{Z_c}{\tanh(\gamma L)} \tag{2.4.1}$$

and the multiplication of the two expressions gives

$$Z_c = \sqrt{Z_{in,sc}\, Z_{in,oc}} \tag{2.4.2}$$

This was already evident in eq. (2.2.3), observing that a measurement in short-circuit conditions rules out the shunt terms leaving only those of the series impedance, and conversely an open-circuit measurement reduces to nearly zero the flowing current and thus the voltage drop on the series terms, leaving only the shunt admittance terms.

This method for estimating the characteristic impedance is really adopted in practice, especially for low and medium frequency applications, when the measurements of the short-circuit and open-circuit input impedance values may be performed at a single or few frequency values using ordinary instrumentation, such as oscilloscopes and current probes.

Besides reasoning on the accuracy of the measurement method and the propagation of uncertainty, it is evident that both $Z_{in,sc}$ and $Z_{in,oc}$ must be in the scale of the used instrument, so neither too small nor too large; we may state that the appropriate length L is close to an odd number of eighths of the test wavelength, or, put in another more practical way, that the measurements are to be taken at somewhat different frequency values to avoid singularities, based on the assumption that the line p.u.l. parameters are not varying appreciably for changes of the test frequency within less than few %.

For the uncertainty $u(Z_{in,sc})$ and $u(Z_{in,oc})$ in the measurement of $Z_{in,sc}$ and $Z_{in,oc}$, with independently distributed errors, the estimated Z_c is characterized by

$$u(Z_c)=\sqrt{\frac{1}{2}\left[u(Z_{in,sc})^2+u(Z_{in,oc})^2\right]} \tag{2.4.3}$$

Also the attenuation factor α and the phase factor β may be estimated from these measurements. Taking the square root of the ratio of $Z_{in,sc}$ and $Z_{in,oc}$ expressions (obtained by setting $Z_t=0$ and $Z_t=\infty$ in eq. (2.3.12) above, respectively) and replacing it with the $\tanh(\gamma L)$ term

$$\sqrt{Z_{in,sc}/Z_{in,oc}}=\tanh(\gamma L)=\frac{1-\exp(-2\gamma L)}{1+\exp(-2\gamma L)} \tag{2.4.4}$$

it is possible to isolate the $\exp(-2\gamma L)$ term; changing the sign in the exponent and correspondingly reversing the fraction gives:

$$\exp(2\gamma L)=\frac{1+\sqrt{Z_{in,sc}/Z_{in,oc}}}{1-\sqrt{Z_{in,sc}/Z_{in,oc}}} \tag{2.4.5}$$

Then, taking the log and separating the real and imaginary part of γ,

$$\alpha=\frac{1}{2L}\ln\left|\frac{1+\sqrt{Z_{in,sc}/Z_{in,oc}}}{1-\sqrt{Z_{in,sc}/Z_{in,oc}}}\right| \tag{2.4.6}$$

$$\beta=\frac{1}{2L}\left\{\arg\left[\frac{1+\sqrt{Z_{in,sc}/Z_{in,oc}}}{1-\sqrt{Z_{in,sc}/Z_{in,oc}}}\right]+2n\pi\right\} \tag{2.4.7}$$

where the $2n\pi$, n integer, impedes to determine β uniquely, but rather a series of values that differ by π/L.

When the phase factor β has been determined, also the phase velocity is known by applying eq. (2.1.20).

It appears that for the correct determination of α, a piece of line is needed, with a length that causes a loss of some dB as a minimum; on the contrary, the unambiguous determination of β requires a piece of line so short that the discrimination of π/L is possible (put in another way, since an estimate of β may be independently determined by means of the simple relationship $\beta=2\pi/\lambda$, the value of λ in the transmission line needs to be known with enough accuracy to discriminate which value of n above is the most appropriate).

From the observation of line resonances another simplified method for the determination of the characteristic impedance may be derived [169], sec. A1.1 (the standard indicates it as a mean characteristic impedance and stems from an assumed constant permittivity, and thus capacitance): it needs the determination of the frequency difference Δf between two resonance minima or maxima and the value of capacitance C of the line sample under test (e.g. cable sample). The characteristic impedance is thus

$$Z_c = \frac{1}{2\Delta f\, C} \tag{2.4.8}$$

For an accurate estimation of the frequency difference the average of several oscillating half periods may be used, as done in Example 10.3; for too a long cable sample the resonance frequency will be too low, creating problems of frequency resolution to the measuring equipment (e.g. frequency range, IF filter bandwidth and number of points).

2.5 Propagation and reflection

The behavior of reflection terms is now considered when they travel the line back and forth, reflected from the two ends, and overlap each other with a variable result depending on the phase (or time) relationship and the nature of the discontinuity at each end. Two time-domain signals are considered that are typical of the study of transmission lines and of the time-domain reflectometry (see TDR and TDT in sec. 11.10 and, in particular, 11.10.5): the step and impulse signals are quite handy in the Laplace domain, may be generated with enough accuracy and excite transmission line response in a way to suitably investigate discontinuities and related reflections.

2.5.1 Propagation and reflection of a step signal

Reflections are considered in a time-domain perspective, rather than just quantifying the amount of sinusoidal signal that is transmitted or reflected at some interface (e.g. between the source and the line, or between the load and the line). During propagation, partially reflected waves overlap traveling back and forth along the line and all form the signal transmitted from the source to the load. Some experiments may be found in sec. 11.10.5.

The propagation along the line may be described in steps:

1. the signal $E(t)$, e.g. a step voltage, is applied at the source port (port 1), and depending on the mutual relationship between the source impedance and the line characteristic impedance, a fraction of it, $V_0(t)$, is applied to line input;

2. the signal propagates as $V_0^+(t)$ along the line and after 1τ appears, attenuated, at the load port (port 2), where a fraction of it is passed through to the load, becoming $V_1^L(t)$, and a complementary fraction, $V_1^-(t)$, is reflected back to the source;

3. after another 1τ time interval, at 2τ from the initial time instant, the attenuated copy of $V_1^-(t)$ appears at port 1, where again it is partially reflected back into the line as $V_2^+(t)$ or passed back to the source as $V_2^S(t)$, and a change of sign will be seen; the observed change depends on the phase relationship between the original applied step and the reflected copy getting back from the line;

4. the term $V_2^+(t)$ propagates again along the line in the load direction, undergoing further attenuation and arriving at the load after another 1τ time interval, that is 3τ since the beginning; the same described at step 2 repeats with $V_2^+(t)$ term, reaching port 2 and partially transferring to the load as $V_3^L(t)$ or being reflected back into the line as $V_3^-(t)$; ... and so on.

This propagation/reflection mechanism applies successive reflected copies every 2τ at the two ports, at the source and at the load, with the latter delayed by 1τ, that is physically the time required for the first version of the applied signal to go from the source to the load (propagation time). The steps described above are graphically organized in the "reflection diagram", also called "bounce diagram" or "lattice diagram", shown in Figure 2.5.1.

To support the diagram with numeric values that help the reader in following the reflected terms, we focus on a specific problem where a 5 V step voltage is applied to a $50\,\Omega$ line (e.g. a coax cable) that is terminated at the other end on a mismatched load, with $Z_L = 100\,\Omega$. The coax line is assumed to have a 0.66 velocity factor k (0.66 times the light velocity), so that $v = 198\,\mathrm{m/\mu s}$, or in other words $50.5\,\mathrm{ps/cm} = 5.05\,\mathrm{ns/m}$ (rounded to $5\,\mathrm{ns}$ for simplicity).

Due to variable phase relationships of the reflected copies of the applied signal, while they are traveling back and forth along the line, the instantaneous signal may differ from the original copy: it will reach a steady state and a stable wave-shape when the reflection terms extinguish after having traveled many times the line in both directions (the number of times depending on the line attenuation and the amount of signal reflected at the source and load interfaces). This is the first source of distortion of a signal transmitted along a line, whose integrity is affected by the reflections of its own copies combining during the propagation towards the load.

It is recalled that one trip along the line takes the propagation delay time τ, that is in principle a constant, so that all the components of the signal are delayed by the same amount τ; in real cases, being the phase velocity slightly variable with frequency, the propagation delay varies correspondingly of the same amount. This is a second-order source of distortion, particularly relevant with fast steep signals, whose spectrum occupies an extended frequency range.

Now it is possible to write down the equations for the model above, to be used as a quantitative reference for measurements of line reflections and signal distortion.

The voltage at the source port (port 1, longitudinal position $x = 0$) at $t = 0\,\tau$ is:

$$V_0(0) = E\,\frac{Z_c}{Z_S + Z_c} \tag{2.5.1}$$

having indicated with the capital letter V the phasor of the voltage or its Laplace transform.

Arriving at the other end (the load port, port 2, at the position $x = L$) there is partial reflection and transmission, so that:

$$V_1^-(L) = V_0(0)\,e^{-\alpha L}\Gamma_L \tag{2.5.2}$$

$$V_1^+(L) = V_0(0)\, e^{-\alpha L}(1+\Gamma_L) \tag{2.5.3}$$

with

$$\Gamma_L = \frac{Z_L - Z_c}{Z_L + Z_c} \qquad \Gamma_S = \frac{Z_S - Z_c}{Z_S + Z_c} \tag{2.5.4}$$

The reflected wave $V_1^-(L)$ travels back, undergoing further attenuation and arrives at the source port as $V_2^-(0)$, where it is reflected back into the line as a new progressive wave $V_2^+(0)$:

$$V_2^-(0) = V_1^-(L)e^{-\alpha L} = V_0(0)\, \left[e^{-2\alpha L}\Gamma_L\right] \tag{2.5.5}$$

$$V_2^+(0) = V_1^-(L)\, e^{-\alpha L}\Gamma_S = V_0(0)\, \left[e^{-2\alpha L}\Gamma_L \Gamma_S\right] \tag{2.5.6}$$

The voltage term transmitted back to the source at the time instant 2τ sums together with the pre-existing source voltage and the total voltage at the source port becomes

$$V_2(0) = V_0(0) + V_2^-(0) + V_2^+(0) = E\, \frac{Z_c}{Z_S + Z_c}\, \left[1 + e^{-2\alpha L}\Gamma_L(1+\Gamma_S)\right] \tag{2.5.7}$$

It is worth noting that superposing at each port the incident and the reflected term gives the same result — as expected — of applying directly the known formula for the transmission coefficient, in this case $(1+\Gamma_S)$.

The reflected term $V_2^+(0)$ arrives at the load port at 3τ, after undergoing the attenuation of one line trip:

$$V_3^+(L) = V_2^+(0)\, e^{-\alpha L} \tag{2.5.8}$$

$$V_3^-(L) = V_2^+(0)\, e^{-\alpha L}\Gamma_L \tag{2.5.9}$$

At the load port the voltage is given by summing the pre-existing voltage and the two terms above:

$$\begin{aligned} V_3(L) &= V_1^+(L) + V_3^-(L) + V_3^+(L) = \\ &= V_0(0)\, e^{-\alpha L}(1+\Gamma_L) + V_0(0)\, \left[e^{-2\alpha L}\Gamma_L\Gamma_S\right]\, e^{-\alpha L}(1+\Gamma_L) = \\ &= E\, \frac{Z_c}{Z_S + Z_c}\, e^{-\alpha L}(1+\Gamma_L)\, \left[1 + e^{-2\alpha L}\Gamma_L\Gamma_S\right] \end{aligned} \tag{2.5.10}$$

Repeating the mechanism for one complete trip more, so reflecting back to the source arriving there at 4τ and then reflecting back to the load, we obtain the following expressions for the net voltages at the source and the load at 4τ and 5τ time instants, respectively:

$$
\begin{aligned}
V_4(0) &= V_2(0)+V_4^-(0)+V_4^+(0) = V_2(0)+V_3^-(L)e^{-\alpha L}+V_3^-(L)e^{-\alpha L}\Gamma_S = \\
&= E\,\frac{Z_c}{Z_S+Z_c}\left[1+e^{-2\alpha L}\Gamma_L(1+\Gamma_S)+e^{-2\alpha L}\Gamma_L\Gamma_S e^{-2\alpha L}\Gamma_L(1+\Gamma_S)\right] = \\
&= E\,\frac{Z_c}{Z_S+Z_c}\left[1+(1+\Gamma_S)(e^{-2\alpha L}\Gamma_L)+\Gamma_S(1+\Gamma_S)(e^{-2\alpha L}\Gamma_L)^2\right]
\end{aligned}
\tag{2.5.11}
$$

$$
\begin{aligned}
V_5(L) &= V_3(L)+V_5^-(L)+V_5^+(L) = V_3(L)+V_4^+(0)e^{-\alpha L}\Gamma_L+V_4^+(0)e^{-\alpha L} = \\
&= E\,\frac{Z_c}{Z_S+Z_c}\,e^{-\alpha L}(1+\Gamma_L)\left[1+e^{-2\alpha L}\Gamma_L\Gamma_S\right]+ \\
&+E\,\frac{Z_0}{Z_S+Z_0}\left[e^{-2\alpha L}\Gamma_L\Gamma_S\right]\,e^{-\alpha L}\Gamma_L e^{-2\alpha L}\Gamma_S(1+\Gamma_L) = \\
&= E\,\frac{Z_c}{Z_S+Z_c}e^{-\alpha L}(1+\Gamma_L)\left[1+e^{-2\alpha L}\Gamma_L\Gamma_S+(e^{-2\alpha L}\Gamma_L\Gamma_S)^2\right]
\end{aligned}
\tag{2.5.12}
$$

Figure 2.5.1 shows the lattice diagram with the reflected parts at various time instants that are multiples of the propagation time $\tau = L/v$ (bounces proceed diagonally with the time flowing from top to bottom); the results of an example with a source impedance $Z_S = 25\,\Omega$, line characteristic impedance $Z_0 = 50\,\Omega$ and a load impedance $Z_L = 150\,\Omega$ are reported, having assumed no line attenuation.

The expressions in square brackets above are the sum of several terms in the form of a geometric series, whose limit is known; going to the limit for infinite reflections, the closed form expressions for the voltages at the source and the load are the same as physically reasonable, because at $t = +\infty$ with vanished transients the longitudinal line voltage drop is zero:

$$
V(0) = V(L) = E\,\frac{Z_c}{Z_S+Z_c}\,e^{-\alpha L}\,\frac{1+\Gamma_L}{1-e^{-2\alpha L}\Gamma_S\Gamma_L}
\tag{2.5.13}
$$

As additional check, the expressions may be simplified by replacing Γ_L and Γ_S with the corresponding fractions of source, line and load impedance and assuming the line attenuation zero ($\alpha = 0$, $\Gamma_S = (Z_S - Z_c)/(Z_S + Z_c)$ and $\Gamma_L = (Z_L - Z_c)/(Z_L + Z_c)$), so that $V(0) = V(L) = Z_L/(Z_S + Z_L)$.

What is important to observe as a conclusion is that when the source and the load resistance are on the same side with respect to the line characteristic impedance (that is either $Z_S, Z_L < Z_c$ or $Z_S, Z_L > Z_c$), we will observe a series of voltage steps of the same sign, but of decreasing amplitude, one above the other as in a staircase with decreasing step height to reach the steady state value; on the contrary, when

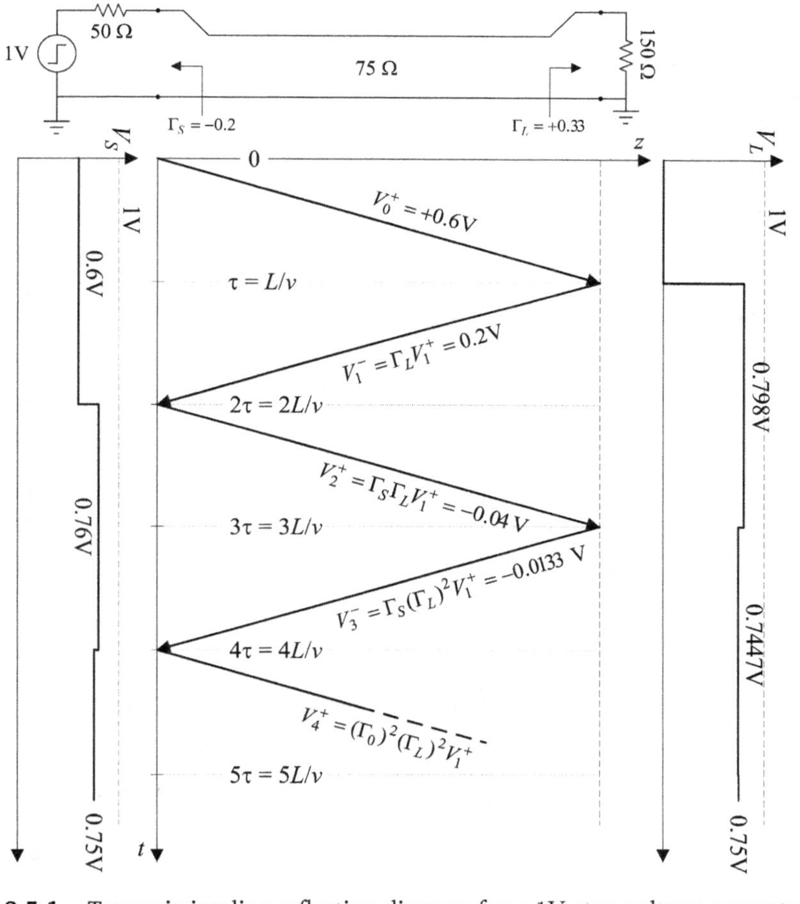

Figure 2.5.1 – Transmission line reflection diagram for a 1V step voltage generator with internal impedance $Z_S = 25\,\Omega$ driving a transmission line with characteristic impedance $Z_0 = 50\,\Omega$ and no attenuation, terminated on a load $Z_L = 150\,\Omega$. The step voltages developing through time at each bounce of reflected waves are plotted at the two sides, at the source and the load ports, respectively (at steady state with no line voltage drop the two voltages are expected equal and given by the simple voltage divider rule $Z_L/(Z_S + Z_L)$).

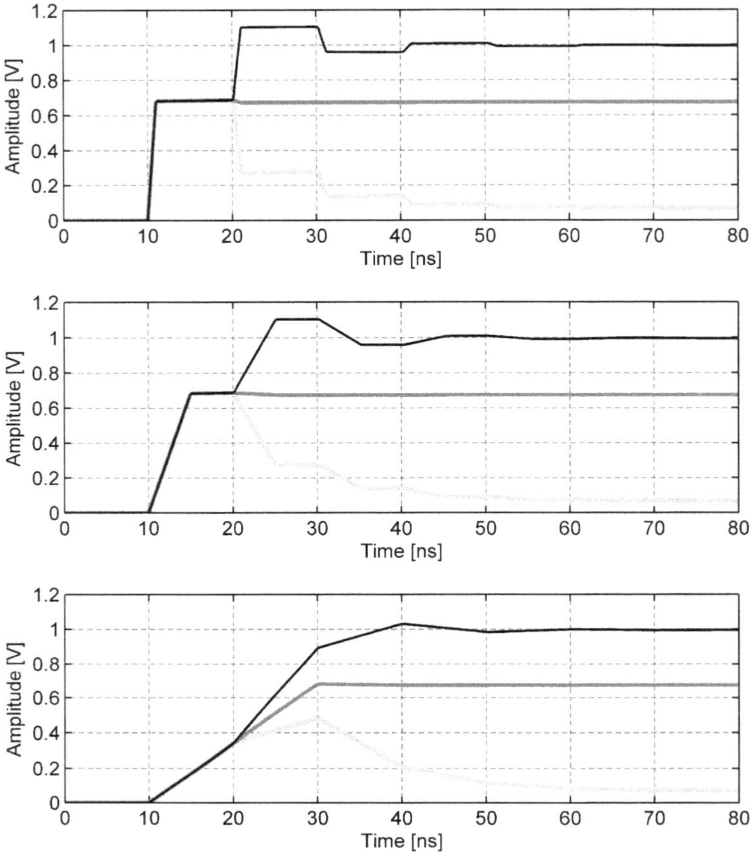

Figure 2.5.2 – Step voltage with three rise time values (1 ns top, 5 ns middle and 20 ns bottom) applied to 1 m RG 58 coax line (4.8 ns propagation time) in three loading conditions (open circuit, black, matched 50 Ω, dark gray, and short circuit, light gray).

the the two terminations are on opposite sides with respect to the line characteristic impedance (that is either $Z_S < Z_c$ & $Z_L > Z_c$ or $Z_S > Z_c$ & $Z_L < Z_c$), there will appear ringing, with reflected terms of alternating sign, so summing and subtracting with the pre-existing voltage and reaching the steady state value with a decreasing oscillatory behavior. It is underlined that it is said "resistance" on purpose; the more general case of "impedance", so including reactance, is more complex and is considered below in sec. 2.6.

When the rise time becomes much longer than the transmission line delay τ, the reflections get "lost" in the slow transition region and appear as fluctuations and small jumps of the rising edge. Regardless of the rise time and rising slope, the amplitude of the reflection terms is the same in all cases because it depends on the line matching conditions. From a practical viewpoint, it is possible to distinguish between rising edges that make the reflections visible (that we may call "fast transients" or "high frequency transients") and those that mask line reflections (similarly called "slow transients", or "low-frequency transients"). The former are needed to make the reflection effects (overshoot, oscillations, etc.) become apparent. We may say that a particular circuit is operating at low frequency if the initial rising transient is still developing when the reflection terms are back; considering a transient duration t_r, this condition is equivalent to $t_r > 5\tau$ approximately, but there exist other criteria all more or less equivalent that depend on the application, the signal and transient shape, the logic thresholds — if any —, and the tolerated amount of reflections.

The same rationale may be applied to sinusoidal signals, and other signals in general. For slow sines with a period much longer than the line propagation delay τ, again, reflections are not visible; a sinusoidal signal has a smaller slope change than a step signal, where the rising edge excites reflections and the final steady value put them in evidence, so that, in general, reflections and their effects are less visible. The effects that a sine signal with a shorter period will undergo are not visible in amplitude, for the reflected terms are all nearly of the same amplitude, but the phase relationship and the phase delay become important: the delay along the line becomes a relevant fraction of the period and so the phase shift of the reflected terms, that when superposing to the original signal may affect the result significantly.

For example, let's consider a line terminated with a $100\,\Omega$ resistor and fed with a $1\,\mathrm{V}$ sine; at low frequency the voltage delivered at the load is $1\,\mathrm{V}$ and the current flowing is $10\,\mathrm{mA}$; voltage and current waves are in phase along the whole line thanks to the resistive load. Let's increase the frequency until the line delay is a quarter of a period: the reflected terms will get back to the source with a $180°$ displacement, thus the voltage wave is reducing at the input side, and at the same time the load sees a larger voltage, twice the previous one, and the two voltage signals at the source and at the load are not in phase any longer. The load current is simply given by the load peak voltage ($2\,\mathrm{V}$) divided by the $100\,\Omega$ load, giving $20\,\mathrm{mA}$. The input voltage will be half the original value and the current twice, fulfilling power conservation through the line (assumed lossless, or with negligible losses). The overall behavior is that of a 1:2 transformer, as it is known indeed: *quarter-wave transformer* (see eq. (2.3.15)).

Be the load resistance much larger, like for an open-circuit termination, voltage and current amplitude would have undergone a dramatic change: the input current is maximized as if the line is in short circuit and the voltage at the load side that is open circuited swings to the maximum allowed by the line losses, to infinite in an ideal case. This is a well known behavior of line resonance, as it appears from eq. (2.3.12), with the terminating load $Z_t = \infty$, so that the input impedance is infinite for the zeros of $\tanh(\gamma L)$, that occur every π.

2.5.2 Propagation and reflection of an impulse signal

We have considered so far a test signal (the step function) that lasts longer than the round-trip time of the transmission line (i.e. twice the propagation time τ), so that the first reflection term and its overlapping to the original signal can be seen. When on the contrary applying fast signals with a short time duration, they will undergo the same reflections but there will be no overlapping, as when a pulse is sent down the line to identify the location of any impedance mismatch (e.g. defect, variation, disconnection) by means of its own reflections and their time of arrival (this is called time domain reflectometry). The test signal in this case is selected purposely short, to have clearly separable reflected components: a pulse is preferred for its reduced time axis occupation and thus to enhance temporal (and spatial) resolution. The energy required to generate a pulse is also proportionally lower, while spectrum occupation at the highest frequencies is nearly similar to a step signal, being determined by the rise time. One thing, however, should be considered: rise and fall edges shall be as symmetric as possible to avoid spectral aberrations.

Of course, once the line velocity is known, propagation time and propagated distance are immediately related one to the other; if, however, the line has a strong attenuation or its p.u.l. parameters are changing widely with frequency, too a short pulse may be undesirable for its extended bandwidth and frequency-axis occupation, with components undergoing different propagation times and thus being reconstructed with a phase shift at the return, and unavoidable distortion (see sec. 3.3.1).

It is interesting to observe that a pulse shorter than the round-trip time applied to the transmission line will not see as input impedance anything else than the characteristic impedance, because when the first copy (the "original") is applied to the line, there are yet no reflected components, and all the concepts used to reason on stationary waves (sec. 2.5.3), input impedance equation (2.3.12) in sec. 2.3, and composition of reflected components are not applicable. The pulse does not "see" the other end of the line, that might be terminated on a short or an open circuit with no difference; only after 2τ seconds the difference would be dramatically perceivable, as shown in Figure 2.5.3 around $10\,\mathrm{ns}$ for the two terminating conditions. The slightly smaller pulses visible at $15\,\mathrm{ns}$ are the copies reflected back by the far end undergoing phase reversal: the voltage is reversed in short-circuit conditions, the current in open-circuit conditions. The same applies to the rise time of a step signal, which, if steep enough and shorter than the round-trip time, will see again only the characteristic impedance and will not be affected by reflections that occur subsequently.

This simple reasoning clarifies why when using the impulse response mode in Vector Network Analyzers, it is said that the location of line discontinuities and imperfections is spotted out quite efficiently, but not their type (i.e. capacitive, inductive): the low frequency "tail" of the step signal is necessary for this.

The design of a pulse generator takes benefit of such a condition and the generator will operate almost always in matched load conditions. This because at high frequency (for short pulses and fast rise times) the value of the characteristic impedance is constant and real; only at low frequency the influence of r and g is relevant and pulses of long time duration might suffer from slight impedance mismatch.

If we consider the more general case of an electronic component like a logic gate sending signals over a PCB, with a trace long enough to ensure a round-trip time of about 5 ns (let's say that we have a trace length of about 30 cm), most of the fast logic families switching in one or few ns will not see the end of the line during transitions. The use of matching resistors in series with the driven line aims at stopping further reflections, as we note briefly when talking of near-end crosstalk terms in sec. 8.3.4.4, and at limiting the effect of the line input capacitance; the latter has a negative impact on the stability of the driver circuit (quite critical when using operational amplifiers) and causes a peaky current absorption at signal transitions, with possible slew rate

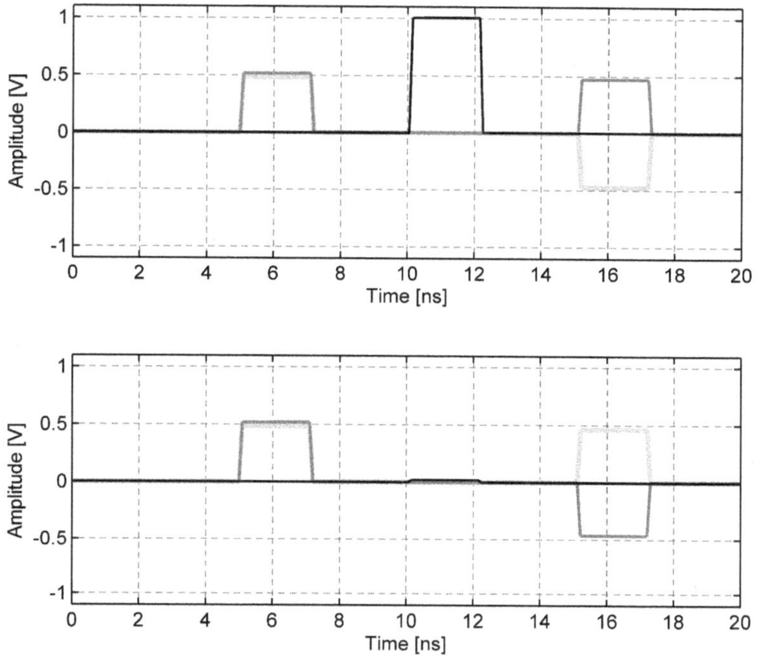

Figure 2.5.3 – Short pulse (100 ps rise and fall time, 1 ns time duration, 5 ns initial delay) applied to 1 m RG 58 coax line (4.8 ns propagation time) terminated on open circuit (top) and short circuit (bottom): line input voltage (dark gray), input line current (light gray, multiplied by 50), far end voltage (black). The short circuit condition is 1 Ω to make the small jump in the far end voltage visible.

limitation and distortion. Also for this the adoption of series resistors is particularly relevant.

2.5.3 Standing wave pattern and VSWR

Let's take again the phasor voltage equation (2.1.10), that may be reformulated as

$$
\begin{aligned}
V(z) &= V_1 \left[\exp(-\gamma z) + \frac{V_2}{V_1} \exp(+\gamma z) \right] = \\
&= V_1 \left[\exp(-\gamma z) + \Gamma_t \exp(-2\gamma L) \exp(+\gamma z) \right] = \\
&= V_1 \exp(-\gamma L) \left[\exp(+\gamma x) + \Gamma_t \exp(-\gamma x) \right]
\end{aligned}
\tag{2.5.14}
$$

where $x = L - z$. The phasor current equation may be soon determined dividing by Z_0.

When two waves of identical frequency travel in opposite directions, the interference phenomenon called *standing waves* occurs. In the extreme case of a lossless ($\alpha = 0$) line with maximum reflection at the load termination (open circuited, thus $\Gamma_t = 1$), the magnitude of the voltage produced by the composition of the two waves is

$$|V(x)| = |2V_1 \exp(-\gamma L)|\,|\cos(\beta x)| \tag{2.5.15}$$

The magnitude oscillates thus between minima (zero valued) and maxima; the separation of consecutive minima (or maxima) is $\beta(x_{i+1} - x_i) = \pi$ or $x_{i+1} - x_i = \lambda/2$; the minima are located at $\beta x_i = \pi/2 + n\pi$, n integer, and the maxima $\pi/2$ apart.

The instantaneous voltage at the points of minimum is always zero, while at the points of maximum oscillates in time at the same frequency of the signal and spans to the maximum level.

The most common characterization of the standing wave voltage pattern is the VSWR (Voltage Standing Wave Ratio), defined as

$$\text{VSWR} = \frac{|V(x)|_{\text{max}}}{|V(x)|_{\text{min}}} = \frac{1+|\Gamma_t|}{1-|\Gamma_t|} \tag{2.5.16}$$

If the line has losses ($\alpha > 0$) the expression above is modified as

$$\text{VSWR} = \frac{|V(x)|_{\text{max}}}{|V(x)|_{\text{min}}} = \frac{1+|\Gamma_t|\exp(-2\alpha x)}{1-|\Gamma_t|\exp(-2\alpha x)} \tag{2.5.17}$$

and implies that the VSWR is not a constant, since minima and maxima are distributed along the line and they have thus a different value of the position x. Practically speaking, if the attenuation is not so relevant ($\alpha/\beta \ll 1$) with terminating impedance values at the source and load side that do not produce too large reflection coefficients (let's say less than 0.5), the determination of the VSWR as the ratio of adjacent maximum and minimum is accurate enough.

The reflection coefficient Γ_t may be calculated back from the VSWR as

$$\Gamma_t = \frac{\text{VSWR}-1}{\text{VSWR}+1} \tag{2.5.18}$$

Several problems and adverse effects may be put in relationship with the observation of a standing wave pattern and large VSWR values:

- the voltage at the points of maximum may exceed or be too close to the voltage rating of some of the connected equipment or the cable itself, thus causing voltage breakdown and contributing to losses;
- similarly the current at the points of maximum may exceed or be too close to the cable rating, thus causing local heating;
- because of the presence of an impedance mismatch (at the output and then proceeding backward causing a variation of the input impedance with frequency), the power-transfer efficiency from the source will vary with frequency.

Referring to the last two points, it is underlined that thus in real cases cable attenuation itself is not the only factor that affects and determines power loss (and cable heating in extreme cases), but also the impedance mismatch at the ends, and the resulting increase in voltage and current waveforms due to standing waves, shall be accounted for. Such increase of voltage and current intensity raises dielectric losses and ohmic losses, roughly proportional to the square of the rms values of voltage and current, respectively.

To this aim an expression that takes into account both cable attenuation (measured as insertion loss, see sec. 10.2) and cable mismatch is available, giving an estimate of total cable losses in dB:

$$K_{tot-loss} = 10\log_{10}\left[\frac{\alpha^2 - |\Gamma|^2}{\alpha(1-|\Gamma|^2)}\right] \tag{2.5.19}$$

where α is the cable attenuation measured in dB for power and Γ is the generic reflection coefficient, indicated above as Γ_t.

In the previous Chapter at sec. 1.2.2.1 the beneficial effect of attenuators for impedance matching was considered: in many cases, when the load is expect to vary, such as for the input impedance of an antenna used over a wide frequency range, the use of moderate attenuation (e.g. 6 or 10 dB) ensures a high reliability of the results, in terms of controlled standing waves and reflection coefficients. For power applications this solution cannot clearly be implemented easily due to large dissipated power and overheating.

2.6 Line termination on arbitrary impedance

We have seen how a step voltage propagates along the transmission line and how the reflection phenomenon develops; the amount of reflection, and thus the final shape that the voltage at the source and load ports will take, depends on the degree of matching of the connected impedance.

It is interesting to consider the various cases of a transmission line terminated on an open circuit, a short circuit, a pure resistance, a pure capacitor, a pure inductor, and their combinations, including transmission line sections. The intention is to have a range of configurations and cases with a description of the related waveforms when excited by a step signal, so to recognize them as fingerprints and classify new cases more easily; for an accurate determination of the unknown terminating impedance, of course, there will be the need of numeric techniques and detailed analysis of signals (e.g. spotting out small details that indicate deviations from expected behavior and high-order response, and at the same time smoothing experimental curves to remove noise and artifacts and to fit easily with the expected behavior).

A thorough overview of the response of a variably terminated line is shown in [268]: the acronyms TDR (Time-Domain Reflectometry or Reflection) and TDT (Time-Domain Transmission) are used to indicate that the subject of the virtual measurement is reflection (i.e. input impedance) or transmission (i.e. propagation). TDR and TDT are considered in sec. 6.5.1 and 11.10: ideally a step is coupled into the input port of the line and both the reflected and transmitted signals are probed at input and output ports, respectively; when comparing time- and frequency-domain measurements, TDR and TDT are put in relationship with return loss and insertion loss.

The graphs appearing in Figure 2.6.1 are taken from the Picosecond's application note [268] and cover possible line loading schemes, however limited to a few reactive components and thus leaving aside very complex circuits.

In the following two special extreme cases (line terminated in a capacitive load and in an inductive load) are treated more in detail.

2.6.1 Line terminated in a capacitive load

The line is considered fed with a matched source ($Z_S = Z_c$) on the left (source side) and terminated with a capacitor C at the other end (load side). The applied signal is a step function. We know that the reflection coefficient at the source is 0 (and thus there will be no other reflection back to the load after the first term getting back from the load to the source) and that at the load requires a more complex expression: when considering a single frequency (phasor solution), it will be a complex number where $Z_L = -jX_C = 1/(j\omega C)$, but for a time-domain solution we need to resort to the Laplace transform, so that

$$\Gamma_L = \frac{\dfrac{1}{sC} - Z_c}{\dfrac{1}{sC} + Z_c} = \frac{1 - s\tau_C}{1 + s\tau_C} \tag{2.6.1}$$

where $\tau_C = Z_c C$ is a sort of time constant of the circuit, considering Z_c as a resistor. The applied signal $e(t)$ enters the line as $v_1(t) = \Gamma_S e(t) = \frac{1}{2} e(t)$ and travels along the line: after 1τ it reaches the load and the resulting load voltage $v_2(t)$ is

$$v_2(t) = v_1(t - \tau)\,(1 + \Gamma_L) \tag{2.6.2}$$

and in the Laplace domain, after simplifications,

$$V_2(s) = \frac{1}{2} E(s)\, e^{-s\tau}\,(1 + \Gamma_L(s)) \tag{2.6.3}$$

If a step function $E\,u(t)$ is applied as input signal, then

$$\begin{aligned} V_2(s) &= \frac{E}{2}\frac{1}{s}\,\frac{1 + s\tau_C + 1 - s\tau_C}{1 + s\tau_C}\, e^{-s\tau} = E\,\frac{1}{(1 + s\tau_C)\,s}\, e^{-s\tau} = \\ &= \left[\frac{1}{s} - \frac{1}{s + 1/\tau_C}\right] E\, e^{-s\tau} \end{aligned} \tag{2.6.4}$$

Inverse transformation gives the time domain waveform of the load voltage as

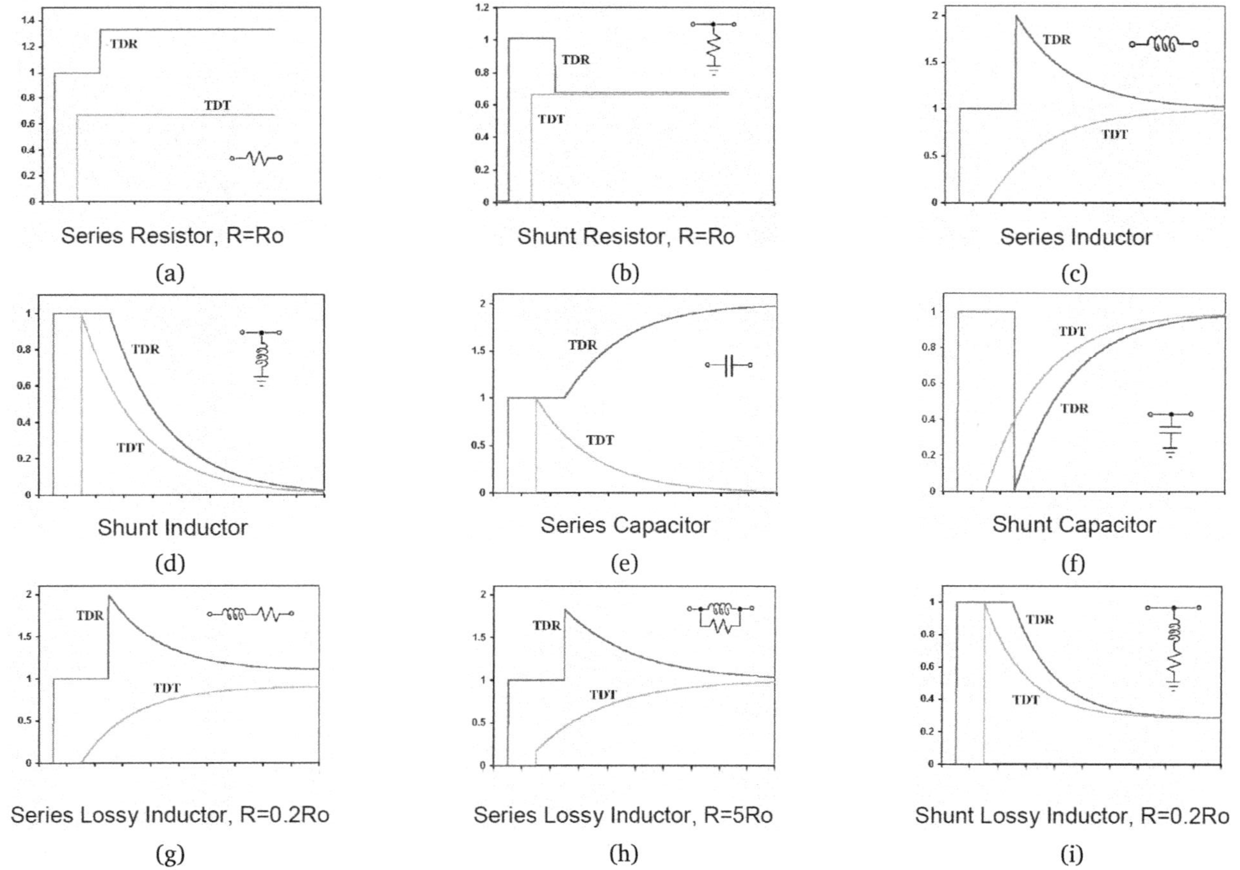

Figure 2.6.1 – Step response of various loading scheme of a transmission line: TDR and TDT responses are shown in dark and light gray, respectively; the reflection term is always lagging with respect to the transmission response because of the two-way delay (Reproduced with permission from Picosecond, Application Note AN-15, Copyright 2004 [268]).

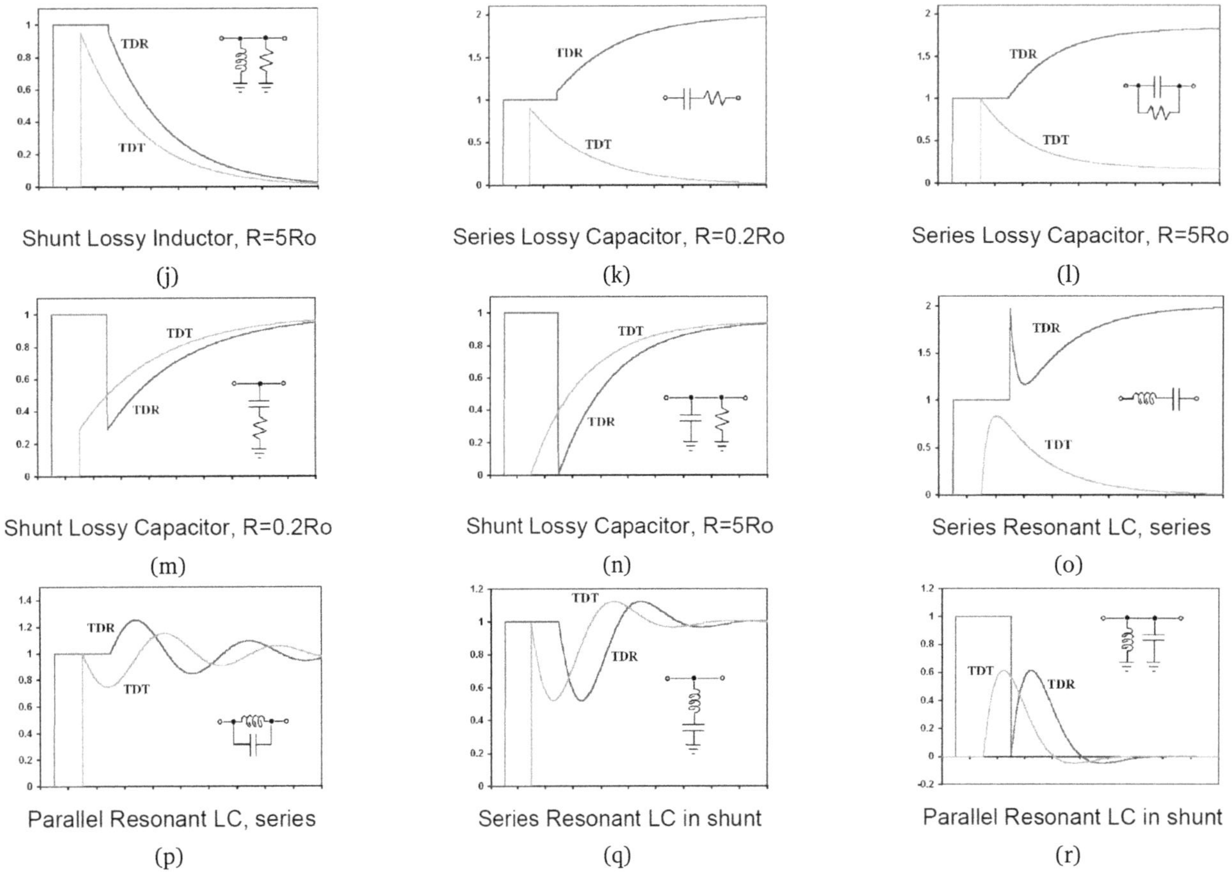

Shunt Lossy Inductor, R=5Ro (j)

Series Lossy Capacitor, R=0.2Ro (k)

Series Lossy Capacitor, R=5Ro (l)

Shunt Lossy Capacitor, R=0.2Ro (m)

Shunt Lossy Capacitor, R=5Ro (n)

Series Resonant LC, series (o)

Parallel Resonant LC, series (p)

Series Resonant LC in shunt (q)

Parallel Resonant LC in shunt (r)

Figure 2.6.1 – Step response of various loading scheme of a transmission line: TDR and TDT responses are shown in dark and light gray, respectively; the reflection term is always lagging with respect to the transmission response because of the two-way delay (Reproduced with permission from Picosecond, Application Note AN-15, Copyright 2004 [268]).

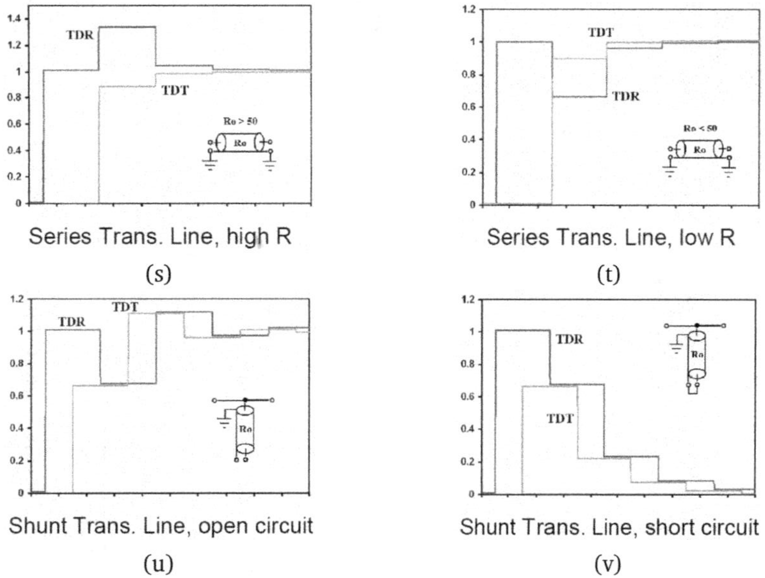

Figure 2.6.1 – Step response of various loading scheme of a transmission line: TDR and TDT responses are shown in dark and light gray, respectively; the reflection term is always lagging with respect to the transmission response because of the two-way delay (Reproduced with permission from Picosecond, Application Note AN-15, Copyright 2004 [268]).

$$v_2(t) = E\,u(t-\tau)\,\left[1 - e^{-(t-\tau)/\tau_C}\right] \tag{2.6.5}$$

The voltage reflected back to the source at time $t=\tau$ reaches port 1 at $t=2\tau$ and as per eq. (2.5.10) the resulting voltage $V_1(s)$ in the Laplace domain is

$$V_1(s) = \frac{E}{2}\frac{1}{s}\left[1+\left(1+\frac{1}{2}\right)\Gamma_L(s)\,e^{-s\tau}\right] = E\,\frac{5-s\tau_C}{(1+s\tau_C)\,s}\,e^{-s\tau} \tag{2.6.6}$$

and the inverse transform will give again the time response waveform.

If the pure capacitor is replaced by a parallel connected R/C circuit (such as when the capacitor represents the stray capacitance of a resistor, or a high-value resistor in parallel takes account of the capacitor $\tan\delta$), then the expressions above substantially hold, provided that the time constant is changed to $\tau_L = RZ_c/(R+Z_c)\,C$. The recalculation of the reflection coefficient Γ_L in eq. (2.6.1) gives

$$\Gamma_L = \frac{\dfrac{R}{1+sRC} - Z_c}{\dfrac{R}{1+sRC} + Z_c} = \frac{R - Z_c - sRZ_cC}{R + Z_c + sRZ_cC} = \frac{\dfrac{R-Z_c}{R+Z_c} - s\tau_C}{1+s\tau_C} \tag{2.6.7}$$

$$\begin{aligned} V_2(s) &= \frac{1}{2}E(s)\,e^{-s\tau}\,(1+\Gamma_L(s)) = \frac{1}{2}E(s)\,\frac{1+s\tau_C+\dfrac{R-Z_c}{R+Z_c}-s\tau_C}{1+s\tau_C}\,e^{-s\tau} = \\ &= \frac{R}{R+Z_c}\,\frac{1}{1+s\tau_C}\,E(s)\,e^{-s\tau} = \frac{1}{CZ_c}\,\frac{\tau_C}{1+s\tau_C}\,E(s)\,e^{-s\tau} \end{aligned} \tag{2.6.8}$$

If a step function $E\,u(t)$ is applied as input signal, then

$$V_2(s) = E\frac{1}{s}\,\frac{R}{R+Z_c}\,\frac{1}{1+s\tau_C}\,e^{-s\tau} = E\,\frac{R}{R+Z_c}\left[\frac{1}{s} - \frac{\tau_C}{1+s\tau_C}\right] \tag{2.6.9}$$

Inverse transformation gives the time-domain waveform of the voltage across the load:

$$v_2(t) = E\,u(t-\tau)\left[\frac{R}{R+Z_c} - \frac{R}{R+Z_c}e^{-(t-\tau)/\tau_C}\right] \tag{2.6.10}$$

Verification of the expressions for limit cases:

- let us assume $R=Z_c$ and $C=0$, so obtaining a perfectly matched load with no inductive component: the first term in square brackets becomes $\frac{1}{2}$ (and we are indeed applying to the line the voltage $\frac{E}{2}$) and the second term vanishes: correctly the voltage reaching the load without reflections is half the source voltage;
- let us now take either $R=0$ or $C=0$: eq. (2.6.10) becomes identically zero, since no voltage develops across a short circuit;

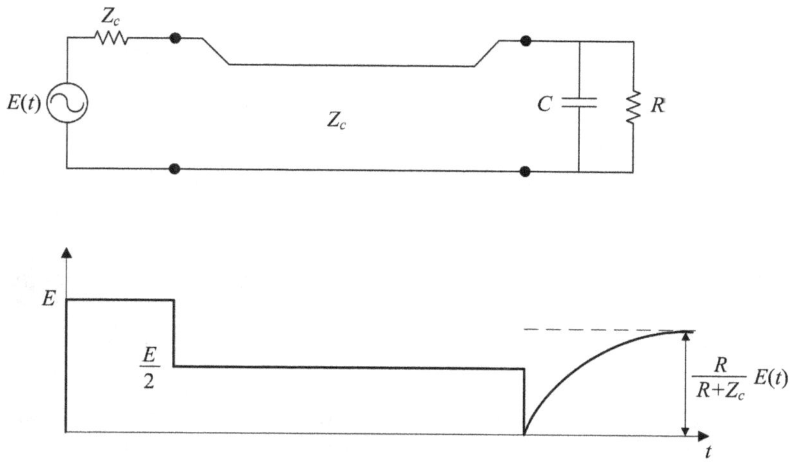

Figure 2.6.2 – Time response of the line terminated on a capacitive/resistive load.

- for the general case $R>0$ and $C>0$ we may observe that the capacitive part is not relevant for low frequency components (provided R is large enough), and this appears from the reflection coefficient Γ_L in eq. (2.6.7), that for $s \simeq 0$ becomes correctly $\Gamma_L = \frac{R-Z_c}{R+Z_c}$; in the first time interval close to the transition edge characterized by high frequency components, the loading impedance is virtually zero, a nearly total mismatching occurs with -1 reflection and it is "testified" by the second term in eq. (2.6.10), that disappears after a few time constants.

2.6.2 Line terminated on inductive load

The capacitor at the load side is now replaced by an inductor L (see Figure 2.6.3), leaving the line match-terminated at the source side ($Z_S = Z_c$). The applied signal is a step function. We know that the reflection coefficient at the source is 0 (and thus there will be no other reflection back to the load after the first reflected term from the load); at the load when considering a single frequency (phasor solution), it will be a complex number with $Z_L = jX_L = j\omega L$, expressed once more with the Laplace transform method:

$$\Gamma_L = \frac{sL-Z_c}{sL+Z_c} = -\frac{1-s\tau_L}{1+s\tau_L} \tag{2.6.11}$$

where $\tau_L = L/Z_c$ is a sort of time constant of the circuit, considering Z_c as a resistor. The applied signal $e(t)$ enters the line as $v_1(t) = \Gamma_S e(t) = \frac{1}{2}e(t)$ and travels along the line: after 1τ it reaches the load and the resulting load voltage $v_2(t)$ is

$$v_2(t) = v_1(t-\tau)\,(1+\Gamma_L) \tag{2.6.12}$$

and in the Laplace domain, after simplifications,

$$V_2(s) = \frac{1}{2} E(s)\, e^{-s\tau}\,(1+\Gamma_L(s)) \tag{2.6.13}$$

If a step function $E\,u(t)$ is applied as input signal, then

$$V_2(s) = \frac{E}{2}\frac{1}{s}\frac{1+s\tau_L-1+s\tau_L}{1+s\tau_L}\,e^{-s\tau} = E\,\frac{\tau_L}{1+s\tau_L}\,e^{-s\tau} \tag{2.6.14}$$

Inverse transformation gives the time-domain waveform of the load voltage:

$$v_2(t) = E\,u(t-\tau)\,e^{-(t-\tau)/\tau_L} \tag{2.6.15}$$

If the pure inductor is replaced by a series connected R-L circuit (such as when the inductor represents the stray inductance of a resistor, or a low value resistor in series takes account of the inductor ohmic resistance and losses), then the expressions above substantially hold, provided that the time constant is changed to $\tau_L = L/(R+Z_c)$. The recalculation of the reflection coefficient Γ_L in eq. (2.6.11) gives

$$\Gamma_L = \frac{sL+R-Z_c}{sL+R+Z_c} = \frac{1+s\tau_L - \dfrac{2Z_c}{R+Z_c}}{1+s\tau_L} = 1 - \frac{2Z_c}{R+Z_c}\frac{1}{1+s\tau_L} \tag{2.6.16}$$

$$\begin{aligned} V_2(s) &= \frac{1}{2}E(s)\,e^{-s\tau}\,(1+\Gamma_L(s)) = E(s)\,e^{-s\tau}\left(1-\frac{Z_c}{R+Z_c}\frac{1}{1+s\tau_L}\right) = \\ &= E(s)\,e^{-s\tau}\left(1-\frac{Z_c}{L}\frac{\tau_L}{1+s\tau_L}\right) \end{aligned} \tag{2.6.17}$$

If a step function $E\,u(t)$ is applied as input signal, then

$$\begin{aligned} V_2(s) &= E\frac{1}{s}\left(1-\frac{Z_c}{L}\frac{\tau_L}{1+s\tau_L}\right)e^{-s\tau} = \\ &= E\,e^{-s\tau}\left[\frac{1}{s}-\frac{Z_c}{R+Z_c}\left(\frac{1}{s}-\frac{\tau_L}{1+s\tau_L}\right)\right] = \\ &= E\,e^{-s\tau}\left[\frac{1}{s}\frac{R}{R+Z_c}+\frac{\tau_L}{1+s\tau_L}\frac{Z_c}{R+Z_c}\right] \end{aligned} \tag{2.6.18}$$

Inverse transformation gives again the time domain waveform of the load voltage:

$$v_2(t) = E\,u(t-\tau)\left[\frac{R}{R+Z_c}+\frac{Z_c}{R+Z_c}e^{-(t-\tau)/\tau_L}\right] \tag{2.6.19}$$

Verification of the expressions for limit cases:

- let us assume $R = Z_c$ and $L = 0$, so a perfectly matched load with no inductive component: the first term in square brackets becomes $\frac{1}{2}$ (and again we are ap-

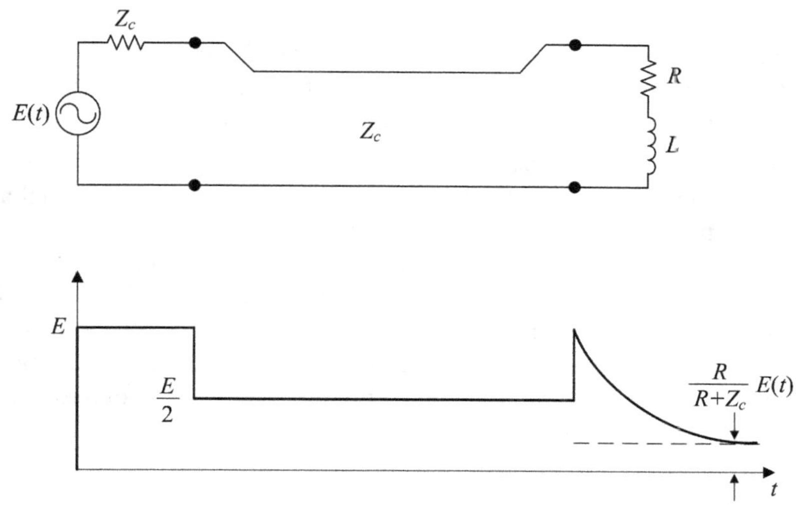

Figure 2.6.3 – Time response of the line terminated on a inductive/resistive load.

plying the voltage $\frac{E}{2}$ to the line) and the second term vanishes, so correctly the voltage reaching the load without reflections is half of the source voltage;

- let us now take $R = 0$ and $L > 0$: eq. (2.6.19) becomes the same as eq. (2.6.15), because the first term vanishes and the coefficient of the second one simplifies to $\frac{1}{2}$;
- for the general case $R > 0$ and $L > 0$ we may observe that the inductive part is not relevant for low frequency components (provided R is large enough), and this appears from the reflection coefficient Γ_L in eq. (2.6.16), that for $s \simeq 0$ becomes correctly $\Gamma_L = \dfrac{R - Z_c}{R + Z_c}$; in the first time interval, close to the edge of the transition characterized by high frequency components, the loading impedance is virtually infinite, so a nearly total mismatching occurs with $+1$ reflection, and it is "testified" by the second term in eq. (2.6.19), that disappears after a few time constants.

2.6.3 Simulated examples

The expressions just derived for the two cases of capacitively and inductively terminated line are now verified using a circuit simulator (PSpice®[5]).

[5] PSpice is available for free at several websites as student version with a limited capability in terms of number of components, e.g.
`http://www.engr.uky.edu/~cathey/pspice061301.html`,
`https://sites.google.com/site/eeelabeul/download-software/pspice-9-1-student-version`.

The considered case is a 1 m RG 58 coax cable fed by a square wave generator (1 ns rise time) and terminated circuits (shunt R//C or series R-L circuit). The experiment is prepared to have time constants faster and slower than the rise time: the set of values considered for simulations are reported in Figure 2.6.4.

The length of the coax cable was selected to have propagation time longer than the signal rise time and the selected time constants, so to have a distinct signal at the far end, without superposition of reflected terms. With a matched near end at the source no reflections should occur in any case, but the generator output impedance might be not exactly 50 Ω during the whole transition at signal edges. Observing that time constant values are no longer than 25 ns, and assuming 50 ps/cm for the propagation speed, a 1 m cable with one-way 50 ns propagation time allows to separate the transient response at signal edge and the effect of multiple reflections due to unavoidable mismatches with a convenient time-scale of the oscilloscope.

2.7 Line discontinuities and non-ideality

So far the considered transmission lines were homogeneous and close to ideal. Real transmission lines, besides attenuation and losses and some dispersion of parameters at high frequency, may feature discontinuities for several reasons. For example, a line may be a combination of different coax sections joined by interconnects, such as connectors, adapters and so on, maybe using different types of coax cables, so with a change of line parameters, even if the final nominal characteristic impedance is the same (i.e. 50 Ω). For PCB planar lines signal traces may change slightly their size, or the characteristic impedance may be influenced by the proximity of other lines, especially when going to the pins of a high density connector (see Chapter 6). Of course discontinuities are recognized as such depending on instrumentation settings and the range of the analysis: maximum frequency and number of points for Vector Network Analyzers, sampling time for oscilloscopes, etc..

2.7.1 Mismatch of two consecutive line sections

If we can neglect line attenuation, then each line section is loaded by the characteristic impedance of the other line section, that except for very low frequencies is a real number and the resulting reflection and transmission coefficients are real numbers too. Assuming two line sections, line 1 and line 2, featuring a different characteristic impedance, Z_{c1} and Z_{c2}, the reflection at the interface between the two cannot be solved and removed; the possible solutions are to match the source at the other end of line 1 and the load at the other end of line 2, so to extinguish reflections created by the inner mismatch, and the use of a matching attenuator at the interface (see sec. 1.2 and 1.2.2.1), improving the reflection coefficient seen from each side, at the cost of reducing signal amplitude. When using an attenuator with the function of impedance matching pad, the overall line attenuation is increased, of course, but the individual reflection coefficients of line 1 entering the attenuator and of the attenuator feeding line 2 are reduced. When considering more in detail the errors introduced by the line mismatch at either ends (that for VNA measurements will be defined as "port match

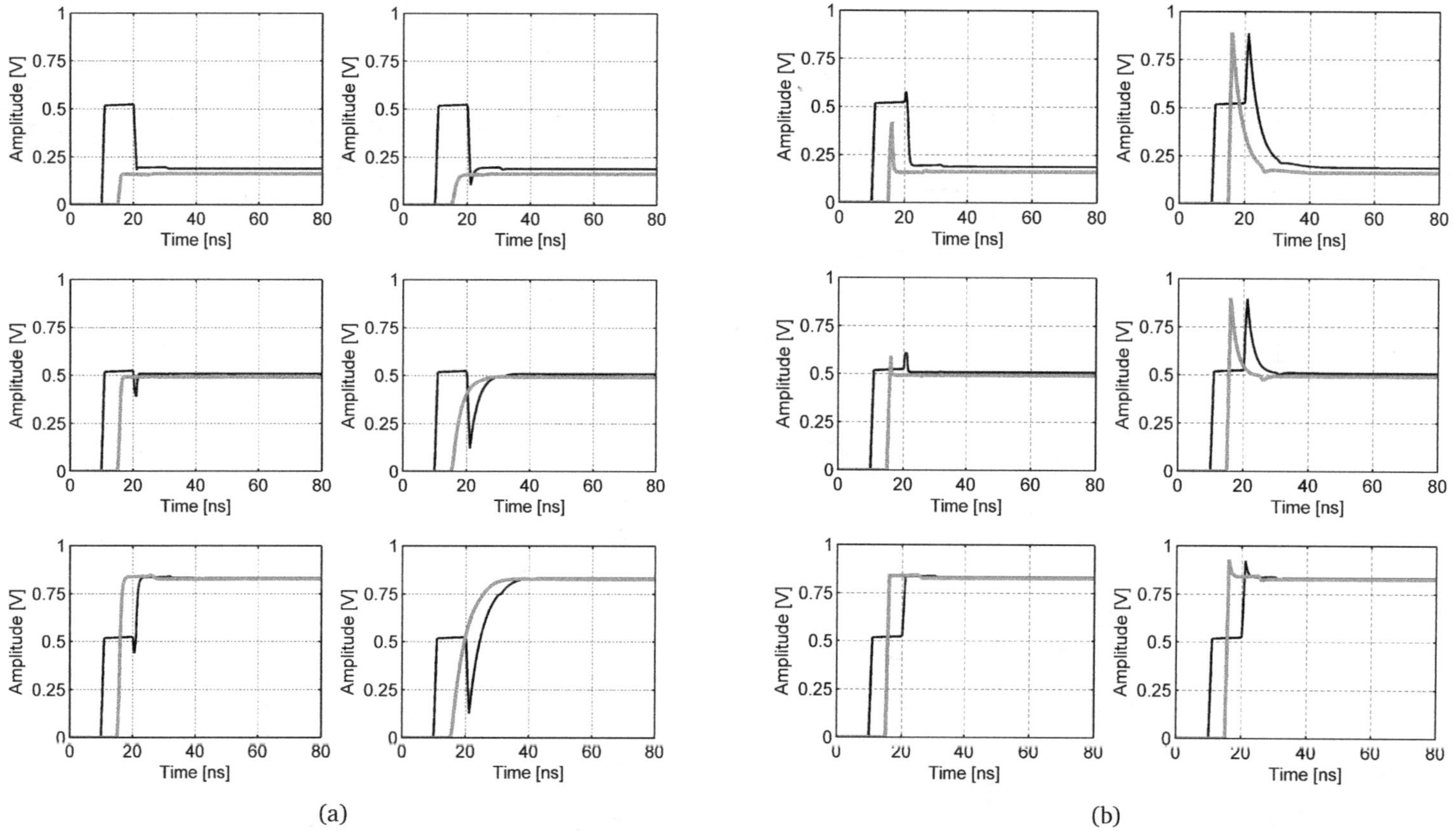

Figure 2.6.4 – Test of capacitive and inductive termination of a transmission line with 1 ns rise time and 4.8 ns propagation time: (a) resistive-capacitive R/C termination 10 Ω/10 pF, 10 Ω/100 pF, 50 Ω/10 pF, 50 Ω/100 pF, 250 Ω/10 pF, 250 Ω/100 pF (left to right, top to bottom); (b) resistive-inductive R-L termination 10 Ω – 20 nH, 10 Ω – 200 nH, 50 Ω – 20 nH, 50 Ω – 200 nH, 250 Ω – 20 nH, 250 Ω – 200 nH (left to right, top to bottom).

error", or in other words source match and load match error, see sec. 11.11.3.2), an attenuator may be used to decouple port match errors and to spot out the VNA port match errors in the observed ripple of the insertion loss curve; a 10 or 20 dB high performance attenuator ruling out line mismatches and mutual interaction is often used for VNA verification as shown in sec. 11.12.2.

Mismatch between cable sections and between cables and connectors will be considered in Chapter 10, when relating reflection coefficients, line propagation and scattering parameters, and in Chapter 11, when running examples both in frequency and time domain. The reader is thus invited to have a look at sec. 10.2.3.2 and 11.10.5, where experimental results are discussed.

When considering two supposedly equal cables joined by an interconnect (e.g. a "barrel", that is a female-female short straight connection, assuming that cables are male terminated as usual), in order to evaluate any mismatch we shall go to a much finer level of detail, adopting a more complete model of characteristic impedance, that includes losses. Time domain measurements are quite indicated, using a time resolution and rise time of the excitation signal that are compatible with the length of the interconnect (as said at the end of sec. 2.5.1, a "fast transient" is used): we will see in Chapter 11 that frequency range and time resolution are intimately connected and roughly reciprocal, with some minor influence due to the selection of the frequency-to-time transformation processing. In Figure 11.10.11(b) for example, the female-female extension added to the setup creates a second smaller peak due to the transition from the cable section, the two having slightly different characteristic impedance.

2.7.2 Branching

It is quite common during measurements, and most of all when distributing signals across PCBs, to be tempted (or compelled) to branch a line to connect two (or more) loads. Some questions arise immediately: is this a relevant discontinuity? may I do this and with which signal degradation?

There are in practice two types of branching: one consists of two different transmission lines departing from the same common point (like for a T connection with coax cables, or a PCB bus line fed at a mid-point); the other configuration consists of loads connected along a distributing line with no significant branching of a second transmission line (let's say that we have a short distance between the line and the load terminals) and we may call it "tapping". When branching is achieved by means of a power splitter or some sort of transformer, impedance matching is preserved with minor deviations that may affect accuracy, due first to the device working principle (e.g. resistive divider, built on planar lines, etc.), and then to parasitics.

The case of two branching lines may be treated formally by the same approach that is used to derive the equations of the quarter-wavelength resonating stub (see also sec. 6.1.4.3 for considerations on PCB vias). When unavoidable for practical reasons, the circuit may include series resistors to better match to the input impedance of the branching lines, avoiding extreme situations when moving between resonances and anti-resonances. It is a solution that works to transmit a signal that shall plainly be received correctly, but is not applicable for measurement or power applications. The

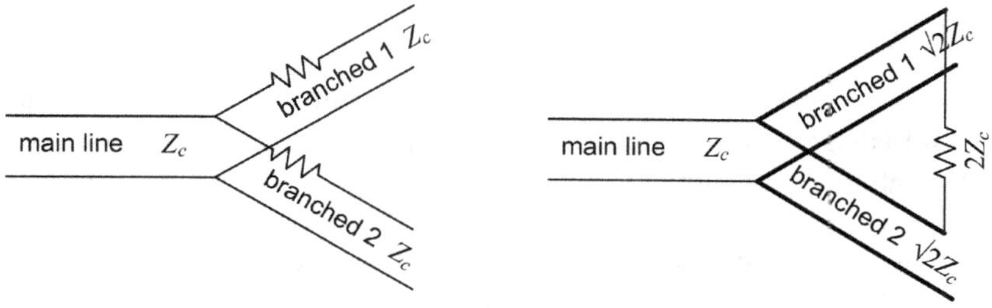

Figure 2.7.1 – Examples of branching of two transmission lines from a main one: (left) series impedance matching, (right) Wilkinson power divider.

Wilkinson power divider is another example of two identical branched lines from the main line: impedance matching is achieved by setting the characteristic impedance of the two branched lines to $\sqrt{2}Z_c$ (Z_c being the characteristic impedance of the main line) and connecting their two far ends together by a $2Z_c$ resistor (see Figure 2.7.1).

Loading elements along the line may be treated by a combination of methods: if the loading capacitance is relevant, then the per-unit-length capacitance of the specific line section may be increased correspondingly (this is what was done for traveling-wave amplifiers, where the stray capacitance of the branched semiconductor or tube was trimmed to a slightly lower constant value that for each section was included in the calculations of the line response). Saying that the loading inductance is the relevant parameter implies that these elements are significantly loading the line, to the extreme of a near short-circuit condition, and this would be definitely of no practical use; in any case such a configuration shall be analyzed with a circuit simulator.

A useful application of a branched short line in short-circuit condition (shorted stub, or simply "stub") is that of canceling by reflection a step function (such as that available in oscilloscopes as the calibration waveform) and generate an almost rectangular impulse: the tapped step waveform propagates along the stub and is reflected back with opposite polarity subtracting from the main step waveform defining the end of the impulse; the width of the impulse will be approximately twice the propagation time along the stub. Of course the impedance mismatch will cause some attenuation and most important reflections that will alter slightly the rising and falling edges of the impulse (see sec. 2.5.2). This is incidentally what was suggested by Gans and Nahman to physically limit the duration of step-like functions to ease the calculation of the Fourier transform, as explained in sec. 3.1.4.

There are several elements on-board PCBs that because of their dimensions may be seen still as a loading element, but at frequencies that are becoming commonplace they behave as branched lines: device and connector pins, vias, pads are all examples of elements prevalently capacitive but with a self resonance frequency in the order of ten to twenty MHz. They are reviewed in Chapter 6 at sec. 6.1.3 and 6.1.4, and in particular at sec. 6.1.4.3.

2.7.3 Lossy lines and dispersion

A lossy line is a line for which attenuation is relevant and the signal amplitude at the other end is significantly reduced. However, by amplification or simple numeric correction, such an inconvenience would be soon corrected, were it frequency independent (thus resulting in a simple matter of a scale factor), that is, in other words, if the effect of such attenuation is a proportional reduction of the signal waveshape without distorting it.

Distortion arises from the different attenuation that the various signal components undergo, depending on their frequency: usually, the higher the frequency the higher the attenuation. Thus in reality the overall signal amplitude is less affected than its individual high-frequency components: the first "victims" are the high-frequency parts of a signal, namely signal edges (resulting in rise-time degradation) and some drooping for short square pulses. Furthermore, excessive losses at high-frequency modify the resulting line impedance connected to the transmitter and receiver, affecting also impedance matching and further deformation at signal edges.

An increased rise time and the appearance of ringing and oscillations have the drawback of interfering with fast bit patterns, where the bit duration is comparable to the time duration of these effects: the increased rise time and the "tails" of superimposed ringing affect thus the adjacent symbol, resulting in increased inter-symbol interference (ISI). This phenomenon is usually evaluated and tested by constructing the eye diagram (or eye pattern) of a data stream (see sec. 3.3.3).

The reasons for line losses and line attenuation may be many, starting from the physical behavior of the line itself and then including also the mutual effect of nearby lines:

- the simplest effect is the increased resistance of line conductors due to skin effect (see sec. 1.1.2);
- at higher frequency dielectric losses add another relevant resistive component (see sec. 1.1.1.2), causing dielectric and cable overheating in extreme cases;
- depending on the type of line (coaxial, with braided or solid shield, or pair, possibly twisted, or PCB line) radiation losses may be significant, increasing with frequency; this term becomes relevant at very high frequency, with wavelength comparable to the line geometry, and is represented by an additional frequency-dependent resistive component;
- the amount of radiation defines also the coupling to nearby lines, that is responsible for both signal reduction due to crosstalk (see sec. 5.1.6.3 and 8.3.4) and for energy loss; it is usually true that the energy lost for radiative coupling is not so relevant and that this term is significant only for the electromagnetic emissions (EMI) of the line; on the contrary, coupling to nearby lines may have a dramatic effect on signal reduction and on impedance discontinuity, especially if such coupling occurs only for a portion of the line length, thus causing impedance change within the line itself (see sec. 2.5 and 2.7.1).

2.7.3.1 Skin effect and current distribution

The combined effect of skin effect and coupling from nearby lines (also through the shared return plane) is the most relevant cause of conduction line losses: skin effect goes with the square root of frequency and for small conductors (e.g. signal cables and PCB transmission lines) an appreciable change of resistance is observed starting from some MHz; a tenfold increase of resistance is thus reached when the frequency increases by two orders of magnitude, that is nearly 1 GHz. For much faster signals, such as high-speed serial buses, with frequency spectra extending up to several GHz the variation of conductor resistance and conduction losses may be dramatic.

2.7.3.2 Dielectric conduction and dielectric losses

An ideal dielectric (such as the insulating layer between conductors in a cable or the PCB substrate) would behave like a pure capacitor at all frequencies: in reality, first, some current leakage may occur due to bulk conduction, that is a resistive behavior appearing at low frequency and even at dc; then, also dielectric losses due to dissipation at high frequency shall be taken into account.

Dielectric conduction (or leakage) is due to the interaction of the applied electric field (i.e. external voltage) with the dielectric molecules, resulting in small leakage current in some cases already at dc and increasing with frequency: even if losses are not dramatic and no overheating is observed, this term may be large enough to make the characteristic impedance change slightly, affecting impedance matching.

Dielectric losses on the contrary are due to the orientation of the dipoles in the dielectric under a rapidly changing electric field (see sec. 1.1.1.2); the mobility of the dipoles changes with frequency and the overall effect is the combination of different mechanisms that are described by a dissipation factor (or $\tan\delta$) as a function of frequency. Usually, materials with the smaller permittivity will have also smaller dissipation factors. Moreover, dissipation is also influenced by other environmental conditions, in particular moisture (see sec. 6.1.5).

2.7.3.3 Characteristic impedance, line attenuation and propagation velocity

We have already seen in sec. 2.2 that the characteristic impedance has quite a large imaginary part at low frequency that vanishes above the critical frequency at which the longitudinal resistance can be neglected if compared with the longitudinal inductive reactance. And we commented that such a large imaginary part is of no relevance for impedance matching conditions, since it occurs at low frequency where matching is not so relevant. Of course, a line with large conduction losses has a critical frequency that is brought to the high frequency range, extending at the same time the frequency range over which impedance mismatching occurs.

Line attenuation α was defined in eq. (2.1.12) and related to line parameters responsible for losses, i.e. longitudinal resistance and shunt conductance:

$$\alpha = \frac{1}{2}\left(\frac{r}{Z_c} + gZ_c\right) \tag{2.7.1}$$

As soon as we recognize that the terms r and g are frequency dependent (because of skin effect and dielectric losses), the line attenuation becomes also frequency dependent. The first term accounts for conduction losses, approximately increasing with square root of frequency, while the second term stands for dielectric losses, with a more complex frequency dependency.

$$\alpha_{cond.} = \frac{1}{2}\frac{r}{Z_c} \qquad \alpha_{diel.} = \frac{1}{2} g Z_c = \frac{1}{2} 2\pi f \tan\delta \, c \frac{\sqrt{\varepsilon_r}}{v\,c} = \pi f \tan\delta \frac{\sqrt{\varepsilon_r}}{v} \tag{2.7.2}$$

Dielectric losses have an explicit linear behavior with frequency, but $\tan\delta$ and ε_r add further dependency, the former increasing with frequency (but not for all frequency intervals, see sec. 6.1.2), while the latter usually decreases slightly. The exact behavior of the overall line losses and attenuation depends on many factors: for cables theory and standards indicate a range of expressions as shown in sec. 5.3.1 for coaxial cables and in sec. 5.5.3 for shielded pairs; the frequency establishes also the mode of propagation from TEM, to quasi-TEM, to higher modes, that, if for coaxial structures are well defined and occurring at quite high frequency (see sec. 5.3.1), are extremely relevant when considering planar transmission lines on PCBs (see sec. 6.3).

Including resistive effects of line losses and disregarding other effects (such as inductance and capacitance change with frequency), eq. (2.1.15) is obtained where the phase constant is larger (and the line is slower). Changes of line velocity with frequency clearly delay signal components differently depending on their frequency: this is called *dispersion* and line behavior is such that delay is larger at lower frequency. This is probably in contrast with a first intuitive interpretation, based also on the evident increase of rise and fall times, especially when the frequency and speed of signals are increased. The reason is simply that the effect of rise time degradation is caused by the attenuation of high frequency components of the signal, that is much more relevant than the effect of dispersion, even if it is originally due to line loss.

Dispersion is not so relevant if the transmitted signal has a narrow band occupation (e.g. a modulated bandpass signal); on the contrary base-band signals, such as digital signals with an approximate square pulse shape, are quite exposed to such phenomenon, being their band occupation quite extended, due to the fast rise and fall times and the short time duration of pulses.

From a measurement viewpoint, line dispersion can be spotted out by measuring the line group delay as shown in sec. 3.3.1.

3

Signals and Transforms

Signals and their characterization in time and frequency domains are considered, using as far as possible standard definitions (just recalled when necessary) and commonly accepted terminology. As far as possible, the attention is on the practical implications and how to verify that the procedure is correct, including statistical characterization of results.

A good, exhaustive and commonly accepted set of definitions of the various components and parts of a signal, of steady or transient nature, is quite hard to find; it is demonstrated when thinking in terms of signal analysis, communications or control system engineering, for which various parts and characteristics of signals assume different relevance and naming ambiguity is also possible (let's think to the concept of "bandwidth"). Practically speaking, we will rely on the general engineering knowledge; definitions and distinctions will be made when there is the risk of confusion and misunderstanding.

Continuous time expressions are preferred to discrete time expressions whenever introducing a topic for clarity and didactic purpose; if the implementation requires so, discrete time expressions, as well as digital implementation details, are also included. Confirmation of theoretical statements and assumptions may be attained by simulation or experiments, the latter nowadays accompanied almost always by some post-processing.

3.1 Time-Frequency transforms

Two time-frequency transforms have probably the widest application in signal analysis and will be used in the following: the Fourier transform (or Discrete Fourier Transform, DFT) and the Chirp Z-transform (CZT). A basic restriction that is apparent for the DFT/IDFT pair is the tradeoff between simultaneous resolution in time and frequency domains. This is usually expressed in terms of suitable measures of bandwidth B in Hz and time T in seconds, such that $T \times B \geq 1$. All real-world signals are evidently time-limited and as a consequence cannot be truly band-limited, even if in practice they may be assumed so, with the spectrum components negligibly small beyond some frequency value: measuring equipment is intrinsically band-limited. When jointly processing signals in time and frequency such a limitation becomes evident and truly limits the resulting time and frequency resolution; depending on the transformation kernel and the characteristics of the analyzed signal the equality in the relationship above may hold.

3.1.1 Fourier transform

Quite complete references for Fourier transform theory exist and report mathematical insights [48, 238, 256, 311], offering as well a more intuitive approach [229]. DFT/IDFT implementations are widely available in many math libraries and programming environments, both as Fast Fourier Transform (FFT) and recursive Fourier transform, called Goertzel's algorithm[1], that allows to limit efficiently the computation only to the few needed components with a proportional speed-up. For many implementation details reference is made to Matlab® for a matter of convenience, with other mathematical packages perfectly equivalent.

The DFT transform pair is

$$X[k] = \sum_{n=0}^{N-1} x[n] e^{-j\frac{2\pi}{N}kn} \qquad x[n] = \sum_{k=0}^{N-1} X[k] e^{+j\frac{2\pi}{N}kn} \tag{3.1.1}$$

with the same number N of samples in time (index n) and frequency (index k) domains.

3.1.2 Chirp z-transform

This transform (also known as the "fractional Fourier transform") replaces the unitary exponential Fourier kernel with an exponential term, whose amplitude varies while turning around the origin. It was introduced by Bluestein [42] and then in less than one year treated extensively by Rabiner, Schafer and Rader [277]. Its origin and the form in which it is written belong to the z-transform

---

[1] The original paper is G. Goertzel, "An Algorithm for the Evaluation of Finite Trigonometric Series," *American Mathematical Monthly*, Vol. 65, No. 1, Jan. 1958, pp. 34–35. Details are given in [275].

$$X[k] = \sum_{n=0}^{N-1} x[n]\, z_k^{-n} \tag{3.1.2}$$

but with the sample points located on the spiral described by

$$z_k = A\,W^{-k} \qquad k = 0 \ldots M-1 \tag{3.1.3}$$

where M is the integer number of points, and A and W are complex numbers, indicating the starting point and the step, respectively. W, as a complex number, sets the phase rotation at each step (as it is in the Fourier transform), but also if the points spiral in or out with respect to the origin. The equivalent contour in the s-plane is given by

$$s_0 = \frac{1}{T}\ln A \qquad s_k = \frac{1}{T}(\ln A - k\ln W) \tag{3.1.4}$$

The case with $A = 1$, $M = N$ and $W = e^{-j2\pi/N}$ corresponds to the DFT.

This transform adds further degrees of freedom to the DFT with the varying modulus and the possibility of rotating for the desired angle, expressed as fraction of the full turn. In other words, the number of the calculated frequency (or, better, z) data samples does not have to correspond to the original number of time samples N, no restrictions on M or N which then may be even, odd, prime, ..., and the angular spacing given by W is arbitrary with respect to the original periodicity of the time samples. However, since W is raised to an increasing integer exponent, $-k$, precision and round-off errors become soon a problem, even with double precision libraries, if W is too far from unity. So, its usefulness in following arbitrary spirals in and out of the unity circle is quite limited.

The inverse transform (ICZT) is not commonly referenced and in Matlab there is no ICZT function indeed! Historic references [277, 42] do not mention such inverse transform and searching the Web for "inverse", "chirp z", "ICZT" does not bring abundant results. Frickey [130] proposed, and Liu et al. went on this way [224], to take the complex conjugate of the direct transform CZT applied to the complex conjugate of the spectrum to anti-transform; Frickey himself warns the reader that "it is not clear if this will work for the Chirp-Z transform in general; however, it will work for the Chirp-Z that operates on the unit circle."

$$x[n] = [CZT(X^*)]^* \tag{3.1.5}$$

Another complete reference is: R.M. Merserau, "An algorithm for performing an inverse chirp z-transform," *IEEE Transactions on Audio Electroacoustics*, Vol. 22, no. 5, pp. 387-388, 1974. doi: 10.1109/TASSP.1974.1162603 .

3.1.2.1 Use and advantages

When considering circuits and networks, the ability of the CZT to locate poles in terms of center frequency and bandwidth (i.e. factor of merit) is quite useful. Three contours were tested in [277], one consisting exactly of the unit circle (corresponding to a DFT), one spiraling towards the center and one outside the unit circle. For contours far from the investigated poles, CZT spectrum tend to be quite insensitive to parametric variations of poles, such as the bandwidth, and thus gives a more reliable answer, while for close-up contours frequency details are many more, even if cluttered by the ripple between CZT points, due to the $N-1$ zeros of the CZT.

For the examples in [277], where the CZT is tested for the location of zeros, the conclusion is that the algorithm is not useful to locate transmission zeros, because in many cases it would be difficult, not to say impossible, to distinguish between the dips in the spectrum between poles and those caused by complex zeros. Another outcome is that low resolution spectra very often behave better than high resolution ones: they are smoother, with a more stable amplitude and with less artifacts.

However, the potentially limitless ability of spiraling in and out of the unity circle is bogged down by the not-so-limited, yet insufficient, possibilities of processor floating point numbers and math libraries: with a 1.01 modulus, the W term will get to $20\,10^3$ in $M = 1000$ steps, and bringing it to 1.05 takes off to $1.5\,10^{21}$ in the same number of steps! If this on the one hand limits the implementation of the CZT, on the other hand poses the question on the usefulness of too a curled spiraling path: the CZT is most useful as a flexible method to calculate Fourier spectra over a restricted number of bins, as it is required when doing frequency zooming, or symmetrically time zooming starting from frequency domain data (see sec. 11.9 and 11.10).

3.1.2.2 Implementation

Practically speaking, the CZT/ICZT pair is quite attractive for its freedom of choice of frequency and time resolution and number of samples, so quite suitable also for efficient real-time applications. The CZT algorithm is available as Matlab function `czt()` and attention is directed to its correct use and to parameter setting.

Example 3.1: Use of CZT

Use of CZT on a sample signal (the example comes from the Matlab help for the `czt()` function)

```
fs = 1000; f1 = 100; f2 = 150; % in hertz
m = 1024; % number of czt points
w = exp(-j*2*pi*(f2-f1)/(m*fs)); a = exp(j*2*pi*f1/fs); %A and W
% create a filter response for fft and czt test
h = fir1(30,125/500,rectwin(31));
y = fft(h,1000); z = czt(h,m,w,a);
fy = (0:length(y)-1)'*1000/length(y);
fz = ((0:length(z)-1)'*(f2-f1)/length(z)) + f1;
figure(1); plot(fy(1:500),abs(y(1:500))); hold on; plot(fz,abs(z));
axis([1 500 0 1.2]); xlabel('Hz'); ylabel('Magnitude');
```

Results are not shown: the CZT is used to zoom part of the filter response, demonstrating its flexibility to choose time resolution and display spectrum details.

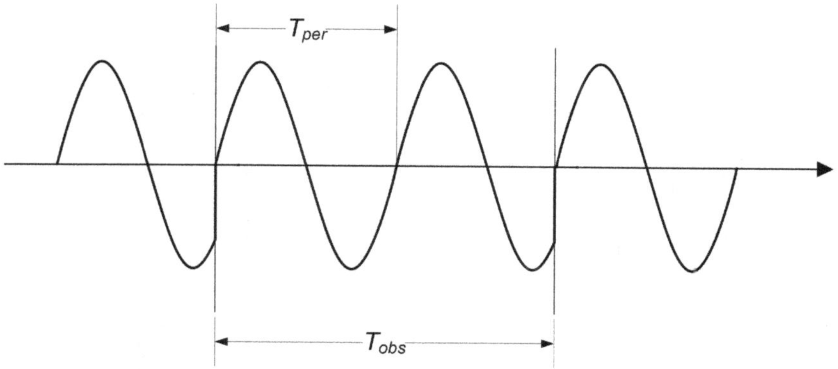

Figure 3.1.1 – Periodic signal and its extension over an observation interval (called also "epoch"), where it is not periodic.

Practical aspects of the use of the CZT/ICZT, of the DFT/IDFT and the required properties of the data vectors are considered in sec. 11.10.

3.1.3 Frequency leakage and windowing

Good references for an introductory, yet comprehensive, treatment of spectral leakage and related countermeasures are [153, 229, 256]. Harris describes the leakage phenomenon in an elegant way: "From the continuum of possible frequencies, only those which coincide with the basis will project onto a single basis vector; all other frequencies will exhibit non zero projections on the entire basis set. This is often referred to as spectral leakage and is the result of processing finite-duration records. Although the amount of leakage is influenced by the sampling period, leakage is not caused by the sampling. An intuitive approach to leakage is the understanding that signals with frequencies other than those of the basis set are not periodic in the observation window." The periodic extension of a signal not commensurate with its natural period exhibits discontinuities at the boundaries of the observation interval; the discontinuities are responsible for spectral contributions (or leakage) over the entire basis set. The form of this discontinuity corresponds to the abrupt step as shown in Figure 3.1.1.

An equivalent explanation may be given by considering the behavior of the *sinc* function that is convolved in frequency with each sample at successive frequency bins, corresponding to the implicit rectangular windowing due to the cut-off of the time domain signal at its boundaries. The *sinc* has a maximum in the origin and zeros at multiples of $1/T$, where T is the time duration of the epoch captured with the rectangular window. If the fundamental and the harmonics of the captured signal do not occur where the *sinc* zeros are (that is to say that they are synchronous with the implicit period T set by the rectangular cut-off), at each frequency f_h there will be non-zero convolution terms that add with arbitrary sign to the main bin located at f_h, thus altering its original amplitude. This is shown in Figure 3.1.2.

Smoothing windows (or equivalently *tapering windows*) are weighting functions applied to data to reduce the spectral leakage associated with finite observation intervals.

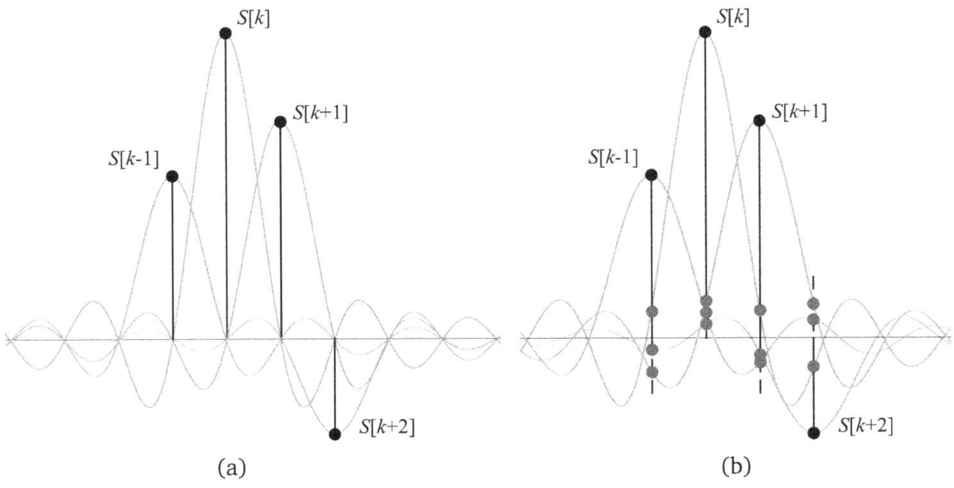

Figure 3.1.2 – Effect of a *sinc* with non-integer period compared to that of the underlying signal: (a) synchronized situation where the *sinc* of each sample do not leak onto the adjacent samples; (b) the observation period (and the *sinc* oscillation period) is increased by 20% and the leaking terms onto adjacent samples are shown as gray dots (for the smallest negative sample $S[k+2]$ this interference amounts to about 30% in absolute value, while for the other larger samples it is less than 20%).

From one viewpoint, the window is applied to data (as a multiplicative weighting function) to reduce the order of the discontinuity at the boundary of the periodic extension where the signal was cut. This is accomplished by matching as many orders of derivative as possible at the boundary of the resulting windowed data and setting them to zero, or nearly so. Thus windowed data are smoothly brought to zero at the boundaries, so that the periodic extension of the data is now continuous up to some high order derivative.

Similarly to the interpretation of frequency leakage with the *sinc* function in frequency domain, the action of smoothing windows may be also interpreted as a uniform reduction of the secondary lobes of the *sinc* function, thus reducing in case of lack of "synchronization" the impact of additional non-zero terms during convolution.

Using windows has impact on many attributes of spectral analysis, including detectability, resolution, dynamic range (see Table 3.1.1).

Concerning the influence on noise level, we observe that the amplitude of the harmonic estimate at a given frequency is biased by the accumulated broad-band noise that falls inside the window bandwidth. In this sense, the window behaves as a filter, gathering contributions over its bandwidth. For the harmonic detection problem, we desire to minimize this accumulated noise and maximize the signal-to-noise ratio, and we accomplish this with small-bandwidth windows. A convenient measure of this bandwidth is the equivalent noise bandwidth (ENBW) (see sec. 4.3.1.5).

Conversely, the interference, or mutual effect, between spectral components is called *frequency leakage*, or *spectral leakage*. Such phenomenon is in reality referred to in some literature as long-range frequency leakage, opposed to the other form of leakage, the short-range frequency leakage, that corresponds to the simpler fact that when the spectrum is represented as samples on a discrete frequency vector, the real underlying signal frequency may lie between two discrete points, thus worsening or biasing the frequency estimate. Radicalizing the definitions for simplicity, the two forms of leakage, despite similar names, have a different nature and behavior and are to be treated differently: smoothing (or tapering) windows are used for the long-range, while interpolation is used for the short-range leakage. It is author's opinion that the true leakage by its proper name be the long-range frequency leakage, that will be shortly addressed, while the short-range leakage is caused by a problem of limitation of the frequency resolution (see sec. 3.1.3.1 below on scalloping loss). However, the two phenomena are in reality inter-related and not so easily separable, because assuming a lack of correspondence between the true signal frequency and the DFT sample points on the frequency axis for the short-frequency leakage to occur, implies that the signal is not periodic in the observation window T, thus leading also to the other form of leakage, the long-range frequency leakage.

Even for the case of a single sinusoidal tone with a definite spectral line (not located at a DFT sample point, or in other words not periodic in the observation window), the leakage from the kernel on the negative-frequency axis biases the kernel on the positive-frequency axis; the sinusoidal tone line spreads around, ending up with a lower peak amplitude and causing an increase of the noise floor due to coherent terms, that are not suppressed by averaging. This bias is the most severe and bothersome for the detection of small signals in the presence of nearby large signals. To reduce the effects of this bias, the window should exhibit low-amplitude sidelobes far from the central main lobe, and the transition from the main lobe to the low sidelobes should be very rapid. One indicator of how well a window suppresses leakage is the peak sidelobe level (relative to the main lobe); alternatively the asymptotic rate of fall-off of these sidelobes may be used.

Another index for characterization of windows is the processing gain (PG), which may be defined as the dc gain of the window, given by the sum of the windows samples. For a rectangular window of N samples this factor is N, the number of terms in the window. For any other window, the gain is smaller, due to the window smoothly going to zero near the boundaries; processing gain represents a bias of spectral amplitudes and must be compensated for.

The use of tapering windows worsens frequency resolution according to the main-lobe width[2], that is usually measured with respect to the rectangular window of time duration T; as a countermeasure, the time duration of the epoch extracted from the signal may be increased proportionally beyond T to compensate for this loss of frequency resolution. Normally, for the most common windows, the main-lobe width at $-3\,\mathrm{dB}$ lies in the range 1.5 to 2.0 times that of the rectangular window, so accounting for a 30% to 50% loss of frequency resolution.

Since the window assigns an effective bandwidth to the spectral lines, we would be interested in the minimum separation between two equal strength lines, such that for arbitrary spectral locations their respective main lobes can be resolved and identified in the final resulting spectrum (we will see in sec. 9.2.3 that this is the operative defi-

---

[2] Just to recall, the rectangular window (i.e. no tapering applied) results in the best frequency resolution, at the cost of the largest frequency leakage.

nition of effective resolution, applied to Spectrum Analyzers and similar equipment). The DFT output points are the *coherent* addition of the spectral components weighted through the window at a given frequency; at the crossover points of the window frequency kernels, the gain from each kernel must be less than 0.5, or in other words the crossover points must occur at the 6 dB points of the windows or below. So, $-6\,\mathrm{dB}$ bandwidth is relevant when evaluating the effective resolution for coherent contribution.

If short time duration tone-like signals must be detected, the non overlapped DFT analysis of the signal might miss the tone if it occurs near the time window boundaries; in general overlapping is good at tracking the dynamic behavior of the spectrum for quasi-stationary slow-varying signals. In this case overlapped sub-sequences must be used and the overlap percentage is usually 50% or 75%, not only for practical reasons, but also to particularly reduce the correlation between the spectral components of successive sub-sequences; this is shown in sec. (4.3.2.3) at eq. ((4.3.57)).

3.1.3.1 Scalloping loss

When evaluating broadband signals populated by many unrelated components, a significant issue is also the *scalloping loss*, that is the attenuation of signal components that lie mid between two bin frequencies; such phenomenon was defined above as short-range frequency leakage. The bin width and thus the bin frequency spacing is set by the chosen transformation period T; at the crossing of the window transforms located at two adjacent bins there is the maximum processing gain reduction and this amounts to $4\,\mathrm{dB}$. The exact expression of the scalloping loss when the sinusoidal tone frequency f_s is not aligned to one of the Fourier spectrum bin frequencies f_k involves the $\mathrm{sinc}(\cdot)$ function (the rectangular window transform) and it is $\mathrm{sinc}^2(\pi(f_s - f_k)T)$, the maximum occurring as expected when the difference between the two frequencies is half the resolution frequency: $|f_s - f_k| = 1/2T$.

Two countermeasures are normally adopted for this phenomenon:

- increase frequency resolution fictitiously by *zero padding* (or zero stuffing), that is adding zero samples to the original sequence increasing the number of samples N involved in the calculation (bringing them always to a power of two for convenience), but keeping the captured signal power unchanged and without any improvement in the signal-to-noise ratio, except the span of the broadband incoherent noise over more frequency bins, but leaving the spectral density (i.e. per unit frequency) unchanged; by the way it is underlined that the number of samples used in the normalization of the DFT calculated spectrum remains the original N, the zero samples bringing no additional contribution; zero padding is best implemented if the necessary zero samples are added symmetrically at the beginning and the end of the signal epoch, thus preserving the phase;

- interpolation of the exact location and height of the component by exploiting the spectral samples in the nearby bins (normally referred to simply as *frequency interpolation*); there exist closed-form expressions for some smoothing windows that give the frequency and amplitude correction terms; interpolation have been accomplished differently in the literature, depending on the used window and the attained degree of approximation (please see sec. 3.1.3.3).

3.1.3.2 Smoothing windows

The most common smoothing windows, that are also found on-board VNAs (and FFT analyzers), are the rectangular (i.e. no windowing), Hann (or von Hann), Hanning, Hamming, Gaussian, Kaiser-Bessel and Dolph-Chebyshev[3] windows.

The Hann window is a cosine window with only one first-order cosine term, described by the following function

$$w[n] = 0.5\left(1 - \cos\left(\frac{2\pi n}{N}\right)\right) \tag{3.1.6}$$

where N is the window length and $-N/2 \leq n < +N/2$. This window has the characteristic of being continuous with a continuous first derivative at the two extremes, so that its discontinuity resides in the second derivative and the spectrum falls off with order 3 versus frequency.

The family of the windows obtained by raising to a power the cosine function, indicated by "$\cos^{\alpha}(\cdot)$", is named Hanning window, of which the Hann window is a particular case for $\alpha = 2$ (the $\cos^2(\cdot)$ term gives a $\cos(\cdot)$ term at twice the frequency, the one appearing in eq. (3.1.6) above).

The Hamming window is a modified Hann window, that changes the relative weight of the cosine function:

$$w[n] = 0.54 - 0.46\cos\left(\frac{2\pi n}{N}\right) \tag{3.1.7}$$

The reason for having selected these exact values of the coefficients is that they are those attaining nearly minimum sidelobe levels; this window represents a special case of the Blackman-Harris, where for $\alpha = 0.53856$ (rounded up to 0.54) a sidelobe level of -43 dB is obtained (the one of the Hamming window).

Needless to say that the rare combination of names (Hann, Hanning and Hamming) generates some confusion!

The Bohman window is obtained by the convolution of two half-duration cosine lobes of the Hann type: the fall-off with frequency is thus of order four.

$$w[n] = \left(1 - \frac{|n|}{N/2}\right)\cos\left(\frac{\pi\,|n|}{N/2}\right) + \frac{1}{\pi}\sin\left(\frac{\pi\,|n|}{N/2}\right) \tag{3.1.8}$$

---

[3] The correct spelling of Chebyshev is quite a frustrating problem: "chebyshev", "chebychev", "tchebychev" etc. are used all quite often. For this reason a look at the original name in Cyrillic gives Пафнутий Львович Чебышёв, that reads more or less "Pafnutii Lvovich Chebyshov" (see Wikipedia `http://en.wikipedia.org/wiki/Pafnuty_Chebyshev`).

The Gaussian window has a Gaussian profile established by

$$w[n] = e^{-\frac{1}{2}\left(\frac{1}{\sigma}\frac{n-N/2}{N/2}\right)^2} \tag{3.1.9}$$

In general, for all windows, the mean-square time duration T_{rms} and the mean-square bandwidth W_{rms} (also called "localizations" of the signal in the time and frequency domain axis, respectively) shall satisfy the relationship $T_{rms}W_{rms} \geq 1/4\pi$, also named *Gabor's inequality*. The peculiar characteristic of the Gaussian wave-shape is that it is the only one for which the equality holds.

As reported in Matlab® documentation, "The Kaiser window is an approximation to the prolate-spheroidal window, for which the ratio of the mainlobe energy to the sidelobe energy is maximized. For a Kaiser window of a particular length, the parameter β controls the sidelobe height. For a given β, the sidelobe height is fixed with respect to window length." Maximizing the mainlobe-to-sidelobe ratio in terms of energy is for sure the reason for some equipment manufacturers of having preferred this window. Moreover, as reported in the original paper by Kaiser [204], a FIR implementation is available, and this is quite important for real-time computationally efficient implementations, so that the Kaiser window was widely used and preferred for implementation on-board Vector Network Analyzers (see sec. 11.9 and 11.10.2)

The Dolph-Chebyshev window specifically addresses the problem of sidelobe attenuation with a flat amplitude of all sidelobes, guaranteed by design below a specified threshold -20α in dB, where α is the window parameter. Its specification originates in the frequency domain and the time domain window $w[n]$ is obtained by inverse DFT:

$$W[k] = (-1)^k \frac{\cos\left[N\ \arccos\left(\beta\cos(\pi k/N)\right)\right]}{\cosh\left[N\, \mathrm{a}\cosh(\beta)\right]} \tag{3.1.10}$$

where $\beta = \cos\left[\frac{1}{N}\, \mathrm{a}\cosh(10^{\alpha})\right]$ and $0 \leq k \leq N-1$.

A synthesis of the main performance indexes of the above windows is shown in Table 3.1.1. A few explanatory notes are necessary: 1) the Hanning window for $\alpha = 2$ is the von Hann window; 2) the side-lobe fall-off of the Dolph-Chebyshev is zero because all side-lobes have the same height; 3) the order of the Kaiser-Bessel [153] includes a multiplicative factor π, that is absent in the Matlab function `kaiserwin()`.

The data reported in Table 3.1.1 are complemented by a graphical description of the time and frequency domain response of the windows, extending the analysis to higher window orders for the Kaiser-Bessel and Dolph-Chebyshev. Calculations have been done using Matlab® and the Signal Processing Toolbox (the product version is not really relevant, because the used functions are quite common and not strongly version dependent). The results in terms of time domain window shape and frequency domain spectrum are shown in Figure 3.1.3, 3.1.4, 3.1.5 and 3.1.6 for an arbitrary number of samples $N = 64$ (tests have been done with $N = 64$ samples for exigencies of graphical representation: a larger number of samples would result in a cluttered spectrum with the zeros so close not to be distinguished). The spectra are normalized for unity dc

Window	Degree	Highest side-lobe [dB]	Side-lobe fall-off [dB/oct.]	Coherent gain	ENBW	-3 dB BW	-6 dB BW	Correlation 75%/50% overlap
Rectangle	–	-13	-6	1.00	1.00	0.89	1.21	0.75/0.50
Hamming	–	-43	-6	0.54	1.36 (1.38)	1.30	1.81	0.707/0.235
Hanning	$\alpha = 1.0$	-23	-12	0.64	1.23	1.20	1.65	0.755/0.318
	$\alpha = 2.0$	-32	-18	0.50	1.50	1.44	2.00	0.659/0.167
	$\alpha = 3.0$	-39	-24	0.42	1.73	1.66	2.32	0.567/0.085
	$\alpha = 4.0$	-47	-30	0.38	1.94	1.86	2.59	0.486/0.043
Bohman	–	-46	-24	0.41	1.79	1.71	2.38	0.545/0.074
Gaussian	$\alpha = 2.5$	-42	-6	0.51	1.39	1.33	1.86	0.677/0.200
	$\alpha = 3.0$	-55	-6	0.43	1.64	1.55	2.18	0.575/0.106
	$\alpha = 3.5$	-69	-6	0.37	1.90	1.79	2.52	0.472/0.049
Kaiser-Bessel	$\beta = 2.0$	-46	-6	0.49	1.50 (1.519)	1.43	1.99 (2.02)	0.657/0.169
	$\beta = 2.5$	-57	-6	0.44	1.65 (1.678)	1.57 (1.594)	2.20 (2.23)	0.595/0.112
	$\beta = 3.0$	-69	-6	0.40	1.80 (1.824)	1.71 (1.729)	2.39 (2.423)	0.539/0.074
	$\beta = 3.5$	-82	-6	0.37	1.93 (1.959)	1.83 (1.855)	2.57 (2.602)	0.488/0.048
	$\beta = 4.0$	—	—	(0.344)	— (2.086)	— (1.972)	— (2.770)	—
	$\beta = 6.0$	—	—	(0.282)	— (2.532)	— (2.387)	— (3.360)	—
Blackman-Harris	3-sample	-61	-6	0.45	1.61	1.56	2.19	0.610/0.126
	4-sample	-92	-6	0.36	2.00 (2.04)	1.90 (1.93)	2.72 (2.70)	0.460/0.038
Dolph-Chebyshev	$\alpha = 2.5$	-50	0	0.53	1.39 (1.41)	1.33 (1.344)	1.85 (1.875)	0.696/0.223
	$\alpha = 3.0$	-60	0	0.48	1.51 (1.537)	1.44 (1.46)	2.01 (2.04)	0.647/0.163
	$\alpha = 3.5$	-70	0	0.45	1.62 (1.65)	1.55 (1.57)	2.17 (2.197)	0.602/0.119
	$\alpha = 4.0$	-80	0	0.42	1.73 (1.76)	1.65 (1.67)	2.31 (2.34)	0.559/0.087
	$\alpha = 5.0$	-100	0	—	— (1.96)	— (1.85)	— (2.60)	—

Table 3.1.1 – Smoothing windows performance [153] verified with numeric calculations (differing values are shown between parentheses).

gain: the original value corresponds to the coherent gain[4] (CG) indicated by Table 3.1.1.

For many windows it may be observed that the first zero of the highest order curve (light gray) occurs at a frequency position that is nearly twice as large as the lowest order curve (black), indicating approximately a mainlobe width proportionally larger.

Quite interesting to consider are the values of Equivalent Noise Bandwidth (ENBW), "-3 dB Bandwidth" and "-6 dB Bandwidth". Harris [153] stresses that "two equal strength main lobes separated in frequency by less than their 3 dB bandwidths will exhibit a single spectral peak and will not be resolved as two distinct lines. The problem with this criterion is that it does not work for the coherent addition we find in the DFT. The DFT output points are the coherent addition of the spectral components weighted through the window at a given frequency. If two kernels are contributing to the coherent summation, the sum at the crossover point (nominally half-way between them) must be smaller than the individual peaks, if the two peaks are to be resolved. Thus at the crossover points of the kernels, the gain from each kernel must be less than 0.5, or the crossover points must occur beyond the 6 dB points of the windows." This is particularly important: it reminds that leakage terms are coherent, so that a 6 dB criterion applies (as already commented at the beginning of this sec. 3.1.3).

The bin is the base width of the frequency representation, that is the sampling pulsation ω_s divided by the number of samples N: this value on the graphs of Figure 3.1.3, 3.1.4, 3.1.5 and 3.1.6 is $\omega_s/32 = 0.031$.

The roll-off for smoothing windows is simply dictated by the type of window as e.g. obtained by some power of the $\cos(\cdot)$ function or else; as it may be noticed smoothing windows have normally a roll-off of 6 dB/octave, with increasing roll-off for the Hanning group for increasing α (as said, the exponent of the $\cos(\cdot)$ term); the Bohman has 24 dB/octave, being obtained as the convolution of two windows featuring 12 dB/octave each; finally, for the Dolph-Chebyshev window expressing roll-off is not meaningful because the side-lobes are all limited by design to the specified level.

Now, what is the criterion for choosing among the different window implementations and parameter values or, in other words, how to decide for the tradeoff e.g. between mainlobe width (and thus resolution) and mainlobe-to-sidelobe amplitude ratio (to reduce spectral leakage)?

Of course, the answer depends on the application: we will see in sec. 11.10.2 that for Vector Network Analyzers window choice converges to having three window types, called "minimum", "normal" and "maximum", referring to the dynamic range given by the mainlobe-to-sidelobe amplitude ratio, and "minimum" corresponds to the rectangular window, so no smoothing applied, but resolution at its maximum; behind this broad classification there "hide" the windows we have just reviewed.

It is evident that side-lobe suppression is particularly effective for Kaiser-Bessel and Dolph-Chebyshev windows, from which the preference of manufacturers for the im-

---

[4] Please, pay always attention to the coherent gain of a smoothing window and to the fact that some versions of Matlab and other mathematical packages/environments might or might not normalize it implicitly; without normalization, the computed smoothed spectra are smaller than the original ones proportionally to the coherent gain.

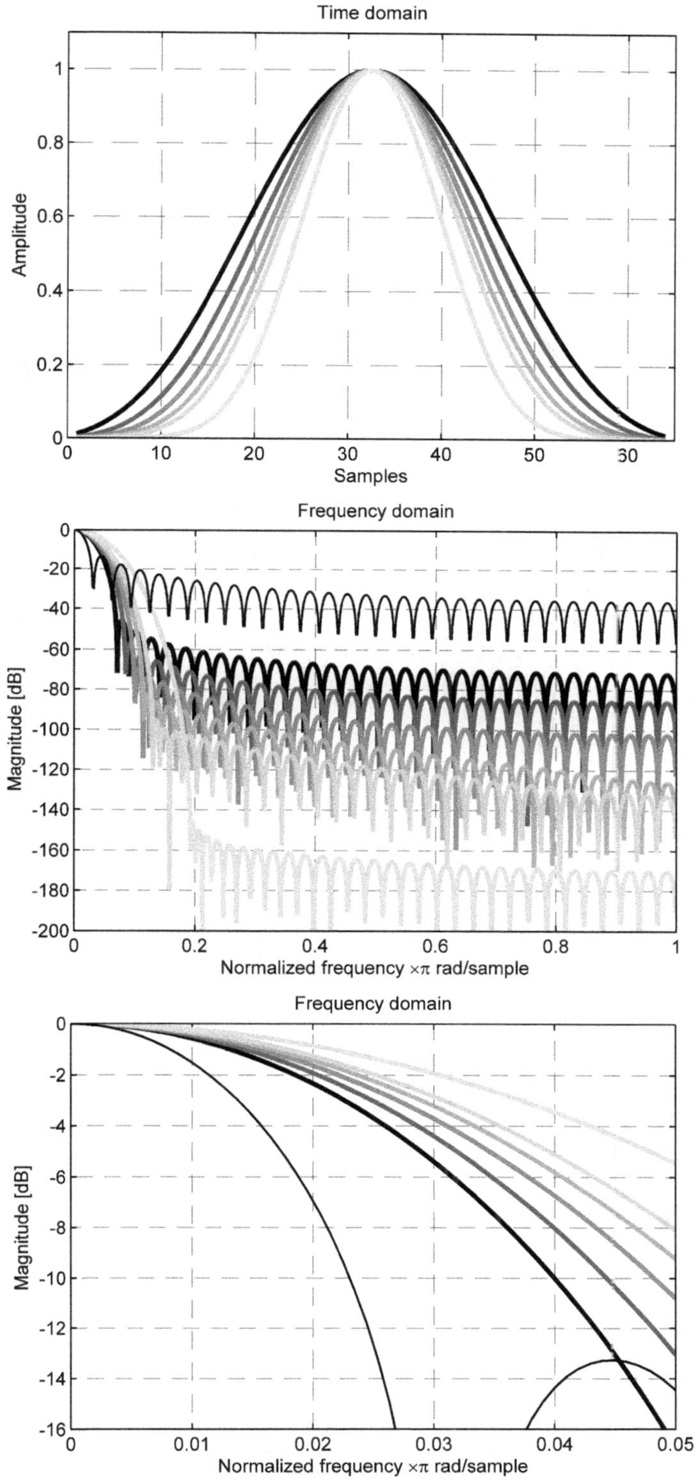

Figure 3.1.3 – Kaiser-Bessel windows: (top) time and (middle, bottom) frequency domain; rectangular (black thin), $\beta = 2.0$, 2.5, 3.0, 4.0, 6.0 (from black to light gray).

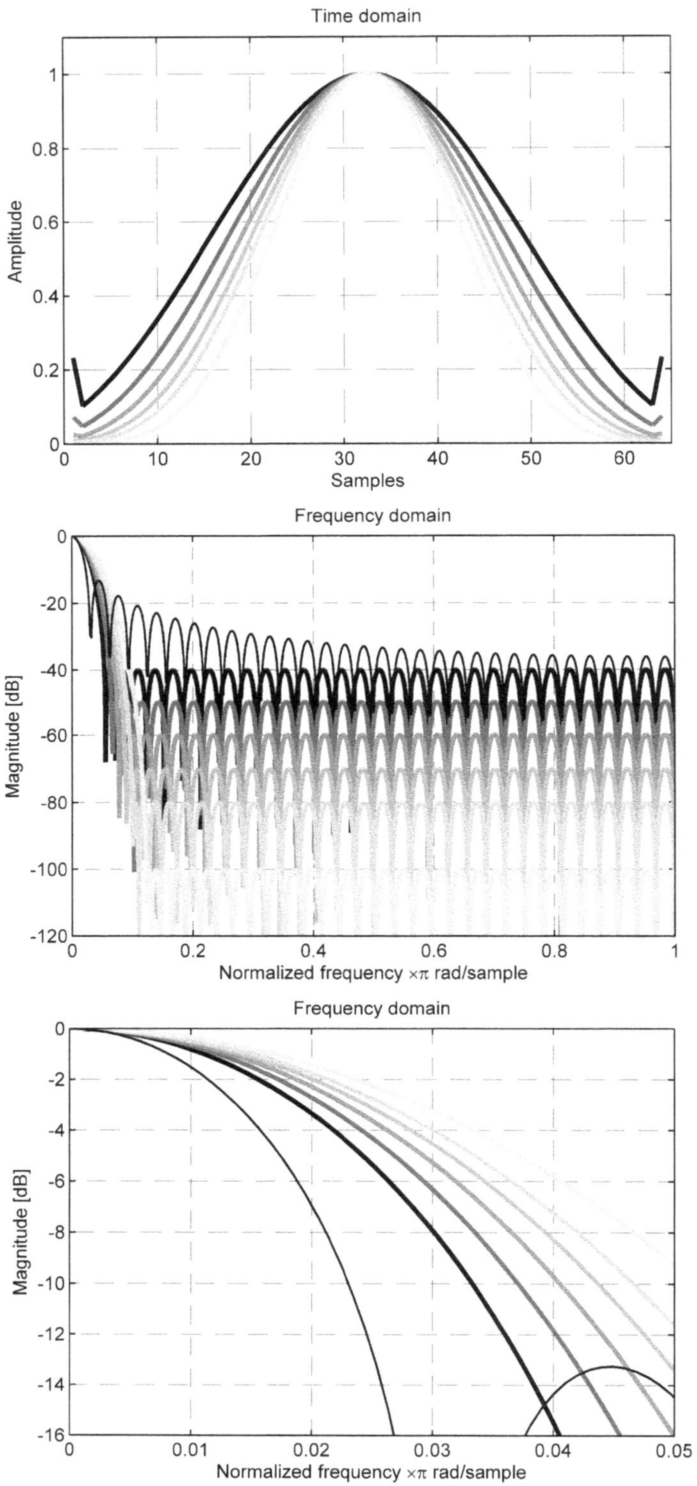

Figure 3.1.4 – Dolph-Chebyshev windows: (top) time and (middle, bottom) frequency domain; rectangular (black thin), $\alpha = 2.0$, 2.5, 3.0, 3.5, 4.0, 6.0 (from black to light gray).

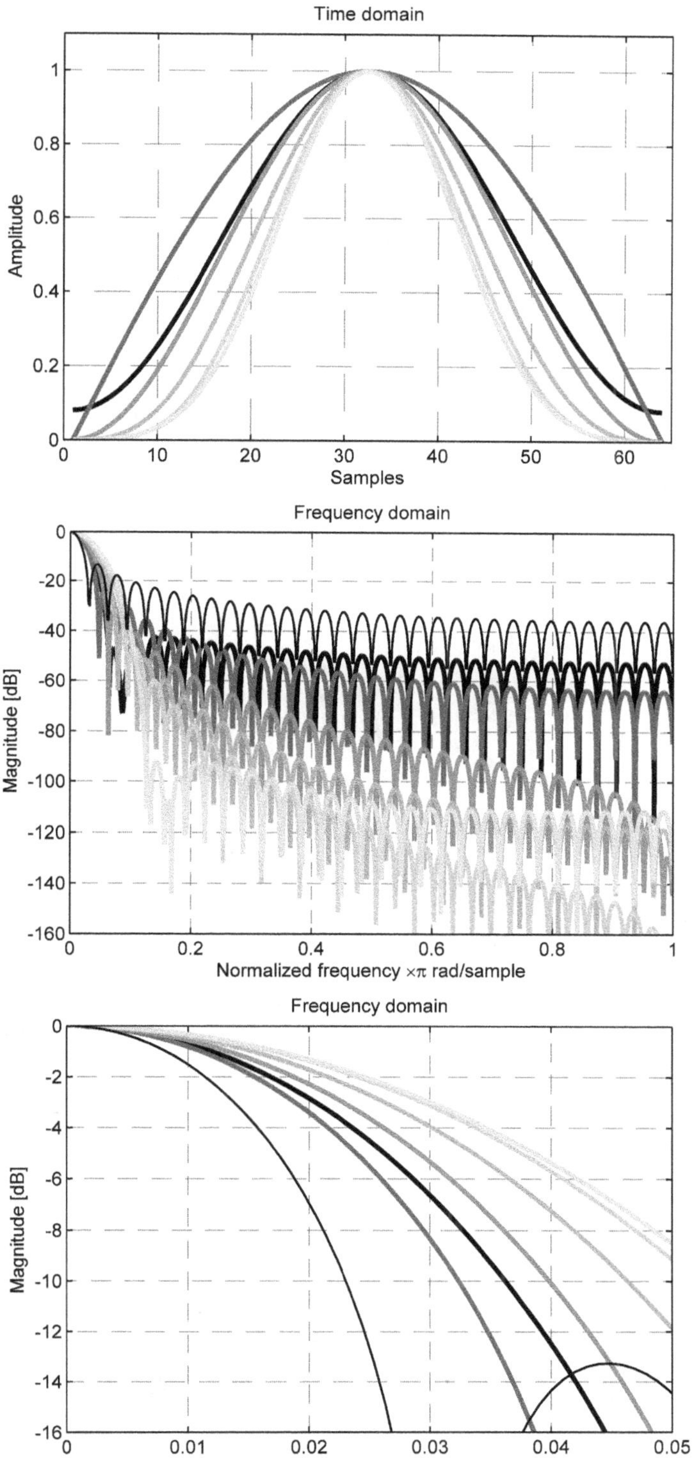

Figure 3.1.5 – (top) time and (middle, bottom) frequency domain; rectangular (thin gray), Hamming (black), Hanning $\alpha = 1.0$, 2.0, 3.0, 4.0 (from drak gray to light gray) and Blackman-Harris order 4 (lightest gray).

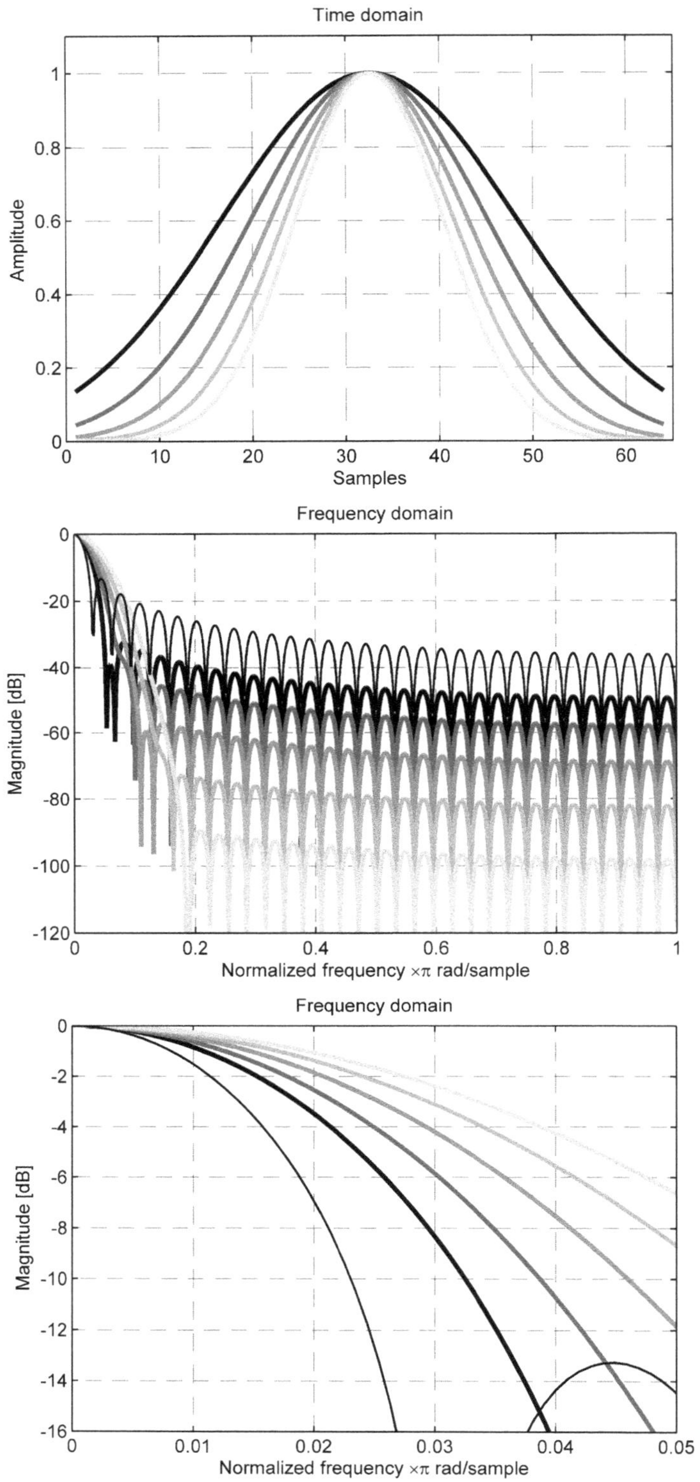

Figure 3.1.6 – Gaussian windows: (top) time and (middle, bottom) frequency domain; rectangular (black thin), $\alpha = 2.0,\ 2.5,\ 3.0,\ 3.5,\ 4.0$ (from black to light gray).

plementation of some of the VNA smoothing windows: in particular, an impressive side-lobe suppression of more than 100 dB is attainable with the Kaiser-Bessel of order 4 and the Dolph-Chebyshev of order 5 (this one corresponding to exactly -100 dB of side-lobe by design). Let's consider this further as an example. By inspection of Table 3.1.1, the D-C window has better performance in terms of main-lobe width, with 1.85 against 1.972 for the -3 dB bandwidth. Although a 6.6% improvement in frequency resolution might seem not so relevant, it is underlined that the main-lobe width values of all windows with 100 dB side-lobe suppression or so are grouped in a restricted interval, not much larger than that 6.6%. The Blackman-Harris of order 4, for example, with nearly 100 dB suppression, has 1.93 of main-lobe width, so the Dolph-Chebyshev is definitely a very good choice for the "maximum" window used in VNAs (see sec. (11.10) and (11.9)), and for very high side-lobe suppression in general!

When time domain or frequency domain time-gated scattering parameters are calculated on VNAs under non-stringent conditions (e.g. the number of points is not too small, there are no heavy discontinuities nearby, etc.), smoothing windows with average side-lobe suppression (e.g. in the range of 50-60 dB) give satisfactory results, while preserving a better time or frequency resolution. Large side-lobe suppression windows, even though available, are thus used rarely.

An overall evaluation of windows effects and suitability for VNAs in both low-pass and band-pass mode is reported in sec. 11.10.6, where the impact on spatial resolution and the estimated peak height is evaluated.

3.1.3.3 Interpolation of frequency and amplitude

As anticipated, the problem of scalloping loss (see sec. 3.1.3.1) affects the accuracy of the amplitude estimation resulting from the Fourier spectrum of signals whose components do not fall exactly on the Fourier frequency bins for the chosen observation period T and number of sample points N (this phenomenon was called short-range frequency leakage to use a common definition). The problem extends to the determination of the exact frequency of the signal component, bracketed by the two adjacent frequency bins f_k and f_{k+1}, and then in some cases to phase correction. The correction is accomplished by applying a fractional change using an opportune coefficient, determined by means of various interpolation techniques.

At a first glance interpolation might be linear between the two samples immediately at the left and the right of the unknown desired signal component. A higher order interpolation is much desirable (e.g. quadratic) to take into account the convexity of the bell-shaped peak, necessitating thus three (or more) points. Interpolation finds its application in accurate frequency determination (e.g. when tracking slowly varying signals, such as the mains frequency of 50 or 60 Hz, or RF oscillator frequency) and amplitude estimation (e.g. when accurate harmonic distortion measurements are needed, such as in Power Quality studies).

One of the first interpolation techniques [195], after simplifying the $\sin(\cdot)$ function with its own argument for small angle values, provides a linear determination of the frequency correction δ starting from the ratio of the left and right bin samples:

$$\alpha = \frac{|S[k+1]|}{|S[k]|} \qquad \delta = \frac{\alpha}{1+\alpha} \qquad f_s = (k+\delta)df \tag{3.1.11}$$

The amplitude is then corrected by making reference to either the left or the right term, choosing the largest one to improve accuracy:

$$|A_s| = \begin{cases} \dfrac{2\pi\delta}{N}\dfrac{|S[k]|}{\sin(\pi\delta)} \\ \\ \dfrac{2\pi(1-\delta)}{N}\dfrac{|S[k+1]|}{\sin(\pi(1-\delta))} \end{cases} \tag{3.1.12}$$

And the phase may be obtained as

$$\varphi_s = \begin{cases} \angle S[k] - a\delta + \dfrac{\pi}{2} \\ \\ \angle S[k+1] - a(\delta-1) + \dfrac{\pi}{2} \end{cases} \qquad \text{with} \quad a = \pi\frac{N-1}{N} \tag{3.1.13}$$

A few years later the first extensions to tapered signals (i.e. using smoothing windows, please see next section 3.1.3.2) began to appear, often focusing on the most widely used windows, such as the Hanning [138].

For the Hanning window (for which the resulting spectrum is analytically known), the fractional frequency correction δ[5] is given by

$$\delta = \frac{2\alpha-1}{\alpha+1} \tag{3.1.14}$$

and the corrected complex amplitude (thus including the phase information) is

$$A_s = \begin{cases} \dfrac{2\pi\delta(1-\delta)}{\sin(\pi\delta)}\, e^{-i\pi\delta}(1+\delta)\, S[k] \\ \\ \dfrac{2\pi\delta(1-\delta)}{\sin(\pi\delta)}\, e^{-i\pi\delta}(\delta-2)\, S[k+1] \end{cases} \tag{3.1.15}$$

The use of three-sample interpolation to improve the accuracy of the correction is described in [10] again for the Hanning window. A new ratio α is introduced, that takes into account all the three spectrum samples:

$$\alpha = \frac{|S[k]| + |S[k-1]|}{|S[k]| + |S[k+1]|} \tag{3.1.16}$$

The frequency correction for the rectangular and the Hanning windows is

[5] The author in [138] indicates the fractional frequency correction as x_m.

$$\delta_R = s_{\delta R} \frac{|S[k+1]| + |S[k-1]|}{2\,|S[k]| + ||S[k+1]| - |S[k-1]||} \tag{3.1.17}$$

with $s_{\delta R} = \text{sign}(|S[k+1]| - |S[k-1]|)$, and

$$\delta_H = 2\frac{|S[k+1]| - |S[k-1]|}{|S[k-1]| + 2\,|S[k]| + |S[k+1]|} \tag{3.1.18}$$

A set of expressions is also proposed in [10] for estimation with the Hanning window when going beyond the three-point interpolation, that is five- and seven-point interpolation. The three-point interpolation expression is reported below, while the other more complex ones may be found in [10]:

$$A_{3H} = \frac{\pi\delta}{\sin(\pi\delta)} \frac{(1-\delta^2)(4-\delta^2)}{3} \left(|S[k-1]| + 2\,|S[k]| + |S[k+1]|\right) \tag{3.1.19}$$

Finally, in [112] the problem of interpolation is solved numerically, applying the results to the class of Dolph-Chebyshev and Kaiser-Bessel windows. The author, after a quick review of two-point and three-point interpolation techniques (see above) and underlying that the latter is applicable only to windows of the $\cos^n(\cdot)$ type, presents a numeric solution of general applicability using approximating polynomials:

- the ratio α (renamed R in the paper) of the amplitude of spectrum terms keeps unchanged as reported in eq. (3.1.11) and (3.1.16);
- $\delta = P_\delta(\alpha)$ is determined by a polynomial approximation of order L;
- the correction of amplitude is applied in the case of both two-point and three-point interpolations, using the following expressions, that take the coefficients of the window transform $W[k]$[6].

$$P_{2A} = \frac{|W[0]|}{|W[\delta]|} \qquad P_{3A} = \frac{\dfrac{|W[-1]| + 2\,|W[0]| + |W[+1]|}{|W[-1-\delta]| + 2\,|W[-\delta]| + |W[1-\delta]|}}{2\left(1 + \dfrac{|W[1]|}{|W[0]|}\right)} \tag{3.1.20}$$

The window transform is in principle defined only for integer values of its index argument k, so that writing e.g. $|W[-\delta]|$ does not make any sense, being δ a small number between 0 and 1. The author underlines that dense resampling of the frequency axis near the origin of the window ($\omega = 0$) is necessary and in the code example given in his Appendix he uses a 64 density factor. For such calculation DFT or CZT may be used.

As for the correction coefficients above, the corrected amplitude is:

[6] In C. Offelli and D. Petri, "Interpolation techniques for real-time multifrequency waveform analysis," *IEEE Trans. on Instrumentation and Measurement*, vol. 39, no. 1, pp. 106–111, Feb. 1990, it was noticed that the interpolation coefficients may be calculated based on the DFT bins of the taper window rather than those of the signal.

$$A_s = \begin{cases} P_{2A}\ |S[k]| & \text{for two} - \text{point interp.} \\ \\ P_{3A}\ (|S[k-1]| + 2\,|S[k]| + |S[k+1]|) & \text{for three} - \text{point interp.} \end{cases} \tag{3.1.21}$$

Finally, two general and useful reviews of frequency estimation methods are:

- J. Schoukens, R. Pintelon and H. Van Hamme, "The Interpolated Fast Fourier Transform: a Comparative Study," *IEEE Transactions on Instrumentation and Measurement*, Vol. 41, no. 2, April 1992, pp. 226-232; doi: 10.1109/19.137352
- I. Santamarida, C. Pantaleon and J. Ibanez, "A comparative study of high-accuracy frequency estimation methods," *Mechanical Systems and Signal Processing*, Vol. 14, no. 5, 2000, pp. 819-834; doi: 10.1.1.32.4945.

3.1.4 Infinite time duration signals

In Chapter 2 we saw that impulse and step signals are quite relevant in describing propagation along transmission lines, represent two primitive functions for the determinations of transfer functions, and will be used extensively in time-domain reflectometry (TDR) and transmission (TDT). For TDR and TDT being implemented directly in time domain (e.g. see sec. 6.5.1 and 10.4.4) or by frequency domain measurements (see sec. 11.10), we clarify a little what is behind the transformation of limited and infinite time duration signals, the impulse being an example of the former and the step signal of the latter.

When applying various forms of discrete Fourier transform (e.g. numeric transforms, such as DFT and FFT), the periodization of the signal implies that it is assumed of finite time duration (see Figure 3.1.7(a) and (b)); the chosen observation period is shorter or equal to the signal time duration and the periodization requires that no discontinuities occur when joining in a cyclic fashion the beginning and the end of such observation period. Even if in general equal values (and equal derivatives) are required for no artifacts due to signal truncation and leakage to appear (see sec. 3.1.3), for the considered signals this translates into a "smooth return to zero" requisite. This is addressed by multiplicative tapering windows: depending on the window the resulting signal after multiplication is smoothly brought to zero at its extremes together with a certain number of its derivatives, of course at the expense of some waveshape deformation and attenuation (the latter is compensated by taking into account the window gain). While this is satisfactory for small discontinuities and imperfections (e.g. caused by approximate estimate of signal period, timing inaccuracy, etc.), for step signal the straightforward application of tapering windows brings to unavoidable errors and underestimations.

Since step-like signals are quite an important class of signals used for time-domain reflectometry and VNA low-pass processing mode, special procedures have been proposed for their transformation, preserving characteristics such as dc value and a dense enough frequency resolution of the resulting transform. The first attempt[7] tried to remove the discontinuity by differentiating the step-like signal before transformation, leading thus the problem back to working with impulse-like signals only, but losing dc value information and increasing noise remarkably. This is also the approach used by LeCroy for their SPARQ software[8]: before differentiating waveforms are de-noised first by averaging them and then using a proprietary wavelet-based software.

Nicolson [252] then proposed to drive the step $f(t)$ back to zero after the initial transition (that brings the high frequency information) by subtracting a ramp signal $r(t)$ of the same height of the step $r(T)=f(T)$, with the resulting signal linearly approaches zero, so with still a discontinuity in the first derivative. Even more drastically Gans and Nahman[9] brought to zero the step-like signal by subtracting a delayed copy $\tilde{f}(t)=f(t-T)$: this is the method for cheaply generating a short impulse at the output of a step generator using the reversed reflection from a shorted stub of adequate length; so, despite being rough from a mathematical point of view (discontinuities are all there), the method could be attractive for direct hardware implementation. Although procedurally different, it was demonstrated by Waldmeyer[10] that the three methods lead substantially to the same result (if evaluated on the harmonic frequencies of the base time duration T) and that they all aim at solving (or reduce) the distortion related to the truncation error.

A straightforward comparison of the resulting spectra of the last two methods highlights that the frequency components are complementary, the former resulting in even-order components, while the latter in odd components (and uneven separation on the frequency axis between dc and the first one). Following this observation it was proposed to overlap the two, obtaining a denser frequency representation[11].

What Waldmeyer concludes is that, first, the three methods somehow convert a step-like signal to an impulse-like one and unavoidably disturb the dc value, and, second, they are all equivalently giving as solution the transform of the truncated step-like signal $F(k)$ corrected by an additive term $f(t=T)/(1-\exp(-j2\pi k/N))$. He clearly states that this is quite a generally applicable formula, in that for an impulse-like signal (and any signal that returns to zero at the end of the time interval) the added corrective term is zero.

It was however with Cormack and Binder [90] that a thorough presentation of the problem and a quantitative assessment were offered: they consider all the mentioned techniques, together with a more elegant signal analytical extension, and including also the basic windowing approach. Given the step signal that they call $f_2(t)$, the ex-

---

7 H.A. Samulon. "Spectral analysis of transient response curves," *Proc. of IRE*, vol. 39, pp. 175-186. 1951.

8 A. Blankman, "LeCroy SPARQ S-Parameter Measurement Methodology," Technical Brief, Rev 3, July 2011.

9 W.L. Gans and N.S. Nahman, "Continuous and discrete Fourier transforms of step-like waveforms," *IEEE Transactions on Instrumentation and Measurement*, Vol. 31, June 1982, pp. 97-101. doi: 10.1109/TIM.1982.6312529

10 J. Waldmeyer, "Fast Fourier transform for step-like functions: The synthesis of three apparently different methods," *IEEE Transactions on Instrumentation and Measurement*, Vol. 29, no. 3, March 1980, pp. 36-39. doi: 10.1109/TIM.1980.4314858.

11 A.M. Shaarawi and S.M. Riad, "Computing the complete FFT of a step-like waveform," *IEEE Transactions on Instrumentation and Measurement*, Vol. 35, no. 1, Jan. 1986, pp. 91-92. doi: 10.1109/TIM.1986.6499064.

tension is nothing more than pairing a preceding and following signal segment $f_1(t)$ and $f_3(t)$ (called "precursor and continuation functions") that are joined to the existing signal. minimizing the discontinuity in value and derivatives and possibly going smoothly to zero when the observation window is going to end. To this aim the authors propose precursor and continuation functions based on a constant (that is a delayed step) and an exponential function whose three coefficients are determined based on continuity of the signal and the first two derivatives at the joining point (e.g. $t=0$ and $t=T$). The paper contains a thorough interesting discussion on how to bring the extension function segments beyond T back into the $[0, T]$ interval (time aliasing), obtaining two new functions $g_1(t)$ and $g_3(t)$, called "compensation functions", for which Nicolson ramp is a special case without exponential tapering. Pragmatically the authors warn against the possibility that an unwise selection of the end of the observation window T might occur at a point in the signal with a false local slope that brings to an ever increasing solution, instead of a gently decaying exponential; in this case T shall be moved slightly until the characteristics of the signal allow a solution with the desired behavior (this is shown exaggeratedly in Figure 3.1.7(c)). It is shown that the Fourier transform of the compensated signal defined over $[0, T]$ is the same of the original extended signal except for $1/T$ scaling, if the other frequency aliasing terms are neglected (that can be done if the number of points is large).

In [92] the same Waldmeyer's expression for spectral correction is considered,

$$F[k]=T\left[\mathrm{DFT}\{f[n]\}+\frac{f[N-1]}{N(1-e^{-j2\pi k/N}}\right] \tag{3.1.22}$$

deriving also an equivalent $\mathrm{sinc}^2(\cdot)$ correction term for Cormack's EF-FFT method.

Once all function terms are defined over the $[0, T]$ interval, they show also how a multiplicative tapering window may be defined that is equivalent to the additive precursor and continuation compensation functions. They don't give a general solution that would be quite complex, but solve the problem for the specific case of a Hanning window (see sec. 3.1.3.2): the resulting compensation functions are zero for the precursor and the known $\cos(\cdot)$ downward shape for the continuation term over the added T interval $[T, 2T]$; although the tapering to zero is achieved, the authors comment that the resulting signal shape evaluated over the continuation time interval beyond T is not close to the expected physical behavior, even prolonging the time interval after T. It is not absolutely clear if this occurs for all tapering windows in general and if there are better choices, but the example proposed in [90] comparing the use of extensions and Hanning window show a remarkable difference with the Hanning estimated response nearly 2 dB lower than an expected value of 33 dB; bringing the fast rising part of the signal (featuring the high frequency content) at the center of the Hanning window reduces the error to about 0.5 dB. The authors comment this as a confirmation of the "undesirable unequal weight given to each sample in the data set by a conventional window", that is the difference in the compensation function for the precursor (zero) and the extension (the Hanning slope). This rings a bell regarding the complaint for not centering the tapering window on the high-frequency characteristics of signal, as pointed out when discussing VNA time gating (see sec. 11.9).

Needless to say, frequency resolution may be further improved if the length of the continuation function is much longer than the original data size: Cormack, Blair and McMullin [91] show a few examples of 2- and 16 fold increase, with and without decaying exponential (i.e. using a flat continuation function or a tapered exponential).

It may be said that EF-FFT by Cormack et al. sounds very accurate and flexible, although there is always the choice of a direct DFT or CZT implementation including the correction for step-like waveforms, using the theoretically correct Waldmeyer's formula (reported also in [92]). The satisfactory performance of the latter is confirmed by Nicolson [251] in his pioneering work, where the discrepancies with respect to theoretically and frequency-domain proofed examples is really a fraction of dB (mostly smaller than 0.1 dB). The use of impulse-like signals, of course, solves the problem at the source, but brings along another problem, that of limited spectral power that excites each component with an intensity much smaller than that of the step-like signal with a similar rise time.

Let's consider now the example waveform in [92] and shown in Figure 3.1.8: a step function with a superimposed sinusoidal ringing with exponential decay.

$$x(t) = \left\{U_S + U_0 e^{-a(t-t_s)} \sin(2\pi f_s (t-t_s))\right\} u(t-t_s) \tag{3.1.23}$$

The time duration is $t = 0 \ldots T$, $U_S = 1$, $U_0 = 0$, $a = 12.5/(N-1)$, $f_s = 75/(N-1)$. The total number of points is N (=256 in the original example, but we consider also 4096) and the relevant parameter for the on-going discussion is the step time position t_s ($= 200/(N-1)$ in the original example, and expressed here as a fraction of the time duration T).

The hand-calculated Fourier spectrum is compared with that resulting from direct application of DFT using smoothing windows and applying also Waldmeyer's correction C. Time domain vectors are expressed as a function of $n = 0 \ldots N-1$ (or $t = 0 \ldots (N-1)\Delta t$, with $\Delta t = T/N$) and frequency domain vectors are expressed as a function of $k = 0 \ldots N-1$ (or $f_k = 0 \ldots (N-1)\Delta f$, with $\Delta f = 1/(NT)$); the original notation [92] is partly discrete and partly continuous, expressing frequency with pulsation $\omega = 2\pi f$.

$$X(f_k) = U_S \frac{e^{-j\omega_k t_s}}{j\omega_k} + U_0 2\pi f_k \frac{e^{-j\omega_k t_s}}{(a + j\omega_k)^2 + (2\pi f_s)^2} \tag{3.1.24}$$

The details of calculations are shown in Example 3.2 reporting the Matlab scripts.

The correction to be applied to the DFT calculated spectrum [92], that for a matter of publication date we call "Waldmeyer's correction", is given by

$$C(f_k) = \frac{x(N-1)}{N\left[1 - e^{-j2\pi k/N}\right]} \tag{3.1.25}$$

After comparing the results shown in Figure 3.1.9 and paying attention to the two relevant parameters, that is the number of points N and the position of the step rising edge t_s, it is possible to draw the following conclusions:

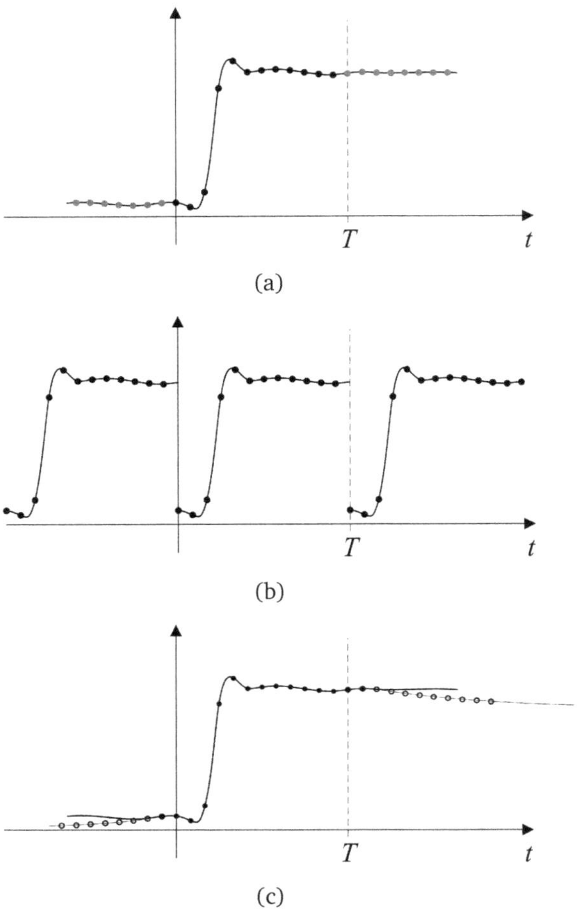

Figure 3.1.7 – Step-like signal transformation: (a) example of step-like signal with samples outside the established $[0, T]$ interval; (b) truncation and periodization of DFT calculation; (c) application of Cormack and Binder method with precursor (left) and continuation (right) sections; a few samples are added before 0 and after T to taper derivatives locally.

- Waldmeyer's correction shall be applied to the original $X(f_k)$ by summing vector C; for the two vectors real and imaginary parts are synchronized and give the correct amplitude profile, while this is not true for the smoothed versions;

- spectra obtained using smoothing windows are moderately correct, but cannot be further corrected;

- Waldmeyer's correction corrects the amplitude at the resonance and the curve slope after that; it is far more effective if the number of points is large;

- by aligning the transient edge at the center of the window (i.e. $t_s = 0.5T$), not only correction C performs slightly better, but all spectra obtained with the direct method (e.g. windowed spectra) are closer to the correct one and the error in general is acceptable, even if still noticeable;

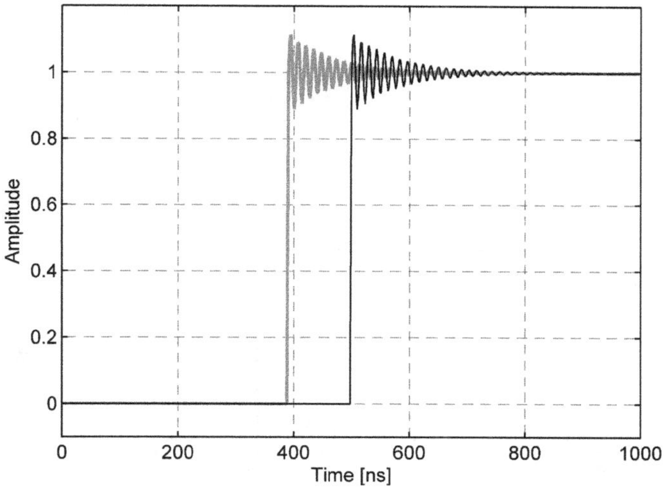

Figure 3.1.8 – Step-like waveform with ringing for testing of DFT, windowing and Waldmeyer's correction: the gray curve is the one appearing in [92], the black one aligned at the center will be used to demonstrate the relevance of centering the time window. $T = 1$ μs was set arbitrarily.

- this latter remark supports the need to correctly align the window during time gating of VNA data, as said above and considered later in sec. 11.9.

Example 3.2: Performance of DFT and Waldmeyer's correction for step-like signal

Let's take a closer look at the implementation of the expressions, especially for a few changes to the normalization by T in `XT1` and `XT2` expression, with respect to the original one.

```
N=4096; n=0:1:N-1; T=1e-6; t=n/N*T; Us=1; U0=0.125;
a=12.5/(N-1); fr=75/(N-1);
NS=round(0.39*N); ts=NS*T/N; ns=round(ts/T*N);
u_0=[ones(1,N)]; u_ns=[zeros(1,ns) ones(1,N-ns)];
x=(Us + U0.*exp(-a*(n-ns)).*sin(2*pi*fr*(n-ns))).*u_ns;
%% True spectrum (Cormack)
wn=2*pi*n/T;
XT1 = Us.*exp(-1i*wn*ts)./(1i*wn*T);
XT2 = U0.*(2*pi*fr*N).*exp(-1i*wn*ts)./((a*N+1i*wn*T).^2+(2*pi*fr*N).^2);
XT = (XT1+XT2);
%% Basic FFT
X0=1/N*fft(x); X0(1)=X0(1)*0.5; X0=X0(1:N/2); f=wn(1:N/2)/(2*pi);
%% Windowed FFT
wh=hanning(length(x)); X1H=1/N*fft(x.*wh');
X1H(1)=X1H(1)*0.5; X1H=X1H(1:N/2);
wdc80=chebwin(length(x),80); X2dc80=1/N*fft(x.*wdc80');
X2dc80(1)=X2dc80(1)*0.5; X2dc80=X2dc80(1:N/2);
%% Walmeyer's correction
C=x(N)./(N.*(1-exp(-1i*2*pi*n/N))); C=C(1:N/2);
```

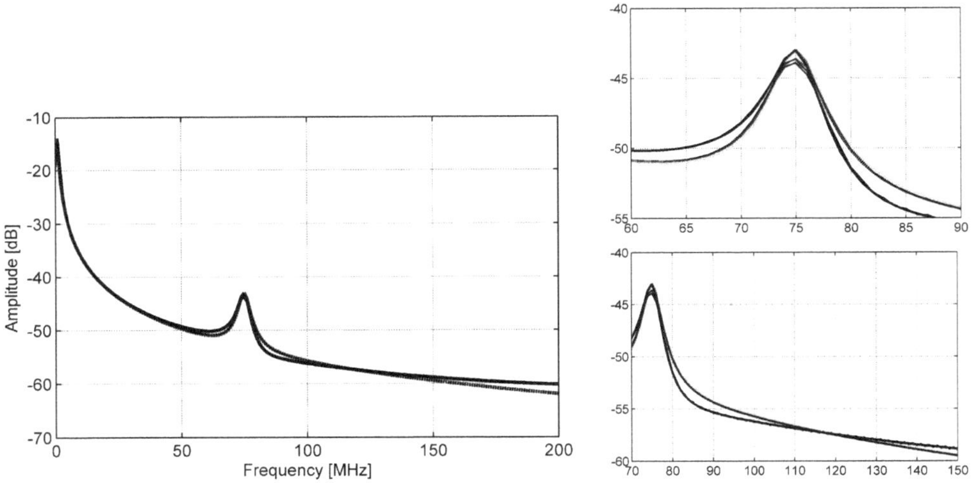

(a)

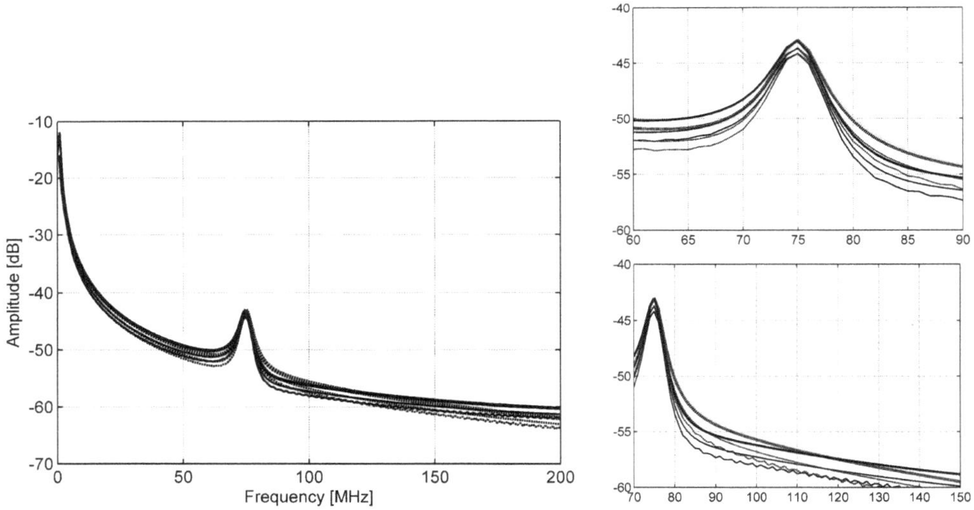

(b)

Figure 3.1.9 – Spectra for the two considered step-like waveforms for (a) transient at the original position $t_s = 0.39T$ and (b) transient aligned at center position $t_s = 0.5T$; each figure reports the magnitude of hand-calculated $X(f_k)$ (thick gray), two DFT spectra with Hanning and Dolph-Chebishev 80 dB windows (thin black) and Waldmeyer's correction applied to non-windowed DFT (thick black). Solid lines are obtained with $N = 512$ (as in the original example) and dashed lines with $N = 4096$.

3.2 Rise time and bandwidth

3.2.1 Rise time

In simple terms, considering a signal step function between two voltage levels V_- and V_+, the rise time is the time taken by the signal to change from a specified low state (or low level, V_L) to a specified high state (or high level, V_H). With the word "specified" we intend that these levels are known, as well as any related tolerance and criterion to assess that the signal has reached the high state; they are for example established by a standard, an internal procedure, or are typical of a class of digital signals (such as e.g. TTL logic levels). Think for instance to an oscillating behavior, with the signal crossing the high level threshold upward, but then crossing again in the opposite direction going farther away. A signal may be reaching the desired level in a stable way, remaining in a tolerance band around or above the level, or oscillating far from the crossed threshold line, outside the tolerance band; this is exactly how the settling time of operational amplifiers is defined.

Typically these low and high threshold pairs (V_L, V_H) in electronics and signal analysis are taken as $(10\%, 90\%)$ or $(20\%, 80\%)$, depending on the type of signal and the electrical context, or undergo more detailed specifications, such as for the input and output signals for some old logic families.

The rise time may be also defined going beyond the measurement of a time interval between two crossing points, that is yet the preferred method, that can be implemented in a very straightforward way with oscilloscopes and visually evaluating the waveshape: definitions based on bandwidth and impulse response give almost equivalent results, as it is shown in the following.

3.2.1.1 Rise time based on threshold crossing (10-90% and 20-80%)

Crossing a pair of thresholds may be defined as a suitable criterion, identifying the low and the high level so to mask out both superimposed noise and any waveform deformation when detaching from the lower pedestal and when reaching the upper level. Deformation, however, occurs for many reasons related to the behavior of the electronic circuits, such as variable gain, slew rate limitation, saturation, variable output impedance, ringing, etc.

The most diffused threshold pair is 10% / 90% and the corresponding rise-time is indicated as t_{10-90}. When a large amount of ringing and reflections is present (also because of the wide frequency spectrum of signals), a safe identification of the beginning and the end of the rising edge is performed relying on more robust thresholds, 20% and 80%, and the rise-time is indicated as t_{20-80}.

3.2.1.2 Rise time based on slope during transition

The rise time may be also related to a slope concept, while the signal is moving from the lower to the upper level: the center slope value ($t_{ctr.sl.}$) is captured at 50% of amplitude between the two steady levels; the identification of the maximum slope value

($t_{max.sl.}$) on the contrary necessitates some post-processing (not only the mathematical operations for its quantification from the measured data, but also some filtering and smoothing) and corresponds to the steepest slope during the transition.

The advantage of the center slope definition is that the measurement is taken right midway between the two levels, so in the best position for rejection of noise and ringing. However, the measurement of the slope, and then the calculation of the corresponding rise-time value, requires some post-processing, being not readily available on all oscilloscopes:

$$t_{ctr.sl.} = \frac{\mathrm{V}_H - \mathrm{V}_L}{slope(50\%)} \tag{3.2.1}$$

The maximum slope rise time is calculated in the point of the rising edge where the steepest slope is located, usually close to the beginning:

$$t_{max.sl.} = \frac{\mathrm{V}_H - \mathrm{V}_L}{slope(\max)} \tag{3.2.2}$$

For simple first-order low-pass circuits (RC and RL circuits) the maximum slope method gives as result the time constant of the circuit, that occurs exactly at the beginning.

3.2.1.3 Rise time based on bandwidth

Rise time may be also based on an agreed bandwidth, such as -3 dB, -6 dB or the Equivalent Noise Bandwidth, ENBW. For the first two bandwidths the definition of rise time is quite operative: first, find in the signal spectrum the frequency at which the amplitude is 3 dB or 6 dB below the dc value; then, the rise time is determined with inverse proportionality using π at the denominator as the scaling factor (some authors propose instead using a multiplicative factor of 0.3 to 0.45 at the numerator, as discussed later at the end of sec. 3.2.4).

$$t_{-3\,\mathrm{dB}} = \frac{1}{\pi B_{-3\,\mathrm{dB}}} \qquad t_{-6\,\mathrm{dB}} = \frac{1}{\pi B_{-6\,\mathrm{dB}}} \qquad t_{\mathrm{ENBW}} = \frac{1}{\pi B_{\mathrm{ENBW}}} \tag{3.2.3}$$

It is recalled that the Equivalent Noise Bandwidth criterion is indicated by some authors as the “rms” criterion.

The reason for using such definition is that it allows to pass rapidly from time to frequency domain, and back. For example the EN 61188-1-2 standard [76] reports in its Table 1 the rise times of most common logic families, evidently using this criterion, even if not explicitly said.

3.2.1.4 Rise time based on impulse response

In this case the analyzed signal is assumed as the result of an ideal step function applied to a linear system with impulse response $h(t)$, which may be determined as the first derivative of the step response. Of course such operation is correct in theory, but has the serious drawback of amplifying the noise and high frequency disturbance, so that filtering and smoothing is nearly unavoidable; in reality, the relevance of this definition is more theoretical than practical. The rise time is defined as the standard deviation of such impulse response multiplied by a scaling factor $\sqrt{2\pi}$.

$$\sigma^2 = \int_{-\infty}^{+\infty} t^2 \frac{h(t)}{H(0)}\,\mathrm{d}t - \left[\int_{-\infty}^{+\infty} t\,\frac{h(t)}{H(0)}\,\mathrm{d}t\right]^2 \tag{3.2.4}$$

$$t_\sigma = \sqrt{2\pi}\,\sigma \tag{3.2.5}$$

where $H(0)$ is the dc gain of the system (the dc value of the Fourier transform of $h(t)$).

3.2.2 Rise time through LTI systems

Let's consider a linear time invariant (LTI) system, such as a linear electric circuit, e.g. a RC low-pass filter (or something more complex than that). This system has an impulse response $h(t)$, a corresponding frequency response $H(f)$, that for simplicity has a low-pass behavior, thus with a bandwidth B that corresponds to the upper -3 dB cut-off frequency f_h, being the lower one $f_l = 0$. This assumption holds quite well because we are focusing mostly on measuring instruments and devices that have all a more or less extended low-pass response (multimeters, oscilloscopes, their probes, etc.). Again for simplicity we subdivide broadly low-pass system responses into first-order (one pole) system, Gaussian response system and second-order (two poles, either real or conjugate complex) system, identified by the following subscripts: "I", "g", "II".

Let's apply an ideal step function to each of the systems above and determine the output response and its properties.

3.2.2.1 First-order low-pass system

For this step response, the time domain relationship is:

$$V(t) = V\left(1 - e^{-t/\tau}\right) \tag{3.2.6}$$

and solving for t

$$t = -\tau\,\ln\left(1 - \frac{V(t)}{V}\right) \tag{3.2.7}$$

The bandwidth intended as the -3 dB cut-off frequency is related to the time constant by $B = 1/(2\pi\tau)$.

The transient output response is made of three time intervals:

- $[0,\, t_L]$ for the output voltage to go from V_- (assumed 0 in practice for all logic families) to the lower threshold level V_L,
- $[t_L,\, t_H]$ for the output voltage to go from V_L to the lower threshold level V_H),

- $[t_H, t_F]$ for the output voltage to go from V_H to the final level $V_F = V_+$).

Taking the (10%, 90%) definition of rise time, we have

$$t_L = -\tau \ln(1-0.1) = 0.1054\,\tau \text{ and } t_H = -\tau \ln(1-0.9) = 2.3026\,\tau,$$

so that the rise time is $t_{r,10-90\%} = 2.1972\,\tau$ and the relationship with the bandwidth is $B = 0.3497/\tau$. Conversely, taking the (20%, 80%) definition of rise time, we have $t_L = -\tau \ln(1-0.2) = 0.2231\,\tau$ and $t_H = -\tau \ln(1-0.8) = 1.6094\,\tau$, so that the rise time is $t_{r,20-80\%} = 1.3863\,\tau$ and the relationship with bandwidth is $B = 0.2206/\tau$.

3.2.2.2 Gaussian response system

For a Gaussian response low-pass system [12], with a frequency response $|H(\omega)| = e^{-\omega^2/\sigma^2}$, a bandwidth $B_{-3\,\mathrm{dB}} = \frac{\sigma}{2\pi}\sqrt{\frac{3\ln 10}{20}} \simeq 0.0935\,\sigma$, and a corresponding time-domain impulse response given by the inverse Fourier transform of the frequency response $h(t) = \frac{\sigma}{2\sqrt{\pi}}\, e^{-\sigma^2 t^2/4}$, the output step response is calculated with the convolution integral:

$$V(t) = V\,\frac{1}{2}\left[1 + \operatorname{erf}\left(\frac{\sigma t}{2}\right)\right] \tag{3.2.8}$$

Again, for the three time intervals $[0, t_L]$, $[t_L, t_H]$ and $[t_H, t_F]$ we can calculate the corresponding time instants by comparing the output voltage with the selected levels. Taking the (10%, 90%) definition of rise time, we have $t_{10} = -1.8122/\sigma$ and $t_{90} = +1.8122/\sigma$, obtained by solving the equations $0.1 = \frac{1}{2}\left[1 + \operatorname{erf}(\sigma t_{10}/2)\right]$ and $0.9 = \frac{1}{2}\left[1 + \operatorname{erf}(\sigma t_{90}/2)\right]$, both solved for the argument of the $\operatorname{erf}(\cdot)$ equal to ± 0.9062. Once multiplied by 2 (as required for the argument of the $\operatorname{erf}(\cdot)$), the total rise time is the difference $t_{90} - t_{10}$, that is twice t_{90}: $t_{r,10-90\%} = 3.6244/\sigma$ and the relationship with the bandwidth is $B_{-3\,\mathrm{dB}} = 0.3389/\tau$.

Conversely, taking the (20%, 80%) definition of rise time, we have $t_L = -1.1902/\sigma$ and $t_H = +1.1902/\sigma$, so that the rise time is $t_{r,20-80\%} = 2.3804/\sigma$ and the relationship with the bandwidth is $B_{-3\,\mathrm{dB}} = 0.2226/\tau$.

During these calculations we have had evidence that a Gaussian response system is non-causal and its impulse response is in principle non-zero over the entire time interval from $-\infty$ to $+\infty$. Strictly speaking it is thus a non-realizable system, except that truncation to a convenient time duration may be operated and then a time shift by the same amount is necessary, to have the beginning of the left tail of the impulse response coincide with $t = 0$. Truncation may be operated when the impulse response value is sufficiently small to be negligible; the Gaussian curve may be characterized by points

[12] An attractive feature that justifies the use of the Gaussian response is that it is the filter with the minimum group delay (see sec. 3.3.1 and 3.3.1); from the point of view of the step transient response, a Gaussian system minimizes the rise time fulfilling the "no overshoot" condition.

corresponding to a know fraction of the peak value, taken equal to unity: the half maximum (0.5) and the tenth of maximum (0.1) points are 1.1774σ and 2.1460σ, respectively. Considering a threshold of 1%, that is a hundredth of the maximum value, we obtain 3.0349σ, only 41% larger than the "tenth of maximum" value, testifying the rapid decay of the Gaussian response. Going to 0.1% the truncation shall be shifted backward to 3.7169σ.

This approach lends itself to digital implementation, using a Finite Impulse Response (FIR) filter with the desired response. Yet, for analog implementation there is no feasible way to implement truncation and time-shift unless expensive and cumbersome delay lines are used. Rather, by exploiting the Central Limit Theorem, the cascaded connection of a large number of simple first-order low-pass filters gives an overall Gaussian response[13]. Alternatively, high-order linear phase analog filters have responses very close to the Gaussian one: Bessel filters ensure the linear-phase response by construction and with a moderate order they ensure a very good approximation of the Gaussian response; this is treated in more detail in Example 3.3 and Table 3.2.2.

Example 3.3: Considerations on the approximation of the Gaussian response

A Gaussian response is not readily available, but can only be approximated with analog or digital solutions:

- for analog implementation there are two options:
  - a cascaded connection of N first-order low-pass cells leads to an approximately Gaussian response thanks to the Central Limit Theorem: see Figure 3.2.1 for a comparison of the impulse and frequency response; it is evident that the filter rise-time remains too slow, detaching well from the base line at zero, but slowing down already at abut 20% (the equivalence with the Gaussian filter consists of equal time constants, so invoking eq. (3.2.15) below, the low-pass filter time constant is reduced proportionally to the square root of the number of stages, to result in the same overall time constant; the time constant is used directly with the `zpk()` function);
  - a linear phase filter, such as a Bessel or elliptic architecture, is able to fit the desired Gaussian response even better with a moderate filter order, while keeping the phase response linear (see Figure 3.2.1); performance is much better and especially for filter orders of 3 and 4 (the numeric values stating the performance are reported in Table 3.2.2, together with the other filters (similarly to the low-pass filter, the equivalence consisted of a bandwidth increased proportionally to the square root of filter order, to be fed to the `besself()` function).
- the digital implementation may be achieved with a FIR filter that matches the Gaussian response, without worrying about the linearity of the phase response, usually ensured by the FIR architecture and in any case achieved *in extremis* by using forward and backward time filtering (see Matlab function `filtfilt()`); it is not a major issue once truncation and time shifting of the Gaussian response is anyway accepted.

[13] This statement is based on the same rationale used to demonstrate the effect of cascaded LTI blocks on the rise time of the propagating signal in sec. 3.2.4.

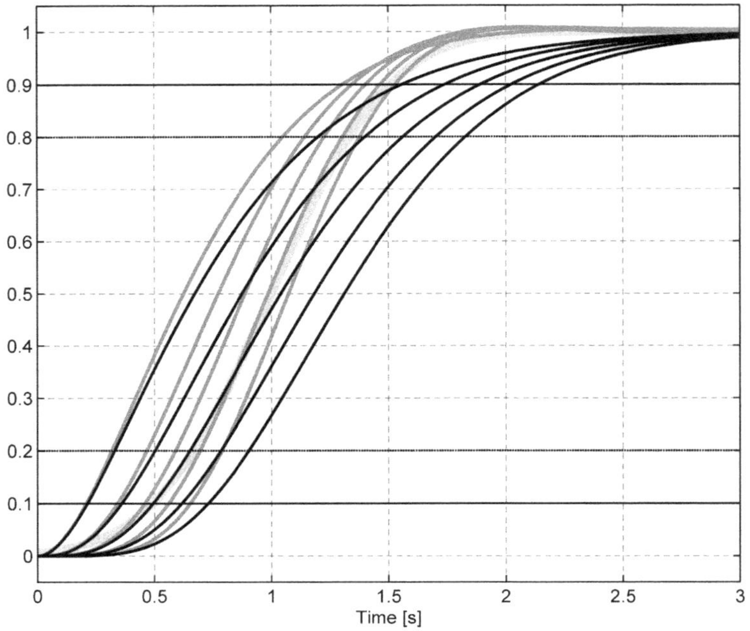

(a)

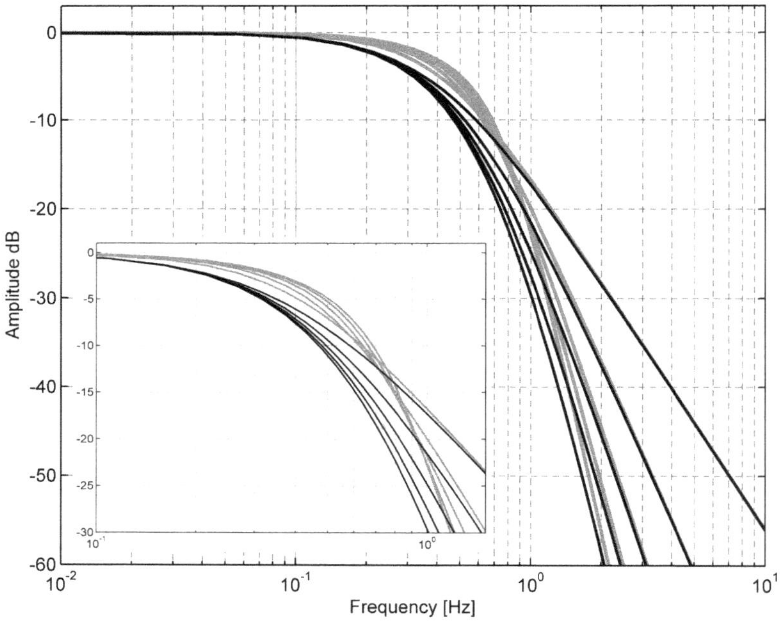

(b)

Figure 3.2.1 – Comparison of cascaded single-pole low-pass filters (black curves, for order 2 to 6) and Bessel filters (gray curves, for order 2 to 6), to simulate Gaussian response (light gray thick curve): (a) time domain step response, (b) frequency response. (The order for single-pole low-pass filters corresponds to the number of filter sections, i.e. the number of poles; the Bessel filter order corresponds to the number of complex pole pairs.)

3.2.2.3 Second-order low-pass system

For a second-order low-pass system the transfer function is characterized by the natural resonance frequency ω_n and the damping factor ζ (that by the way is inversely proportional to the factor of merit Q as $\xi = 1/2Q$):

$$H(s = j\omega) = 1/\left[\omega_n^2 + 2\zeta\omega_n s + s^2\right] \tag{3.2.9}$$

For the step response, the following cases are distinguished: critically damped system ($\xi = 1$), under-damped system ($\xi < 1$) and over-damped system ($\xi > 1$).

$$y_{cd}(t) = 1 - e^{-\omega_n t}\,(1 + \omega_n t) \tag{3.2.10}$$

$$y_{ud}(t) = 1 - \frac{e^{-\xi\omega_n t}}{\sqrt{1-\xi^2}}\sin(\omega_d t + \theta) \tag{3.2.11}$$

$$y_{od}(t) = 1 - \frac{1}{2\omega_n\sqrt{\xi^2-1}}\left(s_2 e^{s_1 t} - s_1 e^{s_2 t}\right) \tag{3.2.12}$$

where $\theta = \arctan\left(\sqrt{1-\xi^2}/\xi\right)$ and s_1 and s_2 indicate two real poles ($s_{1,2} = -\xi \pm \sqrt{\xi^2 - 1}$, when the system is overdamped); $\omega_d = \omega_n\sqrt{1-\xi^2}$ is the damped natural frequency (defined for $\xi < 1$, that is when system response is oscillatory).

A sketch of the typical step response of a second-order system in the three conditions (underdamped, critically damped and over-damped) is shown in Figure 3.2.2.

The determination of the (10%, 90%) and (20%, 80%) rise times is quite complex, not only because there are three different equations describing system response for varying ξ, but also because they are trascendental equations; a numeric approach has been followed then, and the results are summarized in Table 3.2.1.

Some further considerations follow that are useful to interpret measurement results.

For the under-damped case the overshoot (or better the maximum overshoot, that occurs in the first half period and corresponds to the peak value V_p minus the steady state value V, once the transient response has vanished) in its fractional expression is

$$\text{overshoot}\% = 100\left(\frac{-\xi\pi}{\sqrt{1-\xi^2}}\right) \tag{3.2.13}$$

The "peak time" t_p, that is the time measured from the detachment of the waveform from the lower level at which the peak value occurs, is

$$t_p = \frac{\pi}{\omega_n\sqrt{1-\xi^2}} \tag{3.2.14}$$

These two equations relate the internal system parameters ξ and ω_n to quantities that are easily measurable: the peak value corresponds to the oscilloscope peak value

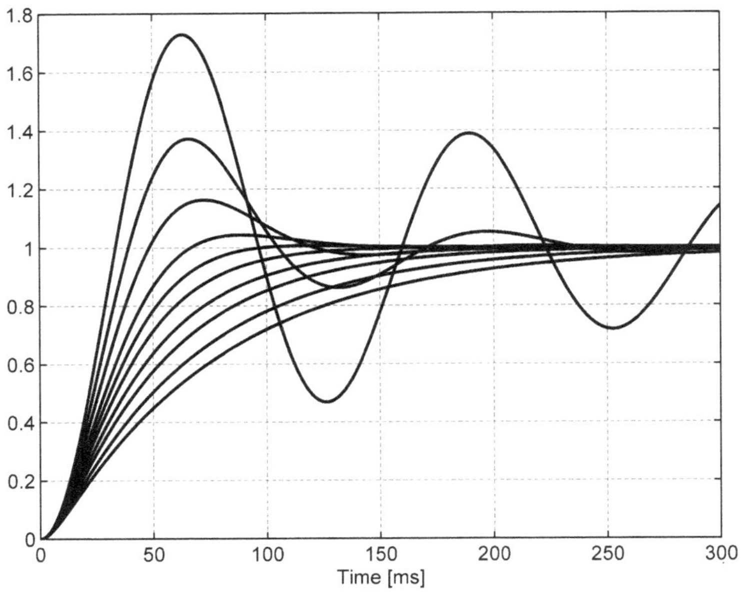

Figure 3.2.2 – Step response of a second-order system with varying damping: $\xi =$ 0.1, 0.3, 0.5, 0.7, 0.85 (under-damped), $\xi = 1$ (critically damped) and $\xi = 1.2$, 1.4, 1.7, 2.0 (over-damped).

marker, subtracting the steady state value (another marker may be used or directly the applied step amplitude V); the peak time is again measured by the same peak marker with respect to the trigger instant. Eq. (3.2.13) is then solved for ξ, and ω_n is calculated using eq. (3.2.14).

In reality, if the system is not too damped (that is if ξ is smaller than 1, but not too much), the oscillations of the transient response are visible and allow the determination of ω_d; so, another expression in ξ and ω_n is available, that is useful to improve the accuracy of the estimate (the system of equations is now over-determinate).

If the system is over-damped, then the peak value and peak time cannot be defined and the waveform has in general less features that may be recognized and used. The two poles are real; if ξ is sufficiently large, one of the two poles is much faster than the other one and its transient response term vanishes rapidly, so that the overall time response is quite that of a first-order system with the dominant pole.

Also the case of a critically damped system ($\xi = 1$) can be treated mathematically to support the identification of its internal parameters (in this case only ω_n remains, provided that the assumption that $\xi = 1$ is valid).

In any case it is always possible to set up a system of equations relating time and amplitude of some points of the transient response captured by markers and solve the system with a least mean square or similar technique.

Step response type	ξ	$t_{r,10-90\%}$	$t_{r,20-80\%}$
	0.1	1.105	0.789
	0.2	1.205	0.861
	0.3	1.322	0.945
	0.4	1.464	1.045
under-damped	0.5	1.638	1.163
	0.6	1.853	1.304
	0.7	2.146	1.485
	0.8	2.464	1.672
	0.9	2.877	1.904
critically damped	1.0	3.349	2.166
	1.1	3.850	2.451
	1.2	4.356	2.749
over-damped	1.3	4.857	3.054
	1.5	5.836	3.666
	1.7	6.792	4.272
	2.0	8.197	5.166

Table 3.2.1 – Rise time values for the second-order low-pass system for varying ξ and normalized by ω_n; other values of ξ can be obtained by interpolation.

3.2.3 Equivalence of rise-time definitions

The rise time definitions given in sec. 3.2.1 and the analysis of the step response of three basic LTI systems considered in sec. 3.2.2 are combined and the result is shown in Table 3.2.2 (taken from [199], but with corrections and integrations).

A few points are unclear in the original table:

- even taking t_σ as a reference for all the three considered LTI systems, the values of t_σ are different for the three systems, and cannot be taken all equal to 1.0;
- the original table shows values under columns headed "$F_{3\,\mathrm{dB}}$" and "F_{RMS}" with obvious meaning; our notation uses "-3 dB" and "ENBW" and "B" replaces "F"; yet, it is not immediately clear if these values represent the frequencies or the respective rise time values, calculated for example with the $1/(\pi F)$ relationship; for this reason they are recalculated using the Matlab script in Example 3.5.

Once t_σ is given the role of reference rise time (any other rise time definition would work), the other definitions are evaluated, giving a picture of the variability of rise time under different methods of calculations and different dynamic system responses.

The usefulness of this table may be explained with the following considerations, exemplified then in Example 3.4:

Impulse response	t_σ	t_{10-90}	t_{20-80}	$t_{ctr.sl.}$	$t_{max.sl.}$	$B_{-3\,\mathrm{dB}}$	$t_{-3\,\mathrm{dB}}$	$B_{-6\,\mathrm{dB}}$	$t_{-6\,\mathrm{dB}}$	ENBW	t_{ENBW}
One pole, $\tau=0.399$	0.615	0.877	0.553	1.330	0.399	0.398	0.800	0.689	0.462	1.348	0.236
Two poles, crit. damped $\omega_n=0.282$	0.513	0.947	0.612	1.164	0.767	0.363	0.878	0.563	0.565	0.984	0.324
Gaussian, $1/\sigma=0.281$	0.439	1.019	0.672	0.996	1.00	0.332	0.959	0.470	0.677	0.804	0.396
Bessel, var. order $n=2$	0.625	1.095	0.727	1.266	0.992	0.312	1.021	0.453	0.703	0.810	0.393
Bessel, var. order $n=3$	0.636	0.998	0.672	1.077	0.988	0.346	0.921	0.478	0.666	0.834	0.382
Bessel, var. order $n=4$	0.652	0.940	0.634	0.987	0.952	0.370	0.860	0.506	0.629	0.869	0.367

Table 3.2.2 – Rise-time values for different LTI dynamic systems: single real pole, two coincident real poles and Gaussian response, together with an approximation of the Gaussian response obtained with a Bessel filter with two, three and four pole pairs. (values calculated numerically with an accuracy of 0.1 ms, i.e. 0.01%)

- it is possible to convert between different rise-time definitions, by first normalizing the given rise-time by t_σ and then multiplying by the coefficient that appears in Table 3.2.2; see Example 3.4-1) and -2);
- if conversion between different systems is desired, addressing the problem of what the same rise time would be changing system response, the information on time constant in the first column of Table 3.2.2 is used: the multiplying coefficient is given by the ratio of time constants; see Example 3.4-3);
- the rise time value based on the $-3\,\mathrm{dB}$ bandwidth for a single-pole system coincides with the time constant, deriving it directly from the definition of $-3\,\mathrm{dB}$ bandwidth itself (the difference of 0.001 between the two values is a check for the overall accuracy of calculations); the other two bandwidth-based definitions of rise time follow and the larger the bandwidth the smaller the rise time (so going from $B_{-3\,\mathrm{dB}}$ to ENBW); the higher-order systems with a steeper roll-off have smaller ENBW values and correspondingly larger ENBW-defined rise time values;
- the Bessel implementation of the Gaussian filter fits quite well for $n=2$ and $n=3$; simply cascading separated first-order low-pass sections would give too a low response with the rise time approaching the Gaussian response too slowly with the number of sections.

Example 3.4: Comparison and verification of rise-time definitions for different dynamic systems and impulse responses (see Table 3.2.2)

1) Let's take a single-pole RC circuit with time constant $\tau=10\,\mathrm{ms}$ and we want to evaluate the 10-90% rise time: we know from Table 3.2.2 that $t_{max-sl}=\tau$, thus $t_{10-90}=0.877\,t_\sigma=0.877/0.399\,\tau=2.2\tau=22\,\mathrm{ms}$, and this is quite a known result.

2) Analogously, for a two-pole system with real coincident poles (critically damped), for the same time constant $\tau=10\,\mathrm{ms}$, the 10-90% rise time would be $t_{10-90}=0.947\,t_\sigma=0.947/0.363\,\tau=2.6\tau=26\,\mathrm{ms}$, as expected longer than the single-pole circuit, but not that much.

3) Now think of passing from the rise-time value for the single-pole system to the value for the same rise-time definition for the other two systems; normalization in Table 3.2.2 is such that the time constants of the dynamic systems are adjusted for unity t_σ; in this case the $t_{10-90}=0.877\,t_\sigma=22\,\mathrm{ms}$ above in 1) becomes $t_{10-90}(\mathrm{two-pole})=0.947\,t_\sigma=23.8\,\mathrm{ms}$ and $t_{10-90}(\mathrm{Gaussian})=1.02\,t_\sigma=25.6\,\mathrm{ms}$ for the two-pole and the Gaussian system, respectively.

Now, using Matlab, the three dynamic systems are first, modeled, and then the step response is simulated, calculating the corresponding rise-times, based on their definitions; the results are shown in Figure 3.2.3 and numeric values are given only if they differ from those reported in Table 3.2.2. The developed code is shown in Example 3.5 (alike lines are omitted to a compact piece of code).

Example 3.5: Matlab code for step-response simulation and rise-time calculation (Table 3.2.2)

```
%Step response of LTI systems and rise-time
dt=1e-3; t=0:dt:20; N=length(t); delta=0.0003;
tau1=0.399; sys1=zpk([],-1/tau1,1/tau1); %1st order
tau2=0.282; sys2=zpk([],[-1/tau2 -1/tau2],1/(tau2)^2); %2nd order, crit.  damped
sig3=1/0.281; trunc3=3.0349*sig3; n3=find(abs(t-trunc3)<dt,1,'first');
t3=[-t(n3:-1:1) t(2:1:n3)];
h3=sig3/(2*sqrt(pi))*exp(-sig3^2/4*t3.^2);
HF3=fft(h3); NFFT3=floor(length(h3)/2); HF3=HF3(1:NFFT3); HF3=HF3/HF3(1);
df3=1/(max(t3)-min(t3)); f3=0:df3:(NFFT3-1)*df3;
sys3=frd(HF3,f3,'FrequencyUnit','Hz');
%Calculation of impulse response
h1=impulse(sys1,t); h2=impulse(sys2,t);
%Calculation of step response
y1=step(sys1,t); y2=step(sys2,t); y3=0.5*(1+erf(sig3*t3/2));
%y3=step(sys3,t); %Time response of FRD model cannot be simulated!
%Calculation of rise-time performance
%t_sigma
t_sig1=sqrt(2*pi)*std(h1); t_sig2=sqrt(2*pi)*std(h2); t_sig3=sqrt(2*pi)*std(h3);
%10% = 0.1, 90% = 0.9
t_1090_1=t(abs(y1-0.9)<delta)-t(abs(y1-0.1)<delta); ...
%20% = 0.2, 80% = 0.8
t_2080_1=t(abs(y1-0.8)<delta)-t(abs(y1-0.2)<delta); ...
%slope
dy1=(y1(2:N)-y1(1:N-1))/dt; dy2=(y2(2:N)-y2(1:N-1))/dt; dy3=(y3(2:N)-y3(1:N-1))/dt;
%center slope
n10=find(abs(y1-0.1)<delta,1,'last'); ...
n90=find(abs(y1-0.9)<delta,1,'first'); ...
t_sl50_1=(1-0)/dy1((n10+n90)/2); ...
%max slope
t_slmax_1=(1-0)/max(dy1); t_slmax_2=(1-0)/max(dy2); t_slmax_3=(1-0)/max(dy3);
%Bandwidth -3dB
B3_1=bandwidth(sys1,-3)/(2*pi); B3_2=bandwidth(sys2,-3)/(2*pi); ...
t_B3_1=1/(pi*B3_1); t_B3_2=1/(pi*B3_2); t_B3_3=1/(pi*B3_3);
%Bandwidth -6dB
B6_1=bandwidth(sys1,-6)/(2*pi); B6_2=bandwidth(sys2,-6)/(2*pi); ...
t_B6_1=1/(pi*B6_1); t_B6_2=1/(pi*B6_2); t_B6_3=1/(pi*B6_3);
%ENBW
w=linspace(0.01,100,300); dw=w(2)-w(1);
[H1,P1,W1]=bode(sys1,w); [H2,P2,W2]=bode(sys2,w); [H3,P3,W3]=bode(sys3,w);
ENBW_1=sum((H1(1,1,:)*dw).^2)/H1(1,1,1)^2; ...

t_ENBW_1=1/(pi*ENBW_1); t_ENBW_2=1/(pi*ENBW_2); t_ENBW_3=1/(pi*ENBW_3);
```

3.2.4 Cascaded LTI systems and rise time propagation

If we have N linear time invariant systems in a cascaded connection with an input signal entering from the left and propagating to the right, the signal propagates following the rule of convolution of each impulse response and we will finally see that the output is the result of the overall convolution of the block impulse response. The total rise time, with a demonstration based on the Central Limit Theorem, results in the square root of the sum of the square of the single rise times.

$$t_{r,tot} = \sqrt{t_{r1}^2 + t_{r2}^2 + \cdots + t_{rN}^2} \tag{3.2.15}$$

This relationship is extremely useful and recalled by many, but its proof is lost in the mist of the '50s: the original demonstration is attributed to Henry Wallman in a paper

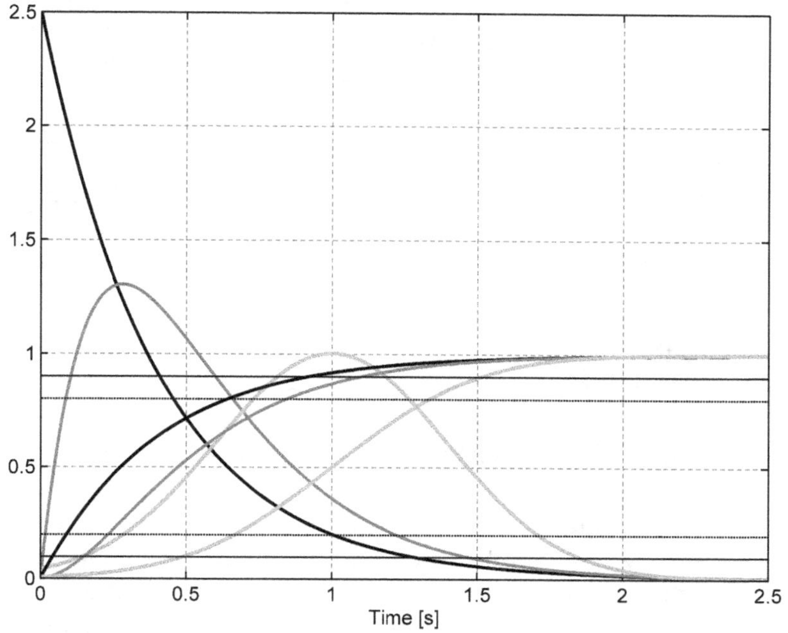

Figure 3.2.3 – Step response of the three dynamic systems of Table 3.2.2

written for the Applied Mathematics Symposium in 1950[14], but the one-page paper simply states that "the step-function response of cascaded networks, each free of overshoot, tends to error-function integral." A more detailed examination appears in the book by Valley and Wallman published two years before [325], but no demonstration is given. On the contrary, in Johnson and Graham [199], Appendix B, the rationale of the proof is clearly stated: it is based simply on the observation that the incoming step signal convolves with the impulse response of the network; similarly convolution occurs between the probability density functions of two summed random variables and in this case their variances add; then, the square of the dispersion of the edge of the step signal and of the impulse response add in the same way. If the same definition of rise time and associated variance is used, then the proof holds and eq. (3.2.15) keeps valid; the important thing is to adopt the same definition of rise time for all the cascaded networks: 10-to-90%, 20-to-80%, central slope, maximum slope, etc.

From eq. (3.2.15) the formula for the compensation of the rise time of measuring blocks $t_{m.bl.}$ under linearity assumption may be readily derived: the DUT rise time may be estimated by rms subtracting $t_{m.bl.}$ from the measured rise time t_{meas}.

$$t_{DUT} = \sqrt{t_{meas}^2 - t_{m.bl.}^2} \tag{3.2.16}$$

[14] H. Wallman, "Transient response and the central limit theorem of probability", Proceedings of Symposium in Applied Mathematics, Vol. 2, pp. 91, 1950.

No. of stages N	Overshoot %	Rel. rise time	Exp. rise time eq. (3.2.15)
1	4.30	1.00	1.00
2	6.25	1.32	1.41
4	8.40	1.69	2.00
6	10.00	1.95	2.45

Table 3.2.3 – Rise time and overshoot for cascaded two-pole tuned networks

Whilst many think that the expression (3.2.15) is exact and may be applied in all cases, "it is actually a statement of trend in the limit" [325]: it is nevertheless quite accurate and already for $N = 2$ the error is only 10%!

It was underlined that this rule holds for networks without overshoot. It is thus interesting and useful to consider what happens in case of overshoot in the step response of our cascaded networks: if the overshoot is small, below e.g. 2%, then the overall overshoot grows extremely slowly or does not grow at all when the number of cascaded stages is increased; if the overshoot is more evident, e.g. between 5 and 10%, then the overshoot increases approximately as the square root of the number of stages, but the rise time much less. For this latter case Valley and Wallman give tabular data for cascaded networks with original overshoot of 4.3% (Table 3.2.3)[15].

Correspondingly, Valley and Wallman give the relationship that we have already seen between the rise time t_r (in the 10-to-90% definition) and the $-3\,\mathrm{dB}$ bandwidth $B_{-3\,\mathrm{dB}}$ for a network "without excessive overshoot":

$$t_r = \frac{0.35 \div 0.45}{B_{-3\,\mathrm{dB}}} \tag{3.2.17}$$

The considered case of "not excessive overshoot", that might be assumed less than 10% with some approximation, have practical usefulness covering the case of many measuring amplifiers and oscilloscope probes.

3.2.5 Signal bandwidth

The most intuitive definition of bandwidth of a signal is the highest frequency component that is significant; then, to understand what "significant" really means and implies, we may attack the problem from different viewpoints.

If we take an ideal square wave and we begin to select the components of its spectrum, starting from the dc component (or 0th harmonic), then adding the fundamental and the higher order components one by one, two facts are observed: the top of the re-

[15] For further details and examples on the step-function response of multistage amplifiers Valley and Wallman indicate two references:
H.E. Kalhnann, R.E. Spencer and C.P. Singer, "Transient Response," *Proceedings of the IRE*, Vol. 33, pp. 169-195, 1945.
A.V. Bedford and G.L. Fredendall, "Transient Response of Multistage Video-frequency Amplifiers," *Proceedings of the IRE*, Vol. 27, pp. 277-284, 1939.

sulting signal gets flatter resembling that of the square wave and the rising slope gets steeper, reducing correspondingly the rise-time. When a threshold is fixed for either the amplitude or the rise time, then the corresponding "significant" harmonic order can be identified.

The problem with an ideal square wave is that it has infinite bandwidth and zero rise time; the components of the Fourier spectrum have an amplitude inversely proportional to their harmonic order, that is they drop off with a $1/f$ behavior. On the contrary, for a real waveform that resembles a square wave, when again approaching the problem of identifying its bandwidth, we might ask ourselves when (i.e. for which component subtracted progressively from the reconstructed waveform) such a waveform begins to depart substantially from the reference ideal square wave of identical period. Again we need to fix a threshold and a rule might be that the power associated to the single component is equal or less to the 50% of the corresponding square wave component (going from power to amplitude, this means that the amplitude is 70% of the corresponding square-wave component). Or we can decide that the critical frequency is the one above which the components of the real waveform start to drop off faster than $1/f$ (and this for a nearly trapezoidal wave-shape occurs at a frequency that is the reciprocal of the rise time, as shown in the next section and commented later on in sec. (9.2.13)).

Practically speaking, bandwidth is a concept related to the occupied frequency axis with respect to some characteristic or some use of the signal: we have already seen the -3 dB bandwidth, as a general means to establish the frequency interval where a signal has the majority of its power concentrated, or for dynamic systems the frequency interval for which an input signal is not substantially attenuated. The -6 dB bandwidth has quite a similar meaning, adding a delta interval to the previously determined -3 dB frequency interval, the extension of which depends heavily on the roll-off, that is the rate of decrease of signal spectrum or system response versus frequency (usually measured in dB/decade or dB/octave). Choosing one criterion or the other depends on the application: we have seen in sec. 3.1.3.2 that the -6 dB criterion is recalled when coherent composition of signals is to be accounted for and a 6 dB attenuation of signal amplitude corresponds to 0.5; conversely a 3 dB attenuation corresponds to 0.707 and we may observe also that 0.707 is the factor that comes into play when two sinusoidal tones of the same intensity are added in quadrature, that is assuming a random phase relationship, or similarly when the Power Spectral Densities of two uncorrelated noise sources are summed together. The Equivalent Noise Bandwidth ENBW is another form of characterization of the bandwidth occupation of a dynamic system, including the roll-off and the side-lobes in the stop-band: we have considered it as a performance index of smoothing windows and for rise time definition (see sec. 3.2.17).

3.2.6 Rectangular and trapezoidal pulse train

A rectangular pulse train of amplitude A, with period T and rectangle duration τ (that corresponds to a square wave of duty cycle τ/T) is shown in Figure 3.2.4(a). The coefficients of the Fourier series are

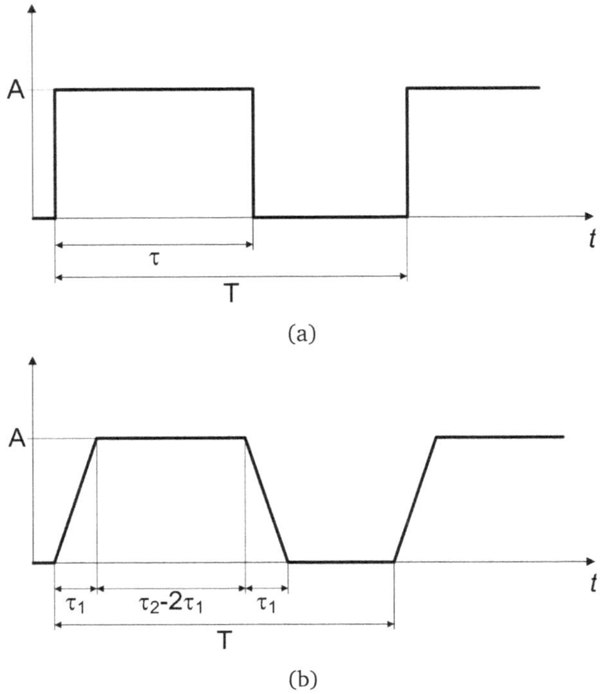

Figure 3.2.4 – (a) Rectangular pulse train and (b) Trapezoidal pulse train.

$$c_n = \frac{A\tau}{T}\frac{\sin(\pi n\tau/T)}{\pi n\tau/T} = \frac{A\tau}{T}\,\mathrm{sinc}(\pi n\tau/T) \tag{3.2.18}$$

with T appearing in both fractions, in order to have a more elegant and compact expression, that involves the *sinc* function.

The integral of the square of the first lobe gives the power that is contained in the first components of the spectrum between 0 and $1/\tau$:
$\int_0^{1/\tau} |R(f)|^2\, df = 0.92A^2\tau$ (the result being obtained by numerical integration).
The total power of the rectangular pulse is $A^2\tau$ by integration in the time domain over the entire period T (and use of the Parseval's theorem). Thus, the spectral width of the rectangular pulse encompasses the 92% of the power, leaving outside in the high order components only the 8%.

Let's consider a trapezoidal pulse train waveform as a more realistic representation of a digital base-band signal (see Figure 3.2.4(b)): the former time interval τ, used in the rectangular pulse train example for the active part of the wave, becomes τ_2 and the rise and fall times of the impulse flanks have time duration τ_1. The amplitude is always A and the period T. It is easy to see that the trapezoidal waveform is obtained by the convolution of a rectangular pulse $r_1(t)$ of width τ_1 with a narrower rectangular pulse $r_2(t)$ of width τ_2. Thus the Fourier transform of the trapezoidal pulse train is given by the products of the two transforms $R_1(f)$ and $R_2(f)$.

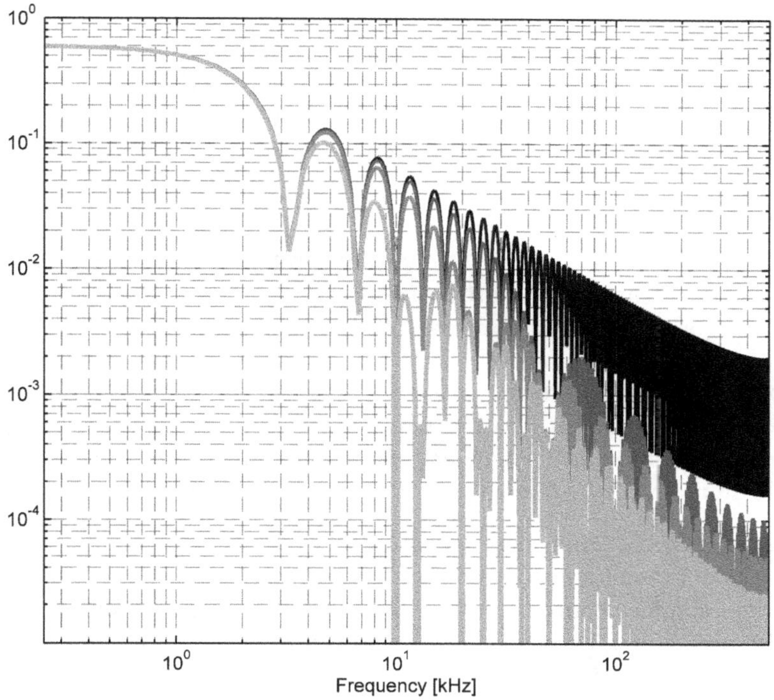

Figure 3.2.5 – Spectra of a rectangular pulse (black) and a trapezoidal pulse with varying rise time: 2%, 4% and 8% of the period in three gray tones. In this example the period is $T = 1$ ms, the duty cycle 30%, the amplitude is $A = 1$ and 3:1 zero-padding is used to visually improve the calculated spectrum.

It is instructive to study what is the frequency extension of the spectrum for different values of the rise time τ_1 (expressed as % of the period T) and to evaluate the bandwidth in terms of energy and of faithful reproduction of the pulse rising and falling edges (see Figure 3.2.5). The behavior may be quite different depending on the amount of rise time: any practical quick rule to estimate the bandwidth of digital signals based solely on the clock frequency must assume something on the rise time. If we agree that many clock waveforms have a rise time between 4% and 12% of the period, then adopting the average value of 8% leads to an estimate of the bandwidth based on eq. (3.2.17), that ranges between $0.35/0.08\,f_{ck} = 4.37 f_{ck}$ to $0.45/0.08\,f_{ck} = 5.62 f_{ck}$, so on average 5 times f_{ck}.

The other parameter that is relevant to the spectrum shape is the duty cycle, that is the τ/T ratio. A square wave of 1 kHz period is studied for varying duty cycle (0.5, 0.4, 0.25, 0.125 and 0.0625) for two cases: slow rise time, that is 2%, and fast rise time, that is 0.2% (not as fast as normally available waveforms from signal generators, but enough to see the difference on the frequency scale we have chosen). The two cases are reported in Figure 3.2.6. The spectrum is flat at the lowest frequency, so that the first value on the displayed frequency axis corresponds to the dc value: it is simply $A\tau/T$, that for $A = 1$ corresponds to the duty cycle itself. The fundamental, even if not immediately visible in the figures, is by eq. (3.2.18) A/π, in our case 0.3183. The other thing that is worth noting is that the smaller the duty cycle (and the shorter the time duration of the mark portion of the period), the farther the first dip in the spectrum, followed by the first lobe: the frequency location of the first dip (or zero) is the inverse of the mark duration, that is $1/\tau$. Whatever the duty cycle, all spectra stay beneath the same envelope, touching it with their lobe tips. The effect of the rise time is signified by a dip in the envelope, that for the 2% case (20 µs rise time) is correctly located at its reciprocal, that is 50 kHz, and moves to 500 kHz when the rise time is reduced by an order of magnitude to 0.2%.

Going back to consider in more detail the waveform edge, the trapezoidal waveform is still only an approximation of a real digital signal (e.g. a clock signal). Real signals may have one (or a combination) of the following characteristics:

- smooth corners, both when detaching from the baseline or joining the upper level, like after the action of a real-pole low-pass filter; this is an appreciable characteristic when reducing electromagnetic emissions and interference (EMI), while on the contrary the next characteristic is to be avoided;
- oscillations when joining the upper level (called "ringing") due to more complex dynamics of the interposed linear system, represented by the presence of at least a pair of complex conjugate poles (indicating a 2nd-order block); the frequency of the oscillations is related to the natural frequency of the pole pair and the amplitude of the first oscillation (called "overshoot"), as well as the decay of the amplitude of the successive oscillations to the damping of the pole pair; the damping may be influenced not only by the resistive components of the circuit, but also by the losses in the cables and reactive components;
- the rising slope may not be led back to a linear behavior, because of a change of the electrical characteristics of the signal generator (e.g. a logic gate) while switching from the lower state to the upper state; it is a very complex case, that highly depends on the internal structure of each gate and its technology: while switching with a steep slope on a RC load, the current absorbed by the capacitor is larger at the beginning and this might saturate the driving gate, that adds its higher internal resistance and slows down the slope, that in turn becomes faster only when the absorbed current is low enough to bring the gate back from saturation.

The effect on the spectrum of the characteristics considered above is reported in the following figures.

Figure 3.2.7 shows the reduction of the high frequency components and the progressive reduction of the occupied bandwidth for increased smoothing, obtained by increasing the length (the number of "taps") of a Finite Impulse Response (FIR) filter implementing a moving average. When the rising edge loses its sharp corners at the beginning and the end, the spectrum amplitude begins reducing at about 200 kHz (dark gray curve with respect to black curve). When the tapering of the rising edge is improved by increased smoothing, even if the rise time is not increasing dramatically (the 10-90% rise time of the original waveform is 15 µs, and that of the last curve slightly less than 20 µs), the occupied bandwidth shrinks and the spectrum amplitude starts decreasing at 100 kHz (at 50 kHz for maximum smoothing).

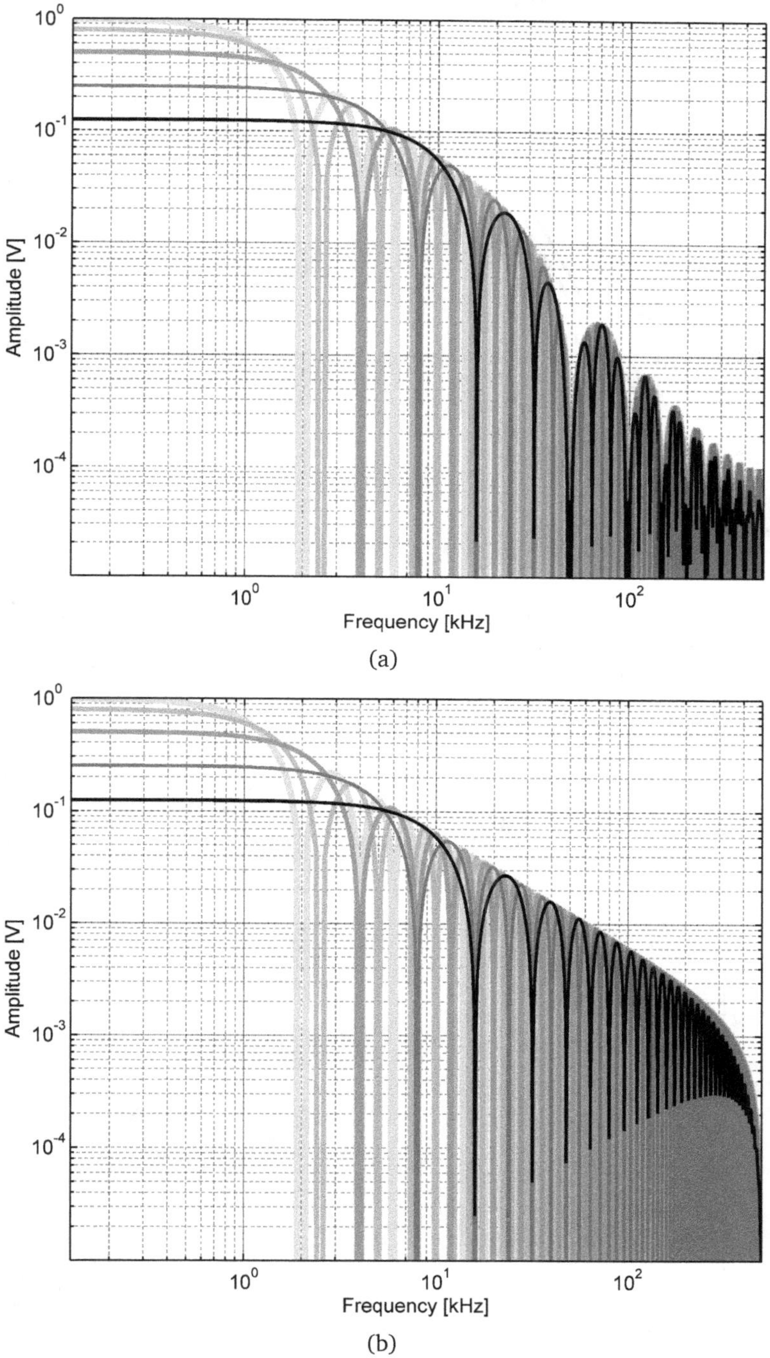

Figure 3.2.6 – Spectra of a rectangular pulse with (a) 2% and (b) 0.2% rise-time for varying duty cycle (0.5, 0.4, 0.25, 0.125 and 0.0625 from light gray to black). In this example the period is $T = 1\,\mathrm{ms}$, the signal excursion is 0-1 V and 3:1 zero-padding is used to visually improve the calculated spectrum (250 Hz resolution).

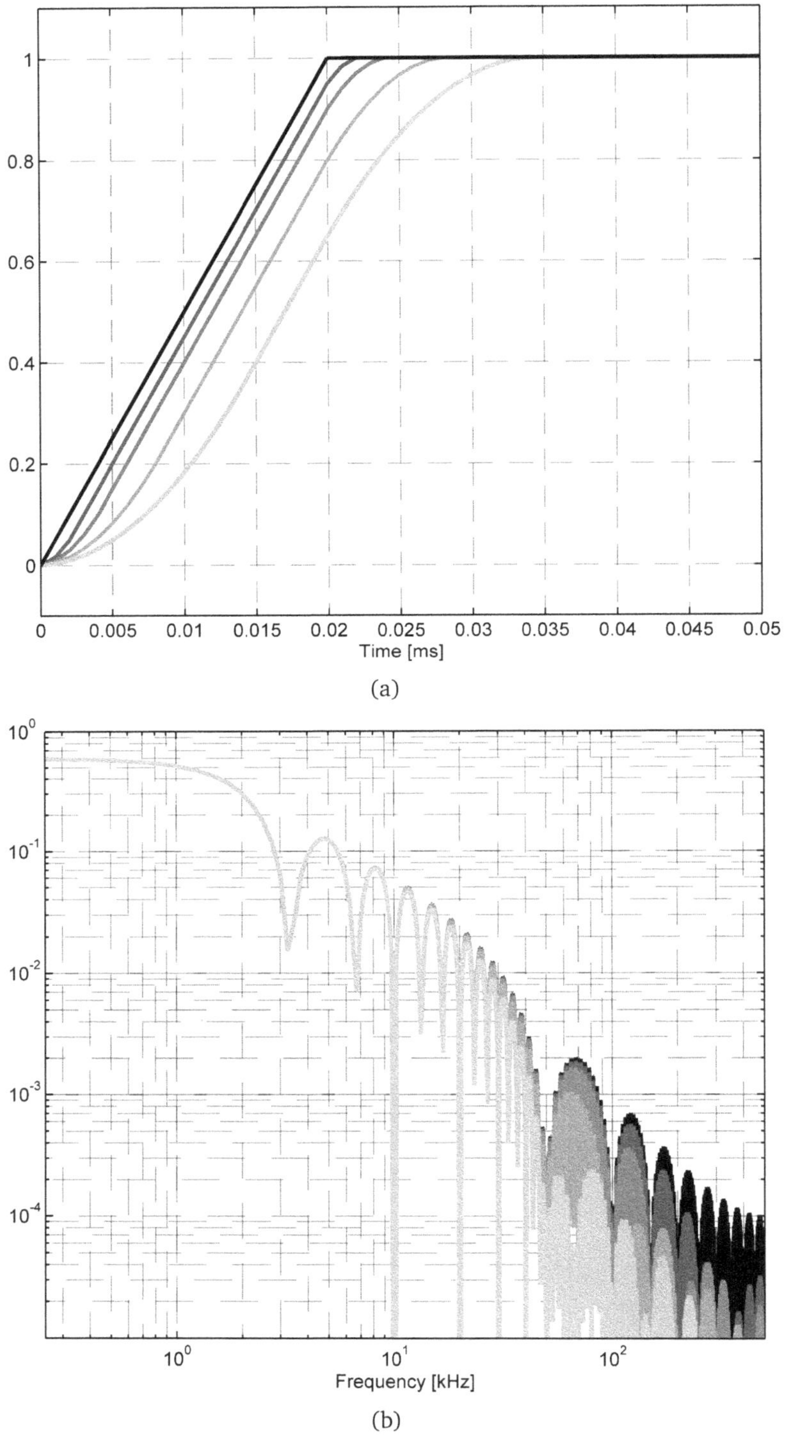

Figure 3.2.7 – Effect of the smoothing of corners obtained with a moving average filter of increasing order: (a) rising edge in time domain, (b) resulting spectra.

Similarly, the slew rate limitation also smooths the edges and reduces the bandwidth. It is the case of a line driver (e.g. unity buffer or operational amplifier) feeding a cable with a square waveform.

Oscillations due to multiple reflections across an unmatched line (see sec. 2.5.1) or in case of crosstalk (see sec. 5.1.6 and 8.3.4) may occur anywhere along the signal waveform, including the waveform edge. Being oscillations caused by a repetitive phenomenon when reflections and standing waves occur, the effect on the spectrum is a local increase around the oscillation frequency. Of course, reducing the bandwidth and smoothing sharp edges helps limiting reflections and crosstalk and makes in general the transmission line electrically shorter.

3.2.7 Base-band modulations

Quite often, looking into the specifications of some transmission protocol, the question arises on how the bit rate may be transformed into bandwidth, to understand how much of the physical channel the modulation is using and up to which frequency we have to extend our calculations and prepare for measurements.

Base-band modulations may be described at a first approximation by recalling the trapezoidal pulse train and its spectrum. However, it shall be remembered that the signals used in reality are random signals and that channel coding may occur before sending the symbols over the physical channel. Base-band modulations are used in a variety of existing protocols that are often named altogether as *serial data links*; examples of such protocols are RS 232, Ethernet, USB, SATA, etc.

A general model valid in all cases is composed of a source of information, whose output is made of single binary digits or symbols and the digital message is an ordered sequence of symbols; the symbols have their own statistical distribution and correlation, but when evaluating the spectral properties of the modulation, the first attempts and the general theory assume that the information is uniformly distributed and uncorrelated at the origin.

For binary sources, the bit stream is passed to a code block generator, that organizes the stream of information in bit blocks (or code words) of length suitable for successive channel coding and transmission. Channel coding is implemented by assigning each block to a channel symbol (or signal), i.e. one of the electric signals that are sent over the physical channel. Why the need to pack bits together in code blocks? it is not mandatory, nor always implemented, but M-ary codes allow to transmit more information at the same signaling rate, i.e. with the same rate of symbol transmission. As an example, M-ary codes may be implemented at the physical channel level in various ways, such as using different amplitudes for the same symbol (Pulse Amplitude Modulation), different phases of the same sinusoidal signal (Phase Shift keying in its various implementations, diffusely adopted for modem modulations), different frequency values of the same sinusoidal signal (Frequency Shift Keying, as used in robust codes for satellite transmission, safety related signaling, etc.).

At this stage it is also possible to implement channel coding selecting carefully channel symbols to suppress or modify undesirable characteristics of the so formed encoded message before its transmission: dc component removal, scrambling, breaking of repetitive sequences are all examples of shaping of the encoded sequence to better suite channel characteristics and to optimize spectral properties.

Using more complex symbols featuring M levels does not mean that we are obtaining a neat proportional advantage of M over the "2" of binary codes: first, information is quantified by entropy not by the amount of levels and it goes roughly with $\log_2(\cdot)$, besides accounting for many other details such as correlation at the source, correlation artificially put to shape the spectrum and then correlation added by channel coding (see below); second, the average transmitted power shall be brought equal to have a meaningful comparison; third, the capacity of the physical channel is limited and this limitation is multiple: the maximum frequency depends on the channel bandwidth, the type of modulation and the spectrum shape that can travel through the channel depend also on dispersion and distortion, and the amount of channel coding and error control codes added to the message establish the amount of payload per second and the band occupation, altogether to fulfill the Shannon theorem on channel capacity; fourth, as a consequence, for a given transmitted power and channel noise, the obtained error probability at the receiver positioned at the other end of the channel is a complex tradeoff among all these factors.

These concepts and many more, together with useful examples and reference data, may be found in several textbooks; the ones I am familiar with from my university days are Carlson [54] and Proakis [274].

In base-band modulations the elementary symbol has approximately the form of a square (or trapezoidal) pulse and uses two levels ($M = 2$). When the signaling rate is large, pulse shaping is indispensable to keep the overall bandwidth under control: to this aim raised-cosine and Gaussian shaping is usually adopted.

3.3 Channel propagation

The channel used to propagate the signal (i.e. the connecting cables or the PCB traces, including launchers, connectors, couplers, and so on) exerts its influence on the signal rise time and in general on the signal waveform, as anticipated at the end of Chapter 2 when speaking of dispersion and group delay distortion. An ideal cable has a constant phase and group delay in the frequency interval of interest and lets the signal pass through undistorted. Only the loading of the cable capacitance may have effect on the signal generator and its slew rate.

Going further and considering the non-ideal characteristics of the channel through which the signal is transmitted and propagates, we may say that when the attenuation varies with frequency (see e.g. a measured Insertion Loss like in Figure 10.2.6), its bandwidth gets smaller and the rise time increases. In reality also the characteristic impedance may be variable as a function of frequency, as it happens in printed circuit boards where the dielectric permittivity of the substrate is not constant: this phenomenon takes similarly the name of "dispersion" and will be considered in detail in Chapter 6.

The rise-time definition and interpretation were discussed in sec. 3.2.1 and 3.2.1.3; a direct way of getting the rise time from frequency domain data is to base its definition on bandwidth. When propagating through a cable Johnson and Graham [199] propose to find first the corner frequency (or "knee frequency") at which the cable attenuation is 3.3 dB and then the 10-90% rise time is defined as $t_{10-90\%} = 0.5/f_{3.3\,\mathrm{dB}}$, in agreement with the $-3\,\mathrm{dB}$ definition given above (see Table 3.2.2). This method works well, provided that the degradation with frequency is smooth and dispersion is not too relevant, as it happens sometimes with transmission lines built on PCBs (see sec. 6.3 and 6.3.1).

What happens when a limited bandwidth signal is sent over a limited bandwidth channel, or, in other words, what is the resulting rise time? The same applies when measuring a signal with an oscilloscope probe that has limited bandwidth and one is interested in separating the influence of the probe itself from the information regarding the rise time of the measured signal. Composition of signal and system rise time was considered in sec. 3.2.2: the square root composition of the sum of squares is generally used, so that there is always an increase of the resulting rise time, not too evident if the system rise time is only about 50% shorter (i.e. system response is approximately twice as faster).

Practically speaking, there are two options when dealing with the influence of system rise time and its compensation:

- if we are doing a measurement, the original signal rise time can be estimated quite well, even if the probe (or measuring equipment) has a limited rise time of the same order or even slightly worse;
- if we are interested in sending an undistorted waveform through a channel, then the latter shall have a rise time that is less than 50% of the signal rise time to keep its influence under a sensible threshold of 10%; it shall be underlined that this is not the only condition for an undistorted transmission of signals across a channel, and that other factors such as group and phase delay (see sec. 3.3.1), as well as attenuation and dispersion (see Chapter 6 for the various types of planar transmission lines and sec. 10.2.1) shall be considered.

3.3.1 Group delay and phase response

The phase response of a transmission channel or system is simply the argument $\phi(f) = \angle H(f)$ of the complex frequency response $H(f)$. Modern instruments that sweep the assigned frequency interval have the option of plotting the phase response as well as the *group delay*. The group delay is the delay of the envelope amplitude of the sinusoidal components of a signal; conversely, the phase delay measures the delay of the phase and not of the envelope amplitude. Both are calculated versus frequency (or more "classically" versus pulsation ω): the phase delay is the amount of delay obtained by dividing the phase response by pulsation; the group delay corresponds to the first derivative of the phase response.

$$\tau_\phi = \frac{\phi(\omega)}{\omega} \qquad \tau_g = \frac{\partial\phi(\omega)}{\partial\omega} \tag{3.3.1}$$

For a linear system (such as the LTI considered so far) the two quantities are constant with frequency and equal.

Going back to the definitions of velocities for a transmission line (see sec. 2.2), the velocities may be readily defined as the inverse of the corresponding delay quantity, so that we have the phase velocity and the group velocity (line velocity was introduced at the end of sec. 2.1). The importance of these two quantities may be seen when considering a modulated sinusoidal signal $s(t)$ that propagates through the line:

$$s(t) = m(t)\cos(\omega_0 t) \tag{3.3.2}$$

where $m(t)$ is the modulating signal and ω_0 is the carrier frequency.

The fact that $m(t)$ is assumed slowly varying reflects in the two symmetric spectra $\frac{1}{2}M(\omega-\omega_0)$ and $\frac{1}{2}M(\omega+\omega_0)$ that are concentrated around ω_0 or quasi monochromatic. To see the effect of line propagation, the propagated signal $v(t)$ shall be calculated by taking the inverse Fourier transform of the spectrum of $S(\omega)$ multiplied by the line response $e^{-j\gamma(\omega)L}$. Considering only the positive frequency axis and after some simplifications that involve approximating the propagation constant with a Taylor expansion of two terms (first order) for the phase constant $\beta(\omega)$ and only one term (order zero) for the attenuation $\alpha(\omega)$ (thus assumed constant and frequency independent):

$$\begin{aligned} v(t) &= \Re\left\{\frac{1}{2\pi}\int_0^{+\infty} M(\omega-\omega_0)e^{-j\gamma(\omega)L}e^{j\omega t}d\omega\right\} = \\ &= \Re\left\{e^{j(\omega_0 t-\beta(\omega_0)L)}e^{-\alpha(\omega_0)L}\frac{1}{2\pi}\int_0^{+\infty} M(\omega-\omega_0)e^{-j\beta'(\omega_0)(\omega-\omega_0)L}e^{j(\omega-\omega_0)t}d\omega\right\} \\ &= m(t-\beta'(\omega_0)L)e^{-\alpha(\omega_0)L}\cos(\omega_0 t-\beta(\omega_0)L) \end{aligned} \tag{3.3.3}$$

This expression tells us that the modulating signal moves with a different velocity with respect to the sinusoidal carrier, the former given by the inverse of the derivative of the phase constant and the latter by the phase constant itself, and we recognize the group velocity and the phase velocity, respectively. Of course this is an interpretation based on the mathematical separation of the two parts, while the signal that propagates along the line is one thing only.

A variation of line velocity with frequency will cause dispersion of frequency components that form the signal applied to the line, thus causing distortion of the received signal at the other end: some components will arrive later, the phase relationship between them will be altered and the shape of the received signal will differ from the original one, despite the total rms value is the same[16]. Both velocities (and the corresponding time delays) if varying with frequency indicate distortion. However, information is carried by the envelope and group quantities assume a particular relevance.

The usefulness of using time-delay quantities is first of all that the phase response curves may be de-trended and the linear dependency on frequency removed (see sec. 7.1.1.7 where de-trending is achieved by subtraction of line delay and initial phase shift): as a result the scale on which the delay curve is displayed is optimized and any deviation is more visible and can be recognized using a flat horizontal ideal curve as reference, rather than a sloped curve versus frequency[17].

Second, the delay is the quantity that interest us, because it is the one directly affecting the signals and in more direct connection with channel velocity and line length.

Third, the use of the group delay in particular amplifies deviations at high frequency, where they are more likely to occur, being obtained by means of a derivative. Let's consider a phase response versus frequency with some oscillations, with a maximum peak-to-peak variation $\phi_{pp,\max}$ evaluated as the difference between adjacent highs and lows. The phase delay is directly proportional and will be varying of the same quantity; the group delay will show larger oscillations being the first derivative (see eq. (3.3.1)) and will be more susceptible to fast changes of the phase response, even if they remain within the assigned maximum peak-to-peak variation.

In Figure 3.3.1 a RG 8 coaxial cable was tested measuring insertion loss by a VNA two-port measurement (see Chapter 11); phase unwrapping, phase delay and group delay are calculated separately from real and imaginary parts of S_{21}.

3.3.2 Impedance mismatching and reflections

Another characteristic of the propagation channel that may lead to signal distortion is the impedance mismatch of the source and/or the load with respect to the line characteristic impedance, causing not only a small amplitude deviation, but, most important, reflections. The effect of reflections is twofold: corruption by overlapping of reflection terms to the original signal and crosstalk to adjacent lines. Crosstalk will be thoroughly considered in Chapter 8 with some practical considerations and applications anticipated in Chapter 6, when considering planar transmission lines fabricated on printed circuit boards. Let's focus on a single channel (i.e. transmission line) and limit the analysis mismatching and reflections on the same channel.

We have heard often the term "controlled impedance" and "impedance matching" for several applications, analog and digital, where a transmitter and a receiver shall be

16 With no too a formal demonstration it is sufficient to think to a sinusoid at a given fundamental frequency f_1 with superimposed a third harmonic $f_3 = 3f_1$ with one of two possible phase shifts, zero degrees or half a period of the third harmonic component, that is 60° at the fundamental; we know that the resulting signal is quite different: a sort of a cowboy hat in the first case (and this technique is used to enhance the rms of the signal without increasing its peak value, for instance in electric drives) and a peaky signal with increasing slope on its flanks in the second case.

17 An example used by Agilent to discuss phase response and delay response shows a constant phase versus frequency with some oscillations; this is clearly not realistic and exaggerated, being all phase responses more or less linear with frequency.

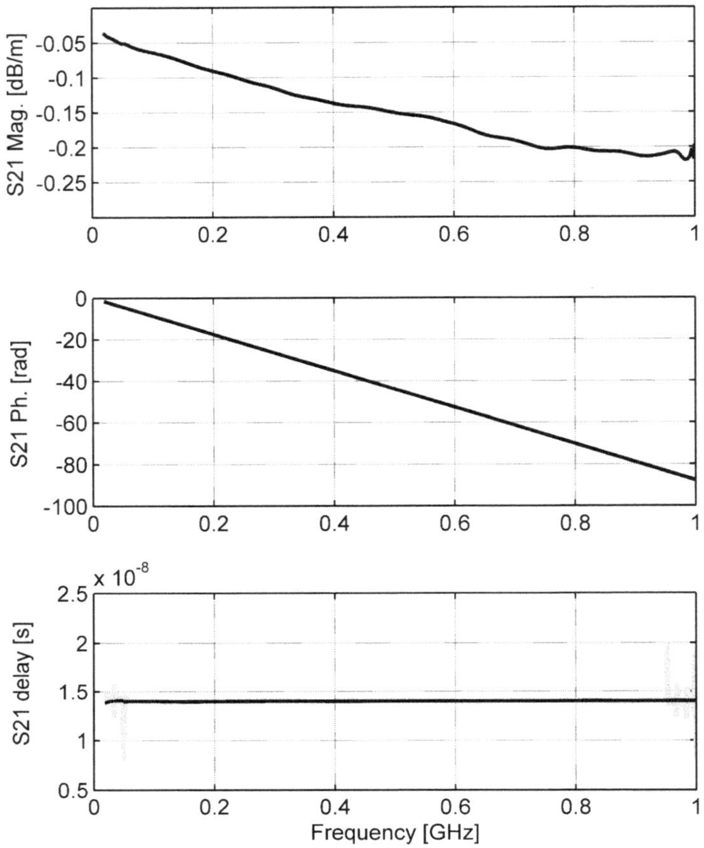

Figure 3.3.1 – Measurement of S_{21} for 3.5 m of RG8 cable: magnitude, phase (unwrapped) and phase and group delay. Since the cable is a linear device the phase delay (black) and group delay (gray) correspond; the noise at the beginning and the end of the group delay trace is due to the calculation using phase differentiation.

coupled through a channel. The extent up to which impedance matching shall be ensured depends on the application, on signal characteristics and on relevant phenomena subject to specification. So, for steep rising/falling edges the bandwidth is more extended and the chance higher that non-ideality of cables and connectors, as well as elements located on the PCB, turn out in significant impedance mismatch. From this the saying "not to use more bandwidth than strictly necessary", in other words taking care of smoothing front edges and corners by limiting the speed of transceivers and intervening on pulse shaping in extreme cases.

Impedance mismatch and the consequential reduction of transmitted signal and signal corruption due to superposition of reflected terms were reviewed in Chapter 2: we know how to estimate reflection and transmission coefficients, the shape and behavior of reflections for typical signals (pulses and step signals) and that several circuit techniques may be used to enhance impedance matching.

One possibility used in coaxial lines and especially for temporary test setups is to add attenuators that were already thoroughly reviewed in sec. 1.2 and 1.2.2.1; impedance matching is achieved at the expense of some signal loss due to the inserted attenuation.

When considering the electrical interaction of the signal transmitter with the physical channel in digital circuits, a commonly adopted solution is the insertion of matching resistors, that is small series resistors that shall ensure a minimum driving resistance (the transmitter internal impedance is usually lower than that of the channel) and reduce also the capacitive current at signal edges charging and discharging the channel capacitance. We shouldn't forget that digital signals are to be evaluated first in time domain; the peaky current at the steep signal edges wouldn't be noted by a frequency domain analysis. The criteria for resistor calculation can be found in a few application notes for fast logic circuits, such as ECL and LVDS [41].

The selection of adequate tools for correctly analyzing such phenomena is far from trivial: line losses are a frequency-domain concept, impedance and line attenuation may be effectively measured as return loss and insertion loss (see sec. 10.2), but transmitter "impedance" might change during the transition from the low to the high state and the development of the rising edge, and the interaction of reflections with the applied and received signal is better displayed by means of eye diagrams. Time domain reflectometry (discussed in sec. 6.5 for planar transmission lines) is quite effective in spotting out impedance mismatches by means of the related reflections, by leading them back to the point of origin along the channel. We will see in sec. 11.8 and 11.10 how to turn a frequency-domain equipment (the VNA) into a pseudo time domain instrument, suffering some limitations due to signal processing and mainly leakage phenomena, but affording a much more favorable signal-to-noise ratio. If then the exigency of identifying channel properties and its parameters, or equivalent circuit representation (e.g. see sec. 6.5.1.3), for later modeling and simulation is considered, it is quite evident that there is no best method or instrument, either in frequency or time domain, but only preferred ones, and that a combination of various techniques is always a necessity.

Going back to the original question about the tolerable mismatching and the conditions to meet, it is thus evident that for simple applications a direct evaluation of signal bandwidth, channel physical dimensions and electrical properties, and known signal-to-noise criteria can work: in these conditions connectors are almost always non-relevant and the attention is focused on cables; satisfactory termination at both ends is achieved by series compensation (or series-parallel for faster logic families) and the tolerable mismatching is a direct consequence of a direct estimate of the required signal-to-noise-and-distortion ratio.

Conversely, for large signaling rates, with complex pulse shapes and reconstruction techniques inside the receiver, as well as equalization and advanced filtering at the channel interface, the resulting quality of service (in terms e.g. of bit error rate) is far from trivial and can be treated and correctly addressed by more articulated approaches: modeling of the modulation and its statistical properties, channel coding, equalization and receiver decoding, with the inclusion of the transmitter, channel and receiver electrical models, possibly integrating experimental results. Several simulation tools can host all these elements, once a common description is achieved: wave solvers can output results for physical elements such as the channel and its components in terms of scattering parameters; they are also able to simulate eye diagrams, but cannot implement complex signal patterns and logic.

On the other hand, also the measurement of the channel characteristics is more complex: not only the necessary spatial resolution is smaller and requires more bandwidth, but also accurate calibration and low uncertainty are necessary. Additionally, it's not impedance of a single channel any longer to be measured, but rather the mutual influence with surrounding elements and other channels: line balancing, common-to-differential conversion, crosstalk (near end and far end), etc., all reviewed in Chapter 8 and 10 and considered in sec. 6.4.3 for practical aspects in PCB fabrication.

3.3.3 Eye diagram

Eye diagrams (or eye patterns) are obtained by recording data streams using a specific point of the signal to trigger. Let's assume base-band channel symbols that might be imagined corresponding to a +1 or -1 rectangular pulse of time duration T. A data stream is sent over the channel and captured by an oscilloscope; it is customary to call this test data stream a PRBS (pseudo-random binary sequence) data stream[18]. Triggering on the rising or the falling edge using a persistence option aligns all captured symbols at the same time reference and allows to display them overlapped: the instantaneous values, including aberrations, deformations, inter-symbol interference, are displayed for each captured symbol and the information on their statistics is given graphically by the density of the symbol traces created by superposition or numerically by histograms and statistics. The eye diagram gives thus a combined representation in graphical form of relative probability and spanned values for each symbol versus the horizontal time axis.

The eye diagram displays also any variability due to instability of the source or variable delay of spectrum components, that affect not only the shape of each symbol taken singly, but also in relationship with the symbols that precede and unavoidably shape the spectrum. This is usually all-comprehensively called time jitter and can be readily evaluated by observing the dispersion of single symbol traces at relevant points of the graph, e.g. used for symbol discrimination and decision (usually symbol edges are the most susceptible to time jitter and they in turn influence the timing for the identification of the optimal decision point).

The acquired symbols shall be evaluated if they are discernible by a hypothetical receiver, establishing if the distance between them (in the sense of the decision criterion at the receiver) is adequate to ensure a low error probability. Without going into the details of the decision models and the estimation of errors and bad classifications [54, 274], some graphical properties are considered when evaluating an eye diagram: vertical opening, horizontal opening, width between crossovers, and the variability or statistical consistency of these parameters with respect to the incoming acquired traces. An example of eye diagram is shown in Figure 3.3.2.

[18] The reason for this is that tests of bit error rate (BER) or frame error rate (FER) are performed with long data stream that shall be as random as possible not to bias the resulting statistics.

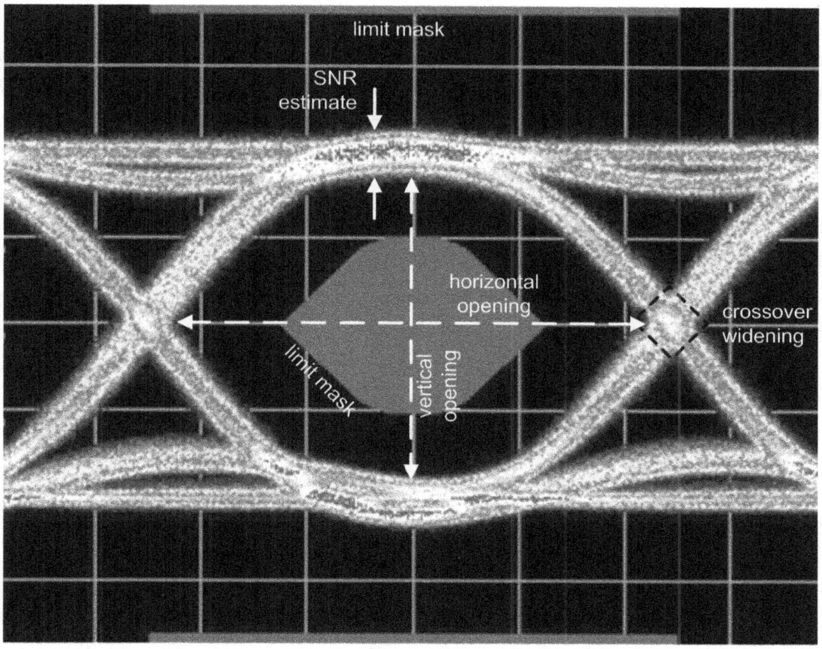

Figure 3.3.2 – Example of eye diagram with a limit mask area (dark gray) and some of its relevant characteristics.

Eye diagrams and their properties are often evaluated against acceptability masks that set the minimum vertical and horizontal opening, limit the skewness of rising and falling slopes and the dispersion of the crossover points, thus setting a limit to the time jitter. These limits are set by the adopted protocol or standard; for the same protocol, even changing the data rate or the type of connecting cable, or its length, the limits are deemed to change. Once the transfer function of the physical channel (including interfaces and connected equipment) is known by measurements or numeric simulation, then the eye diagram can be simulated by feeding the transfer function block with synthetic data streams, such as predetermined pseudo-random sequence or recorded real sequences.

3.3.3.1 Properties of eye diagrams

The *vertical opening* indicates how the symbols are separated and distinguishable in the best decision time, in principle at the center of the symbol time interval T, that is far from rising and falling slopes where transitions are. The measure is taken between the extreme closest points of the two distributions on each side, usually considering the 3σ points on the two distributions, thus ensuring a very high confidence (under normality assumption, this corresponds to 99.7% of confidence). It is interesting to observe that the EN 60512-25-6 [74] standard indicates hastily the quantification of the openings at 50% of the nominal voltages, without going into the details of how to select such voltages accurately, which influence they have, the fact that the slopes and crossings cover an area and have probabilistic behavior, etc. However, since it is not absolutely sure that the widest separation between symbols occurs always at the center of symbol time interval, the analysis of the vertical opening helps in identifying also the best decision time. Worsening of the vertical opening may occur because of Inter-Symbol Interference (ISI): let us think of crosstalk or reflection terms badly overlapping amid each symbol, affecting the vertical opening and thus the distance between symbols and the noise immunity at the decision point. The solution will be the removal or reduction of such terms, and if not possible, the optimization of the decision point by shifting it along the time axis.

The *horizontal opening* is analogously taken as the time difference between points featuring an amplitude that is halfway the two steady values of logic "0" and "1"; as before the usual measure is between 3σ points of the distributions and this might be slightly above or below the mid axis of ideally crossing symbols. In principle the horizontal opening is very close to the symbol time interval T, but some factors may affect it:

- slow rising and falling slopes have influence not only at the symbol edges, but also inside the symbol time interval and any skewness affects the horizontal opening and reduces the time interval available to set the decision point;
- duty cycle variations, symbol time T variations and in general time jitter also affect the horizontal opening, because the crossovers are not all located in a precise point in time, but create a cloud, with the horizontal opening measured at best between the internal sides of the cloud.

The reasons for fluctuations or variations of duty cycle and symbol time may be many:

- stability of transmission clock, with issues related to systematic and random variations considered in sec. 3.5 when talking about frequency stability and phase noise;
- unpredictable variations of the group delay and in general of the physical channel: for radio channels the propagation path is typically subject to fading and other parametric variations, especially with moving transmitter or receiver, and with environmental variability;
- limited capability of the transmitter output stage going into saturation or slew rate limitation when driving rising or falling edges, thus deviating also from the impedance matching condition; the reason may be a damaged channel, aging of materials, critical stray parameters, or crosstalk-induced voltages affecting the instantaneous voltage, and the absorbed current, during multiple transitions on different channels;
- crosstalk-induced voltages, this time directly overlapping to the symbol waveforms on the physical channel, affecting their symmetry and waveshape.

Crossovers (or eye crossings) occur when different symbols are overlapped: in the example above, two binary digits have two symbols that are a positive and a negative square pulse (this is a baseband signal with a non-return to zero coding); assuming the two levels at $+1$ and -1, crossings occur in the middle at the 0 line. Any misalignment or skewness in the rising and falling edges, as well as in the time duration of the respective symbols, will shift the ideal crossing point up or down, and also will move it to the left or the right of the ideal crossing time. The steeper the slope of rising and falling edges, the larger the susceptibility to timing errors. In general, eye crossing characterization gives an indication of duty-cycle distortion or pulse symmetry problems. Eye crossings feature a 2-D distribution in amplitude and time with respect to the identified ideal center: mean value and dispersion, as well as a complete histogram or a graphical display representation, quantify the spread and center of gravity of the cloud of crossing points; usually the peak-to-peak and rms value of *time jitter* are used.

The time jitter of symbols affects the span and stability of crossovers. In reality the presence of ISI at the margins of the symbol time interval, even if not affecting heavily the choice of the decision point, causes a large span of crossovers, bringing rising and falling symbol edges quite far apart (this phenomenon was already mentioned when talking about crosstalk-induced voltages).

When building a eye diagram by overlapping many captured symbols, we collect also information on the probability distribution; in modern instrumentation, the probability is also indicated quite effectively by false colors, creating a live histogram of the collected symbols. Each parameter (vertical opening, horizontal opening, crossovers, time jitter) is also statistically characterized with sample average and dispersion.

When making a test, these parameters and properties shall be compared to limits, that are given in a numerical or graphical form: the former are limit values for average and dispersion quantities (such as the rms and the peak-to-peak jitter), the latter are polygonal areas, also named masks, overlapped to the collected symbol traces.

Until now it was implicitly assumed that the source of data to be displayed as eye diagram is coming from a single channel. When more than one channel is available, it is quite useful to display them overlapped when synchronization is the objective of our analysis: if two groups of lines are clearly distinguishable in the eye diagram, it means that the two channels have slightly different timings and that the difference is deterministic: this is the so called “time skew”, usually caused by a different physical length of the two lines, or unusual loading or coupling on one of the two, or (quite rarely) as a result of particular combination of crosstalk. All these effects are well explained in the Anritsu App. Note cited below.

3.3.3.2 Simulation of eye diagrams

Eye diagrams are simply obtained by feeding a time domain block bearing the channel impulse response with a data sequence of known symbol time T and realigning the so obtained transmitted symbols every T. The channel impulse response is recovered from measurements in various ways: a simple low-pass assumption sided by identification of the poles, or time constants, that fit approximately experimental data; a more complex rational approximation of the channel frequency response, possibly including non-linear effects due to e.g. skin effect and losses, as it is normally done for planar transmission lines (see sec. 6.5.1 and, in particular, paragraph 6.5.1.3 dealing with line parameters fitting for SPP method); direct measurement of time-domain response, e.g. by TDT.

In Matlab several functions of the Communication Toolbox can create digital streams (`commsrc.pattern()` for example) and then add various amounts of jitter (e.g using the previous `commsrc.pattern()`) and noise (`awgn()` function applied last).

Quite a thorough analysis may be accomplished using EyeScope, that has an interactive GUI to select the signals to analyze and which measures we want to display, and then print: vertical and horizontal opening, the time at which they occur, rise and fall times, signal-to-noise ratio that roughly corresponds to the "thickness" of horizontal parts of the eye diagram, various measures of jitter (peak-to-peak, rms, random and deterministic).

Theoretical foundations for modeling and analyzing jitter may be found in the following references:

- Anritsu, "Understanding Eye Pattern Measurements," *Application Note 11410-00533*, Rev. A, 2010-03;
- Ou N. et al., "Jitter Models for the Design and Test of Gbps-Speed Serial Interconnects," *IEEE Design & Test of Computers*, Vol. 21, July-August 2004, pp. 302-313;
- R. Stephens, "Jitter Analysis: The dual-Dirac Model, RJ/DJ, and Q-Scale," *Agilent Whitepaper 5989-3206EN*, 2005-06;
- Agilent, "71501D Eye-Diagram Analysis User's Guide," 70874-90023, 2002-05.

Stephens in particular explains very accurately the dual-Dirac jitter model, considered also in `commonscope.eyediagram()` help of Matlab Communication toolbox.

3.4 Distortion

Signal distortion occurs when the signal (assumed a pure sinusoid for simplicity) is fed to a non-linear circuit and there appear components at the output, that are harmonically related to the fundamental. This is usually demonstrated by taking a representation of the non-linear circuit in terms of an approximating polynomial

$$y(t) = a_1 x(t) + a_2 x^2(t) + a_3 x^3(t) + \cdots \tag{3.4.1}$$

where the instantaneous values of the input signal $x(t)$ are affected not only by amplification/attenuation that is linear (indicated by the coefficient a_1), but also by a deviation from the linear behavior, approximated by higher order terms.

If the input signal is a sinusoid $x(t) = A\,\cos(\omega t)$, with $\omega = 2\pi f$, the resulting additional terms are sinusoidal terms at integer multiples of the frequency f and this is demonstrated by considering duplication and bisection formulas for trigonometric functions.

$$
\begin{aligned}
y(t) &= a_1 A\cos(\omega t) + a_2 A^2\cos^2(\omega t) + a_3 A^3\cos^3(\omega t) = \\
&= a_1 A\cos(\omega t) + \frac{a_2 A^2}{2}\left[1+\cos(2\omega t)\right] + \frac{a_3 A^3}{4}\left[3\cos(2\omega t)+\cos(3\omega t)\right] = \\
&= \frac{a_2 A^2}{2} + \left[a_1 A + \frac{3a_3 A^3}{4}\right]\cos(\omega t) + \\
&+ \left[\frac{a_2 A^2}{2}\right]\cos(2\omega t) + \left[\frac{a_3 A^3}{4}\right]\cos(3\omega t)
\end{aligned}
\tag{3.4.2}
$$

It is clear that several components appear as a result of the assumed third order non linearity: a dc term, a fundamental frequency term (that affects the nominal linear input-output relationship) and additional terms at higher frequency, namely the second and third harmonic of the input signal.

Even order harmonics will disappear in the output if the circuit is balanced and fully differential. Unavoidable asymmetry will yield some amount of even-order distortion.

3.4.1 Gain compression

The small signal gain that we have identified as the nominal input-output relationship (a_1) is affected by a third-order term. The non-linearity is of the compressive type, that is the slope of the linear input-output relationship reduces for larger signals (the third-order term has a negative coefficient a_3 that reduces the gain). This is the so-called *gain compression*. When the amplitude of the input signal is such that the resulting output is 1 dB below the value expected by the linear a_1 relationship alone, we say that the signal has reached the 1 dB compression point $A_{1\,\mathrm{dB}}$.

Considering amplitude, and not power, the distortion equation may be written as

$$
20\ \log\left|a_1 A + \frac{3a_3 A^3}{4}\right| = 20\ \log a_1 - 1 \tag{3.4.3}
$$

that gives

$$
A_{1\,\mathrm{dB}} = 1.09\sqrt{\left|\frac{a_1}{a_3}\right|} \tag{3.4.4}
$$

It is quite customary to talk about 1 dB compression for RF products, where signals are measured in power terms, so expressed in dBm: for an equivalent expression for the 1 dB compression point “10” replaces “20”.

$$
A_{1\,\mathrm{dB}} = 1.029\sqrt{\left|\frac{a_1}{a_3}\right|} \tag{3.4.5}
$$

It is underlined that in all expressions regarding the 1 dB compression point (eq. (3.4.4) and (3.4.5)) the $A_{1\,\mathrm{dB}}$ value is specified in linear scale, not in dB.

3.4.2 Desensitization

If two signals of much different amplitude are applied to the circuit, the stronger one reduces the small-signal gain that the other one should "see": this phenomenon is called *desensitization*.[19]

If the input signal is composed of two tones at different frequencies $x(t) = A_1\cos(\omega_1 t) + A_2\cos(\omega_2 t)$, with the first one taken as the weak one that is of interest, the resulting output for the ω_1 component alone in the presence of a third-order distortion is

$$y_1(t) = \left[a_1 A_1 + \frac{3}{4}a_3 A_1^3 + \frac{3}{2}a_3 A_1 A_2^2\right]\cos(\omega_1 t) \tag{3.4.6}$$

that for $A_1 \ll A_2$ reduces to

$$y_1(t) = \left[a_1 + \frac{3}{2}a_3 A_2^2\right] A_1\cos(\omega_1 t) \tag{3.4.7}$$

where the second term within square brackets is negative for compressive input-output relationships and results in gain reduction, also when the incumbent component A_2 is not at the level typical of signal compression.

3.4.3 Cross-modulation

Considering again two tones at two different frequencies $x(t) = A_1\cos(\omega_1 t) + A_2\cos(\omega_2 t)$ with much different amplitude, if the amplitude of the stronger one A_2 is not constant but subject to some modulation, the weak one at ω_1 also exhibits a modulated amplitude caused by cross-modulation. Let's assume that the modulation of A_2 is of the sinusoidal type with modulation index m and modulation frequency ω_m: $A_2(1 + m\cos(\omega_m t))$. The resulting output for the ω_1 component alone in the presence of a third-order distortion is

$$y_1(t) = \left[a_1 A_1 + \frac{3}{2}a_3 A_1 A_2^2\left(1 + \frac{m^2}{2} + \frac{m^2}{2}\cos(2\omega_m t) + 2m\cos(\omega_m t)\right)\right]\cos(\omega_1 t) \tag{3.4.8}$$

and the component exhibits an amplitude modulation of A_1 at ω_m and $2\omega_m$ that was not in the original signal.

[19] A distinction is necessary: the same word "desensitization" has been used for two different concepts; please see sec. 9.2.13 where "pulse desensitization" is due to the limited rise time of the Spectrum Analyzer with respect to the fast front and short duration of the incoming pulse.

Typical examples are the measurement of weak radiated electromagnetic field emissions in the presence of strong narrowband radio sources, such as FM broadcast radios, or the measurement of many channels in parallel within the same frequency-multiplexed physical channel (such as telephone and data network or cable TV).

3.4.4 Inter-modulation

When feeding multiple tones to a non-linear circuit, the output contains not only harmonic components of each tone present in the original signal, but also components that are not harmonically related to a single tone, whilst they are the result of sum and difference of integer multiples of two input tones. When the sum of two tones is fed to a non-linear circuit, the terms with exponent of 2 or higher of the polynomial approximation mix the two tones, by the implicit multiplication resulting from squaring (or raising to the third or higher power) the sum of two terms.

If a two-tone input signal $x(t) = A_1 \cos(\omega_1 t) + A_2 \cos(\omega_2 t)$ is fed to our third-order non-linearity, the resulting terms are powers of exponent 2 and 3 of the binomial:

$$\begin{aligned} y(t) &= a_1 \left[A_2 \cos(\omega_2 t) + A_1 \cos(\omega_1 t)\right] + a_2 \left[A_2 \cos(\omega_2 t) + A_1 \cos(\omega_1 t)\right]^2 + \\ &+ a_3 \left[A_2 \cos(\omega_2 t) + A_1 \cos(\omega_1 t)\right]^3 \end{aligned} \tag{3.4.9}$$

This gives rise to dc and harmonic terms when raising to the second and third power the single terms and originates inter-harmonics (or inter-modulation terms), when the product of the two terms of each binomial is considered:

$$\begin{aligned} \omega_1 \pm \omega_2 &: \quad a_2 A_1 A_2 \cos(\omega_1 + \omega_2)t + a_2 A_1 A_2 \cos(\omega_1 - \omega_2)t \\ 2\omega_1 \pm \omega_2 &: \quad \frac{3a_3 A_1^2 A_2}{4} \cos(2\omega_1 + \omega_2)t + \frac{3a_3 A_1^2 A_2}{4} \cos(2\omega_1 - \omega_2)t \\ 2\omega_2 \pm \omega_1 &: \quad \frac{3a_3 A_2^2 A_1}{4} \cos(2\omega_2 + \omega_1)t + \frac{3a_3 A_2^2 A_1}{4} \cos(2\omega_2 - \omega_1)t \end{aligned} \tag{3.4.10}$$

The terms at the two fundamental frequencies are:

$$\begin{aligned} \omega_1 &: \left[a_1 A_1 + \frac{3}{4} a_3 A_1^3 + \frac{3}{2} a_3 A_1 A_2^2\right] \cos(\omega_1 t) \\ \omega_2 &: \left[a_1 A_2 + \frac{3}{4} a_3 A_2^3 + \frac{3}{2} a_3 A_2 A_1^2\right] \cos(\omega_2 t) \end{aligned} \tag{3.4.11}$$

The relevant observed phenomena are two: as before the original (and true) amplitude of the two tones at the fundamental is affected by cross-modulation effects, but also third-order inter-modulation products appear. The latter are quite relevant, not only because they populate the spectrum with components that are extraneous to the original signal, but in particular when the difference between ω_1 and ω_2 is small (as in the case of adjacent channels, or quite close beating frequencies): the third-order products appear in vicinity of the original ω_1 and ω_2 components, and distinguishing and separating them is possible only adopting quite a narrow bandwidth and frequency resolution. Of course, if the interest is on the measurement of channel power (see sec. 9.3.2), such extraneous components cannot be separated and affect the accuracy of the result. In general, mixing tones through a potentially non-linear circuit, such as an amplifier or active filter, leads to inter-modulation distortion products, that affect adjacent channels.

This phenomenon is quite relevant and therefore a whole set of metrics was derived to quantify this kind of distortion. Inter-modulation products, as any other distortion product, are measured with respect to the original signal, or carrier, and are quantified in dBc, that is "dB below carrier". It is the case also of distortion products in Analog-to-Digital converters: the spurious free dynamic range (SFDR) is the largest distortion product with respect to the applied tone taken as the reference carrier.

By observing the equations above showing the fundamental terms and the third-order inter-modulation terms, it is recognized that as the amplitude A_1 or A_2 of the tones increases, the desired response increases linearly, while the term affecting the amplitude at the fundamental increases with the third power, as well as the amplitude of the third-order inter-modulation products. On a log scale the slope of the desired fundamental term is 1, while all the others have 3; the ,fundamental however, starts from a larger value because the nominal gain a_1 is larger than the coefficients of second- and third-order distortion a_2 and a_3. This means that the two curves will cross each other at some far point for a large enough amplitude of the input tone: this point is called third-order intercept and is indicated as IP3. This point may be determined on the basis of the sole non-linearity coefficients.

Let's consider two equal tones, with $A_1 = A_2 = A$:

$$\begin{aligned} y(t) &= \left[a_1 + \frac{9}{4}a_3A^2\right] A\cos(\omega_1 t) + \left[a_1 + \frac{9}{4}a_3A^2\right] A\cos(\omega_2 t) + \\ &+ \frac{3}{4}a_3A^3\cos((2\omega_1 - \omega_2)t) + \frac{3}{4}a_3A^3\cos((2\omega_2 - \omega_1)t) \end{aligned} \tag{3.4.12}$$

Assuming sensibly that $a_1 \gg 9a_3A^2/4$ to carry out simplified calculations, the input level A at which the inter-modulation products have the same amplitude of the components at the fundamental frequencies is given by

$$|a_1|\, A_{\mathrm{IP3}} = \frac{3}{4}\, |a_3|\, A_{\mathrm{IP3}}^3 \tag{3.4.13}$$

so that the amplitude on the abscissa of the intercept of the IP3 point is

$$A_{\mathrm{IP3}} = \sqrt{\frac{3}{4}\frac{|a_1|}{|a_3|}} \tag{3.4.14}$$

Of course, when approaching the IP3 level it is possible that the made assumption no longer holds, the effective gain is lower and higher-order inter-modulation terms begin to appear. So, more correctly the third-order intercept may be determined by making as many measurements as possible of the input-output relationship for increasing input amplitude and then extrapolating the intercept.

By comparing eq. (3.4.4) and (3.4.14), it is possible to derive a relationship between the two quantities:

$$\frac{A_{1\,\mathrm{dB}}}{A_{\mathrm{IP3}}} = \frac{\sqrt{0.145}}{\sqrt{4/3}} = -9.64\,\mathrm{dB} \tag{3.4.15}$$

3.4.4.1 Measurement of the third-order inter-modulation intercept

A direct measurement is possible: by applying two tones of equal amplitude A_{in} at frequencies ω_1 and ω_2 and recording the output amplitude at the two frequencies A_{ω_1,ω_2} and at the third-order inter-modulation products A_{IM3}, the ratio of the two is

$$\frac{A_{\omega_1,\omega_2}}{A_{\mathrm{IM3}}} = \frac{|a_1|\,A_{in}}{\frac{3}{4}\,|a_3|\,A_{in}^3} = \frac{4}{3}\frac{|a_1|}{|a_3|}\frac{1}{A_{in}^2} \tag{3.4.16}$$

and recalling eq. (3.4.14) above, we obtain

$$\frac{A_{\omega_1,\omega_2}}{A_{\mathrm{IM3}}} = \frac{A_{\mathrm{IP3}}^2}{A_{in}^2} \tag{3.4.17}$$

Example 3.6: Effect of large distortion terms on smaller measured amplitudes

Once the inter-modulation distortion products and the intercept point for third-order inter-modulation IP3 have been defined, what is the effective influence of larger signals on the measurement of the amplitude of small signals and the related accuracy? Given a $A_{s,in} = 10\,\mu\mathrm{V}$ signal entering the spectrum analyzer, we want to estimate the amount of corruption caused by an $A_{int,in} = 10\,\mathrm{mV}$ interfering signal.
The IP3 point of the mixer is at 0 dBm, that is $A_{\mathrm{IP3}} = 224\,\mathrm{mV}$. Let's assume that there is no desensitization and cross-modulation, so that the gain for the signal is the same as for the inter-modulation terms. Eq. (3.4.17) gives

$$\frac{A_{s,out}}{A_{\mathrm{IM3}}} = \frac{A_{\mathrm{IP3}}^2}{A_{int,in}^2} = \frac{(224\,\mathrm{mV})^2}{(10\,\mathrm{mV})^2} = 501.8 = 54\,\mathrm{dB} \tag{3.4.18}$$

If the IP3 point of the mixer is lowered to $-10\,\mathrm{dBm}$, that is $A_{\mathrm{IP3}} = 70.7\,\mathrm{mV}$, the recalculated effect is 34 dB, as it might have been evident observing that a reduction by an order of magnitude of IP3 corresponds to a reduction by an order of magnitude of the signal-to-distortion ratio, i.e. 20 dB, if amplitude, and not power, is considered.

Example 3.7: Maximum input level for given limit of inter-modulation terms

It is desired to determine the maximum input level in a two-tone test for which the inter-modulation products do not exceed the Spectrum Analyzer noise floor, so that they are, practically speaking, "invisible". The terms regarding the output on the left side of eq. (3.4.17) are expressed in terms of the input power terms on the right side:

$$P_{out} - P_{\mathrm{IM3},out} = 2(P_{\mathrm{IP3}} - P_{in}) \tag{3.4.19}$$

Since the gain affects P_{in} leading to P_{out}, as well as the equivalent input-referred inter-modulation level $P_{\mathrm{IM3},in}$ leading to $P_{\mathrm{IM3},out}$, it simplifies at the left-hand side of the equation:

$$P_{in} - P_{\mathrm{IM3},in} = 2(P_{\mathrm{IP3}} - P_{in}) \tag{3.4.20}$$

$$P_{in} = \frac{2P_{\mathrm{IP3}} + P_{\mathrm{IM3},in}}{3} \tag{3.4.21}$$

The maximum input signal level at which the inter-modulation products are equal to the noise floor F (also called "displayed noise level") is

$$P_{in,max} = \frac{2P_{\mathrm{IP3}} + F}{3} \qquad F = -174\,\mathrm{dBm} + \mathrm{NF} + 10\log_{10} B \tag{3.4.22}$$

and NF indicates the noise figure (that is the amount of amplification of the various internal noise sources provided by the spectrum analyzer).

3.4.5 Multiple harmonic components of the input signal

The eq. (3.4.1) is again considered, assuming non-linear terms of order two and three, this time feeding the system with a sinusoidal input signal that has its own harmonic components, namely the second and the third to keep things simple: $x(t) = A_1\cos(\omega t) + A_2\cos(2\omega t) + A_3\cos(3\omega t)$, with $\omega = 2\pi f$.

$$\begin{aligned} y(t) &= a_1\left[A_1\cos(\omega t) + A_2\cos(2\omega t) + A_3\cos(3\omega t)\right] + \\ &+ a_2\left[A_1\cos(\omega t) + A_2\cos(2\omega t) + A_3\cos(3\omega t)\right]^2 + \\ &+ a_3\left[A_1\cos(\omega t) + A_2\cos(2\omega t) + A_3\cos(3\omega t)\right]^3 \end{aligned} \tag{3.4.23}$$

The calculation of the resulting output spectrum has been carried out with symbolic manipulation, using the Symbolic Toolbox in Matlab®, and grouping and simplifying the resulting terms by hand[20]. The resulting terms are summarized here below:

[20] First the symbolic expression for $y(t)$ was built, expanded with `expand()` and then the correct expression with $\cos(\cdot)$ terms at multiples of the fundamental was selected from the set of results made available by the function `simple()`: the desired result is given by the operation `combine(sincos)`. To facilitate the management of each step and the grouping of relevant terms, the harmonic terms at 2ω and 3ω were written as $\omega 2$ and $\omega 3$, i.e. independent symbolic variables to avoid unwanted grouping and simplifications, and then recombined manually.

$$
\begin{aligned}
0:&\quad \frac{1}{2}a_2(A_1^2+A_2^2+A_3^2)+\frac{3}{2}a_3(\frac{1}{2}A_1^2A_2+A_1A_2A_3)\\
\omega:&\quad a_1A_1+a_2(A_1A_2+A_3A_2)+\frac{3}{4}a_3(A_1^3+A_1^2A_3+A_2^2A_3+2A_1A_2^2+2A_1A_3^2)\\
2\omega:&\quad a_1A_2+\frac{1}{2}a_2(A_1^2+2A_1A_3)+\frac{3}{2}a_3(\frac{1}{2}A_2^3+A_1^2A_2+A_3^2A_2+A_1A_2A_3)\\
3\omega:&\quad a_1A_3+a_2A_1A_2+\frac{1}{4}a_3(A_1^3+3A_3^3+3A_1^2A_3+3A_2^2A_3+3A_1A_2^2)\\
4\omega:&\quad a_2(A_1A_3+\frac{1}{2}A_2^2)+\frac{3}{4}a_3(A_1^2A_2+A_3^2A_2+2A_1A_2A_3)\\
5\omega:&\quad a_2A_2A_3+\frac{3}{4}a_3(A_1^2A_3+A_3^2A_1+A_2^2A_1)\\
6\omega:&\quad \frac{1}{2}a_2A_3^2+\frac{1}{4}a_3(A_2^3+6A_1A_2A_3)\\
7\omega:&\quad \frac{3}{4}a_3(A_1A_3^2+A_2^2A_3)\\
8\omega:&\quad \frac{3}{4}a_3A_2A_3^2\\
9\omega:&\quad \frac{1}{4}a_3A_3^3
\end{aligned}
\qquad (3.4.24)
$$

Provided that the sinusoidal signal of fundamental ω has large enough second and third harmonics, so that the single terms are all distinguishable from the noise floor, the system can be solved in a_1, a_2 and a_3 with the following considerations:

- it is known that a_1 shall be unity and it may be used as an input or to check the correctness of the solution;

- taking the three equations in ω, 2ω and 3ω the system has a unique solution; adding a fourth equation, e.g. for 4ω, the system is over-determined and a least mean square method may be used; however, at 4ω there will appear new harmonic terms not taken into account in our model, as explained below;

- higher order expressions cannot be used because the harmonics of the external signal have been neglected starting from 4ω;

- there will be one solution for each power level reaching the mixer, i.e. each value of the input attenuator, progressively going to zero;

- the reading for the largest attenuation (e.g. 30 dB) is to be retained as the measurement of the external distortion of the fed sinusoidal input, despite a larger noise floor due to the large attenuation; for best results the RBW is kept narrow enough to have a noise floor down to by 20 dB minimum, as observed in sec. 9.4.1 and Figure 9.4.1.

Example 3.8: Determination of third-order distortion products

Let's consider the first three harmonic components A_1, A_2 and A_3 applied to a mixer, for which non-linear terms of order two and three are considered; the resulting spectrum components at the mixer output displayed by the Spectrum Analyzer H_1, H_2 and H_3 may be written as:

$$\begin{aligned} H_1: &\ a_1A_1 + a_2A_1A_2 + \frac{3}{4}a_3A_1^3 \\ H_2: &\ a_1A_2 + \frac{1}{2}a_2A_1^2 + \frac{3}{2}a_3A_1^2A_2 \\ H_3: &\ a_1A_3 + a_2A_1A_2 + \frac{1}{4}a_3A_1^3 \end{aligned}$$

Setting the input attenuator of the Spectrum Analyzer at Att.=30 dB and bringing the mixer far from its distorting operating points (see sec. 9.2.11 and 9.4), the following values are measured for the applied sinusoidal signal: $A_1 = -1.12\,\mathrm{dBm}$ (or $0.773\,\mathrm{mW}$, or $124.31\,\mathrm{mV}$), $A_2 = -40.15\,\mathrm{dBm}$ (or $96.61\,\mathrm{nW}$, or $1.39\,\mathrm{mV}$) and $A_3 = -48.54\,\mathrm{dBm}$ (or $14.00\,\mathrm{nW}$, or $0.529\,\mathrm{mV}$). Reducing the input attenuation to 0 dB, the mixer is brought full into its distortion region and we want to estimate a third-order approximation of its non-linear input-output function, determining a_1, a_2 and a_3.
To this aim the following was read on the Spectrum Analyzer display: $H_1 = 120.10\,\mathrm{mV}$, $H_2 = 2.17\,\mathrm{mV}$ and $H_3 = 0.733\,\mathrm{mV}$.
Once the so determined A_1, A_2 and A_3 values are entered in the equation system $\mathbf{H} = \mathbf{Qa}$ above, simple inversion gives the solution; in this case $a_1 = 0.9662$, $a_2 = 0.0028$, $a_3 = 0.0105$.

3.5 Frequency stability, phase noise and time jitter

Clocks and frequency sources are used in a wide variety of applications (metrology, telecommunications, electronics, etc.) and they are implemented in different technologies, featuring a wide range of performance and cost. Regarding performance, the need was felt already in the '60s to characterize them, aiming at least at two main purposes: to allow for meaningful comparisons between similar devices developed by different laboratories, or between different devices in a given application; to access application performance in terms of the measured oscillator frequency stability [266, 291]. Many papers have been written on the topic that is now extended to all the digital devices and formats in use, not only for commercial telecommunication and electronic products (think of serial high-speed links in network applications, short range serial/parallel paths inside an electronic board, clock references and digital synthesis, etc.), but also for measuring instruments and metrology (in particular digital synthesis and Phase Locked Loop technique, even inside our spectrum analyzer). The IEEE Std. 1139 [174, 175], that collects definitions and test methods, has been issued a few times in the last 30 years and may be considered a very useful reference for stable reliable information, solving the problems of sometimes inconsistent and confusing notation of the many papers and technical notes on the subject.

Given a sinusoidal signal with the argument of the sine indicated by $\phi(t)$, the instantaneous frequency is given by

$$\nu(t) = \frac{1}{2\pi}\frac{\mathrm{d}\phi}{\mathrm{d}t} \tag{3.5.1}$$

The nominal frequency of the signal is indicated by ν_0 and is the average value of ν. If the frequency is constant, then the phase increases linearly and we have the well known expression of the sinusoidal signal with amplitude V_0 and initial phase ϕ_0:

$$V(t) = V_0 \sin(2\pi\nu_0 t + \phi_0) \tag{3.5.2}$$

The initial phase may be taken equal to zero without loss of generality.

The ideal sinusoidal signal is now affected by noise that causes phase perturbation $\phi(t)$ and amplitude perturbation $\varepsilon(t)$: in both cases we assume that the two perturbations are small with respect to the nominal values, so that $\varepsilon(t)/V_0 \ll 1$ and $\varphi(t)/\nu_0 \ll 1$.

$$V(t) = [V_0 + \varepsilon(t)]\sin[2\pi\nu_0 t + \phi(t)] \tag{3.5.3}$$

By recalling eq. (3.5.1) the frequency fluctuation $\delta\nu$ around ν_0 due to the phase fluctuation (or phase deviation) $\phi(t)$ is

$$\delta\nu = \frac{1}{2\pi}\frac{\mathrm{d}\phi}{\mathrm{d}t} \tag{3.5.4}$$

and the fractional frequency fluctuation is

$$y = \frac{\delta\nu}{\nu_0} \tag{3.5.5}$$

Similarly, for phase instability defined by the instantaneous phase deviation $\phi(t)$ above, we may define a fractional phase fluctuation

$$x = \frac{\delta\phi}{2\pi\nu_0} \tag{3.5.6}$$

It is clear that the two quantities, $x(t)$ and $y(t)$, are related as

$$y(t) = \frac{\mathrm{d}x(t)}{\mathrm{d}t} \tag{3.5.7}$$

Time quantities, when varying randomly around a reference point, are said subject to "jitter", or "jittering". Phase and time jitter may be intuitive. For a definition we may recall the standard ANSI C63.14 (1998): phase jitter, "the phenomenon, from causes known or unknown, that results in a relative shifting in the phase of the signal; the shifting in phase may appear to be random, cyclic, or both"; time jitter, "a measure of the uncertainty of the repetitive position of a time mark", but related to periodic waveforms, it may be also identified with "time-related, abrupt, spurious variations in the duration of any specified, related time interval." In the following the concept of jitter will be related to other better defined time and frequency quantities.

Once the fundamentals of the model were laid down, it was already evident many years ago that the problem offered itself for two viewpoints and treatments: two kinds of parameters or quantities were needed in order to meet requirements of different applications, namely spectral parameters (related to the spread of signal energy in the Fourier frequency spectrum) and time parameters (allowing assessment of the stability over a given time interval). For the former spectral densities of phase and frequency fluctuations are considered, for the latter the statistical distribution and the variance of frequency and time quantities. A comprehensive discussion and a unifying theory with experimental evidence was proposed in [102].

3.5.1 Spectral densities in the Fourier frequency domain

The fluctuation of phase and frequency does not change the total power of the periodic signal, but alters its distribution in frequency, that is the power spectral density; in general, for a periodic signal that has impulsive delta functions in its line spectrum located at harmonic frequencies, phase and frequency fluctuation spreads the power of these components in their neighborhood. Several power spectral density functions may be identified based on fluctuation quantities above, such as one for phase fluctuation ϕ, one for the frequency fluctuation $\delta\nu$ and one for the fractional frequency fluctuation y, all defined as the mean square value of the fluctuation divided by the resolution bandwidth used when measuring or calculating it. The IEEE Std. 1139 [175] first defines the spectral density for the normalized frequency fluctuation, the normalized amplitude fluctuation and phase fluctuation:

$$S_y(f) = \frac{1}{B}\, y_{rms}^2(f) \qquad S_a(f) = \frac{1}{B}\, a_{rms}^2(f) \qquad S_\phi(f) = \frac{1}{B}\, \phi_{rms}^2(f) \tag{3.5.8}$$

with the three terms on the right indicating mean squared quantities evaluated over the chosen resolution bandwidth B, provided that the Fourier representation is such that the three quantities are approximately constant over the band B (considering a discrete frequency implementation, they are the sum of the squares of the frequency bins falling inside the band B).

The relationships between the power spectral densities may be exemplified as:

$$S_\phi(f) = \frac{\nu_0^2}{f^2}\, S_y(f) = \frac{1}{f}\, S_{\delta\nu}(f) \tag{3.5.9}$$

Please note that $S_\phi(f)$ is expressed in $\mathrm{rad}^2/\mathrm{Hz}$, while $S_{\delta\nu}(f)$ is expressed in $\mathrm{Hz}^2/\mathrm{Hz}$.

The phase jitter in rms terms may be derived from $S_\varphi(f)$ as the square root of the integral over the frequency interval $[f_1, f_2]$:

$$\phi_{jitter,rms} = \sqrt{\int_{f_1}^{f_2} S_\phi(f)\, \mathrm{d}f} \tag{3.5.10}$$

Such a band-pass integral has been referred to as Rutman's band-pass variance, when calculated on octaves of band related to the sampling time adopted for the Allan variance analysis:

$$\beta_y^2(\tau_k) = \int_{1/4\tau_k}^{1/2\tau_k} S_y(f)\,\mathrm{d}f \tag{3.5.11}$$

Conversely, the lowest part of the spectrum below f_1 (but above the lowest frequency of the spectrum f_0), gives an indication of phase wander, subject to diverge when negative power-law noise terms are present:

$$\phi_{wander,rms} = \sqrt{\int_{f_0}^{f_1} S_\phi(f)\,\mathrm{d}f} \tag{3.5.12}$$

Let's consider the sinusoidal signal spectrum $S_v(f)$, that in the absence of any noise and fluctuation would be a single delta located at ν_0; the power P per unit bandwidth is given by $S_v(f)/R_0$, where R_0 is the reference resistor, usually taken as $50\,\Omega$.

To interpret the shape of the spectrum of a sinusoidal signal subject to phase instability, we may say that any deterministic variations of considerable value will cause a modulation of the signal and creates sidebands around the center frequency ν_0 (called *carrier*), while small random fluctuations have the effect of broadening the spectrum around the center frequency ν_0, that would be otherwise a single line.

If we choose to express the spectrum distribution as the fraction of power in a given frequency interval with respect to the carrier power, the resulting function is expressed in terms of the difference of f and ν_0. Such a function shall increase while approaching ν_0 on either side. It may be shown that this function is Lorentzian and thus avoids any singularity in the origin (at ν_0 it cannot go to infinite, even if it is inversely proportional to f^2). This is the first definition of the $\mathcal{L}(f)$ function, with the specificity of taking the single side-band (SSB) power over 1 Hz interval with respect to the carrier power. This definition however does not keep separate AM and FM terms, leading to ambiguous results, and is not applicable when the mean square value of ϕ exceeds $0.1\,\mathrm{rad}^2$. For this reason the IEEE Std. 1139 now defines $\mathcal{L}(f)$ in terms of $S_\phi(f)$, simply saying that they are the same except for a factor of 2, due to the fact that $\mathcal{L}(f)$ works on the SSB components, thus leaving out half of the power. A simplified expression of $\mathcal{L}(f)$ that neglects any noise floor and $1/f$ noise component is

$$\mathcal{L}(f) = \frac{1}{\pi}\frac{\pi\nu_0^2 c}{(\pi\nu_0^2 c)^2 + f^2} \tag{3.5.13}$$

where c is a scalar to adjust the shape of the curve.

The noise far from the carrier will always be some negative number below $0\,\mathrm{dBc/Hz}$, indicating that the noise components are below the carrier. However going towards the 1 Hz point and beyond it, closer to the carrier frequency it is easy to see that it is not forbidden nor wrong to observe positive values, that indicate that the noise is above the carrier. This shall be interpreted correctly, since we are operating below 1 Hz and thus with a frequency bandwidth that is smaller than 1 Hz: the total power integrated over a 1 Hz bandwidth will always be unity.

It is interesting to observe another parameter, the width of the spectral line at ν_0 defined as the half-power width, or $-3\,\mathrm{dB}$ width: $f_{-3\mathrm{dB}} = 2\pi\nu_0^2 c$.

Regarding the effectiveness of frequency-domain spectral representations (spectral densities) compared to time-domain quantities that are reviewed in the next sec. 3.5.2, it is underlined that using spectral densities is always correct and conveys a larger amount of information, provided that the estimation is correctly performed with respect to known phenomena of spectral leakage, poor resolution, data correlation, resulting all in bias and variance of spectral estimators (see sec. 3.1.3). The contrast is clear: time-domain estimates are broad-band quantities, while spectral densities give a more accurate representation of details. In this regard Percival [266] states that "Allan variance nor the band-pass variance is a meaningful measure of frequency stability until after a parallel spectral analysis has been performed."

3.5.1.1 Power-law noise classification

Spectral densities due to random noise of all high stability frequency standards can be modeled by the power-law model, where spectral densities vary as a power of f. When considering the behavior of the power spectral density curves of noise, it is first of all to be recalled that some components may be of the discrete-frequency type and caused by external disturbance, e.g. due to the supply frequency and its harmonics: these components are of course excluded from this classification. The selected power terms to express spectral densities are identified based on practice, observations and physical explanations.

$$S_y(f) = \sum_{\alpha=-2}^{+2} h_\alpha f^\alpha \qquad S_\phi(f) = \sum_{\beta=0}^{+4} h_\beta f^\beta \tag{3.5.14}$$

In the two expressions above, of course, the h coefficients are different for $S_\phi(f)$ and $S_y(f)$; in the following those of $S_y(f)$ are retained for further discussion, based on the definitions of the IEEE Std. 1139 [175]. It is recalled that both models in series of power terms are truncated and put to 0 for frequency f above the upper limit that the IEEE Std. 1139 indicates with f_h and that corresponds to the cutoff frequency of an infinitely sharp low-pass filter.

So, frequency instability due to broadband phase random noise is in general classified as power-law noise terms, distinguishing the following categories:

- Random Walk FM: it's a noise very close to the carrier and thus difficult to measure; $S_y(f)$ is proportional to $1/f^4$; it is usually due to very slow environmental conditions acting on the signal source;
- Flicker FM: it is usually related to the physical resonance of the oscillator and to the electronic circuits employed in the signal source; $S_y(f)$ is proportional

to $1/f^3$; if the oscillator is subject to considerable drift, this noise term may be masked;

- White noise: due to the random walk of phase, it is quite a common noise term; $S_y(f)$ is proportional to $1/f^2$; it usually tends to reduce slightly with time and then it becomes a flicker FM noise for longer observation periods;
- Flicker PM: this noise shows as a flicker modulation of the phase; $S_y(f)$ is proportional to $1/f$;
- White PM: with nearly the same origin of the flicker PM it is flat with frequency, so $S_y(f)$ is proportional to $1/f^0$.

The identification of frequency behavior and the interpretation in terms of separate noise mechanisms with a well definite frequency model are quite a useful method for treating noise processes using a frequency-domain representation; after considering time-domain quantities in the next sec. 3.5.2, the mutual relationships between the considered quantities will be summarized in sec. 3.5.4.

3.5.2 Dispersion and variance in time domain

The frequency is subject to fluctuations that we have expressed by means of the fractional change $y(t)$ and a first intuitive measure of the amount of fluctuation is the total spread or, better, its variance. Despite its simplicity, the practical usefulness of the variance is much limited, since it diverges for all real oscillators. The sample variance is used instead: in general, N samples of time duration τ_0 are taken, each spaced T. A common choice that leads to the Allan variance (see sec. 4.1.5 for an introduction) is to have $T = \tau_0$ (or, *zero dead-time*) and $N = 2$ (that is a direct difference between adjacent samples); the biasing was estimated by Barnes [314] in real experiments, in particular considering that frequency counters and other instruments have a non-zero dead-time between successive samples. Then, the modified Allan variance [27] goes beyond this definition and groups n successive samples for the calculation of an ensemble difference, that is then used to calculate the overall variance.

Data may be affected by drift and other deterministic errors; after such components have been accounted for (see [11], and [27], sec. III.E), the Allan and modified Allan variance may be calculated, however, especially for long-term observations, this operation is subject to a significant uncertainty due to non-stationarity of systematic components, the need for some data fitting and de-trending, and finally "experimenter's plausibility check" [167]. As noted by Howe [167], the weak side of Allan variances from a statistical point of view is the reduced number of statistical samples (see sec. 4.1.5), or in other words a non-perfect exploitation of the available original data samples, that may run short quickly if long observation times are required. To this aim he proposed a periodic repetition of the original data, based on the observation that the modified Allan variance goes to zero for long averaging times and based on the initial necessity to shift the time reference with which the average and difference behind the calculation of the modified Allan variance. This method leads to the calculation of an alternative measure of variance, called "Totdev", based on the modified Allan variance calculated on a virtual data sequence that is three times longer than the original one. As pointed out by the author, "Totdev is the current IEEE-recommended statistical test of oscillator frequency stability at long-term averaging times, namely, those τ values beyond 10% of the length of the whole data run." ([175], eq. (A.25), (A.26)).

The ITU-T standard G.810 [193] establishes the definitions of the quantities involved in the evaluation of frequency instability in time domain. On the contrary, the IEEE Std. 1139 "attacks" at its appendix A.3 soon with the two-sample deviation and Allan deviation. The content of this section is the result of a balanced comparison between the two formulations, looking especially for mutual confirmations and identifying changes of notation and different assumptions or interpretations. The notation of the ITU-T standard is changed to avoid confusion with the IEEE one used above: for each new quantity the original ITU-T notation is however indicated.

The instantaneous phase model is given by the following expression, that takes into account constant offset, drift terms and completely random phase deviation[21]:

$$\varphi(t) = \varphi_0 + 2\pi\nu_0(1+y_0)t + \pi D\nu_0 t^2 + \phi(t) \tag{3.5.15}$$

where φ_0 is the initial phase, ν_0 is the nominal frequency, the third term takes into account drift behavior (mainly due to aging effect and temperature influence, if applicable) with D the linear fractional frequency drift rate, and $\phi(t)$ is the random phase component (called before "phase fluctuation").

The *time error function* is the time error of a clock, with respect to a frequency standard, that is the difference between the time of that clock and the frequency standard one: $x(t) = T(t) - T_{ref}(t)$.

So, with eq. (3.5.15) above, assuming a stable reference clock used to evaluate the source under test, the time error function is given by

$$x(t) = x_0 + y_0 t + \frac{D}{2}t^2 + \frac{\phi(t)}{2\pi\nu_0} \tag{3.5.16}$$

The *time interval error* (TIE) *function* is the difference between the measure of a time interval as provided by a clock and the measure of the same time interval as provided by a reference clock:

$\mathrm{TIE}(t;\tau) = [T(t+\tau) - T(t)] - [T_{ref}(t+\tau) - T_{ref}(t)] = x(t+\tau) - x(t)$.

The *time deviation* (TDEV or σ_x) is a measure of the expected time variation of a signal as a function of the integration time and is expressed as

$$\mathrm{TDEV}(n\tau_0) = \sqrt{\frac{1}{6n^2(N-3n+1)} \sum_{j=1}^{N-3n+1} \left[\sum_{i=j}^{n+j-1} x_{i+2n} - 2x_{i+n} + x_i \right]^2} \tag{3.5.17}$$

[21] Please, take care to and distinguish between φ and ϕ, the latter corresponding to the instantaneous phase of the IEEE Std. 1139 and reviewed in sec. 3.5.1.

where x_i are the error samples; N is the total number of samples; τ_0 is the time error sampling interval; τ is the integration time; n is the number of sampling intervals within the integration time $\tau = n\tau_0$.

Eq. (3.5.17) above is a modified two-sample Allan deviation that we have already seen at sec. 4.1.5. The IEEE Std. 1139 [175] and the ITU-T Std. G.810 [193] make both use of Allan deviation (ADEV) and modified Allan deviation (MDEV) to quantify time-domain fluctuations, addressing in this way two problems: i) any constant time offset has no influence, so that we are focusing on random fluctuations and not on systematic errors, and ii) these two quantities may be put in direct relationship with the frequency-domain spectral densities, succeeding in the interpretation of slopes, as already done for the power-law model of spectral densities.

The ITU-T Std. G.810 presents both the Allan deviation and the modified Allan deviation[22], together with their relationships with the phase power spectral density $S_\phi(f)$.

$$\mathrm{ADEV}(\tau = n\tau_0) = \sqrt{\frac{1}{2n^2\tau_0^2(N-2n)} \sum_{i=1}^{N-2n} (x_{i+2n} - 2x_{i+n} + x_i)^2} \tag{3.5.18}$$

$$\mathrm{ADEV}(\tau) = \sqrt{\frac{2}{(\pi\nu_0\tau)^2} \int_0^B S_\phi(f)\, \sin^4(\pi\tau f)\, df} \tag{3.5.19}$$

where the integral is taken over the measurement bandwidth B. Eq. (3.5.18) corresponds to eq. (A.20) given in IEEE Std. 1391.

The Allan deviation is not influenced by constant frequency offset, but systematic effects (like night/day periodic drift) may heavily influence the results; if the frequency drift dominates, then the slope of the Allan deviation is 1, linear with τ.

It has been shown in the literature that the Allan deviation discriminates all the power-law noise types, except Flicker PM and White PM. To correct this adverse effect the modified Allan deviation was conceived [12, 291].

$$\mathrm{MDEV}(\tau = n\tau_0) = \sqrt{\frac{1}{2n^4\tau_0^2(N-3n+1)} \sum_{j=1}^{N-3n+1} \left[\sum_{i=j}^{n+j-1} x_{i+2n} - 2x_{i+n} + x_i \right]^2} \tag{3.5.20}$$

with $n = 1, 2, \ldots \left\lfloor \frac{N}{3} \right\rfloor$;

$$\mathrm{MDEV}(\tau) = \sqrt{\frac{2}{(\pi\nu_0 n^2\tau)^2} \int_0^B S_\phi(f)\, \frac{\sin^6(\pi n\tau_0 f)}{\sin^2(\pi\tau_0 f)}\, df} \tag{3.5.21}$$

22 The ITU-T Std. G.810 uses ADEV and MDEV symbols for Allan deviation and modified Allan deviation, omitting the quantity on which they are calculated (in our examples x, but it might be e.g. y). It is quite common a more compact notation using capital sigma, of the kind $\Sigma_y^2(\tau)$ to indicate the Allan deviation of y; adding “mod” indicates the modified Allan deviation $\Sigma_{y,mod}^2(\tau)$.

Eq. (3.5.20) corresponds to eq. (A.23) given in IEEE Std. 1391. The MDEV converges for all the most relevant noise sources affecting timing signals and oscillators, and distinguishes the various power-law terms above. Opposite to ADEV, MDEV may depend on the sampling time τ_0, in particular when White PM noise dominates. Again systematic effects may affect the results.

Interestingly the IEEE Std. 1391 reports a relationship between x and y deviation:

$$\mathrm{ADEV}_x(\tau) = \frac{\tau}{\sqrt{3}} \mathrm{MDEV}_y(\tau) \tag{3.5.22}$$

For long integration times τ, around 10% or more of the total data points ($\tau \geq 0.1 N \tau_0$), the Allan deviation is potentially erroneous and biased due to an intrinsic insensitivity to odd noise processes, that is odd with respect to the midpoint of the array of data points, e.g. when a noise process in x has a monotone behavior going from up on the left to low on the right, or viceversa.

3.5.3 Sinusoidal modulation and linear drift as systematic effects

It is quite common that the oscillator frequency may fluctuate on a periodic basis, that we assume for simplicity a pure sinusoidal modulation of frequency F_m over a span of $\pm\Delta\nu_0$. An example is the change of room temperature during night and day, or between working days and weekends, or an external disturbance due to magnetic field at the supply frequency, in turn modulated by the overall load, that might depend on the use of heating and conditioning, lighting, etc.

In the frequency domain we have simply $y(t) = \dfrac{\Delta\nu_0}{\nu_0} \sin(2\pi F_m t)$ and the corresponding power spectral density has a spectral line located at F_m:

$$S_y(f) = \frac{1}{2} \left(\frac{\Delta\nu_0}{\nu_0}\right)^2 \delta(f - F_m) \tag{3.5.23}$$

The Allan variance and modified Allan variance corresponding to ADEV and MDEV, considering eq. (3.5.19) and (3.5.21), become

$$\Sigma_y^2(\tau) = \left(\frac{\Delta\nu_0}{\nu_0}\right)^2 \frac{\sin^4(\pi\tau F_m)}{(\pi\tau F_m)^2} \tag{3.5.24}$$

$$\Sigma_{y,mod}^2(\tau) = \left(\frac{\Delta\nu_0}{\nu_0}\right)^2 \frac{\sin^6(\pi n\tau_0 F_m)}{n^2(\pi n\tau_0 F_m)\sin^2(\pi\tau_0 F_m)} \tag{3.5.25}$$

It may be easily seen that both quantities are zero when the numerator is zero, and this occurs when the modulation period and the sampling time are equal: thus, it is always advisable to perform calculations for multiples of the mains fundamental period, that is 20 ms or 16.66 ms. On the contrary, when sampling occurs at points that are half the modulation period (or an odd multiple of the half), the effect is maximum.

If the frequency is subject to a linear drift of d Hz/s, e.g. due to the progressive warm-up of the equipment, the two Allan deviations diverge linearly with τ and are thus of no use. For this reason Percival [266] underlines that in his example data are "de-trended": the procedure is straightforward by linear fitting in the least square sense a first fraction of data (e.g. 50%) and then using this estimate to remove the drift (assumed linear) from the remaining data. However, Percival underlines that data de-trending, removing drift from the observations, may be more effective in frequency domain, rather than in time domain. Analogously, he supports the frequency-domain method, when unevenly sampled data are to be considered, for which the Allan and modified Allan deviations cannot be easily estimated directly. This is also highlighted in the IEEE Std. 1139 [175], as reported briefly at the end of sec. 4.1.5.

Some other measures of frequency stability have been considered and proposed to solve both this problem and in general the behavior of the various forms of Allan deviation with respect to specific power-law orders, such as -3 and -4, relevant for very long time intervals.

3.5.4 Conversion between frequency and time domain quantities

In this section the problem of correctly relating time- and frequency-domain quantities is considered: there have been discussions on which of the two approaches (analysis in time or frequency domain) is the most accurate and gives the soundest and most reliable results, and at which extent the results obtained with the two approaches are exchangeable and comparable. Of course, the answer is not definitive and simple, depending on the process that we analyze: there are some known extreme pathological examples mentioned at the end of the section.

The basic relationships between time- and frequency-domain quantities are thus reviewed (with as much reference as possible to standards), with some additional comments on their suitability and any known ambiguity or inaccuracy.

The phase jitter is a scalar time-domain quantity, rigorously defined starting from the power spectral density of the phase fluctuation $S_\phi(f)$ as

$$\varphi_{rms} = \sqrt{\int_B S_\phi(f)\,\mathrm{d}f} \tag{3.5.26}$$

and the time jitter is correspondingly

$$t_{rms} = \frac{\varphi_{rms}}{2\pi\nu_0} \tag{3.5.27}$$

This section aims also at completing the power-law model of sec. 3.5.1.1, by finding the relationships between frequency-domain and time-domain quantities presented so far to characterize noise processes, in relationship to the generation of periodic signals.

We recall first of all eq. (3.5.14)

$$S_y(f) = \sum_{\alpha=-2}^{+2} h_\alpha f^\alpha \qquad S_\phi(f) = \sum_{\beta=0}^{+4} h_\beta f^\beta \tag{3.5.28}$$

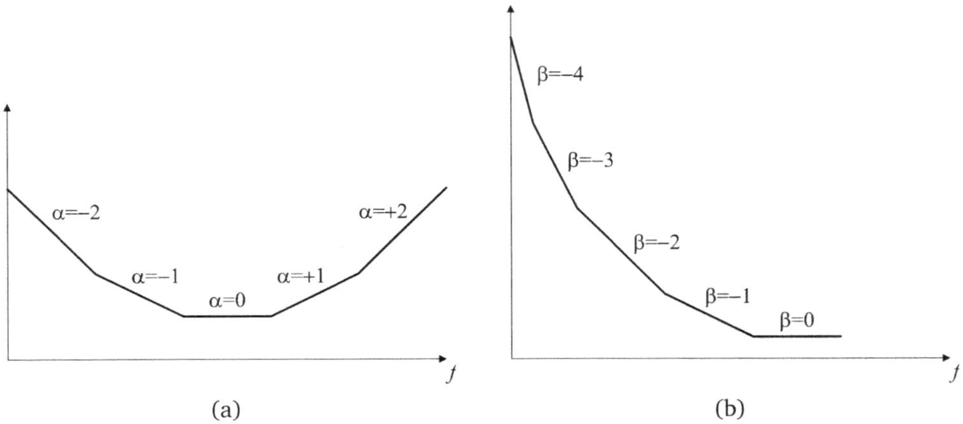

Figure 3.5.1 – Sketch of power-law profiles for (a) $S_y(f)$ and (b) $S_\phi(f)$ plotted on log-scaled axes.

underlying that α and β exponents are dedicated to the PSD of y and ϕ. Knowing the relationship between the two original quantities, for the same noise process the exponent values correspond as $\beta = \alpha - 2$.

For the five power-law noise processes considered (Random Walk FM, Flicker FM, White noise, Flicker PM and White PM), the corresponding values of α is -2, -1, 0, $+1$ and $+2$ respectively, while β starts from -4 up to 0.

From the relationship between $S_y(f)$ and $S_\phi(f)$ expressed in eq. (3.5.9) it is possible to derive a relationship between the two sums of f powers:

$$S_y(f) = \sum_{\alpha=-2}^{+2} h_\alpha f^\alpha = \frac{\nu_0^2}{f^2} S_\phi(f) = \nu_0^2 \sum_{\alpha=-2}^{+2} h_\alpha f^{\alpha-2} = \nu_0^2 \sum_{\beta=0}^{+4} h_\alpha f^\beta \tag{3.5.29}$$

that explains the use of α and β exponents and a relationship between the two sets of h coefficients:

$$h_{\beta=\alpha-2} = \nu_0^2 h_\alpha \tag{3.5.30}$$

Graphically the two expressions for $S_y(f)$ and $S_\phi(f)$ correspond to two piecewise linear profiles on log axes, as shown in Figure 3.5.1.

Now the relationship between $S_y(f)$ and $\Sigma_y(\tau)$ is considered, confining all the complexity of dead time between measurements, sampling time and integration time in a "transfer function" $\Psi(f)$:

$$\Sigma_y(\tau) = \sqrt{\int_0^\infty S_y(f)\,|\Psi(f)|^2\,df} \tag{3.5.31}$$

For a two-sample deviation, corresponding to the original Allan deviation, the transfer function takes the form

$$|\Psi(f)|^2 = 2\frac{\sin^4(\pi\tau f)}{(\pi\tau f)^2} \tag{3.5.32}$$

and assuming the five-term power-law expression for $S_y(f)$, eq. (3 5.31) becomes the formidable expression below

$$\Sigma_y^2(\tau) = h_{-2}\frac{(2\pi)^2}{6}\tau + h_{-1}2\ln 2 + h_0\frac{1}{2\tau} + h_1\frac{1.038 + 3\ln(2\pi f_h\tau)}{(2\pi\tau)^2} + h_2\frac{3f_h}{(2\pi\tau)^2} \tag{3.5.33}$$

with f_h corresponding to the cut-off frequency of an infinitely sharp low-pass filter that unambiguously defines the bandwidth for the estimation. The f_h value is subject to the condition that $2\pi f_h\tau \gg 1$.

Recalling eq. (3.5.21), a similar expression may be attained to calculate the modified two-sample Allan deviation starting from $S_y(f)$. The "transfer function" $\Psi(f)$ takes the form

$$|\Psi(f)|^2 = \frac{2}{n^4}\frac{\sin^6(\pi\tau f)}{(\pi\tau_0 f)^2\sin^2(\pi\tau_0 f)} \tag{3.5.34}$$

and, again under the assumption of the five-term power-law expression for $S_y(f)$, eq. (3.5.31) takes the form

$$\Sigma^2_{y,mod}(\tau) = h_{-2}0.824\frac{(2\pi)^2}{6}\tau + h_{-1}\frac{27}{20}\ln 2 + h_0\frac{1}{4\tau} + h_1\frac{0.084}{\tau^2} + h_2\frac{3f_h\tau_0}{(2\pi)^2\tau^3} \tag{3.5.35}$$

If the used scale M is considered for a given sampling time t_s, then an averaging time interval $\tau = M\,t_s$ may be defined. Because the Allan variance acts as an approximate high-pass spectral analyzer on phase modulation (PM) noise and an approximate octave bandpass analyzer on frequency modulation (FM) noise, linear regions in the log-log "sigma-tau" plot are associated with regions of power-law behavior f^β in the PM spectrum $S_x(f)$, which maps to Σ_x^2 by the formula

$$\Sigma_x^2(\tau) = \frac{8}{\tau^2}\int_0^{+\infty}\sin^4(\pi f\tau)\,S_x(f)\mathrm{d}f \tag{3.5.36}$$

The identification of power-law components of $S_x(f)$ from those of $\Sigma_x^2(\tau)$ is an example of a parametric inversion of eq. (3.5.36) from Allan variance to spectrum. There is no problem with this practice if the actual spectrum is known to have the desired parametric form, i.e. under the commonly used assumption of power-law spectral density: see, for example, [266], section IV, where a generally adopted approach called "pilot analysis" is presented. The pilot analysis is based on the interpretation of the sequence of Allan variance values for sampling times distributed in a geometric fashion ($\Sigma_y^2(\tau_k)$ and $\Sigma_y^2(\tau_{2k})$) as the band-pass integral of the spectral density function, over an octave of frequency, between one half and one quarter of the sampling time τ_k (that is $1/2\tau_k$ and $1/4\tau_k$). However, Percival underlines the so obtained poor frequency resolution and the fundamental correlation between the two versions of the Allan variance $\Sigma_y^2(\tau_k)$ and $\Sigma_y^2(\tau_{2k})$.

The band-pass integral of the spectral density is often referred to as Rutman's band-pass variance:

$$\beta_y^2(\tau_k) = \int_{1/4\tau_k}^{1/2\tau_k} S_y(f)\mathrm{d}f \tag{3.5.37}$$

On the other hand, claims of non-parametric inversion formulas have been published. In [139] it is said that the Allan variance does not always determine a unique PM spectrum, but the ambiguity occurs for log-periodic spectral modulations, that do not arise from any known physical theory and are thus of limited practical significance.

To summarize, the Allan variance, in general, does not completely characterize the covariance properties of the noise process. Although the Allan variance statistic remains useful for revealing broad spectral trends, the extraction of spectral details by this technique is difficult, if not impossible in some cases.

4

Statistic and Random Processes

Probability and stochastic processes are essential to model measured phenomena, results and the measuring instrument itself.

As for the previous chapter dealing with signals in a deterministic perspective, but already introducing statistical elements, the present chapter aims at presenting statistic and probabilistic concepts that will support our experiments: probabilities and distributions, confidence intervals, correlations and power spectral density functions, etc. Continuous time expressions may be preferred for clarity and reader's ease, but their digital discrete time counterparts are necessary for computer-based post-processing of results, for the evaluation of their consistency and finally to display them in suitable graphical form.

Definitions and basic concepts are briefly reviewed, but without the thoroughness and completeness of a textbook: the intention is to offer a repository of the most relevant elements of theory, without repeating the work done in other good references that treat exhaustively theory and examples, such as Montgomery and Runger [248], Drake [111], Papoulis [262], Carlson [54], Proakis [274].

4.1 Statistic and probability

We recall briefly some definitions of statistics that will be used in the following. For a more complete treatment the reader is redirected to [247] and [262], that represent very complete and comprehensive references for probability theory and statistics.

4.1.1 Expectations and statistical moments

The expected value noted by $E\{\cdot\}$ is an average operation weighted by the likelihood of occurrence of the argument, or of the contained variables, that is to say weighted by the Probability Distribution Function (PDF) indicated by $p(x)$.

Considering a random variable x or a stationary random process $x(t)$ (for this, please, see sec. 4.2), its moments are defined as the expected values of increasing powers of x, so that the order of the moment corresponds to the exponent of x:

$$\mu_k = E\left\{x^k\right\} = \int_{-\infty}^{+\infty} x^k p(x)\, \mathrm{d}x \tag{4.1.1}$$

having used for brevity x as argument of the integral (should have been X, to distinguish it from the variable x).

Thus, the moment of order 0, μ_0, is simply equal to 1 because it is the integral of the PDF; the moment of order 1, μ_1, gives the mean value; the moment of order 2 gives the mean square value, indicated by ψ^2. For the latter and for higher order moment it is often convenient to express them as central moments, that is to calculate them about the mean value; the usual notation uses an appended "c".

$$\mu_k^c = E\left\{(x-\mu)^k\right\} = \int_{-\infty}^{+\infty} (x-\mu)^k p(x)\, \mathrm{d}x \tag{4.1.2}$$

The central moment of order 2 becomes thus the variance, indicated normally by σ^2. Descriptive statistics say that "variance" attempts to capture (or "model") in a single value the extent to which data vary from some basis; the chosen basis here is the mean value. The square root of variance is dispersion, which is the expected difference each sample has from the base value.

Another parameter normally considered a signal property rather than a statistical property is the *crest factor*: it is defined as the ratio between the peak value of a signal and its rms. Its use is for the definition of the input scale of data-acquisition systems to avoid over-range or to decide on the dynamic range of a receiver or transmitter to maintain linearity and avoid significant distortion. If we consider a Normally distributed noise as input signal, we can derive a relationship between its statistical properties and the crest factor. In principle, a Normal distribution doesn't exclude the occurrence (very rare, but possible) of huge values located at the extremes of the distribution and above any threshold that may be selected.

Prob. %	Crest factor
4.6	2.0
1.0	2.6
0.37	3.0
0.1	3.3
0.01	3.9
0.006	4.0
0.001	4.4
0.0001	4.9

Table 4.1.1 – Probability of exceedance of the assumed crest factor for Normal distribution

4.1.2 Time averages

When a random variable is characterized by a time axis being the result of the measurement of a random process, then the question may be posed of the use of time averages, noted by $\langle x \rangle$, rather than ensemble averages (in sec. 4.2 this is further considered).

$$\langle x \rangle = \frac{1}{T} \int_0^T x(t)\,\mathrm{d}t \tag{4.1.3}$$

4.1.3 Statistical distributions

Some statistical distributions assume particular importance for our problems.

4.1.3.1 Uniform distribution

The uniform distribution is often used to indicate the maximum uncertainty on the expected outcome, that is only bounded between the two extremes a and b. This distribution is e.g. used to describe the quantization error of Analog-to-Digital converters (ADCs), where $a = -q/2$ and $b = +q/2$, with q being the resolution of the quantizer, corresponding to the ADC full scale FS, divided by the number of quantization levels, that is 2^n with n the number of bits. In the general case the mean value of a r.v. with uniform distribution is $(a+b)/2$ and the dispersion is $(b-a)^2/12$, that for an ADC correspond to 0 and $q^2/12$, respectively. Another example of a uniform distribution is related to the expression "the error of the instrument is $\pm e\,\%$", that refers to a definition of the error in the classical sense with no knowledge about its distribution, except its boundaries, normally assumed symmetric.

4.1.3.2 Normal, or Gaussian, distribution

The Normal, or Gaussian, distribution is well known and describes satisfactorily many natural phenomena, first of all thermal noise, in general affecting all electrical and electronic systems and measurements. Its distribution is

$$p(x) = \frac{1}{\sqrt{2\pi}} e^{-(x-\mu)^2/2\sigma^2} \tag{4.1.4}$$

and is sometimes manipulated by a change of variable $z = (x - \mu)/\sigma$ to have a normalized zero-mean r.v. with PDF and CDF values tabulated in many books. What is sometimes overlooked is that a truly Gaussian distribution has arbitrarily large outcomes associated to a non-null probability, characteristic that is lost each time the original phenomenon is recorded in a digital system or passes through an electrical or electronic system, that has bounded outputs; what we obtain is a truncated Gaussian distribution. In general truncation of small noise signals by data acquisition system is not an issue because of the much larger scale setting with respect to span and dispersion. When setting the scale the operator shall be aware of the probability of missing a reading because of an out-of-scale value that saturates the measurement system: this is often addressed by considering the crest factor and the associated probability (see before sec. 4.1.1 and Table 4.1.1); values between 4.0 and 4.9 of crest factor give extremely low probabilities of missing a sample.

Example 4.1: Matlab code for confidence intervals of a Gaussian distribution

The following code simply calculates the value of the coverage factor value k that multiplying the standard deviation σ defines the bounds of the confidence interval for a given assigned confidence level α. With quite a practical approach the code is executed many times changing the argument until the desired confidence level is attained. The `sqrt(2)` term is justified by the fact that the `erf()` function is defined in Matlab for variance equal to 1/2.
For symmetric confidence intervals:

```
>> 1-erf(k/sqrt(2))
```

For asymmetric confidence intervals:

```
>> (1-erf(k/sqrt(2)))/2
```

The *Central Limit Theorem* states that, given n r.v. x_i of arbitrary distribution (subject however to some constraints), the PDF of the r.v. z that is the sum of the said r.v. has a Gaussian distribution with mean value the sum of the mean values and variance the sum of variances. If the sum is normalized by the number of variables n and the x_i r.v. share the same PDF (e.g. are all outcomes of the same experiment as in the case when we average successive readings of the same variable to obtain a more precise measurement), then z is an unbiased estimator of the population mean μ_x and the mean square error of this estimator goes to zero with increasing n, being its variance $\sigma_z^2 = \sigma_x^2/n$. Thus, the variance of z is known, and so the error of the *sample mean estimator*, once the variance of the population σ_x^2 is known.

When inspecting sampled data organized in a histogram, an approximate bell shape is immediately associated to a Gaussian distribution. However, real data distributed symmetrically around the center of the distribution may fit better other distributions than the Normal one; a distinction is made between short-tail, moderate-tail and long-tail distributions, indicating with the term "tail" the extremes of the distribution at the left and the right of the central body that go to zero. When the tails go to zero very fast, like if the distribution is truncated, we may guess that the distribution is approximately uniform, or maybe trapezoidal, or that it is a heavily truncated version of other symmetric distributions, such as the Gaussian itself. A moderate-tail distribution is the Gaussian itself, when the small probability associated to the farthest points in the tails is neglected. Examples of long-tail distributions, where the probability of the tails cannot be disregarded, are the Cauchy $p(x) = \frac{1}{\pi(1+x^2)}$ and the Laplace $p(x) = \frac{1}{2b} \exp\left(-\frac{|x-\mu|}{b}\right)$ distribution.

4.1.3.3 Rayleigh distribution

Given a two-dimensional Gaussian distribution with two independent variables x and y of zero mean and same standard deviation σ, the random variable that represents the modulus or length of this two-dimensional vector, $z = \sqrt{x^2+y^2}$, has a Rayleigh distribution (see sec. 4.3.2.3 for a practical example). The distribution is

$$p(z) = \frac{z}{\sigma^2}\, e^{-z^2/2\sigma^2} \tag{4.1.5}$$

The mean value is $\sigma\sqrt{\frac{\pi}{2}} = 1.25\sigma$ and the standard deviation is $\sigma\sqrt{2-\frac{\pi}{2}} = 0.655\sigma$, where σ is the standard deviation of the two underlying Gaussian distributions.

This distribution is the first one of a series of unsymmetrical distributions, that are generally used to describe the behavior of combined Gaussian distributions and the effects on the overall intensity related to attenuation and propagation.

When the summed squared Gaussian components are three, the distribution is called Maxwell. Both the Rayleigh and the Maxwell may be considered a variant of the Chi-square distribution, when the considered variable is the square root of the sum of the squares, rather than the sum of the squares, of the individual normally-distributed input variables.

4.1.3.4 Nakagami-Rice distribution

A vector formed by two Gaussian random variables with zero mean has a Rayleigh distribution; if such a vector is moved around a point with a fixed value a, then the resulting sum z has a Nakagami-Rice distribution

$$p(z) = \frac{z}{\sigma^2}\, e^{-(z^2+a^2)/2\sigma^2}\, I_0(az/\sigma^2) \tag{4.1.6}$$

where I_0 is a modified Bessel function of the first kind and of zero order.

Of course the distribution depends on two parameters, a and σ, or better $\sigma\sqrt{2}$, that is the root mean square amplitude of the Rayleigh distributed random vector; details for the application to radiofrequency propagation and scattering are given in [194].

4.1.3.5 Log-normal distribution

If we consider a random variable x whose logarithm has a Normal distribution, then the r.v. x has a log-normal distribution, given by

$$p(x) = \frac{1}{\sigma\sqrt{2\pi}}\frac{1}{x}\exp\left(-\frac{1}{2}\left(\frac{\ln x - \mu}{\sigma}\right)\right) \tag{4.1.7}$$

where μ and σ are the mean and standard deviation not of the r.v. x, but of its logarithm, so that the corresponding quantities for x (that we call m for the mean value and v for the standard deviation) are: mean value $m = \exp\left(\mu + \sigma^2/2\right)$ and standard deviation $v = \exp\left(\mu + \sigma^2/2\right)\sqrt{\exp(\sigma^2) - 1}$. Conversely, given m and v of the x r.v., the resulting mean value and standard deviation of the log-normal distribution are: $\mu = \ln\left(\frac{m}{\sqrt{1 + v/m^2}}\right)$ and $\sigma = \sqrt{\ln\left(1 + v/m^2\right)}$.

The reciprocal of a r.v. with log-normal distribution also has log-normal distribution.

To conclude, as highlighted in [194], it can be considered that a log-normal distribution of a variable means that the numerical values are the result of the action of numerous causes of slight individual importance which are multiplicative, as opposed to the Gaussian distribution that by the Central Limit Theorem is the result of the action by summation.

Considering a frequency domain equipment that manipulates and expresses measured quantities in dB (such as a Spectrum Analyzer), the error affecting the output results from the combination of the many error terms associated to the uncertainty of internal blocks (such as attenuator, mixer, IF processing). These blocks are each characterized by a gain or attenuation and the individual error terms are combined by multiplication; if their dB-expressed counterparts are considered, then the combination is additive. For the output error resulting from the former situation (thus considering the original quantities before any representation in dB) a log-normal distribution may be assumed. The usual approach however is to consider the dB-expressed quantities as the reference ones and apply to them any reasoning on uncertainty and associated PDF: the additive nature of the relationship leads to a final Gaussian PDF postulated for the output error, that is the one generally used to evaluate propagation of uncertainty.

4.1.3.6 Chi-square distribution

The Chi-square distribution refers to a variable indicated by χ_n^2 that is the sum of n squared Gaussian variables. The PDF takes the form

$$p(\chi_n^2) = \frac{1}{2^{n/2}\Gamma(n/2)}(\chi_n^2)^{(n/2-1)}\,e^{-\chi_n^2/2} \tag{4.1.8}$$

with Γ indicating the Gamma function. Its integral, the CDF, is the so-called “chi-square distribution with n degrees of freedom”. The mean value and the variance of χ_n^2 are equal to n and $2n$, respectively.

4.1.3.7 *F*-distribution

When considering the ratio of two r.v. y_1 and y_2 both with Chi-square distribution and each normalized by its number of degrees of freedom n_1 and n_2, the resulting r.x. $x = \dfrac{y_1/n_1}{y_2/n_2}$ has a F distribution, with a PDF as:

$$p(x) = C\, x^{(n_1/2-1)} \left(1 + \frac{n_1}{n_2} x\right)^{-\frac{n_1+n_2}{2}} \tag{4.1.9}$$

where C is a constant that depends on the beta function and on the degrees of freedom: $C = \left(\dfrac{n_1}{n_2}\right)^{n_1/2} \dfrac{1}{B\left(\dfrac{n_1}{2}, \dfrac{n_2}{2}\right)}$. It may be observed that in spite of n_1 and n_2 being usually positive integers (because they are degrees of freedom), this is not strictly required, being sufficient that they are positive real numbers.
The mean value of x is $\dfrac{n_2}{n_2 - 2}$ with two peculiar characteristics: it does not depend on n_1 and n_2 shall be greater than 2 (the opposite would make y_2 a very poor variable, by the way).
Similarly, for the variance to be defined it is required that $n_2 > 4$: the variance is then $\dfrac{2n_2^2(n_1+n_2-2)}{n_1(n_2-2)^2(n_2-4)}$.
The cumulative distribution function CDF requires the calculation of the so-called "incomplete Beta function". The F distribution owes its "F" to Fischer (Sir Ronald Fischer), but was first introduced by G. Snedecor.

4.1.3.8 Student's *t* distribution

Another distribution that is mostly useful with problems with a low number of degrees of freedom (such as when a limited number of samples are available to make an estimate) is the "t" distribution, called — as known — Student's distribution after the pseudonym used by W.S. Gossett, when he published his work while working at the Guinness brewery[1].

$$p(x) = \frac{1}{\sqrt{\nu\pi}} \frac{1}{\left(1 + \dfrac{x^2}{\nu}\right)^{\frac{\nu+1}{2}}} \frac{\Gamma\left(\dfrac{\nu+1}{2}\right)}{\Gamma\left(\dfrac{\nu}{2}\right)} \tag{4.1.10}$$

where ν is the number of the degrees of freedom and $\Gamma(\cdot)$ is the Gamma function.

---

[1] "Student" [William Sealy Gosset], "The probable error of a mean," *Biometrika*, Vol. 6 n. 1, pp. 1-25, March 1908.
Please have a look at the quite complete and insightful Wikipedia page at: `http://en.wikipedia.org/wiki/Student%27s_t-distribution`.

This family of curves depends on the ν parameter and for $\nu \to +\infty$ the distribution approaches the Normal distribution. The importance of this distribution is in describing the behavior of small data sets for phenomena with original Normal distribution: the t variable obtained by the following transformation

$$t = \frac{\bar{x} - \mu}{s/\sqrt{n}} \tag{4.1.11}$$

with $\bar{x}$ the sample mean of a random sample of size n (n elements) from a Normal distribution with population mean μ and s the sample standard deviation, has a Student's t distribution with $\nu = n-1$ degrees of freedom. It is evident that it is not possible to operate with less than two elements in the sample and the interest is however in small samples, because for tens of elements the two distributions (Student's t and Gaussian) are already not easily distinguishable.

So the t distribution is useful to describe the sample mean estimator when the population mean is known (or a very good estimate of it may be obtained) and the population standard deviation is unknown. When considering the sample mean estimator with an unknown variance of the population, the best replacement for it is the sample variance estimated from the available samples; when the number of samples is very low it may deviate largely from that of the population. The t distribution is thus used to express the confidence interval for low n values[2].

4.1.4 Statistical sampling and estimates

The estimation of moments based on a sample gives the so-called "sample mean" and "sample variance": the former, noted $\bar{x}$, is simply the arithmetic mean of all the N items of the sample and leads to a consistent estimate of the mean value μ of the entire population; the latter, noted s^2, is a biased estimate of the variance of the population, unless the normalization factor is brought to $N-1$ instead of N (the reason is that s^2, as a central moment, is evaluated with respect to the known base value, that is the sample mean $\bar{x}$, and not the mean of the population μ, thus losing one degree of freedom).

$$\bar{x} = \frac{1}{N}\sum_{i=1}^{N} x_i \qquad s^2 = \frac{1}{N-1}\sum_{i=1}^{N}(x_i - \bar{x})^2 \tag{4.1.12}$$

Sample averages may be obtained by means of ensemble averages (that is "expectation") as above, or by means of time averages, when the values come from a process that shall be stationary and ergodic (see sec. 4.2.2). Thus time averages over the available record may replace ensemble averages.

For clarity we recall briefly what "consistent" means and some other terms used for estimation of statistical variables.

---

[2] It was proposed in an early draft of the EN 50121-2 standard, still at the working document stage, to evaluate the compliance to limits of traction line emissions (caused mainly by pantograph arcing and thus intermittent) with a minimum number of recordings. It is still kept in the EN 55011 standard to evaluate compliance of production from testing on a sample set.

The *bias error* b is a constant difference between the estimate $\hat{\theta}$ and the — supposed known — true value θ^*; the word "error" may be dropped, using simply bias and leaving the term error for random errors. The *random error* is defined by the expected dispersion of the estimate around the true value.

$$b_\theta = E\left\{\hat{\theta}\right\} - \theta^* = \lim_{N\to\infty} \frac{1}{N} \sum_{i=1}^{N} \hat{\theta}_i - \theta^* \quad \sigma_\theta = \sqrt{\lim_{N\to\infty} \frac{1}{N} \sum_{i=1}^{N} (\hat{\theta}_i - E\{\theta\})^2} \qquad (4.1.13)$$

It is evident that the sample mean has zero bias and its random error is the standard deviation, or the sample dispersion as a replacing estimate.

In some cases it is more informative to use the fractional counterparts of the bias error and random error with respect to the true value, that is $\varepsilon_b = b_\theta/\theta^*$ and $\varepsilon_r = \sigma_\theta/\theta^*$.

An *unbiased* estimator features a zero bias error. An estimator with a mean square error lower than any other estimators is said *efficient*. A *consistent* estimator features a zero bias error and a random error going to zero with increasing probability, as the number N of observations increases.

Depending on the underlying distribution, the best estimator may vary, so that the sample mean might not be the best choice always. As put in the NIST Handbook for Statistical Methods [254], "the optimal (unbiased and most precise) estimator for location of the center of a distribution is heavily dependent on the tail length of the distribution. The common choice of taking N observations and using the calculated sample mean as the best estimate for the center of the distribution is a good choice for the normal distribution (moderate tailed), a poor choice for the uniform distribution (short tailed), and a horrible choice for the Cauchy distribution (long tailed)." (see sec. 4.1.3 above for an introduction to probability distributions). The best estimator for the uniform distribution is the simple mid-range value, obtained by averaging the two extreme values of the interval a and b ($(a+b)/2$): because of the many deviations from a clean vertical cut at the extremes of the distribution, this estimator is still subject to some inaccuracy and may be supported by some robust estimate of the two extreme values. For the Cauchy distribution, instead, the median is the best estimator of the center location. A quick check to identify the distribution is of course the inspection of a suitable plot of the data.

PDFs may be estimated from observations by the histogram method: let's consider again a set of N data values x_n ($n = 0, 1 \ldots N$) obtained from the recorded signal $x(t)$, stationary and with mean value μ. It is straightforward to estimate the PDF of x_n by choosing a set of intervals of width w (or w_i for generality, and thus different for each i), defined as $d_i = a + iw$, where a is the lower bound of the range of x values and i is an integer $i = 0, 1, \ldots K$, with K indicating the number of classes (or bins) chosen for the $[a, b]$ interval, so that $K = (b-a)/w$. If the number of data values falling within a single interval $[d_i, d_{i+1}]$ is indicated as N_i, such that $N = \sum N_i$, then a histogram can be built and the estimate of PDF p (and of its integral P, the Cumulative Distribution Function, CDF) may be computed as: $p_i = N_i/N$ and $P_i = \sum_{k=0}^{i} N_k/N$.

The common practice is to assign the obtained relative probability value to the center of the bin, located in X_i and embracing a range of values $[X_i - \Delta x/2,\ X_i + \Delta x/2]$. If the true underlying PDF has no change of slope within each bin (i.e. its first derivative is constant), then the average of all the ideal points falling inside the i-th bin gives effectively the true PDF value p_i^*; on the contrary, if there is a change of slope of the original PDF over a single bin, this means simply that the chosen width Δx is too large and that a smaller one would give better results. The resulting relative bias error of the PDF estimate depends on the higher order derivatives of the PDF and a first approximation is [31]

$$\varepsilon_b(x) = \frac{(\Delta x)^2}{24} \frac{p''(x)}{p(x)} \tag{4.1.14}$$

While the sample mean and the sample variance (or dispersion) are quite intuitive and "natural" ways to express the center of gravity and the dispersion of sampled data, depending on the nature of the sample data there are other estimators that reveal a better robustness in practice. Data may result from a set of raw measurements performed by the same instrument and same person, or by including variability due to environmental factors, drift and aging, or may include operator's influence, or being extended to completely different laboratories, thus undergoing a change of paradigm. This is quite common when making inter-laboratory comparisons, testing the applicability and the consistency of procedures and methods of calibration or measurements. It is quite common that even renowned laboratories adopting certified procedures and high-performance instruments may result in dispersed data and outliers: examples are Round Robin verifications of Open Area Test Site attenuation, Vector Network Analyzer comparisons (such as those made by ANAMET[3]), noise power measurement verifications, etc.. Other indicators of the center of gravity and dispersion turn out to be more robust and to deliver more consistent results, especially when the data sample size is not large: in particular the median is quite robust in the presence of outliers, and the median absolute deviation (MAD) is a good measure of dispersion, as well as the inter-quartile range. The median is the middle value of a list of ordered samples, arranged e.g. in ascending order; when the number of samples is even, there is only one point in the middle, so that the result of the median is taken as the midpoint of the two middle values (i.e. their mean).

This approach may be extended to complex values (e.g. those representing scattering parameters, considered in Chapter 10), by taking the absolute value, i.e. the distance between each sample and the sought median (or spatial median, as it is usually called, because we are considering a 2-D problem). The median is determined as that value minimizing the sum of all distances from the sample data (by the way the sample mean is nothing more than the value that minimizes the Euclidean distance of the data samples in the mean square sense).

[3] Automatic Network Analyzer METrology group, founded in 1993 by the UK's National Physical Laboratory (NPL); an interest group gathering people and organizations "involved in RF, microwave and millimeter-wave measurements, including network analysis" (`http://www.npl.co.uk/networks/electromagnetics-network/anamet-the-rf-and-microwave-metrology-club`).

4.1.5 Allan variance

The Allan variance may be falsely classified among these other statistical estimators (as it is done), but it is a variance (or dispersion) operator that is applied to time series, so that it can be considered also a purely statistical index if time average and ensemble average can be exchanged. In reality, it is widely used to characterize time series and fluctuation of frequency-related phenomena, as shown in sec. 3.5.2.

The first role of Allan variance is fairly conventional: to provide a measure of variation, performing it at a variable sampling period (that is time elapsed between two adjacent samples, or readouts). The Allan deviation over the sample values is a general measure of frequency stability at the reference sampling rate; extending beyond adjacent samples (overlapped Allan variance), we can get an Allan variance for (synthetic) slower sampling rates.

More interestingly the Allan variance assists in the analysis of residual noise and clock stability at short and long time intervals, enabling the identification of the underlying noise behavior[4]: it was introduced by Allan [11] in 1966 and then underwent an intense activity of analysis and reworking (see e.g. [27, 167, 266, 175]). In frequency measurements of noise-like processes (see sec. 3.5.1 and 3.5.1.1 on phase noise and its power law terms) five different types of noise are normally defined, based on the behavior of their spectra and Allan variance: white noise phase modulation, flicker noise phase modulation, white noise frequency modulation, flicker noise frequency modulation, and random walk frequency modulation.

The Allan variance has a strong, but not so obvious, relationship with the power spectral density, that is normally used to characterize the correlation properties and frequency distribution of such processes (see sec. 4.3). A log-log plot of Allan variance versus sample period produces approximate straight lines of different slopes in four of the five cases above. A different (more complex) form called "modified Allan deviation" [12] can distinguish between the remaining two cases[5].

The Allan variance is built on a running difference calculated on the acquired samples and may be defined as a "zero dead-time" (no delay) two-sample variance. From an operational point of view, the computation of the Allan variance is faster than classic variance: Allan variance is based on the value of the previous sample (note that an N-element array implies only $N-1$ difference values). Basically

$$\Sigma_y^2 = \frac{1}{2}\left\langle (\bar{y}_{k+1} - \bar{y}_k)^2 \right\rangle = \frac{1}{2(N-1)} \sum_{k=1}^{N-1} (\bar{y}_{k+1} - \bar{y}_k)^2 \tag{4.1.15}$$

where the infinite time average $\langle\cdot\rangle$ is to be replaced by a finite summation of N samples. The value "2" in the denominator is intended to produce the same result as the classical variance for white noise.

---

4 Spectrum based techniques are recognized as more efficient, but the Allan variance (and its various modified versions) are still considered a valid and useful tool; see sec. 3.5.4.

5 The modified Allan variance appeared in 1981 in the cited paper by Allan and Barnes and in a paper in the same session by J.J. Snyder, "An ultra-high resolution frequency meter," 35th Annual Symposium on Frequency Control, Fort Monmouth, NJ, 1981, pp. 464-469.

To keep the meaning and usefulness of the Allan variance[6], the quantity y (that in sec. 3.5 is denoted as "fractional frequency fluctuation", when speaking of frequency stability) is related to the original quantity x, that is the "fractional phase fluctuation" (but stands also for any direct time measurement of fluctuation of periodicity):

$$\bar{y}_k = \frac{x_{k+1} - x_k}{\tau} \tag{4.1.16}$$

where τ is the time difference between the original samples x and in a two-term difference establishes also the time location of quantities $\bar{y}_k$.

By combining the equations above, it is possible to obtain the also known expression with the triplet of x quantities:

$$\Sigma_y^2 = \frac{1}{2(N-1)\tau^2} \sum_{k=1}^{N-1} (x_{k+2} - 2x_{k+1} + x_k)^2 \tag{4.1.17}$$

To correctly calculate the Allan variance, samples shall be uniformly distributed with no dead time, that otherwise would in general bias the result. In [284], sec. 5.16, the problem of unevenly spaced data is briefly considered: Riley examines the problem of satellite clocks polled on Mondays, Wednesdays and Fridays, thus featuring 2.33 average sample time, while the sampling is operated on a 2-2-3 days basis; the obvious interpolation of data samples before processing gives wrong results, while Riley suggests to convert and process the data in the frequency domain, keeping their time tags (that shall be always stored next to the sample values). For further reading he reports two quite complete references: P. Tavella and M. Leonardi, "Noise Characterization of Irregularly Spaced Data," 12th European Frequency and Time Forum, pp. 209-214, March 1998; C. Hackman and T.E. Parker, "Noise Analysis of Unevenly Spaced Time Series Data," *Metrologia*, Vol. 33, pp. 457-466, 1996.

If the difference between samples is extended in time, to far away samples with respect to the k instant, while keeping always the triplet form, we obtain the overlapped estimate of the Allan variance. The reason of the term "overlapped" is that now the leftmost x term refers to a time instant that goes behind the center term, referring to earlier time instants. Deciding a time interval $\tau' = m\tau$, the so-obtained Allan variance expression is calculated on slightly less samples, that is $N - 2m$ from the original N:

$$\Sigma_y^2 = \frac{1}{2(N-2m)\tau^2} \sum_{k=1}^{N-2m} (x_{k+2m} - 2x_{k+m} + x_k)^2 \tag{4.1.18}$$

There exists also a modified version, called *modified Allan variance*, that collects m differences of y taken at τ' time distance (thus in the overlapped Allan variance sense) and then takes the average; so, $2m$ samples at τ fundamental sampling time are needed to calculate the difference terms as in the overlapped Allan variance, but additional m samples are needed to shift m times by τ and to calculate the average over m

[6] The Allan variance was introduced to measure clock stability extending the analysis to long time intervals; for this reason the original quantities that often, non to say always, referred to in the literature are the fractional frequency and phase fluctuations.

terms: from this the fact that the modified Allan variance needs $3m$ samples windows and thus reduces slightly the usability of the available data samples.

As clearly reported in [284], the modified Allan variance may be written as

$$\Sigma_y^2 = \frac{1}{2m^4(N-3m+2)} \sum_{j=1}^{N-3m+2} \left\{ \sum_{i=j}^{j+m-1} \left[\sum_{k=i}^{i+m-1} (y_{k+m} - y_k)^2 \right] \right\} \tag{4.1.19}$$

$$\Sigma_y^2 = \frac{1}{2m^2(N-3m+1)\tau^2} \sum_{j=1}^{N-3m+1} \left\{ \sum_{i=j}^{j+m-1} (x_{i+2m} - 2x_{i+m} + x_i)^2 \right\} \tag{4.1.20}$$

Many other variances and techniques have stemmed from the original Allan variance and are briefly reported in sec. 3.5.

The problem of establishing the confidence intervals, and thus the uncertainty, of the so obtained variance estimators is quite complex: the NIST handbook [284] and the IEEE Std. 1139 [175], Annex E, consider both systematic instabilities and common pitfalls and external factors, as well as the uncertainty for the remaining random error. For the non-overlapped Allan deviation the confidence interval, assuming a Gaussian distribution that is valid when M is large ($M \geq 10$), is defined as

$$U_\alpha = \Sigma_y(\tau)\, \kappa_\alpha \sqrt{M} \tag{4.1.21}$$

where κ_α is a constant that changes with the type of noise falling in the estimated Allan dispersion (for 1σ confidence intervals $\kappa_2 = 0.99$, $\kappa_1 = 0.99$, $\kappa_0 = 0.87$, $\kappa_{-1} = 0.77$, $\kappa_{-2} = 0.75$).

More generally, abandoning the simplifying assumption of a Gaussian distribution, the Chi-square distribution gives a more reliable picture of the asymmetric distribution: the degrees of freedom are shown in the standard, taken from D.A. Howe, D.W. Allan and J.A. Barnes, "Properties of Signal Sources and Measurement Methods," 35th Annual Symposium on Frequency Control, pp. 669-717, May 1981. Further contributions have been given by Lesage et al. [218], who have analyzed other variance terms: they show that the uncertainty of the modified Allan variance is of the same order of the original Allan variance, even if slightly larger (their calculations were developed only for white phase noise ($\alpha = 2$) and white frequency noise ($\alpha = 0$)).

4.2 Random processes: definitions and stationarity

A random process (or stochastic process) is a collection of random values (random variables) that represent how a system evolves through time, giving such values as output. What we have when the sequence of values of some output is recorded as the time passes is a *time series*. Such a system is the perfectly specular image of a perfectly deterministic system with its own output variables to which the indeterminacy of randomness has been added: this does not mean that the new output will be highly unpredictable, or will be subject to a kind of weird behavior, but simply that its values will be characterized by a probability distribution, that may also have time dependency, that is that may vary as long as the process develops through time.

The latter is an extreme case that may be encountered often in the real world, such as for example the noise of a device that is warming up. In general, we will focus on processes with well behaved statistics, that represent already a useful tool for modeling and representing the behavior of many real systems.

Additionally, from a practical point of view the mentioned "time" in many cases might be discrete rather than continuous, such as when the observed system is e.g. a stock market index, a performance index of some quality measure (number of defects, time between two failures, etc.), a quantity that is measured, acquired or stored at given instants of time only. In many situations the underlying process is still continuous in time, but we do not have access to it or it is better to reduce the degrees of freedom and the amount of stored and analyzed information to some discrete time instants. When considering electrical and electronic systems and signals, the knowledge of the underlying physical principles tells us that the phenomena are continuous time (down to a sufficiently small scale, of course). From this, we may say that with a little flexibility continuous and discrete time are not mutually exclusive and that they may be interchanged as appropriate.

4.2.1 Basic definitions

For any experiment in which the outcomes develop through time and cannot be predicted within a reasonable error due to unavoidable experimental errors (e.g. noise and measurement uncertainty) we talk of a random process. A time history of data points recorded at some time during an experiment is only one of the possible realizations; the collection of all such realizations defines completely the random process. Being the number of collectibles and the time allowed for the experiment finite, we are bounded to work with a limited number of realizations to characterize the process or to characterize it with *a priori* knowledge, based on what we know about its statistical properties and inner mechanisms. Sometimes it may be convenient to fit a general known statistical model to the available realizations.

Examples of random processes may be a sinusoidal signal generated by an oscillator whose fundamental frequency is bound to vary because of random internal noise and to drift because of temperature variations and aging; the mains voltage spectrum whose change in time (e.g. the amplitude of fundamental and harmonics) is due to external causes, such as the varying operating conditions for the loads connected to the same supply network that are unpredictable and not under experimenter's control; additive noise alone, in principle, transforms all signals in random processes for which there are no two really identical recordings.

A random process is thus identified by a set of time recordings (time history) in the form of signals, whose time duration, amplitude resolution and frequency resolution are established by the way they were measured and recorded. Several characteristics of course may be due to the influence of test setup and the adopted measurement method: we will see that besides the well-known parameters of time-domain recording (sampling time, observation window, etc.), when recording with a more or less direct measurement in frequency domain, the dynamic nature of several phenomena shall be considered. For example, using frequency sweeping and envelope detection, several parameters and spectrum analyzer settings may bias irreversibly the results (see Chapter 9). Internal noise and distortion are another source of artifacts.

4.2.2 Stationarity and ergodicity

A random process is said stationary if its ensemble statistics do not depend on time; this gives us the certainty that statistics calculated on a data sample do not depend on time. Of course depending on which statistics may or do not vary with time, a random process may be more or less strictly stationary, with a range of degrees of stationarity. Regarding stationarity, a process is classified as [32]:

- first-order stationary, if its first-order probability density function remains equal regardless of any shift in time, so that the average value is constant;
- second-order stationary, if its second-order probability density function does not vary over any time shift applied to both values;
- strict sense stationary (SSS), if all its statistical properties do not vary over any applied time shift;
- wide-sense stationary (WSS) to relax the requirements of SSS, since it becomes evident that the strict requirements of a SSS process are more than that is often necessary in order to adequately approximate calculations on random processes; for a WSS process the average value is constant and the correlation function is defined only by the time-shift.

The fact that some statistics do not depend on time does not imply that ensemble and time statistics coincide. If they do, we say that the process is *ergodic*. Given a stationary process, we are going to define weakly ergodic and strongly ergodic processes. A process is said *weakly ergodic* if time average for mean and covariance computation give the same result as ensemble averages; if all other statistical properties of the process are calculable through time averages, then the process is *strongly ergodic*.

To interpret the two definitions with a closer look to measurements and metrology jargon, we may say that stationarity indicates that the process statistics are independent on the initial time instant and that repeated measurements at different instants of time are consistent (repeatability). Ergodicity states that ensemble and time-domain statistics may be interchanged and that the results obtained with one of the two methods may be reproduced also with the other one.

Besides the direct knowledge of the underlying physical mechanisms, from which stationarity may be postulated, suitable tests exist to evaluate the stationarity of sampled random data [31, 32]. It may be assumed for simplicity that any non-stationarity relevant for our problems is revealed by a time trend of the mean square value of data; by the way, it is possible to find a non-stationary random process with constant mean square value, even if such a process is unusual in practice and of no use for our problem. By extension, testing that some other statistical indexes of the time series (such as the mean or higher order moments) is constant over time gives an indication of its stationarity. Similarly, in frequency domain the spectral properties of the time series may be tested for similarity when calculated on two (or more) different segments of data. The analysis of the behavior of Fourier spectrum of the series (or, better, of its log representation[7]) may similarly shed some light on the time behavior of spectral properties: taken some specific time instants and frequency values that are relevant to or representative of our problem, the assessment of stationarity of the original time series is achieved by means of a two-factor analysis of the variance calculated for the time instants, for the frequency values and for the interaction, taking them altogether[8].

Functions implementing tests for stationarity are available in some mathematical programming environments and packages, mostly applied to economy and econometrics, so that spectrum based methods are often not included.

4.2.3 Other characteristics

A few other properties that may be useful when manipulating frequency and time domain expressions are collected altogether and discussed in the following.

4.2.3.1 Parseval's theorem

Parseval's theorem is often used to pass from time to frequency domains and viceversa; the involved quantity is the energy of the process, that shall feature finite energy, or in other words, shall be square-integrable. For periodic signals energy is replaced by power over a given time interval T.

The continuous-time version that expresses the energy E using Fourier transform with a bilateral notation extending symmetrically to negative time or negative frequency is

$$E = \int_{-\infty}^{+\infty} |x(t)|^2 \, \mathrm{d}t = \int_{-\infty}^{+\infty} |X(\omega)|^2 \, \mathrm{d}\omega \tag{4.2.1}$$

This expression is also called Rayleigh's theorem, identifying with "Parseval's theorem" the version for periodic signals that involves power, rather than energy. From a continuous time perspective, the Parseval's theorem for periodic signals makes use of a time integral over the period T and an infinite summation of the Fourier series terms that define the line spectrum of the periodic signal:

$$P = \frac{1}{T} \int_{T} |x(t)|^2 \, \mathrm{d}t = \sum_{k=-\infty}^{+\infty} |c_k|^2 \tag{4.2.2}$$

---

7 G.M. Jenkins and M.B. Priestley, "The spectral analysis of time-series," *Journal of the Royal Statistical Society*, Series B, Vol. 19, 1957, pp. 1-12.

U. Grenander and M. Rosenblatt, *Statistical Analysis of Stationary Time Series*, John Wiley & Sons, New York, 1957.

8 M.B. Priestley and T. Subba Rao, "A Test for Non-Stationarity of Time-Series," *Journal of the Royal Statistical Society*, Series B, Vol. 31, 1969, pp. 140-149.

where the time integral is over a period T no matter the exact limits of integration and the c_k indicate the complex coefficients of the Fourier series.

If considering discrete-time signals and conversely discrete-frequency transform, for a full digital implementation, Parseval's theorem is still valid and takes a form that is more similar to the continuous time version:

$$P = \sum_{n=0}^{N-1} |x[n]|^2 = \sum_{k=0}^{N-1} |X[k]|^2 \tag{4.2.3}$$

where $x[n]$ and $X[k]$ are the two discrete sequences for the signal and its DFT.

4.2.3.2 Convolution and multiplication

Convolution and multiplication form a pair of dual operators, in that they may be exchanged when moving between time and frequency domains.

Convolution between two signals $u(t)$ and $v(t)$ is indicated as $u*v$ and corresponds to the following operation

$$z = u*v = \int_{-\infty}^{+\infty} u(\lambda)v(t-\lambda)d\lambda \tag{4.2.4}$$

and analogously for discrete sequences $u[n]$ and $v[n]$ of length N we have

$$z = u*v = \sum_{-(N-1)}^{N-1} u[p]v[n-p] \tag{4.2.5}$$

Going to the transform domain the convolution between signals corresponds to the product of transform, that is

$$Z(\omega) = U(\omega)V(\omega) \tag{4.2.6}$$

and analogously

$$Z[k] = U[k]V[k] \tag{4.2.7}$$

with obvious meaning of capital letters (Fourier transforms) of signals above.

Convolution is primarily involved in the calculation of the time response of a linear system, for which the output is given by the convolution of the input signal with the system impulse response. Conversely, in the frequency domain convolution occurs when, for example, applying a smoothing window to a signal.

The duality convolution-multiplication is extremely useful in some demonstrations and theorems, such as when determining the correlation function of the output of a linear time-invariant system (see sec. 4.3.1 below).

By the way convolution is not limited only to the relationship between time-domain signals and their frequency-domain transforms. Convolution is also used in probability, for example to express the resulting PDF of a random variable that is the sum of two other random variables; the role of the transforms to multiply in the other domain is assumed by the characteristic function, that however will not be used and is not further considered.

As a last comment, please note that, despite the general similarity, the convolution integral and the cross-correlation function are different (have a look simply at the sign of the dummy variable).

4.2.3.3 Additive White Gaussian Noise

White Gaussian-distributed noise is often used as the reference random process to test and verify many relationships. The choice is reasonable when considering noise processes in real situations, because they originate from large-number statistics, ensured quite often by the so-called "change of scale": e.g. noise in semiconductors due to electrons and holes collisions at the nanoscale brings to a Gaussian distributed process, when measured at a larger scale, by invoking the Central Limit Theorem. It may happen however that particular forms of noise have power-law model (see e.g. sec. 3.5.1.1), but they can be converted to and from a white noise process after multiplication by a suitable power of frequency f, while keeping the original statistical distribution. The effect of parametric disturbance, such as temperature and other environmental factors, may not fall in any of these categories; in this case, special care shall be taken, but often the process may be considered weakly non-stationary, leaning on the different time-scales of the phenomena (temperature is slowly varying, our electrical signals much faster[9]).

Derived processes after mathematical manipulations (such as the intensity of a baseband noise, or the intensity of a band-pass noise) may feature different statistical distributions, e.g. Rayleigh or Chi-square when working with the squares of Gaussian-distributed components, but may remain white, that is uncorrelated.

4.3 Auto-correlation and Power Spectral Density

Definitions and considerations open this section, establishing the basic properties for random processes evaluation.

Given the process $x(t)$, the auto-correlation function is defined in a general way as

$$R_{xx}(t_1,t_2) = E\left\{x(t_1)\,x(t_2)\right\} = \iint\limits_{-\infty}^{+\infty} x(t_1)\,x(t_2)\,f(x_1,x_2)\,\mathrm{d}x_1\,\mathrm{d}x_2 \tag{4.3.1}$$

[9] An exception is maybe a long sweep made with the spectrum analyzer, while the internal noise changes during unavoidable temperature fluctuations (see sec. 9.2.10.2).

The average power of $x(t)$ is the value $E\left\{x^2(t)\right\} = R_{xx}(t,t)$; the auto-covariance is given by the auto-correlation function with the average value removed:
$C(t_1,t_2) = C(t_1,t_2) - \mu(t_1)\mu(t_2)$.
The cross-correlation of two random processes $x(t)$ and $y(t)$ is $R_{xy}(t_1,t_2)$:

$$\begin{aligned} R_{xy}(t_1,t_2) &= E\left\{x(t_1)\,y^*(t_2)\right\} = E\left\{x^*(t_1)\,y(t_2)\right\} = \\ &= \iint_{-\infty}^{+\infty} x(t_1)\,y^*(t_2)\,f(x_1,y_2)\,\mathrm{d}x_1\,\mathrm{d}y_2 \end{aligned} \tag{4.3.2}$$

Here are some definitions:

- two processes $x(t)$ and $y(t)$ are mutually orthogonal if $R_{xy}(t_1,t_2) = 0$ for every t_1 and t_2; they are uncorrelated if $C_{xy}(t_1,t_2) = 0$ for every t_1 and t_2;
- a process $x(t)$ is a white noise if values $x(t_1)$ and $x(t_2)$ are uncorrelated; if these two variables are also independent, then the process is strictly white noise.

The Fourier transform of the auto-correlation function of a random process $x(t)$ corresponds to the Autospectral Density Function, ADF, also called Power Spectral Density, PSD, $S_{xx}(f)$.

$$S_{xx}(f) = \int_{-\infty}^{+\infty} R_{xx}(\tau)\,e^{-j2\pi f\tau}\mathrm{d}\tau \tag{4.3.3}$$

Analogously, the Fourier transform of the cross-correlation function between two random processes $x(t)$ and $y(t)$ is called Cross-Spectral Density Function, CDF[10], or simply Cross-Power Spectral Density, $S_{xy}(f)$.

$$S_{xy}(f) = \int_{-\infty}^{+\infty} R_{xy}(\tau)\,e^{-j2\pi f\tau}\mathrm{d}\tau \tag{4.3.4}$$

The ADF functions are real-valued even non-negative functions, while the CDF is a complex-valued function. All these functions are defined for a frequency interval that extends from $-\infty$ to $+\infty$: remembering that they are even, it is always possible to define a so-called one-sided PSD, defined only for non-negative frequency values, so closer to reality, that is usually indicated by the letter "G":

$$G_{xx}(f) = 2S_{xx}(f) \qquad \text{for } f \geq 0 \tag{4.3.5}$$

A sufficient condition for their Fourier transformation is that the integrals of their absolute values are finite, and this holds for quite a lot of correlation functions, except some pathological cases, such as the correlation of an infinite sinusoidal signal, whose integral of the absolute value does not converge (but is anyway bounded).

[10] Rarely used and misleading, because it is confused with the Cumulative Distribution Function.

These relationships are often called Wiener-Khintchin relations[11].

4.3.1 Random signals and Linear Time Invariant systems

Let's consider the input process $x(t)$ to a Linear Time Invariant (LTI) system (such as a filter or an amplifier) and the output process $y(t)$, and for simplicity let's focus on stationary processes, so to write the auto-correlation functions as $R_{xx}(\tau)$, $R_{yy}(\tau)$ and $R_{xy}(\tau)$, with $\tau = t_1 - t_2$, and correspondingly the PSDs, $S_{xx}(f)$, $S_{yy}(f)$ and $S_{xy}(f)$. Under the assumption of stationarity and ergodicity[12], the latter are related to each other through the LTI frequency response $H(f)$ as

$$S_{xy}(f) = H(f)\, S_{xx}(f) \qquad S_{yy}(f) = |H(f)|^2\, S_{xx}(f) \tag{4.3.6}$$

so that it is easy to calculate the output mean square value

$$\psi_y^2 = \overline{y^2} = \int_{-\infty}^{+\infty} S_{yy}(f)\, \mathrm{d}f = \int_{-\infty}^{+\infty} |H(f)|^2\, S_{xx}(f)\, \mathrm{d}f \tag{4.3.7}$$

or for the one-sided PSD functions $G_{xx}(f)$ and $G_{yy}(f)$

$$\overline{y^2} = \int_{0}^{+\infty} G_{yy}(f)\, \mathrm{d}f = \int_{0}^{+\infty} |H(f)|^2\, G_{xx}(f)\, \mathrm{d}f \tag{4.3.8}$$

As a reminder, in the following notation $S_{xx}(f)$ indicates the two-sided PSD that extends from $-\infty$ to $+\infty$ on the frequency axis, while $G_{xx}(f)$ indicates the one-sided PSD, defined only for the positive frequency axis. $G_{xx}(f)$ is thus in direct relationship with frequency domain measurements. Last, it is useful to recall that $\overline{y^2} = R_{yy}(0)$.

The LTI action (e.g. for a filter) is to shape the PSD of the input random process. Assuming an uncorrelated white noise with a flat PSD $S_{xx}(f)$ as input, the memory that is in the LTI time constants reflects into the spectrum of the output noise $S_{yy}(f)$. As it is shown in the basic example of sec. 4.3.1.4, the former white uncorrelated input noise becomes "colored" and has an auto-correlation that is no longer a Dirac delta, but shows the same time constant of the LTI system. We will shortly see, when talking about the Equivalent Noise Bandwidth in sec. 4.3.1.5, that also the total rms output noise is related to the extension of the LTI frequency response.

Immediate applications of these equations are, first, the estimation of the equivalent input noise spectrum of an amplifier by dividing the measured output spectrum by its

[11] The relationship between the autospectrum and the autocorrelation functions was demonstrated independently by Norbert Wiener and Alexander Khintchin in the '30s and is thus usually called Wiener-Khintchin relation. Please have a look at the basic information for first reference reported in Wikipedia: `http://en.wikipedia.org/wiki/Wiener%E2%80%93Khinchin_theorem`

[12] Ergodicity is necessary in the proof when exchanging the integrals (time domain averages) with the expectation operator.

gain (or frequency response) and, second, the identification of a frequency response by manipulating the autospectrum and the cross-spectrum of input and output.

4.3.1.1 Thermal noise

This phenomenon for metallic resistors was studied by Johnson and by Nyquist in the late '20s, who set down the basic theory still in use today. A resistor is a source of pure thermal noise, called also "Johnson noise", that depends on its value R and the temperature T (measured in Kelvin degrees). The random motion of electrons in the metal is the source of the noise and the open-circuit noise voltage has a Gaussian distribution (compatible with the Central Limit Theorem applied to the many collision events) with zero mean and a variance of $\frac{2(\pi kT)^2 R}{3h}$, where k is the Boltzmann constant ($1.37\,10^{-23}$J/deg) and h is the Planck constant ($6.62\,10^{-34}$Js).

The complete power spectral density expression was shown to be

$$S_v(f) = \frac{2Rh\,|f|}{e^{h|f|/kT} - 1} \tag{4.3.9}$$

but a handy simplification that holds for the lower portion of the frequency interval leads to

$$S_v(f) \simeq 2RkT\left[1 - \frac{h\,|f|}{2kT}\right] \simeq 2RkT \tag{4.3.10}$$

The fraction inside square brackets is much smaller than unity and may be ruled out for the commonly used frequency values: it is equal to 1 at a temperature of 300 °K when the frequency is 12400 GHz!

So, thermal noise has both a Gaussian distribution and a constant PSD, from which the usual definition "white Gaussian noise", abbreviated in AWGN, where the leading "A" stands for "additive", i.e. non-parametric (see sec. 4.2.3.3). Rigorously speaking thermal noise is not the only noise manifestation that fits the definition of AWGN; shot noise (see below) fits it as well, except for the impulse in the spectrum origin at dc.

If the noise is expressed in terms of total rms value, then also the bandwidth B adopted for the quantification shall be accounted for.

$$\eta_{th.n.} = 4kTR \qquad \psi_{th.n.} = \sqrt{4kTRB} \tag{4.3.11}$$

This definition of η is compatible with a one-sided approach that uses only positive frequencies, as in real systems. For calculations involving the Fourier transform that require functions with even symmetry in the frequency domain, thus using positive and negative frequency values, the value in eq. (4.3.11) shall be divided by 2.

4.3.1.2 Shot noise

Shot noise is generated by the random emission of electrons or by the random passage of electrons and holes across a potential barrier. Among the few demonstrations available in the literature, the one presented in [239] is rigorous enough, yet keeping clear and intuitive; it considers the auto-correlation function rather than simply the root mean square value, in order to demonstrate that also this noise is flat (i.e. white) except for the non-null average value in the origin. Let's consider a current $i(t)$ of impulsive nature made of n charge carriers passing a given section in the unity of time, so that we can always take a short time interval δt, where only one electron of charge q (for holes the reasoning is perfectly identical) is or is not present, and the current there is exactly $q/\delta t$ or 0. The average current value is $\bar{I} = nq$. The auto-correlation function (given by the multiplicative superposition of a $i(t)$ time piece and its time lagged copy $i(t+\tau)$) may be computed for:

1) time lag $\tau = 0$, readily obtaining the root mean square

$$R(0) = \lim_{T\to\infty} \frac{1}{T} \int_{-T/2}^{+T/2} i^2(t)\,\mathrm{d}t = \lim_{T\to\infty} \frac{1}{T} \left[N \left(\frac{q}{\Delta t} \right)^2 \Delta t \right] = n \frac{q^2}{\Delta t} = q\bar{I} \tag{4.3.12}$$

where $N = nT$ is the total number of charges passing in the time window T, each giving a $q/\delta t$ current term; for $\delta t \to 0$ the term corresponds to an impulsive value in the origin of the auto-correlation function, that gives correspondingly a flat spectral density;

2) time lag $\tau \neq 0$, where the probability of two overlapping pulses (one in $i(t)$ and the other in $i(t+\tau)$) may be computed: the probability of a pulse in $i(t)$ over a time interval δt is $n\,\delta t$; if $i(t)$ and $i(t+\tau)$ are independent, then the probability of an overlapping pulse is $(n\,\delta t)^2$, and this is to be considered for the total number of time intervals in T, that is $T/\delta t$; as observed before each pulse carries $q/\delta t$ current, so that

$$R(\tau) = \lim_{T\to\infty} \frac{1}{T} \left[(n\Delta t)^2 \frac{T}{\Delta t} \left(\frac{q}{\Delta t} \right)^2 \Delta t \right] = (nq)^2 = \bar{I}^2 \tag{4.3.13}$$

This term is responsible for the impulse at the origin of the spectrum.

The overall auto-correlation function is thus given by the sum of the two terms, that is the impulse in the origin with area $q\bar{I}$ and the constant term $\bar{I}^2$. The two-sided spectral density of the current $i(t)$ comes directly from the Fourier transform of the auto-correlation function, resulting in a constant for the delta in the time origin at $\tau = 0$ and a delta in the frequency origin (dc, or $f = 0$) with area equal to the constant term $\bar{I}^2$.

To conclude, the mean-square shot noise current in the frequency band B that does not contain dc is thus given by

$$\overline{i_{sh}^2} = 2q\bar{I}\,B \tag{4.3.14}$$

where the "2" is added for having considered a two-side spectral density for the calculation with the positive-frequency band B. For clarity it is recalled that $q = 1.622\,10^{-19}$ C is the electron charge and $\bar{I}$ is the dc bias current.

Shot noise is a white noise and its PDF is Poisson distributed, rather than Gaussian.

4.3.1.3 Flicker noise

Flicker noise is a noise that has a spectral density proportional to $1/f^{\beta}$, with β approximately 1. Several explanations of its origin have been proposed, but it is still an ill-understood phenomenon [245]. In resistive materials, its origin seems to be caused by a fluctuation of the mobility of the free charge carriers. For semiconductors several mechanisms may take place depending on the physical properties of the device, construction and technology.

For bipolar devices it has been associated with the surface properties of a semiconductor device and has been tentatively separated into surface noise and leakage noise [127]. The former is attributed to the fluctuating occupancy of the surface states: the general energy-band description used in teaching semiconductor properties is developed, by assuming a symmetrical and periodic electron potential function; at the surface the usual distribution of energy states is altered and this happens across an infinitesimal layer, shorter than the electron mean path. The surface states are similar to bulk traps: they capture holes and electrons and keep them in the trap, until they are released or they recombine. Several models have been proposed[13] that differ mainly in the assumed number of levels. The considerable concentration of holes and electrons near the surface implies that sometimes the surface conductivity is larger than the bulk conductivity (despite the somewhat reduced mobility of carriers through the region, whose thickness however is comparable to or smaller than their free mean path) and an appreciable amount of current can flow through the surface space-charge layer. The random change of carriers concentration in these surface states produces a fluctuation of surface conductance, and hence a leakage current fluctuation; surface conductance is usually voltage dependent and is proportional to the applied voltage amplitude. The $1/f$ spectral density assumption for flicker noise was shown to hold down to extremely low frequencies, for some devices and systems as low as one cycle per month, where the noise merges with the natural drift. It is underlined that assuming $\beta = 1$, i.e. assuming the true "$1/f$" curve slope, is only an approximation and that devices of common use exist with significant deviations from it, such as the OPA27 and OPA37 operational amplifiers, whose input noise voltage characterization is shown in Figure 4.3.1. The PSD of the equivalent input noise shown in Figure 4.3.1(b) might seem different from the original one of the OPA37: the amplifier has a corner frequency that is about an order of magnitude lower, but several flat thermal noise terms that increase noise floor exert their influence (resistors in the biasing, gain and protection network) and shift the crossing point between the original $1/f$ curve and the horizontal axis of

---

[13] L. Tamm, "Uber eine mogliche Art der Elektronenbindung an Kristalloberflächen," *Physik. Z. Sowjetunion*, vol. 1, pp. 733-746, June, 1932.
J. Bardeen, "Surface States and Rectification at a Metal Semiconductor Contact," *Physical Review*, vol. 71, pp. 717-727, May 15, 1947.
W. H. Brattain and J. Bardeen, "Surface Properties of Germanium," *Bell System Technical Journal*, vol. 32, pp. 1-41, January, 1953.
J. Bardeen and S. R. Morrison, "Surface Barriers and Surface Conductance," *Physica*, Vol. 20, pp. 873-884, Nov. 1954.
J. Bardeen, R. E. Coovert, S. R. Morrison, J.R. Schrieffer and R. Sun, "Surface Conductance and the Field Effect on Germanium," *Physical Review*, vol. 104, pp. 47-51, Oct. 1, 1956.
F. Seitz, *The Modern Theory of Solids*, pp. 395-406, McGraw-Hill, New York, 1940.
R.H. Kingston, *Semiconductor Surface Physics*, University of Pennsylvania Press, Philadelphia, 1957.

the high frequency noise floor to a lower frequency; if the noise floor of approximately $5-6\,\mathrm{nV}/\sqrt{\mathrm{Hz}}$ is transposed into the original OPA37 noise spectrum (Figure 4.3.1(a)) for comparison, the new crossing point is right at 1 Hz.

Considering briefly also MOS devices, flicker noise is mainly generated by tunneling effects in the surface oxide layer of the material. For many different devices it may be also generated by the imperfect contact between two conducting materials, but, in this case, it is called “contact noise”.

Flicker noise is modeled by a noise current source

$$\overline{i^2_{n,flk.}} = \frac{K_{flk} I^m}{f^n} B \tag{4.3.15}$$

where K_{flk} is the flicker noise coefficient, and m and n indicate the dependency on current and frequency.

Flicker is also relevant when describing frequency stability (see sec. 3.5), especially for very stable frequency sources, so to exclude the relevance of other noise mechanisms[14].

From a very general point of view, curiously enough, 1/*f* noise is present in nature in unexpected places, e.g. speed of ocean currents, the flow on Japanese expressways, the yearly flow of Nile as measured over the last 2000 years, the loudness vs. time of a piece of classical music, as noticed in [165].

Flicker noise is also named “pink noise”, especially in the audio field, identifying it with the simple property that its PSD decreases 3 dB per octave with increasing frequency.

4.3.1.4 Thermal noise through a low-pass filter

Now let’s consider two basic elements we have already seen: a thermal noise source with a constant flat PSD and a low-pass filter built with a RC circuit; moreover, the noise source happens to be the same resistor R of the low-pass filter. The Thevenin equivalent of the resistor gives a noise voltage source v_n in series with a noiseless resistor R; the voltage source has a positive frequency spectrum $V_n = 4kTR$ [W/Hz].

The RC filter is a first-order low-pass filter with the frequency response

$$|H(f)|^2 = \frac{1}{1+(f/B)^2} \tag{4.3.16}$$

where $B = 1/2\pi RC$ is the -3 dB bandwidth that corresponds to the filter cut-off frequency. The output bilateral PSD function is

$$S_y(f) = \frac{2kTR}{1+(f/B)^2} \tag{4.3.17}$$

and its inverse transform, the auto-correlation function, is

[14] E. Rubiola and V. Giordano, “On the 1/f Frequency Noise in Ultra-Stable Quartz Oscillators,” *IEEE Transactions on Ultrasonics, Ferroelectrics, and Frequency Control*, Vol. 54, No. 1, Jan. 2007, pp. 15-22. doi: 10.1109/TUFFC.2007.207.

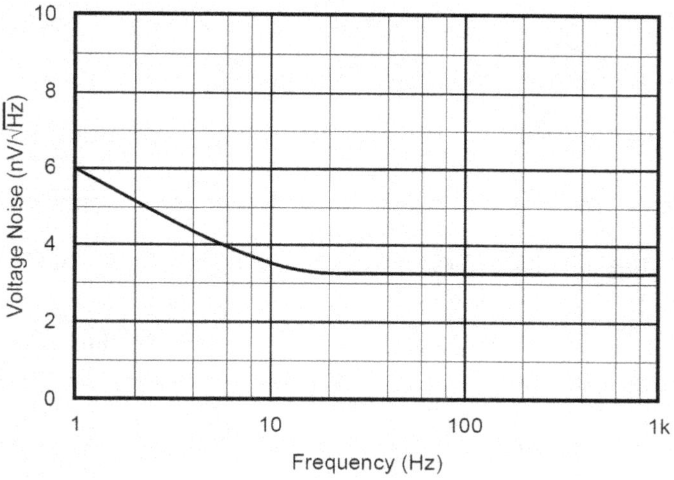

(a)

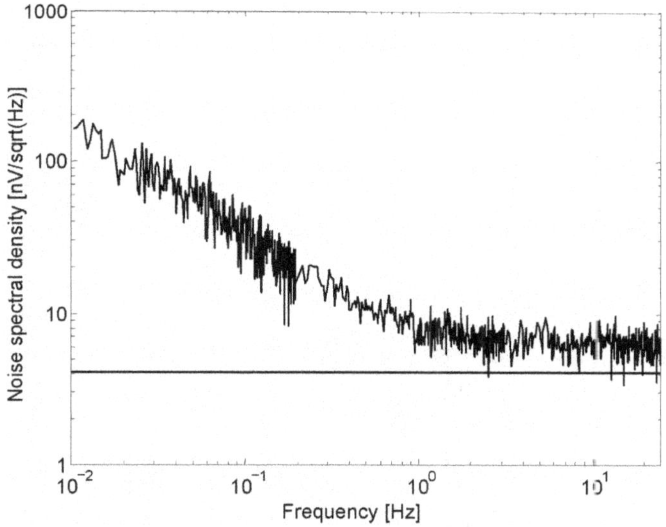

(b)

Figure 4.3.1 – Flicker noise slope lower than 1 for the input noise voltage of the OPA37 Operational Amplifier (Texas Instruments, OPA37 datasheet, Aug. 2005), confirmed by direct measurement of the total equivalent input noise of a magnetic field probe designed around this OA [233]. The slope resulting from the original input noise data in (a) is about $\beta = 0.3$ despite the limitation of the frequency range to only 1 Hz; the measurement of a differential amplifier counting two devices and a few resistors accounts for $\beta = 0.71$. (Courtesy Texas Instruments)

$$R_{yy}(\tau) = 2kTR\,\pi B\, e^{-2\pi B|\tau|} = \frac{kT}{C}\, e^{-|\tau|/RC} \tag{4.3.18}$$

that shows that there is appreciable correlation over a time interval that is proportional to the filter time constant, but with a mean square value of the output that doesn't depend on R, that is the source of the thermal noise!

$$\psi_y^2 = R_{yy}(0) = \frac{kT}{C} \tag{4.3.19}$$

The explanation resides in the particular problem we have selected, where the resistor value influences both the amount of thermal noise and the cut-off frequency of the filter (using B hides the fact that there is a dependency on R): the larger the resistor value the higher the noise, but the stronger the cut-off with a lower cut-off frequency.

It suffices to calculate the Equivalent Noise Bandwidth of the filter (see sec. 4.3.1.5),

$$\text{ENBW} = \int_0^{+\infty} \frac{1}{1+(f/B)^2}\,\mathrm{d}f = \frac{\pi}{2}B = \frac{1}{4RC} \tag{4.3.20}$$

and observe that by definition of ENBW the mean square value at the output is given by the noise power density multiplied by ENBW:

$$\psi_y^2 = \eta_{th.n.}\text{ENBW} = 4kTR\frac{1}{4RC} = \frac{kT}{C} \tag{4.3.21}$$

This demonstration has used heavily the definitions that appear in sec. 4.3.1.5 below regarding the Equivalent Noise Bandwidth.

4.3.1.5 Noise bandwidth of a filter

The Equivalent Noise Bandwidth B_n (also indicated as ENBW) of a real system with transfer function $H(f)$ (see above at the beginning of sec. 4.3.1) and with a gain given by $|H(f_0)|$, is the bandwidth of an ideal filter with box-like shape and with the same gain, which yields the same level of output power as the real system [166, 239]. The noise bandwidth of a filter can be defined for low-pass and band-pass filters. If white noise voltage $v(t)$ with constant flat PSD $S_{vv}(f) = \eta$ is applied to the input of the filter, the mean-square output voltage is given by

$$\overline{v_{o1}^2} = \int_0^{+\infty} S_{vv}(f)\,|H(f)|^2\,\mathrm{d}f = K_v \int_0^{+\infty} |H(f)|^2\,\mathrm{d}f \tag{4.3.22}$$

Now let the white noise be applied to an ideal low-pass filter having a gain of H_0 for $0 \le f \le B_n$ and a gain of 0 elsewhere, where B_n is the ENBW we are looking for. The mean-square output voltage is given by

Order	Real poles		Butterworth
1	$1.571\,f_c$	$1.571\,B_{-3\,\mathrm{dB}}$	$1.571\,B_{-3\,\mathrm{dB}}$
2	$0.785\,f_c$	$1.220\,B_{-3\,\mathrm{dB}}$	$1.111\,B_{-3\,\mathrm{dB}}$
3	$0.589\,f_c$	$1.155\,B_{-3\,\mathrm{dB}}$	$1.042\,B_{-3\,\mathrm{dB}}$
4	$0.491\,f_c$	$1.129\,B_{-3\,\mathrm{dB}}$	$1.026\,B_{-3\,\mathrm{dB}}$
5	$0.420\,f_c$	$1.114\,B_{-3\,\mathrm{dB}}$	$1.017\,B_{-3\,\mathrm{dB}}$

Table 4.3.1 – Noise bandwidth of low-pass filters: for real-pole filters the ENBW is given in terms of the corner frequency f_c or the $-3\,\mathrm{dB}$ bandwidth $B_{-3\,\mathrm{dB}}$; for Butterworth the ENBW is given in terms of the $-3\,\mathrm{dB}$ bandwidth $B_{-3\,\mathrm{dB}}$ only [239].

$$\overline{v_{o2}^2} = \int_0^{B_n} S_{vv}(f) H_0^2 \,\mathrm{d}f = K_v H_0^2 B_n \tag{4.3.23}$$

Now B_n is to be determined to make the two mean square values equal:

$$B_n = \frac{1}{H_0^2} \int_0^{+\infty} |H(f)|^2 \,\mathrm{d}f \tag{4.3.24}$$

and B_n may be then defined the equivalent noise bandwidth of the filter. The noise bandwidth of a band-pass filter is defined in a similar way simply changing the limits of integration.

Two classes of filters are considered: one with N real poles all with the same cut-off frequency and the other is a Butterworth of order N. The equivalent noise bandwidth is reported in Table 4.3.1 for N ranging from 1 to 5; f_0 indicates the pole frequency and B_{-3dB} is the frequency at which the gain is $-3\,\mathrm{dB}$ the static gain H_0 (or center band gain for bandpass filters).

4.3.1.6 Equivalent input noise and noise through a LTI system

The use of the model above allows to reason on the concept of Equivalent Input Noise (EIN), a concept largely applied in radiofrequency and electronics, e.g. for instruments and active devices. The application of the EIN concept is straightforward: measuring (or quantifying in other ways) noise at the output of an electronic system, identified as the random process $y(t)$, $n(t)$ is the EIN that going through the LTI causes exactly this same output noise $y(t)$. In reality the definition is largely applied to noise power (identified by mean square value ψ_y^2 or noise PSD $G_{yy}(f)$), because talking of the equality of single realizations of a random process is a nonsense. Noise power is divided by the LTI gain H_0 (the static gain $H(0)$ for low-pass systems or center-band gain $H(f_0)$ for band-pass systems); the reason behind the normalization for the LTI gain is that the interest in noise is for the operating frequency range of the electronic system, normally characterized by an almost constant frequency response $H(f)$.

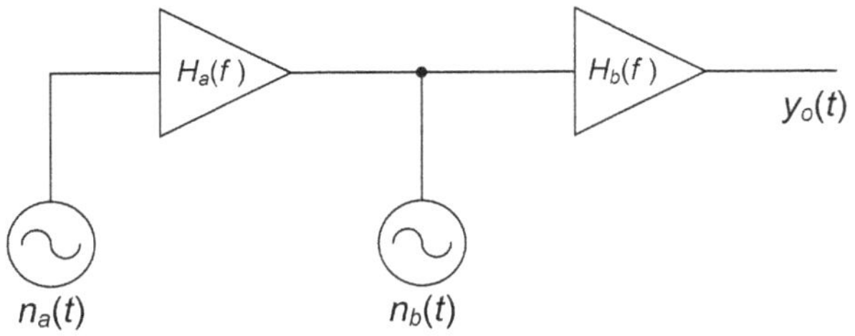

Figure 4.3.2 – Two-stage electronic system featuring two blocks with frequency response $H_a(f)$ and $H_b(f)$ and noise sources identified with their time-domain signals.

$$\psi_n^2 = \psi_y^2 / H_0^2 \tag{4.3.25}$$

Noise sources are in reality many inside a complex electronic equipment; let's try a simplification in order to better visualize the concept. An amplifier is made of two stages as in Figure 4.3.2, one input stage a and one output stage b, each with its own internal noise sources (e.g. resistors, semiconductors, supply noise, etc.). For each block an equivalent noise source is put at its input side and summed with the signal getting out from previous block and fed into it. This choice has two underlying assumptions: a linear summing model is assumed for the propagation of noise and we are again applying the concept of equivalent input noise for each building block. The former is normally accepted in many cases, especially for thermal noise sources inside electronic systems and justified by the fact that the noise amplitude is so small not to change appreciably the operating point of electronic devices; parametric noise is also possible and can be modeled with more sophisticated approaches [32], but it is beyond our scope. The latter is simply justified saying that, first, we will show that the technique is sound and easy to implement at high level (for an entire electronic system); going down into the low-level details, the proliferation of smaller blocks and equivalent noise sources shall stop and it's more sensible to take a direct measurement of a portion of the system identified as black box.

Given the measured output noise $n_o(t)$, assumed it is zero mean, with an estimated mean square value $\psi_{n,o}^2$ over the operating bandwidth B and a constant power spectral density $\eta_o = \psi_{n,o}^2 / B$, the EIN $n_b(t)$ before block b is described by $\eta_b = \eta_o / |H_{b,0}|^2$.

Assuming AWGN Gaussian white noise for all blocks , the input block a transfers its EIN $n_a(t)$ to its output with a constant PSD $\eta_{a,o} = \eta_a |H_{a,0}|^2$ over its operating bandwidth B_a, then falling off with the same roll-off of the $H_a(f)$ frequency response. The output noise of block a sums to the EIN of block b, $n_b(t)$, resulting in an equivalent PSD given by $\eta_{b,i} = \eta_{a,o} + \eta_b = \eta_a |H_{a,0}|^2 + \eta_b$. At the output of block b, that is the output of our system, the output noise has a constant PSD of

$$\eta_o = \eta_{b,o} = \eta_{b,i} |H_{b,0}|^2 = \eta_a |H_{a,0}|^2 |H_{b,0}|^2 + \eta_b |H_{b,0}|^2 \tag{4.3.26}$$

From the point of view of the voltage spectral density values (often used to characterize operational amplifiers and other electronic devices) eq. (4.3.26) becomes a root-mean-square sum of the terms.

It derives from eq. (4.3.26) that if the output block b has a unity gain and all the gain required for the application is concentrated in the input block, then its internal noise is the most relevant and influences the output noise; of course, under the simplifying assumption of equal or similar EIN values for each block a and b. In general, in the attempt to optimize the apportionment of the gain between the blocks minimizing the output noise, under the constraint that the product of the two gains $H_{a,0}$ and $H_{b,0}$ shall be constant and equal to the total gain H_0, gains are chosen nearly equal[15].

Combination of voltage and current noise sources When it comes to consider both voltage and current sources, then the equivalent resistance and impedance connected to the noise sources shall be accounted for. The use of equivalent voltage and current noise sources is beyond the scope of this book and is taken as a matter of fact, owed to the different noise mechanisms in semiconductor devices that we have briefly reviewed a few pages ago. Noise current sources are transformed into equivalent voltage sources by multiplying for the equivalent loading resistance, that is the total resistance "seen" from the noise current source terminals: it is a common practice in Operational Amplifiers to multiply for the parallel of the resistors insisting on either of the OA input terminals, namely feedback resistor, input source resistance, other gain resistor, and offset balancing resistor. Some OAs are preferable to connect to low resistance sources (e.g. bipolar OAs with very low voltage noise spectra and input noise spectra in the nA range), while FET OA shall be selected with high impedance sources.

4.3.1.7 Signal-to-noise ratio

We can attack the problem of propagation of noise through LTI systems from a different point of view, that of the signal-to-noise ratio (SNR) and how the system response affects it. The signal-to-noise ratio is defined as the ratio of power of signal $s(t)$ and noise $n(t)$ over bandwidth B, or even the entire frequency range from 0 to infinite. At the input the SNR is:

$$\mathrm{SNR}_i = \frac{\psi_s^2}{\psi_n^2} = \frac{P_s}{P_n} \tag{4.3.27}$$

The origin of the noise may be outside or inside the system, and for the latter we have adopted the abstraction of bringing it back to the input in terms of equivalent input noise (see sec. 4.3.1.6 above). Thus, $s(t)$ and $n(t)$ are summed together and combined into $x(t)$ before entering the system; we assume that $s(t)$ and $n(t)$ are incoherent and that there is no net contribution to the cross-product that has a zero average ($E\left[s(t)\,n(t)\right] = 0$) and does not appear in the system output $y(t)$ when calculating expectation or average.

The power of the two processes going to the output is affected by the LTI frequency response; assuming that the LTI filter is conceived to have a flat gain g over the signal bandwidth and that its Equivalent Noise Bandwidth (ENBW) B_n is known, then

---

[15] Of course this is over-simplifying and many design trade-offs occur, regarding the choice of devices for bandwidth, EIN, input current, open loop gain, etc.. But this sneaks in the kingdom of electronic circuit design, that may be found cleverly treated among others by Horowitz [165].

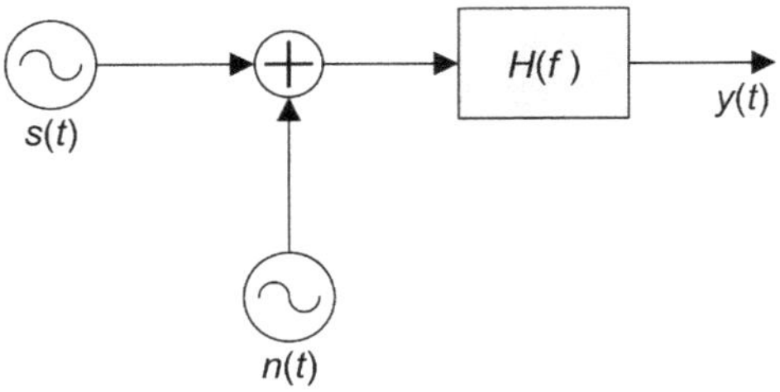

Figure 4.3.3 – Linear Time Invariant system with signal and noise source.

$$\text{SNR}_o = \frac{g\,P_s}{\eta B_n} = \frac{P_s^o}{P_n^o} \tag{4.3.28}$$

As exemplified in Table 4.3.1, the ENBW B_n is always larger than the $-3\,\text{dB}$ bandwidth, because the out-of-band filter response (made of the roll-off portion and the stop-band) is to be included in the computation; so the wisest behavior to maximize SNR_o is to, first, choose the smallest bandwidth B compatible with the signal bandwidth and, second, to select the highest filter roll-off (i.e. number of poles and filter type), as technically attainable, or reasonable.

Since signal and noise cannot be really separated, but are accessible only at the LTI output, the successive measurement of the mean square value of the output, with and without the signal $s(t)$ applied, gives

$$\begin{aligned} P'_y &= P_s^o + P_n^o \qquad \text{with} \\ P''_y &= P_n^o \qquad\qquad \text{without} \end{aligned} \tag{4.3.29}$$

and the SNR_o may be calculated readily as $P'_y/P''_y - 1$. The same is applicable when performing the rms measurement of a steady signal and aiming at correcting for the internal noise contribution: the rms of the tone is readily obtained by quadrature-subtracting the estimate noise rms (corresponding to P''_y) from the total rms (corresponding to P'_y) (see sec. 9.2.12).

4.3.1.8 Equivalent Noise Resistance, Equivalent Noise Temperature and Noise Figure

Some definitions are reviewed that are mostly useful for rapid calculations of equivalence between different systems and devices and to interpret datasheets. The section ends with comments on the behavior of cascaded blocks and passive two-port devices. Quite a thorough treatment may be found in [87], Chap. 8, by Adamson.

Equivalent Noise Resistance The EIN of a device or system, assumed with a flat PSD, may be expressed in terms of Equivalent Noise Resistance (ENR) determining the resistance value R_{eq} that has the same PSD; eq. (4.3.11) is used to this aim. As reference at room temperature (300 °K) two $50\,\Omega$ and $1\,\text{k}\Omega$ resistors have $0.8284\,\mu\text{pW/Hz}$ and $16.57\,\mu\text{pW/Hz}$, respectively, or as voltages $0.91\,\text{nV}/\sqrt{\text{Hz}}$ and $4.07\,\text{nV}/\sqrt{\text{Hz}}$.

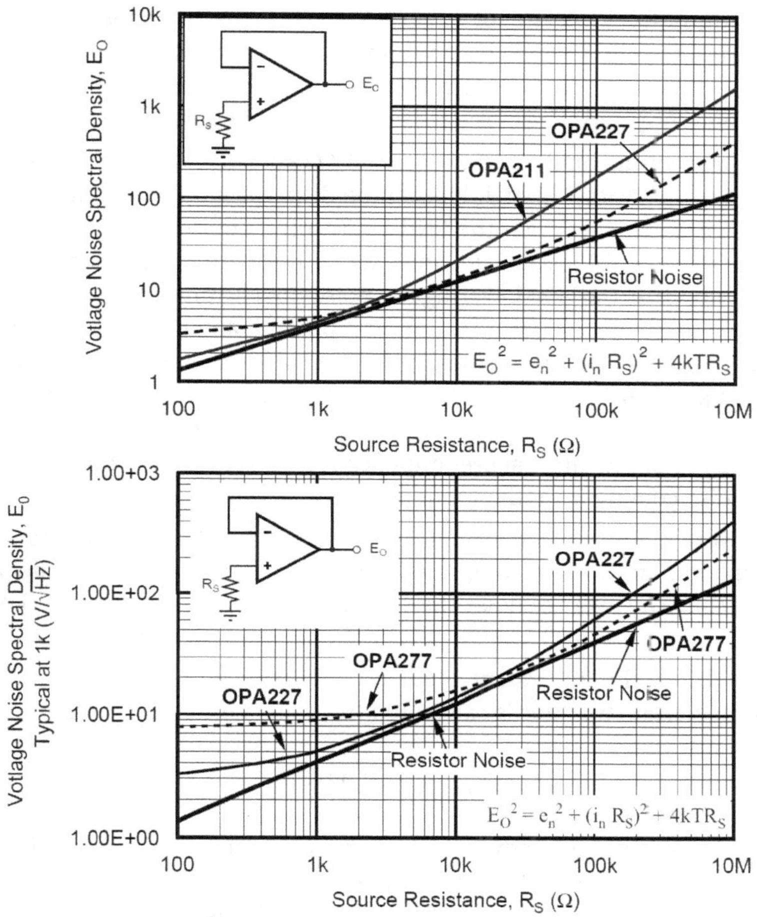

Figure 4.3.4 – Examples of Equivalent Input Noise graphs compared to the source resistance noise: (above) comparison of OPA211 and OPA227, (below) comparison of OPA277 and OPA227. (Courtesy Texas Instruments)

As shown in sec. 4.3.1.6, it is customary for Operational Amplifiers to express the total EIN by comparing it to the noise brought in by the internal resistance of the signal source, that sets a lower limit to OpAmp performance measured at its output.

In Figure 4.3.4 there are reported the EIN curves normally shown in OpAmp datasheets for comparison between different devices and with reference resistors[16]. The OPA211 with a very low voltage EIN is quite effective when the source resistance (and the values of the other resistors of the circuit) is low and doesn't enhance the current EIN term; the OPA227 on the contrary is more effective when the source resistance is large and the OPA277 even more. The tangent points between each OpAmp curve and the R_S noise line move from about $1\,\mathrm{k}\Omega$ for the OPA211 to $9\,\mathrm{k}\Omega$ for the OPA227 and up to about $16\,\mathrm{k}\Omega$ for the OPA277.

---

[16] References for operational amplifiers of this example: Texas Instruments, OPA211 datasheet, Sept. 2007; Texas Instruments, OPA227 datasheet, Jan. 2005; Texas Instruments, OPA277 datasheet, Apr. 2005.

The expression reported in the lower right part of the two graphs is the synthesis of the three terms contributing to the noise measured at the OpAmp output, that is the voltage EIN e_n, the current EIN i_n (multiplied by the resistance seen at the terminals, in this case only R_S) and the source thermal noise (due to resistance R_S).

Considering a spectrum analyzer and $R_{\text{ref}} = 50\,\Omega$ resistance value taken as reference, it is possible to quantify its EIN highlighting the disproportion with respect to the reference resistance value: the larger the ENR the worse its noise performance[17]. In Figure 4.3.5 the performance of a spectrum analyzer is considered, including or excluding its internal pre-amplifier (largely lowering the noise floor). It is now evident that the concept of the EIN is an abstraction that does not point at any noise source located at the input, that the noise sources are many, distributed along the SA (in particular in the mixer) and that the use of a pre-amplifier before some of the noisiest internal blocks effectively reduces the EIN. If the noise would have been generated at the very input of the SA, the use of the internal pre-amplifier would have been ineffective, amplifying signal and noise coming from the input in the same way.

Three tests were made, starting from a 10 dB setting for the internal attenuator without pre-amplifier, and then turning on the pre-amplifier and finally reducing the attenuator to 0 dB. The average noise floor in the three cases, as shown in the rightmost part of the figure, is $-116.03\,\text{dBm}$, $-125.64\,\text{dBm}$ and $-135.91\,\text{dBm}$, respectively.

From eq. (4.3.11), the equivalent input resistance in terms of noise power density ENR of the spectrum analyzer is calculated from the measured power

$$P_{n.} = 4kTB\,\text{ENR} \tag{4.3.30}$$

where B is the resolution bandwidth adopted for the measurement (1 kHz) and the temperature T is the internal one, that the spectrum analyzer diagnostic system says to be $42\,°\text{C}$. In the foregoing analysis the uncertainty is taken into account of assuming that the temperature of all components relevant to the measured noise is $42\,°\text{C}$; a lower bound of $30\,°\text{C}$ is also considered and the corresponding ENR is calculated.

The resulting ENR is obtained by transforming the dB quantities into linear quantities and then squaring and dividing for $4kTB$:

- Att=10 dB, Pre-Amp. no: the measured average power of the noise floor is $0.0025\,\text{pW}$ and $\text{ENR} = 144.6\,\Omega$; at $30\,°\text{C}$ the resulting ENR is $150.4\,\Omega$;
- Att=0 dB, Pre-Amp. no: the measured average power of the noise floor is $0.00027\,\text{pW}$ and $\text{ENR} = 15.8\,\Omega$; at $30\,°\text{C}$ the resulting ENR is $16.4\,\Omega$;
- Att=0 dB, Pre-Amp. yes: the measured average power of the noise floor is $0.000026\,\text{pW}$ and $\text{ENR} = 1.49\,\Omega$; at $30\,°\text{C}$ the resulting ENR is $1.54\,\Omega$.

[17] Alternatively, the noise disproportion is quantified in power dB of the ENR ratio $10\log_{10}((\text{ENR} - R_{\text{ref}})/R_{\text{ref}})$ and is called Excess Noise Ratio, whose abbreviation is unfortunately ENR. It may be defined similarly for the Equivalent Noise Temperature considered in the next paragraph.

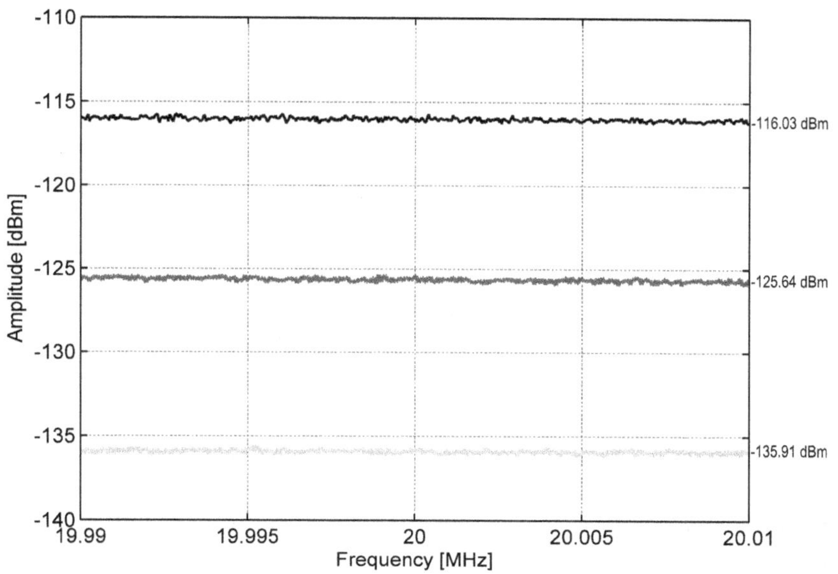

Figure 4.3.5 – Noise floor of a Spectrum Analyzer with and without the internal pre-amplifier: 10 dB internal attenuation, no pre-amplifier (black curve), 10 dB internal attenuation with pre-amplifier (dark gray curve), 0 dB internal attenuation with pre-amplifier (light gray curve).

It is evident that reducing the internal attenuation (in this case from 10 to 0 dB) or switching on the pre-amplifier largely lowers the internal noise floor and the equivalent noise resistance; the two options are nearly equivalent, each giving an order of magnitude reduction.

Last, a comment that is valid for any definition of noise in "equivalent" terms: when measuring noise power in mismatched conditions (or not perfectly matched conditions) the reflection coefficient that occurs at the port shall be considered, so that, in terms of power, a factor $(1-|\Gamma|^2)$ shall be accounted for.

Equivalent (or effective) Noise Temperature Another parameter that is subject to change in eq. (4.3.11) and is suitable to establish an "equivalence" is temperature: given a reference resistance value R e.g. of $50\,\Omega$, a device or a system has an Equivalent Noise Temperature (ENT) T_{eq} that has the same PSD or, if measured at the output, gives the same mean square value.

The ENT concept may be applied also to non-thermal noise, as in the case of another typical noise source of semiconductors and tubes, the shot noise. Shot noise is a white incoherent noise that has no temperature dependency, but rather depends on the average current I biasing the device (see sec. 4.3.1.2). The mean square shot noise current is given by

$$\overline{i^2_{n,sh.}} = 2qIB \tag{4.3.31}$$

so that the equivalent shot noise voltage across the reference resistor R is $2qI\,BR^2$ and comparison with thermal noise for the same resistor gives an equivalent temperature

$$T_{eq,sh} = \frac{qIR}{2kT} \tag{4.3.32}$$

where T is the real ambient temperature the resistor "sees". It comes that very noisy devices, such as shot noise source, may have a very large ENT that however does not correspond to the physical temperature.

If the noise process is not incoherent and the PSD is not flat, it might be misleading to give a unique ENT value without any other information: it is always possible to associate an ENT value, provided that the frequency interval is narrow enough that the PSD is nearly constant, or it is clearly stated that the original PSD is not constant and the process is not white.

Noise Figure Finally, a parameter that is used mostly for low-noise and RF semiconductor devices and circuits is the Noise Factor (NF) and the Noise Figure, as its expression in dB, defined in several ways. The preferred definition is based on the signal-to-noise ratio SNR, so that the noise factor is the ratio of the SNR at the input over the SNR at the output. If a system adds no noise to the input signal, then the SNR at the input and the output have the same value because both the signal and the noise coming from the preceding block or device has undergone the same amplification; the resulting noise factor is unity and the noise figure is 0 dB.

Assuming white noise behavior for the input source (with power P_n) and for internal sources, it is not difficult to transform the definition of NF based on the ratio of signal-to-noise ratios, into one based on equivalent temperature or resistance. The internal noise power P_{int} is transformed into an equivalent input noise power $P_{eq,int}$ that undergoes the same amplification of the input signal and input noise.

$$\mathrm{NF} = \frac{\mathrm{SNR}_{in}}{\mathrm{SNR}_{out}} = \frac{\dfrac{P_s}{P_n}}{\dfrac{G\,P_s}{G\,P_n + P_{int}}} = \frac{\dfrac{P_s}{P_n}}{\dfrac{G\,P_s}{G\,(P_n + P_{eq,int})}} = 1 + \frac{P_{eq,int}}{P_n} \tag{4.3.33}$$

Now the ratio between the input-referred internally generated noise power and the externally fed input power may be transformed into equivalent temperature or equivalent resistance, assuming thermal noise behavior:

$$\mathrm{NF} = 1 + \frac{\tilde{T}_{int}}{\tilde{T}_S} = 1 + \frac{\tilde{R}_{int}}{\tilde{R}_S} \tag{4.3.34}$$

$\tilde{R}_S$ is not the resistance of the signal source, but its equivalent resistance in terms of the generated noise ENR (so, including practically also the physical source resistance), and similarly the temperature term is ENT, not indicating the physical temperature of the equipment or the environment, as explained previously in this section.

Example 4.2: Determination of noise floor and sensitivity

Let's try to calculate the sensitivity of a RF receiver using the information of its noise figure. The source of signal is characterized by the signal power P_S and noise power due to the internal resistance P_N, both expressed per unit bandwidth. By definition of noise figure

$$\mathrm{NF} = \frac{\mathrm{SNR}_{in}}{\mathrm{SNR}_{out}} = \frac{P_s/P_n}{\mathrm{SNR}_{out}} \tag{4.3.35}$$

so that, in turn,

$$P_s = P_n \times \mathrm{NF} \times \mathrm{SNR}_{out} \tag{4.3.36}$$

and over the bandwidth B the total signal power is

$$P_s = P_n \times \mathrm{NF} \times \mathrm{SNR}_{out} \times B \tag{4.3.37}$$

Expressing all quantities in dB of power and making some simplifications

$$P_s[\mathrm{dBm}] = P_n[\mathrm{dBm/Hz}] \times B[\mathrm{Hz}] + \mathrm{NF}[\mathrm{dB}] + \mathrm{SNR}_{out}[\mathrm{dB}] \tag{4.3.38}$$

Assigning SNR_{out} the value of the minimum required SNR_{min}, it is possible to determine the minimum input power that is thus detectable.

Before proceeding, the noise power is estimated, by applying the assumption of perfect matching, that consists of equal resistance of the signal source and receiver input:

$$P_n = \frac{4kT\,R_S}{4R_{in}} = kT = -174\,\mathrm{dBm/Hz} \tag{4.3.39}$$

at room temperature. Thus

$$P_s[\mathrm{dBm}] = P_n[\mathrm{dBm/Hz}] \times \mathrm{B}[\mathrm{Hz}] + \mathrm{NF}[\mathrm{dB}] + \mathrm{SNR}_{out}[\mathrm{dB}] \tag{4.3.40}$$

$$P_s[\mathrm{dBm}] = -174\,[\mathrm{dBm/Hz}] + 10\log \mathrm{B}[\mathrm{Hz}] + NF[\mathrm{dB}] + \mathrm{SNR}_{min}\,[\mathrm{dB}] \tag{4.3.41}$$

4.3.1.9 Cascaded blocks and noise performance

What happens when cascading several blocks, either active or passive from the noise viewpoint? Let's assume that each block has a noise factor NF_i, or equivalently a noise temperature T_i, and a gain (or attenuation) G_i (all expressed in linear quantities): the overall, or total, noise factor (and as a consequence the noise figure), as seen at the output of the last block, is given by

$$\mathrm{NF}_{tot} = \mathrm{NF}_1 + \frac{\mathrm{NF}_2 - 1}{G_1} + \frac{\mathrm{NF}_3 - 1}{G_1 G_2} + \cdots + \frac{\mathrm{NF}_n - 1}{G_1 G_2 \ldots G_{n-1}} \tag{4.3.42}$$

or, equivalently, the overall noise temperature is

$$T_{tot} = T_1 + \frac{T_2}{G_1} + \frac{T_3}{G_1 G_2} + \cdots + \frac{T_n}{G_1 G_2 \ldots G_{n-1}} \tag{4.3.43}$$

where n is the total number of cascaded blocks and $i = 1,\ 2, \ldots n$.

It is evident that if G_1 is large, the noise from the first block will dominate and the other contributions will be small or even negligible. To this aim, a factor of merit is defined and sometimes used, called "noise measure", or "equivalent noise measure":

$$\mathrm{NM} = \frac{\mathrm{NF}-1}{1-\dfrac{1}{G}} \tag{4.3.44}$$

A low value of NM indicates a good noise performance, when weighted against the device (or block) gain and its effect on the overall system of cascaded blocks.

When the number of cascaded blocks is reduced as in normal situations and the more practical case of a noise source connected to the measurement instrument (e.g. power meter or spectrum analyzer) is considered, two characteristics may then be further considered taking into account the behavior at high frequency, i.e. reflection and transmission properties. We distinguish thus the effects of mismatches at input and output of the connecting block assumed as a generic two-port network (so the return loss at the input and output port, S_{11} and S_{22}) and the effects of its attenuation, seen as the insertion loss S_{21}.

Regarding the effect of mismatch it is obvious to consider the noise signal as any other signal that undergoes reflection, so that only a fraction of it is measurable. However, the input impedance of the attached two-port network influences the behavior of the noise source in the measuring instrument itself, where correlation of reflected noise terms in either direction occurs: to this aim the use of isolating devices has become common to normalize the behavior (once the isolating devices are characterized).

Straightforwardly the effect of the attenuation of the two-port passive network insertion loss parameter is simply that of reducing the noise power term reaching the measuring instrument and shall be compensated for.

Of course, any passive network, is a source of noise by its own temperature and the noise is mostly of the thermal (or Johnson) type (see sec. 4.3.1.1); taking into account the effective transmission (or "gain") of the network the noise temperature will be multiplied by $(1-S_{21})$.

4.3.2 Correlation Function and Power Spectral Density estimate

In this section the various methods for the estimation of the Power Spectral Density (distinguishing between Auto-Spectrum and Cross-Spectrum density functions) are considered. The subject is quite articulated and several background topics are simply recalled and not dealt with in detail to keep the section compact. For some of them reference is made directly to the previous sections 4.3 with the basic relationships and 3.1.1 on the Fourier transform, including problems of frequency leakage and windowing. Additional references reported in the Bibliography are [31, 32, 48, 229, 256, 275]. Further references may be found e.g. in [32], chap. 3:

- G.C. Carter, C.H. Knapp and A.H. Nuttall, "Estimation of the Magnitude-Squared Coherence via Overlapped Fast Fourier Transform Processing," *IEEE Transactions on Audio and Electroacoustics*, Vol. 21, No. 4, p. 337, Aug. 1973; doi: 10.1109/TAU.1973.1162496;

- L.D. Enochson, "Digital Techniques in Data Analysis," *Noise Control Engineering*, Vol. 9, No. 2, p. 138, Nov./Dec. 1977;
- R.K. Otnes and L.D. Enochson, *Applied Time Series Analysis*, Wiley, New York, 1978.

4.3.2.1 Correlation function

The usual definition of sample correlation function (calculated from a single data record $x(t)$ for auto-correlation $\hat{R}_{xx}(\tau)$ and similarly from $x(t)$ and $y(t)$ for cross-correlation $\hat{R}_{xy}(\tau)$) is used to evaluate expected value and variance of the estimate:

- the expected value $\hat{R}_{xx}(\tau)$ is effectively the cross-correlation function $R_{xx}(\tau)$, so the estimate is unbiased;
- the variance the expression calculated in [31], sec. 8.4, assuming jointly Gaussian processes $x(t)$ and $y(t)$, is

$$\sigma^2\left[\hat{R}_{xx}(0)\right]=\frac{2}{T}\int_{-\infty}^{+\infty}R_{xx}(\xi)\,\mathrm{d}\xi \tag{4.3.45}$$

$$\sigma^2\left[\hat{R}_{xx}(\tau\gg 0)\right]=\frac{1}{T}\int_{-\infty}^{+\infty}R_{xx}(\xi)\,\mathrm{d}\xi \tag{4.3.46}$$

Assuming a bandwidth-limited Gaussian white noise over a bandwidth B for sampled data records of length T, the mean value and variance of the estimated auto-correlation function are as follows:

$$\varepsilon\left[\hat{R}_{xx}(\tau)\right]=\varepsilon\left[\psi_x^2\right]=\frac{1}{\sqrt{2BT}}\sqrt{1+\frac{R_{xx}^2(0)}{R_{xx}^2(\tau)}} \tag{4.3.47}$$

and for $\tau=0$ it simplifies to

$$\varepsilon\left[\hat{R}_{xx}(0)\right]=\varepsilon\left[\psi_x^2\right]=\frac{1}{\sqrt{BT}} \tag{4.3.48}$$

The variance is then

$$\sigma^2\left[\hat{R}_{xx}(\tau)\right]=\frac{1}{2BT}\left[R_{xx}^2(0)+R_{xx}^2(\tau)\right] \tag{4.3.49}$$

Considering the cross-correlation function, similar expressions may be found:

$$\varepsilon\left[\hat{R}_{xy}(\tau)\right]=\frac{1}{\sqrt{2BT}}\sqrt{1+\frac{R_{xx}(0)R_{yy}(0)}{R_{xy}^2(\tau)}} \tag{4.3.50}$$

$$\sigma^2\left[\hat{R}_{xy}(\tau)\right]=\frac{1}{2BT}\left[R_{xx}(0)R_{yy}(0)+R_{xy}^2(\tau)\right] \tag{4.3.51}$$

These expressions hold for conditions on B and T that are quite easy to fulfill: $T > 10\,|\tau|$ and $BT > 5$.

The auto- and cross-correlation functions are also analyzed to find peak locations τ', useful for example to evaluate the time delay between two data records related to different propagation paths. Under the assumption of bandwidth limited white Gaussian noise in [31] the statistics of τ' are derived: the mean value is zero showing that there is no bias of the estimate; the variance $\sigma^2\,[\tau']$ is given by

$$\sigma^2\,[\tau'] = \frac{0.866}{(\pi B)^2}\,\varepsilon^2\left[\hat{R}_{xx}(\tau)\right] \tag{4.3.52}$$

where $\varepsilon^2\left[\hat{R}_{xx}(\tau)\right]$ is the normalized mean square error, that is equal to the variance over the square of the expected mean value.

The interval of confidence associated with the estimate of a peak position will be determined by the correct multiple of the standard deviation (square root of eq. (4.3.52)), in order to ensure the assigned probability (e.g. $1.96\,\sigma$ for 95% probability).

4.3.2.2 Power Spectral Density from the Correlation function

A first method for the determination of the Power Spectral Density spectrum (PSD), or auto-spectrum, $S_{xx}(f)$ is based on its own definition as the Fourier Transform of the auto-correlation function $R_{xx}(\tau)$ (see at the beginning of sec. 4.3 above). By the adopted notation it is implicitly assumed that the random process that we consider is stationary. Analogously for the cross-correlation $R_{xy}(\tau)$ between two random processes $x(t)$ and $y(t)$, the Cross Spectral Density, or cross-spectrum, is the Fourier Transform of $R_{xy}(\tau)$.

By observing that correlation functions are always even functions of τ, it follows that the desired spectral densities are given by the real part only of the Fourier Transform. By the duality of auto-correlation and autospectrum as a Fourier Transform pair, it follows that the integral over one domain corresponds to the value in the origin in the other domain; so, the integral over the frequency axis of the autospectrum gives the value in the origin of the auto-correlation, i.e. the mean square value (that in turn is the sum of the variance and of the square of the mean value).

$$\psi_x^2 = \int_0^{+\infty} G_{xx}(f)\,\mathrm{d}f = R_{xx}(0) \tag{4.3.53}$$

$$G_{xx}(0) = \int_{-\infty}^{+\infty} R_{xx}(\tau)\,\mathrm{d}\tau \tag{4.3.54}$$

Since the auto-correlation function is biased by the constant term corresponding to the square of mean value, it results that its Fourier Transform (the autospectrum) in frequency domain shows a delta function at the origin ($f = 0$) with area equal to μ_x^2.

4.3.2.3 Power Spectral Density via Fourier Transform of data record $x(t)$

Without demonstrating it [31, 32, 208, 311], we simply state that given a random process $x(t)$, for which we have J records of time length T indexed by index $j = 0 \ldots J-1$, the autospectrum is given by the expectation of the modulus squared as shown below.

$$S_{xx}(f) = \lim_{T\to\infty} \frac{1}{T} E\left\{ |X_j(f,T)|^2 \right\} \quad (4.3.55)$$

A first implementation is the method commonly referred to as the *Periodogram*; it owes its name to the capacity of spotting out hidden periodicity of signals as commented by Schuster[18] at the very beginning of the XX century (see [311], sec. 2.2.1).

$$S_{xx}(f) = \frac{1}{N} \left| \sum_{n=0}^{N-1} x[n] e^{-j\frac{2\pi}{N}n} \right|^2 = \frac{1}{N} |X[k]|^2 \quad (4.3.56)$$

Using the squared modulus of the DFT, the resulting kernel is not the rectangular window typical of implicit windowing over the observation epoch, but a triangle, whose transform is given by the convolution of the spectra of two rectangular windows.

The periodogram as an estimator is to be characterized by error bias and variance . The periodogram is an unbiased estimator, in that it goes to zero for an increasing number of samples N. Regarding the bias Stoica and Moses [311] point out the effect of the main lobe and of the sidelobes, ending in the so called frequency leakage (see sec. 3.1.3). Its variance, on the contrary, is the main problem and it does not reduce with N, keeping on fluctuating around the true PSD with erratic (noise-like) behavior: from this the periodogram is an inconsistent spectral estimator.

Bartlett[19] introduced the idea of averaging the single squared transforms, while keeping the rectangular implicit window, thus maximizing the frequency resolution, without solving the problem of frequency leakage.

Welch[20] further modified the Bartlett method using overlapped data segments (or epochs) and applying windowing, offering lower variance than the Bartlett method. The Welch method, despite the non-dramatic reduction of the variance of the estimate and the existence of many other methods for spectral estimation, is quite popular in that it may be implemented easily and is applicable to signal spectra available in absolute value only, as those resulting from heterodyne measurements (e.g. using a Spectrum Analyzer) or by analog filtering with a conveniently small resolution (see sec. 4.3.2.4 later on).

---

[18] A. Schuster, "The periodogram of magnetic declination as obtained from the records of the Greenwich Observatory during the years 1871-1895," *Transactions of Cambridge Philosophical Society*, Vol. 18, pp. 107-135.

[19] M.S. Bartlett, "Smoothing periodograms for time series with continuous spectra," *Nature*, Vol. 161, pp. 686-687.

[20] P. D. Welch, "The use of fast Fourier transform for the estimation of power spectra: A method based on time averaging over short, modified periodograms," *IEEE Transactions on Audio and Electroacoustics*, Vol. 15, pp. 70-73, June 1967.

By adjusting number of records, chosen window and frequency resolution, consistency of spectral estimate may be slightly improved and optimized for a specific case.

In general methods for spectrum estimation may be divided primarily in parametric and non-parametric, the FFT-based methods belonging to the latter category. Besides improvements of the FFT-based technique, using various types of averaging and smoothing windows, an important step is to leave the conventional Fourier transform approach, based on the implicit assumption that the data that cannot be observed outside the window are zero (that is a non-demonstrated and probably unrealistic assumption). If the user has some knowledge of the system or signal he/she is studying, then such *a priori* information may be conveyed in the selection of a better model, than starting from the complete absence of clues about the model structure: auto-regressive (AR) methods such as Burg and Yule-Walker algorithms are based on a dynamic system interpretation, for which the order may be set starting from some assumptions or iteratively, based on the goodness of the results. This part goes under the so-called "model order selection". Similarly, auto-regressive moving-average (ARMA) models offer a more complex and flexible representation of the studied dynamic model. The case of sinusoids in white noise, considered by Pisarenko decomposition and generalized by Prony's method for damped arbitrary sinusoids, covers quite well the most common signal analysis problems and may be seen as a special ARMA process [208]. All these methods go under the name of "spectrum analysis" or "spectrum estimation" and may be found treated in depth and from different viewpoints in many reference textbooks [32, 208, 256, 275, 311], the most complete and accurate probably being the one written by Stoica and Moses [311].

Remaining on simple, easy to implement and easy to control techniques, the most attractive technique available to achieve a better estimation accuracy is *averaging* (the Welch method): Fourier spectra are computed on signal records, possibly using overlapping to increase their number and time resolution, and then averaged exploiting the benefits of sample estimates. In some cases smoothing windows are not available, as when averaging is done by the spectrum analyzer itself: further post-processing on external computer is then still possible.

Incoherent averaging The most common form of averaging in eq. (4.3.55) is applied to the successive squared spectra to implement the expectation operator; this operation is called *incoherent averaging*, is performed without pha se information and the time location of each spectrum is free of constraints. It is known that the resulting signal-to-noise ratio of the spectral representation improves with the number of samples J along the time axis, having used the term "noise" to identify any incoherent component superimposed to the signal or added by the Fourier mathematical operations. What is improving is the variance of the spectral noise components, resulting in a flatter spectrum noise floor, not the ratio between signal power and the average noise power [290]. In other words, averaging of successive Fourier transforms is increasing the *processing gain*.

A single transform has already an *inner processing gain* that is due to the behavior of each frequency bin as a bandpass filter, narrower as the number of points N increases and with increasing signal-to-noise ratio, that goes with $20\,\log_{10}(\sqrt{N})$. This conclusion is valid for a number of transform points "larger than about 20-30" [229] and for a signal not overwhelmed by noise; to simplify, there is an increase by 3 dB of the amplitude SNR for each doubling of the number of frequency points N.

Let's consider the average of successive Fourier spectra over time, so moving between spectra rather than along the frequency axis of a single spectrum: the average noise floor is still there, but we have an improvement in terms of reduction of noise floor variance, that may be called *integration processing gain*. If the signal-to-noise ratio is measured between the peak of the sinusoidal tone and the largest noise floor value, reducing noise variance has impact on the signal-to-noise power ratio. The reduction of the variance goes with J, so that a twice larger number of spectra to be averaged gives a halved variance and a 30% smaller standard deviation. For independent spectra (i.e. without correlation for example due to overlap), the integration processing gain is given in dB by $10\log_{10}(\sqrt{J})$. Please see Figure 4.3.6 for an example.

Frequency Domain Equipment, such as Spectrum Analyzer, operates always in incoherent mode: see Figure 9.3.1 in Chapter 9 for a demonstration of noise floor smoothing by averaging. It is again observed that what incoherent averaging reduces is the variance of the noise floor profile, not the variance of noise, that strictly speaking corresponds to its mean square value and thus its energy (see sec. 4.3).

Effect of overlap and smoothing windows When overlap is used, it is known that there is correlation between spectra of partially overlapping windows and the amount of correlation depends on the used smoothing windows [153] (see the rightmost column of Table 3.1.1, that reports correlation coefficients for 75% and 50% overlap for the selected smoothing window). This has impact on the resulting SNR that for the most common overlap percentages of 50 and 75% is:

$$\mathrm{SNR}_{incoh} = \begin{cases} \frac{1}{J}\left[1+2c_{0.5}^2\right] - \frac{2}{J^2}\left[c_{0.5}^2\right] & ovl = 0.5 \\ \\ \frac{1}{J}\left[1+2c_{0.75}^2+2c_{0.5}^2\right] - \frac{2}{J^2}\left[c_{0.75}^2+c_{0.5}^2\right] & ovl = 0.75 \end{cases} \tag{4.3.57}$$

This relationship holds for the vast majority of smoothing windows and the omitted case for 0.25 overlap is never really relevant [153].

All this is said for the sinusoidal tone of interest located exactly on a frequency bin. If the tone frequency is intermediate between two bins, the expression for the inner processing gain is still valid, but it is "shifted downwards a little", so that the resulting signal-to-noise ratio is somewhat smaller due to *scalloping loss* (see sec. 3.1.3.1).

Coherent averaging When the time information (or phase information) is available and time records may be synchronized, then *coherent averaging* may be used. The meaning of the word "synchronization" implies that a periodic signal is to be estimated and that a reference frame for the phase can be identified, such as when triggering on a fundamental periodic waveform with oscilloscopes and data acquisition systems. In such cases averaging may be done in time domain on the synchronized (i.e. time-aligned) waveforms (or epochs) or in frequency domain, aligning the phase of the

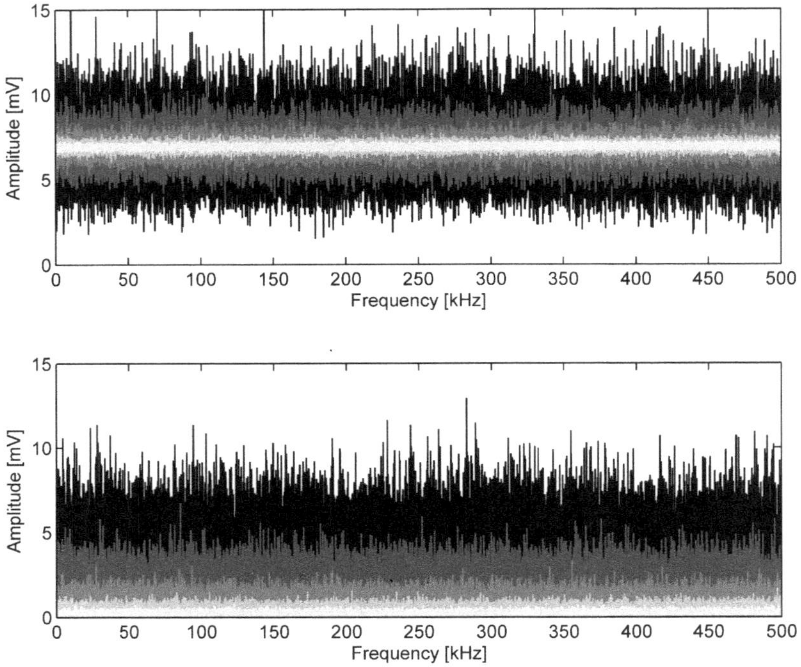

Figure 4.3.6 – Example of incoherent (above) and coherent (below) averaging of successive spectra of a normally distributed white noise; five values of $J = 4$, 16, 64, 256, 1024 were tested: incoherent averaging reduces the variance of the noise profile converging to the noise variance, while coherent averaging progressively reduces the noise floor.

fundamental component of each epoch. If not properly aligned, the vectorial sum of sinusoidal terms with different, and random, phase relationships results in a reduction of the amplitude of the output, as in the case of a random noise components, for which averaging is thought. For independent samples, the reduction of noise power is proportional to the number of averaged samples J, while the amplitude of the coherent components keeps constant (if synchronization is provided). There is a true reduction of the average noise floor and the signal-to-noise power ratio improvement expressed in dB is $10\ \log_{10}(J)$.

Averaging operator Averaging is performed in real-time using some forms of the so-called running average filter. The idea is that a vector $U[k]$ of memory cells is used to store the components (or frequency bins) of the resulting spectrum and a similar vector of cells represents the new freshly computed or measured spectrum $U'[k]$; the running average algorithm at the time instant t_i replaces each value in $U_i[k]$ with a value resulting from the average of the last $i-1$ passed values and the new one:

$$U_i[k] = \frac{(i-1)U_{i-1}[k] + U'[k]}{i} \tag{4.3.58}$$

The multiplication by $(i-1)$ of course recovers the running sum from the average stored in $U_{i-1}[k]$.

A more general algorithm is obtained if an aging factor is included: the resulting algorithm is called exponential average, because the aging factor α gives way to a geometric series of the type $(1-\alpha)^i$, weighting each sample, while it moves farther away in the past from the present instant of time.

$$U_i[k] = \frac{(1-\alpha)(i-1)U_{i-1}[k] + \alpha U'[k]}{i} \tag{4.3.59}$$

It is evident that the two values $\alpha = 0$ and $\alpha = 1$ bring to the extreme cases of no innovation and no average, respectively. Normal values used in practice are in the range $0.5 \ldots 0.9$.

The flattening of noise floor with the reduction of its variance is given by

$$\text{SNR}_{exp.av.} = 10 \log_{10} \frac{\alpha}{2-\alpha} \tag{4.3.60}$$

that for $\alpha = 0.5,\ 0.6,\ 0.7,\ 0.8,\ 0.9$ gives the following values: $\text{SNR}_{exp.av.} = -4.77,\ -3.68,\ -2.69,\ -1.76,\ -0.87\,\text{dB}$.

Statistical properties of the noise floor Let's consider the probability distribution of the noise floor: the external noise $n_e(t)$ is assumed white with a Normal distribution and it can be demonstrated that also the real and imaginary parts of its Fourier transform $N_e(f)$ are normally distributed and that they are independent. The resulting amplitude, or absolute value, $|N_e(f)|$ is not normally distributed any longer, but has Rayleigh PDF (see sec. 4.1.3), resulting from two Normal distributions in quadrature and the relationship $|N_e(f)| = \sqrt{N_{e,re}^2(f) + N_{e,im}^2(f)}$.

The internal noise of the Fourier transform depends on the precision of the mathematical representation adopted by the processor and consists of the combined effect of round-offs in calculations, round-offs in the sin and cos tables, etc.. Proakis and Manolakis [275], sec. 6.4, present an estimation of quantization errors in case of fixed-point processors, using a b bit representation, for a N-point DFT: including scaling and its impact in worsening the signal-to-noise ratio, the resulting intrinsic power signal-to-noise ratio of fixed-point DFT is

$$\text{SNR} = \frac{2^{2b}}{N^2} \tag{4.3.61}$$

The numeric noise of Fourier algorithms is white with Normal distribution, except for very low frequency where the effect of scaling results in "snow drift" noise, so slightly increasing when frequency is going to zero [225]. Usually, when using high-precision arithmetic, numeric noise is orders of magnitude smaller than the external noise and doesn't change the overall behavior remarkably, so that these considerations are generally equally applicable to evaluations performed directly in the frequency domain and to those resulting from the computation of the spectrum from time domain recordings. The Rayleigh PDF of the modulus of two orthogonal Gaussian-distributed signals is shown in Figure 4.3.7.

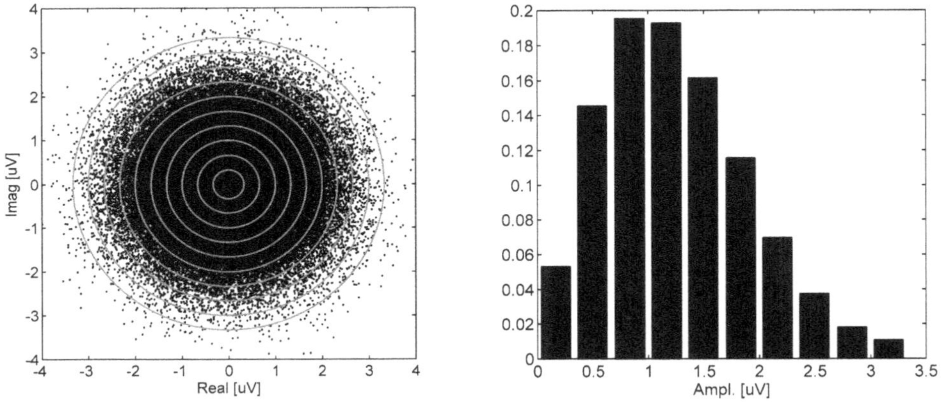

(a) (b)

Figure 4.3.7 – Rayleigh distribution of magnitude resulting from the sum of two Gaussian random variables in quadrature (Real and Imag): (a) isolevel curves in a 2-D representation, (b) histogram of resulting intensity.

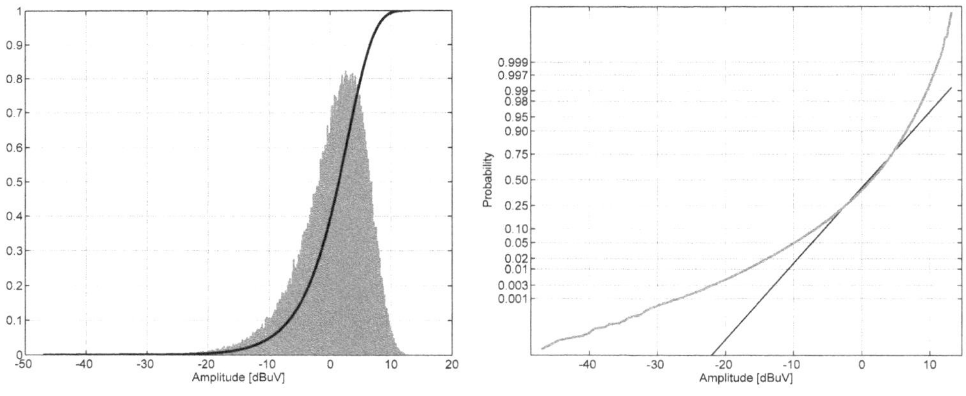

(a) (b)

Figure 4.3.8 – dB-transformed data resulting from the sum of two Gaussian random variables in quadrature (Real and Imag): (a) probability density function, PDF, (scaled) and cumulative density function, CDF (b) plot for graphical verification of Gaussianity assumption.

If a dB transformation is applied to the resulting Rayleigh-distributed intensity (i.e. taking $20\log_{10}(\cdot)$ of it), the small positive values near the origin are stretched and the new PDF gets a tail towards negative dB values, possibly Gaussian-like. Yet, it only resembles a Gaussian distribution because the left and right tails have a much different aspect (see Figure 4.3.8(a)). The graphical verification shown in Figure 4.3.8(b) confirms that the distribution may only look Gaussian at the center around 0 dB, but the tails are the real problem of this distribution; further tests of the goodness-of-fit with respect to a Gaussian distribution performed with the Kolmogorov-Smirnov[21] criterion all fail, even with a relaxed level of significance assigned to the positive hypothesis.

It is true however that adding further noise sources, even of modest amplitude, reduces heavily the negative tail, resulting in a much more symmetrical distribution, that better resembles a truncated Gaussian.

The negative tail appearing. when the log transformation is applied, biases towards lower values the sample mean estimator and is thus responsible for the lower response of the RMS detector of the Spectrum Analyzer (see sec. 9.1.9.2 and 9.2.7), that we know "flattens" the noise floor. On the contrary, for true undistorted power measurements the use of a linear scale is strongly advisable (see sec. 9.3.2).

Error of the estimate Going back to eq. (4.3.55), starting from a Gaussian distributed signal, the real and imaginary parts of the Fourier transform for each component are both Gaussian random variables and their squared modulus (sum of the squares of two independent Gaussian variables) has a chi-square distribution:

$$\frac{\hat{G}_{xx}(f)}{G_{xx}(f)} = \frac{\chi_2^2}{2} \tag{4.3.62}$$

that is a chi-square variable with $n = 2$ degrees of freedom. The square root is conversely Rayleigh distributed.

It is evident that since the mean and variance of such a random variable are n and $2n$, respectively (see sec. 4.1.3.6), the estimate is heavily biased and thus inconsistent. The normalized standard error, given by the ratio of standard deviation and mean, is

$$\varepsilon\left[\frac{\hat{G}_{xx}(f)}{G_{xx}(f)}\right] = \frac{\sqrt{2n}}{n} = \sqrt{\frac{2}{n}} \tag{4.3.63}$$

Averaging J different successive independent squared Fourier spectra is thus mandatory and can reduce the error, with each spectrum adding two degrees of freedom:

$$\varepsilon\left[\frac{\hat{G}_{xx}(f)}{G_{xx}(f)}\right]_J = \sqrt{\frac{2}{2J}} = \frac{1}{\sqrt{J}} \tag{4.3.64}$$

[21] The Kolmogorov-Smirnov criterion is used to decide if a sample comes from a population with a specific distribution and is based on the empirical distribution function (ECDF). The comparison of the theoretical CDF being tested and the ECDF built on the collected samples is done with the maximum distance criterion. Please see also sec. 4.1.4 before in this chapter.

It is apparent that this expression indicates the known linear reduction of noise floor variance with the increased number of averages (see sec. 4.3.2.3 and in particular the paragraph on “incoherent averaging”).

Now, to pave the way for the next section, we observe that first, the frequency resolution of a Fourier transform can be seen as the bandwidth $B = \Delta f = 1/T$ of an equivalent band-pass filter, with T the time duration of each epoch over which the Fourier spectrum $X_j(f, T)$ is calculated, and that, second, the J averaged Fourier spectra span over a total time duration that is $T_{tot} = JT$. Thus eq. (4.3.64) may be written as

$$\varepsilon\left[\frac{\hat{G}_{xx}(f)}{G_{xx}(f)}\right]_J = \frac{1}{\sqrt{BT_{tot}}} \qquad (4.3.65)$$

and the same will be found for eq. (4.3.73), describing the normalized error, when PSD estimation is done by analog filtering.

4.3.2.4 Power Spectral Density by Analog Filtering

A straightforward approach to PSD estimation based on analog techniques is the use of a band-pass filter with bandwidth B_{bp}, followed by rms measurement (either performed with a high performance multimeter or oscilloscope, or with a Spectrum Analyzer, that is described in Chapter 9). The output shall be normalized by B_{bp}. When using digital instrumentation the rms measurement is performed with special code for squaring, time integration and averaging, as it is for the most modern instruments: integration time and sampling time influence behavior and accuracy of rms estimate.

$$\Psi_T(f, B_{bp}) = \frac{1}{T\, B_{bp}} \int x^2(t, f, B_{bp})\,\mathrm{d}t \qquad (4.3.66)$$

$$\hat{G}_{xx}(f, B_{bp}) = \lim_{T\to+\infty} \Psi_T(f, B_{bp}) \qquad (4.3.67)$$

$$G_{xx}(f) = \lim_{B_{bp}\to 0} \hat{G}_{xx}(f, B_{bp}) \qquad (4.3.68)$$

Using an infinitesimally small bandwidth B_{bp} is not practical, so for a finite B_{bp} value the estimate above is a biased estimate of the PSD and the bias is shown to be [32]

$$b\left[\hat{G}_{xx}(f, B_{bp})\right] = \frac{B_{bp}^2}{24}\, G''_{xx}(f) \qquad (4.3.69)$$

similarly to the estimate error for probability density function given in sec. (4.1.4).

This relationship is an over-estimate of the bias error term for most real cases. It is worth underlying that the bias error acts always to decrease the dynamic range of the PSD estimate, so that “peaks are under-estimated and valleys are not so deep”. In the case of a second-order system or a system with a sharp resonance peak, it is shown

[32] that the bias error around the peak depends on the ratio with the half-power resonance bandwidth, B_r, roughly equal to $2\xi f_r$, where ξ is the damping factor and f_r is the resonance frequency.

$$b\left[\hat{G}_{xx}(f, B_{bp})\right] = -\frac{1}{3}\left(\frac{B_{bp}}{B_r}\right)^2 \tag{4.3.70}$$

The variance of the estimate may be calculated assuming that the filtered signal over each B_{bp} frequency interval behaves like a bandwidth limited Gaussian white noise, and it is reasonable if B_{bp} is sufficiently small.

$$\sigma^2\left[\hat{G}_{xx}(f, B_{bp})\right] = \frac{G^2_{xx}(f)}{B_{bp}T} \tag{4.3.71}$$

The total mean square error of the PSD estimate is the sum of the variance in eq. (4.3.71) and the square of the bias in eq. (4.3.69):

$$E\left[(\hat{G}_{xx}(f, B_{bp}) - G_{xx}(f))^2\right] = \frac{G^2_{xx}(f)}{B_{bp}T} + \left[\frac{B^2_{bp}}{24}G''_{xx}(f)\right]^2 \tag{4.3.72}$$

and its normalized version is

$$\varepsilon^2\left[\hat{G}_{xx}(f, B_{bp})\right] = E\left[\frac{(\hat{G}_{xx}(f, B_{bp}) - G_{xx}(f))^2}{(G_{xx}(f))^2}\right] = \frac{1}{B_{bp}T} + \frac{B^4_{bp}}{576}\left[\frac{G''_{xx}(f)}{G_{xx}(f)}\right]^2 \tag{4.3.73}$$

It may be observed that B_{bp} appears both at the denominator and at the numerator and thus there are conflicting exigencies: a small B_{bp} reduces bias, but a large one is desirable to reduce the random part of the error. Normally PSDs of real signals exhibit peaks with large second derivative values and desired information is often concentrated around them, and this exacerbate the bias problem even more.

We may assume in general that the noise source has broadband characteristics, so that B_{bp} may be set to the most appropriate value over a wide range and the second derivative of ADF should be small if sharp peaks and steep slopes are absent. When considering broadband noise-like signals, steep slopes may occur in some cases: at very low frequency when measuring semiconductors because of flicker noise component; at higher frequency peaks and humps may appear if the measurement is contaminated by narrow-band external signals (such as the mains fundamental and its harmonics) or one of the elements of the measurement chain or the device itself exhibit a resonance or undesired phase shift. There are for example Operation Amplifiers with visible humps in the PSD profile of the equivalent input noise usually due to some peaking at high frequency of the closed-loop OA frequency response.

If a digital implementation is preferred, the estimate of the one-sided ADF is computed via Discrete Fourier Transform (DFT) of the input signal and ensemble averaging over J records each of length T (so that frequency resolution is set to $df = 1/T$ and the total time duration is $T_{tot} = JT$). It is easy to demonstrate (see [32], p. 284-5) that the normalized random error is the same seen for the analog implementation, resulting in the same expression (4.3.64) reported above.

The use of DFT is of course subject to its own requisites and criteria, discussed in sec. 3.1.1, 3.1.3 and 3.1.3.2), that fall under the following terms: frequency resolution, frequency leakage, scalloping and scalloping losses, non-stationarity effect, etc..

4.3.2.5 Equivalence of autospectra

When two PSDs must be compared for equality, we revert to a statistical evaluation, where the confidence intervals of the two PSDs are compared.

An estimate $\hat{G}(f)$ of an original reference PSD named $G(f)$ has an approximate normal distribution of its frequency bins if the number of averages N_r used to derive the estimate was large enough (let's say ten or some tens). Neglecting the bias of the estimate, the mean value and the variance of the estimate will be

$$E\left\{\hat{G}(f)\right\}=G(f) \qquad \mathrm{var}\left\{\hat{G}(f)\right\}=\frac{1}{N}G^2(f) \tag{4.3.74}$$

Then the $(1-\alpha)$ confidence interval that brackets $G(f)$ is defined as

$$\hat{G}(f)\left(1-z_{\alpha/2}\frac{1}{\sqrt{N_r}}\right)\leq G(f)\leq \hat{G}(f)\left(1+z_{\alpha/2}\frac{1}{\sqrt{N_r}}\right) \tag{4.3.75}$$

where $z_{\alpha/2}$ is the $100\,\alpha/2$ percentage point of the Normal distribution.

A logarithmic transformation may be operated on the PSD estimate $\hat{Y}(f)=\log(\hat{G}(f))$, so that the $\hat{Y}(f)$ distribution is closer to a normal distribution than the original $\hat{G}(f)$, without the frequency dependence for the variance of each frequency bin.

$$E\left\{\hat{Y}(f)\right\}=\log(G(f)) \qquad \mathrm{var}\left\{\hat{Y}(f)\right\}=\frac{1}{N_r} \tag{4.3.76}$$

In this case the $(1-\alpha)$ confidence interval that brackets $\hat{Y}(f)$ is

$$\log(\hat{G}(f))-z_{\alpha/2}\frac{1}{\sqrt{N_r}}\leq \log(G(f))\leq \log(\hat{G}(f))+z_{\alpha/2}\frac{1}{\sqrt{N_r}} \tag{4.3.77}$$

The test of equivalence of two PSD estimates, $\hat{G}_1(f)$ and $\hat{G}_2(f)$, based on the same frequency resolution and number of bins n_b (identified by the frequency vector F), but obtained for generality with a different number of time records, N_{r1} and N_{r2}, respectively, is then performed as follows.

The $g(F)=\log\left[\hat{G}_1(F)/\hat{G}_2(F)\right]$ variable of the same size of the number of frequency bins has a zero mean ($E\left[g\right]=0$) and $\mathrm{var}\left[g\right]=1/N_{r1}+1/N_{r2}$. The statistic "normalized sum of square"

$$X^2=\frac{1}{\mathrm{var}\left[g\right]}\sum g^2(F_i) \tag{4.3.78}$$

has a Chi-square distribution with N_F degrees of freedom and $X^2 \simeq \chi^2_n$. A Chi-square goodness-of-fit test may be performed and the two estimates are said equivalent if $X^2 \leq \chi^2_{n,\alpha}$, having identified a satisfactory level of significance α for the test.

4.3.2.6 Use of cross-correlation

This method uses the parallel measurement of DUT signal through two different measurement channels (that do not need to be identical in performance or characteristics, but "approximately similar" is beneficial). It is based on the determination of the cross-spectral density function and some interesting properties, that may be exploited in the post-processing. It is well described in [290], where the focus is on the mathematical foundations and the statistic behavior of quantities, and where experimental results are presented and commented as well. The method is reported of being introduced for the first time in 1952 for the determination of the angular size of stellar radio sources[22]; again, in the same field of application, it was used for interferometric measurements[23], and then in the sixties began to be applied to radiometry[24] and thermometry; more recently it features wide application to the accurate evaluation and comparison of noise sources and in particular noise behavior of semiconductors, especially as a diagnostic tool.

The problem is clearly described in [290]: we have a signal source (the DUT) featuring low signal level (e.g. noise) requiring to remove (or to reduce) the influence of amplifiers and measuring instrument. Two amplification/filtering chains are setup (called instrument A and instrument B), each fed with the DUT signal $c(t)$ and adding to it their own internal noise, $a(t)$ and $b(t)$. We assume to be able to measure at the same time the received composite signals $x(t) = c(t) + a(t)$ and $y(t) = c(t) + b(t)$, e.g. by using a network analyzer, a signal receiver or more simply a dual-channel FFT analyzer or oscilloscope, of course with much different performances related to the internal noise floor, dynamic range and sensitivity. From time domain relationships is immediate to pass to frequency domain with signals replaced by respective transforms indicated by capital letters. Since the DUT signal $c(t)$ is common to the two measured signals, the remaining components assumed to be random and independent[25], it is intuitive that the expectation of the cross-spectrum converges to the autospectrum of $c(t)$, i.e. $S_{cc}(f)$; of course, assuming ergodic stationary signals, the expectation is replaced by time average of the acquired M records:

[22] R. Hanbury Brown, R.C. Jennison and M.K. Das Gupta, "Apparent angular sizes of discrete radio sources," *Nature*, Vol. 170, 1952, pp. 1061-1063

[23] R. Hanbury Brown and R.Q. Twiss, "A test of a new type of stellar interferometer on Sirius," *Nature*, Vol. 178, 1956, pp. 1046-1048.

[24] C. M. Allred, "A precision noise spectral density comparator," *Journal of Research of National Bureau of Standards*, 66C, 1962, pp. 323-330.

[25] Correlated noise components between the two instruments may exist, such as crosstalk and environmental induced phenomena, due to e.g. magnetic field, temperature, etc.

$$\begin{aligned}\langle S_{yx}\rangle_M &= \frac{1}{T}\langle YX^*\rangle_M = \frac{1}{T}\langle (C+A)(C+B)^*\rangle_M = \\ &= \frac{1}{T}\left[\langle CC^*\rangle_M + \langle CB^*\rangle_M + \langle AC^*\rangle_M + \langle AB^*\rangle_M\right] = \\ &S_{cc} + O(\sqrt{1/M})\end{aligned} \tag{4.3.79}$$

The cross-terms rule out to zero in principle with an infinite time history and their power decreases proportional to the square root of M (see sec. 4.3.2.3 and 9.3). This power represents the minimum measurable level of the DUT signal.

Setting the DUT signal to zero, $c(t)=0$, the resulting (and measurable) cross-spectrum shall tend to zero; the estimation $\hat{S}_{xy}$ made by averaging successive cross-spectra $S_{xy,m}$ (one for each value of the index $m=1\ldots M$) has a Rayleigh distribution with $2M$ degrees of freedom, with bias (average value) and variance as

$$E\left\{\hat{S}_{xy}\right\} = \sqrt{\frac{\pi}{4M}} = \frac{0.866}{\sqrt{M}} \qquad \operatorname{var}\left\{\hat{S}_{xy}\right\} = \frac{1}{M}\left(1-\frac{\pi}{4}\right) = \frac{0.215}{M} \tag{4.3.80}$$

It is interesting to note that the ratio between dispersion and average value is independent of M:

$$\frac{\operatorname{dev}\left\{\hat{S}_{xy}\right\}}{E\left\{\hat{S}_{xy}\right\}} = 0.523$$

Other quantities of interest are the sum and difference of average and dispersion, normalized by the average, that identify the height of the band around the mean value where the noise power is likely to be located ($\pm 1\sigma$ confidence interval); the two bounds expressed in dB of power are $10\ \log_{10}(1-\text{dev}/\text{avg}) = -3.21$ dB and $10\ \log_{10}(1+\text{dev}/\text{avg}) = +1.83$ dB. For increasing M we observe the band reducing in extension and moving to lower power levels. When considering the combined effect of the DUT noise and the instrumentation noise, at some large M the noise band is lower enough that the DUT is prevailing and the total measured noise does not reduce any longer.

Taking the measurement of either channel $x(t)$ or $y(t)$, the estimate of the auto-spectrum (or power spectral density) is considered to be obtained via Fourier transform (see sec. 4.3.2.3), so that amplitude at each frequency point results from the sum of the squares of the real and imaginary components of the Fourier transform $X(f)$ or $Y(f)$; the sum is normalized by the observation interval T (i.e. the record length), correctly giving a power spectral density as result. The auto-spectrum $S_{xx}(f)$ or $S_{yy}(f)$ is thus χ^2 distributed with two degrees of freedom for each time record of index m and $2M$ degrees of freedom in total when doing the sample estimate. This time the deviation over average ratio decreases with the square root of M.

Term	avg	var	PDF
$B'A'$, $B''A''$, $B''A'$, $B'A''$	0	$\frac{1}{4}$	Normal
$B'C'$, $B''C''$, $B''C'$, $B'C''$, $C'A'$, $C''A''$, $C''A'$, $C'A''$	0	$\frac{\kappa^2}{4}$	Normal
$C'^2+C''^2$	κ^2	κ^4	χ^2 $\nu=2$

Table 4.3.2 – Terms involved in the estimation of the cross-spectral density S_{yx}.

$$\frac{\operatorname{dev}\left\{\hat{S}_{xx}\right\}}{E\left\{\hat{S}_{xx}\right\}}=\frac{\operatorname{dev}\left\{\hat{S}_{yy}\right\}}{E\left\{\hat{S}_{yy}\right\}}=\frac{1}{\sqrt{M}}$$

Apportioning S_{yx} into the single components and re-arranging them in a different order, it is possible to isolate groups of terms with similar behavior and characteristics (see [290] for details). S_{yx} is given by the expectation of the product YX^*, where "*" indicates complex conjugate. The transform of each signal A, B, C, as well as X and Y, may be subdivided into real and imaginary parts, indicated by $'$ and $''$, respectively.

$$\begin{aligned}S_{yx} &= \frac{1}{T}E\{YX^*\}=\cdots \\ &= \frac{1}{T}E\{(B'A'+B''A''+B'C'+B''C''+C'A'+C''A''+C'^2+C''^2) \\ &\quad + j\,(B''A'-B'A''+B''C'-B'C''+C''A'-C'A'')\}\end{aligned} \tag{4.3.81}$$

For simplicity the variances of noise transforms A and B are taken equal to unity, and thus the real and imaginary parts have a variance of $1/2$ (by conservation of energy); for the wanted signal C its variance is taken equal to κ^2 (i.e. measured against the two noise sources). The usefulness of the procedure reveals of course when $\kappa^2 \ll 1$, that is when the desired signal (e.g. the DUT noise) is masked by the instruments noise. A table like in [290] summarizes the result in terms of average and variance and related distribution (see Table 4.3.2).

Thus it is easy to group the terms in order to represent the real part of the unwanted noise, the imaginary part of the unwanted noise and the desired terms, that, it is underlined, are real.

Now, it is possible to examine the estimators we may apply to the quantities in eq. (4.3.81) and to identify the best one in terms of biasing and consistency.

- Absolute value $\left|\hat{S}_{yx}\right|$. All dual channel analyzers have the option of displaying the absolute value (intensity), but this estimator brings in also the imaginary terms that contain half the power of the background noise of used instrument chain and no information on C (the desired DUT signal); we know that increasing M the background noise reduces and sooner or later it drops below the κ^2 power level of the C signal.

- Real part $\Re\left[\hat{S}_{yx}\right]$. We see immediately a 3 dB improvement of the signal-to-noise ratio; the main operative drawback is that this quantity may be negative and thus cannot be expressed in dB. For large M the average and variance of the estimator can be reconstructed from the average and variance of the composing terms (see Table 4.3.2):

$$E\left\{\Re\left[\hat{S}_{yx}\right]\right\}=\kappa^2 \qquad \operatorname{var}\left\{\Re\left[\hat{S}_{yx}\right]\right\}=\frac{1+2\kappa^2+2\kappa^4}{2M}$$

- Averaging the absolute value of the real part $\left|\Re\left[\hat{S}_{yx}\right]\right|$. This estimator corrects the problem of negative values that are folded up onto the positive half-plane. The resulting PDF has an increase at the origin and low positive values with respect to the original Gaussian distribution.

- Averaging after discarding negative values of the real part $\Re\left[\hat{S}_{yx}\right]_+$. Negative values are not folded, but discarded; the resulting PDF is more similar to the original one, but is stretched vertically to larger values of probability to account for the reduced number of samples (only the positive are retained, thus occurrences in each bin are normalized for a smaller number).

- Averaging the maximum after replacing with zero the negative values of the real part $\max\left\{\Re\left[\hat{S}_{yx}\right]_{0+}\right\}$. Negative values are replaced with zeros and the maximum is taken, so that the result is either a positive value or zero; this implies that there is a concentration of components with zero value (those corresponding to original negative values); the resulting PDF is thus the original one, except for a Dirac in the origin, that counts all the zero results.

The best estimator is then selected based on its bias and variance: Rubiola and Vernotte [290] state clearly that "having accepted that an estimator suitable to logarithmic plot is positive, thus inevitably biased, the best choice is the estimator that exhibits the lowest variance and the lowest bias." The process of selection discards the estimator based on the absolute value focusing on those taking the real part $\Re\left[\hat{S}_{yx}\right]$. Then they proceed reasoning on the resulting distribution when negative values are treated in different ways, observing that "the naif approach of just discarding the negative values before averaging turns out to be the worst choice among the considered estimators." Then they finally demonstrate that the $\max\left\{\Re\left[\hat{S}_{yx}\right]_{0+}\right\}$ has the lowest bias.

5

Cables

This Chapter is the first of three considering the main elements of test setups and is based upon principles and material for transmission lines contained in Chapter 2. Cables are the most straightforward interconnecting means and are complemented by planar transmission lines on Printed Circuit Boards considered in the next Chapter 6. Coaxial cables and pairs are considered, discussing materials and behavior, including theoretical works regarding shielding properties and referring to the applicable standards. Standards are not only useful and indispensable references that establish relevant and minimal performances, but shed some light on the measurement methods to verify cable performance and how to read and interpret datasheets.

5.1 Cable performance and characteristics

In this section we review parameters and performance indexes that are normally used to characterize cables and that are referred to in the normal engineering practice and applicable standards. Of course the term "cable" encompasses cables used for signal, measurement, data network and telecommunication applications, leaving outside power cables, for which the considered standards are not even applicable.

Standards for cabling systems in use in Information Technology applications (namely network and data connections) have converged on a classification of the environment in which the cable operates, that uses four letters to classify mechanical (M), ingress (I), climatic and chemical (C) and electromagnetic (E) agents, resulting in the now famous acronym "MICE". So, a cable is the result of a tradeoff of several characteristics, that make it suitable for its target application and environment, rated for the four mentioned agents with an eye on cost, weight, usability, etc. The focus in the following is on electromagnetic characteristics, considering marginally those strictly electrical, such as the voltage withstanding capability, the rated voltage and current, etc.. Detailed conditions for letter "E" as reported in the EN 50173-1 standard [58] simply recall the EMC test levels adopted in the most common electromagnetic environments: IT, residential, industrial. The first one is not an official designation; we want to indicate the discrepancy between the EMC immunity standard for IT equipment (EN 55024 [68]), that differs from and is in some tests less stringent than the recent versions of the generic application EMC standard for the residential environment (EN 61000-6-3 [79]). For the tests whose results are more influenced by cable performance, we simply recall i) the radiated electromagnetic field disturbance (ranging from 80 to 2700 MHz with a field intensity between 1 and 10 V/m), ii) the conducted radiofrequency disturbance (ranging from 150 kHz to 80 MHz with amplitude of 3 or 10 Vrms) and iii) the electrical fast transients (with amplitudes from 500 V to 2000 V peak and a fast rise-time of only 5 ns). In the other environments lines are subject to induction, and based on this assumption common mode disturbance is usually required in standards: from this the importance of shielding and transfer impedance (see sec. 5.1.7).

5.1.1 DC resistance and resistance unbalance

This parameter is simply related to the wire gauge of internal conductors and the quality of production in keeping the dc resistance values balanced for the two conductors in a pair and in the whole multi-pair cable. Even if not explicitly specified, the dc resistance parameter may regard also the cable screen, in particular if it is either part of the circuit (i.e. it is used as return conductor) or it may be subject to intense coupled current (e.g. extraneous supply current due to conductive coupling). To this aim, i.e. ensuring good electric contact and a minimum longitudinal dc resistance, aluminum foil screens are backed up by a longitudinal equipotentializing wire.

Many standards specify such a basic measurement, but there are no other details for its execution. The resistance value to measure is not extremely low and the general care of 4-wire measurements for moderately low resistance may be adopted: this technique is described in many multimeter manuals and consists of using two terminals to apply the test current (usually dc, but also an ac signal may be used for better rejection of offsets and Seebeck "battery effects") and two high-impedance terminals for voltage reading; the latter may be positioned exactly where the sample to measure is connected, minimizing the influence of leads and connecting elements.

When measuring resistance unbalance between pairs and between conductors in the same pair, resolution and sensitivity of the instrument and of the method shall be correspondingly increased: tight fabrication tolerances and controlled material properties suggest small unbalance values in the order of a fraction of % to a few %; for cable samples of moderate length and with low dc resistance values, a sensitivity of a hundred μΩ is usually necessary.

5.1.2 Inductance and capacitance terms

The main relationships that rule inductance and capacitance terms have been briefly reviewed in Chapter 1 at sec. 1.1.3 and 1.1.4.

5.1.2.1 Inductance

The theoretical expressions for the determination of self- and mutual inductance (see sec. 1.1.3) depend on the natural logarithm of the ratio of distance and radius and on the magnetic properties of materials; with no ferromagnetic materials the value is determined by the conductor geometry alone, i.e. conductor cross-section and thickness of the insulation, that establish the separation of conductors.

Practically speaking, for low and medium frequency applications with receivers connected to the cable line featuring hundreds of ohm of input impedance, inductance terms and inductive voltage drop are less relevant. At higher frequency, for RF applications, a standardized guaranteed characteristic impedance is required, as for coaxial cables and multi-pairs used in high-bit-rate networks. Inductance and capacitance are thus inversely proportional and are in general less variable because the geometry is well controlled, with the cable undergoing qualification with the performance indexes described in the following.

The use of linear non-ferromagnetic materials is always highly advisable, not to say mandatory, but skin effect, and also proximity effect in a less manner, cause anyway slight changes to the geometry (see sec. 1.1.3.2 and 1.1.3.3): both in the inner conductors and in the cable shield/armor the current distribution changes with frequency and so does the resulting inductance. If for a well-designed commercial cable the overall change is within the declared tolerance, for precision airlines even very modest changes to inductance and characteristic impedance are relevant, as pointed out at the end of sec. 11.12.3.3. For connectors, with a more complex shape, such effects may be more pronounced and significant changes of inductance may occur.

5.1.2.2 Capacitance

The theoretical expressions for the determination of self- and mutual capacitance are much similar to those for inductance (see sec. 1.1.4): both depend on the natural logarithm of the ratio of distance and radius and depend also on the presence of other conductors and their potential. Capacitance terms, however, are more susceptible to the electrical properties of used materials, with varying permittivity and dielectric losses; moreover, the gradient of the electric field may be quite large near the conductor surface, where the insulating sheath is located, locally affecting dielectric behavior.

In multi-pair cables mutual capacitance terms are many, from mutual capacitance between the two conductors in a pair, C_{12}, and self-capacitance of each of them towards cable shield (and internal shields), C_{10} and C_{20} (see Figure 5.1.1), to the many mu-

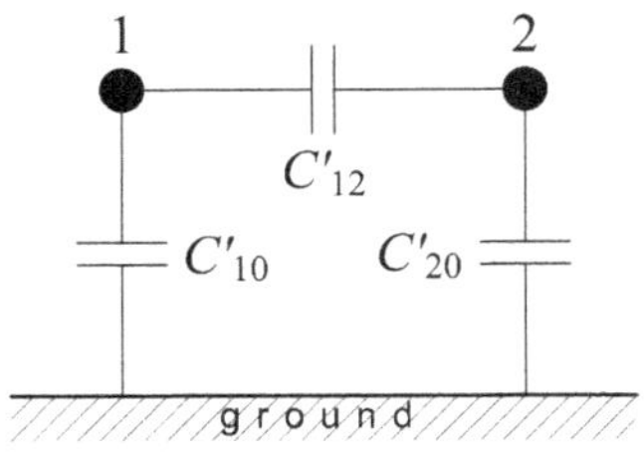

Figure 5.1.1 – Diagram of the capacitive terms of a pair; multiple pairs may be added multiplying the number of reciprocal terms; the "ground" may be a real ground (or grounded object) for unshielded pairs or the cable shield, or pair shield.

tual capacitance terms between different pairs, C_{ij}. The measured (or measurable) mutual capacitance $C_{m,12}$ "sees" the physical mutual capacitance C_{12} and the two self-capacitance terms in series, so that $C_{m,12} = C_{12} + C_{10} \| C_{20} = C_{12} + 0.5\, C_{10}$. Self terms are normally larger and thus any imbalance translates in a non-negligible capacitance difference, directly increasing the common-to-differential mode transformation (see sec. 10.4.1); practical values of capacitance unbalance are again in the order of a fraction of % to a few %.

For pairs in unbalanced conditions, and as a general method, the measurement of the capacitance terms may be executed as follows [63]: measure the capacitance of conductor 1 with respect to conductor 2 and all other conductors, shields, ground, all bonded together (name this value C_1', with $C_1' = C_{10} + C_{12}$); measure similarly capacitance of conductor 2 with respect to conductor 1 and all other conductors, shields, ground, all bonded together (name this value C_2', and again $C_2' = C_{20} + C_{12}$); measure the capacitance of conductor 1 and 2 bonded together with respect to all other conductors, shields, ground, all bonded together (name this value C_3', and again $C_3' = C_{10} + C_{20}$). Then the mutual capacitance between conductors 1 and 2 is

$$C_{m,12} = \frac{C_1' + C_2'}{2} - \frac{C_3'}{4} - \frac{(C_1' - C_2')^2}{4C_3'} \tag{5.1.1}$$

The IEC 60096-1 [169], sec. 12.1.2, gives a simpler formula for the mutual capacitance between inner conductors without the third term above:

$$C_{m,12}^{\mathrm{IEC}} = \frac{2(C_1' + C_2') - C_3'}{4} \tag{5.1.2}$$

By the way the mutual cable capacitance may be used for a first estimate of the "mean" characteristic impedance (as in IEC 60096-1 [169], sec. A1.1), once the resonance frequency of the cable sample is known, as seen in sec. 2.4, eq. (2.4.8).

As an additional term, the unbalanced capacitance to earth (or capacitance unbalance to earth) C_u, may be readily determined from the above quantities as

$$C_u = C_1' - C_2' \tag{5.1.3}$$

The standard EN 50289-1-5 [63] recommends that measurements are carried out at a test frequency between 500 and 2000 Hz using a RCL bridge; there is no minimum length of cable sample and required accuracy is 1%. The reason for selecting these test frequencies is on one side avoiding frequency values too close to the mains frequency and to avoid the resonance of the cable sample (that might be an entire coil sampled in production and thus even some km long), but also to limit the longitudinal contribution of the cable inductive reactance (that would mask part of the transverse capacitive reactance under measurement). Such modest test frequencies might be compatible with the limited exigencies and instrumentation capability for those applications like serial-data links at low signaling rate, telephone connections, etc., but certainly leave behind quite a relevant and extended portion of the frequency interval.

It is possible to perform the same measurement at the minimum VNA frequency (e.g. 10 kHz) and then verify with some test frequency points in a convenient interval (such as one or two octaves) if the longitudinal inductive reactance is playing a role or not. To this aim the measured capacitive reactance shall halve for each octave step, if it is not and the reduction is larger, the residual error is due to the correspondingly increasing inductive reactance.

Capacitance terms are very important because they explain pair imbalance and common-to-differential transformation, that are then quantified in a proper way by the many signal integrity parameters that we review in the following. They receive attention in all those applications where signal bandwidth is quite low, but cable length is considerable, from hundreds of meters to many km. Crosstalk is directly proportional to mutual capacitance and the isolation of selected pairs dedicated to special functions (see sec. 5.1.6.3) is subject to the availability of pairs with a particularly low mutual capacitance: there are specific requirements for capacitance imbalance and maximum mutual capacitance in terms of guaranteed maximum and confidence interval.

Moreover, a low value of the equivalent input capacitance of a pair is always desirable when considering that not only it increases the cable velocity and characteristic impedance, but keeps lower the current taken from the source when driven with steep edges: when applying a step signal with a significantly large rate-of-rise, the input current is quite close to the short circuit current and this may have impact on the linearity of the driving source and on its speed in recovering from saturation.

5.1.3 Return Loss and Insertion Loss

These two terms correspond to input impedance and attenuation, and are described in Chapter 10, when considering scattering parameters and their interpretation. Their measurement is not difficult (see sec. 10.2.3), unless the frequency range extends down in the kHz range or up to several GHz and beyond: in the first case the expected attenuation is so small that a long cable coil is needed to obtain meaningful results; in the second case, even with modest cable length the accuracy may be influenced by measurement noise and also connectors and cable-connector joints become relevant. Concepts of reflection and attenuation, use of scattering parameters and correction of setup non-ideality will be addressed in particular in Chapter 11.

Practically speaking insertion loss gives the attenuation of the cable under ideal circumstances and underestimates cable losses, that are also caused by standing wave phenomena and the local increase of voltage and current intensity (see sec. 2.5.3).

In sec. 4.5.6 of IEC 61935-1 [171] the temperature coefficient of the insertion loss is broadly classified for "screened" and "unscreened" cables: 0.2 %/°C for screened cables and 0.4 %/°C below 250 MHz, 0.6 %/°C above for unscreened cabling. The physical explanation, nor the context of such values (e.g. experimental, statistical, worst-case), is not given.

The required accuracy for field test equipment is specified at sec. 6 of IEC 61935-1 for return and insertion loss: for category II and III equipment values between 2 and 3 dB are required for return loss at the reference frequency of 100 MHz, about 40% higher at 250 MHz, while for insertion loss values between 1.2 and 1.5 dB are required at 100 MHz, 50% higher at 250 MHz; specifications for category IV equipment are slightly more strict, ranging between 2.4 and 3.2 dB for return loss and 1.2 to 2.5 dB for insertion loss this time at 100, 250 and 600 MHz.

In the old IEC 61935-1 (2005) reference was made to the ISO/IEC 11801 standard for insertion- and return-loss measurements, while in the new version that supersedes the old one by 2012 the reference has been updated to the EN 50173-1. However, the latter is much less detailed regarding measurement methods and advised practice.

5.1.4 Propagation delay, delay skew and cable velocity

The definition of propagation delay and cable velocity may be found in Chapter 2: they are inter-related by cable length, and thus when expressed with p.u.l. parameters they are one the reciprocal of the other. Of course the concept of delay skew applies only when there is more than one signal conductor, so to multi-pair cables.

The determination of propagation delay τ can be readily accomplished by means of a complete (i.e. giving complex numbers and not only the intensity) S_{21} measurement: the phase constant β divided by $2\pi f$ is the propagation delay. The uncertainty (expressed for a coverage factor $k = 1.96$ corresponding to 95% probability of the confidence interval under Gaussian distribution assumption) required by the IEC 61935-1 [171] is fair, with a limit of 2.5 ns for measurements in the range 0-60 ns and 5 ns for up to 600 ns. In reality, to complicate things, the standard at its sec. 6 specifies the characteristics for field test equipment: for propagation delay and delay skew the required "accuracy" is $27\,\mathrm{ns}$ and $10\,\mathrm{ns}$, respectively, so quite more permissive than the uncertainty values above and leaving the doubt to the reader whether uncertainty and accuracy are used with same or similar meaning, or not.

Considering a multi-pair cable, any difference in p.u.l. inductance and capacitance terms of each pair translates into a slightly different delay value (and delay skew as a consequence, relevant if the pairs are used to carry simultaneous signals). From this it is evident that attention shall be given to varying mutual capacitance when using pairs in a cable and that high speed data transmission requires carefully selected pairs for balance and symmetry. Delay skew is not only responsible of mode conversion (see sec. 10.4.1), but also of attenuation and reduced bandwidth for differential signals, as demonstrated in sec. 8.2.5.

5.1.5 Rise time

The rise time of a system has been extensively considered in sec. 3.2, regarding definition with respect to bandwidth, propagation through cascaded linear time-invariant (LTI) systems and consequences on time-domain response to pulse, step or square waves. A cable, as all other circuits and systems, is characterized by its own rise time, that is mainly due to input capacitive effect and bandwidth. The former simply indicates the capacitive loading on the signal source when the cable is driven by the fast portion of the signal. If the rise time were determined only by cable bandwidth (maximum usable frequency), it would be the same for any length; on the contrary, looking at the data shown in Table 5.1.1, dependency on cable length is evident (the rise time is approximately proportional to the square of length). In reality, the concept of bandwidth is still valid, but bandwidth shall be determined as the frequency at which attenuation is 3 dB and this depends on length; moreover, with an approximate reasoning, attenuation itself at high frequency does not increases linearly with frequency but slightly less, nearly with the square root of frequency. So, doubling the length doubles the attenuation, but the −3 dB frequency doesn't recede proportionally, because we need an even smaller frequency such that the new attenuation correctly matches, and the reduction of frequency goes approximately with the square.

This rationale on attenuation explains why the three cables in Table 5.1.1 give such different values of rise time, not because of their capacitance (that is almost the same) as in a RC lumped circuit, but due to attenuation (and the three cables are ordered for decreasing attenuation).

Length	**10-90% rise time [ns]**		
[feet]	**RG-174**	**RG-58**	**RG-8**
1	0.004	0.002	0.0002
2	0.014	0.006	0.001
3	0.032	0.014	0.002
4	0.056	0.024	0.004
5	0.088	0.038	0.006
10	0.35	0.15	0.025
20	1.4	0.61	0.10
50	8.8	3.8	0.64

Table 5.1.1 – Experimental values of the 10-90% rise time for commercial coax cables [199]; rise time increases proportionally to the square of length.

5.1.6 Crosstalk

For a definition of crosstalk, please, make reference to the beginning of Chapter 8 and to sec. 8.3.4 in particular. The general reference for this section is the EN 61935-1 standard [171]; some reference values and limits are also reported in the EN 50288-

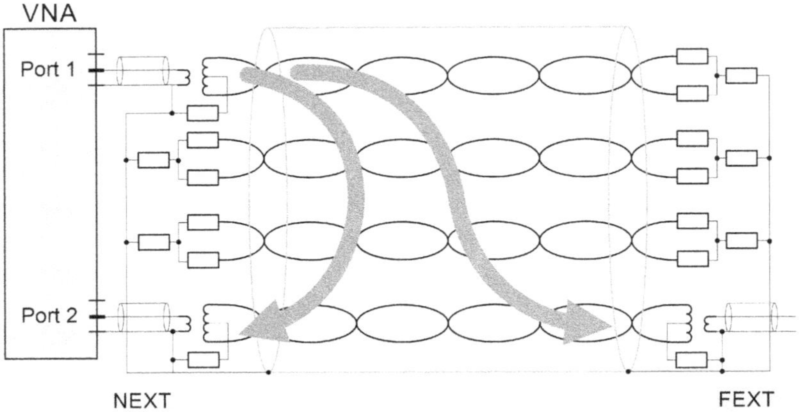

Figure 5.1.2 – Scheme of the NEXT or FEXT measurement (please, note the use of baluns between single-ended VNA and differential lines; center-tap baluns are used with correct termination on the common-mode impedance, see sec. 10.4.2).

2-1 standard [60] up to 100 MHz, in the EN 50288-10-1 standard [62] up to 500 MHz and in the EN 50288-9-1 standard [61] up to 1000 MHz.

5.1.6.1 Near-end crosstalk (NEXT) and power sum NEXT

In Chapter 8 we define crosstalk as a voltage appearing at a victim line k in relationship to the voltage/current flowing in a source line i due to electromagnetic coupling between the two. We have also seen that the so-formed crosstalk may be referred to either of the two ends of the victim line, and we identify with near-end crosstalk the one measured at the same side of the generator connected to the source line.

From the definition it is clear that the S_{21} parameter can quantify the coupling, once the VNA signal of one of the two ports is applied to the source line (the VNA is the mentioned generator, see Chapter 11) and the other port is connected to the near-end of the victim line (near-end crosstalk, or NEXT), at the same side where the generator is connected (see Figure 5.1.2). The standard suggests the use of baluns (also shown in Figure 5.1.2) that may be found with satisfactory performance up to a few GHz; otherwise a VNA with the balanced-port option may be used. The schemes for the measurement of the relevant scattering parameters are shown in sec. 10.4.3.

$$\mathrm{NEXT}_{i,k} = -20\log_{10}|S_{21\,i,k}| \tag{5.1.4}$$

having noted with $S_{21\,i,k}$ the coupling between line i and k, measured as an insertion loss of the type S_{21}.

The uncertainty required by the IEC 61935-1 [171] for multi-conductor pairs is 1 dB up to 100 MHz, 1.2 dB at 250 MHz and 2 dB at 1000 MHz, values well above those established for VNA calibration errors (see sec. 11.5 and 11.11.1), but that shall account for instrumentation and setup uncertainty. Additionally, the same standard at sec. 6 for field test equipment specifies an even larger required “accuracy” indicating values between 2 and 5 dB at various frequencies and for various equipment categories.

We immediately observe that line attenuation influences the measured NEXT. It is sometimes required to give an equivalent NEXT, after the effect of the attenuation on the victim line k is removed, and this is called Attenuation-to-Crosstalk Ratio (ACR) and is expressed as the difference of the measured NEXT and the Insertion Loss of the victim pair IL_k.

$$\mathrm{ACRN}_{i,k} = \mathrm{NEXT}_{i,k} - \mathrm{IL}_k \tag{5.1.5}$$

The power sum NEXT (PSNEXT) term aims at summing all the contributions on the victim line by all the pairs of a multi-pair cable and so it is applicable only if the total number of pairs is larger than two. In this case, given the source line with index i and the victim line with index k, the sum of the power terms it is extended to all the possible source lines with $i = 1 \ldots N$, but $i \neq k$. Correctly, the sum is made on the crosstalk terms expressed as power, and not as voltage.

$$\mathrm{PSNEXT}_k = -10 \log_{10} \left(\sum_{i=1,\, i \neq k}^{n} 10^{\mathrm{NEXT}_{i,k}/10} \right) \tag{5.1.6}$$

The standard exemplifies the number of required measurements for the case of four pairs: with the source line in the first pair, two NEXT measurements shall be performed for each conductor of the other pairs, for a total of six measurements; then the measurements are repeated when the source line is the other conductor, still in the first pair. The total number of measurements is thus twelve. Insertion loss IL may be measured as a true S_{21} measurement for each line separately.

5.1.6.2 Far-end crosstalk (FEXT) and power sum FEXT

For the far-end crosstalk we have a similar measurement method (see Figure 5.1.2): the VNA signal of one of the two ports is applied to the source line and the other port is connected to the far-end of the victim line, opposite to the side where the generator is connected. For this reason the standard says that the FEXT measurement is applicable only to a laboratory environment, while the NEXT can be performed also on installed cabling: the two ends of each cable are physically farther apart without the possibility of connecting the two ports of the same VNA. However, a spectrum analyzer and RF generator may be used instead, synchronized via the 10 MHz input and trigger input connected by means of cables of the same length and thus featuring the same time delay, provided that also the internal trigger delay is taken into account and compensated, if different for the two instruments. Compensation may be easily achieved by changing the length of the trigger-feeding cables.

$$\mathrm{FEXT}_{i,k} = -20 \log_{10} |S_{21\,i,k}| \tag{5.1.7}$$

An Attenuation-to-Crosstalk Ratio (ACR) for the far-end crosstalk ACRF may be defined in the same way as done for the near-end crosstalk above:

$$\mathrm{ACRF}_{i,k} = \mathrm{FEXT}_{i,k} - \mathrm{IL}_k \tag{5.1.8}$$

The power sum FEXT (PSFEXT) term is analogous to the just seen PSNEXT, summing all the contributions on the victim line by all the pairs of a multi-pair cable. Again, the sum is made on the crosstalk terms expressed as power, not voltage.

$$\mathrm{PSFEXT}_k = -10 \log_{10} \left(\sum_{i=1,\, i \neq k}^{n} 10^{\mathrm{FEXT}_{i,k}/10} \right) \tag{5.1.9}$$

5.1.6.3 Alien (or exogeneous) crosstalk, near-end and far-end

When considering lines belonging to different cables and thus separated by their own screens (when present, and we may assume it is the normal policy) and by an increased distance, the standards talk of alien crosstalk, or exogeneous crosstalk.

The definitions are trivial, because they correspond exactly to NEXT and FEXT, but source and victim lines belong to different cables, instead of simply different pairs within the same cable (see Figure 5.1.3). This is the usual situation in industrial applications when considering cable segregation: cable-to-cable coupling is far more common in the engineering practice and related design, than pair-to-pair crosstalk, that is cable manufacturer's competence, and it is normally accepted as per datasheet and used as input to our engineering problems. For such applications with the term crosstalk it is intended the alien crosstalk, because it is the quantity that is under designer's or installer's control, modifying cable layout, distance, routing, segregation, and optimizing termination and earthing.

The number of required measurements is quite large: given two cables containing each 4 pairs (typical of data transmission applications), there are 16 pair combinations, with each pair of the first cable combining with all the pairs (one at a turn) in the other cable; because of the need of measuring both near- and far-end alien crosstalk terms, the number of measurements rises to 32; since the first cable was assumed as the disturbing cable since the beginning, reversing the roles, with the second cable as the disturbing cable, measurements shall be repeated for a total of 64.

There is a certain degree of variability of the resulting alien crosstalk to be considered when source and victim cables are in their final installation: presence of conductive parts surrounding the cables (e.g. concrete cable-duct or metallic cable-tray), grounding of cable shields, etc. To this aim the standard EN 50174-1 [59] in Appendix D states the following: "It is however difficult to measure the common mode crosstalk loss of installed cabling or of a cable in laboratory conditions. The measured crosstalk loss depends very much on the position of the cable with respect to earth, how the unused pairs and cable screen (if any) are connected and the common mode terminating impedance with respect to earth that is provided by the attached equipment. In the case of installed cabling the common mode crosstalk loss depends very much on the earth reference point that is used and that differ for the transmitter and receiver."

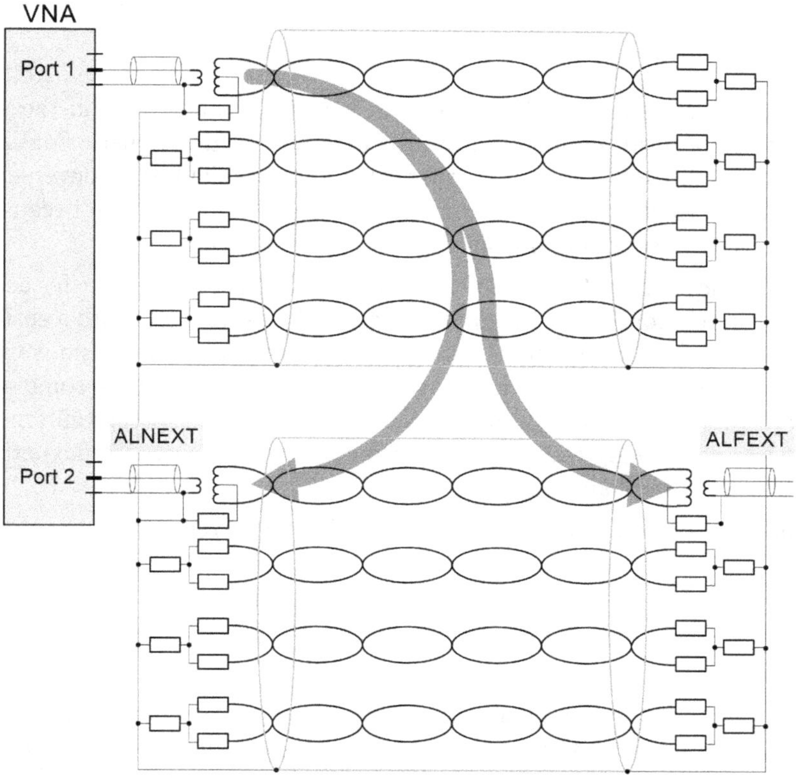

Figure 5.1.3 – Scheme of the alien crosstalk measurement (ANEXT or AFEXT).

One final comment on the configuration of the pairs belonging to the same cable, considered by standards for information technology and data communication applications: in many industrial applications all the lines connected to a given equipment or apparatus are housed in the same cable for a matter of robustness, ease of installation, certification of the application. When the use of each pair is much different, so that a significant disturbance might be coupled because of internal crosstalk, pairs are distributed on opposite sides of the cable and very often separated by internal shields; this not only ensures a low degree of electromagnetic coupling, but eases the safety case when worst-case failures and situation shall be considered, such as possible insulation failure and leakage.

5.1.7 Transfer impedance and transfer admittance

The transfer impedance Z_t is in general the most relevant parameter that identifies the goodness of a shielded cable (e.g. a coaxial cable or a shielded pair) with respect to external disturbance in the form of electromagnetic field hitting the external surface and causing voltage across the inner conductor(s). The parameter links this voltage to the current that is flowing in the cable shield, from this the use of the term "impedance" and the measuring unity of Ω.

For a fraction $\mathrm{d}x$ of the total cable length, the voltage term is

$$dV = Z_t I_{sh} dx \tag{5.1.10}$$

where Z_t is the transfer impedance and I_{sh} is the current flowing in the shield on the internal surface, so originating from the external electromagnetic field and passing through the shield. It is evident that for a perfect shield at a conveniently high frequency the current is confined completely on the external surface because of skin effect, and the corresponding transfer impedance term is zero.

At high frequency it is more convenient to make reference to the transfer admittance Y_t, that puts in relationship an external voltage on the cable shield with a current flowing in the inner signal conductors; in this case the coupling mechanism is capacitive, transversely through the dielectric. In this section, when examining some of the approaches to quantify the transfer impedance of a cable, the transfer admittance will be considered as well, being the coupling mechanisms and the electromagnetic models quite similar and complementary.

In general, by diffusion, even a solid shield will have current leaking to the internal surface given by

$$Z_t = R_{dc} \frac{\theta\, t_{sh}}{\sinh(\theta t_{sh})} \tag{5.1.11}$$

where θ is related to the skin effect and to the skin depth δ by $\theta = (1+j)/\delta$, and t_{sh} is the shield thickness [295, 320]; R_{dc} has the obvious meaning of dc resistance. This is the so-called Schelkunoff basic formulation for a solid shield (tube).

When a braided shield or a non-ideal foil shield are considered with respect to a better solid shield, or double braided, or foiled and braided shield, the Z_t term is larger because of a larger dc resistance, but this would explain only a small increase that might be compensated for by increasing the shield thickness. At higher frequency the larger Z_t values are on the contrary due to non-ideality of wire pattern or of foil characteristics: optical coverage, non uniform distribution and defects, such as holes and cracks, oxidation, etc. For these reasons the transfer impedance curve versus frequency can show many deviations from an initial approximate inductive behavior linear with frequency.

When deviating from an ideal balanced symmetric cable featuring a solid shield, magnetic field coupling to the inner conductor(s) may be represented by an equivalent mutual inductance M, that sums to the resistance, that is the real part of Z_t. The equivalent inductance model is able to predict Z_t behavior satisfactorily for the vast majority of cables in the medium frequency range, that is from the frequency value where the dc resistance cannot explain any longer the increase of coupled voltage to the inner conductor, up to a maximum frequency at which the combined effect of transverse coupling capacitance shall be accounted for, that normally occurs at frequencies above a hundred MHz. The mutual inductance term includes both distributed non-ideality of the shield braid (due to how the shield is wound and small distributed defects, such as oxidation and braid displacement due to friction) and local defects (such as hole in the shield or imperfect termination on connector shell).

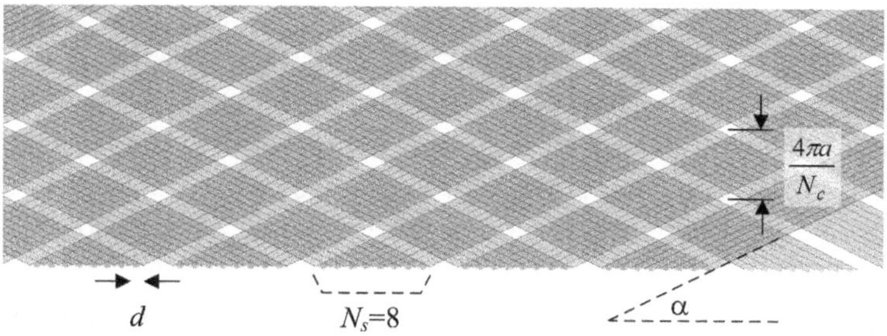

Figure 5.1.4 – Cable braid pattern and most relevant quantities.

We now review the development of such models, suitable for braided shields and solid shields with defects.

5.1.7.1 Vance's model (1975)

Vance [326] accounts for the micro-holes that occur in a braid of diameter D ($a = D/2$ is the shield radius), due to overlapping N_c carriers, each made of N_s single wires (strands) of diameter d, with a pitch angle α, that is referred to cable axis (see Figure 5.1.4). Some factors are derived describing geometrical performance of the cable: the filling factor F and the optical coverage K.

$$F = \frac{N_c N_s d}{4\pi a\,\cos(\alpha)} \tag{5.1.12}$$

$$K = 2F - F^2 \tag{5.1.13}$$

The IEC 60096-1 standard [169], sec. 3.2.2, reports a similar expression for the filling factor (indicated there by K_f):

$$K_f = \frac{N_c N_s d}{2\pi \bar{D}} \left(1 + \frac{\pi^2 \bar{D}^2}{L^2}\right) \tag{5.1.14}$$

where $\bar{D}$ is the average braid diameter as above, but corrected for $2d$, as it was proposed later by Tyni (see sec. 5.1.7.2 below) and L is the braid lay length.

The number of holes in the braid per unit length N_h is

$$N_h = \frac{4\pi a\,\cos(\alpha)\,\sin(\alpha)}{N_s^2 d^2} \tag{5.1.15}$$

Taking into account the amount of copper effectively woven around the ideal cylinder of the shield, for the dc resistance we obtain

$$R_{dc} = \frac{4}{\pi d^2 N_s N_c \sigma\,\cos(\alpha)} \tag{5.1.16}$$

to be used directly in Schelkunoff's formula (5.1.11), after replacing shield thickness t_{sh} with an equivalent quantity $0.67d/\sqrt{\cos\alpha}$. The formula is the one that also appears in Kley's paper on modeling and optimization of single-braid cable shields [212]. There is no 100% certain source of information that ensures the correctness without typos of F and K factors expressions, due to the age of papers, the low-quality scanning and the fact that some publications are not publicly accessible to cross-check expressions.

For the so-formed apertures of rhombic shape distributed along the braided shield, after defining the eccentricity $e=\sqrt{1-(w/l)^2}$ (with w the width along the minor axis and l the length along the major axis) of the ellipse circumscribing the rhombus formed by two carriers with opposite inclination of $+\alpha$ and $-\alpha$, it is possible to calculate an equivalent mutual inductance that describes the amount of electromagnetic field that couples through these apertures. Vance proposes an expression taken from his references (Kaden and Markuvitz)[1] for the inductance and capacitance related to a hole or aperture in the shield:

$$C_{12}=\nu\frac{pC_1C_2}{4\pi^2a^2\varepsilon} \qquad M_{12}=\nu\frac{\mu m}{4\pi^2a^2} \tag{5.1.17}$$

where $a=D/2$ is the shield radius, μ and ε refer to the inner dielectric, ν is the number of apertures per unit length, and p and m are the electrical and magnetic polarizabilities of the aperture. The terms C_1 and C_2 indicate respectively the distributed capacitance of the cable (that is the capacitance between the inner conductor and the shield) and of the shield towards the environment (that is for a line with the shield as inner conductor surrounded by other conducting parts including earth).

Of course, still having to determine p and m, the problem of determining the two coupling capacitance and inductance terms is not solved.

The magnetic polarizability is determined by Vance in the plane tangent to the shield for the two ellipse axes:

$$m_l=\frac{\pi l^3}{24}\frac{e^2}{K(e)-E(e)} \qquad m_w=\frac{\pi l^3}{24}\frac{(1-e^2)\,e^2}{E(e)-(1-e^2)\,K(e)} \tag{5.1.18}$$

with $K(e)$ and $E(e)$ indicating the complete elliptic integrals of the first and second kind. On this Kley [212] comments that since Vance's formulas are approximated, there is no need of being more accurate than 10% in determining the elliptic integrals, that may be approximated by a $\cos(\cdot)$ function.

The electric polarizability is reported by Vance with reference to [247] as

$$p=\frac{\pi l^3}{24}\frac{1-e^2}{E(e)} \tag{5.1.19}$$

[1] From Vance's references:
H. Kaden, *Wirbelströme und Schirmung in der Nachrichtentechnik*, Springer-Verlag, Berlin, 1959;
N. Marcuvitz, *Waveguide Handbook*, MIT Radiation Laboratory Series, vol. 10, McGraw-Hill, New York, 1951 [231].

The calculated expressions for the resulting mutual inductance and mutual capacitance terms M_{12} and C_{12} take into account if the pitch angle with which the wires are woven is smaller or larger than 45°.

$$M_{12} = \begin{cases} \frac{\pi\mu_0}{6N_c}(1-K)^{1.5}\frac{e^2}{E(e)-(1-e^2)K(e)} & \text{for } \alpha < 45^\circ \\ \\ \frac{\pi\mu_0}{6N_c}(1-K)^{1.5}\frac{e^2/\sqrt{1-e^2}}{K(e)-E(e)} & \text{for } \alpha > 45^\circ \end{cases} \tag{5.1.20}$$

$$C_{12} = \begin{cases} \frac{\pi C_1 C_2}{6\varepsilon N_c}(1-K)^{1.5}\frac{1}{E(e)} & \text{for } \alpha < 45^\circ \\ \\ \frac{\pi C_1 C_2}{6\varepsilon N_c}(1-K)^{1.5}\frac{\sqrt{1-e^2}}{E(e)} & \text{for } \alpha > 45^\circ \end{cases} \tag{5.1.21}$$

where K is the optical coverage (the notation of the original Vance's paper is kept, even if this term is much similar to the one used for the elliptic integrals); K is defined above in eq. (5.1.13). The terms C_1 and C_2 are the same distributed capacitance terms defined above.

Vance offers an alternative expression for e that depends on the inclination angle α:

$$e = \begin{cases} \sqrt{1-\tan^2\alpha} & \text{for } \alpha < 45^\circ \\ \\ \sqrt{1-\cot^2\alpha} & \text{for } \alpha > 45^\circ \end{cases} \tag{5.1.22}$$

M_{12} adds to the Schelkunoff's expression (5.1.11) and gives the usual expression of transfer impedance

$$Z_t = R_{dc} + j\omega M_{12} \tag{5.1.23}$$

and the mutual capacitance term determines the transfer admittance

$$Y_t = j\omega C_{12} \tag{5.1.24}$$

5.1.7.2 Tyni's porpoising term for Vance's expression (1976)

The model was improved by noting that the current that flows along the shield does this following the weaving pattern and so in a helical shape following the wire strand, when it goes up and down the crossing strands, thus staying alternatively on the inside and outside of the shield; this mechanism is said "porpoising". This of course worsens the separation of the two surfaces (the inner and the outer) of the shield and increases the transfer impedance; this behavior was modeled by Tyni [323], who added a third term to the transfer impedance expression (5.1.23).

$$Z_t = R_{dc} + j\omega M_h \pm j\omega M_b \tag{5.1.25}$$

where M_h is the *hole inductance* already considered by Vance and M_b is the *braid inductance* given by the porpoising effect. The $\pm$ sign indicates the in-phase and out-of-phase terms contributed by the various inductive loops; as noted by Sali [294] for braid angles $\alpha < 45°$ the braid inductance opposes the hole inductance. The expressions reported for Tyni's inductance terms are:

$$M_{b,\mathrm{Tyni}} = \frac{-\mu_0 h}{4\pi D}\left(1 - \tan^2(\alpha)\right) \tag{5.1.26}$$

$$M_{h,\mathrm{Tyni}} = \frac{2\mu_0 N}{\pi \cos(\alpha)}\left(\frac{w}{\pi D}\right)^2 \exp\left(-\frac{\pi d}{w} - 2\right) \tag{5.1.27}$$

$$h = \frac{2d}{1 + w/d} \tag{5.1.28}$$

The term h indicates the effective distance between the two layers of braid (indicated in [93] as "radial spindle separation"), D is the mean diameter of the braid, d is the single wire diameter and w is the mean distance between two carriers (strands of wires), indicated in Sali's paper by the letter b. D is calculated starting from the diameter over dielectric D_0 (that is the internal diameter of the braid) and the braid wire diameter d, as $D = D_0 + 2d$. This diameter was later corrected by Katakis by adding h. A relationship for α may also be derived:

$$\tan\alpha = \frac{\pi(D + h)}{l} \tag{5.1.29}$$

that allows to determine the lay length l once the braid angle α is known, or better viceversa to determine accurately α by measuring the lay length.

It is briefly noted that eq. (5.1.27) was wrongly reported in Sali's publication [294], eq. (A.4), without the exponent "2" and including the term "α" in the exponential function, that in another Sali's publication [2] is absent. Cudd and Benson [93], on the contrary, are a much more accurate source of information.

5.1.7.3 Sali's correction to Tyni's formulation (1991)

Sali [294] underlines the completeness of Tyni's model and indicates it as a useful tool for cable braid optimization by balancing accurately the values of cable geometrical parameters, by compensating braid and hole inductance terms, resulting in a lower transfer impedance. However, Sali criticizes Tyni's model for the estimate of the flux linkage between the inner and outer layers (that we referred to above as the porpoising effect), that shows some discrepancy for cables with optimized braid, and proposes

[2] S. Sali, "Error Analysis of Crosstalk between Communication Cables," WSEAS International Conference, Izmir, Turkey, September 13-16, 2004, ISBN 960-8457-02-5.

a better estimate of the flux linkage area and of the parameter h, that otherwise following the previous expression revealed to overestimate the inductance.

Sali's notation in the following expressions is commented to match that of previous expressions: the inclination angle θ is α, the outer braid diameter D_m is D. The new porpoising inductance by Sali is

$$M_{b,Sali} = -\frac{\mu_0}{2\pi D}\frac{1-\tan^2(\theta/2)-\tan(\theta)}{\cos(\theta)}(A_s\sin(\theta)+A_n)\frac{N_h}{N} \tag{5.1.30}$$

where the various new parameters are related to the geometrical parameters as:

$$A_n = \frac{d}{8}(4-\pi) \qquad A_s = (Z_x - Q_x)(Q_y - d) + k_v(Q_x - Z_x) \tag{5.1.31}$$

$$v = \left(\frac{2\pi D}{N^2 d^2}\right) K_f \sin(2\theta) \qquad Q_y = h - A_y \tag{5.1.32}$$

$$k_v = A_y\left[1-\sqrt{\frac{R}{L_x}}\right] \qquad L_x = \frac{b}{2} + nd - \frac{d}{\sin(2\theta)} \tag{5.1.33}$$

$$R = b - Z_x + \frac{d}{\sin\theta} \qquad Q_x = R - \frac{b}{2} - Z_x \tag{5.1.34}$$

$$Z_x = L_x + A_x \tag{5.1.35}$$

$$A_y = d\sin\theta + b \qquad A_x = \frac{d}{2\sin\theta} + \frac{b}{2} \tag{5.1.36}$$

Sali does not introduce any expression for the determination of the hole inductance M_h, and rather confirms explicitly the use of the Tyni's expression.

The outer diameter of the braid D is assigned the value $D = D_0 + 2d$, with D_0 the cable diameter up to the outer surface of the dielectric cylinder and d as above the wire diameter (sometimes the practical correction is adopted, $D = D_0 + 2.25d$). Sali (or Katakis, as reported by Cudd and Benson [93][3]) proposes a more correct $D = D_0 + 2d + h$ with the value of h, that indicates the mean distance between layers, determined by solving the two equations

$$\tan(\theta) = \frac{D+2d+h}{l} \tag{5.1.37}$$

$$\frac{2\pi}{N}h^2\csc(\theta) = h\left((1-N_s)d + \frac{2l}{N}\sin(\theta)\right) - 2d^2 = 0 \tag{5.1.38}$$

[3] Both Sali and Katakis were Ph.D. students of the University of Sheffield.

where l is the lay factor (and not the major axis of the ellipse connected to the rhombic holes of the braid).

5.1.7.4 Kley's matching of experimental results (1992-1993)

Kley[4] takes another direction, noting on the one hand approximations and lack of complete modeling of some phenomena and fitting on the other hand experimental data, determining in this way corrective values for coefficients and exponents. In [212], sec. IV.A, regarding the hole inductance M_h (theoretically determined for a tube with negligible thickness) and some attenuation caused by the non-zero thickness (called "chimney effect"), the author himself admits that "All attempts of the author to analytically determine the chimney effect were defeated, because they were too far away from the reality of the measurements."

Solid shield Kley starts from a solid shield of thickness t with circular holes of radius r; he indicates the hole inductance with the symbol M_{LL} and the inductance related to eddy current in the hole walls, that he calls skin inductance, as L_{SL}. Regarding eddy current effect, he underlines that normally the overall impedance related to such effect is made of a real and an imaginary part of approximately the same value: he thus quantifies losses due to the eddy current phenomenon, and related to the resistive part, as a way to determine the inductance, for which no formulation is available. Kley proposes an expression for resistance and then equates it to inductive reactance:

$$\omega L_{SL} \simeq R_L = \frac{r}{4\pi\sigma\delta r_0^2} \qquad (5.1.39)$$

where the dependency on frequency other than the one of the reactance is hidden in the skin depth δ. The underlying assumption is that the skin depths is much smaller than the hole radius $\delta \ll r_h$. Of course, being the magnetic field over the hole aperture neither longitudinal nor uniform, as assumed in the model, there will be the need to complement the expression, by including the shield thickness and weighting it with respect to the hole size.

The overall transfer impedance of the solid tube with circular holes along its length is

$$Z_t = Z_R + \frac{k}{L} j\omega M_{LL} + \frac{k}{L}(1+j)\omega L_{SL} \qquad (5.1.40)$$

where the term $(1+j)$ accounts for the said resistive-inductive behavior of the hole.

This expression was subject to experimental confirmation based on several different samples and curve fitting. Kley was able to determine corrections to the used expression for M_{LL}: the average radius shall take into account the wall thickness and thus shall be increased by half of it, and this was already pointed out by Sali above; a multiplicative reduction factor of 0.875 was introduced, to take into account the

[4] Kley's work on cable shielding and transfer impedance began during his Ph.D., whose dissertation was published in 1991: T. Kley, "Optimierte Kabelschirme - Theorie und Messung", Diss. ETH No. 9354, Zurich, Switzerland, 1991.

curvature of the shield; the exponential approximation of the more correct elliptic integrals needs a coefficient 1.84 to match the experimental results. Similarly, for L_{SL} he proposes the same corrections to account for shield thickness and for the exponent that is brought to 2.3. The resulting error with respect to the experimental data is larger for L_{SL} term (about $\pm 30\%$) than for M_{LL} (about $\pm 15\%$), simply because the fitting is operated on the overall Z_t and the term M_{LL} is much larger and has thus a correspondingly larger weight in the curve fitting process.

The two expressions proposed by Kley are much more practical than the previous ones, so that they are nowadays used often by numeric simulators:

$$M_{LL} = \mu_0 \frac{0.875 r_h^3}{3\pi^2 (r_0 + d/2)^2} e^{-1.84 d/r_h} \tag{5.1.41}$$

$$L_{SL} = \frac{d\, r_h}{2\sigma\delta (r_0 + d/2)^2} e^{-2.3 d/r_h} \tag{5.1.42}$$

Analogously, for the transfer admittance he defines a coupling capacitance taken from Kaden

$$C_t = -K_Y C_1 C_2 \tag{5.1.43}$$

where K_Y is the coupling factor, C_1 is the p.u.l. capacitance of the cable (i.e. between the inner conductors and the shield) and C_2 is the p.u.l. capacitance between the shield and the external environment. This formula was declared valid only for $\varepsilon_{r1} = \varepsilon_{r2} = 1$ by Kaden, that is for air dielectric inside the tube and outside it. In reality, by calculations reported in his Ph.D. dissertation Kley extends the validity to other dielectrics; by citing Halme, the "chimney effect" is taken into account by the already mentioned 0.875 factor. Since the used dielectrics are all with very low losses, the real part of the transfer admittance is absolutely negligible.

By curve fitting of experimental data taken for several geometries, Kley proposes a formula for the coupling capacitance of a hole in a perfect circular tube

$$C_t = -C_{10} C_{20} \frac{\varepsilon_{r1}\varepsilon_{r2}}{\sqrt{\varepsilon_{r,i} + \varepsilon_{r,o}}} \frac{0.875\, r_h^3}{6\pi^2 \varepsilon_0 (r_0 + d/2)^2} e^{-2.3 d/r_h} \tag{5.1.44}$$

The coupling capacitance involves quantities that describe elements outside the cable (e.g. the dielectric properties of the external environment), while the hole inductance expression uses only quantities related to the shield, provided that the proximity effect with other external conductors can be ignored and that the current distribution in the shield is symmetric.

Braided shield Kley considers then a braided shield, using the modeling approaches reviewed above, but aiming at simplifying them and then introducing corrective factors by matching experimental results. The parameters that characterize the braid are inherited from the past models (but with different notation) and some new assignments are made: the average braid diameter already given by Sali becomes $D = D_0 + 2.5d$

(with D_0 the diameter of the cable including the dielectric, but excluding the braid and with d the diameter of a single braid wire); the filling factor is defined by taking the total area of the braid wires depending on their inclination and comparing it with the area of the spanned by the shield, so the minimal filling factor is $G_0 = N_c N_s d / 2\pi D$ and the filling factor is $G = G_0 / \cos\alpha$ (Kley uses m and n to indicate N_c and N_s, respectively); finally, the optical coverage is defined from the filling factor as $B = G(2-G)$.

Using Tyni's Z_t eq. (5.1.25), Kley separates differently the three terms:

$$Z_t = Z_R + j\omega L_T + (1+j)\omega L_S \tag{5.1.45}$$

where Z_R is the basic Schelkunoff's expression for the solid tube accounting for the whole transfer impedance, not only for the resistive part, that is however preponderant; the two terms L_T and L_S are the coupling and skin inductance, respectively, using the same notation of Kley's paper. These two terms are split into constituents, by making reference to past formulations and adjusting them to experimental evidence.

The coupling inductance L_T is formed by the hole inductance M_L and by the braid inductance M_G.

The hole inductance M_L is determined based on Vance's work by approximating the elliptic integrals with $(2-\cos\alpha)$ and introducing the already seen corrective factors coming from experimental observations.

$$M_L = \mu_0 \frac{0.875(2-\cos\alpha)}{6m} (1-G)^3 e^{-\tau_H} \tag{5.1.46}$$

where $\tau_H = 9.6G\sqrt[3]{B^2 d/D}$ is the relaxation exponent for the chimney effect.

The braid inductance M_G is related solely to the geometry of the braid and is due to the mutual effect of the two parts of the braid with angle α and $-\alpha$ around each hole:

$$M_G = -\mu_0 \cos 2\alpha \tag{5.1.47}$$

Finally, the skin inductance L_S is also decomposed into two terms, one due to the effect of the field penetrating the shield and causing eddy current effects on the hole walls ("chimney effect") and one due again to the mutual effect of the two halves of conductors around the holes created in the braid:

$$\omega L_S = \frac{1}{\pi\sigma\delta}\left[\frac{1}{D_L} + \frac{1}{D_G}\right] \tag{5.1.48}$$

where the two diameters are fictitious and used only to obtain the correct output from the model. They are determined with some reasoning on the expected behavior and with respect to known formulations, to obtain an approximate expression:

$$\frac{1}{D_L} \sim (1-G)\, e^{-\tau_E} \cos\alpha \qquad \frac{1}{D_G} \sim -\cos 2\alpha \tag{5.1.49}$$

The analysis of the experimental results, by taking the difference between the measured coupling inductance L_T and the calculated hole inductance M_L, brings to a more complex formulation of braid inductance M_G. Dependency on G_0 arises and the more complete expression is

$$M_G = -\mu_0 \frac{0.11d}{2\pi D G_0} \cos(2k_1\alpha) \tag{5.1.50}$$

where $k_1 = \frac{\pi}{4}\left[\frac{2}{3}G_0 + \frac{\pi}{10}\right]^{-1}$.

On the contrary the braid inductance does not depend on the contact resistance between individual wires and surface conditions (e.g. oxidized). This means that two fictitious diameters can be separately defined:

$$\frac{1}{D_L} \sim 10\pi G_0^2 \frac{\cos\alpha}{D_m}(1-G)\,e^{-\tau_E} \qquad \frac{1}{D_G} \sim -\frac{3.3}{2\pi D_m G_0}\cos(2k_2\alpha) \tag{5.1.51}$$

where $k_2 = \frac{\pi}{4}\left[\frac{2}{3}G_0 + \frac{3}{8}\right]^2$, $\tau_E = 1.25\tau_H = 12G\sqrt[3]{B^2 d/D}$, D and G_0 were defined at the beginning.

Kley presents thus a set of expressions adjusted to match measurements and theory for the expression of the transfer impedance of braided shield cables.

$$Z_t = Z_R + j\omega L_T + (1+j)\omega L_S \tag{5.1.52}$$

where

$$Z_R = \frac{1}{\sigma G_0 \cos\alpha}\,\frac{2}{\pi^2 D d}\,\frac{d_R(1+j)/\delta}{\sinh\left(d_R(1+j)/\delta\right)} \tag{5.1.53}$$

$$L_T = \frac{\mu_0}{m}\left[0.875\frac{\pi}{6}\left(2-\cos\alpha(1-G)^3 e^{-\tau_H}\right) - \frac{0.11}{n}\cos(2k_1\alpha)\right] \tag{5.1.54}$$

$$\omega L_S = \frac{1}{\pi\sigma\delta}\frac{1}{D}\left[10\pi G_0^2\cos\alpha(1-G)e^{-\tau_E} - \frac{3.3}{2\pi G_0}\cos(2k_2\alpha)\right] \tag{5.1.55}$$

Analogously, in order to determine the transfer admittance in a very similar way, the coupling capacitance is determined by matching the experimental results and thus introducing corrective factors: K_Y has the usual 0.875 corrective factor for the shield curvature, the elliptic integrals used by Vance are approximated with $\cos\alpha$ and the exponent τ_E for the chimney effect is corrected to be primarily dependent on G.

$$C_t = -C_{10}C_{20}\frac{\varepsilon_{r1}\varepsilon_{r2}}{\sqrt{\varepsilon_{r,i}+\varepsilon_{r,o}}}K_Y \tag{5.1.56}$$

with

$$K_Y = 0.875\frac{\pi\cos\alpha}{6\varepsilon_0 N_c}(1-G)^3 e^{-\tau_E} \tag{5.1.57}$$

and τ_E as defined above. Kley underlines that such formulation for transfer admittance is in reality valid for a rotational symmetric three-conductor system; in general it is valid if the cable is approximately coaxial, with its diameter small compared to the distance between the shield and other environment elements that determine the capacitance term C_2.

5.1.7.5 Conclusions

The behavior of the so-defined Z_t versus frequency is what interests us more, in order to explain the many curves that are seen in datasheets and literature and to critically assess cable performance. We have already said that at low frequency Z_t is real and its magnitude is R_{dc}, with a slight increase for a sufficiently large frequency due to skin effect. Then, when inductive reactive terms prevail, Z_t begins to increase with a slope of 20 dB/decade, as implied by the $j\omega$ multiplicative factor, provided that the mutual inductance term is constant.

Two peculiar behaviors are worth of consideration: first, the curve shows a drop in magnitude before the inductive part begins, and, second, Tyni noticed first that for some very good cable with large optical coverage the inductive part of Z_t increases less than linearly with frequency and precisely as $\sqrt{f}$.

Demoulin [103] proposed a further modification of the porpoising inductance, so that it accounts for this behavior:

$$M_p' = M_p \left(1 + \frac{b}{\sqrt{j\omega}}\right) \tag{5.1.58}$$

with b a parameter that depends on cable characteristics, but was not better specified.

The phase of Z_t detaches from 0 at dc and then goes towards 90° when the simpler Vance's model is considered, whereas it may oscillate between -90° and +90° for Tyni's model that includes the porpoising inductance (5.1.26). The modified porpoising inductance (5.1.58) is seen in experimental data as a change of magnitude, rather than a specific change of phase.

Considering practical factors and deviations from nominal parameters (on which Z_t predictions are calculated), Cudd, Benson and Sitch [93] report: variations of braid geometry due to "pull down" occurring to the braid wires during cable manufacture; spindles occasionally may miss a wire and braid coverage factor as well as the number of wires deviate from nominal.

Moreover, the approach proposed in their paper [93] is that of calculating minimum and maximum Z_t for a given model (e.g. Tyni model, including Sali's correction, indicated as Katakis's correction), when geometrical factors are varied around nominal values for their declared tolerance; the largest part of measured Z_t values is below the calculated maximum value, except for a few singular, and probably critical, cases.

For miniature a sub-miniature cables (e.g. RG 174) variations of Z_t are non-linear with frequency; observing that all cables had silver-plated braid and citing some supporting

IEC works [93], the proposed conclusion is that a significant current crosses braid-wire contacts thanks to the very low contact resistance.

5.1.8 Transverse and longitudinal conversion loss

The unbalance of a balanced line (a shielded pair) may be physically explained by mainly considering the difference in the self capacitance terms of each pair conductor towards the shield. From an operative viewpoint the resulting effect is the conversion of common-mode signals into differential-mode ones, so that the cable and the connected equipment is more exposed to common-mode disturbance, that in principle would be of no harm. The quantification of the overall cable performance with respect to common-to-differential mode transformation (and viceversa) is addressed by defining the Transfer Conversion Loss (TCL) and Longitudinal Conversion Loss (LCL) on one hand, and Transfer Conversion Transmission Loss (TCTL) and Longitudinal Conversion Transmission Loss (LCTL) on the other hand. The pairs of indexes refer to the near-end and far-end transformation, respectively; the two terms of each pair, in reality, are reciprocal, so only one index per pair is relevant, e.g. TCL and TCTL. Definitions in the IEC 61935-1 [171] are not very clear; a new version of the standard is about to be published. From the definition based on common-to-differential mode transformation and from the distinction of near and far ends, cable performance indexes may be put in relationship with mixed-mode S parameters (see sec. 10.4 and 10.4.3) as follows: $\mathrm{LCL} = S_{dc11}$ (or S_{dc22}), $\mathrm{TCL} = S_{cd11}$ (or S_{cd22}), $\mathrm{LCTL} = S_{dc12}$ (or S_{dc21}), $\mathrm{TCTL} = S_{cd12}$ (or S_{cd21}).

Already 15 years ago the problem of correct quantification of LCL (or TCL) was addressed as far as measurement setup and correct separation of components by means of an LCL (or TCL) probe are concerned[5].

Limits for these quantities have been under review both at IEC (new release of IEC 61935-1) and at IEEE (e.g. the 802.3 working groups). For evaluation and comparison of these measured characteristics, attention has been given to accuracy, or uncertainty, in particular because in many cases measurements may be performed on site, under non ideal conditions and possibly with less performing instruments. The expression for measurement error based on eq. (52) of IEC 61935-1 and written using the notation that is detailed in sec. 11.5 is

$$\delta \mathrm{IL} = E_{TF} \frac{1 - E_{SF} E_{LF}}{(1 - S_{dd11} E_{SF})(1 - S_{dd22} E_{LF}) - S_{dd11} S_{dd22} E_{SF} E_{LF}} \qquad (5.1.59)$$

and is based on a complete 12-term error model for the VNA (or field tester), as analyzed in a recent work [6].

---

[5] I.P. MacFarlane, "A probe for the measurement of electrical unbalance of networks and devices," *IEEE Transactions on Electromagnetic Compatibility*, vol. 41, no. 1, pp. 3-14, Feb. 1999.
D. Schwarzbeck, "Calibration of LCL-Probes," Schwarzbeck Mess-Elektronik, App. Note, [online] www.schwarzbeck.com/appnotes/LCLCAL1.PDF (last accessed Aug. 2015).

[6] Y.S. Meng, A.C. Patel, Y. Shan and H.N. Pandya, "Error Model for Insertion Loss Measurement in Testing of Balanced Twisted-Pair Cabling: A Complete Expression and Its Simplification," XXXIth URSI General Assembly and Scientific Symposium, Aug. 16-23, 2014, Beijing, China. doi: 10.1109/URSI-GASS.2014.6928988

5.1.9 Screening factor

The screening (or shielding, or reduction) factor defines the shielding properties of the cable in an equivalent way, aiming at quantifying the amount of voltage appearing on the internal conductors for a given longitudinal voltage E applied externally to the cable shield. The external longitudinal voltage E is also accompanied of course by a flowing shield current, that is the one supporting the definition of transfer impedance (see sec. 5.1.7). The use of the screening factor concept is quite diffused for low-frequency applications, especially when the supply voltage might be applied to the cable shield during worst-case or fault situations, posing the problem of electrical safety of equipment connected to the inner signal conductors and possibly people safety. Suitable for large systems and applications extended in space, the use of the screening factor is extended also to the quantification of interference for low frequency signals, usually up to some kHz or tens of kHz.

The screening factor is simply defined as the ratio of the external and internal voltage. The flowing of screen current is the cause of induction to the inner conductors and depends heavily on the amount of current penetrating the external armor thickness appearing at the inner armor surface, as well as the use of additional underlying conductive shields. The penetration is ruled by the skin effect and the skin depth (see sec. 1.1.2 and eq. (1.1.3)), more pronounced for ferromagnetic materials (the magnetic permeability is at the denominator, inside the square root); ferromagnetic materials, such as many types of steel, are used in reality for their mechanical robustness and the reduced penetration is added value: in reality, being these materials subject to changes of their permeability and even to saturation, they need a thorough characterization at different levels of the flowing current, i.e. of applied external longitudinal voltage.

When measuring the screening factor, the experimental setup shall be designed in order to address the following issues: i) the induction on the measuring wires, not belonging to the cable sample under test, shall be minimized, not to get wrong values, only partly given by the voltage effectively appearing on inner conductors; ii) the longitudinal voltage and the screen flowing current shall be considered at once, thus sizing correspondingly the feeding system (because of the need of testing the nonlinearity of the cable armor if made of ferrous materials, longitudinal voltages up to 1 kV per km may be necessary). Especially when the cable sample is short ($1.5\,\mathrm{m}$, as per VDE standard 0472 [331], see Figure 5.1.5), the resistance of bonding connections and joints shall be quite low, so that usually large cross sections and accurately prepared mating surfaces are used, not to impact remarkably on the resulting accuracy: $25\,\mathrm{mm}^2$ to $50\,\mathrm{mm}^2$ cross section and the use of continuous soldered joints are indicated.

To address the first point, the VDE standard (and other regional standards, especially for railway applications) indicates the need to have the two wires connecting the far end of the cable sample following the same path: one wire, identified as current line C, carries the test current that recloses on the cable screen when the longitudinal voltage is applied; the other wire, called measurement signal line S, brings the far end voltage measurement to the voltmeter located at the near end. The suggested path is rectangular at a distance of $400\,\mathrm{mm}$, so to simulate the expected inductance of the earth loop, that is around $2\,\mu\mathrm{H/m}$. The maximum coupling between C and S lines is obtained by either using a multi-wire cable, with one inner wire assigned to S and the

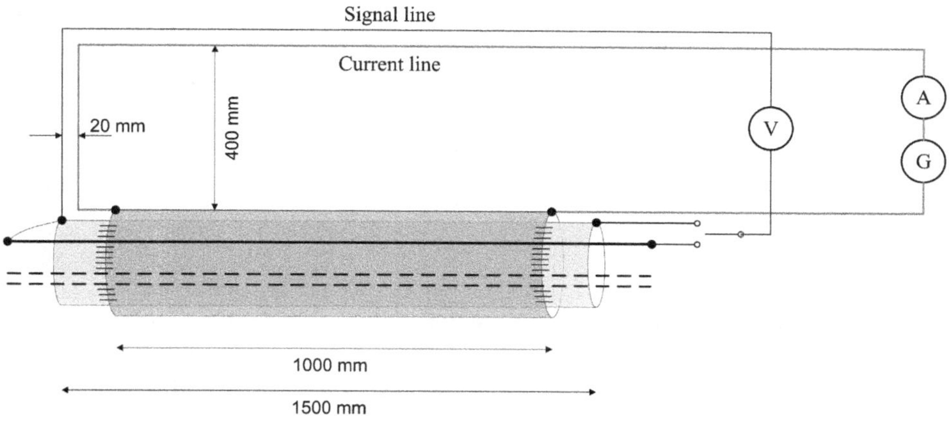

Figure 5.1.5 – Test setup for measuring the screening factor (VDE 0472 [331]): "G" is the current generator, "A" is the amperometer for current measurement and "V" is the voltmeter for voltage measurement.

other to C, or for very large currents using two flat bars for C sandwiched around the signal wire S. The VDE standard indicates $20\,\mathrm{mm}$ separation, that is usually however intended as maximum separation.

The current line C is connected to the cable armor and screen by soldering as continuously as possible on their surface a copper conductor (or strap) with suggested cross section of $25\,\mathrm{mm}^2$. The inner conductor subject to test is connected at the near end side to the signal line S and bonded to the cable armor/screen at the other end. Thus, the measured induced voltage on the inner conductors corresponds to a longitudinal voltage, having used the far end of the cable sample as the potential reference for both the armor (for the external inducing longitudinal voltage) and for inner conductor (for the induced longitudinal voltage).

The screening factor, also named "reduction factor", is given by the ratio of the two voltages measured with the switch connected to the inner victim conductor and to the outer shield; the former represents the induction and the latter is the longitudinal voltage drop across the external shielding.

5.1.10 Shielding effectiveness or attenuation

This is not a uniquely defined property nor an intrinsic cable parameter, so that its value depends on the exact definition adopted and on the test setup. Normally it is assigned the meaning of either the ratio of the radiated power (captured by some means, as with the outer cylinder of the triaxial method) and the power fed into the cable, or the ratio of the inner shield current and the outer shield current, where the outer portion of shield current is responsible for radiation to the outside, as well as it is true the opposite, that is the inner portion is responsible for noise pickup from external disturbance.

However, measured shielding effectiveness normally shows a good correlation with values derived from transfer impedance measurements: the reason is that transfer impedance measurements are normally done with low impedance values and the driving line is short-circuited with the shield under measurement, so that the measurement result minimizes the contribution of the capacitive transfer admittance; at low frequency this is correct and the shielding effectiveness is then influenced by the inductive part of the transfer impedance.

5.2 Cable materials

Materials and techniques applied to cables are reviewed to establish a common basis of knowledge and terminology. The aim is not only to give quantitative data for the electrical characteristics of materials, but also to give a quick synopsis of commercial types, common choices and other definitions.

When talking of material properties, a great deal of approximation is to be accepted and taken in due consideration. If we consider for example relative permittivity ε_r and dielectric losses (e.g. $\tan\delta$) of insulating dielectric, the same material may have different values for many reasons: the values are objectively different because measured at a different temperature or frequency, but also they are subject to the expected variability of measured quantities due to uncertainty, repeatability and reproducibility, and to the adoption of different measurement methods. Moreover, the reported values are often scarcely defined in terms of measurement conditions, so that their applicability is even more vague; we must remember that in many cases the sources of information are leaflets and tiny datasheets, presentations and web pages, ... whatever our expectations and opinions for the latter. The same may be repeated for PCB materials that are reviewed in Chapter 6, even if for PCB a set of standards establishes quite well the measurement methods for laminates and the descriptive values of dielectric materials are often used to accurately design planar transmission lines and other devices.

In general cable production is standardized for several low-level details that are relevant to cable performance: for coaxial cables the EN 50290-2-1 [66] at sec. 6 reports material constants, standardized wires for both inner conductor (either stranded or solid) and braided shield. The construction of the latter is well considered and construction constants are given for the stranding factor and the braid angle, crossed with other performance factors (such as voltage gradient) and overall weight and size, filling factor and outer and mean diameter estimators, to cite a few. The overall size and tolerance for coax cables with nominal characteristic impedance of $50\,\Omega$, $75\,\Omega$ and $93\,\Omega$ are specified in Table 13 of the standard for various different dielectric materials, such as solid polyethylene, cellular polyethylene, polytetrafluoroethylene and cellular fluorinated ethylenepropylene. Also the external sheath has received attention in Table 19, where PVC is included (excluded from the list of dielectric materials).

Current carrying capability is standardized for both coaxial cables and balanced pairs.

5.2.1 Inner insulator (dielectric)

There is potentially a wide variety of dielectric materials: some are preferred for mechanical properties, during cable production or its use over time, some for stability, small losses, etc. For signal cables a few selected dielectrics are used for the largest part of applications: polyethylene, PTFE or Teflon, solid and foamed, and a few others. For some high-voltage applications semi-conductive (sometimes even called "semiconductor") coatings are used sometimes to reduce field gradient and accumulated charge; such coatings are obtained e.g. with black-carbon fibers or metallic particles and the properties of the compound may be anisotropic and non-linear. In large power RF applications, where the potential difference and the electric field may be quite high, many coaxial cables have a hollow cylindrical inner conductor, that achieve a combination of advantages: it is lighter than a solid conductor and carries the same amount of RF current, that distributes around the periphery because of skin effect; it keeps the electric field lower in proximity of its surface thanks to the larger curvature radius, avoiding thus dielectric breakdown.

The most relevant properties of dielectric materials were reviewed at the beginning in sec. 1.1.2. When considering cables as elements of real-world applications, not only the variability with frequency (in terms of reduction of the relative permittivity and increase of loss tangent), but also that with temperature is of concern: attenuation of a long cable run may be different in different times of the day or seasons, using a cable in a long tunnel with a significant longitudinal temperature gradient may result in apparent differences between otherwise identical cable samples.

Anisotropy might be relevant also for cables as it is for planar structures, such as PCBs (see sec. 6.1.2.1 and in general sec. 6.3): coaxial cables have a circular symmetry of the electric field aligned along the radial direction; pairs and multi-pair cables have the electric field in even-mode and odd-mode aligned in quite different directions.

5.2.2 Shield

The function of the shield (or screen) of a cable is to reduce the coupling of external disturbance and to improve the confinement of the electromagnetic field, thus improving the coupling between conductors (or to the shield itself) and stabilizing the self-capacitance terms.

Conductive shields exist of the braided type (for flexibility), in foil (usually aluminum, applied with a thin insulating layer), or solid. Combinations are possible of the first two, in the so-called double braided or shield-and-foil cables. When using foiled shield, the equipotential connection along the cable for bonding and earthing is ensured by a thin wire connected in parallel to the shield (the equipotentializing wire increases also the shield cross section reducing the longitudinal dc resistance, see sec. 5.1.1).

In general braided shields are preferable to foiled shields, but the material and the adopted process to ensure good electric contact between wires is of paramount importance: oxidation may occur, worse in case of penetration of moisture, and when chemical substances are released by the external cable jacket, during the last phases of its polymerization. When oxide covers the surface of shield wires, at crossings the electric contact is in general not guaranteed any more: for tinned copper the tin oxide is easily removed by friction during cable movements, but silvered copper is a more valid solution since silver oxide is conductive (it was already commented in sec. 5.1.7.5 above); the use of bare copper is absolutely to be avoided (some old coaxial cables, such as RG8, have copper shield and it is easy to see the first cm at the extremities of a cable sample, compromised by air and moisture leaking along the braid). When oxidation occurs, not only the optical coverage worsens, but "microphone noise" comes into play: the cable movement shifts oxidized and clean portions of the wire surface, changing the potential barrier and the contact resistance during the movement. A reduction of the optical coverage is relevant both for emissions and immunity, but only at high frequency (when speaking of transfer impedance in sec. 5.1.7, the frequency range at which this property of the cable gets significant is in the hundreds of MHz). The porpoising terms are also affected by oxidized braid wires, that modify the flow of current across the braid.

For very high-frequency applications in the GHz range, it is quite common to use solid-shield cables (called rigid or semi-rigid cables), especially for long cable runs: when handling large power levels, the reason is a better confinement of the electromagnetic field and robustness for cable bending; for precision application, cable geometry is more constant, as are characteristic impedance and transmission properties.

The superiority of solid shield with respect to other shields may immediately be seen when observing that the definition of transfer impedance starts from the calculable transfer impedance of a solid tube, to which defects in form of holes are added, obtaining higher transfer impedance values and thus poorer screening (see sec. 5.1.7).

5.2.3 External sheath and armour

The external sheath has a protection function with respect to mechanical action, chemical agents, environmental factors. Rubber, PVC and Polyethylene are the most common, but other materials may be used, depending on the application and desired characteristics: UV resistant, temperature class, flame retardant, halogen free, etc.

The external sheath does not influence the electrical characteristics of the cable, that are bounded to the region within the metallic shield, unless the cable as a whole is part of a larger electric system (e.g. when the cable is tested by the triaxial method (see sec. 5.6.1) or the line injection method (see sec. 5.6.2)). In these cases the dielectric permittivity and the thickness of the sheath shall be known with some accuracy, because they have a direct influence on the matching of the external transmission line. It is however wise to allow for some discrepancy of its electrical properties, or adapt the method to cope with some amount of mismatch, and thus reflections. When crosstalk to nearby lines is of concern, given the usually adopted approximations and the air surrounding the cable, the influence of the cable jacket is negligible.

The consistency of jacket and its mechanical properties (e.g. rigidity, smoothness or roughness and grip, are quite relevant when "connectorizing" the cable and to ensure that it keeps in place inside the connector, transmitting the least stress onto the latter when subject to stress, flexure, vibration. It is not by chance that there are high-quality cables for stable and repeatable microwave measurements that are extremely flexible and with a rough external coating, to better mate firmly in place when headed with suitable connectors (this is again reviewed in sec. 5.7.1.1 at the end of this chapter).

In the early days of cable production, not wisely selected plasticizers were responsible of various kinds of contamination: diffusing through the braid into the dielectric, they altered dielectric constant and dielectric losses, increasing the overall power loss and the temperature of the cable; for some shield materials corrosion and oxidation was also a concern, worsening remarkably shielding characteristics and transfer impedance, and giving rise to microphonic noise (see previous section 5.2.2).

The set of IEC/CENELEC standards EN 50290-2-XX, for example, report the requirements for various cable sheath materials.

When besides an insulating protecting sheath the cable has metallic armour, for low and medium frequency the latter may improve shielding performance. Coaxial cables are almost never armoured and armoured solutions are for heavy use and installation in harsh environments, where low-frequency performance may be relevant especially for protection from induction at supply frequency and stray current pick-up. The relevant performance figure is thus the screening factor (see sec. 5.1.9) and test methods shall ensure that possible non-linearity of the external armour and its influence on the screening performance are taken in the due account: the reason is that armours are nowadays almost always made of steel; to test them an intense current (or, equivalently, longitudinal voltage) shall be applied, of course at the supply frequency.

5.2.4 Other characteristics

When the influence of temperature is evaluated from a broad perspective, a combination of mechanisms is to be considered, for both short-term and long-term phenomena: increasing the temperature causes higher losses in the dielectric, but also dielectric expansion, with the possibility of contamination, deformation, mechanical interference with the shield and strain at connector interface; additionally, temperature accelerate oxidation of the shield wires worsening transfer impedance and microphonic effect, especially in the presence of adverse environmental conditions and questionable plastic materials; the overall warm-up of the cable modifies thus its geometry and as a consequence electrical characteristics, with or without the possibility of restoring original conditions when cooling down again. During higher temperature phases bending and torsion very likely cause permanent deformation and mechanical damage. Cables used outdoor under the sun are warm enough to be highly exposed to this kind of damage: pinched or crushed with feet, left connected to a piece of equipment that is moved around, crushed under equipment moved on a table, etc. This is particularly relevant for permanent installations in harsh locations (e.g. on a transmission tower, at an industrial plant, etc.) and also when performing tests outside under the sun (e.g. test of electromagnetic emissions of various kinds of vehicles to be done on site; verification, aiming, calibration of large antennas and transmission systems; etc.).

Movement and bending, even if not causing permanent or significant damage or deformation, will influence cable performance because of unavoidable movement of the inner conductor and slight dielectric elastic deformation, in particular at the joint with the terminal connectors, where most of mechanical stress is applied (see sec. 5.7.1.1, and in particular sec. 5.7.1.2): the effect of the eccentricity of inner conductor caused by cable movement and bending can be transmitted to connector center pin, that tilts, worsening by far the effect and the consequences.

Cable uniformity may be not only compromised by mechanical stress or warming during its use, but may be already compromised at the factory during production: in old cables it was found repeatedly that the extrusion process caused periodic variations of the geometry, thus tuning the cable for resonance in the range of one to several GHz, worsening VSWR significantly.

Bad cable connections and, in particular, cable-connector interaction are probably the first cause of low-quality, weird or even completely wrong measurements and are sometimes quite difficult to tackle.

Headed cables may be purchased complete of the desired length and this is the preferred choice when dealing with high-quality microwave elements, quite expensive and that shall deliver accurate and hard-to-obtain performances. But there are many situations in which preparation and assembling shall be done in-house, or at least is not part of a large well-monitored and controlled production. When preparing the cable peeling and cutting to dimension the various parts the degrees of freedom are many: more or less applied force, more or less sharp cuts, tolerances on lengths of naked inner conductor and removed jacket and dielectric, correct "combing" and repositioning of the excess braided shield for crimping, etc.. There are many instructions, more or less complete available from cable and connector manufacturers, at amateurs and associations sites, etc., as well as manuals and even videos: they are helpful, but often trivial and missing the important steps where you risk to go wrong. For example nearly nobody tells clearly if the pin shall be flushed against the dielectric surface or not; doing so avoids that a gap with air instead of dielectric is created before entering the pin, but complicates soldering (see sec. 7.1.1.3 for such discontinuity). When soldering the inner pin (or pins), the correct amount of inner conductor shall go in the pin cup and soldering shall be fast enough not to damage the dielectric; to this aim a large peak temperature on the cup may be used with a larger soldering iron relying on thermal inertia and dissipation of the whole cable sample.

5.3 Coaxial cable

The coaxial cable has a symmetrical compact structure, with a high degree of coupling between the inner conductor and the external shield. It exists as simple coaxial cable (in brief "coax") and double-shielded coax; the double shield has almost always the function of improving the electromagnetic characteristics of the first shield (namely the longitudinal resistance and the optical coverage) and it is thus at the same potential and intimately electrically bonded to the first one. Flexible, semi-rigid and rigid cables are usually distinguished, while under the term flexible there is a wide range of cables with different diameters, rigidity, weight, etc..

After sec. 5.1 reporting theoretical methods and expressions, this section gives indications of the relevant properties and useful expressions. Various deviations from the uniform coaxial structure are considered, addressing for example partially filled cables, eccentricity and in general sensitivity to geometry (some formulas will be equally applicable to connectors). The same will be done for single wires and multiple wires in the successive sections 5.4 and 5.5.

5.3.1 Theoretical expressions

A coaxial cable has a perfectly calculable geometry with circular symmetry and thus there exist quite known expressions for its capacitance and inductance, showing their dependence on the inner and outer radii, sometimes indicated as a and b, but here for clarity referred to with d and D:

$$C = \frac{55.556\varepsilon_r}{\ln(D/d)} \quad [\mathrm{pF/m}] \tag{5.3.1}$$

$$L = 200\,\ln(D/d) \quad [\mathrm{nH/m}] \tag{5.3.2}$$

The dominant mode of propagation is the TEM mode, giving handy expressions for the characteristic impedance and phase velocity (or propagation velocity):

$$Z_c = \frac{60}{\sqrt{\varepsilon_r}} \ln\left(\frac{D}{d}\right) \quad [\Omega] \tag{5.3.3}$$

The phase velocity may be calculated from the per-unit-length values of inductance and capacitance given above; cables with slightly larger characteristic impedance (usually associated to lower capacitance) will be faster, with a velocity factor in the range of 0.75-0.85, whereas other coaxial cables have generally a velocity factor of 0.65-0.75, mostly 0.66. The factor "60" is the result of simplifying the characteristic impedance of vacuum[7]; the more complete theoretical expression is [278, 332]:

$$Z_c = \frac{Z_0}{2\pi\sqrt{\varepsilon_r}} \ln\left(\frac{D}{d}\right) \tag{5.3.4}$$

For the attenuation terms analogously [332]:

$$\alpha_c = \left(\frac{0.014272\sqrt{f}}{Z_c}\right)\left(\frac{1}{d} + \frac{1}{D}\right) \qquad \alpha_d = 0.091207\, f\, \sqrt{\varepsilon_r}\, \tan\delta \tag{5.3.5}$$

It is noted that dielectric losses in α_d do not depend on cable dimensions.

There are some "golden" values for D/d ratio established nearly 80 years ago (already considered at the end of sec. 2.2): 2.718 optimizes breakdown voltage, 1.65 power transfer and 3.59 attenuation (conduction loss term α_c). The use of different dielectric materials of course affects slightly these optimal values.

It is customary to see small coax cables featuring higher attenuation (see sec. 5.3.3 for measurement results on cable samples), leading thus to prefer large cables. Large cables are not only heavier and lack of flexibility, but might also propagate higher modes. There are various solutions for the identification of these modes, but we are interested in usable closed-form expressions, so the two simple expressions of λ_c given in [332] are sufficient for our purposes.

---

[7] Sladek [305] reports 59.95860±0.00006, while EURAMET [123] gives 59.93916 assuming the air permittivity $\varepsilon_r = 1.000645$. Of course coefficients in eq. (5.3.1) and (5.3.2) shall be correspondingly adjusted.

$$\lambda_{c,\mathrm{TE11}} = 2\pi r_m \left[1 - \frac{1}{6}\left(\frac{t}{2r_m}\right)^2 - \frac{7}{120}\left(\frac{t}{2r_m}\right)^4 - \ldots\right] \tag{5.3.6}$$

where $r_m = (D+d)/2$ and $t = D-d$, and

$$\lambda_c = \frac{\pi(D+d)}{2} \tag{5.3.7}$$

The latter corresponds exactly to the circumference in between the inner conductor and the external shield, with a diameter that is the average of the two.

The IEEE Std. 287 [172], Table 1, reports the pairs "inside diameter of outer conductor" / "theoretical limit in GHz for TE11/H11 mode", assuming air dielectric: 14.29 mm / 9.5 GHz, 7.00 mm / 19.4 GHz, 3.50 mm / 38.8 GHz, 2.92 mm / 46.5 GHz, 2.40 mm / 56.5 GHz, 1.85 mm / 73.3 GHz, 1.00 mm / 135.7 GHz, where we recognize the classification of precision coaxial connectors (see Chapter 7)[8].

Example 5.1: Determination of the critical frequency for standard coax cables

Let's consider some standard gauge coax cables and apply eq. (5.3.7) above. The considered cables are RG 316, RG 58, RG 213 and LMR 400, the latter a low-loss cable of the size of RG 213. The critical frequency results from the cited equation, once the velocity of propagation is known, that is the velocity factor k of the cable.

$$f_c = \frac{c\,k}{\lambda_c}$$

The diameters of inner conductor and shield for the three cables are: 0.02" (0.508 mm), 0.036" (0.91 mm), (2.25 mm), 0.108" (2.74 mm) and 0.079" (2.01 mm), 0.139" (3.53 mm), (7.6 mm), (8.0 mm) in the order the cables were listed above. The velocity factors are: 0.7, 0.66, 0.66 and 0.85. Thus the critical frequencies are: 50, 28.4, 12.8 and 15.1 GHz.

What may be observed is that, in principle, a smaller cable is preferable to push farther the critical frequency of the first higher mode, but we know that the higher losses would make this cable in any case useless at such frequency. Low-loss cables tend to have a larger inner conductor and slightly larger velocity factor, and these two factors partially compensate; however a larger inner conductor leads always to a lower critical frequency.

The example is concluded by applying the same consideration to a connector, such as a 3.5 mm type (air-filled SMA, see sec. 7.1.1): the internal and external diameters are 1.52 and 3.5 mm and the velocity factor is approximately 1 because of the air dielectric, so that the resulting critical frequency is about 38 GHz (the rated maximum frequency is however 33 GHz [304], page 7). From this two considerations for connectors: smaller connectors are needed for the higher frequencies of modern spectrum analyzers and VNAs; the air-filled types have higher critical frequencies than the dielectric counterparts.

[8] A thorough pioneering work reporting carefully the electrical parameters of precision air-lines for the required accuracy and level of detail of the time is: R.E. Nelson and M.R. Coryell, "Electrical Parameters of Precision, Coaxial, Air-Dielectric Transmission Lines," National Bureau of Standards Monograph 96, June 30, 1966.

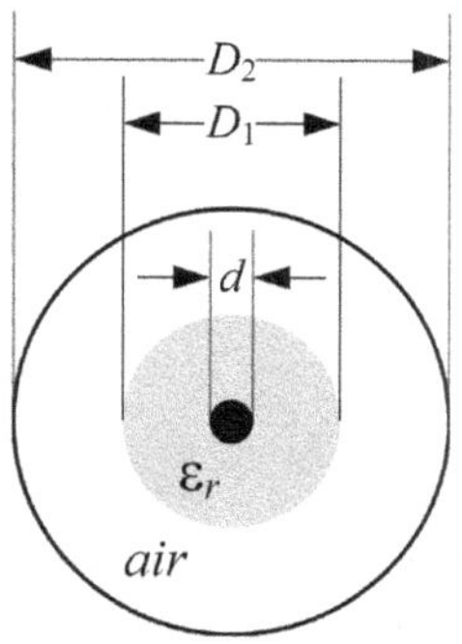

Figure 5.3.1 – Sketch of the partially filled coaxial cable

Quite interestingly, another non-ideality of coax cables is that they may introduce some distortion of about tens of ppm measured with respect to the applied sinusoidal tone [13] (please, see sec. 7.4.3 for a similar phenomenon occurring in connectors, Passive Intermodulation). This was observed with braided shield coax cables, in particular for braids made of metals other than copper (e.g. steel, aluminum, or with nickel plating).

To be practical, it is sensible not to expect a high stability and accuracy of characteristic impedance because of cable production tolerances, change of geometry and dimensions due to bending and other mechanical or environmental stress, including temperature effects, and influence of connectors. Regarding temperature effect, dielectric permittivity may change by less than a tenth of % per °C (see sec. 1.1.1.2 for dielectrics in general and sec. 6.1.2.2 for materials used in PCBs) and its square root in the expression of the characteristic impedance varies by about one half. Also dimensional changes are possible with temperature, not so much for expansion of metallic parts, as for softening of the insulating material and deformation; deformation occurs as crush and ovalization, and eccentricity of inner conductor (see sec. 5.3.1.3 below).

5.3.1.1 Partially filled coax-cable (1)

This is a particular case of a reduced dielectric thickness that leaves room to air, e.g. to reduce dielectric losses or when tooling the cable at its end near the connector. So the usefulness of these expressions is to understand how much dielectric is possible to remove before there are significant changes. As shown in Figure 5.3.1, the diameter of the dielectric sleeve around the inner conductor of diameter d is indicated by D_1; the outer diameter that corresponds to the external shield surrounding the three concentric regions of air, dielectric sleeve and inner conductor, is D_2.

$$Z_c = \frac{Z_0}{2\pi} \ln\left(\frac{D_2}{d}\right) \sqrt{\frac{\varepsilon_r \ln\left(\frac{D_2}{D_1}\right) + 1 \ln\left(\frac{D_1}{d}\right)}{\varepsilon_r \ln\left(\frac{D_2}{d}\right)}} \tag{5.3.8}$$

The expression may be calculated by stacking air and dielectric regions and the corresponding capacitance expressions for cylindrical symmetry. The expression above is shown in Ragan [278], p. 168, while in Wadell [332], p. 54, there is a typing mistake[9]: the denominator of the first fraction inside square root is written as d rather than D_1. The formula was verified by removing the dielectric sleeve and verifying that eq. (5.3.3) reappears; please, note that the dielectric sleeve is removed when we put $D_1 = d$, not zero; conversely, when removing the air putting $D_1 = D_2$, the square root reduces correctly to $1/\sqrt{\varepsilon_r}$.

5.3.1.2 Partially filled coax-cable (2)

Of much more use and interest is the case of a discontinuous dielectric that is mostly air and shall accommodate for supporting beads (washers) of dielectric material: this configuration is normally encountered in many situations, such as old low-loss semi-rigid cables, the support for the center pin of connectors (see sec. 7.1.1) and the triaxial and quadriaxial setups for transfer impedance measurements (see sec. 5.6.1).

Ragan [278] distinguishes between low and high frequency:

- in the first case the effect of beads is that of filling the otherwise air dielectric, and thus their effect may be accounted for with a slightly increased equivalent permittivity, given simply by the weighted average of air and bead material permittivities; at low frequency this simplifying approach is valid because the wavelength is much longer than bead size and spacing, and one shall account only for the overall "loading" effect;
- at high frequency, when wavelength shortens and becomes comparable with bead spacing, a more careful design is necessary, backed up by some guidelines for bead spacing.

In general, when adding beads, dielectric breakdown might be of concern if very large power levels are applied to the line, as it was e.g. for radar applications; the cable diameter, as well as the inner conductor, are larger in these cases allowing for larger potential differences and lowering the electric field intensity at surfaces. However, the weak point of bead supported cables was the unavoidable air gap between the inner conductor and the bead surface placed around it; at this interface the electric field concentrates inversely proportional to the relative permittivity of the added beads. Nowadays foam filled cables have replaced air-beads cables with a more uniform distribution of the electric field and solving the problem of air interfaces, since foam is deposited to adhere to the inner conductor.

When frequency is large enough, sections containing the beads behave like short mismatched sections along the line, presenting a slightly lower characteristic impedance than that of the all-air cable. Two phenomena shall be accounted for: a reflection coefficient shall be considered, that depends on dielectric permittivity of bead material and on bead thickness (visual interpolation of Figure 4.11 in [278] gives the following approximate linear expression for the reflection coefficient: $\Gamma = 1 + L\sqrt{\varepsilon_r}/2\lambda$, where ε_r is the relative permittivity of bead material and L its thickness); the array of beads along the line may trigger resonance phenomena caused by bead periodicity.

---

[9] The original publication cited by Wadell is: T. Hatsuda, "Computation of Impedance of Partially Filled and Slotted Coaxial Line," *IEEE Transactions on Microwave Theory and Techniques*, Vol. 15, n. 11, Nov. 1967, pp. 643-644. However, Ragan 20 years before gives the correct expression in his publication.

The latter might not be an adverse effect if the line is going to be used at a specific frequency, such as in radio transmission and radar applications; in these cases bead spacing may be optimized to also compensate for the former effect, that is reflection at each bead spacer. The quarter-wavelength tuning indicated in Ragan[10] requires that spacing (including bead thickness) is $\lambda/4$; Lawson introduced the $3\lambda/4$ - $\lambda/4$ spacing, that performs much better than the simple $\lambda/4$ method around the center frequency used for tuning, but worsens departing from it by about 20%. In practical cases, where fabrication tolerances and line losses shall be considered, tuning is not so accurate and curves tend to smooth.

In this direction Reed[11] proposed a batch tuning of beads, grouping them by a convenient number (in his case eight), each group tuned to a slightly different wavelength, and then the groups spaced randomly: a little optimization of geometry and of the chosen wavelength allowed to broaden considerably the bandwidth of the cable.

Of course, the other way round solution is to make the bead sections "transparent" from characteristic impedance viewpoint, and this may be achieved by under-cutting the inner conductor, so that the slightly smaller diameter compensates for the increased permittivity. However, besides the fabrication difficulties, this solution exacerbates the problem of voltage breakdown even more, having reduced the inner conductor diameter, that is in the opposite direction of the good recipe for high voltage breakdown withstanding capability.

5.3.1.3 Eccentric round coaxial cable

This is another structure that is not purposely used in practice nowadays, but it is extremely useful to understand the influence of fabrication tolerances and deformation when cutting and tooling cable ends for termination. The original formulas are derived from Wadell [332], but because of the relevance in describing small deviations from circularity and their impact on accuracy and uncertainty, this problem has then been considered quite thoroughly by Leuchtmann and Rüfenacht [219, 220], as reported also by Lewandowski [222].

For inner center conductor eccentric only in one direction (see Figure 5.3.2(a)), measured as a distance between conductor and shield centers, that is the fraction a of the external radius $D/2$:

$$Z_c = \frac{Z_0}{2\pi\sqrt{\varepsilon_r}} \operatorname{arccos}\left[\frac{D}{2d}(1-a^2)+\frac{d}{2D}\right] \tag{5.3.9}$$

For eccentricity in both directions (let's say horizontal and vertical, see Figure 5.3.2(b)), measured as fractions a and b of the external diameter, again between the centers of the inner conductor and the external shield:

---

[10] The references cited by Ragan are:
J.L. Lawson, "Design and Test of Concentric Transmission Lines," RL Report, No. 141, July 14, 1941;
E.U. Condon, "Low-loss Coaxial Cables for Micro-Waves," Research Report R-94293-E, Westinghouse Research Laboratories, Apr. 17, 1941.

[11] J. Reed, "Broad-band Bead Spacing," MIT Bachelor's Thesis, Jan. 1943 (again in Ragan [278]).

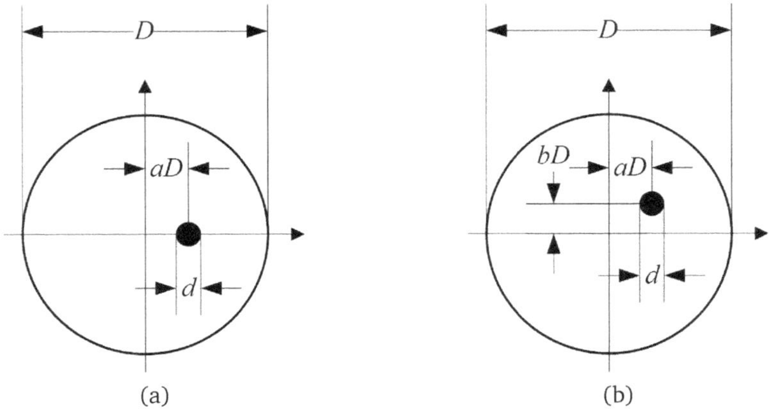

Figure 5.3.2 – Eccentric coaxial cable: (a) eccentricity along one axis, (b) eccentricity along two axes

$$Z_c = \frac{Z_0}{2\pi\sqrt{\varepsilon_r}} \ln\left[\frac{D}{d}\left(1 - \frac{e^2k^2}{(k^2-1)}\right)\right] \tag{5.3.10}$$

where

$$e = \frac{b}{a+1}, \quad k = 2\frac{a+1}{d}.$$

This expression given by Wadell corresponds quite well to eq. (C.6) in Lewandowski, obtained under the more than acceptable assumption that $2e/(D-d)$ is small, i.e. for small eccentricity. Equations contain a quadratic term of the eccentricity (that is smaller than 1), so that for good cables the effect is negligible.

Leuchtmann and Rüfenacht studied not only eccentricity, but also deviation from circularity of the inner conductor, by modeling it with a harmonic function of increasing order. The cited report [220] deals thoroughly with eccentricity and other geometry variations that are reviewed in the next section: that reference is also a useful reference to cross-check own calculations and models.

5.3.1.4 Very accurate coaxial cable model

The complete model proposed by Daywitt [94] (and resumed by Ridler in [87], Chap. 9, and by METAS in their VNA Tools II manual [339]) is reported for completeness. This model is used extensively to characterize airlines when the required accuracy is high. The model is similar to that considered at the beginning of this section, accounting for a double correction of all the four per-unit-length parameters, taking into account skin effect and correction for the effect of finite thickness, for principal mode only (factor F_0). The expressions are reported here for completeness, but can be found in the cited references.

A few differences were noted between Daywitt's original work and VNA Tools II: the material conductivity is no longer constant and varies between a dc value σ_{dc} and a decreasing value in ac that goes with the square root of frequency in GHz; this correction is however not further commented in [339], but the form suggests a dependency like skin depth, that is however already explicitly included in the d_0 expression.

Of course the dc (or low-frequency) part of expressions is missing, so that the resistance R' shall be completed by its dc value. It is worth underlying also that the effect of conductivity is included by means of skin depth and that assuming infinite conductivity puts to zero skin depth and thus the quantity d_0; the remaining terms are thus $L' = L_0$ and $C' = C_0$ of an ideal lossless line.

$$\sigma = \sigma_{dc} - \sigma_{hf}\sqrt{f}\,[\mathrm{GHz}] \qquad k = \omega\sqrt{\mu\varepsilon} \tag{5.3.11}$$

$$d_0 = \frac{\sqrt{\dfrac{2}{\sigma\mu\omega}}\left(1+\dfrac{D}{d}\right)}{4D\ln\left(\dfrac{D}{d}\right)} \tag{5.3.12}$$

$$F_0 = \frac{\left(\dfrac{D}{d}\right)^2 - 1}{2\ln\left(\dfrac{D}{d}\right)} - \frac{\dfrac{D}{d}\ln\left(\dfrac{D}{d}\right)}{\dfrac{D}{d}+1} - \frac{1}{2}\left(\frac{D}{d}+1\right) \tag{5.3.13}$$

$$C_0 = \frac{2\pi\varepsilon}{\ln\left(\dfrac{D}{d}\right)} \qquad L_0 = \frac{\mu\ln\left(\dfrac{D}{d}\right)}{2\pi} \tag{5.3.14}$$

$$R' = 2\omega L_0 d_0\left(1-\frac{k^2d^2F_0^2}{2}\right) \qquad L' = L_0\left(1+2d_0\left(1-\frac{k^2d^2F_0^2}{2}\right)\right) \tag{5.3.15}$$

$$G' = \omega C_0 d_0 k^2 d^2 F_0 \qquad C' = C_0(1+d_0k^2d^2F_0) \tag{5.3.16}$$

$$Z' = R' + j\omega L' \quad Y' = G' + j\omega C' \quad \gamma = \sqrt{Z'Y'} \quad Z = \sqrt{\frac{Z'}{Y'}} \tag{5.3.17}$$

5.3.2 Sensitivity to geometry variations

Some manipulations of expressions appearing in the previous section and calculation of differentials with respect to geometrical parameters allow to derive some simple expressions that predict the sensitivity of cable quantities (characteristic impedance and capacitance) from small changes to cable geometry and parameters. The considered variability regards: variations of diameter of the inner conductor (d) and outer shield (D), eccentricity, electrical conductivity.

The following expressions may be extended to precision coaxial connectors, that share with cables a cylindrical geometry. Additionally, connectors may suffer of specific mechanical misalignment, eccentricity, skewness, when mating that are not covered by these expressions.

5.3.2.1 Diametrical tolerances

Directly from expression (5.3.4) of characteristic impedance of an ideal coaxial cable, differentiating with respect to d or D, we obtain a linear relationship that links the resulting characteristic impedance change ΔZ_c to the considered diameter change:

$$\Delta Z_c = \frac{Z_0}{2\pi\sqrt{\varepsilon_r}}\Delta d \qquad \Delta Z_c = \frac{Z_0}{2\pi\sqrt{\varepsilon_r}}\Delta D \tag{5.3.18}$$

Such changes of characteristic impedance may be expressed also as fractional changes:

$$\frac{\Delta Z_c}{Z_c} \simeq \frac{1}{\ln\dfrac{D}{d}}\left(\frac{\Delta D}{D} - \frac{\Delta d}{d}\right) \tag{5.3.19}$$

Diameter is thus an extremely important parameter to the aim of ensuring the designed or nominal characteristic impedance.

The variation of diameter may also occur, not only because of fabrication tolerance (static), but also during temperature changes: using different metals for the inner conductor and the outer shield is not advisable, because for a precision airline structure this would imply on average a change of 0.0001 in VSWR every $10\,°\mathrm{C}$ (not large, but can be further reduced by using same or similar metals, such as various copper alloys).

5.3.2.2 Eccentricity

If eccentricity is considered, we observe that an increase of distance between the inner conductor and the external shield towards e.g. the left is partially compensated by the corresponding reduced distance occurring at the right; it is intuitive thus that the expressions will bear a square of the eccentricity e.

$$\Delta Z_c = \frac{Z_0}{2\pi\sqrt{\varepsilon_r}}\ln\left(1 - \frac{4e^2}{D^2 - d^2}\right) \tag{5.3.20}$$

Eccentricity is well tolerated with respect to ΔZ_c, or alternatively VSWR: assuming a high precision airline for which the limit on VSWR is only 1.001, $0.1\,\mathrm{mm}$ of eccentricity are allowed for a $7\,\mathrm{mm}$ diameter.

5.3.2.3 Electrical conductivity

The electrical conductivity of metals is large, but not infinite, and this influences the longitudinal resistance and skin effect most of all. We have already reviewed variability in conductivity for alloys, plating and similar manufacturing solutions. In [305] calculation of the effects of internal resistance and inductance and some simplifications after series expansion give the following approximate formula for the sensitivity to metal conductivity

$$\Delta Z_c = \frac{302}{\sqrt{\sigma f}} \left(\frac{1}{d} + \frac{1}{D}\right)(1-j) \tag{5.3.21}$$

Lewandowski [222] gives a comprehensive result for the effect of geometrical and conductivity variations altogether on resistance per unit length and, as a consequence, on characteristic impedance and propagation constant. The reason for considering both electrical conductivity and cable geometry, is that changing the latter not only affect inductance and capacitance, but also longitudinal resistance, and hence losses.

$$\Delta R = -\frac{1}{2}\Delta\sigma - \frac{d}{D+d}\Delta D - \frac{D}{D+d}\Delta d \tag{5.3.22}$$

We know that characteristic impedance at low frequency undergoes significant variations due to the relevance of resistive terms (longitudinal resistance and shunt conductance), but, at the same, time at low frequency impedance mismatch is less relevant from a practical point of view. However, when considering precision coaxial lines, such as assurance airlines, and precision connectors, even this phenomenon shall be accounted for. Inversion of equation above gives the critical frequency as a function of the electrical conductivity for some limit A on ΔZ_c:

$$f^* = \frac{0.001824}{\sigma A^2} \left(\frac{1}{d} + \frac{1}{D}\right)^2 \; [\text{MHz}] \tag{5.3.23}$$

At high frequency the limit is represented by the mode cutoff frequency that was determined for TE_{11} mode with eq. (5.3.7) in sec. 5.3.1.

5.3.2.4 Effect on cable reflection and transmission

Considering the scattering parameters that describe the cable section, namely return loss S_{11} for reflections and insertion loss S_{21} for transmission, Lewandowski [222] gives two synthetic expressions that account for all the reviewed variations, distinguishing the effect on characteristic impedance and on longitudinal resistance, clarifying thus the effect on reflection and transmission of ordinary geometry and conductivity variations. These expressions are given for the analysis of connectors, but having assumed a cylindrical structure they are quite extendable to coaxial cables; noting also the low values of S_{21} for short lengths (as pointed out in sec. 7.1.1.5 for connectors), they are more relevant if applied to cables a hundred times longer.

$$\Delta S_{11} = \frac{1}{2}\frac{\Delta Z_c}{Z_c}\left(1 - e^{-2\gamma L}\right) \tag{5.3.24}$$

$$\Delta S_{21} = e^{-2\gamma L}\left[1 - (1+j)\alpha L \left(\frac{\Delta R}{R} - \frac{\Delta Z_c}{Z_c}\right)\right] \tag{5.3.25}$$

	RG 213	RG 58	RG 316
D [mm]	7.25	3.51	1.2
d [mm]	2.25	0.91	0.03
ΔD [µm]	72	35	12
Δd [µm]	22	9	3
$\max\left(\frac{\Delta Z_c}{Z_c}\right)$ %	1.7	1.5	1.4
ΔS_{11} %	0.85	0.75	0.7
$\max\left(\frac{\Delta R_c}{R_c}\right)$ %	0.5	0.5	0.5
ΔS_{21} %/1 m	1 @ 100 MHz 3 @1 GHz	1.5 @ 100 MHz 4.7 @1 GHz	4 @ 100 MHz 12 @1 GHz

Table 5.3.1 – Variation of return loss S_{11} and insertion loss S_{21} of coaxial cables for common geometry and conductivity variation (1%). For cables with an overall length in the order of meters, the impact on S_{21} is relevant (S_{21} is expressed per 1 m length) and of the same order of magnitude of S_{11}. Smaller cables have larger attenuation and a correspondingly larger ΔS_{21}.

Assuming typical cable fabrication tolerances, it is possible to estimate the effect on the final ΔS_{11} and ΔS_{21}, as shown in Table 5.3.1. When we consider precision airlines (see sec. 11.12.3.3), the observed dimensional variability with primary metrology methods will be quite low around a fraction of or few µm.

5.3.2.5 Periodic diameter changes and induced resonances

A phenomenon that sometimes happens (and is reported to have happened in the past [278], page 254) due to low quality fabrication is the appearance of standing waves and considerable drops in the transmission properties. The reason is a periodic fluctuation of dielectric diameter, of centering of the inner conductor and lack of circularity, that may be caused by the fabrication process (e.g. irregular drawing) or improper storage (e.g. improper coiling, excessive stacking, high temperatures). Depending on the type of cable, its diameter and production process the wavelength that triggers resonance may range between several tens of cm down to a few cm, so covering the frequency range between approximately 1 and 10 GHz.

The standard IEC 60096-1 [169], sec. A2.1 and following, reports a method for the determination of characteristic impedance where it is underlined the need of defining a local characteristic impedance if the "cable is not uniform longitudinally", without clarifying if they intend slight deviations as just considered or simply interconnected different sections. The method is said not to be applicable for periodicity shorter than 2.5 m. In general, suitable measurement methods shall identify any resonance frequency of Z_c by adopting a small enough frequency resolution and sensitivity for the vertical scale; probably too a long cable sample containing several variations delivers obfuscated Z_c resonances due to unavoidable shifts and deviations of periodicity.

5.3.3 Measured coax cable performance and reference data

Performance are measured and reported in datasheets for commercial cables, however limiting to only average values of cable attenuation, sometimes simply stating compliance with a limit. Differently, the complete description of cable response necessitates the S_{21} insertion loss curve. Similarly, characteristic impedance is never measured directly (S_{11} return loss), but only stated as a nominal value. This information is less commonly reported in datasheets, but of course may be reconstructed from insertion loss curve (see sec. 10.1.3). The measurement of S_{11} and S_{21} quantities is treated in more detail in sec. 10.2.3, where measured curves are also reported for some cable samples as example. Similarly, sample measurements are reported in Chapter 11 to support the analysis of the various settings and modes of the Vector Network Analyzer: see for example sec. 11.10.6.

In the following, measurement results are reported that appear in the literature or available from the web, useful to support design, evaluation, analysis, without the need of repeating tests and experiments.

5.3.3.1 Attenuation

Attenuation is quite a complex matter when it comes with a complete connection link, including connectors at the extremities, thus influenced by connector characteristics and how it was mounted (soldered? crimped? by professionals? new or old? etc.). To have an idea of the variability of performance of "identical" complete cables, see for example Figure 10.2.6, where three 30 m samples of RG316 cable headed with SMA connectors are shown, soldered and crimped by the undersigned 10 years ago; they were used outdoor a few times a year and do not show any unusual dispersion or weird behavior, even if one of them has unusual noise in S_{11} trace.

The rated attenuation of the cable alone in good nominal conditions is considered, as it can be found in datasheets and manuals. Attenuation is normalized to length, but there is no general agreement on the measurement units, so it is usually expressed in [dB/m], [dB/100 m], [dB/ft], sometimes not expressly indicated, but recognizable if the average reference figures are known. For convenience of accuracy of the representation attenuation in Table 5.3.2 is expressed in dB/100 m.

Atten. [dB/100 m]	**Test frequency**			
Cable type	**100 MHz**	**1.2 GHz**	**2.5 GHz**	**6 GHz**
RG 58	20.3	69.2	105.6	169.2
RG 213	9.2	33.1	49.9	93.8
LMR 400	4.9	15.7	22.3	35.4

Table 5.3.2 – Reference values of attenuation in [dB/100 m] for some coaxial cables.

There are several methods to determine cable attenuation: see sec. 2.1.1 for an introduction and sec. 6.5.1 for an application to PCB planar transmission lines. These methods are quite comprehensively presented in IEC 60096-1 [169], Appendix A4, and are reported in the following.

Half-power frequency When performing a swept measurement of input impedance (e.g. reflection coefficient or S parameter S_{11}) with the cable sample in short- or open-circuit conditions, using a small enough frequency resolution at each line resonance, it is possible to determine also two frequency points at the left and right of the resonance frequency (f_r^- and f_r^+), at which the power is reduced to one half. The attenuation in dB is then determined including the propagation velocity v:

$$\alpha = 8.686 \frac{\pi}{v} (f_r^+ - f_r^-) \tag{5.3.26}$$

The method is indicated by the standard for a test frequency of 200 MHz, so implying an interval of validity in the hundreds MHz range.

Admittance circle with sliding short-circuit An alternative method suggested for higher frequencies is based on plotting the input admittance when the cable under test is terminated in a short section of coaxial line with a moveable short circuit. The plot of measured values for different positions of the short circuit at a given frequency is a circle in the admittance plane and cuts the real axis in two points: the distance between these two points is used to determine the attenuation. Yet, the IEC 60096-1 does not go farther and there are no details that allow to apply the method. It is briefly observed that the span of the circle is related to a reactive behavior, instead, and that, recalling eq. (2.1.12), attenuation may be determined by taking the center of the circle (that corresponds to Y_c, as demonstrated in eq. (2.3.12) for Z_c by taking $Z_t = 0$); then multiplying by r and g is necessary and they should be separately measured.

Substitution method This method is based on a setup able to compensate and keep constant the effects of impedance mismatch using attenuators at each end of the cable sample: the attenuation is simply the difference between two measurements of S_{21} (or received voltage/power if using a spectrum analyzer), with and without the cable sample. With cable sample removed, attenuators connected back-to-back are suitable for VNA calibration or adjustment of the received level in the spectrum analyzer.

A more careful evaluation of terms that make the overall insertion loss reveals that besides cable sample attenuation, also reflection attenuation shall be considered: reflection attenuation is simply the loss of signal due to reflection at the cable-attenuator interface and is given by $20\log_{10}\left(\frac{1}{1-|\Gamma|^2}\right)$, where Γ is the cable-attenuator reflection coefficient. If the calibration of the setup is done by joining the two pads with a short section of the same cable then reflection influence will be duly taken into account and compensated during calibration itself. The standard cites also "interaction attenuation" without giving any further detail nor explanation.

5.3.3.2 Cable losses

We already said (see sec. 2.5.3) that cable losses are determined by both attenuation (defined and quantified for a matched terminated cable) and additional losses for standing waves and VSWR. The expression of the overall losses $K_{tot-loss}$ (see eq. (2.5.19)) was given as depending on cable attenuation α in linear units (not in dB) and on reflection coefficient Γ for mismatching at cable ends. It is instructive then to evaluate cable losses for increasing cable mismatch, in order to identify in power applications for which degree of mismatching cable attenuation is really relevant.

In Figure 5.3.3 two cables are considered, RG 213 representing a commonly used and available coax cable for RF applications and LMR 400, that is more expensive and slightly more challenging to handle; for comparison, RG 58 is added, as the reference for cheap inadequate cable. Assuming a value of VSWR = 1.5 (that is a reflection coefficient $\Gamma = 0.2$), the total cable loss at 100 MHz for the three selected cables is 0.054, 0.097 and 0.216, as expected not so different from the cable attenuation itself reported in Table 5.3.2; increasing the VSWR to 4 (and the reflection coefficient goes to $\Gamma = 0.5$) the total losses become 0.105, 0.168 and 0.373. We observe two facts: the increase of losses is larger for the high quality LMR 400 cable than for the other two, even if its losses keep always the smallest (under large VSWR, high-quality cables are more affected, and other characteristics, such as power handling capability, become important); considering the effect of frequency, instead, losses are much larger for the cheaper RG 58 cable, that at 6 GHz for VSWR = 1.5 features 1.8, while the other two cables 0.96 and 0.377.

Going to the evaluation of power losses in real applications, let's consider immunity testing for electromagnetic field, using a biconical antenna at 100 MHz fed with $100\ \mathrm{W}$ and featuring – as known – a large VSWR[12]. Using 10 m of RG 213 cable (really quite a common choice!) the total cable losses in dB are in the range of 0.31-0.33 dB/m, thus 3.1-3.3 dB in total, that is 51-53%, and the total power loss is thus around $50\ \mathrm{W}$! In this example the VSWR was exaggerated a little, because values of 7 or 8 for the VSWR are encountered around $20-40\ \mathrm{MHz}$, while at $100\ \mathrm{MHz}$ we are probably around 3 or 4, that gives 0.15-0.19 dB/m and thus 30-36% of power loss. In any case the capability of dissipating power in the range of tens of Watts is a characteristic that shall not be forgotten and overlooked, for the cable and cable-to-connector joints!

Now it is clear that exigencies of radio transmitters, handling several hundreds of Watts and kW, are for extremely good matching and power handling capability. Since quite often more than one antenna is fed at a time (e.g. for radio base stations covering different spatial sectors), the power splitter has also the function of improving the VSWR seen by cable and connectors, decoupling antenna input mismatch.

[12] For this type of antennas the most troublesome issue in using them for immunity tests is the highly reactive behavior in the low frequency range, that is between 20 and 100 MHz, where various types of resonances may occur and the VSWR may reach values as large 7 or 8. This has impact not only on the just reviewed cable losses, but also from the point of view of voltage stress on the feeding amplifier.

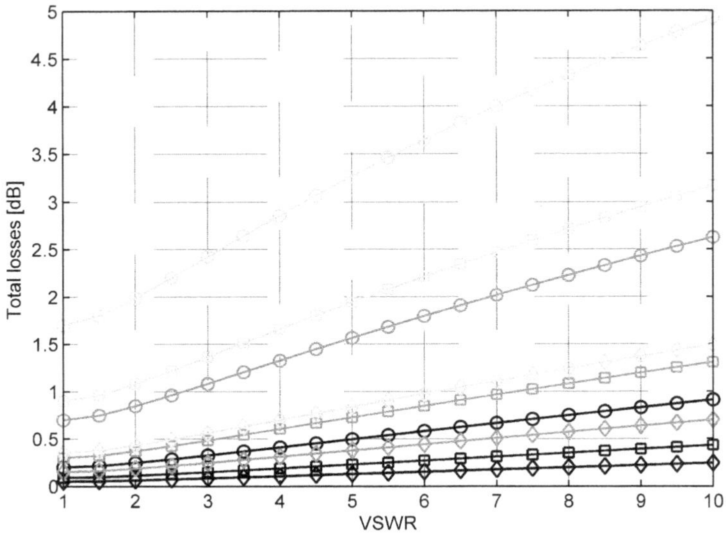

Figure 5.3.3 – Estimated total cable losses as per eq. (2.5.19) for coaxial cables as a function of cable mismatch: three values of frequency, 100 MHz (black), 1 GHz (dark gray), 6 GHz (light gray); three cables, LMR 400 (diamond), RG 213 (square), RG 58 (circle).

5.4 Single wires

5.4.1 Wire facing a corner and a C-shaped ground plane

This section collects expressions for conductor geometries characterized by a single wire with a return path that is not a flat ground plane, such as an angular profile and a C-shaped profile, so similar to extraneous structure that are used sometimes as or become accidentally the return path.

5.4.1.1 Wire facing a corner

This case covers all configurations where a cable or conductor is run close to the angular portion of a structural element, such as along one of the corners of a girder. The geometry is simplified and the angular element is assumed to have side D and the wire diameter is d, with the wire standing approximately midway in the center of the diagonals. The medium has dielectric permittivity ε_r. The solution was given by Wheeler [333] and reported by Wadell [332], sec. 3.9.1.

$$Z_c = \frac{\zeta_0}{4\pi\sqrt{\varepsilon_r}} \ln\left\{\left(\frac{D}{d}\right)^2 + \sqrt{\left[\left(\frac{D}{d}\right)^2 - 1\right]^2 + \left(\frac{D}{d}\right)^2 - 1}\right\} \tag{5.4.1}$$

config.	shape	s	n	m	c
(a)	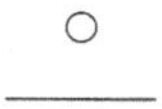	2	1	1	2
(b)	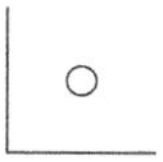	$\sqrt{2}$	2	2	1
(c)		$\frac{4}{\pi}$	2	4	2
(d)	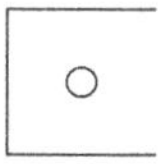	$\frac{4}{\pi}\tanh\frac{\pi}{2}$	3	2	$\frac{4}{9}$
(e)	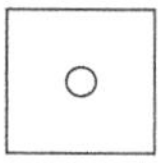	$\frac{2}{\sqrt{\pi}}\left[\Gamma\left(\frac{5}{4}\right)\right]^2$	4	8	1
(f)	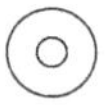	1	∞	1	0

Table 5.4.1 – Shapes and coefficients for the wire in a polygonal shield configuration [333], Fig. 2: (a) one plane, (b) corner, (c) parallel planes, (d) channel, (e) square cylinder, (f) circular cylinder.

5.4.1.2 Wire in a polygonal shield

The work by Wheeler [333] cited in the previous section offers the extension of the characteristic impedance formulation to surrounding sheets of various forms (polygon shields), accommodating for the most common geometries. The formula is general and contains coefficients that shall be assigned values for a specific geometry:

$$Z_c = \frac{Z_0}{2\pi\sqrt{\varepsilon_r}}\frac{1}{m}\ln\left\{1+\frac{1}{2}s^m(S^m-1)+\sqrt{\left[\frac{1}{2}s^m(S^m-1)\right]^2+c(S^m-1)}\right\} \tag{5.4.2}$$

Three coefficients s, m and c are assigned for each shape (see Table 5.4.1) and S is a shape factor defined as

$$S = \frac{D}{d} \tag{5.4.3}$$

where d is the inner conductor diameter and D is the diameter of the circle that would touch all the planes of the surrounding shield.

5.4.1.3 Wire in a C-shaped channel

The trough or C-shaped channel has the geometry of a square with one side missing; the wire of diameter d is approximately in the center of the diagonals with a distance D from the walls. The medium has dielectric permittivity ε_r. The solution is given by [333] and reported by Wadell [332], sec. 3.2.6.

$$\begin{aligned} Z_c &= \frac{Z_0}{4\pi\sqrt{\varepsilon_r}} \ln\left\{1+\left[\frac{1}{2}\left(\frac{4}{\pi}\tanh\frac{\pi}{2}\right)^2\left(\left(\frac{D}{d}\right)^2-1\right)\right]+\right. \\ &\left.\sqrt{\left\{\frac{1}{2}\left(\frac{4}{\pi}\tanh\frac{\pi}{2}\right)^2\left[\left(\frac{D}{d}\right)^2-1\right]\right\}^2+\frac{4}{9}\left[\left(\frac{D}{d}\right)^2-1\right]}\right\} \end{aligned} \tag{5.4.4}$$

5.4.2 Wire over/above dielectric layers and ground planes

This is a general case where a naked wire of diameter d is laid down in contact over a dielectric of thickness h or above it at the height H with air in the middle; the dielectric is lined up by a ground reference plane on the other side. These configurations are encountered for example when building a custom transmission line on a PCB or shielded lines, windings and coils (for the case of ground planes over and beneath the wire and dielectric, please, see sec. 5.4.2.3 for wire-in-a-slab configuration).

5.4.2.1 Wire over a dielectric layer and ground plane

The characteristic impedance for wire over dielectric, so in physical contact with it, is:

$$Z_c = \frac{Z_0}{2\pi\sqrt{\varepsilon_{re}}}\,\mathrm{acosh}\left(\frac{2h}{d}\right) = \frac{Z_0}{2\pi\sqrt{\varepsilon_{re}}}\ln\left(\frac{4h}{d}\right) \tag{5.4.5}$$

where h is the height of the dielectric layer including wire radius: h is the distance between the conductor center and the base of the dielectric layer. The expression is accurate for $h \gg d$. The value of the effective relative permittivity ε_{re} is to be computed with an approach similar to that for the microstrip equivalent width w_{eq} (see sec. 6.3.1.1). The microstrip eq. (6.3.2) is used with the hot trace width w replaced by an equivalent wire diameter d_{eq}, calculated from the wire diameter d; to this aim the formula given by Wadell in his sec. 3.9.3 is obtained by polynomial interpolation of data appearing in the original publication by Cohn [85], Figure 4 ($x = t/w$).

$$\frac{d}{d_{eq}} = 0.5008 + 1.0235x - 1.0230x^2 + 1.1564x^3 - 0.4749x^4 \tag{5.4.6}$$

In Cohn the equivalence was given with an opposite purpose, that of relating a rectangular conductor to a circular one; the unknowns are two, t and w (the thickness and width of the rectangular trace, respectively, in the notation used in sec. 6.3.1), giving – as known – different values of characteristic impedance, even if keeping their ratio constant, as soon as the change is significant with respect to the thickness of the dielectric substrate h (see sec. 6.3.1 on microstrips lines on printed circuit boards). The formula given by Cohn in graphical form is ascribed to Flammer[13].

5.4.2.2 Wire above a dielectric layer and ground plane

For the case with a $h-t$ layer of air in between, the characteristic impedance is

$$Z_c = \frac{Z_0}{2\pi\sqrt{\varepsilon_{r,\text{eff}}}} \operatorname{acosh}\left(\frac{1-u^2}{2v} + \frac{v}{2}\right) \tag{5.4.7}$$

where

$$u = \frac{1}{(4h/d)^2 - 1} \qquad v = \frac{2}{\dfrac{8h}{d} - \dfrac{d}{2h}} \tag{5.4.8}$$

and the effective relative permittivity is given by

$$\varepsilon_{r,\text{eff}} = \frac{\ln\left(\dfrac{4h}{d}\right)}{\ln\left(\dfrac{4(h-t)}{d} + \dfrac{4t}{d\,\varepsilon_r}\right)} \tag{5.4.9}$$

valid for quasi-TEM propagation.

5.4.2.3 Wire in a slab (sandwich of dielectric layer and ground planes)

The wire is surrounded by the dielectric layer (immersed into it) and that layer is lined up with ground planes both above and below. The dielectric layer thickness is D with permittivity ε_r and the diameter of the wire in the middle is d; the thickness of ground planes is not relevant.

The characteristic impedance is given by the following expression accurate to better than 1% for $D/d > 2$.

$$Z_c = \frac{Z_0}{2\pi\sqrt{\varepsilon_r}} \ln\left(\frac{4D}{\pi d}\right) \tag{5.4.10}$$

When the wire diameter d is large and approaches D, the more accurate equations given by Wheeler [334] are needed[14]:

$$Z_c = \frac{15}{\sqrt{\varepsilon_r}} \ln\left[1 + \frac{g_1 g_2}{2} + \sqrt{\left(\frac{g_1 g_2}{2}\right)^2 + 2g_2}\right] \tag{5.4.11}$$

[13] C. Flammer, “Equivalent radii of thin cylindrical antennas with arbitrary cross section,” *Stanford Research Institute Technical Report*, March 15, 1950.

[14] Formulas reported by Wadell [332], sec. 3.9.1, have a typo in the definition of g_2.

where

$$g_1 = \left(\frac{4}{\pi}\right)^4 \qquad g_2 = \left(1+\frac{2g}{w}\right)^4 - 1 \tag{5.4.12}$$

5.5 Paired and multi-conductor lines

This section contains expressions for various conductor grouping and assemblies: purposely built lines to carry signals, like digital transmission lines, or lines used for supply distribution or control signals, characterized also high-frequency signal propagation.

5.5.1 Parallel wires

This is a broad category of geometries and materials, embracing parallel flat cables, control cables for industrial automation and supply distribution lines, usually modeled and evaluated from different points of view, depending on application and context.

The characteristic impedance for two parallel conductors of diameter d at distance D in a homogeneous dielectric of relative permittivity ε_r was found expressed as

$$Z_c = \frac{Z_0}{\pi\sqrt{\varepsilon_r}} \operatorname{acosh}\left(\frac{D}{d}\right) \tag{5.5.1}$$

or

$$Z_c = \frac{Z_0}{2\pi\sqrt{\varepsilon_r}} \operatorname{acosh}\left(\frac{2D^2-d^2}{d^2}\right) \tag{5.5.2}$$

both declared valid for $d \ll D$. The two expressions are reported by Wadell [332], sec. 3.3.1, as belonging to Ramo-Whinnery the former, and W. Hilberg the latter[15]. They are not clearly declared equivalent, but the reader is induced in believing them equivalent under a wide range of d and D values; it is apparent, however, that they are different by simplifying the latter for $d \ll D$ and observing that "2" is inside the $\operatorname{acosh}(\,)$ function in one case and at the denominator outside it in the other case. Further searches confirm that the former is correct and may be found in many textbooks, while the latter found no confirmation.

Conduction losses with few substitutions in the original formula may be written as

$$\alpha_c = \frac{D}{2\sqrt{D^2-d^2}} \sqrt{\frac{f\varepsilon}{\sigma}} \operatorname{acosh}\left(\frac{D}{d}\right) \tag{5.5.3}$$

The factor α_c is in neper/m.

[15] S. Ramo and J.R. Whinnery, *Fields and Waves in Modern Radio*, Wiley, London, 1944.
W. Hilberg, *Electrical Characteristics of Transmission Lines*, Artech House, Norwood, MA, 1979.

5.5.2 Twisted pair

The concept is simple (twisting two parallel conductors), but the expressions describing the electromagnetic behavior are more complex than previous ones. This wire arrangement has long attracted a lot of attention because of several advantages [52]:

- it is mechanically convenient with minimum separation between the two conductors ensured by the twisting;
- coupling of external disturbance of the magnetic type is almost canceled, if differential signals are considered (the magnetic field induction common mode component remains unchanged with respect to simply parallel wires);
- the line inductance is minimized, thus reducing longitudinal voltage drop and pushing at higher frequency any resonance in combination with parasitics and connected capacitance.

Twisted pairs may be completely filled with dielectric or partially filled with air, leaving dielectric only for the insulating sheath around conductors. In any case, even if the cable is completely filled, the filler has normally dielectric properties different from those of the insulating sheath.

The expression for the characteristic impedance is that of two parallel wires, replacing the relative permittivity with the effective relative permittivity:

$$Z_c = \frac{Z_0}{\pi\sqrt{\varepsilon_{r,\mathrm{eff}}}}\,\mathrm{acosh}\left(\frac{D}{d}\right) \tag{5.5.4}$$

$$\varepsilon_{r,\mathrm{eff}} = 1 + q(\varepsilon_r - 1) \tag{5.5.5}$$

$$q = 0.25 + 0.0004\,\theta^2 \qquad \theta = \arctan(T\pi D) \tag{5.5.6}$$

having indicated with θ the pitch angle of the twist and with T the number of twists per unit length, e.g. the #twists/m.

However, considering the use of twisted pairs when reduction of line inductance is desired (even for short lengths, when it is not really behaving like a transmission line), it is also interesting and convenient to describe twisted pair behavior in terms of its total equivalent inductance L and capacitance C, from which the characteristic impedance may be derived as classical $Z_c = \sqrt{L/C}$.

$$L = \frac{\mu_0}{\pi}\,\mathrm{acosh}\left(\frac{D}{d}\right) \tag{5.5.7}$$

$$C = C_1 + C_2 + C_3 \tag{5.5.8}$$

$$C_1 = \int_a^b \frac{\varepsilon_0}{D + (1/\varepsilon_r - 1)\sqrt{D^2 - x^2} - \sqrt{d^2 - x^2}/\varepsilon_r}\,\mathrm{d}x$$

$$C_2 = \frac{\pi \varepsilon_0}{\arccos\left(\dfrac{D}{d}\right)} \qquad C_3 = \int_a^b \frac{\varepsilon_0}{D - \sqrt{d^2 - x^2}} \, dx$$

When ε_r is set to unity, the expressions may be used to calculate the capacitance of the air dielectric; otherwise ε_r should be set to the dielectric permittivity of the sheath material, or to an equivalent permittivity value that takes into account the geometry, the filling material or the air. Wadell [332], page 70, reports that when tested against experimental results, expressions are accurate within the tolerance of the measurement method and geometrical variations of the insulating sheath.

Regarding the value of the pitch angle θ, it is observed that it has little influence in the above formulation, that is however approximate and derived from that of parallel wires. In general pitch-angle value is set to have a compact and mechanically stable arrangement, but too large values around 40-45° and above may subject the wire to excessive stress and increase too much the overall length and weight. In multi-pair cables different pairs may have slightly different pitch angles in order to improve mechanical assembling and to reduce crosstalk between adjacent pairs.

5.5.3 Shielded pair

A shielded pair is shown in Figure 5.5.1. The formulation of Ramo, Whinnery and Van Duzer [279] gives the following characteristic impedance for even- and odd-modes valid for $d \ll D$ and $d \ll s$, the latter not so easy to meet in reality:

$$Z_{c,e} = \frac{Z_0}{\pi\sqrt{\varepsilon_r}} \left\{ \ln\left[\frac{2s}{d}\,\frac{1-(s/D)^2}{1+(s/D)^2}\right] - \frac{\left[1+4(s/d)^2\right]\left[1-4(s/D)^2\right]}{16(s/d)^4} \right\} \tag{5.5.9}$$

$$Z_{c,o} = \frac{Z_0}{2\pi\sqrt{\varepsilon_r}} \ln\left[\frac{s/d}{2(s/D)^2}\left(1-(s/D)^4\right)\right] \tag{5.5.10}$$

The attenuation terms for conduction and dielectric losses are:

$$\alpha_c = \frac{R}{2Z_0} \qquad \alpha_d = \frac{\pi\sqrt{\varepsilon_r \mu_r}}{\lambda_0}\,\frac{\varepsilon_r''}{\varepsilon_r} \tag{5.5.11}$$

where ε_r'' indicates the imaginary part of the complex relative permittivity.

The EN 50290-2-1 standard [66], at sec. 10.1, reports a more complex and articulated expression, that takes into account two different frequency dependencies, without separating explicitly conduction and dielectric losses (recognizable by inspecting the parameters that appear in the expressions) and leaving the expressions of each coefficient without clear explanation and incomplete.

$$\alpha = A + B\sqrt{f} + Cf \tag{5.5.12}$$

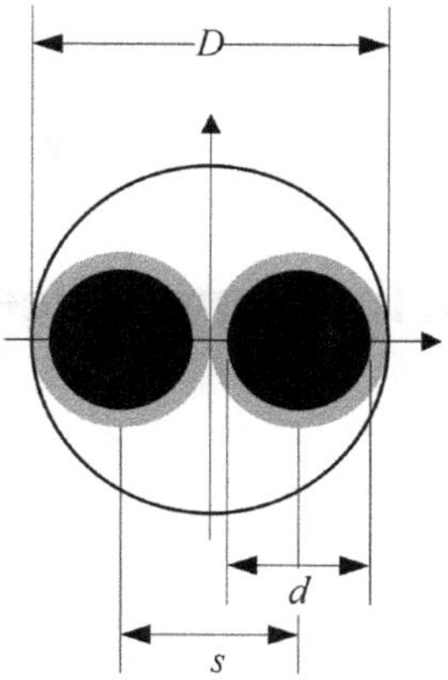

Figure 5.5.1 – Shielded pair (and twisted shield pair) geometry.

5.5.4 Shielded twisted pair

Let's consider a shielded pair whose inner conductors separated by s and each of diameter d , as shown in Figure 5.5.1, are twisted. Wadell [332] clearly says that there are no closed form expressions or analysis presenting expressions valid specifically for shielded twisted pair and that reference is made mostly to the equations for shielded pair. Their degree of approximation worsens and their validity is questionable for separation approaching conductor diameter (thin conductor sheath) and for cable diameter approaching twice the wire diameter (Wadell says "d approaching D", that is clearly a mistake). A reasonable approach that he proposes is to use twisted-pair equations for the odd mode (see sec. 5.5.2 above) and coax equations for even mode (see sec. 5.3)[16]. A similar result was obtained by Levin [221], who for the characteristic impedance of the odd-mode $Z_{c,o}$ finds half the value of a pair without the surrounding shield:

$$Z_{c,o} = \frac{Z_0}{2\pi} \ln\left(\frac{s}{d}\right) \tag{5.5.13}$$

The radius $R = D/2$ of the shield has no influence. Levin does not give explicitly the even-mode characteristic impedance, but his "electrodynamic wave impedance terms" when selecting one of the pair conductors and the shield give $Z_{c,e}$:

$$Z_{c,e} = \frac{Z_0}{2\pi} \ln\left(\frac{R}{d}\right) \tag{5.5.14}$$

Calculations were derived assuming that the internal conductors and their separation are small with respect to the shield radius.

---

[16] In K.P. Marx, "Propagation Modes, Equivalent Circuits and Characteristic Terminators for Multiconductor Transmission Lines with Inhomogeneous Dielectric," *IEEE Transactions on Microwave Theory and Techniques*, Vol. 21, n. 7, July 1973, pp. 450-457, propagation for shielded pair is comprehensively presented.

Many shielded twisted pairs may have small deviations from the ideal twisted pair because of i) eccentricity with respect to the cylindrical shield axis and ii) loosening of the two twisted conductors (so that separation s increases).

5.6 Measuring transfer impedance and admittance

This section complements sec. 5.1 collecting and discussing methods for the quantification of cable screening properties (namely transfer impedance and transfer admittance). The reference standard is the IEC 60096-1 [169], undergoing several editions and showing the synthesis of several discussions that have appeared also in the literature (almost always in *IEEE Transactions on Electromagnetic Compatibility* and related symposium proceedings). Some methods are not yet part of the IEC 60096-1, but possess equally good or even better performance of standardized methods.

The first method appeared as preliminary activity of the Sub-Committee 46A of IEC in 1973 (Fowler and Simons, as recalled in [147]), as the pioneering work at Boeing by Knowles and Olson [213] (published between 1970 and 1974), and other related works by Bridges, Miller, Madle, Oakley, Salt[17]; activities went on vigorously at the beginning of the '80s with multiple research directions and cooperation of the well known actors, Crawford, Eicher, Fowler, Halme, Sztenkuti, forming the ground for the many works to come in the '90s, the most relevant of which put forward by Broydé, Demoulin, Hoeft, etc.

Let's assume an electromagnetic field incident on the victim cable (the cable under test, CUT) and that the cable is not strongly coupled with such field, with cable diameter that is small compared to cable length, electromagnetic field wavelength and distance of the field source. "As the field at the surface of the screen is directly related to density of surface current and surface charge, the coupling may be assigned either to the total field or to the surface current and charge densities. Consequently we may simulate the coupling into the cable by reproducing through any means the surface currents and charges on the screen. Because we assume a cable of a small diameter, we may neglect higher modes." [147] There has been thus convergence on some methods, preferable for simplicity, for the covered frequency range, the minimum or maximum limits of sample length, or for intrinsic accuracy[18]:

- *triaxial method* (the oldest method proposed already by the SC46A in the '70s), subject to variants that require either a short-circuit termination or a matched termination; it is probably the best known method, yet with intrinsic limitation

---

[17] H. Salt, "The surface transfer impedance of coaxial cables," Proc. of the IEEE EMC Symposium, Seattle, WA, USA July 23-25, 1968, pp. 198-209.
R.J. Oakley, "Surface transfer impedance measurements – A practical aid to communication cable shielding design," Proc. of the IEEE 18th Intern. Wire and Cable Symposium, Atlantic City, NJ, Dec. 1969.
J.E. Bridges and D.A. Miller, "Standard EMC cable parameter measurements," Proc. of IEEE Southeastern Electromagnetic Compatibility Symposium, Ashbury Park, NJ, USA, June 17-19, 1969, pp. 264-269.
P.J. Madle, "Attenuation testing of shielded cables and packages," TRW Inter-Office Correspondence 69-7231.11-25, Oct. 1969.

[18] Agreement between results of different methods is always an issue, leading to discrepancies of some dB in many cases, ruled out by the comment "What is 6 dB between friends?" made by the Chairman of SC46A/WG1, as reported by Halme [147].

of sample length related to the maximum applicable frequency, before standing waves begin to appear;

- *quadriaxial method* (same age of triaxial method), proposed to overcome unwanted coupling to the external environment of the triaxial method and to have a more uniform current distribution on the cable shield, assured by better matching;
- *line injection method*, appearing in the late '80s first in the Swiss PTT publications and then adopted by IEC in 1990, normally ascribed to Eicher, is a simple method, suitable also for a time domain implementation (Fowler [128, 129]); it is quite useful in spotting out defects in braids and terminations;
- *reverberation or mode stirred chamber method*, illuminating the cable sample with an impinging electromagnetic field, virtually with no upper limit frequency, is probably the newest method, beginning to appear in the '80s (probably the first pioneering work is the one by Frankel[19]); it has received much interest in order to fix some initial discrepancies and to fully trace it back to the other methods, it is an expensive method for the need of a reverberating chamber and RF amplification, but it has virtually no upper frequency limit, while the low frequency limit is determined by the size of the chamber;
- *current injection method*, conceived and then proposed in the late '70s, appearing in the work documents of the IEC Technical Committee 46 and then published by members of this group and other experts in the late '80s and early '90s; current is applied into a triaxial structure following rigorously the definition of transfer impedance; differences in the results are determined by the current measuring point in the structure, i.e. on the positive terminal of the generator, on the return or around the cable under test before it enters the triaxial structure by means of current probes (each method has its drawbacks regarding susceptibility to leakage current, radiation losses, sensitivity, etc.) [103].

The most relevant methods [148] are reviewed in the next sections.

5.6.1 Triaxial method

The triaxial method is sometimes classified as a closed transmission line method or as a method of coupling with longitudinally homogeneous structures. The basic characteristic is that the driving signal is coupled onto the shield of the cable under test (line 2) by building around it another coaxial transmission line (line 1), of which the cable shield is the hot conductor and the return is along the outer cylinder (see Figure 5.6.1). Each transmission line has its own characteristic impedance and propagation constant

---

[19] S. Frankel, "Cable Shielding Effectiveness Testing Terminal Response of Braided-Shield Cables to External Monochromatic Electromagnetic Fields," *IEEE Transactions on Electromagnetic Compatibility*, vol. 16, No. 1, Feb. 1974, pp. 4-16. doi: 10.1109/TEMC.1974.303317

with obvious meaning of the subscripts. Since in the definition of Transfer Impedance in eq. (5.1.10) the shield current was assumed uniform and constant, care shall be taken of all resonances and propagation phenomena, so that impedance matching at the two ends shall receive special attention.

5.6.1.1 Considerations on design and construction

The so-formed external additional line 1 has its own characteristic impedance Z_{c1}, for matching to the driving source in order to maximize the applied power. It is evident that depending on the diameter of the cable under test d, the resulting characteristic impedance of the outer line Z_{c1} is subject to change, and thus the diameter of the outer conductor D shall be varied accordingly to keep the outer line matched. Being acceptable an approximate impedance match, the test fixture is usually built with a set of external diameters D_k to accommodate for the most common cables; usually a set of brass or copper tubes of 26, 40 and 60 mm diameter is predisposed to cover the widest range of mostly used coaxial cables.

The dielectric of the outer line 1 is air, except for the supporting dielectric disks (of thickness s), that hold the cable under test in the correct position along the center axis; the resulting characteristic impedance is given by the known formula for a coaxial line (see eq. (5.3.3)), using average relative permittivity. The average relative permittivity takes into account the percentage of space occupied by supporting disks and their dielectric permittivity; if they are not completely solid, but lightened by holes, then the volume permittivity will be lower. Given the total length l, the contribution of the disks is $\frac{ns}{l}\varepsilon_{r,disk}$, where n is the number of used disks. The average permittivity is

$$\varepsilon_{r,1,av} = \frac{ns\varepsilon_{r,disk} + (l-ns)\varepsilon_{r,air}}{l} \tag{5.6.1}$$

A second perturbing element is the cable sheath and its thickness t; it represents a coaxial capacitor C_{sh} in series to the remaining part of line 1 C_{air} with $\varepsilon_r = 1$:

$$C_{sh} = 55.556 \cdot \varepsilon_r \cdot \log\left(\frac{d}{d+2t}\right) \qquad C_{air} = 55.556 \cdot 1 \cdot \log\left(\frac{d+2t}{D}\right) \tag{5.6.2}$$

The fractional change of $C_{air,0}$ (the one without the plastic sheath, calculated with respect to cable screen) by the series connection of C_{air} and C_{sh} may be as large as 10%, sometimes even larger, and is much more relevant than the previous contribution of dielectric disks. The two phenomena act in opposite direction, but the effect of the cable sheath is preponderant and the two only partially compensate. Correspondingly, designing line 2 to match the $50\,\Omega$ target value of characteristic impedance with air as dielectric will lead to a slightly larger characteristic impedance.

The triaxial structure (shown in Figure 5.6.1) is suitable also for configurations in short circuit or in open circuit, rather than only matched-impedance configuration. When in short circuit the current is maximized in favor of the measurement of transfer impedance Z_t; in open circuit it is the transfer admittance Y_t that comes out. In matched conditions both exert their influence on the circuit [148]. For ease of compar-

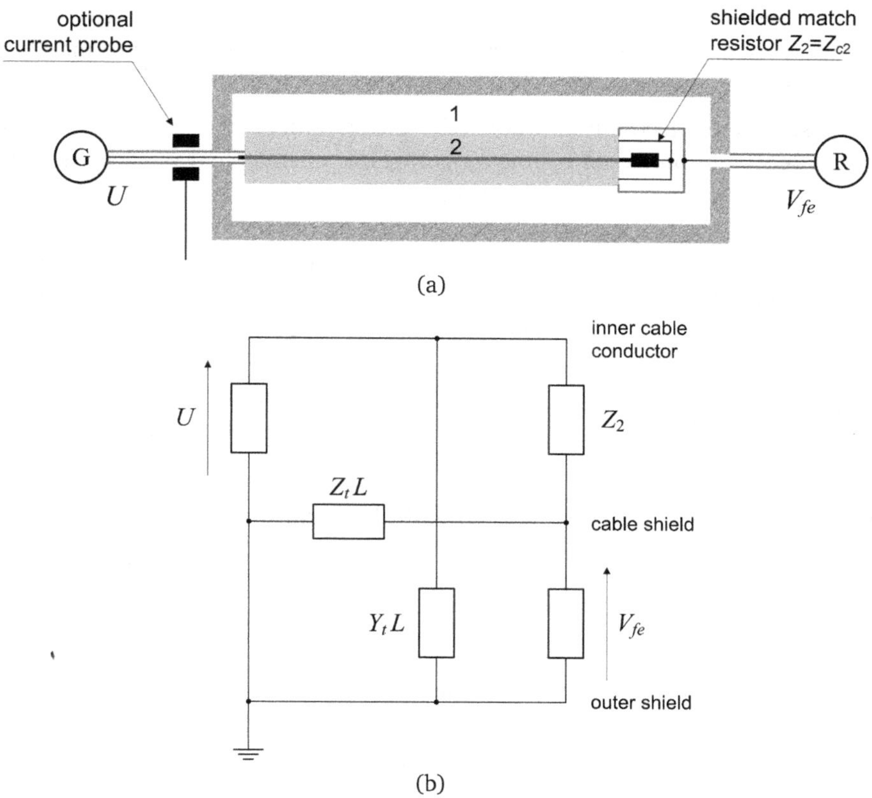

Figure 5.6.1 – Triaxial measurement method: (a) test fixture, (b) equivalent circuit.

ison another quantity is used in place of the admittance Y_t: it is the capacitive coupling impedance $Z_f = Z_{c1} Z_{c2} Y_t$.

For practical reasons the best choice to give a ground reference to the circuit is to ground the outer cylinder, that would be in any case very difficult to leave floating, due to its larger size and parasitic capacitance coupling to the external environment. It can thus be located on a ground plane and connected to the cold (or grounded) side of the VNA connector. One additional consideration is that also the other connector of the VNA is to be grounded; only some VNAs have independent floating port connectors (this is one of the cases where they are really useful!) and if using a spectrum analyzer with a RF generator, then both connectors are always bonded and grounded. It comes at hand to have one side of the outer cylinder short circuited on its internal conductor (that is the shield of the CUT), because in this way we are giving the same ground potential to that side of the CUT shield. However, keeping the outer line 1 short circuited at one end prevents any remarkable influence of capacitive coupling on that side of the line.

It is observed at the end of the next section that the triaxial structure and its equations lend themselves to exchanging the injecting line and the CUT (namely line 1 and line 2), using the CUT to apply a signal in matched conditions (the source is connected at one end, a matched termination at the other end) and measuring the voltage appearing across the other transmission line 2. It is obvious that the two conditions cannot be applied at the same time: so, either the outer cylinder is grounded, or it is connected to the receiver.

Since the short circuit condition allows the measurement of the transfer impedance alone, the receiver connected to line 2 operates in mismatched conditions; the measurement shall be carried out by taking the envelope of the measured voltage, neglecting the zeros given by the standing waves forming along the structure. Short structures will have too a high resonance frequency, so that the peaks of received voltage are too separated to give an accurate envelope. The same can be said for measurements performed in open circuit conditions, isolating the transfer admittance.

Given the equivalent circuit of the coupled transmission lines appearing in Figure 5.6.1 (b, c, d), the set of equations to derive screening terms and to relate them to measurable quantities are derived in the next sections [49, 148].

5.6.1.2 Basic near-end and far-end crosstalk equations for low frequency

The near-end and far-end crosstalk voltages appearing at the two ends of line 2 terminated on the own characteristic impedance Z_{c2} are:

$$V_{ne} = V_2(0) = \frac{L}{2} Z_t I_{s0} \frac{1-\exp\left(-(\gamma_1+\gamma_2)L\right)}{\gamma_1+\gamma_2} \tag{5.6.3}$$

$$V_{fe} = V_2(L) = -\frac{L}{2} Z_t I_{s0} \frac{1-\exp\left(-(\gamma_1-\gamma_2)L\right)}{\gamma_1-\gamma_2} \exp\left(-\gamma_2 L\right) \tag{5.6.4}$$

It is also possible to express the two voltages as a function of the applied test voltage, that does not require an additional current measurement; to this aim Halme's notation is used [148], adding explicitly the effect of transfer admittance through the capacitive coupling impedance $Z_f = Z_{c1} Z_{c2} Y_t$.

$$V_{ne} = V_2(0) = \frac{Z_f + Z_t}{\sqrt{Z_{c1} Z_{c2}}} \frac{L}{2} \frac{1-\exp\left(-(\gamma_1+\gamma_2)L\right)}{\gamma_1+\gamma_2} \tag{5.6.5}$$

$$V_{fe} = V_2(L) = -\frac{Z_f - Z_t}{\sqrt{Z_{c1} Z_{c2}}} \frac{L}{2} \frac{1-\exp\left(-(\gamma_1-\gamma_2)L\right)}{\gamma_1-\gamma_2} \exp\left(-\gamma_2 L\right) \tag{5.6.6}$$

If Z_f is negligible, as usually is especially at low frequency, the two expressions correspond in amplitude (neglecting line losses) and differ in phase. The expressions may be seen as the composition of terms related to the cable under test (those with subscript "2"), to its screening efficiency (the first numerator with Z_f and Z_t) and to the setup (the terms with the subscript "1").

Evaluating eq. (5.6.3) and (5.6.4) at low frequency, the products $\gamma_1 L$ and $\gamma_2 L$ are both much smaller than unity and the exponential functions may be well approximated by first order terms only, giving

$$V_{ne,LF} \cong \frac{1}{2} Z_t I_{s0} L \qquad V_{fe,LF} \cong -\frac{1}{2} Z_t I_{s0} L \tag{5.6.7}$$

whose difference brings to the quantification of the screening factor. This is true of course for a uniform shield current I_{s0}, so up to a maximum frequency limit dictated by test setup length L and line 1 velocity.

When the inducing line 2 is terminated on a short circuit to maximize screen current I_{s0}, at low frequency we may say that the current is given by the source voltage loaded by the source impedance, made equal to the characteristic impedance Z_{c2}.

The two expressions containing the propagation constants may be indicated using Halme's notation

$$S_{ne} = \frac{1-\exp\left(-(\gamma_1+\gamma_2)L\right)}{\gamma_1+\gamma_2} \qquad S_{fe} = \frac{1-\exp\left(-(\gamma_1-\gamma_2)L\right)}{\gamma_1-\gamma_2}\exp\left(-\gamma_2 L\right) \tag{5.6.8}$$

and they are responsible for the modulation of the envelope at high frequency, when the first phase rotation becomes relevant, that is when a half-period rotation is made:

$$f_{n,hf} = \frac{c}{\pi L\left|\sqrt{\varepsilon_{r1}}+\sqrt{\varepsilon_{r2}}\right|} \qquad f_{f,hf} = \frac{c}{\pi L\left|\sqrt{\varepsilon_{r1}}-\sqrt{\varepsilon_{r2}}\right|} \tag{5.6.9}$$

It is worth observing that frequency f and length L are interchangeable, that is fixing a frequency value there is a critical length above which resonances[20] appear and S factors begin rotating and reducing their intensity (20 dB/decade).

The cut-off frequency for far-end quantities $f_{f,hf}$ contains the difference of square roots of permittivities (that is the difference of reciprocals of velocities): having $\varepsilon_{r1} = \varepsilon_{r2}$ brings the cut-off frequency to infinite and the far-end voltage becomes a smooth curve much more suitable than near-end one for analysis. Of course, only an approximation of the equality condition is possible due to, first, different materials (the outer cylinder has mostly air as dielectric, even if plastic supports exert their influence, see sec. 5.6.1.1), and then practical tolerances and approximations.

Thus, at high frequency the expressions containing exponential terms cause both attenuation and phase rotation, so there is a reduction in the measured near-end and far-end voltages, as well as repeated zeros and local maxima. We may state that the useful frequency interval for the two expressions extends up to where the two terms are within an arbitrary threshold e.g. 10% (i.e. 1 dB) from the low frequency value. As shown in Figure 5.6.2, the useful frequency interval of the far-end curve extends always more than that of the near-end curve, so that it is advised "side" for the measurement in all cases.

So far matched conditions at both ends of the two concentric coaxial lines have been assumed: for practical reasons and also — as explained at the beginning — to spot out either of the two transfer impedance and admittance terms, mismatching and reflection may occur at either end. This has the effect of mixing the near- and far-end terms, having no infinite absorbing effect at line ends any longer. The two equations

[20] We call "resonance" in this case not a properly said resonance of the triaxial setup, rather a singularity of the sum and difference of coupling terms.

(5.6.5) and (5.6.6) generate mixed end voltages, that are distinguished by appending a "r" for reflection:

$$V_{ne,r} = V_{ne} + \Gamma V_{fe} e^{-\gamma_2 L} \qquad V_{fe,r} = V_{fe} + \Gamma V_{ne} e^{-\gamma_2 L} \tag{5.6.10}$$

Halme et al. state that with much different near- and far-end coupling terms, even for moderate reflection, the result may be highly variable and erroneous, because one of the reflected terms is absolutely not negligible. However, in the extreme case of e.g. short circuit at the near end, V_{ne}, will be negligibly small and despite the complete reflection the V_{fe} term is perfectly measurable and correct.

In reality, the short-circuit configuration on the outer line is well used because of practical convenience and to configure the setup in a known mismatch condition. Breitenbach et al. [49] report the resulting equation (their eq. (3-8)) with a slightly different notation and using the "reversed" configuration, that sees the signal applied to the cable under test and the voltage measured at either end of the outer transmission line, that becomes line 2. In synthesis the ratio of the far-end voltage (the short circuit is located at the near end) and the applied voltage is an oscillating function where a third term appears: besides the two velocities given by the sum and difference of line velocities, the presence of the exponential term that describes the reflection along the whole length of the setup introduces a third term that depends on the velocity of waves on the measuring line (line 2) traveling back and forth the line length, so with a factor of 2 more. The measured waveform has periodic maxima that are retained to identify the envelope as the quantification of cable screening:

$$\left|\frac{V_{fe}}{V_1}\right|_{\max} = \frac{c}{\omega Z_{c1}} \left| \frac{Z_t - Z_f}{\sqrt{\varepsilon_{r1}} - \sqrt{\varepsilon_{r2}}} + \frac{Z_t + Z_f}{\sqrt{\varepsilon_{r1}} + \sqrt{\varepsilon_{r2}}} \right| \tag{5.6.11}$$

Up to the first cut-off frequency Z_f is negligible compared to Z_t that may be thus easily determined.

In the next section there is a brief review of Kley's expressions to address the more general problem of the influence of the terminating conditions.

5.6.1.3 Kley's notation for mismatched conditions

To distinguish the effect of terminal conditions regarding mismatch (i.e. matched termination, short circuit and open circuit), Kley's notation [211] is far more complete and suitable for reasoning on different termination combinations; only a few changes are made to improve readability and interpretation of expressions and their symmetry. Kley's scheme has one side identified by the longitudinal position $x = 0$ and the opposite side by $x = l$, suggesting the idea that they stand for near end and far end, respectively, whereas the generator is connected at $x = l$ and this defines the near-end side. So we will use "n" and "f" to indicate generator side and the opposite one. Finally, Kley distinguishes between the characteristic impedance of the two transmission lines, using "1" for the inner CUT line; the terminating impedances, assumed of resistive nature, are originally indicated with R_0 and R_1 for t.l. 1 and R_2 and R_3 for t.l. 2, with R_0 and R_2 at the far-end, opposite to the generator, and R_1 being the internal

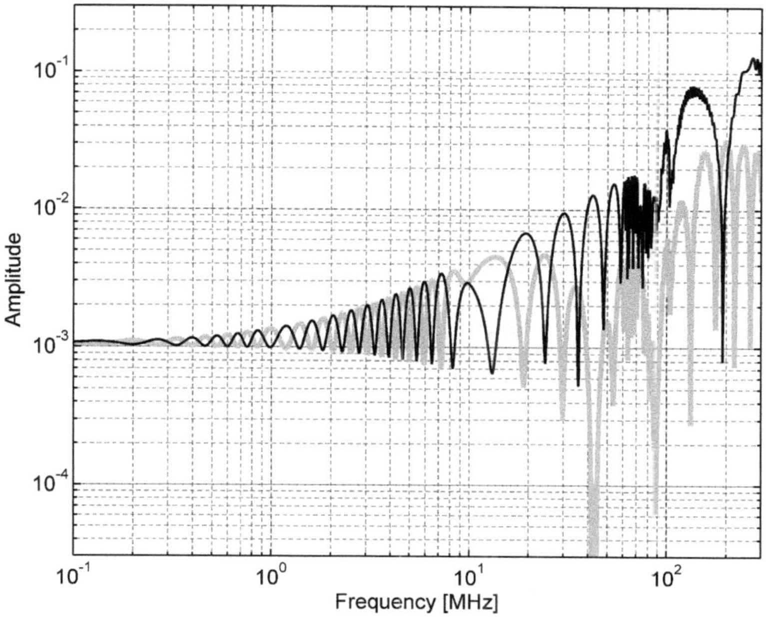

Figure 5.6.2 – Typical near-end (gray) and far-end (black) voltages normalized to the input voltage for a 8.5 mm diameter cable single braid with low optical coverage (about 0.88): the far-end voltage has slightly wider oscillations than the near-end voltage; the curve envelopes are flat at low frequency, then increasing with Z_t, and at about 30 MHz the larger Y_t (that we might expect due to the low optical coverage) begins dominating.

source resistance. It is however preferable to explicit "n" and "f" like in "1n" that is R_1, and "1f" that is R_0.

$$V_{2,ne} = Vl \frac{R_{2f}}{(R_{1n}+R_{1f})(R_{2n}+R_{2f})} [Z_t g_n + Y_t Z_{c1} Z_{c2} h_n] \tag{5.6.12}$$

$$V_{2,fe} = -Vl \frac{R_{2n}}{(R_{1n}+R_{1f})(R_{2n}+R_{2f})} [Z_t g_f + Y_t Z_{c1} Z_{c2} h_f] \tag{5.6.13}$$

where g and h are corrective coefficients to account for reflections and standing waves:

$$\begin{aligned} g_f &= \frac{1}{D}\left\{\left[\gamma_1 \frac{R_{1f}}{Z_{c1}} + \gamma_2 \frac{R_{2n}}{Z_{c2}}\right][\cosh(\gamma_1 l) - \cosh(\gamma_2 l)] + \gamma_1 \sinh(\gamma_1 l) + \right. \\ &\left. -\gamma_2 \sinh(\gamma_2 l) + \frac{R_{1f} R_{2n}}{Z_{c1} Z_{c2}} [\gamma_2 \sinh(\gamma_1 l) - \gamma_1 \sinh(\gamma_2 l)]\right\} \end{aligned} \tag{5.6.14}$$

$$\begin{aligned} h_f &= \frac{1}{D}\left\{\left[\gamma_2 \frac{R_{1f}}{Z_{c1}} + \gamma_1 \frac{R_{2n}}{Z_{c2}}\right][\cosh(\gamma_1 l) - \cosh(\gamma_2 l)] + \gamma_2 \sinh(\gamma_1 l) + \right. \\ &\left. -\gamma_1 \sinh(\gamma_2 l) + \frac{R_{1f} R_{2n}}{Z_{c1} Z_{c2}} [\gamma_1 \sinh(\gamma_1 l) - \gamma_2 \sinh(\gamma_2 l)]\right\} \end{aligned} \tag{5.6.15}$$

$$\begin{aligned} g_n &= \frac{1}{D}\left\{\left[\gamma_1\frac{R_{1f}}{Z_{c1}}-\gamma_2\frac{R_{2f}}{Z_{c2}}\right][1-\cosh(\gamma_1 l)\cosh(\gamma_2 l)]+\right.\\ &-\left[\gamma_1\frac{R_{1f}R_{2f}}{Z_{c1}Z_{c2}}-\gamma_2\right]\cosh(\gamma_1 l)\sinh(\gamma_2 l)+\\ &-\left[\gamma_1-\gamma_2\frac{R_{1f}R_{2f}}{Z_{c1}Z_{c2}}\right]\sinh(\gamma_1 l)\cosh(\gamma_2 l)+\\ &\left.-\left[\gamma_1\frac{R_{2f}}{Z_{c2}}-\gamma_2\frac{R_{1f}}{Z_{c1}}\right]\sinh(\gamma_1 l)\sinh(\gamma_2 l)\right\} \end{aligned} \tag{5.6.16}$$

$$\begin{aligned} h_n &= \frac{1}{D}\left\{\left[\gamma_2\frac{R_{1f}}{Z_{c1}}-\gamma_1\frac{R_{2f}}{Z_{c2}}\right][1-\cosh(\gamma_1 l)\cosh(\gamma_2 l)]+\right.\\ &-\left[\gamma_2\frac{R_{1f}R_{2f}}{Z_{c1}Z_{c2}}-\gamma_1\right]\cosh(\gamma_1 l)\sinh(\gamma_2 l)+\\ &-\left[\gamma_2-\gamma_1\frac{R_{1f}R_{2f}}{Z_{c1}Z_{c2}}\right]\sinh(\gamma_1 l)\cosh(\gamma_2 l)+\\ &\left.-\left[\gamma_2\frac{R_{2f}}{Z_{c2}}-\gamma_1\frac{R_{1f}}{Z_{c1}}\right]\sinh(\gamma_1 l)\sinh(\gamma_2 l)\right\} \end{aligned} \tag{5.6.17}$$

$$\begin{aligned} D &= (\gamma_1^2-\gamma_2^2)l\left\{\cosh(\gamma_1 l)+\frac{\sinh(\gamma_1 l)}{R_{1f}+R_{1n}}\left[Z_{c1}+\frac{R_{1f}R_{1n}}{Z_{c1}}\right]\right\}\\ &\left\{\cosh(\gamma_2 l)+\frac{\sinh(\gamma_2 l)}{R_{2f}+R_{2n}}\left[Z_{c2}+\frac{R_{2f}R_{2n}}{Z_{c2}}\right]\right\} \end{aligned} \tag{5.6.18}$$

Equations may be analyzed for cases of matched impedance (R terms equal to characteristic impedance), short circuit (some R terms equal to zero) and open circuit (some R terms brought to infinite) conditions. Kley observes that these three conditions may be used to measure Z_t and Y_t with different frequency ranges of validity: the matched condition has the widest range of validity and gives the combination $Z_t \pm Y_t Z_{c1} Z_{c2}$, the short-circuit condition gives readily Z_t and the open-circuit condition Y_t.

5.6.1.4 Measurement methods

The IEC 60096-1 standard [169], including amendment 2 [170], reports measurement methods for the low-frequency resisitive-inductive transfer impedance measurement and for transfer admittance at higher frequency, applying the signal both directly and through matching to the outer line characteristic impedance, extending the frequency range and improving the accuracy by using baluns (common-mode ferrite rings). Some interesting considerations may be found in a two-part paper written by Mund and Schmid for Wire & Cable Technology International[21].

Transfer impedance - Feeding through a resistance The generator is connected through a resistor R set to 1.4 times the characteristic impedance of the outer line; the length of the coupled inner and outer lines is l. Indicating with U_1 the applied voltage measured before the said resistance and with U_2 the voltage measured at the end of the cable under test by a matched coupled instrument, the transfer impedance is

$$|Z_t| = \frac{2R}{l}\frac{U_2}{U_1}F' \tag{5.6.19}$$

where F' is a correction factor for the frequency response of the system given by

$$F' = \frac{x\,(1-n^2)\,\sqrt{\cos^2 x + m^2 \sin^2 x}}{\sqrt{n^2(\cos x - \cos nx)^2 + (\sin x - n \sin nx)^2}} \tag{5.6.20}$$

where $m = Z_1/R$ is the ratio of the characteristic impedance of the outer line to the feeding resistance (set to 1.4 if possible) $n = \lambda_1/\lambda_2$ but is better defined directly as the ratio of propagation velocities, $n = v_2/v_1$, $x = \dfrac{2\pi l}{\lambda_1}$.

Transfer impedance - Direct feeding If the outer line is fed directly the voltage U_1 is the voltage directly applied to the line, larger than what was applied by using the series connected resistor R and this turns out useful to have larger voltages available, e.g. when the expected transfer impedance is low.

$$|Z_t| = 2Z_1\frac{2\pi}{l}\frac{U_2}{U_1}F'' \tag{5.6.21}$$

where F'' is a correction factor for the frequency response of the system given by

$$F'' = \frac{(1-n^2)\,\sin x}{\sqrt{n^2(\cos x - \cos nx)^2 + (\sin x - n \sin nx)^2}} \tag{5.6.22}$$

where $m = Z_1/R$ is the ratio of the characteristic impedance of the outer line to the feeding resistance, set to 1.4 if possible, $n = \lambda_1/\lambda_2$ but is better defined directly as the ratio of the propagation velocities, $n = v_2/v_1$, $x = \dfrac{2\pi l}{\lambda_1}$.

Transfer admittance - Direct feeding The triaxial setup is surrounded by an additional screen becoming a quadriaxial setup, in order to prevent the energy from radiating out of the system, but also to avoid unwanted disturbance coupling, considering the fairly low expected admittance values. The setup is fed directly from the

[21] B. Mund and T. Schmid, "Measuring EMC of HV Cables & Components with the Triaxial Cell - Part 1, - Part 2," *Wire & Cable Technology International*, Jan. 2012, pp. 88-90 and 116-118.

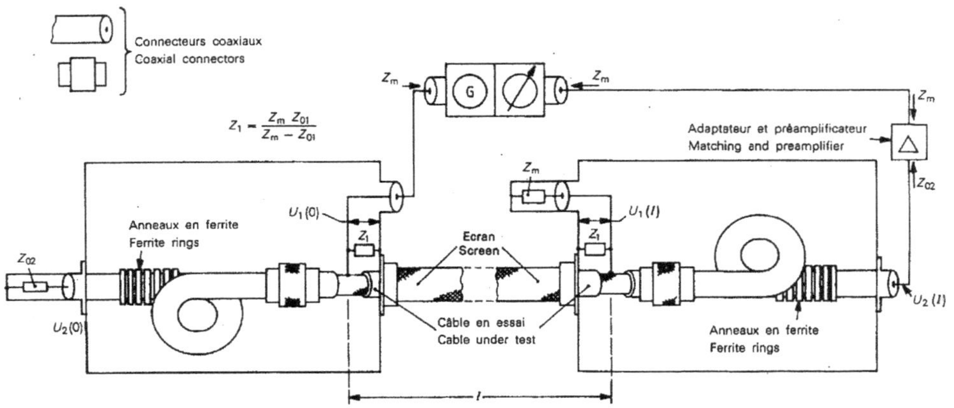

Figure 5.6.3 – Balun lined triaxial setup as per IEC 60096-1, A5.

generator; the length of the setup l shall be short compared to the wavelength λ (e.g. one thirtieth), limiting thus the frequency range: $l = 1\,\mathrm{m}$ is a quite common value and it is suitable up to about 30 MHz.

$$|Y_t| = \frac{2}{Z_2 l} \left|\frac{U_2}{U_1}\right| \frac{f_0}{f_m} \tag{5.6.23}$$

where $f_0 = 30\,\mathrm{MHz}$ and f_m is the measuring frequency.

Transfer impedance and admittance - Balun terminated setup This method is proposed in IEC 60096-1, sec. A5.4, as valid up to 1 GHz. The inner line is terminated at both ends by coaxial baluns made of ferrite rings, avoiding current flowing on the outer surface of the cable sample (see Figure 5.6.3). The outer line is match terminated at best on its characteristic impedance, taking into account the effect of the generator.

$$|Z_t| = \frac{2Z_1}{l} \left|\frac{U_2(l)}{U_1(l)}\right| F''' \tag{5.6.24}$$

By indicating $U_1(l)$ and $U_2(l)$ it is specified that these voltage shall be measured at the far end with respect to the point of injection, simply because reflections and stationary waves along the structure are relevant now that the frequency is allowed to increase up to 1 GHz, so with about 2.5-3 wavelengths contained in a $l = 1\,\mathrm{m}$ long cable sample.

$$F''' = \frac{l\sqrt{(\alpha_1-\alpha_2)^2+(\beta_1-\beta_2)^2}}{\sqrt{\left[1-e^{(\alpha_1-\alpha_2)l}\cos(\beta_1-\beta_2)l\right]^2+\left[e^{(\alpha_1-\alpha_2)l}\sin(\beta_1-\beta_2)l\right]^2}} \tag{5.6.25}$$

where α_1 and α_2 are the attenuation of the outer and inner line, and β_1 and β_2 are the corresponding phase constants.

Attenuation and phase constants may be determined with a separate measurement that characterizes the triaxial setup, e.g. based on the open- and short-circuit combined impedance measurement discussed in sec. 2.4.

5.6.2 Line injection

The method has been described as a modified triaxial method, where the external coaxial transmission line is replaced by a two-wire system (or multi-wire system, if more than one injection wire is used). When one injection wire is used, it is run parallel externally to the cable shield, with which it forms a two-wire transmission line, whose characteristic impedance is matched at best with the impedance of the signal source (usually $50\,\Omega$). The setup has one degree of freedom, that is the side where the source is to be connected, or in other words if the injection is to be at the near end or far end.

The method is quite well described in [117, 159]; Sztenkuti [315] credits the invention to Fowler[22], back in 1971. The injecting line is applied to the external surface of the cable sample and is fed with the driving signal at either end, between the hot trace (the applied conductor) and the cable shield: each end shall be terminated on the characteristics impedance. Practically the injection line is built around the cable sample, following at best the circumference of the cable by means of a flat copper strap, a single wire or a set of wires. The so-formed transmission line includes thus the shield as return plane of an approximate microstrip, with the cable sheath behaving as the inner dielectric. Impedance matching is reached easily by calculating the characteristic impedance of the curved microstrip and adjusting the width of the hot trace: formulas given in sec. 6.3.1 may be used, assuming that the hot trace is narrow enough not to worry about the curvature. For a PVC cable jacket of 1 mm thickness, the expected hot trace width w for $50\,\Omega$ characteristic impedance is $2.5\,\mathrm{mm}$ using an adhesive copper band 5 to 10 hundredths of mm thick, applied to the outer surface of the cable sheath.

Not only characteristic impedance, but also phase velocity, of the injecting line are matched as closely as possible to that of the victim line, that is the cable sample. It is generally believed that microstrips are slower than coaxial lines, but this is true for PCB dielectric with larger permittivity than cable dielectric (nearly twice, for FR4 with respect to polyethylene or PTFE), so that the capacitive term is prevalent. The PVC of the cable sheath (with a relative permittivity lower than 3 and normally in the range 2.7-2.9) and the moderate width of the hot trace compensate a lot for capacitive effects, so that the effective relative permittivity ε_{re} ranges between 2.2 and 2.3 for the considered values of sheath thickness and permittivity. The resulting velocity factor is $1/\sqrt{\varepsilon_{re}}$, so around exactly 0.66, as for most polyethylene dielectric cables. Reduction of ε_{re} to accommodate for faster cables with velocity factor around 0.7 may be attained by increasing slightly the thickness of the hot trace (the one applied externally onto the cable sheath) up to e.g. some tenths of mm.

It is important that common mode coupling with external circuits is avoided and that the driving current is all flowing in the injecting line, so that ferrite rings that increase the common mode impedance are applied to the feeders of the injecting line and on the two sections outside the coupled central section.

---

[22] E.P. Fowler, "On the Interference Immunity of Coaxial Cables", Proc. IEEE Region 8 Convention, Lausanne, Switzerland, Oct. 1971, par. B11

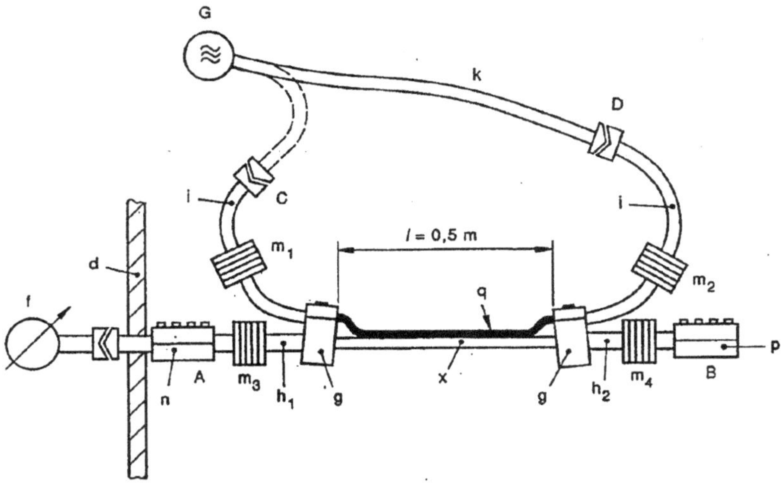

Figure 5.6.4 – Setup of the line injection measurement method as in IEC 96 [170]: m_1 to m_4 are decoupling ferrite rings; h_1 and h_2 additional screening with metallic tubes; x and q are the CUT and the injecting line respectively.

It is customary to measure the voltage on the inner conductor at the opposite side from where the driving signal is applied, thus performing a far-end crosstalk measurement. The method lends itself also to specular measurement of the near-end type, by exchanging the driving point; since near-end and far-end crosstalk terms result from the difference and the sum of the inductive and capacitive terms, respectively, their separate measurement allows to distinguish the constituents of the transfer impedance. It is important to remember that since RF instrumentation (generators, spectrum analyzers, VNAs) are single-ended grounded equipment, when injecting and measuring at opposite sides, there is an unwanted equipotential bonding between the reference grounds, creating an additional shunting path. There are various solutions: the injected current may be forced into the system using the already mentioned ferrite rings increasing the common-mode impedance (as indicated in the test setup of IEC 96, shown in Figure 5.6.4), an isolation transformer may be used to separate instrumentation grounds at the risk of high-frequency resonance, or measurements may be done using a fiber-optic connected power meter (of course without phase information).

As pointed out by Eicher and Boillot [117], the weak point of the method, but also its strength, is that the injection line illuminates only a part of cable shield, covering a fraction of the circumference, so that more than one measurement is necessary to adequately cover it entirely, spotting out possible inhomogeneities and defects. Under ideal conditions of a uniform shield all measurements should give the same result.

Measurement of Z_t is performed with the following sequence:

- calibration of feeding and measuring parts by short circuiting near and far end;
- measurement of insertion loss of the injection circuit A_i: the generator and receiver (or the two VNA ports) are connected across the injection line;

- measurement of insertion loss of the cable under test A_c: the generator and receiver (or the two VNA ports) are connected across the c.u.t.;
- measure the near-end transmission A_{ne} injecting and measuring the c.u.t. voltage at the same side, with both terminals at the far-end side match terminated;
- measure the far-end transmission A_{fe} injecting at the near end and measuring the c.u.t. voltage at the far end, with the unused terminals match terminated;
- all four A measurements shall be corrected for the insertion loss measured during calibration; then the quantity A_t may be calculated for either near-end or far-end measurements: $A_{t,ne} = A_{ne} - A_i/2 - A_c/2$, $A_{t,fe} = A_{fe} - A_i/2 - A_c/2$;
- Z_t is determined as

  $$Z_{t,n} = \frac{2}{l}\sqrt{Z_{c,i}Z_{c,c}}\,\mathrm{ENV}(T_n)\,[1+\pi f l] \text{ or}$$

  $$Z_{t,n} = \frac{2}{l}\sqrt{Z_{c,i}Z_{c,c}}\,\mathrm{ENV}(T_n)\left[1+\pi f l\,\frac{v_i - v_c}{v_i + v_c}\right],$$

  where ENV stands for "envelope", v_i and v_c are the propagation velocities in the injection and c.u.t., respectively, and T are the linear equivalents of the measured transmissions in dB, that is $T_n = 10^{-A_{t,ne}/20}$ and $T_f = 10^{-A_{t,fe}/20}$.

Last, the question arises if the results obtained with the line injection method are consistent and similar to those obtained with other methods, such as the triaxial method, in particular observing the most relevant difference, that is the inhomogeneity of the former with respect to the theoretical circular symmetry of the latter. Hoeft [159] states this saying "same results" and verifies it by comparison with results obtained with the quadriaxial method applied to a RG 213 sample of 1.12 m length. Additional measurements were made also on shorter samples, 0.5 and 0.3 m long.

The transfer impedance obtained with the far-end line injection method matches in the low frequency range the measured dc resistance and has no sign of high frequency resonances indicating any supervening limitation (long line effects are barely visible above 500 MHz), increasing almost linearly with frequency as expected; some deviation from the expected linear behavior is observed in the medium frequency range (between 300 kHz and 50 MHz).

Conversely, when using the near-end line injection method, resonances appear and they are quite comparable with those observed when the quadriaxial method is applied. After concluding that the line injection method is as good as other consolidated and accepted methods in their frequency range of validity and that can be used at much higher frequency, the authors try to explain the observed slight deviation from linearity and determine if it is due to the method itself. The first explanation for such behavior is that porpoising occurs on the braided shield of the RG 213 cable sample. To verify this the authors [159] built a cable with solid shield (no porpoising) and one defect, responsible for the measured transfer impedance: the measured curve showed no deviation from the expected linear behavior, so that the observed phenomenon was due to the cable sample itself and not to the line injection method, that results in a reliable and flexible method, substantially in agreement with the other accepted methods. The same conclusion is supported by Eicher and Boillot [117], who, commenting one of their references, indicate a maximum deviation of ±6 dB between the two methods.

5.7 Results on repeatability and non-linearity

When measuring or verifying the characteristics of DUTs and setup elements, the laboratory environment is normally assumed. Changing the environment for outdoor applications or applying harsh environments and accelerated aging in simulated conditions will impose further restrictions or steer the choice towards other solutions, largely relying on datasheet information: materials shall be suitable for temperature excursion, but their performance will be unavoidably influenced and largely varying; a combination of countermeasures and provisions will be needed to protect the setup from environmental agents (protection with gel or plaster against humidity, thermal shunting by conductive elements or thermal insulation, etc.); movement, vibration and shaking are particularly challenging, also because there is lack of complete information about cable behavior, and cable-connector combination (setup elements shall be hold in place and protected against transmission of mechanical stress and mechanical resonances with the due care; flexible cables shall be preferred to rigid ones, that transmit quite efficiently stress and vibrations and are in general heavier).

Cables are specified for commercial applications (following the standards reviewed before in this chapter), but a complete standard for laboratory applications similar to the IEEE Std. 287 for connectors is missing; the IEC 61935-1 and IEC 60096-1 standards are quite all-comprehensive for cable performance, but are not responsible to indicate limits and recommended values for various applications. In general cables have less variability than connectors and a denomination such as the RG classification gives enough confidence in their performance. Characteristic impedance and attenuation are clearly defined in this way as they appear in datasheets, but transfer impedance, stability with aging and temperature, and in general repeatability are almost always neglected. Cable characteristics have changed remarkably since the '70s, so that the lifetime of the results in scientific publications has shortened correspondingly. Additionally, the need of evaluating characteristics such as transfer impedance or mode conversion was felt only with the progress in EMC and fast data connections.

Transfer impedance and admittance were dealt with in sec. 5.1.9 and basic properties such as characteristic impedance, attenuation and propagation velocity were considered in sec. 2.4 and 5.3.3 for coaxial cables, and are easily extendable to other types of cables, here the attention is on repeatability and distortion, normally not considered by manufacturers datasheets. A few reliable data and references regarding these characteristics are reported below, aiming to guide the reader in an overall estimation of the phenomena and their significance.

5.7.1 Repeatability of Return Loss and Insertion Loss

Excluding evident loss of performance due to various defects, such as oxidation of braid and microphonic effect, non-linear behavior causing passive intermodulation (see sec. 5.7.2), or also lack of production control with out-of-control tolerance, such as the repetitive variation of cable diameter while extruding (see sec. 5.3.2.5), the remaining variability may be analyzed theoretically by applying the considerations and expressions reported above in sec. 5.3.2. To author's knowledge references to measured repeatability and variability of cable characteristics during normal use are very few. Lewandowski in his Ph.D. thesis reports a few measurement results that are however a combination of the effects of cable flexure on the cable itself and end connections, that he cannot exclude. Similarly, some tests performed by the author to exacerbate the problem, clearly identify the impact on connector of cable-induced mechanical stress as the most relevant issue.

5.7.1.1 Weak cable/connector interaction, Lewandowski (2004)

Lewandowski [222] presents several results while discussing the main topic of his Ph.D. thesis, that is modeling of VNA uncertainty. Fig. 6.7 in [222] reports the repeatability of 16 measurements performed on a 1.85 mm coaxial male offset short in the form of a curve of sample dispersion of the magnitude of the reflection coefficient versus distance (estimated from time domain measurements assuming a speed of light propagation). A stochastic model was used also (one of the versions of NIST StatistiCal[23]) where the applied perturbations were based on the microscope analysis of the conditions of the connectors, observing non-repeatability of the connector joint at the mating plane, flexing of the center conductor and supporting bead (occurring at about 3 mm away from the mating plane) and bending of the connector socket fingers (occurring at about 1.1 mm away from the mating plane); so it is evident that it is not the repeatability of the cable itself and its characteristics that is evaluated, but the consequential impact of cable flexure on terminal connections. Two peaks of dispersion of the reflection curve located mainly at two distances that correspond to the two ends of the cable were identified, the far end at about 700-800 mm and the near end at about 50 mm; the latter was the largest (about 45 ppm against 25 ppm) and the most interesting, for which they proceed to zoom discovering several peaks at 55, 30 and 5 mm. The points at 30 and 5 mm distance from the VNA port plane are already within the adapter that connects the cable to the VNA and, as expected, this is a very critical point, where the stress exerted by cable flexure (or bending) concentrates and applies: tilting of the inner pin, as well as flexure of adapter parts are possible, as it is observed in practice. Especially when using semi-rigid cables and in general heavier low-loss cables together with small connectors such as SMA, 3.5 mm and 2.92 mm, mechanical stress and its influence on setup stability and repeatability is always a concern: using a stabilizing collar (e.g. metallic band or tape with resin compatible with the cable sheath material) around the entry point into the connector is always a good idea when looking for home-made patches; flexible high-quality microwave cables and cable assemblies may be used instead, to ensure the least transmitted mechanical stress and a firm retention at connectors.

[23] StatistiCal™ is the newest VNA calibration software package where much of the work appearing in [222] is reflected; available online: http://www.nist.gov/ctl/rf-technology/related-software.cfm.

The values reported in Table 5.7.2 are quite small and correspond probably to the best dispersion attainable, the involved NIST staff taking all precautions in terms of methods and components to minimize the impact of cable flexure.

5.7.1.2 Extreme cable flexure and connector response

Measurements were made by recording the variability of return loss and insertion loss under extreme flexure with weak connectors, such as commercial standard-grade SMA. It was said that the entry point of the cable into the terminal connector shall be robust and that an additional collar may be useful, but the mechanical stress created by cable flexure applies all at the SMA mating plane, towards the VNA port. In this case there are few chances to improve mechanical mating, unless connectors are changed reverting to more robust connector types, such as type N, or laboratory precision grade connectors (see sec. 7.1.1). Of course, it is always possible to limit cable movement by proper setup.

For the results shown in Figure 5.7.1, significant flexure was applied to a semi-rigid cable (LMR 200 type) terminated by SMA; bending geometries were 90° and U-shape bend with curvature radius of about 15 cm with a cable length of 1 m. The curves of S_{11} and S_{21} errors (calculated by difference with respect to a reference S_{11} and S_{21} measurement at rest) show a periodic variation that is clearly visible in the S_{21} curve, repeating every about 400 MHz. This value is compatible with the length of the cable (exactly 95 cm) and the velocity factor of about 0.7. The repetitive maxima are clearly visible as predicted by Juroschek [202], whose method for interpretation of connector defects in S_{11} and S_{21} is discussed in sec. 7.1.1.7: as predicted, the height of maxima increases with frequency, even if the behavior is not monotonic.

On the contrary, flexible high-quality cables, headed with suitable high-performance stable connectors, are the correct choice for extremely accurate measurements or when environmental factors or long mission times require so. Nevertheless, the overall setup and the way cables are held and kept in place (e.g. by using clamps, wire straps, etc.) play a fundamental role, especially for the response to vibrations and shock (as underlined in sec. 5.7.2 below). As an example Phaseflex® by Gore is quite often used for microwave accurate test setups, even if in [198] its use is suggested also for long service times, in order to reduce maintenance and replacement: performance results are very good, not so much in static conditions compared with brand new samples of competitors, as after relevant mechanical stress was applied. Repeated flexure and accelerated aging were applied while measuring insertion loss in magnitude and phase; the latter, in particular, showed to be particularly selective for the compared cable samples, whilst it is not a so often measured performance. The effect of variability of the phase response was additionally analyzed as for its effect on the accuracy of VNA repeated measurements [269]: two successive return loss measurements of a low-loss DUT (1.85 mm male-female adapter) are made, the second one with a significant flexure of the used cable extension; the resulting difference is -50 dB on average, the maximum where some deviation of measured return loss is appreciable being -45 dB. Similarly, tests performed on the insertion loss curves show some difference, that — as anticipated — is not so dramatic in amplitude (± 0.008 dB maximum up to 50 GHz),

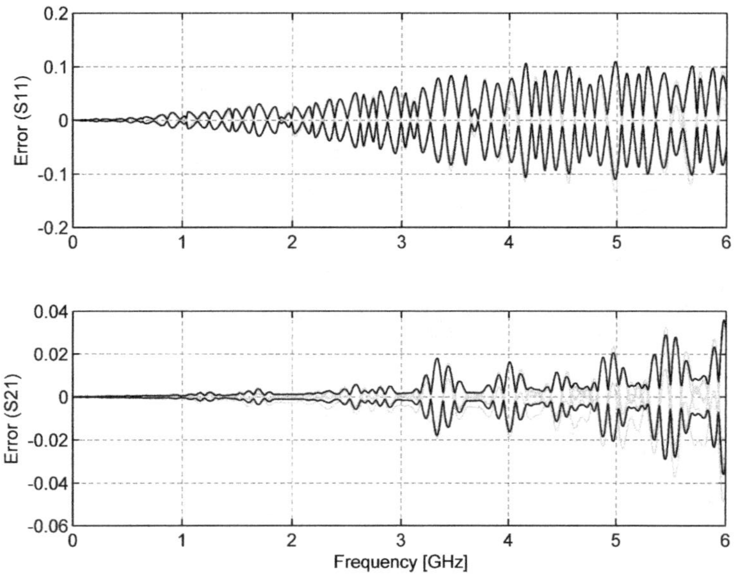

Table 5.7.1 – Return loss S_{11} and insertion loss S_{21} variability due to cable flexure transmitted onto SMA connectors.

Connector	5 GHz	15 GHz	25 GHz	35 GHz	45 GHz	55 GHz
7.6 mm male offset-short	$\pm 0.16 + j \pm 0.1$	$\pm 0.2 + j \pm 0.2$	$\pm 0.25 + j \pm 0.35$	$\pm 0.3 + j \pm 0.4$	$\pm 0.35 + j \pm 0.55$	$\pm 0.4 + j \pm 0.7$
5.4 mm male offset-open	$\pm 0.08 + j \pm 0.1$	$\pm 0.1 + j \pm 0.12$	$\pm 0.12 + j \pm 0.16$	$\pm 0.15 + j \pm 0.25$	$\pm 0.18 + j \pm 0.3$	$\pm 0.2 + j \pm 0.35$
matched load	$\pm 0.01 + j \pm 0.01$	$\pm 0.002 + j \pm 0.02$	$\pm 0.025 + j \pm 0.03$	$\pm 0.05 + j \pm 0.05$	$\pm 0.06 + j \pm 0.06$	$\pm 0.09 + j \pm 0.1$

Table 5.7.2 – Repeatability of the reflection coefficient for random cable flexure under different terminations (7.6 mm male offset-short, 5.4 mm male offset-open and matched load) reported by Lewandowski [222] (16 repeated measurements): the displayed values are in % for the real and imaginary part (the latter always the largest one) and correspond to the maximum span observed in the original figures, being impossible to numerically calculate the dispersion.

but more significant in phase ($0.5°$ every $10\,\mathrm{GHz}$). These values are quite close to those extensively reported for connector repeatability and cable-connector repeatability and probably represents the best values attainable.

5.7.2 Passive intermodulation (PIM)

In the attempt of identifying the elements of the test setup that may be a source of intermodulation products, cables, like connectors in sec. 7.4.3, have been included in the investigation. There are many concomitant mechanisms, but there is general agreement that passive intermodulation (PIM) occurs at surface contact: for cables the culprit is mainly the screen braid and the moderate oxide layer that might occur between braid wires; cables do not contain ferromagnetic materials and are protected by the outer sheath, that may however contaminate the braid accelerating oxidation; moreover, if dirt and pollutant may leak in, it is more likely at the interface with connectors. PIM is treated thoroughly in N.B. Carvalho and J.C. Pedro, *Intermodulation Distortion in Microwave and Wireless Circuits*, Norwood, MA, Artech House, 2003.

Sample results obtained by Hati, Nelson and Howe[24] regard amplitude-modulation (AM) and phase-modulation (PM) noise.

Regarding AM noise semi-rigid cables are largely better than flexible cables, with a noise contribution to the system that is nearly negligible and lower than that of flexible cables by about 30 dB. Regarding PM noise, semi-rigid cables behave better especially at low vibration frequency (up to 100 Hz), while semi-rigid and flexible cable have similar PM noise at higher frequency; it is briefly noted that in the curves reported in that paper the semi-rigid cable suffers from resonance effects, increasing PM noise around 10 Hz. This is perfectly understood since, mechanically speaking, the semi-rigid cable setup has a larger factor of merit, as said before. So, the optimal configuration may be a balanced combination of semi-rigid and flexible cables, adding standoffs and fastening methods and in particular dampening materials, such as stiro-foam, plaster and clay-like agents.

[24] A. Hati, C.W. Nelson and D.A. Howe, "Vibration-Induced PM and AM Noise in Microwave Components," *IEEE Transactions on Ultrasonics, Ferroelectrics, and Frequency Control*, vol. 56, no. 10, Oct. 2009, pp. 2050-2059.

6

Printed Circuit Boards and planar transmission lines

6.1 PCB geometry and its elements

A PCB is a sandwiched structure of different layers of copper (or in general conductive material) separated by insulating layers of dielectric material[1]. Any pattern can be drawn on each copper layer, but in general we will talk about pads, vias and traces: pads are small areas of copper used to solder Surface Mount Technology (SMT) components and to make signals accessible to test probes; vias were largely used in the past for through-hole components and now they are mainly used to pass from one layer to the other, e.g. also to equipotentialize two ground or supply copper areas, and still to connect some through-hole components; traces are used to connect different components and other elements (e.g. pads and vias) and in general to distribute supply, ground and signals. Traces may be thin or large depending on several factors: current rating, mechanical robustness and the ability to cool connected components behaving like heatsinks, the implementation of transmission lines with known and controlled characteristics, etc..

---

[1] There are some small deviations, such as Aluminum substrates for increased dissipation, but a thin insulating layer is there too.

Some knowledge of PCB terms and elements is assumed and the description will consider briefly the employed materials and the standard dimensions commonly available, that influence our design choices and accuracy for transmission line realization. Some terms and uses are specific to PCB technology and fabrication and they shall be taken into account when ordering a PCB, designing and finalizing PCB layout files and selecting fabrication options.

The conductive layers are often called "copper layers", while the dielectric layers are simply "layers". Both have their standard thickness that is reviewed later on in sec. 6.1.1. Copper layers come in standard thickness of $35\,\mu m$, but also thinner and thicker ones are available, such as $18\,\mu m$ and $70\,\mu m$; rather than the thickness, the weight of copper per unit of area is sometime specified, expressed in oz/ft^2, ounces per square feet: $70\,\mu m$ corresponds to $20\,oz/ft^2$, while the $35\,\mu m$ to 10. Dielectric layers come in different thickness and materials depending if they are the outer layers or the inner ones, because the PCB structure is built symmetrically starting from the mid plane (where the core is located) using pre-preg layers ("pre-preg" means pre-impregnated) lined up with copper sheets, and then growing up mirror-like with new layers. Depending on the chosen dielectric materials, only some thickness values may be available, and this has impact on the design of the transmission line structures: simply speaking the thinner the dielectric layer the smaller the width of the hot trace to achieve the same characteristic impedance; a larger permittivity has the same effect.

Vias are divided between vias and micro-vias, when the diameter falls below some threshold value, normally around $0.4-0.5\,mm$. Vias may be of the *through* type if they cross orthogonally the whole layer structure of the PCB putting in connection the two outer copper layers, but there exist also *blind vias* (if they end up onto one of the inner copper layers) or *buried vias* (if they remain confined inside the PCB, going from one inner copper layer to another one). With the increasing miniaturization of components and the tight tolerances for trace width and separation, vias have become a burden, with an embarrassingly large size with respect to the other elements, so wasting board space and somehow constraining the design, from which the need for micro-vias.

A PCB is thus obtained piling up different assembled layers that are already processed, so that each has its own traces, planes and holes for vias. The reason for using pre-preg layers is that each glues to the other layers, once the so-formed PCB goes in the oven. It is important to remember that traces laid down on pre-preg layers will sink for their thickness (or part of it) into the softened pre-preg while it is partially melting: this implies that first, these traces do not add up fully in the calculation of the overall board thickness, but, more important, the assumed substrate thickness for calculation of transmission lines will not be preserved. Practically speaking with a typical trace thickness t of $35\,\mu m$ (or $1.4\,mils$) and with a substrate thickness h that in multi-layer boards may range between $200\,\mu m$ and $400\,\mu m$, the error on the ratio may be between 5 and 15%, reduced to one half if choosing $18\,\mu m$ of copper thickness. We know that this has an impact on microstrip structures, and less on stripline structures, that become only slightly offset. Conversely, very large traces and entire planes do not sink in the pre-preg substrate. For this reason, copper pour is also useful to preserve more accurately board geometry through the fabrication process. True microstrips (not loaded or covered versions) are placed on outer layers, thus avoiding this type of problems.

Holes for vias are made at various steps of the process: if all vias are through-hole from top layer to bottom layer they might be even drilled at the end of the process; however, nowadays with SMD technology vias mainly route signals between layers and thus blind vias and buried vias (see sec. 6.1.4) are quite common and the production process has been adapted to this, drilling each pre-preg layer before assembling the PCB. The copper layer arrangement may be different depending also on the number of layers in the board: outer layers are commonly signal layers to connect to the top and bottom components, but of course contain also the power supply connections to components themselves and a certain amount of ground plane, so that they may be thought of as "mixed" layers; then, correct power supply distribution requires internal layers for distribution of supply rails, and depending on the type of board we will see unipolar logic supplies and also symmetric voltages, positive and negative, with respect to ground. The schemes for the layer stack-up and its use are many and the designer is absolutely free of choosing his/her own; a general reference is the IPC standard 317 [176] covering configurations up to 8 layers. Depending on the required board functionality, supply layers may be replaced by ground layers, e.g. for a test board for passive components, or when the number of planar transmission lines is large and signal integrity of high-speed lines is more important.

The outer copper layers, that are in contact with the environment and will host the components to be soldered, undergo several treatments: pads that are used to solder component or connector pins may be gold plated (e.g. ENIG, flask gold, or similar), tin plated, or simply predisposed with a cream layer for manual, oven or wave soldering; the areas of the PCB where the solder shall not adhere are protected by a solder mask (see sec. 6.1.5), that is effective also for the general protection of the PCB and for the electrical insulation of traces; the inner part of vias is metalized as well as the annular ring around, also called "pad". It is important to understand the sequence of PCB fabrication steps because this not only sheds some light on tolerances, but also indicates what is possible and impossible when asking for special adjustments, or to skip one process step, etc. Large companies with long experience in PCB design and production know this very well and control quite accurately such degrees of freedom, but semi-pros in this particular field should always collect documentation and ask even trivial questions on details that might be hastily dismissed under an all-catching term or definition.

Printed circuit boards are a complex assembly of materials and laminates with different properties, such as glass temperature, thermal expansion coefficient, viscosity, besides electrical properties: these mechanical properties exert their influence on the final product. Some good books for an overview of PCB fabrication, especially to have an introduction to terminology and common practice, as well as relationships with circuit design and CAD, are:

- R.H. Clark, *Printed Circuit Engineering*, Van Nostrand, 1989;
- R.S. Khandpur, *Printed Circuit Boards: Design, Fabrication, Assembly and Testing*, Tata - McGraw Hill, 2005;
- C.A. Harper, *High Performance Printed Circuit Boards*, McGraw Hill, 2000.

Details, tips and experimental data can be often found in papers and presentations at components and packaging conferences or specialized journals[2], rather than in manufacturers' leaflets and datasheets, that usually stick to standardized procedures and report the required.

6.1.1 PCB dimensions and tolerances

The thickness of complete standard PCBs is around $1.55 - 1.6\,\mathrm{mm}$, to match the gauge of edge connectors, even when the $70\,\mu\mathrm{m}$ copper thickness is chosen; however if a tight fit of the edge connectors is desired it shall be noted that edge connectors (e.g. SMA edge connectors) come with slightly different clearance between upper and lower pins, with more than 0.1 mm tolerance and thus shall be selected carefully depending on the real PCB size. If using vertical PCB mount connectors there is no problem any longer of matching the PCB thickness. Other PCB thickness values are possible, depending on the manufacturer's capacity and laminates availability, as well as different arrangements of signal, power and ground layers and vias in between: several examples appear in the literature [105, 300], both seen externally as a finished object, and as parts of it, cross-sections and SEM or X-ray photographs, with an overview and many details regarding non-ideality of edges and surfaces, dielectric texture, etc.

When stacking a PCB, the alignment of layers and of pads and vias within is critical: not only tolerances of the initial alignment shall be considered, but also deformation of the dielectric sheets, in terms of shrinkage, growth, warping. Pads and vias may be still in electrical connection, but the useful cross-section is reduced, it is more prone to breakage and does not have the electromagnetic characteristics assumed and established at design time. This type of misalignment is called in general "misregistration". Other fabrication tolerances are reviewed individually when evaluating the sensitivity of planar transmission line expressions to trace thickness, trace width and dielectric thickness; copper roughness is also a further influential factor, especially at high frequency. An interesting work[3] offers a comprehensive and clear presentation dating back to 1974, when not only PCB technology was deemed by larger tolerances, but also the first high speed circuits began to appear. The aim of that work was to statistically evaluate all factors in order to optimize design elements (e.g. drill size and pad size) and process management (e.g. machine maintenance or replacement and concentration of production controls for the most influential parts of the process). A wide range of errors is reviewed, characteristic also of fabrication process at that time, so including e.g. the creation of the artwork and phototools (no CAD programs available!); also the quality of the printing process, the mechanical tooling and the quality

[2] The best known conferences on this broad topic are the IEEE Electronic Components and Technology Conference (ECTC), IEEE Electronics Packaging Technology Conference (EPTC), IEEE Workshop on Signal Propagation on Interconnects, and DesignCon.
An authoritative journal is the IEEE Transactions on Advanced Packaging, that addresses issues related to fabrication, testing, environmental conditions, electric and electromagnetic behavior of devices and boards, sharing the latter topic with other journals, such as IEEE Transactions on Electromagnetic Compatibility.

[3] R. Marcucci and C.F. Tomes, "Registration System Study as Applied to Printed Circuit Boards," *IEEE Transactions on Manufacturing Technology*, Vol. 3, no. 1, June 1974, pp. 26-40.

of materials were not the same as today, but they are quite close, if the tolerance values they show in their Table I are considered: errors in the range of 1 to 2 mils are commonplace, as those assumed in more recent works such as [299].

6.1.2 PCB dielectric materials

The choice of dielectric materials depends on weight, cost, aging, maximum working temperature, mechanical robustness and stiffness, behavior in case of fire (the famous acronym FR4 stands for "flame retardant", it is not indicating any electromagnetic characteristic of the material). Many PCB materials in terms of characteristics of laminates for assembling are standardized; the main standard reference is the IPC set of standards, accepted also for military applications in place of the superseded MIL-STD-275. The IPC-2221 standard at sec. 4 indicates the required characteristics (dielectric strength, nominal dielectric permittivity, mechanical properties and thermal expansion, to cite a few) and cross-links to the IPC-4101 and IPC-4102 standards for laminates and pre-impregnated bonding layers that form the multi-layer PCB structure. Special materials not contemplated in industrial standards are always possible.

We will see that the larger the material permittivity the lower the resulting transmission line characteristic impedance. Also reducing layer thickness brings to a lower characteristic impedance. For small quantities, such as when designing a prototype, the problem is always unit price and availability: the factory processing our PCB design might not have our selected material as a standard option and they will have to buy the minimum quantity; in general PCBs requiring high performance materials are more expensive by a factor of two to three.

What do the high performance dielectrics have that the cheap ones do not possess?

FR4 is an acronym for a wide range of epoxy fiber boards, with a dielectric permittivity ranging approximately between 4 and 5 and with quite large (but most of all not accurately defined) dielectric losses. However, once a specific FR4 has been identified, the range of variability of dielectric permittivity reduces to what is declared by the manufacturer (see for example Table 6.1.1 and Panasonic R1566 laminates, with not so dispersed dielectric permittivity values). The usual behavior of permittivity is a reduction of 10-15% going from 1 MHz to several GHz; since the characteristic impedance depends approximately on the reciprocal of the square root of the permittivity, we might expect an increase of less than 9% of the characteristic impedance (this is considered in more detail in sec. 6.3, where also experimental data are reported). Dielectric losses expressed as complex permittivity, or more commonly as $\tan\delta$, are normally less accurate and for general-purpose low-cost materials information is limited to one or few values, often specified as upper limits. Improved FR4 materials exist and PCB manufacturers normally state if using improved FR4 with tight tolerances or thoroughly characterized, and make datasheets available.

However, when designing a test setup for frequencies of several GHz, the uncertainty due to the many unknowns or ill-specified parameters steers the designer towards better materials, the most common and known manufactured by Isola, Nelco, Rogers, Taconic, just to cite a few. A wide selection of materials with various dielectric permittivity values is available (from simply better glass-epoxy materials, to PTFE with glass, ceramic, etc.), thus giving one more degree of freedom to the design related to the thickness of the dielectric layers and trace width for a given target characteristic impedance of planar transmission lines. The price of these laminates in some cases may be absolutely affordable; keep in mind that the overall price of a PCB is determined not only by the material, but also by the required fabrication options, such as gold plating, micro-vias, etc., and by the quantity.

It is remembered a fundamental weakness of standard dielectric laminates based on fibers with not a tight control on fabric and fiber patterns: considering fiberglass and similar products, the glass weave is impregnated by resin and the two have distinct and different dielectric properties, i.e. a larger dielectric permittivity around 6 but much lower dielectric losses the former, and a lower dielectric permittivity around 3 to 4, but larger dielectric losses the latter; the overall dielectric properties of the laminate are a combination of the two, depending on the respective percentages, but the glass weave is also responsible for anisotropy depending on its orientation and local variations of dielectric properties. One solution sometimes adopted is to use materials with slightly lower glass permittivity, so that the difference of permittivity with the epoxy resin is also smaller. When going to tiny traces for highly integrated transmission lines, the periodic structure of the laminate and its inhomogeneity may exert their influence. As reported in [319], the effect for vias is not perceivable up to 50 GHz, while for planar transmission lines there may be variable resonances depending on orientation (the so called "glass weave effect"); the authors give an expression that accounts for the distribution of the fibers in the horizontal and vertical direction to predict the resulting resonance frequency. The factor of merit may vary significantly with the relative orientation of the planar line and the glass weave pattern, depending on how many ups and downs of the weaved thread occur along the line length: so, the simulated resonances move between 35 and 70 GHz, while the resonance equation would predict also resonances beginning with 12.7 and 24 GHz, that are hardly visible in the simulation results. Common-mode and differential scattering parameters are reported, resulting in a more pronounced effect for the differential insertion loss S_{dd21}. Two practical factors however, reduce the glass weave effect: tolerance in the distribution of glass fibers and resin dielectric losses. A statistical analysis of the effect of a randomly distributed pitch size of the glass fibers show that the resonance peak is reduced by 10 dB; a larger dissipation factor of the resin, from 0.03 to 0.1, of course worsens the S_{dd21} curve by lowering and inclining it, but at the same time reduces the relevance of resonance peaks proportionally. Of course, asking for denser weave materials (such as 1652, 2116, 2113) with respect to coarser weave ones (e.g. 106 and 1080) reduces glass weave effect as well. It is remembered also that wider traces are less susceptible and that skewed or zig zag traces also are quite immune to weave effect: practical experiments performed by Altera confirmed that a reduction of about one order of magnitude in time domain signal skew is attainable for zig zag traces with respect to straight traces.

6.1.2.1 Testing methods and significance of permittivity values

There are several test methods to determine dielectric permittivity of PCB materials for RF and microwave applications. Some of the methods are specified in IPC or IEC standards, others come from best practice. IPC-TM-650, under sec. 2.5.5, lists several standardized methods, for low and high frequency, based on different techniques (e.g. suitable for clad laminates ready for production or requiring stripping of the clad laminate and/or cut a sample to fit the test cell; measurements are made with a variety of capacitance bridges, LCR bridges, Q factor meters and network analyzers).

Description of test methods A first intuitive method is based on the construction of a parallel plate capacitor across one dielectric layer of the board. The method is suitable for low-frequency range measurements carried on with a classic technique, such as capacitance bridge or RCL bridge: the circular shape of electrode disks built on two adjacent copper layers has been determined to reduce fringe electric field and their size is minimized with respect to the required PCB area, while giving measurable capacitance values (see sec. 6.5.1.3). This technique is suitable for frequencies up to about a hundred MHz, but it is most commonly employed up to some MHz with commonly agreed conventional test frequency values of 1 MHz, and also 100 kHz.

A planar transmission line may be also built and its input impedance and resonances exploited to determine capacitance by means of simplified design expressions or regression of more exact formulation: see sec. 6.3 and individual subsections dealing with the various types of planar transmission lines; the preferred line is usually a microstrip, sized e.g. for $50\,\Omega$ characteristic impedance, and this technique works up to several GHz with a minimum frequency established by line length and cross section and usually around a hundred MHz. Additionally, the Microstrip Differential Phase Length (MDPL) method aims at removing many pitfalls and approximations of a single microstrip measurement, but has not yet been accepted and included in IPC standards. However, it is mostly useful to determine the axial components also in the presence of significant anisotropy. In general, when the objective of the substrate material characterization is to improve the accuracy of and optimize the design of some planar transmission line, using the same or a similar line to determine its characteristics is probably a well thought and suitable approach.

More accurate methods are based on the determination of resonances of known configurations: measurement of frequency, as well as the determination of the geometry of the resonating cell are in general more accurate.

A common method is the *Clamped Stripline Resonator* (CSR) working in the GHz range, as specified in the IPC-TM-650 2.5.5.5 [182]. The test is suited to routinely test laminates coming from the production chain; the procedure consists in clamping one laminate sample on each side of a resonator forming a stripline resonator. The test is intended to measure the permittivity in the z axis (that is through the laminate thickness); the laminate has the copper foil removed to perform the test and the surface roughness of the dielectric substrate may be an issue. Since air may be trapped in the surface roughness, the resonator is loosely coupled to reduce the influence of the entrapped air, but this comes at the expense of a stronger electric field in the air-gap with significant components also in the x-y plane, thus giving in reality a mixed permittivity value as result; if the material is anisotropic, the calculated permittivity may be altered by this mutual influence. The presence of the air in any case makes the method to underestimate the “real” permittivity value, even along a unique axis.

Another common test is the *Full Sheet Resonance* (FSR) test, as specified in the IPC-TM-650 2.5.5.6 [183]. The FSR test uses copper-clad laminate to create an open-wall parallel-plate waveguide, establishes a standing wave, measures the resonant peak and determines the permittivity value of the material. The advantages with respect to the previous method are: no potential issues related to the entrapped air, insensitive to the anisotropic nature of the material, the laminate is not destructed to strip off the copper clad. The resonant peak of the standing wave is related to the physical size of the laminate panel and for normal panel size the resonance is located at low frequency, in the hundreds MHz range, and in any case below 1 GHz. It is known that copper roughness, that depends on the deposition process, can have an effect on the effective permittivity at higher frequency and this method fails to cover microwave applications with reliable and accurate data. Moreover, resistive losses in the metallization are not accounted for as well, but by virtue of the same measurement principle at resonance they shouldn't be relevant; however, the change of resistance as well as that of inductance due to skin effect (at one or few GHz the skin depth is already smaller than the metallization thickness) should be considered as a source of error. Simulation results[4] show an amplitude reduction of less than 1% and a resonance frequency shift of some to a ten %, depending on dielectric thickness: going from 2.5 mil to 10 mil the apparent permittivity error drops from nearly 5% to about 1-1.6%. Metal roughness is the other source of error, estimated around 3.6% and 5.1% for 2.5 mil to 10 mil dielectric thickness, so opposite to skin depth behavior. The combined effect of roughness and skin depth makes variations more complex: the overall error including different frequency values may vary between 5 and 10%, or slightly larger.

A test method that is not covered by IPC is the *Split Post Dielectric Resonator* (SPDR) test. A resonator tuned to a specific frequency is the test fixture; the empty resonator is measured first, then the sample is introduced into the resonator and the shift in the resonant frequency is measured. Starting from sample thickness, that shall be measured accurately, permittivity is calculated: it is determined for the x-y plane of the sample only, and for this reason this method shall be used in combination with another method measuring permittivity along z axis. It is in any case very useful to spot out any anisotropy.

Accuracy of test methods The definition and determination of the accuracy of the various measurement methods is a multifaceted problem, that takes into account instrument uncertainty (e.g. with respect to amplitude and to frequency measurement), accurate definition and determination of geometry, impact of fabrication process and its tolerances, etc. Often the accuracy is discussed for one or few methods at a time and results obtained by different works are not perfectly comparable.

In [110] a comparison is made for FR4 material with the help of additional results coming from waveguide-based methods, by which the effect of a slab of dielectric substrate without metalization is evaluated. The results cover several decades of fre-

[4] A. Deutsch, A. Huber, G.V. Kopcsay, B.J. Rubin, R Hemedinge, D. Care, W. Becke, T-M. Winkel and B. Chamberlin, "Accuracy of Dielectric Constant Measurement Using the Full-Sheet-Resonance Technique IPC-TM-650 2.5.5.6," IEEE 11th Topical Meeting on Electrical Performance of Electronic Packaging, Monterey, CA, USA, Oct. 21-23, 2002, pp. 311-314. doi: 10.1109/EPEP.2002.1057939

quency (10 kHz to 10 GHz), but comparisons are done only in the X band: the maximum span of dielectric permittivity values is about 10% and the authors declare an "error bound" of 2% and 5% for real and imaginary parts of the permittivity, based on the accuracy of the instrumentation (VNA) and of the numerical models used to extract dielectric parameters. The authors observe also that fitting over such extended frequency range is achievable adopting a Debye model with more than one relaxation constant ($N = 8$ is proposed and used for the reported results).

In general methods based on transmission line measurement are the least accurate, also because of the necessity of de-embedding setup (and in particular launchers) during calibration, whereas resonance-based methods are intrinsically more accurate and robust. It remains that the physical structure of the setup allows quantification for a preferred axis or a combination of more than one value for anisotropic materials.

Mostly suited test methods for microwave PCB applications Specific applications will benefit of measurement methods with similar characteristics and in general attention shall be given to orientation and polarization of electric field and its interaction with dielectric permittivity and possible anisotropy. Most frequently the electric field will be along the z-axis of the material. An example may be a simple microstrip transmission line circuit: if used at relatively low frequency (less than 1 GHz), then the permittivity value obtained from FSR test are appropriate; if the circuit is used at higher frequencies, the stripline test may be of interest; if the material does not have a rough copper treatment and/or a high degree of anisotropy, this method is appropriate.

If the material uses rough copper and/or has a high degree of anisotropy, then it should be understood that the clamped stripline test will report a lower permittivity value; in this case, it may be best to use the Microstrip Differential Phase Length method [5].

Some microwave PCB applications have E-fields lying in the x-y plane only; for these applications permittivity values obtained with SPDR method would be appropriate.

There are, however, several applications where the electric field has components in all the three axes (in the x-y plane as well as in the z-axis). Most edge coupled applications will fall in this category, and for this type of circuits finding an appropriate and reliable test method is a little more complicated: a combination of methods and of the z-axis and the x-y plane values should be used; which percentage of each will depend on the specific configuration and the amount of coupling between elements; for not too anisotropic substrate materials the error will be anyway not so relevant. Sensitivity analysis is possible by means of finite element programs, by changing one of the two permittivities and evaluating the impact on the final result.

6.1.2.2 Typical PCB dielectric permittivity values

Provided the general inaccuracy of dielectric permittivity and loss tangent values in relation to measurement methods, numerical methods and approximations, fabrica-

[5] N.K. Das, S.M. Voda and D.M. Pozar, "Two Methods for the Measurement of Substrate Dielectric Constant," *IEEE Transactions on Microwave Theory and Techniques*, Vol. 35, no. 7, July 1987, pp. 636-642.

tion tolerances and temperature influence, it may be useful anyway collecting a set of values for the most commonly used PCB materials.

The behavior of ε_r versus frequency of FR4 material reported by IPC [176] is the one considered by many and refers to a glass/epoxy ratio of 40/60, quite common for FR4 boards. The value is 4.7 at 1 kHz and drops linearly to 4.0 at 1 GHz over a log-scale of the frequency. A similar expression is presented in the EN 61188-1-2 standard[6] [76], indicating it as "worldwide average value" of "epoxide/glass laminate": $\varepsilon_r = 4.97 - 0.257 \log_{10}(f/10)$, where f is the frequency in MHz. When calculated at 1 GHz the resulting ε_r is slightly larger, 4.18. The standard says also that the measured $\pm 3\sigma$ variation amounted at ± 0.35 without specifying the type of analysis, the conditions or the number of samples, and then that the variation on a single printed board panel is as high as ± 0.26. The standard recognizes that ε_r values depend also on the resin-to-glass ratio and the results in its Figure 2 (given by the application of the formula above) "takes that into account." The newer version of IPC-2141 [191] reports many more permittivity values for various types of FR4 glass styles and resin contents, as found commercially: the largest ones (type 1500, 7628 and 7629 with about 42-44% of epoxy resin) start from about 4.5 at 1 MHz, for types 2113 and 2116 (with 52-54% resin content) the initial permittivity is about 4.2 and then for types 1080 and 106 (with 62% and 69% of resin) the permittivity lowers to 4.0 and 3.8 respectively; this is in line with the considerations on dielectric permittivity of glass and resin made at the beginning of this section.

The decrease of relative permittivity with frequency is due to the behavior of the epoxy resin, so that higher glass/epoxy ratios will exhibit more stable values versus frequency. And this holds true in general when the used resin is not epoxy, but one of those listed in Table 6.1.1, such as polyimide, polysulfone, etc.. PTFE on the contrary has a quite stable, yet lower, value of the dielectric permittivity and it is often used as a reference material for calibration.

Materials classified by IPC in the Appendix B of IPC-D-317A [176] have a wide range of dielectric permittivity values, from PTFE (brand name Teflon®) in the non-reinforced and glass reinforced versions to various resins (polysulfone, polyetherimide, epoxy, polyimide, cyanate) with different percentages of glass, woven and non-woven. Several permittivity values are reported in Table 6.1.1.

Rogers materials of the RO3XXX family all are quite insensitive to frequency, with the reported dielectric permittivity values constant within about 2% up to 16 GHz. Those of RO4XXX family, on the contrary, have a larger variation versus frequency (they are guaranteed by Rogers up to about 50 GHz, but below 1 GHz the increase of dielectric permittivity is not negligible, by about 3-4%) and a slightly larger sensitivity to temperature (6-8% change for an excursion of 180 °C, between -30 °C and +150 °C).

One last factor that is quite relevant is moisture and the amount of moisture absorbed by the specific material after production and after all treatments and protective coatings have been applied: water permittivity is quite large (see Example 1.1 at the beginning of Chapter 1). How the material is cut and worked and the use of thermal curing and treatments has a great influence on the amount of trapped moisture. Even

[6] The formula here is corrected for a typo appearing in the standard.

coatings that are believed to be durable and tough may trap water when thermally cycled. Solder mask and coatings are reviewed in sec. 6.1.5.

6.1.2.3 Typical PCB dielectric losses

The $\tan\delta$ reported by IPC for FR4 with glass/epoxy ratio of 40/60 has values that range between 0.008 at 1 kHz to 0.022 at 1 MHz; then it remains at this maximum value for a decade (up to 10 MHz), starting to decrease again towards 0.008 at 1 GHz. This identifies a common behavior of this kind of materials, that, however, may be quite different from sample to sample, even towards better values and good performance.

The techniques used to create thin layers of copper may be broadly classified in two groups: electrodeposited copper is made by electrically attracting copper from a bath onto a rotating drum, giving a rough surface that is desirable for adhesion between copper and dielectric, but increases dielectric losses (see sec. 6.1.3.2); rolled copper process casts and mills smoothly the copper onto the dielectric layer, then cold rolled to the final thickness, giving lower dielectric losses. The roughness of the latter is much lower than the former. For tiny metal layer thickness there are conflicting views, about the real relevance of surface roughness (still a small fraction of layer thickness), whose impact on dielectric losses may be estimated in the range of 10 to 30% for the typically thickness (18 and 35 μm).

Dielectric losses are often quantified for a reference planar transmission line, as time or frequency domain quantities, such as attenuation constant, propagation constant, insertion loss or transmission. Details of these methods are reported in sec. 6.5.

6.1.3 Traces

Let's start from trace density and minimum clearance attainable for a board of given dimensions X and Y and number of layers L. Rent's rule cited by Johnson and Graham [199] dates back to the late '70s and states that the average distance between trace axis d (called "trace pitch") for a board populated with N traces is

$$d = \frac{\sqrt{XY}}{N} 2.7L$$

The expression was obtained by observing that when dividing a board into quadrants containing ICs and traces, half of the traces are routed inside the quadrant and half leave the quadrant; further subdividing the quadrant into smaller quadrants, the proportion remains quite unaltered. We may add that a well designed board with well positioned and distributed ICs have probably the large part of the leaving traces terminating inside nearby quadrants. This was observed at the time of sparse logic boards, using logic gates, 8 and 16 bit microprocessors and low-density memory ICs.

At that time vias were all from top to bottom for through-hole components and traces were routed frequently between vias. The board design followed rigid rules and there was often a predetermined grid of via locations: a 100 mils (2.54 mm) separation of vias from center to center allowed one or even two traces running between the vias (called "tracks", so one-track board or two-track board).

Material description	ε_r	$\tan\delta$
(B,I) Non-reinforced PTFE resin (Teflon)	2.1	0.0004-0.003
(I) Woven glass reinforced PTFE: @ 1 MHz	2.55	0.005
(I) Woven glass reinforced PTFE: @ 10 GHz	2.5	0.003
(I) Non-woven glass reinforced PTFE: @ 1 MHz	2.25	0.001
(I) Non-woven glass reinforced PTFE: @ 10 GHz	2.25	0.0015
(B,I) Woven glass / Cyanate resin: @ 1 MHz	3.5-3.7	0.01
(B,I) Woven glass / Polyimide resin: @ 1 MHz	4.5-4.4	0.02-0.025
(I) Woven glass / Epoxy resin: @ 1 MHz	4.7	0.035
(I) Filled PTFE resin (mid/high range): @ 10 GHz	6 / 10	0.003
(N) 4000-6 @ 1 GHz	4.1	0.022
(N) 4000-13 EP @ 1 GHz	3.7	0.009
(R) RO3003	3	0.0011
(R) RO3006	6.15	0.002
(R) RO3010	10.2	0.0022
(R) RO4003	3.38	0.0025
(R) RO4350	3.48	0.0035
(P) R-1566 laminate	4.4-4.7	0.035
(P) R-1551 pre-preg	4.1-4.3	0.035
(T) TLP-5 PTFE-glass @ 10 GHz	2.20	0.0009
(T) TLE-95 PTFE-glass @ 10 GHz	2.95	0.0026
(T) RF-35 PTFE-ceramic-glass @ 1.9 GHz	3.50	0.0018
Alumina	9-10	0.0015
Sapphire (anisotropic), crystal form of Alumina	8.6 hor., 10.55 ver.	0.0015
Gallium Arsenide	12.9	0.001
Indium Phosphide	12.4	0.001

Table 6.1.1 – Table of dielectric materials showing dielectric permittivity, tangent loss, useful temperature range. Information included is derived from: (I) IPC-D-317A [176], App. B; (N) Nelco; (P) Panasonic; (R) Rogers; (B) Bahl [23]; (T) Taconic.

6.1.3.1 Trace capacitance

Traces, when part of a planar transmission line will be considered in detail in sec. 6.3 for various architectures, such as microstrips and striplines, addressing characteristic impedance and propagation. However, traces, as any other copper area, if considered at a sufficiently low frequency, feature a stray capacitance across the PCB dielectric substrate to the supposed underlying reference plane: the capacitance is directly proportional to the area A and inversely proportional to the substrate thickness h; the shape of the area is not really relevant, provided that there is no remarkable disproportion, especially if signals have a wide frequency range. The stray capacitance is

$$C_{area,stray} = \frac{\varepsilon_0 \varepsilon_r A}{h} \tag{6.1.1}$$

Example 6.1: Stray capacitance evaluation

Considering the typical values of a two-layer and four-layer PCB (typically $h = 1.5\,\mathrm{mm}$ and $h = 0.4\,\mathrm{mm}$, respectively), for FR4 material (featuring a relative permittivity approximately between 4.6 and 4.8, decreasing slightly with frequency), the resulting capacitance per unit area ranges between $2.7 - 2.8\,\mathrm{pF/cm^2}$ and $10.2 - 10.6\,\mathrm{pF/cm^2}$. Thus $1\,\mathrm{cm^2}$ of copper area at 1 GHz has a loading reactance of $58\,\Omega$ and $15.3\,\Omega$, respectively. Considering a trace of $w = 1\,\mathrm{mm}$ width, $100\,\mathrm{mm}$ are necessary for $1\,\mathrm{cm^2}$ and at 1 GHz such structure is to be considered already as a combination of distributed inductance and capacitance, so that this estimation is no longer reliable and in reality there will be no such loading on the line. Conversely, if an approximately square area of copper is connected all at the end of a line, then there will be the calculated loading effect. If at the same time we consider the inductance of vias as in eq. (6.1.5) below, then the stray capacitance of the copper around vias can be trimmed to obtain some compensation (see eq. (6.1.3), where the quantity D_{pad} is the size of the copper around the via).

Since the fabrication process is not ideal, traces cannot be assumed ideally straight and cut vertically sharp: irregularities at edges and the consequences of under-etching (a trapezoidal, rather than rectangular, shape) may have some impact when tight accuracy is necessary, especially for tiny dimensions, when such irregularities may influence the behavior of stray capacitance terms of the hot trace considered later in this chapter. As an example, the PCB cross sections shown in [105] at Fig. 38, 43 and 44 have all trapezoidal traces, with either straight or convex sides.

6.1.3.2 Trace roughness

Copper layers are often treated so that the surface is rough enough to promote adhesion to the substrate: the amount of roughness for standard copper foils is in the order of $7 - 8\,\mu\mathrm{m}$[7]and applied to one side only; the reverse-treated foil features the same roughness on both sides; low profile foils have lower roughness, that may be reduced to one half ($3 - 4\,\mu\mathrm{m}$) and one fourth ($1.5 - 2\,\mu\mathrm{m}$). Line losses measured by Hinaga et al. are on average 50% larger for standard foils than for low-profile foils. Line losses (see sec. 6.3 later in this Chapter) are a combination of conduction losses (α_c) and dielectric losses (α_d): both are affected by surface roughness. For α_c a semi-empirical formula was proposed by Hammerstad and Bekkadal [151][8], who modeled the increase with frequency as $\delta\alpha_c = \frac{2}{\pi}\arctan\left(1.4(r/\delta)^2\right)$, where r is the rms value of roughness and δ is the skin depth.

[7] S. Hinaga, M.Y. Koledintseva, K.R. Praveen Anmula and J.L. Drewniak, "Effect of Conductor Surface Roughness upon Measured Loss and Extracted Values of PCB Laminate Material Dissipation Factor," 59th ARFTG Conference Digest, June 7, 2002. doi: 10.1109/ARFTGS.2002.1214679
X. Gu, D.R. Jackson and Ji Chen, "An Analysis of Copper Surface Roughness Effects on Signal Propagation in PCB Traces," Texas Symposium on Wireless and Microwave Circuits and Systems (WMCS), Apr. 4-5, 2013. doi: 10.1109/WMCaS.2013.6563552

[8] Originally conceived by S.P. Morgan, "Effect of surface roughness on eddy current losses at microwave frequencies," *Journal of Applied Physics*, Vol. 20, 1949, pp. 352.

A similar correction was proposed by Groisse et al.[9] using an exponential function $\delta\alpha_c = 1 + \exp\left(-(\delta/2r)^{1.6}\right)$. Both formulations saturate to 2 for increasing frequency, but the Groisse formula is "slower" than Hammerstad-Bekkadal one, so that predicted values are in general equal (at low frequency) or smaller; sometimes this limitation of a maximum doubling of losses at infinite frequency is troublesome and does not match experimental results. On the contrary, Sundstroem[10] proposed a well-grounded method using series expansion in harmonic function terms of surface roughness. The effect on dielectric losses is more complex and usually is estimated by finite element modeling or direct measurement, possibly subtracting the previously estimated conduction losses to isolate it. Guo et al.[11] demonstrate that Hammerstad and Bekkadal expression may in general underestimate attenuation by nearly 30%, using a stripline as reference line with standard roughness of $8\ \mu\mathrm{m}$. Quite interestingly Shlepnev and Nwachukwu[12] demonstrate with the help of very accurate models and measurements of PCB samples that Hammerstad-Bekkadal expression may be adjusted quite satisfactorily, defining precisely the rms roughness δ using the peak-to-valley variation and adding a multiplicative *roughness factor* RF in the form of $(\mathrm{RF}-1)$: for the tested laminates (I-Tera 1080 and 2116 both core and pre-pregs) RF ranged between 2.6 and 2.8, in agreement with the "30% less" observation above.

So far, the discussion has regarded the effect of roughness on attenuation, but nothing was said for the effects on phase constant.

A more accurate approach without the complexity of full 3-D meshing FEM is proposed by Horn *et al.*[13], but no details are given on its implementation: the results look quite accurate, for ten-to-one changes of roughness and three different laminates, predicting both insertion loss and the equivalent permittivity.

By the way, matching correctly the insertion loss in general is not a guarantee that the model is accurate, because conductor losses may be erroneously assigned to dielectric (as additional dielectric losses), matching generally measurement data, but failing for specific conductor profiles. In [50] they underline that this practice was often used, but the corrected loss tangent values are different for different line geometries.

To conclude, roughness may be particularly large for some laminates to promote adhesion and resistance to peeling: *rolled annealed* copper foils are the smoothest, with roughness down to $0.1-0.3\ \mu\mathrm{m}_{\mathrm{rms}}$ (additional treating may be applied to guarantee a minimum roughness, e.g. $0.4-0.6\ \mu\mathrm{m}_{\mathrm{rms}}$); *electro-deposited* foil is produced by deposition of a copper solution on a rotating drum: it is as smooth on the drum side, but

---

[9] S. Groisse, I. Bardi, O. Biro, K. Preis and K.R. Richter, "Parameters of lossy cavity resonators calculated by the finite element method," *IEEE Transactions on Magnetics*, Vol. 32, no. 3, June 1996, pp. 894-897. doi: 10.1109/20.497385

[10] S. Sundstroem, "Stripline Models with Conductor Surface Roughness", Master of Science Thesis, Helsinki University of Technology, Finland, Feb. 2004.

[11] Xichen Guo, D.R. Jackson and Ji Chen, "An Analysis of Copper Surface Roughness Effects on Signal Propagation in PCB Traces," 2013 Texas Symposium on Wireless and Microwave Circuits and Systems (WMCS), Apr. 4-5, 2013, pp. 1-4. doi: 10.1109/WMCaS.2013.6563552

[12] Y. Shlepnev and C. Nwachukwu, "Roughness Characterization for Interconnect Analysis," IEEE International Symposium on Electromagnetic Compatibility, Long Beach, CA, USA, Aug. 2011, pp. 518-523.

[13] A.F. Horn III, J.W. Reynolds, P.A. LaFrance and J.C. Rautio, "Effect of conductor profile on the insertion loss, phase constant, and dispersion in thin high frequency transmission lines," DesignCon 2010, Santa Clara, CA, USA, Feb. 1-4, 2010.

the outer side to be attached to the substrate is in general rougher and controlled by the characteristics of the deposition bath (e.g. $1-3\,\mu m_{rms}$); if *reverse-treating* is chosen, treating is applied to the surface facing the drum, bringing the roughness close to that of the treated rolled annealed. Different types of foils may be used in the same PCB, depending on which core or pre-preg is used. In [50] a thorough presentation of foils and related roughness measurement techniques (stylus profile, optical profiler, scanning electron microscope and photomicrographs) is presented; roughness values and estimated distributions are reported, ranging between 2 and $7\,\mu m_{rms}$. Among the various results they report the characteristic impedance of some lines, showing that smooth rolled annealed copper gives the largest value; the reason is that smooth copper is eroded sooner resulting in faster etching with on average a fraction of mil of over-etching, so the reason is simply that the trace width is slightly smaller. Their results in general seem to support a Hammerstad-Bekkadal expression without requiring any more complicated fitting nor modeling.

6.1.3.3 Guard traces

They are traces at the ground potential laid down between signal traces to reduce crosstalk. When a solid ground plane is already present (as for the most common PCB transmission line structures reviewed soon after), their effect is less pronounced, a large part of the crosstalk being removed by the ground plane itself. Guard traces are always troubling the designer, whether they are useful or, on the contrary, detrimental: an unterminated guard trace is a stub that may resonate in quarter-wavelength and its multiples at a fairly low frequency (see below sec. 6.1.4 and the paragraph on stubs), even if it is perfectly functional for analog electronics and slow digital circuits; terminating it to the ground plane, increases the resonance frequency proportionally to the reduction of the separation between bonding points (vias).

This is evaluated numerically and experimentally by Novak, Eged and Hatvani[14] who show a center trace left floating, then matched terminated at the ends, then grounded at the ends and then finally grounded also at intermediate points. There are some changes of scale of the reported experimental results, a general lack of comments and the results are not evaluated or compared: all this reduces the usefulness of the paper and makes it hard to follow; the paper, however, reports accurately the geometry of the test setup and the Spice code for its simulation.

Bonding guard traces to the ground plane underneath is achieved with vias, that are located along the guard trace with a sufficient density to prevent low-frequency resonances of floating sections. This is a periodic structure (as pin fields of long connectors, and similar arrangements of arrays of vias), that was studied for its impact on transmission line characteristics (see sec. 6.4.3.1); to author's knowledge there has been no attempt to analyze the benefits of breaking of symmetry and reduction of the factor of merit of resonances by randomization of via distribution (like it was proposed years

[14] I. Novak, B. Eged and L. Hatvani, "Measurement by vector-network analyzer and simulation of crosstalk reduction on printed circuit boards with additional center traces," IEEE Instrumentation and Measurement Technology Conference, Irvine, CA, USA, May 18-20, 1993, pp. 269-273.

ago for the distribution of supporting beads in air-dielectric coaxial lines, see at the end of sec. 5.3.1).

Lee *et al.*[15] criticize the use of vias in guard traces because of the drawback of imposing restrictions on routing on the PCB backside; however, it is observed that the ground return plane in proximity of the hot trace of microstrips shall be free of traces. They propose a serpentine guard trace with the aim of increasing the mutual capacitance term, following a simple observation: the mutual inductance (normalized by the self inductance) is in general larger than the mutual capacitance (normalized by the self capacitance), because half of the space surrounding the hot trace is air with a low permittivity; they observe that the difference in time of arrival of even and odd mode is proportional to the difference of these two fractions, and to reduce it the second capacitive term shall increase to compensate the inductive one (see sec. 3.3.1 and 8.3.2). The serpentine increases the mutual capacitance between victim and aggressor lines. Additionally, they show that the length of the individual sections of the serpentine line shall be short enough, to avoid major resonances: the optimum value that minimizes crosstalk looks about twice (or slightly larger) the separation of victim and aggressor lines. The serpentine guard traces not only reduces crosstalk (about $5\,\mathrm{dB/m}$ lower FEXT with respect to a bonded straight guard trace with vias and about $5\,\mathrm{dB/m}$ with respect to no guard traces), but also equalizes propagation times of even and odd modes, thus reducing also time jitter. What they call "conventional" guard trace, that is a trace not frequently bonded to the return plane produces worse jitter due to high-frequency resonances, as expected; in general, the improvement of time jitter is of the same order of that observed for crosstalk.

In general, guard traces load the transmission line they protect, as it is for coplanar waveguides (see sec. 6.3.3), where the surrounding extension of return plane lowers line characteristic impedance. When the guard trace is positioned only at one side (e.g. focusing on the side where crosstalk and interference might come from), the loading is not symmetric and this largely affects common-to-differential mode conversion.

6.1.3.4 Bending and corners

Traces as part of transmission line structures will be fully analyzed in sec. 6.3, later on in this Chapter. Here the attention is focused on a more practical and phenomenological interpretation, deriving simple models for typical situations such as curve bending and corners at changes of direction.

A corner in a trace is normally to be avoided as it represents a discontinuity: at the bend the outer radius is much longer than the inner radius and this is seen as additional capacitance that may be estimated by considering the total trace area in the curve; trimming the excessive area that is not really used for signal flow is called chamfering or mitering (see sec. 6.3.1.3) and reduces significantly the stray capacitance of the bend. This is in fact an option available in PCB CAD programs, where 45° skewed or round bending may be selected.

---

[15] Kyoungho Lee, Hyun-Bae Lee, Hae-Kang Jung, Jae-Yoon Sim and Hong-June Park, "A Serpentine Guard Trace to Reduce the Far-End Crosstalk Voltage and the Crosstalk Induced Timing Jitter of Parallel Microstrip Lines," *IEEE Transactions on Advanced Packaging*, Vol. 31, no. 4, Nov. 2008, pp. 809-817.

Let's consider a trace of width w that is bended sharply at 90°: the corner where the trace change direction is a square of $w \times w$ area, half of which is removed when chamfering at 45° the corner. The excessive capacitance that is removed is that of the capacitor created between the trace and the underlying ground plane:

$$C_{corner,excess} = \frac{1}{2} \frac{\varepsilon_0 \varepsilon_r w^2}{h} \tag{6.1.2}$$

and for a $w = 1\,\mathrm{mm}$ trace with a $h = 1.5\,\mathrm{mm}$ thick FR4 board, the excess capacitance is about $27\,\mathrm{fF}$, getting significant in terms of its reactance only above 10 GHz. We will see that going beyond this low-frequency approximation, such a corner affects the electromagnetic field distribution and propagation (see sec. 6.3.1.3).

6.1.4 Vias

When deciding the best via size, constraints come from the fabrication process (hard constraints regarding minimum and maximum vias that are allowed), from cost-time tradeoff (smaller vias beneath a limit value are called micro-vias, require the use of tiny drill bits, that break and wear more easily and demand extra cost and extra time), from the physical size and tolerance of through-hole components (pins and leads have circular or rectangular shape, have production tolerance and need extra clearance for the solder to flow into the via), and from required electrical characteristics.

Regarding the latter it is easy to say that smaller vias are better for trace routing and have less stray capacitance. Moreover, blind and buried vias are shorter and thus have also considerably less inductance and are less troublesome when seen as stubs and discontinuities of transmission lines.

Before going into the details of estimating their electrical behavior, it is convenient to review the few related standards that may be absolutely optional for small series and laboratory prototypes, but become really relevant, not to say compulsory, when going to production, with automatic component assembly (automatic soldering, pick-and-place, etc.), maybe automatic testing and, finally, certification.

Vias terminate on an annular ring, or pad,[16] that puts in electrical connection the inner plated cylindrical wall and external traces; its size is determined by the plating process characteristics, the via diameter, tooling tolerances and alignment requirements. What happens with CAD software is that all these requirements can be input into the software that automatically generates the right vias, once the nominal diameter is selected and at the end checks the whole design against PCB manufacturer's requirements that are contained in one or more files, normally downloaded from manufacturer's website. A lot of details regarding sizing of the various elements can be found in the three books listed in sec. 6.1 and in [199], sec. 7.1.

The same can be said for trace clearance and width, the separation between various elements, including board edges and mechanical holes, and so on. Clearance between unprotected parts, such as vias and pads, is first required to avoid shorting them when

[16] Standard IPC-6012 "Qualification and performance - Specification for Rigid Printed Boards" specifies annular ring geometry at sec. 3.4.1 and 3.4.2; latest version C, 2010.

soldering components; then, in general, clearance between parts is required to ensure the necessary voltage insulation.

To conclude, larger annular rings and pads are better for soldering, but ask for more board space, have impact on trace routing and are worse for parasitics.

Vias are discontinuities of transmission lines that may be modeled by a lumped capacitance and inductance, that we may call "via parasitics". The effect as discontinuity is appreciable even if vias are not part of the transmission line, but couple to the transmission line under test, behaving like resonating stubs at very high frequency (see next page); the discontinuity is relevant even also as interruption of continuity and uniformity of ground planes, as discussed later in sec. 6.4.3.1.

6.1.4.1 Capacitance

Via capacitance to ground was estimated in [199], relating it to its geometrical dimensions and to the dielectric substrate permittivity:

$$C_{via} = 0.0555 \frac{\varepsilon_r h D_{pad}}{D_{clr} - D_{pad}} \tag{6.1.3}$$

where h is the substrate thickness, D_{pad} is the diameter of the pad around the via and D_{clr} is the diameter corresponding to the clearance for grounded traces and elements around, including the clearance hole through ground planes, that should be the tightest clearance. The coefficient 0.0555 in front of the fraction gives the result in pF if the quantities are expressed in millimeters.

A micro-via with 0.3 mm (12 mils) internal hole and thus about 0.6 mm of annular pad and a clearance of 0.3 mm going through the entire board of 1.6 mm (63 mils) thickness has about 0.83 pF. A larger via of 0.5 mm diameter and 0.9 mm of annular pad and the same clearance features 1.25 pF. On the contrary increasing the clearance around vias from 0.3 mm to 0.5 mm brings the two parasitic capacitance values to 0.50 pF and 0.75 pF. Reducing board thickness from 1.6 mm to 0.4 mm (or similarly for a blind via between two adjacent layers) reduces proportionally the two originally calculated values of parasitic capacitance to 0.21 pF and 0.31 pF.

In [258], based on the observation that for an increasing number of stacked layers vias resemble more and more a coaxial structure, two coaxial line expressions are presented: the leftmost is to be used when the pad capacitance dominates (e.g. many pads at each layer are stacked), the other one is preferred when the via body capacitance dominates.

$$C_{via,coax} = 2\pi\varepsilon_r\varepsilon_0 \frac{h}{\ln\left(\dfrac{D_{clr}}{D_{pad}}\right)} \qquad C_{via,coax} = 2\pi\varepsilon_r\varepsilon_0 \frac{h}{\ln\left(\dfrac{D_{clr}}{D_{via}}\right)} \tag{6.1.4}$$

Other estimates of via capacitance obtained by various methods (e.g. full wave field solver) and expressed in a way that can be reused are considered for comparison:

- reference values in [322] couldn't be used because the dielectric permittivity of the substrate is not given;
- an extensive experimental study is reported in [258], where different geometries and number of layers are considered (results are shown in Table 6.1.2); eq. (6.1.3) gives too small values when the via height is small, while for thick circuits it is quite accurate; in a later work [259] Pajovic reports more extensively the theoretical expressions and extends the method also to differential vias;
- an analytical model is presented in [347], distinguishing again the via body capacitance (called barrel capacitance) and pad capacitance terms, and giving expressions that involve a summation of Bessel and Henkel functions over the mode number (the authors demonstrate that convergence is ensured already at $N=5$); at very high frequency (around 200 GHz) capacitance terms begin to vary and the approach becomes questionable; calculated capacitance values are shown in perfect correspondence with FEM results, but they are way too low with respect to the values shown in Table 6.1.2: at $h=0.51\,\text{mm}$ for slightly smaller pads ($D_{via}=0.2\,\text{mm}$, $D_{pad}=0.51\,\text{mm}$, $D_{clr}=0.86\,\text{mm}$) the resulting capacitance corrected for permittivity is only $35\,\text{fF}$;
- in [215] the general problem of a via through many layers including the effect of connected microstrips and striplines is considered; the results are quite interesting and well in agreement with another reference; despite the lack of closed form expressions, the paper reports many figures with parametric curves; a sample comparison for a similar, but not too similar, geometry ($D_{via}=0.5\,\text{mm}$, $D_{pad}=0.66\,\text{mm}$, $h=0.4\,\text{mm}$, as appearing in Fig. 6, curve 2, there) gives about $0.12\,\text{pF}$.

The effects of the via capacitance for step-like and pulse-like signals, as for any other element (such as pads), are:

- reduced impedance terminating the line during waveform edges, featuring the largest frequency content and resulting in reflections and ringing, localized around steep edges; to determine a suitable limit for parasitic capacitance, we may state that it is required that its reactance is larger than the characteristic impedance by a factor of e.g. five; going to consider signal rise time t_r and its equivalent bandwidth $1/\pi t_r$ for a Gaussian profile, the limit of capacitance value is

$$C_{p,lim}=\frac{t_r}{2.5Z_c}$$

- capacitive loading of line sections causes additional delay for signal propagation: the time necessary for the transient response of RC circuit to vanish is proportional to the time constant, that when a small lumped capacitance C_p is loading a line section with characteristic impedance Z_c is given by Z_cC_p; when driven by a step signal, 2.2 time constants are needed for the step front to rise to 90% of the steady value, so that the added delay is

$$t_{del}=2.2Z_cC_p$$

Via capacitance [pF]	**14 pads** $h = 2.7\,\text{mm}$	**4 pads** $h = 0.51\,\text{mm}$	**3 pads** $h = 0.38\,\text{mm}$
	1.16	*0.64*	*0.62*
$D_{via} = 0.3\,\text{mm}$	1.07	0.66	0.65
$D_{pad} = 0.58\,\text{mm}$	1.22	0.71	0.68
$D_{clr} = 0.84\,\text{mm}$	(1.62)	(0.58)	(0.58)
	[1.33]	[0.25]	[0.19]
			0.59
$D_{via} = 0.3\,\text{mm}$			0.64
$D_{pad} = 0.64\,\text{mm}$			0.69
$D_{clr} = 0.92\,\text{mm}$			(0.54)
			[0.19]
			0.71
$D_{via} = 0.38\,\text{mm}$			0.69
$D_{pad} = 0.66\,\text{mm}$			0.74
$D_{clr} = 0.92\,\text{mm}$			(0.68)
			[0.21]
			0.55
$D_{via} = 0.3\,\text{mm}$			0.57
$D_{pad} = 0.58\,\text{mm}$			0.63
$D_{clr} = 1.06\,\text{mm}$			(0.48)
			[0.10]

Table 6.1.2 – Via capacitance for various geometries [258]: the italic value is the result of TLM modeling, the two underlined values are experimental results obtained with VNA (1 MHz) and RCL bridge (100 kHz), the value between round brackets is given by eq. (6.1.4) right, except the first one that is calculated using eq. (6.1.4) left), the value between square brackets is given by eq. (6.1.3).

In the examples above the resulting C_p for a through-hole via from top to bottom was $0.83\,\text{pF}$. For a line with characteristic impedance $Z_c = 50\,\Omega$ the limit rise time for applied signals is $104\,\text{ps}$; for faster signals the parasitic capacitance needs to be reduced. The additional delay experienced by signals traveling along the line is $91\,\text{ps}$; by itself it is not a problem, provided that there are no synchronization constraints with other signals that are not experiencing the same additional delay; if the delay is common to all the relevant signals, then it does not represent a severe issue, otherwise it's a source of delay skew.

Two remarks are due on the made assumptions:

- a Gaussian profile has been assumed for waveform edge; if a sharper profile, such as a linear one, were assumed, then the bandwidth would be more extended, by up to a factor of two, and the resulting limit value of capacitance would be

$$C_{p,lim} = \frac{t_r}{5Z_c}$$

- the two numbers used to fix the conditions, that is 5 to exclude significant shunting of the characteristic impedance and 2.2 for a satisfactory transient response are one of the possible choices and slightly different values might be used.

If several elements, such as vias, connectors and component pads, are to be connected to such a transmission line routed like a bus, distributing them along the line with a separation corresponding to quite a short time difference, shorter or comparable to the foreseen rise time of applied signals, the effect of each single capacitance will smear out and capacitive terms will load almost uniformly the transmission line, resulting in a lower characteristic impedance. It will be thus necessary to take account of this during design and to choose a slightly narrow trace width to provide the same desired characteristic impedance. In any case the effect of the first one or two elements is always visible from the input terminals, when the signal is applied.

6.1.4.2 Inductance

Via inductance was estimated in [199], with respect to its geometrical dimensions:

$$L_{via} = 0.2h \left[\ln\left(\frac{4h}{D_{via}}\right) + 1 \right] \tag{6.1.5}$$

where h is the substrate thickness and D_{via} is the internal diameter of the via. The coefficient 0.2 in front of the fraction gives the result in nH if the quantities are expressed in millimeters. It is apparent that the length of the via (corresponding to the substrate thickness h) is the most relevant parameter; the via diameter D_{via} does little because it appears inside the logarithm.

Goldfarb and Pucel [136] discuss a different formula for via inductance, examining the effect of a nearby ground plane (the via is specifically the ground via of a microstrip, that thus has a return plane underneath) and concluding with an "imageless" inductance expression that included an empirical correction from "1" to "3/2":

$$L_{via} = \frac{\mu_0}{2\pi} \left[h \ln\left(\frac{h + \sqrt{r^2 + h^2}}{r}\right) + \frac{3}{2}\left(r - \sqrt{r^2 + h^2}\right) \right] \tag{6.1.6}$$

where $r = D_{via}/2$ for compactness.

Using the first formulation by Johnson and Graham, a micro-via with $0.3\,\mathrm{mm}$ $(12\,\mathrm{mils})$ internal hole going through the entire board of $1.6\,\mathrm{mm}(63\,\mathrm{mils})$ thickness has about $1.3\,\mathrm{nH}$. A larger via of $0.5\,\mathrm{mm}$ diameter has a parasitic inductance of $1.14\,\mathrm{nH}$. Conversely, reducing board thickness from $1.6\,\mathrm{mm}$ to $0.4\,\mathrm{mm}$ brings the two parasitic inductance values to $0.21\,\mathrm{nH}$ and $0.17\,\mathrm{nH}$, so much closer and remarkably less than the linear proportionality found for parasitic capacitance.

With Goldfarb and Pucel formulation, instead, the same vias of $0.3\,\mathrm{mm}$ and $0.5\,\mathrm{mm}$ diameter for a board $1.6\,\mathrm{mm}$ thick have $0.54\,\mathrm{nH}$ and $0.41\,\mathrm{nH}$ inductance; reducing board thickness to $0.4\,\mathrm{mm}$, inductance becomes $0.053\,\mathrm{nH}$ and $0.033\,\mathrm{nH}$. Since modern multi-layer boards have always abundance of ground planes, it is sensible to assume that this formulation is closer to practical cases than the Johnson and Graham one.

It is clear that the inductance is the most troublesome parasitic of vias, as it is for connectors and similar devices. There is the possibility however of compensating it by adding some shunt capacitance, possibly at the two ends of the inductive element, so that the resulting pi-cell is trimmed to have $\sqrt{L_p/C_p}$ as equal as possible to the line characteristic impedance. For a through via with the calculated $1.3\,\mathrm{nH}$ parasitic inductance, two capacitors of $0.25\,\mathrm{pF}$ each would be necessary for a $50\,\Omega$ characteristic impedance; they are not so different from the parasitic capacitance of the via, that may be already sufficient and shall be considered in any case to avoid over-compensation. If using the Goldfarb-Pucel estimation, then the necessary capacitance grows and stray terms probably are not sufficient any longer. Of course, such compensation becomes really necessary when the inductive reactance is a significant fraction of the characteristic impedance, that in our example occurs at about 2 GHz. When the parasitic inductance is larger, the benefits of capacitive compensation are clear.

Other estimates of via inductance obtained by various methods (e.g. full wave field solver) and expressed in a way that can be reused are considered for comparison:

- in [322] signal vias surrounded by four grounded vias (e.g. a pattern like in Figure 6.2.1(a), suitable for SMA or smaller connectors) were simulated, with $D_{via} = 160 - 180\,\mu\mathrm{m}$ and anti-pad diameter (thus including clearance around the via) $D_{clr} = 360 - 420\,\mu\mathrm{m}$, and a distance between signal and ground vias $D_{clr} = 385 - 405\,\mu\mathrm{m}$, where the indicated variability refers to the slightly different diameters for different layers of the multi-layer board; the inductance values extracted from the S_{11} measurements are $0.22\,\mathrm{nH}$ and $0.49\,\mathrm{nH}$ for two height values $h = 740 - 1480\,\mu\mathrm{m}$. Calculated inductance values using eq. (6.1.5) and (6.1.6) give $0.57 - 1.3\,\mathrm{nH}$ and $0.22 - 0.63\,\mathrm{nH}$ and $0.63\,\mathrm{nH}$, thus confirming the much better accuracy of eq. (6.1.6), as it will be seen in Example 6.2.

As done for parasitic capacitance, the effects of via inductance for step-like and pulse-like signals are:

- increased longitudinal impedance of the line during waveform edges, which have the largest frequency content, resulting in reflections and a ringing phenomenon localized around steep edges; to determine a suitable limit for the parasitic inductance, we may state that it is required that its reactance is smaller than the characteristic impedance by a factor of, let's say, five; going to consider the signal rise time t_r and its equivalent bandwidth $1/\pi t_r$ for a Gaussian profile, the limit on the inductance value is

$$L_{p,lim} = 0.4\frac{t_r}{Z_c}$$

The two remarks done for capacitance are still valid. A sharper profile of waveform edge would result in a tighter limit of the inductance value by a factor of two, that is

$$L_{p,lim} = 0.2\frac{t_r}{Z_c}$$

6.1.4.3 Resonance and stubs

Vias may be considered not only for their lumped parameter equivalent (in terms of the just analyzed parasitics), but also as a very short "stub" (transmission-line branching element), that may affect the response of the main line.

The effect of a micro-via, 1.6 mm long and with an estimated parasitic capacitance of 0.50 pF, is reported in [45] as causing a 6% reflection with a 50 ps driving signal on a 15" transmission line.

In general, for a stub to be effective and affect signal quality, we need that twice the propagation time through it corresponds to half of the applied frequency, or that its electrical length is a quarter of the applied wavelength [45]: in this case we have a $\lambda/4$ resonance that largely affects signal transmission, pushing the insertion loss in the far negative values (see sec. 10.1.2 and 10.2). Even if the exact velocity of propagation through a via is not known, we may guess that is not so different from a planar transmission line with similar cross-sectional dimensions:

$$f_{\lambda/4} = \frac{1}{4}\frac{v_{via}}{h} \simeq \frac{75}{h\sqrt{\varepsilon_r}} \tag{6.1.7}$$

that gives the resonance frequency in GHz if the via length h is measured in mm; ε_r is the relative permittivity of the PCB material and gives an estimate of the velocity of a generic planar transmission line.

6.1.4.4 Thermals

The so called "thermals", or thermal relief vias, allow the pad and the attached component to be soldered and unsoldered more easily and in a shorter time; when vias are part of large copper areas, such as power supply and ground copper planes, vias with thermals are connected thermally, and electrically, only at the four cardinal points, instead of being continuously connected to the copper around. This may have adverse effect at very high frequency, because it introduces additional inductance.

6.1.4.5 Vias in multi-pin connectors

Whenever possible, it is a good habit to leave clearance between vias of connector pins, such that the ground plane may be poured all over around and return currents may flow - if needed - also between pins rather than running around the whole connector (see sec. 7.1.3.1). The latter case would be like a slot cutting the ground plane and forcing return current to flow farther away increasing the path inductance and the amount of radiated electromagnetic field. To this aim it is quite useful to use some pins to ground the cable bond wires and shield, improving the reduction of crosstalk from two parts of the same connector and creating at the same time an additional cross-connection between the two sides of the connector at the ground potential. This is treated in more detail in Chapter 7 and, in particular, in sec. 7.1.3.

Example 6.2: Estimate of resonance of vias

Let's take two vias as already previously considered with 0.3 mm and 0.5 mm diameter drill; the PCB thickness is 1.6 mm and both go from top to bottom layer. The PCB is a standard FR4 board with $\varepsilon_r = 4.7$. The resonance frequency estimated with eq. (6.1.7) is 21.6 GHz and it is the same whichever the via, only the length h matters. The change of dielectric permittivity accounts for a increase (at 3.3) or reduction (at 6) of about 19% and 13%, respectively.
The resonance frequency may be calculated from the stray capacitance and inductance terms too, estimated with eq. (6.1.3) and (6.1.5), or (6.1.6), respectively: for the two vias the capacitance is $C_{0.3\,\mathrm{mm}} = 0.83\,\mathrm{pF}$ and $C_{0.5\,\mathrm{mm}} = 1.25\,\mathrm{pF}$, and may reduce to 0.50 pF and 0.75 pF if the clearance around the via annular ring is increased; the stray inductance is $L_{0.3\,\mathrm{mm}} = 1.3\,\mathrm{nH}$ and $L_{0.5\,\mathrm{mm}} = 1.14\,\mathrm{nH}$ or $L_{0.3\,\mathrm{mm}} = 0.54\,\mathrm{nH}$ and $L_{0.5\,\mathrm{mm}} = 0.41\,\mathrm{nH}$ with the two formulations. Resonance frequencies are thus 4.84 GHz and 4.22 GHz (or 7.52 GHz and 7.03 GHz with smaller inductance); increasing the clearance around vias, the resonance frequency is increased to 6.24 GHz and 5.44 GHz for the largest inductance estimate, with a more significant effect for smaller vias.
If the board is thinner, the vias are blind or buried, or if full size vias are backdrilled, the overall length reduces: for $h = 0.4\,\mathrm{mm}$ the new resonance frequency with eq. (6.1.7) is proportionally higher, that is four times higher, at 86.4 GHz! Using stray capacitance and inductance recalculated for the new thickness value: $C_{0.3\,\mathrm{mm}} = 0.21\,\mathrm{pF}$ and $C_{0.5\,\mathrm{mm}} = 0.31\,\mathrm{pF}$, and $L_{0.3\,\mathrm{mm}} = 0.21\,\mathrm{nH}$ and $L_{0.5\,\mathrm{mm}} = 0.17\,\mathrm{nH}$; the resulting resonance frequency is 24 GHz and 22 GHz and this is close to what might be seen in practice.
Let's now compare to the measured resonance in [109]: the PCB thickness is 0.1379 inches, that is 3.5 mm, and the dielectric constant is 3.7. The drill size is not given, but by visually comparison with other parameters, it is estimated as 4 mils (0.1 mm); the diameter of the circular pad is 0.3 mm with about 0.15 mm of clearance around (the overall diameter is 0.61 mm, i.e. 0.024 inches). The height of vias is smaller than the PCB thickness because vias were backdrilled; since they connect layers 7 and 9 where the striplines are located, backdrilling from above or below may result in via height reduced to 2.64 mm (from above) or 1.57 mm (from below). The resonance frequency estimated with eq. (6.1.7) for the whole board thickness is 11.1 GHz, while for reduced via height it increases to 14.8 GHz and 24.8 GHz. Using via capacitance and inductance $C_{0.1\,\mathrm{mm}} = 0.81/0.48\,\mathrm{pF}$ and $L_{0.1\,\mathrm{mm}} = 2.99/1.61\,\mathrm{nH}$, the resonance frequency is 3.2/5.73 GHz; but using the corrected inductance formulation in eq. (6.1.6), the inductance becomes $L_{0.1\,\mathrm{mm}} = 1.68/0.84\,\mathrm{nH}$ and the resulting resonance frequency gets to 4.3/7.93 GHz. The measured value in the paper is 8.2 GHz, so that all estimates are more or less in error, useful as rough estimates, rougher for very small elements; it is apparent that the true resonance frequency is quite in between the values estimated with the distributed and the lumped models. In any case, the corrected inductance equation (6.1.6) yields values that are much more reliable. It is not excluded that also surrounding electromagnetically coupled floating vias, as well as multiple ground planes, have some influence on the result.

6.1.5 Solder mask

Solder mask is used to restrict the flow of solder to the desired areas, but it also prevents leakage currents between traces due to impurities during fabrication or later handling and use, and mechanically protects traces; however, it reduces the impedance of lines in the outer layer (microstrips and coupled microstrips), depending on dielectric permittivity of the solder mask itself: the most common value declared by PCB manufacturers is 3.3, but also 2.8 and 4.3 are values that may be found. A further issue is that its thickness is not tightly controlled and this adds another variable to the problem. Materials used for solder mask are many, UV and thermally cured, including various types of resins.

Conformal coating, like solder mask, has the function of protecting PCB from external agents (dust, moisture, chemicals) and is applied after the board has been manufactured and components assembled: it largely improves voltage insulation between the traces and its stability with time and environmental conditions. As for solder mask, coating might cause the increase of the total effective permittivity, thus reducing the characteristic impedance of transmission lines built on outer layers (i.e. microstrips). Broadly speaking, conformal coatings are made of acrylic, epoxy, polyurethane or silicon with approximately known dielectric permittivity, that is 2.2-2.4 for acrylic resins, 3.5 to 4.5 for epoxy resins, 3.3-3.5 for polyurethane and nearly unity for silicon; thickness may be significant, e.g. from one to several tenths of mm.

It was probably to avoid line loading and not to influence characteristic impedance that in [300] it is clearly stated that tested laminates are without solder mask.

It may be guessed, even if it is not absolutely demonstrated, that solder mask is a final operation on the product that is subject to a less tight control than the "core" fabrication process; for conformal coating this is probably even more true, with thickness only approximately controlled by the amount of product poured on the surface. Possible disuniformities, even if not dramatic, may exert their influence on transmission line response, as it is pointed out in [227]: a noticeable phase difference is appreciable at about 10 GHz in the otherwise symmetric S parameters S_{21} and S_{43} (and of course S_{23} and S_{41}) for a symmetric line; a difference in the nuance of solder mask is spotted out and examined under the microscope, indicating a slightly different thickness. Influence on return and insertion loss of microstrips is confirmed by Coonrod [89], who reports a 15-20% higher insertion loss for a RO4350 $0.51\ \mathrm{mm}$ thick laminate. Coonrod himself and others report that for some type of solder mask materials sensitivity to moisture is increased, so that solder mask is not so effective in protecting the PCB as conformal coating might be.

6.2 Test coupons, probes and connections

For experimental testing and in-fixture calibration standards are required that are compatible with and built in the PCB setup: it is thus necessary that the PCB is predisposed for a series of tests by means of purposely realized planar transmission lines (see sec. 6.3 and 6.5) and for the insertion of test probes, thus requiring some space for test probe patterns, either in the through-hole or SMT technology. Several test probe patterns exist and some are standardized, for example by IPC [190, 191].

Test coupons consist of planar transmission lines and other elements with standardized characteristics, that allow to test laminate and substrate properties, as well as to verify the fabrication process, in particular fabrication tolerances. Normally test coupons are required by standards or by agreements between manufacturers and customers, especially when the latter are well-prepared skilled large companies, that lead and drive test methods and standardization activities. Calibration standards realized on-board PCB are often required, for both time- and frequency-domain measurements.

Test probe patterns may be used to install a permanent probing element or to land with a moveable probing head, either manually or automatically (e.g. ATE, Automatic Test Equipment). For fixed probing elements, some are predisposed for the insertion of hand-held probes (e.g. oscilloscope probes), other may be connected directly to a cable with a suitable connector.

Different probe land patterns have evolved through time with remarkably different sizes, coping with increasing frequency ranges and setups and devices of ever smaller size: coaxial edge and board connectors, probe gauges in the mm range for PCBs and then down to tens or hundreds of μm for probing very small board details and down to the wafer level. To simplify, in the following, we classify them as coaxial, large board probes and small wafer probes.

6.2.1 Test coupons and test vehicles

For various reasons simulated structures in all equal to the final product or system may be necessary; test coupons as part of a board, or even stand-alone boards, are thus designed and built.

One reason may be the characterization of the production process in terms of fabrication tolerances, quality of material, uniformity of processing and treating (e.g. solder mask homogeneity and thickness, roughness of trace edges, under-etching, dimensional stability); cross sections may be then also cut and examined with optical or SEM microscopes, X-ray diagnostics, surface scanning systems, and other kinds of gauging and scanning methods.

Most often the attention is on electrical characteristics and performance, again in relationship to production process and its variability, such as e.g. uniformity of dielectric material (including glass weave effects) and its permittivity, alignment of vias and of their parts (e.g. through hole, metalization, pads), surface and profile regularity of traces. In this case the test board shall be equipped with many probing points suitable for connection of the necessary diagnostic tools; in general, during prototyping and testing the number of probing points shall be larger than those remaining during normal production.

When testing pre-production and verifying quality, stability and uniformity, boards may be designed to have small parts equipped with test circuits and structures, very similar to those used for test boards when prototyping. Such predisposition is normally termed "test coupon" or "test area", to distinguish from entire boards designed and build for this, usually called "test vehicles".

In all cases, great care shall be given to probing points and to include additional elements necessary for calibration and verifications, such as calibration standards (e.g. of the Short Open and matched Load type) and additional devices (e.g. disk capacitor for low-frequency determination of the dielectric permittivity, see sec. ????????????).

Whenever the quality and stability of the fabrication process are relevant, in order to ensure the necessary performances (and their uniformity) for the produced devices, well conceived test coupons are highly advisable. There is no particular scheme to follow: the coupon will contain the PCB elements that are more exposed to investigated tolerances and defects and shall allow testing to reveal them with the necessary calibration, time/frequency ranges and probing method. Examples are:

- planar transmission lines with geometry suitable to test the response of the lines themselves (characteristic impedance, attenuation, dispersion, see sec. 6.3 below) and PCB basic characteristics, such as the dielectric permittivity of the various substrates (cores and pre-pregs) and also its anisotropy (e.g. striplines and microstrips have a different electric field distribution and orientation);
- uniformity of substrates using planar transmission lines oriented horizontally and vertically with respect to the glass weave pattern (see sec. 6.5); diagonally oriented lines are normally used to have the least sensitivity to orientation and are clearly recognizable on some test vehicles (see Figure 6.5.1);
- trace width, small return planes and arrays of small vias, to test misregistration and lack of alignment between layers, e.g. measuring common-to-differential conversion (see sec. 6.4.3.2).

6.2.2 Test probes and connectors

Probing of test vehicles, test coupons and normal production boards shall be accommodated by proper design, choosing the right probing points and connection methods depending on several factors: board density, frequency range, measuring equipment, time duration of measurements, available and followed calibration methods. Coaxial connectors are readily connected to vast majority of measuring instruments, but require board estate and may be troublesome in high-density boards, as well as they cannot be used on wafer nor too close to (or directly onto) board elements. Various kinds of RF and microwave probes may be used instead, that may be DIY built or purchased, depending on target size and frequency range.

6.2.2.1 Coaxial connectors

Coaxial connectors may be used to connect to PCBs and other parts of setups with repeatable and robust connections, suitable for commonly used measuring equipment. When planar transmission lines to be tested are built on inner layers, their accessibility is ensured by vias and pads on top (or bottom), that are designed and arranged for minimal impact as discontinuities. For not too a high frequency range and when boards elements are not too small or too short, various types of land patterns for coaxial connectors of the surface mount type can be considered.

Considering Figure 6.2.1, in order to connect to the board with standard connectors instead of probes, mostly used patterns accommodate for SMA or smaller connectors:

- edge connectors are quite known and originate from old RF and microwave tradition; their land patterns are standardized and available in all PCB CAD software; on the cheapest ones care shall be given to the thickness of the PCB and the clearance between upper and lower ground pins, so that the connector fits onto the board edge and is not wiggling loose;

- vertically mounted connectors may be placed anywhere on the PCB and allow a better use and exploitation of the PCB area; standard soldered SMA have the five-pin pattern already shown in Figure 6.2.1(a), but the most useful recently are the snap-in SMA, 3.5 mm, 2.92 mm or 1.85 mm connectors (for example manufactured by Gigaprobes or Molex[17]).

Performances for high-end connectors are the result of a careful design and optimization process as shown in [113] (see Figure 6.2.3): the authors show that not only the geometry of the contact was revised and dimensions reduced, but the big work was on the PCB side for the landing pattern of ground ring to mate with the coaxial shield and to allow for stripline routing to the center pin. The operating frequency range is quite large and extends to 50 GHz, so that even otherwise negligible details are responsible for undesirable responses: the ring of grounded vias is then doubled and the internal ones are pushed closer against the inner pin, that would otherwise be impossible without redesigning the coax geometry. Of course, reducing clearance increases parasitic capacitance and this was addressed by maximizing clearance resulting in non-circular, but scalloped, rings, for the inner pin pad and outer ground ring with vias: eight quite large ground vias (six in the first design) are arranged symmetrically, one of them suppressed to let the stripline in; an additional outer ring of smaller vias rotated by $\pi/16$ is used probably ; no details of the stripline geometry are given and how the unavoidable change of impedance when passing between ground vias was compensated for, but for the small region between the ground ring and the inner pad where the stripline risk to go uncovered, the ground planes on upper and lower layers are extended with a “tongue-shaped” element (that in the paper is called “diving board”).

Quite interesting the dented pad for the inner pin: first, it’s not smooth but modulated by ups and downs, and at the ups there will be nevertheless an increase of coupling capacitance to the inner ring of ground vias; but, second, the number of ups and downs is ten, not eight, so not synchronous with the number and location of ground vias, probably to avoid major resonances. This is a possible explanation of the otherwise strange periodicity. In this sense probably an odd or prime number, such as eleven, would have ensured absolutely independent repetition patterns.

A huge number of simulations were run to perform parametric sensitivity and optimization, resulting on average in a -20 dB return loss at 15 GHz and impedance mismatch at connector in and out interfaces of maximum 10% of 50 Ω when evaluated in TDR for the full frequency range of 50 GHz.

---

[17] http://www.gigaprobes.com/rfconnectors.html (last accessed on Aug. 2015);
http://www.molex.com/molex/products/family?key=precision_connectors&channel=products&chanName=family&pageTitle=Introduction&parentKey=rf_microwave_coax_connectors (last accessed on Aug. 2015).

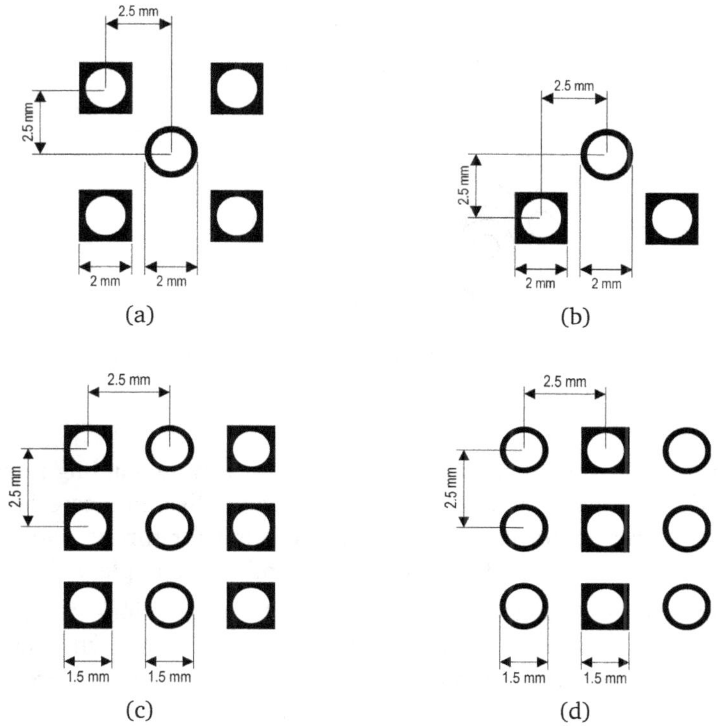

Figure 6.2.1 – PCB probe patterns for test coupons as per IPC-2141 (1996) [190] and EN 61118-1-2 [76]: (a) pattern 1A for SMA connector and single line (finished holes are 1.5 mm diameter), (b) pattern 1B, similar to 1A, for probe tips (again holes with 1.5 mm diameter), (c) multiple single lines, type 2 (smaller finished holes with 1.0 mm diameter), (d) multiple differential lines, type 3 (same finished holes of 1.0 mm diameter as type 2).

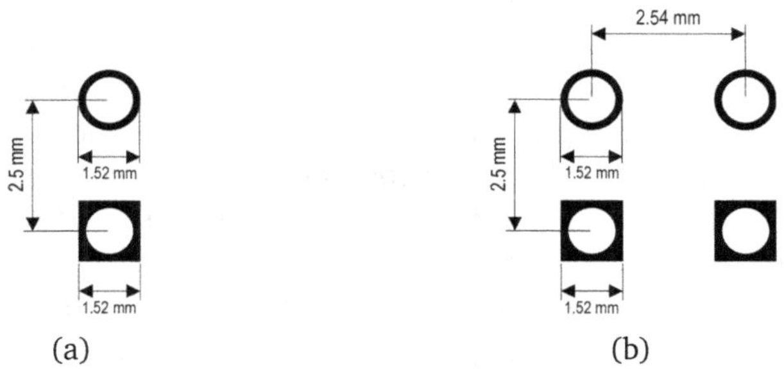

Figure 6.2.2 – PCB probe patterns for test coupons as per IPC-2141 (2004) [191]: (a) pattern "single-ended" (finished holes are 0.91 mm diameter), (b) pattern "differential" (again holes with 0.91 mm diameter).

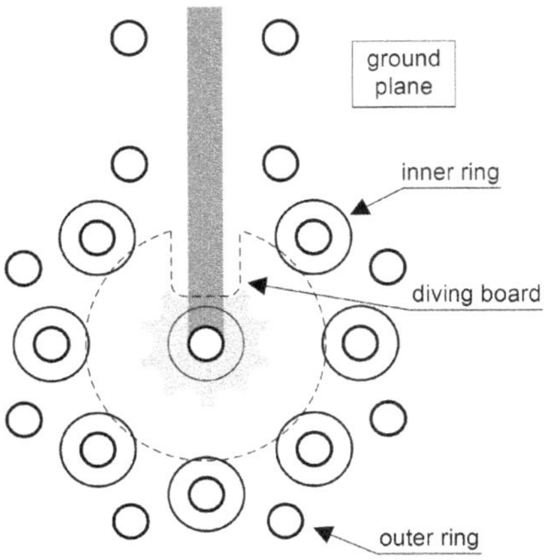

Figure 6.2.3 – Sketch of PCB details for 2.4 mm compression connector pad and signal stripline.

The same criteria of course may be applied even for less demanding interfaces: SMA connectors of SMD or through-hole type may receive the same care in land pattern design as the 2.4 mm connector above, provided that for through-hole components pins are trimmed or backdrilled (see sec. 10.4.3 and the example by Pupalaikis).

One last comment on PCB coaxial connectors with compression-type inner pin: they lend themselves for a direct back-to-back connection (as shown also in the cited [113]), so that de-embedding and connector modeling are possible provided that the available VNA or TDR is able to resolve satisfactorily about 50 ps of connector geometry, so that 15 GHz minimum of frequency range is necessary.

6.2.2.2 Large board probes

The part regarding test probe land patterns and coupons is interesting in order to prepare consistently PCBs for testing. Coupons are specified with one trace length (150 mm minimum for the IPC-2141 1996 version, rounded to imperial units in 2004, so resulting in 152 mm, i.e. 6 in) and with 2.5 mm clearance from other conductive parts nearby. Test probe patterns are for through hole SMA and similar coaxial connectors: spacing is 2.5 mm in the IPC-2141 1996 version, increased to 2.54 mm for the 2004 version (remaining within usual tolerance), so much larger than the microprobe land patterns that will be seen soon after. Their geometry is shown in Figure 6.2.1 and 6.2.2, distinguishing between the two versions of IPC-2141, where a significant change of recommended patterns may be noticed, probably following the speed increase of logic circuits, communication links and planar lines.

Smaller probe patterns revealed necessary with miniaturization on PCBs: they are reviewed with those for wafer measurements altogether and correspond to the largest spacings. An example is given in I. Martinez, “In-situ Calibration and Verification Techniques for the Characterization of Microwave Circuits and Devices”, IEEE International Conference on Microelectronic Test Structures (ICMTS), 2012, where prototype calibration standards were built on thin Alumina substrate.

6.2.2.3 Small wafer probes

Planar transmission lines built on special boards or on wafer are similar to those reviewed in the next section with various dimensions and possibly tighter tolerances. They are also suitable for a much more extended frequency range, extending to several tens of GHz. Probes and probing methods shall adapt in size, tolerance and performance: a measurement station with probes (the most known manufacturers of complete systems, stations without probes, probes or spare parts are Cascade Microtech, Escitec, GGB Industries, Gigaprobes, J microTechnolgy, Semiprobe, Signatone[18]) is necessary when making measurements at tens or hundreds GHz, where small discontinuities are relevant and they cannot be eliminated easily by calibration (no easy-to-find and inexpensive in-fixture calibration standards at these dimensions and for such elements) or time gating (too few data samples and the amount of reflection of such discontinuities is in the order of the measurement target). In these cases the accurate design of test pads is of paramount importance: test pads for station probes are arranged in SG and GSG configurations, that is Signal-Ground and Ground-Signal-Ground patterns. An example of probe tips of the GSG type is shown in [107], Fig. 49. They tend to be standard sized, even if probe tips allow some adjustment in terms of distance, inclination and force when approaching the test pads:

- the pitch, i.e. the distance between probe pin tips and thus pads centers, is standardized in $50\,\mu\text{m}$ increments, such as $150\,\mu\text{m}$, $200\,\mu\text{m}$, $250\,\mu\text{m}$, etc. up to about $500\,\mu\text{m}$ (in [105, 106] there is a good description of probes used in the tests and how the test pads were arranged); for the narrowest pitch also e.g. $30\,\mu\text{m}$ and $75\,\mu\text{m}$ are available, while for the largest ones above $500\,\mu\text{m}$ pitch values vary more coarsely;

- probe tips shall be cleaned, inspected and their planarity verified, often using a target wafer sample; depending on the material used for probe tips, cleaning to remove oxide may be necessary (BeCu is mechanically flexible, hard enough and does not oxidize easily);

- with respect to the target pads probe tips are inclined with a suitable angle: they are flat enough to skate over the pads and vertically enough not to couple too much through the unavoidable stray mutual capacitance, reaching a compromise optimal angle $\theta = \arccos(\varepsilon_{r,\text{eff}-\text{probe}}/\varepsilon_{r,\text{eff}-\text{line}})$, that ensures the correct characteristic impedance of the line (in reality inclination for most probe tips stay within 30° and 45° from horizontal plane and each manufacturer has one or few standardized values); the two dielectric permittivity values refer to that of the planar line connecting probe tips to the probe body and of the planar line subject to measurement; in any case all other lines shall be farther away from probe tips and line with a minimum distance of $0.5-1\,\text{mm}$; there are some patented transmission line technologies and geometries to connect to probe tips, featuring various constructions, such as stripline, coplanar waveguide, etc..

[18] Without representing any endorsement, their sites and product descriptions are believed to be of help for a better understanding and to define characteristics and requirements for a probing setup.

In [106] it is shown how small test pads (16 mil pitch with 8 mil vias arranged in GSG pattern) ensure negligible discontinuity while probing, even touching different portions of the land pads; even a GS pattern of the same size is equally performing; on the contrary, increasing the size of GS pads to 30×30 mil the discontinuity shows with a 3 times larger peak, up to 5 times larger, depending on vias size.

There is a wide range of signal and ground patterns, such as the cited GS and GSG, but also GSSG, GSGSG, etc., connecting to various kinds of single-ended and balanced planar lines with enough reference ground pads for an extended frequency range.

Even when using probe stations for quite high frequency measurements, residual small positioning errors of probe tips may as well affect repeatability of measurements and results. While lifting and lowering probe tips for calibration and measurements, repositioning may be subject to some approximation: when probes are driven manually by an operator for not so frequent measurements that require automatic positioning, marks can be located on pads as a reference for tip positioning; such marks may be small notches that do not affect performance [242], Chapter V. Cascade Microtech gives a short but interesting overview of positioning criteria taking into account the marks of its calibration substrates[19]; it goes without saying that the same criteria may be adopted to mark and size pads and targets in own, maybe application specific calibration standards and in general wafers under test.

Probes shall be calibrated as any other block of the measurement chain: to this aim target substrates with calibration standards (called Impedance Standard Substrates) may be used, where the manufacturer makes every effort to ensure the characteristic impedance up to very demanding measurement frequencies; to this aim very thin lines shall be designed, to reduce moding (that is the propagation of higher modes and consequently deviations from the characteristic impedance definition at the fundamental TEM, or quasi-TEM, mode). In any case, moving from the reference standard substrate to the wafer under measurement will cause unavoidable differences to appear whenever the substrate material and thickness are different. For this reason, and in order to have a wide range of calibration standards, the design and use of on-wafer standards is strongly advisable: modern CAD programs allow a very accurate design and verification by simulation, that by the way couldn't be otherwise; moreover, using built-in standards allows to fully include fabrication technology and launching elements (pads, connections) already in the calibration, removing them quite effectively from measurement results.

The design of calibration standards and calibration itself are backed up by traceability to primary standards realized by renown metrological institutes, such as NIST and NPL. Calibration standards, as will be discussed in sec. 11.5, are "reflect" (opens and shorts, possibly offset) and "line" (transmission lines and matched loads) elements: PCB and wafer calibrations are best performed using TRL, LRM and multiline methods, rather than conventional SOL and similar ones, that are however available on calibration fixtures, or as they are usually called "calibration substrates". Due to the presence of ground points surrounding the signal trace (GSG and similar) the planar lines are most often of the coplanar waveguide type (see sec. 6.3.3); while for PCB

[19] Cascade Microtech, "Proper Infinity probe alignment on various ISSs", Application Note, INFISS-APP-1102, 2002.

technology striplines are possible and perform quite well, better than microstrips and other "open" transmission lines, with wafer technology the latter are almost a forced choice. There are laboratories and manufacturers, such as those cited in this section, that sell impedance standards and calibration substrates [20]. ISSs are equipped with the most common and useful calibration standards (e.g. SOLT and TRL calibration substrates): calibration with ISS is usually a two-tier calibration, having first calibrated the VNA with standard coaxial calibration set at probe connecting cables, and at a second time calibrating the probe with the ISS. For this latter calibration the required error model is more complete than usual 8-term models and Silvonen has long proposed 16-term error models improving the efficiency of calibration algorithms, such as exploiting reciprocity[21] (see sec. 11.5).

6.3 Planar transmission lines on PCB

The approach followed in this section is not that of describing the whole electromagnetic theory and solution methods behind the various planar transmission line structures: they are already comprehensively treated in [88, 108, 332]. On the contrary, the focus is on the selection of the most relevant transmission line structures, on formulas for design and for calculating characterizing quantities, and finally on the evaluation of approximations and on the general agreement of the various formulations applicable to the same problem. It is in fact quite common that in the need of designing a planar transmission line, one of the many on-line calculators and expressions available on the web are used, without an indication of their limitations and degree of approximation. A good synthesis of planar transmission line formulas appears in [23], Chap. 2, and in [88], Chap. 3.

Regarding the accuracy of the expressions that we are going to examine in this section, it is important to stress that:

- substrate dielectric materials may deviate from nominal values of dielectric permittivity and dielectric losses due to fabrication tolerance, PCB post-treatment, temperature; nominal values may be applicable to the broad class of material, subject to change depending on additives characteristic of a specific manufacturer;
- some dielectric materials have a non-negligible level of anisotropy and this has impact on electric field distribution and resulting effective permittivity parameter, that is normally used to take account globally of several effects;
- testing methods recognized and used by manufacturers may not fully agree with limitations of accuracy and frequency range, and material anisotropy;

[20] Cascade Microtech, "Impedance standard substrate datasheet," *Product Note P1215*.

[21] K. Silvonen, K. Dahlberg and T. Kiuru, "16-Term Error Model in Reciprocal Systems," *IEEE Transactions on Microwave Theory and Techniques*, Vol. 60, no. 11, pp. 3551-3558, Nov. 2012. doi: 10.1109/TMTT.2012.2217150

- roughness of the metal layer and dielectric surface may trap air and change not only the overall effective permittivity, but modify locally the electric field distribution and intensity; roughness of copper traces, including etching at the edges, affects current distribution and conduction losses, and also the distribution of electric field where it is already most intense; some roughness of otherwise flat surfaces is anyway necessary to enhance adhesion between the copper foil and the substrate, to produce laminates with adequate mechanical properties;
- the PCB fabrication process has, finally, its own tolerances, for tooling and for meeting nominal values before and after operations, such as oven heating and treatment and pressing of PCB laminates, are performed.

6.3.1 Microstrip

A microstrip is a transmission line made of two parallel flat conductors separated by a dielectric layer (the substrate) of thickness h. The properties of the dielectric layer are ε_r relative permittivity and μ_r relative magnetic permeability (even if for our purpose all the dielectric materials are non-ferromagnetic and thus $\mu_r = 1$). The two conductors are of the same thickness t; conventionally the upper conductor is the "hot" conductor with width w, the lower conductor is the return plane with a width infinite in theory, much larger than w in practice.

The microstrip is quite a common structure in PCBs, since we create a microstrip anytime we have a trace on a layer that is "supported" by a wide plane or even a thick trace beneath at a fixed potential. Multi-layer PCBs are quite suited to host microstrips at different layers: in principle a microstrip should have the hot conductor surrounded by the dielectric layer on one side and air on the other side, but even the case of a trace routed on an internal layer is possible, so with that trace (the hot conductor) completely surrounded by the dielectric material, provided that one of the two sides has no copper, otherwise the so-formed structure would be a stripline (see sec. 6.3.4).

Microstrips are much used on outer layers to directly connect to component pads, especially for small RF and microwave boards. Moreover, they are well suited to measure dielectric permittivity and dielectric loss of the PCB laminate if properly driven and used as resonators, or if design equations are reverse-engineered to determine average values related to measurements.

Microstrips are objects with a very simple structure and with an intuitive distribution of the electric and magnetic field lines, as shown in Figure 6.3.1. Solution methods are not considered and they are quite comprehensively presented in [88, 108].

Before proceeding only a comment on anisotropic materials: if the substrate material is anisotropic with different relative permittivity values ε_{xy} (along the x and y direction parallel to the return plane) and ε_z (along the vertical direction), calculations are done by replacing the relative permittivity ε_r of the material assumed homogeneous with the geometric mean of the two permittivities, that is $\sqrt{\varepsilon_{xy}\varepsilon_z}$. The height (or thickness) of the substrate h is also changed to an equivalent one multiplied by the square root of the ratio of the two permittivities, i.e. $\sqrt{\varepsilon_{xy}/\varepsilon_z}$.

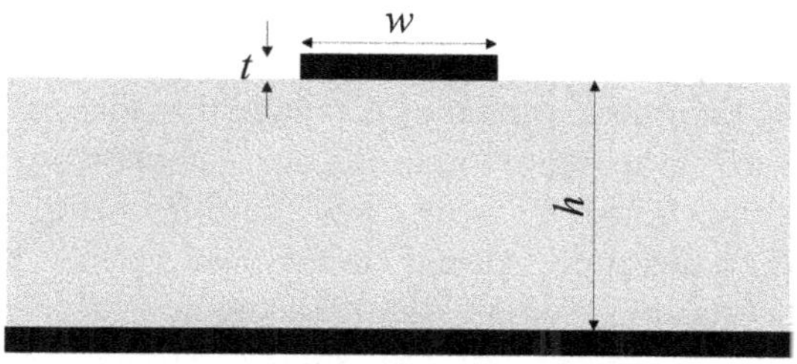

(a)

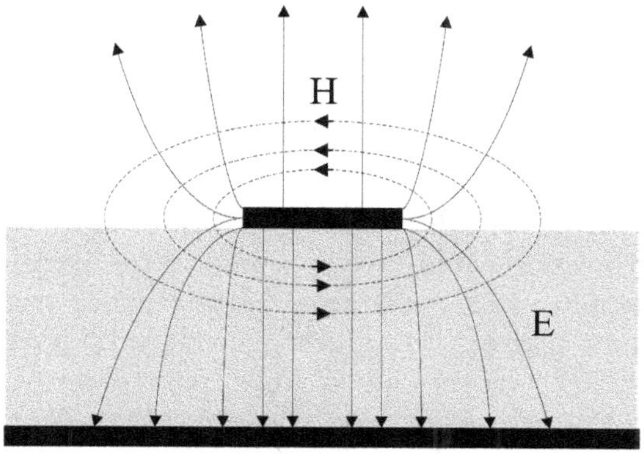

(b)

Figure 6.3.1 – (a) Microstrip line geometry and (b) electric and magnetic field distribution.

6.3.1.1 Characteristic impedance

Microstrip transmission lines feature two different dielectrics, air and substrate. Such asymmetry prevents the application of TEM mode assumption, and also the quasi-TEM mode assumption is not satisfactory at the highest frequencies; superior modes shall then be accounted for and a satisfactory solution is sought with full wave analysis of the structure. Consequently, also the definition of characteristic impedance is liable to change. In a transmission line in TEM mode the selection of integration paths for the electric and magnetic field is not critical and the ratio of the so obtained voltage and current defines the characteristic impedance. Passing to quasi-TEM mode, different portions of space can be recognized with a distribution of electric and magnetic field that need separate integration; voltage and current quantities, and characteristic impedance, are thus no longer uniquely defined. One may wonder thus which is the best definition of characteristic impedance and if this definition holds in all cases or it is subject to conditions and constraints. Jansen and Koster [196] identify three "reasonable" formulations for characteristic impedance:

$$Z_c = \frac{2P}{I^2} \qquad Z_c = \frac{U}{I} \qquad Z_c = \frac{U^2}{2P} \tag{6.3.1}$$

In their work they indicate a frequency dependency that increases going from the first one, to the second and then to the latter, simply because the definition involves conductor voltage, that is varying as a function of the electric field distribution, and as a consequence of the overall microstrip capacitance. Conversely, current (related to magnetic field) is not as sensitive to frequency and supports thus a better definition (the first one). During their discussion they conclude that for microstrip scattering matrices to be valid (unitary) the definition of the characteristic impedance shall include power, so restricting the choice to the first and the third definitions above. Zhu and Wu [348] confirm this conclusion, strongly in favor of the first definition, that is also confirmed by their definition based on 3-D modeling and in agreement with Rautio's definition [281] given a few years before; the small discrepancies between the two 3-D models are due to different launcher discontinuity modeling. It is observed, anticipating the following discussion, that Getsinger's definition of characteristic impedance [134] follows the power-current model (followed by Bhartia and Pramanick [36]), Wheeler [336] was assuming a TEM propagation scheme, Hammerstad and Jensen [152] were applying perturbation to a substantially homogeneous structure, thus assuming TEM mode propagation, and Kaupp [207] was empirically modifying the characteristic impedance expression for the TEM mode of a wire over a ground plane.

There are several expressions that have appeared in the literature for the determination of effective permittivity and characteristic impedance of microstrips, modeling and including geometry, edge effects and dispersion in various ways. Usually the electrostatic problem solution gives the equivalent capacitance terms, leading to the static effective permittivity (low-frequency approximation). The effect of frequency (called "dispersion") is addressed partly by modeling partly by interpolation of numeric or experimental results; from this there are many different corrective terms appearing in the expressions, due to the used interpolation technique and solution method (polynomials, elliptic integrals and so on).

We may distinguish formulations for the calculation of the static effective permittivity $\varepsilon_{\text{eff},0}$ (Bahl and Garg [22] starting from Hammerstad expression [150], Hammerstad and Jensen [152], and Kaupp [207]), for the effect of frequency $\varepsilon_{\text{eff},f}$ (Kirschning and Jansen [210] and Kobayashi [214], all proposing in their original papers a formulation based on the static effective permittivity calculated following Getsinger approach [134]) and for the determination of the characteristic impedance (Bahl and Garg [22], Rogers [286], recalling Hammerstad and Jensen [152], and Kaupp [207]), that is always based on the effective permittivity without showing explicitly frequency [46].

Many equations available for the quantitative description of microstrip behavior are based on the definition of the effective relative permittivity ε_{eff} and the extra width δw (that defines the effective, or equivalent, microstrip width $w_e = w + \delta w$).

Bahl and Garg (or Wheeler) formulation Bahl and Garg [22] took the Hammerstad [150] expressions valid for zero thickness (i.e. neglecting the effect of trace thickness t) and including the t/h ratio that weights trace thickness with respect to substrate thickness. Later reported by other authors, they have undergone some minor changes: Di Paolo [108] includes the effect of conductor thickness t and related width change δw, in [88] the $1.25/\pi$ is replaced by 0.398. Reading Wadell [332], sec. 3.5.1, the formula is ascribed to Wheeler [335] with Schneider's equivalent permittivity expression (tested among others by Bogatin [44]), including also Hammerstad and Bekkadal [151] correction, that is in the form we know and we see below.

$$\varepsilon_{re}(0) = \frac{\varepsilon_r+1}{2} + \frac{\varepsilon_r-1}{2} F - \frac{(\varepsilon_r-1)\,t}{4.6\sqrt{wh}} \tag{6.3.2}$$

$$\delta w = \begin{cases} \dfrac{1.25t}{\pi}\left[1+\ln\left(\dfrac{4\pi w}{t}\right)\right] & \text{for}\, w \le h/2\pi \\ \dfrac{1.25t}{\pi}\left[1+\ln\left(\dfrac{2h}{t}\right)\right] & \text{for}\, w \ge h/2\pi \end{cases} \tag{6.3.3}$$

where

$$F = \begin{cases} \dfrac{1}{\sqrt{1+12h/w}} + 0.04(1-w/h)^2 & \text{for}\, w/h \le 1 \\ \dfrac{1}{\sqrt{1+12h/w}} & \text{for}\, w/h \ge 1 \end{cases} \tag{6.3.4}$$

Limits of such formulation are in terms of t/h and w/h ratio: $t/h \le 0.2$ and $0.1 \le t/h \le 20$. The accuracy is about 1% or better.

It is possible to calculate the microstrip characteristic impedance as [22]:

$$Z_c = \begin{cases} \dfrac{60}{\sqrt{\varepsilon_{re}(0)}} \ln\left[\dfrac{8h}{w_e} + 0.25\dfrac{w_e}{h}\right] & \text{for}\, w \le h \\ \dfrac{120}{\sqrt{\varepsilon_{re}(0)}} \left[\dfrac{w_e}{h} + 1.393 + 0.667\,\ln(\dfrac{w_e}{h} + 1.444)\right]^{-1} & \text{for}\, w \ge h \end{cases} \tag{6.3.5}$$

where the equivalent width $w_e = w + \delta w$ is used to take into account hot trace thickness.

Collin [88] indicates for the relative effective permittivity the expression proposed by Schneider (and further modified by Hammerstad replacing the 10 with a 12 at the numerator), that after manipulation is identical to eq. (6.3.2) above.

Moreover, Collin evaluates numerically the accuracy of the effective relative permittivity as per eq. (6.3.2), by comparing results obtained with different solution methods for different values of relative permittivity (isotropic with $\varepsilon_r = 2$, 6, 10 and anisotropic with $\varepsilon_r = 5.12$ and $\varepsilon_y = 3.4$, that are the characteristics of the boron nitride) and for different w/h ratios ranging from 0.25 to 6. The results confirm that all the considered expressions and solution methods agree: the maximum span occurs normally for $w/h = 4$ and is within 0.6% for $\varepsilon_r = 2$, 1.5% $\varepsilon_r = 6$, 2.4% $\varepsilon_r = 10$ and 0.8% for the anisotropic material; for w/h between 0.25 and 2 that is the most common situation for the desired values characteristic impedance the matching is well within 0.5% for all cases.

Kirschning and Jansen formulation The effect of frequency (dispersion) is taken into account by modifying the relative effective permittivity expression, as proposed by Kirschning and Jansen [210]. They proposed a closed form expression for the effective dielectric constant of single microstrip lines, elaborating Jansen's hybrid mode technique and the related computer program. The formulation is based on Getsinger formula of microstrip dispersion [134], but with a new frequency dependent term.

$$\varepsilon_{\text{eff},f} = \varepsilon_r - \frac{\varepsilon_r - \varepsilon_{\text{eff},0}}{1+D} \tag{6.3.6}$$

where

$$D = D_1 D_2 \left[(0.1844 + D_3 D_4)\, 10h\, f\right]^{1.5763} \tag{6.3.7}$$

$$D_1 = 0.27488 + \frac{w}{h}\left[0.6315 + \frac{0.525}{(1+0.157\, fh)^{20}}\right] - 0.065683 e^{-8.7513 w/h} \tag{6.3.8}$$

$$D_2 = 0.33622\left[1 - e^{-0.03442\varepsilon_r}\right] \tag{6.3.9}$$

$$D_3 = 0.0363 e^{-4.6 w/h}\left[1 - e^{-(fh/3.87)^{4.97}}\right] \tag{6.3.10}$$

$$D_4 = 1 + 2.751\left[1 - e^{(-\varepsilon_r/15.916)^8}\right] \tag{6.3.11}$$

where the term fh is a normalized frequency-thickness with the measuring units GHz cm. Quantities ε_r and ε_{re} have the same meaning used above (see in particular eq. (6.3.2)) and $\varepsilon_{re}(0)$ indicates the quasi-static value of the effective relative permittivity setting $f = 0$, as in eq. (6.3.2) above.

The range of applicability of such formulas is defined as $0.1 \le w/h \le 100$, $1 \le \varepsilon_r \le 20$, $0 \le h/\lambda \le 0.13$, valid up to about 60 GHz.

Rogers formulation Rogers corporation [286] presents expressions that are derived from Hammerstad-Jensen [152] method for both Z_c and static effective permittivity $\varepsilon_{\text{eff},0}$, while the effective permittivity $\varepsilon_{\text{eff},f}$ is calculated using Kirschning-Jansen expressions [210]. Reference to previous work is not complete and does not link expressions with the source; "Getsinger" is mentioned, but in their bibliographic references Bryant and Weiss appear, so that it is intended that the static part from eq. (6.3.12) to (6.3.15) is based on [51]):

$$U_1 = \frac{w}{h} + \frac{1}{\pi}\frac{t}{h}\ \ln\left[1 + \frac{4eh}{t}\ \tanh^2(\sqrt{6.517 w/h})\right] \tag{6.3.12}$$

$$U_r = \frac{w}{h} + \frac{1}{2}\left[\left(U_1 - \frac{w}{h}\right)\left(1 + \frac{1}{\cosh(\sqrt{\varepsilon_r - 1})}\right)\right] \tag{6.3.13}$$

$$A_u = 1 + \frac{1}{49}\ \ln\left[\frac{U_r^4 + \dfrac{U_r^2}{2704}}{U_r^4 + 0.432}\right] + \frac{1}{18.7}\ \ln\left[\left(\frac{U_r}{18.1}\right)^3 + 1\right] \tag{6.3.14}$$

$$B_{er} = 0.564 \left(\frac{\varepsilon_r - 0.9}{\varepsilon_r + 3}\right)^{0.053} \qquad Y = \frac{\varepsilon_r + 1}{2}\frac{\varepsilon_r - 1}{2}\left(1 + \frac{10}{U_r}\right)^{-A_u B_{er}} \tag{6.3.15}$$

$$Z_{01}(x) = \frac{\eta_0}{2\pi} \ln \left[\frac{6 + (2\pi - 6)\, e^{-(30.666/x)^{0.7528}}}{x} + \sqrt{\frac{4}{x^2} + 1}\right] \tag{6.3.16}$$

$$Z_c = \frac{Z_{01}(U_r)}{\sqrt{Y}} \qquad \varepsilon_{\mathrm{eff},0} = \left[Y \left(\frac{Z_{01}(U_1)}{Z_{01}(U_r)}\right)\right]^2 \tag{6.3.17}$$

The influence of the frequency is introduced by using a "filling factor" P in the more accurate formulation given in [210]:

$$P_1 = 0.27488 + \frac{w}{h}\left[0.6315 + 0.525(15.7\, f\, h)^{-20}\right] - 0.065683\, e^{-8.7513 w/h} \tag{6.3.18}$$

$$P_2 = 0.33622\, \left(1 - e^{-0.03442 \varepsilon_r}\right) \tag{6.3.19}$$

$$P_3 = 0.0363\, e^{-4.6 w/h}\left[1 - e^{-(fh/38700)^{4.97}}\right] \tag{6.3.20}$$

$$P_4 = 2.751\, \left[1 - e^{-(\varepsilon_r/15.916)^8}\right] \tag{6.3.21}$$

$$P = P_1 P_2 \left[1000\, f\, h\, (0.1844 + P_3 P_4)\right]^{1.5763} \tag{6.3.22}$$

$$\varepsilon_{\mathrm{eff},f} = \varepsilon_r - \frac{\varepsilon_r - \varepsilon_{\mathrm{eff},0}}{1 + P} \qquad Z_{0,f} = Z_0 \sqrt{\frac{\varepsilon_{\mathrm{eff},0}}{\varepsilon_{\mathrm{eff},f}} \frac{\varepsilon_{\mathrm{eff},f} - 1}{\varepsilon_{\mathrm{eff},0} - 1}} \tag{6.3.23}$$

Kaupp (IPC) formulation IPC [176] proposed some simplified formulas for the estimation of effective relative permittivity. The following was proposed by Kaupp [207] and is based on permittivity values measured (or known) at a single frequency (in this case 25 MHz).

$$\varepsilon_{re} = 0.475 \varepsilon_r + 0.67 \qquad \text{for } 2 < \varepsilon_r < 6 \tag{6.3.24}$$

Characteristic impedance is derived from the formulation for a wire over a ground plane (see sec. 5.4.2.1) by a series of simplifications: the $\mathrm{a\cosh}$ function in the original formulation is replaced by a natural logarithm and then the empirical formula above is replaced for ε_{re}; then going from a round wire to a flat wire, the diameter d is replaced by an equivalent diameter given by $0.8w + t$, with the usual meaning of w as width and t as thickness of the flat conductor, that is the hot trace.

$$Z_c = \frac{87}{\sqrt{1.41+\varepsilon_r}} \ln\left(\frac{5.98h}{0.8w+t}\right) \tag{6.3.25}$$

where the expression for ε_{eff} is included and simplified with other factors. The validity of this expression is declared for $w/h<1$, but then w/h is assigned the range [0.1 ... 3.0] and this justifies the numerical evaluation that follows; regarding dielectric permittivity there are no serious limitations since the declared range is [1...15].

Kobayashi formulation As reported by Collin [88], Kobayashi formulation [214] is the most accurate and "covers the full range of parameter values to be encountered". This is the same conclusion, with some minor remarks, that is reached in [46] (see sensitivity analysis and comparison with experimental data below).

The effective relative permittivity is polynomially fit with frequency, using a corner frequency f_a and an exponent m:

$$\varepsilon_{\text{eff},f} = \varepsilon_r - \frac{\varepsilon_r - \varepsilon_{\text{eff},0}}{1+(f/f_a)^m} \tag{6.3.26}$$

where $\varepsilon_{re}(0)$ indicates as before the quasi-static value of the effective relative permittivity given by eq. (6.3.2) and

$$f_a = \frac{f_b}{0.75+[0.75-0.332/(\varepsilon_r)^{1.73}]\,\frac{w}{h}} \tag{6.3.27}$$

$$f_b = \frac{c}{2\pi h\sqrt{\varepsilon_r - \varepsilon_{re}(0)}} \arctan\left[\varepsilon_r \sqrt{\frac{\varepsilon_{\text{eff},0}-1}{\varepsilon_r - \varepsilon_{\text{eff},0}}}\right] \tag{6.3.28}$$

$$m = \min(m_0 m_c, 2.32) \tag{6.3.29}$$

$$m_0 = 1+\frac{1}{1+\sqrt{w/h}} + 0.32\frac{1}{(1+\sqrt{w/h})^3} \tag{6.3.30}$$

$$m_c = \begin{cases} 1+\dfrac{1.4}{1+w/h}\left(0.15-0.235e^{-0.45f/f_a}\right) & w/h \le 0.7 \\ 1 & w/h > 0.7 \end{cases} \tag{6.3.31}$$

In the expression for f_b c is the speed of light and a simplification often occurs with 47.746 at the numerator.

This formulation, then, takes into account all previous effects, such as trace thickness, by calling eq. (6.3.2) for the $\varepsilon_{\text{eff},0}$ parameter, and introduces the dispersive effect of frequency. The effective permittivity resulting from eq. (6.3.26) shall be used thus in the denominator of eq. (6.3.5). This is confirmed by the definition of effective permittivity with respect to microstrip capacitance with air and with substrate, recalled below in eq. (6.3.34), and by the form of the solution presented in [88], eq. 3.165, where the characteristic impedance is explicitly written as a function of the two capacitance terms: the microstrip capacitance C contains the same dependency on microstrip geometry that appears in eq. (6.3.5).

Attention shall be paid to the units of measure of the original formulation that has been kept unchanged here: h and w are in mm and the frequency is in GHz.

The accuracy claimed in the original paper [214] is better than 0.6% over a very wide range of problem parameters, $0.1 \le w/h \le 10$ and $1 \le \varepsilon_r \le 128$. Collin verified positively the accuracy himself by comparison with numerical results for $0.25 \le w/h \le 6$ and $2 \le \varepsilon_r \le 12$, that in any case covers most configurations.

Capacitance Formulas for microstrip capacitance are also given by IPC and by Bahl, but it is evident that they are obtained as part of solution of the electrostatic problem and calculation of characteristic impedance, and are thus in direct relationship with characteristic impedance expression.

For IPC the capacitance is

$$C = \frac{0.67\,(1.41 + \varepsilon_r)}{\ln\left(\dfrac{5.98h}{0.8w + t}\right)} \quad [\mathrm{pF/in}] \tag{6.3.32}$$

that is readily transformed in [pF/m] dividing the numerator by the conversion factor 39.37, so replacing 0.67 with 0.017.

Similarly, the two factors in square brackets appearing in eq. (6.3.5) above correspond to Bahl's estimation of geometry capacitance, that is with air as dielectric, reported also by Collin [88]:

$$C_{air} = \begin{cases} \dfrac{2\pi\varepsilon_0}{\ln\left(\dfrac{8h}{w} + \dfrac{w}{4h}\right)} & \text{for } w \le h \\[2ex] \varepsilon_0 \left[\dfrac{w}{h} + 1.393 + 0.667 \ln\left(\dfrac{w}{h} + 1.444\right)\right] & \text{for } w \ge h \end{cases} \tag{6.3.33}$$

This term of capacity in air does not consider the effect of the substrate material and is thus to be calculated without applying the correction for anisotropic materials, mentioned at the beginning.

The effect of substrate dielectric is taken into account with the relative effective permittivity ε_{eff}, in order to obtain the microstrip effective capacitance C:

$$\varepsilon_{\text{eff}} = \frac{C}{C_{air}} \tag{6.3.34}$$

Sensitivity of expressions for Z_c Formulations above have different approaches and one may wonder which one is best depending for example on geometry, substrate permittivity and frequency range, or if they agree and results match, and under which conditions; in any case Bahl and Garg formulation has the advantage of simplicity. To this aim a sensitivity analysis was performed with respect to relative permittivity of substrate (see Figure 6.3.2(a)), thickness of substrate (see Figure 6.3.2(b)), thickness of copper trace (see Figure 6.3.3(a)) and width of copper trace (see Figure 6.3.3(b)). When single values are missing[22], it means that expressions are outside their declared interval of validity, that is "if" conditions in previous section are not satisfied.

In the sensitivity analysis also mixed formulations have been tested, that is trying to combine expressions from different authors for the characteristic impedance Z_c (Bahl and Garg, indicated by "b", and Rogers, indicated by "r"), dispersion effect $\varepsilon_{\text{eff},f}$ (Kirschning-Jensen, indicated by "j", and Kobayashi, indicated by "y") and effective static dielectric permittivity $\varepsilon_{\text{eff},0}$ (Bahl and Garg, indicated by "b", Rogers, indicated by "r", and Kobayashi, indicated by "y"). As a brief introductory comment that anticipates conclusions, the overall sensitivity analysis and comparison with measurement results (extensively reported in [46]) confirmed that Bahl and Garg formulation is probably the best one for the characteristic impedance if comparison with available measurements is included, for dispersion effect there is general agreement between the two formulations by Kirschning and Jansen and by Kobayashi (the latter preferable sometimes, especially when compared to experimental data), and last for the static effective permittivity Rogers and Kobayashi formulations are in substantial agreement, slightly better than Bahl and Garg expressions.

Comparison with experimental results (permittivity dispersion) There are various sources of experimental results for permittivity and dispersion effects, to fully test the just reviewed expressions. In Table 6.3.1 there are reported values from four different groups of authors, in some cases picking up more than one configuration, especially when configurations were referring to materials with different static permittivity values (the approximate value is shown between round brackets for each configuration). Complete curves and considerations may be found in [46].

Comparison with experimental results (characteristic impedance) The sources of experimental results for characteristic impedance are much less and those reported are all those that could be found while writing a review work on microstrips [46]. In Table 6.3.2 there are reported values from two authors. Complete curves and considerations may be found in [46].

Considerations on the infinite return plane Regarding the influence of finite width of the return plane w_r, Di Paolo [108], sec. 2.9, reports that it may be neglected whenever $w_r > 3w$, while Wadell [332], p. 111, states that a ratio of 5 has less than 3% effect on characteristic impedance (his original reference is [318]). Of course this effect is influenced by frequency, since at very high frequency the major part of the return current is concentrated beneath the hot conductor, while it spreads more at lower frequency. The microstrip is an unbalanced structure, but if the return plane width is

[22] They are giving rise to a "NaN" (not a number) in the Matlab environment.

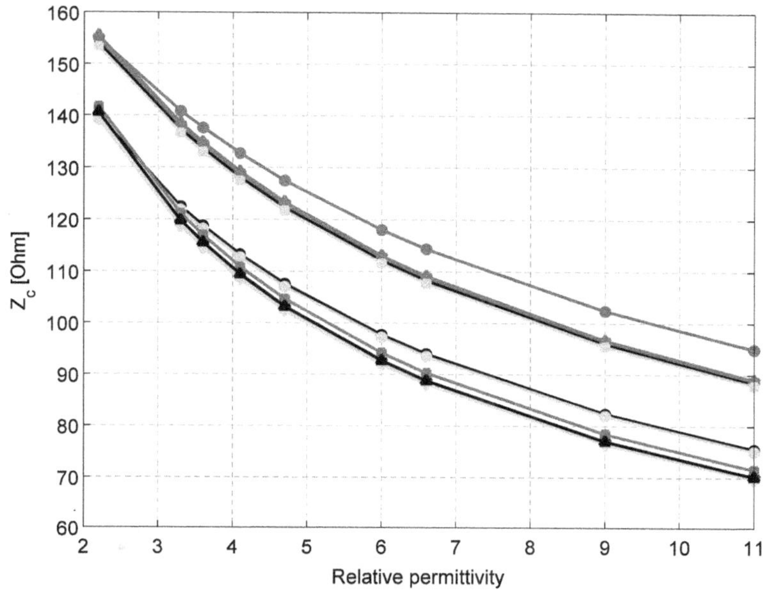

(a)

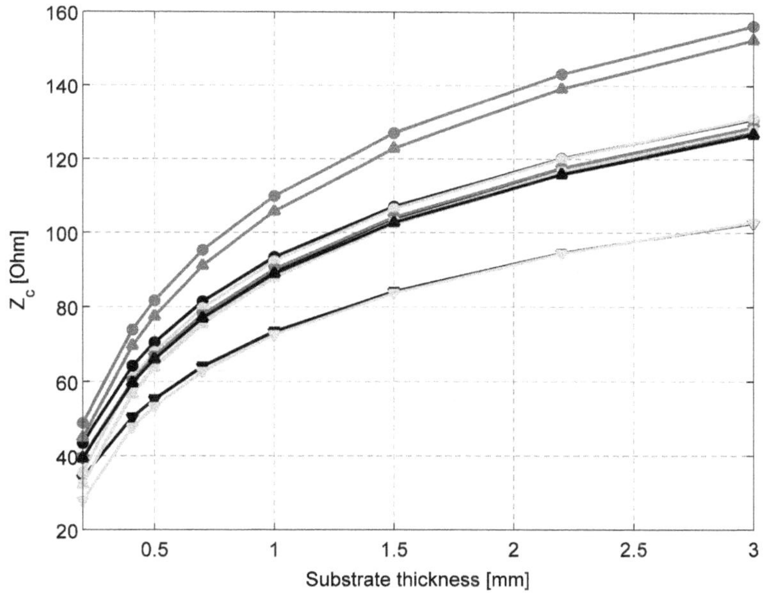

(b)

Figure 6.3.2 – Sensitivity of the expressions for the characteristic impedance of a Microstrip versus (a) substrate relative permittivity ε_r ($w = 0.5\,\text{mm}$, $h = 1.52\,\text{mm}$, $t = 0.035\,\text{mm}$) and (b) substrate thickness h ($w = 0.5\,\text{mm}$, $t = 0.035\,\text{mm}$, $\varepsilon_r = 4.7$): Z_c Bahl (black lines) with $\varepsilon_{\text{eff},0}$ Bahl (triangle), Rogers (square), Kaupp (circle), Kobayashi (reverse triangle); Z_c Rogers (dark gray lines) with $\varepsilon_{\text{eff},0}$ Bahl (triangle), Rogers (square), Kaupp (circle), Kobayashi (reverse triangle); Z_c Kaupp (light gray lines) with $\varepsilon_{\text{eff},0}$ Bahl (triangle), Rogers (square), Kaupp (circle), Kobayashi (reverse triangle).

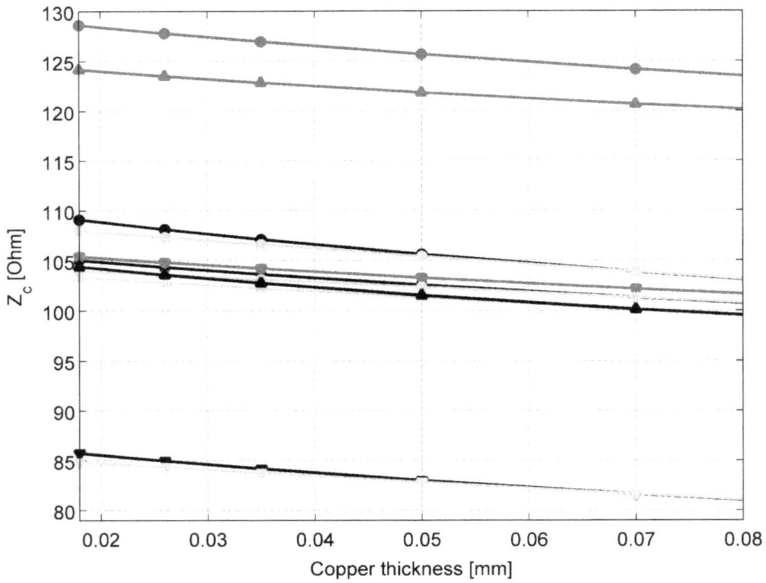

(a)

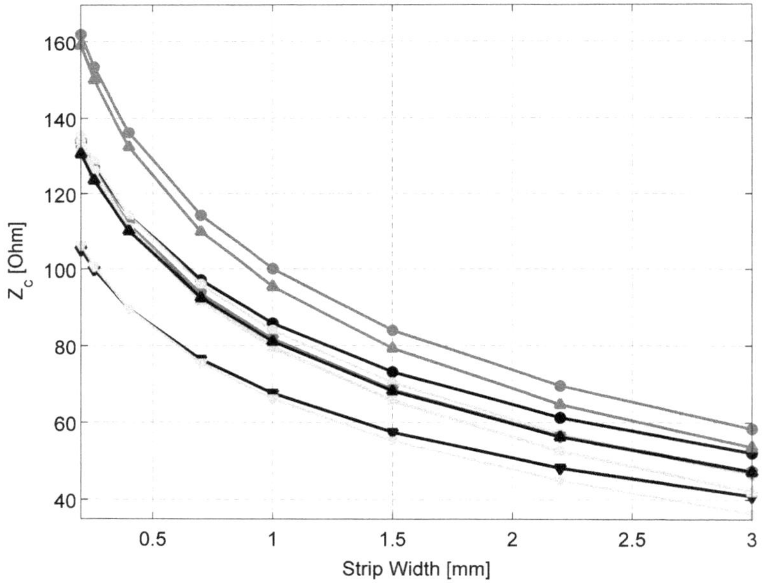

(b)

Figure 6.3.3 – Sensitivity of the expressions for the characteristic impedance of a Microstrip versus (a) trace thickness t ($w = 0.5\,\mathrm{mm}$, $h = 1.52\,\mathrm{mm}$, $\varepsilon_r = 4.7$) and (b) trace width w ($h = 1.52\,\mathrm{mm}$, $t = 0.035\,\mathrm{mm}$, $\varepsilon_r = 4.7$): Bahl & Garg (black), Rogers (dark gray), IPC (light gray)Z_c Bahl (black lines) with $\varepsilon_{\text{eff},0}$ Bahl (triangle), Rogers (square), Kaupp (circle), Kobayashi (reverse triangle); Z_c Rogers (dark gray lines) with $\varepsilon_{\text{eff},0}$ Bahl (triangle), Rogers (square), Kaupp (circle), Kobayashi (reverse triangle); Z_c Kaupp (light gray lines) with $\varepsilon_{\text{eff},0}$ Bahl (triangle), Rogers (square), Kaupp (circle), Kobayashi (reverse triangle).

		yb	yy	yr	jb	jy	jr
Edwards-Owens [116] line 2 ($\varepsilon_r = 9$)	mean (err)	0.545	0.356	0.501	0.284	0.053	0.230
	rms (err)	0.689	0.596	0.663	0.375	0.288	0.342
	mean |err|	0.545	0.417	0.501	0.290	0.235	0.254
Edwards-Owens [116] line 4 ($\varepsilon_r = 8.1$)	mean (err)	-0.012	0.163	0.656	-0.225	-0.030	0.520
	rms (err)	0.153	0.212	0.667	0.267	0.139	0.536
	mean |err|	0.119	0.176	0.656	0.233	0.114	0.520
Edwards-Owens [116] line 7 ($\varepsilon_r = 7$)	mean (err)	-0.764	-0.742	1.001	-0.893	-0.870	0.964
	rms (err)	0.774	0.753	1.016	0.902	0.879	0.977
	mean |err|	0.764	0.742	1.001	0.893	0.870	0.964
Edwards-Owens [116] line 9 ($\varepsilon_r = 6.6$)	mean (err)	1.208	0.601	3.789	1.047	0.419	3.717
	rms (err)	1.230	0.658	3.791	1.068	0.481	3.719
	mean |err|	1.208	0.601	3.789	1.047	0.419	3.717
Deibele-Beyer [101] line A ($\varepsilon_r = 6.6$)	mean (err)	-0.867	-0.710	-0.736	-1.594	-1.411	-1.373
	rms (err)	1.187	1.016	1.045	1.672	1.477	1.438
	mean |err|	0.876	0.733	0.757	1.594	1.411	1.373
Deibele-Beyer [101] line B ($\varepsilon_r = 5.7$)	mean (err)	-1.732	-2.609	-0.270	-1.639	-2.784	0.307
	rms (err)	1.849	2.665	1.309	1.749	2.838	1.111
	mean |err|	1.732	2.609	1.154	1.639	2.784	0.931
Yamashita et al. [344] ($\varepsilon_r = 2.15$)	mean (err)	2.314	1.840	2.214	2.063	1.437	1.930
	rms (err)	2.362	1.963	2.272	2.084	1.481	1.951
	mean |err|	2.314	1.840	2.214	2.063	1.437	1.930
Lee et al. [217] ($\varepsilon_r = 23.5$)	mean (err)	-1.291	-0.136	0.414	*	*	*
	rms (err)	1.444	0.704	0.841	*	*	*
	mean |err|	1.323	0.521	0.600	*	*	*

Table 6.3.1 – Comparison between measured and simulated curves of $\varepsilon_{\text{eff},f}$; acronyms in the column header use the following notation: dispersion calculation $\varepsilon_{\text{eff},f}$ (j=Kirschning-Jensen, y=Kobayashy), followed by effective static permittivity $\varepsilon_{\text{eff},0}$ (b=Bahl-Garg, r=Rogers, y=Kobayashi). Cells marked with * correspond to ε_r values outside the range of validity of Kirschning-Jensen ($\varepsilon_r \leq 20$). Underlined values correspond to the best values (above the median and the mean performance).

		bjb	rjb	bjr	rjr	bjy	rjy	byb	ryb	byr	ryr	byy	ryy
Denlinger [104]	mean (err)	1.406	-3.256	0.920	-3.327	0.934	-3.325	1.303	-3.123	0.836	-3.220	0.849	-3.217
	rms (err)	1.420	4.338	0.948	4.426	0.961	4.424	1.314	4.195	0.860	4.315	0.873	4.311
	mean |err|	1.406	3.389	0.920	3.459	0.934	3.457	1.303	3.262	0.836	3.356	0.849	3.353
Getsinger [134] $w = 0.5''$	mean (err)	0.708	-9.742	0.497	-9.878	0.589	-9.819	0.571	-9.544	0.392	-9.725	0.470	-9.645
	rms (err)	2.872	10.277	2.863	10.424	2.865	10.359	2.795	10.074	2.805	10.267	2.799	10.183
	mean |err|	2.400	9.742	2.432	9.878	2.417	9.819	2.361	9.544	2.406	9.725	2.385	9.645
Getsinger [134] $w = 0.25''$	mean (err)	1.127	-9.113	0.775	-9.281	0.825	-9.257	0.897	-8.775	0.582	-8.998	0.627	-8.967
	rms (err)	3.594	9.399	3.550	9.573	3.554	9.548	3.516	9.064	3.513	9.297	3.512	9.264
	mean |err|	2.618	9.113	2.697	9.281	2.686	9.257	2.624	8.775	2.726	8.998	2.710	8.967
Getsinger [134] $w = 0.1''$	mean (err)	0.286	-10.744	-0.235	-10.985	0.093	-10.833	0.179	-10.583	-0.284	-10.911	0.007	-10.703
	rms (err)	4.110	11.043	4.193	11.296	4.133	11.136	4.082	10.879	4.194	11.224	4.117	11.006
	mean |err|	3.482	10.744	3.653	10.985	3.545	10.833	3.478	10.583	3.663	10.911	3.545	10.703

Table 6.3.2 – Comparison between measured and simulated curves of Z_c; acronyms in the column header use the following notation: Z_c calculation (b=Bahl-Garg and r=Rogers), followed by followed by dispersion calculation $\varepsilon_{\text{eff},f}$ (j=Kirschning-Jensen, y=Kobayashy), followed by effective static permittivity $\varepsilon_{\text{eff},0}$ (b=Bahl-Garg, r=Rogers, y=Kobayashi). Underlined values correspond to the best values (above the median and the mean performance).

reduced down to the width of the hot conductor, then the structure gets balanced and symmetrical. In this case the correct name is "balanced broadside coupled line", and it is no longer a microstrip!

6.3.1.2 Attenuation

Attenuation for signals entering the microstrip may be due to conductor and dielectric losses (for the electromagnetic field confined in the structure) and radiation losses (for the electromagnetic field that "escapes" from the structure, because most of the field lines in the air close back in the structure and are not to be considered always as contributing to radiation losses). Since microstrip is an open transmission line, some of the transmitted energy is not guided through it, but is instead radiated in the surrounding space above dielectric. However, for dielectric materials with high permittivity and convenient shape of conductors, radiation losses are a second-order effect with respect to losses in the structure. For losses we may define attenuation coefficients: α_c for conduction losses and α_d for dielectric losses. These two loss terms may be modeled as a series resistance R and a shunt conductance G such that $\alpha_c = R/2Z_c$ and $\alpha_d = G/2Y_c$ [88].

It is briefly recalled that in most expressions α_c and α_d are in dB per unit length; if not, they will be multiplied by 8.686 (see end of sec. 2.1).

Conduction losses Three sets of expressions have been found for conduction losses of microstrips: the first formulation (formulation A) is presented by Di Paolo [108] and attributed to Pucel, Massé and Hartwig[23]; the other formulation B may be found in [23], Table 2.6; a third one appears in Collin [88], pp. 153-157.

Formulation A.

$$\alpha_c = \frac{R_s}{\zeta h}\frac{8.686}{2\pi}\left[1-\left(\frac{w_e}{4h}\right)^2\right]\left[1+\frac{h}{w_e}+\frac{h}{\pi w_e}\left(\ln\frac{4\pi w}{t}+\frac{t}{w}\right)\right] \quad \text{for } w \le h/2\pi \tag{6.3.35}$$

$$\alpha_c = \frac{R_s}{\zeta h}\frac{8.686}{2\pi}\left[1-\left(\frac{w_e}{4h}\right)^2\right]\left[1+\frac{h}{w_e}+\frac{h}{\pi w_e}\left(\ln\frac{2h}{t}-\frac{t}{h}\right)\right] \quad \text{for } h/2\pi \le w \le 2h \tag{6.3.36}$$

$$\alpha_c = \frac{R_s}{\zeta h}\frac{8.686}{D}\left[\frac{w_e}{h}+\frac{\frac{w_e}{\pi h}}{\frac{w_e}{2h}+0.94}\right]\left[1+\frac{h}{w_e}+\frac{h}{\pi w_e}\left(\ln\frac{2h}{t}-\frac{t}{h}\right)\right] \quad \text{for } w \ge 2h \tag{6.3.37}$$

where R_s is the sheet resistance and

[23] R.A. Pucel, D.J. Massé and C.P. Hartwig, "Losses in microstrip," *IEEE Transactions on Microwave Theory and Techniques*, Vol. 16, no. 6, June 1968, pp. 342-350. doi: 10.1109/TMTT.1968.1126691;
See also "Corrections to: "Losses in microstrip"", *IEEE Transactions on Microwave Theory and Techniques*, Vol. 16, no. 12, Dec. 1968, pp. 1064. doi: 10.1109/TMTT.1968.1126867.

$$D = \left[\frac{w_e}{h} + \frac{2}{\pi}\left(1 + \ln 2\pi\left(\frac{w_e}{2h} + 0.94\right)\right)\right]^2 \tag{6.3.38}$$

Formulation B.

$$\alpha_c = 1.38\,\frac{R_s}{\zeta h}\,\frac{32-(w_e/h)^2}{32+(w_e/h)^2}\,E \qquad \text{for } w \le h \tag{6.3.39}$$

$$\alpha_c = 6.1\,10^{-5}\,\frac{\zeta R_s \varepsilon_{\text{eff}}}{h}\left[\frac{w_e}{h} + \frac{0.667 w_e/h}{w_e/h + 1.444}\right] E \qquad \text{for } w \ge h \tag{6.3.40}$$

where R_s is again the sheet resistance and

$$E = \begin{cases} 1 + \dfrac{h}{w_e}\left[1 + \dfrac{1.25t}{\pi w} + \dfrac{1.25}{\pi}\ln\dfrac{4\pi w}{t}\right] & \text{for } w \le h/2\pi \\[2ex] 1 + \dfrac{h}{w_e}\left[1 - \dfrac{1.25t}{\pi h} + \dfrac{1.25}{\pi}\ln\dfrac{2h}{t}\right] & \text{for } w \ge h/2\pi \end{cases} \tag{6.3.41}$$

Here ε_{eff} is the static effective relative permittivity defined in eq. (6.3.2) at the beginning. The unit of measure of the so obtained conduction loss coefficient α_c is [dB/u.l.], where "u.l." stands for unit of length, that in our case is "meter".

For both sets of expressions the parameter w_e is the effective width, that is extensively used when treating microstrips and other similar devices. In this case the effective width of the hot conductor w_e, again accounting for conductor thickness, replaces the physical width w and takes the following expressions:

$$w_e = \begin{cases} w + \dfrac{t}{\pi}\left[1 + \ln\left(\dfrac{4\pi w}{t}\right)\right] & \text{for } w \le h/2\pi \\[2ex] w + \dfrac{t}{\pi}\left[1 + \ln\left(\dfrac{2h}{t}\right)\right] & \text{for } w \ge h/2\pi \end{cases} \tag{6.3.42}$$

Formulation C.

Another set of equations is found in [88], where the solution is obtained by steps, first identifying the resistance terms R_1 and R_2 for the hot trace and return plane (including skin effect) and then evaluating the current distribution for the return plane beneath the hot trace. The effective width of the return plane is $w + 5.8h$.

$$R_1 \frac{w}{R_m} = LR\left[\frac{1}{\pi} + \frac{1}{\pi^2}\ln\left(\frac{4\pi w}{t}\right)\right] \qquad \text{for } w \le h \tag{6.3.43}$$

$$R_2 \frac{w}{R_m} = \frac{w/h}{w/h + 5.8 + 0.03h/w} \qquad \text{for } 0.1 \leq w/h \leq 10 \tag{6.3.44}$$

where R_m is the skin effect resistance, $R_m = 1/\sqrt{\pi f \mu \sigma}$.

The total attenuation is given by the series composition of the two resistances, thus

$$\alpha_c = 8.686 \frac{R_1 + R_2}{2Z_c} \tag{6.3.45}$$

This set of equations was proposed by Collin as valid for low frequency. In a separate section of his book, dedicated to high-frequency properties of microstrips, when dealing with attenuation, he states clearly “simple formulas giving the attenuation of microstrip lines at high frequencies do not exist”! However, with a numeric example he demonstrates that this low-frequency formula is a valid approximation for narrow strips with w/h ratio smaller than 1-2 and frequency below 10-20 GHz.

Dielectric losses Dielectric losses as reported by Schneider-Glance [296] are calculated simply based on effective ε_{re} and basic ε_r relative dielectric permittivities.

$$\alpha_d = \frac{20\pi}{\ln 10} \frac{1/\varepsilon_{\text{eff}} - 1}{1/\varepsilon_r - 1} \frac{\tan\delta}{\lambda_0} \sqrt{\varepsilon_{\text{eff}}} = \frac{20\pi}{\ln 10} \frac{\varepsilon_r}{\varepsilon_{\text{eff}}} \frac{\varepsilon_r - 1}{\varepsilon_{\text{eff}} - 1} \frac{\tan\delta}{\lambda_0} \tag{6.3.46}$$

In [23], Table 2.6, there is an identical expression, where the effective relative permittivity is a function of frequency; the used quantity F to express it and ε_{re} expression are different from the other F quantity and ε_{eff} expression used for microstrip characteristic impedance (see eq. (6.3.2) and (6.3.4) above).

$$\alpha'_d = 27.29 \frac{1/\varepsilon_{\text{eff}} - 1}{1/\varepsilon_r - 1} \frac{\tan\delta}{\lambda_0} \sqrt{\varepsilon_{\text{eff}}} = \frac{20\pi}{\ln 10} \frac{\varepsilon_r}{\varepsilon_{\text{eff}}} \frac{\varepsilon_r - 1}{\varepsilon_{\text{eff}} - 1} \frac{\tan\delta}{\lambda_0} \tag{6.3.47}$$

$$\varepsilon_{\text{eff},f} = \left[\frac{\sqrt{\varepsilon_r} - \sqrt{\varepsilon_{\text{eff}}}}{1 + 4F^{-1.5}} + \sqrt{\varepsilon_{\text{eff}}} \right] \tag{6.3.48}$$

$$F = \frac{4h\sqrt{\varepsilon_r - 1}}{\lambda_0} \left\{ 0.5 + \left[1 + 2\ln\left(1 + \frac{w}{h}\right) \right]^2 \right\} \tag{6.3.49}$$

It is worth underlining that relative dielectric permittivity values used in these expressions refer to the real part of complex permittivity, with known problems of accuracy related both to the accurate determination of dielectric permittivity and the influence of material losses for increasing frequency, i.e. the imaginary part of the dielectric permittivity (see sec. 6.1.2.1). Again the unit of measure of α_d is [dB/u.l.], with u.l. that stands for “meter” in our case.

6.3.1.3 Typical discontinuities

Several line discontinuities due to lumped components, line terminations or line geometry variations are effectively described in [199], sec. 4.4, where typical behaviors

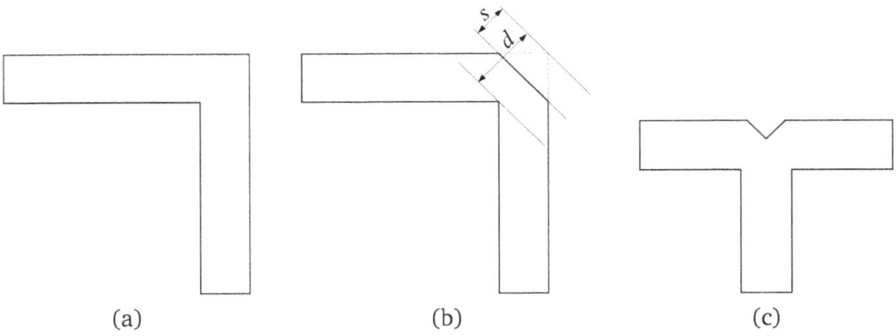

Figure 6.3.4 – Microstrip 90° bend: (a) straight version, (b) chamfered version, (c) T junction a V-shaped slice removed. Observing the chamfered case (b), we define $m = s/d$ as the chamfer ratio, or, multiplied by 100, as the chamfer percentage.

are quickly, but clearly, reviewed. Such phenomena were introduced in sec. 2.6 and will be again considered in Chapter 11, when discussing modes and settings of the Vector Network Analyzer.

Various angles of bending are possible, but the most immediate is the 90° bend: its use is not advisable above let's say 1 GHz due to the associated high reflection coefficient, as well as any bend with larger bending angle beyond 90°. Gradual bending with progressive bends at lower bending angles is a better solution, not to say using circular or elliptic arcs (this is called *mitering* and may be usually set in PCB CAD programs or done manually). Mitering is also called *tapering*, consisting in smoothing sharp changes by using gradual change, following e.g. some exponential or linear curve shape. Another solution is *chamfering* the corner, cutting it at 45° as shown in Figure 6.3.4(b). As said, almost all PCB CAD programs has the miter option with straight chamfering at corners and round connecting arc option. While optimization of the degree of mitering/chamfering requires lengthy numerical simulations, a good enough selection of parameters may be based on commonly used values: when using a round bend the advised bending radius is $r \geq 3w$; when chamfering at 45° (and also for wider angles) the diagonal length of the chamfer is normally set at $l = 1.8w$.

Both a 90° bend and its chamfered version (as shown in Figure 6.3.4(a) and (b)) may be modeled as a lumped T cell with two inductors L in series and a parallel capacitor C connected in their center[24]:

- for the original 90° bend with sharp corners

$$C = 0.001\, h \left[(10.35\varepsilon_r + 2.5)(w/h)^2 + (2.6\varepsilon_r + 5.64)(w/h)\right] \tag{6.3.50}$$

$$L = 0.22\, h \left[1 - 1.35 \exp\left(-0.18(w/h)^{1.39}\right)\right] \tag{6.3.51}$$

[24] M. Kirschning, R.H. Jansen and N.H.L. Koster, "Measurement and Computer-Aided Modeling of Microstrip Discontinuities by an Improved Resonator Method," IEEE MTT-S International Microwave Symposium Digest, May 1983, pp. 495-497.

- for the chamfered 90° bend with 45°-cut corners

$$C = 0.001\,h\left[(3.93\varepsilon_r + 0.62)(w/h)^2 + (7.6\varepsilon_r + 3.8)(w/h)\right] \tag{6.3.52}$$

$$L = 0.44\,h\left[1 - 1.062\,\exp\left(-0.177(w/h)^{0.947}\right)\right] \tag{6.3.53}$$

For h expressed in mm, C and L are then expressed in pF and nH, respectively. These expressions are valid for slim and wide microstrips ($0.2 \le w/h \le 6$) and for most common dielectric permittivity values ($2.3 \le \varepsilon_r \le 10.4$).

Even if the graphical example in Figure 6.3.4(b) was drawn for a 50% chamfering percentage, the amount of chamfering. The ratio $m = 100\,s/d$ may be optimized with respect to the microstrip geometry; Di Paolo [108] cites Douville and James[25] for the following expression of optimal chamfering:

$$m = 52 + 65e^{-1.35w/h} \tag{6.3.54}$$

Another example of mitering is when changing the hot trace width, where sharp 90° corners shall be avoided: at the two open-circuited portions of the wider trace there will occur additional fringing electric field and additional shunt-capacitance associated to it. The width change is purposely designed for example to create a change of the characteristic impedance or to cope with clearance requirements while going to a component pin. The degree of change of the width when moving longitudinally along the trace, or in other words the inclination angle of the skew profile joining the edges of the two trace sections, is not exactly coded, but shall be smoother the larger the width step change: practically speaking the length l of intermediate trace to use to join the two sections is set conservatively as $l \ge 6\,\dfrac{w_1 - w_2}{w_1 + w_2}$. If, as in practice, reduction of trace width is operated close to the destination pin; the effects of the impedance mismatch are combined with those at the pin interface.

When building a T-junction the same problem for the sharp 90° bend occurs: the larger area at the crossing creates additional capacitance; the area may be reduced by smoothly removing a slice of trace area in the crossing on the side on top of the T; the shape of the slice may be a "V" or a trapezoidal shape, as shown in Figure 6.3.4(c).

6.3.1.4 Suspended and inverted microstrip

In this short section there are reported for completeness expressions of characteristic impedance for much less common microstrip arrangements [23]. The common expression for the characteristic impedance is

[25] R.J.P. Douville and D.S. James, "Experimental study of symmetric microstrip bends and their compensation," *IEEE Transactions on Microwave Theory and Techniques*, Vol. 26, no. 3, March 1978, pp. 175-182. doi: 10.1109/TMTT.1978.1129340.

$$Z_c = \frac{60}{\sqrt{\varepsilon_{\text{eff}}}} \ln \left[\frac{f(u)}{u} + \sqrt{1 + \left(\frac{2}{u}\right)^2} \right] \tag{6.3.55}$$

where

$$f(u) = 6 + (2\pi - 6)\, e^{-(30.666/u)^{0.7528}} \tag{6.3.56}$$

For suspended microstrips $u = w/(a+b)$ and for inverted microstrips $u = w/b$.

Attention is to be paid to the fact that the square root of the relative effective permittivity is a symbol, assuming two different expressions for the two arrangements.

For suspended microstrips:

$$\sqrt{\varepsilon_{\text{eff}}} = \frac{1}{1 + \frac{a}{b}\left(a_1 - b_1 \ln \frac{w}{b}\right)\left(\frac{1}{\sqrt{\varepsilon_r}} - 1\right)} \tag{6.3.57}$$

with

$$a_1 = \left(0.8621 - 0.1251 \ln \frac{a}{b}\right)^4 \quad b_1 = \left(0.4986 - 0.1397 \ln \frac{a}{b}\right)^4 \tag{6.3.58}$$

For inverted microstrips:

$$\sqrt{\varepsilon_{\text{eff}}} = 1 + \frac{a}{b}\left(a_1 - b_1 \ln \frac{w}{b}\right)\left(\sqrt{\varepsilon_r} - 1\right) \tag{6.3.59}$$

with

$$a_1 = \left(0.5173 - 0.1515 \ln \frac{a}{b}\right)^2 \quad b_1 = \left(0.3092 - 0.1047 \ln \frac{a}{b}\right)^2 \tag{6.3.60}$$

The declared accuracy is within $\pm 1\%$ for $1 \leq w/b \leq 8$, $0.2 \leq a/b \leq 1$ and $\varepsilon_r \leq 6$. Increasing ε_r to about 10 (i.e. to include alumina) the error increases to about $\pm 2\%$.

6.3.1.5 Practical considerations and conclusions

Various formulations of characteristic impedance of microstrips has been compared at the end of sec. 6.3.1.1, presenting at the same time the results of sensitivity analysis for parameters that are subject to design choices, tradeoffs and optimization.

To summarize the results, we may point out first that the microstrip has a relevant fraction of the electromagnetic field in the air above the hot trace, in particular for small substrate permittivity and low w/h ratios; when the hot trace width increases, electromagnetic field is more in the substrate.

Moreover, with the increase of w/h ratio, characteristic impedance reduces to small values in the range of tens of Ω, while a reduction of trace width increases the characteristic impedance up to a limit value that is only dictated by practical reasons, related to the manufacturing process: minimum trace width ranges often between 6 and 8 mils, depending on manufacturing process, and already at these values the regularity of the trace profile is compromised due to the under-etching; for special PCBs smaller widths ay be achieved.

For low-impedance microstrips, with marked capacitive behavior, most of the electromagnetic field is confined between the hot trace and the return plane beneath: crosstalk with nearby lines is correspondingly reduced. There are two other possibilities to reduce further the impedance and to increase the capacitive component: reducing the substrate thickness (as customary with multi-layer boards) and covering with an insulated top metal shield or a dielectric layer.

6.3.2 Coupled microstrip

A coupled microstrip is made of two microstrips close enough to undergo significant electromagnetic coupling. This circuit is quite commonly used when differential signals of not too wide bandwidth are routed over a PCB: the coupled microstrip is a compact transmission line type that keeps the occupied PCB area conveniently small. When considering application notes and suggested circuit topologies for various logic families and digital transceivers, the coupled microstrip is still the preferred transmission line type. It may be contrasted that those application notes are not so recent and that the cost reduction of multi-layer PCB fabrication and the availability of suitable CAD programs have also paved the way to more complex planar transmission line configurations, such as striplines (see sec. 6.3.4). Striplines are definitely the ultimate choice for signal fidelity.

The dielectric layer has thickness h as for microstrip, with ε_r relative permittivity and μ_r relative magnetic permeability (even if for our purpose all dielectric materials are non-ferromagnetic and thus $\mu_r = 1$). The two “hot” conductors are of same thickness t, but for generality they have two different widths, w_1 and w_2; their separation is indicated by s. The return plane has a sufficiently large width to assume it infinite.

A coupled microstrip is sketched in Figure 6.3.5, together with the typical distribution of electric and magnetic field lines.

6.3.2.1 Characteristic impedance

For symmetric coupled microstrips (for which $w_1 = w_2 = w$) the most straightforward and used approach is that proposed by Garg and Bahl, who go through a quasi-static solution, augmented by correction factors derived from measured values and curve fitting, in order to determine capacitance terms. As shown in Figure 6.3.6, a set of capacitance terms is defined for coupling of each hot conductor to the reference plane and between themselves:

- $C_{f,e}$ is the fringing capacitance of external edges of the hot conductors;
- $C_{f,i}$ is the fringing capacitance of internal edges of the hot conductors;
- C_h is the main capacitance of the capacitor made of hot conductors and underlying reference plane;

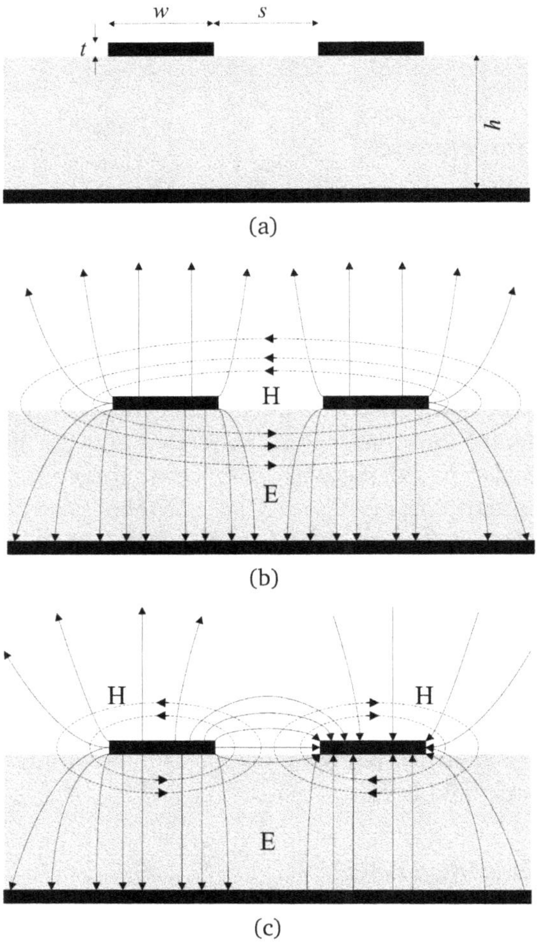

Figure 6.3.5 – Coupled microstrip line geometry and electric and magnetic field distribution: (a) geometry, (b) field distribution with even mode excitation, (c) field distribution with odd mode excitation.

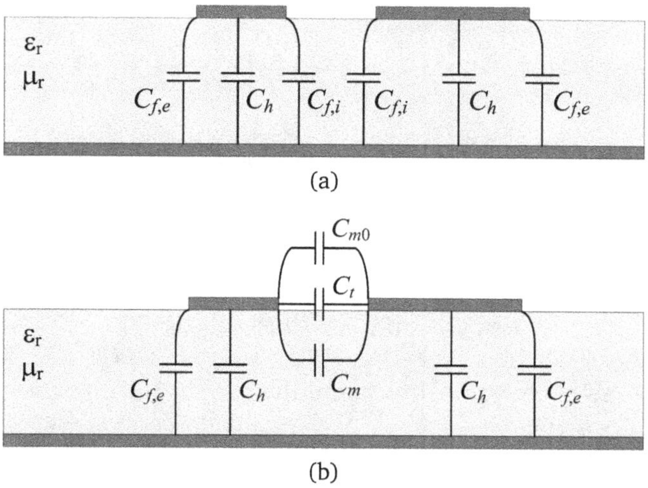

Figure 6.3.6 – Sketch of the capacitance terms of coupled microstrips: (a) even mode, (b) odd mode.

- C_m and C_{m0} are mutual capacitance terms between internal edges of the hot conductors with electric-field lines through the substrate below and through the air above, respectively;
- C_s is the coupling capacitance between facing internal edges of the hot conductors through the air separation of thickness s, considering the edges as the parallel plates of a so-formed capacitor.

Capacitance terms appearing in even- and odd-mode may then be defined as:

$$C_e = C_h + C_{f,e} + C_{f,i} \tag{6.3.61}$$

$$C_o = C_h + C_{f,e} + C_m + C_{m0} + C_t \tag{6.3.62}$$

Some of capacitance terms may be readily quantified by their own definitions:

$$C_h = \varepsilon_0 \varepsilon_r \frac{w}{h} \tag{6.3.63}$$

$$C_t = 2\varepsilon_0 \frac{t}{s} \tag{6.3.64}$$

The fringing capacitance term for the external edges may be determined by taking each hot conductor separately as a normal microstrip and using the same characteristic impedance Z_c and effective relative permittivity ε_{re} of the microstrip shown in eq. (6.3.5) and (6.3.2), respectively.

$$C_{f,e} = \frac{1}{2v_0 Z_c}\sqrt{\varepsilon_{re}} - C_h \tag{6.3.65}$$

where ν_0 is the speed of light in vacuum.

The internal fringing capacitance $C_{f,i}$ is given by[26]

$$2C_{f,i} = C_{f,e} \frac{1}{\left[1 + A\frac{h}{s} \tanh\left(\frac{10s}{h}\right)\right] \sqrt{\frac{\varepsilon_{re}}{\varepsilon_r}}} \tag{6.3.66}$$

where

$$A = \exp\left(-0.1 \exp\left(2.33 - 2.53\frac{w}{h}\right)\right) \tag{6.3.67}$$

and where the used effective relative permittivity ε_{re} is the one given in eq. (6.3.2) for the simple microstrip.

Then, for mutual capacitance terms through the air and the substrate we have

$$C_{m0} = \begin{cases} \dfrac{\varepsilon_0}{\pi} \ln\left(2\dfrac{1+\sqrt{p'}}{1-\sqrt{p'}}\right) & \text{for } 0 \le p^2 \le 0.5 \\[2ex] \dfrac{\pi\varepsilon_0}{\ln\left(2\dfrac{1+\sqrt{p}}{1-\sqrt{p}}\right)} & \text{for } 0.5 \le p^2 \le 1 \end{cases} \tag{6.3.68}$$

where

$$p = \frac{s}{s+2w} \qquad p' = \sqrt{1-p^2} \tag{6.3.69}$$

and the chosen form is such that logarithms replace the notation with elliptic integrals $C_{m0} = \varepsilon_0 K(p')/K(p)$.

The mutual capacitance (also called "coupling capacitance"), taking into consideration effective relative permittivity and fringing capacitance, is

$$C_m = \frac{\varepsilon_0\varepsilon_r}{\pi} \ln\left(\coth\left(\frac{\pi s}{4h}\right)\right) + 0.65 C_f \left(0.02\frac{h}{s}\sqrt{\varepsilon_r} + 1 - \frac{1}{\varepsilon_r^2}\right) \tag{6.3.70}$$

We now recall the two equations (6.3.61) and (6.3.62) written a few paragraphs above for even and odd mode, respectively. The characteristic impedances $Z_{c,e}$, $Z_{c,o}$ and the effective relative dielectric constants $\varepsilon_{\text{eff},e}$, $\varepsilon_{\text{eff},o}$ are then calculated by recalling the two capacitance terms for even and odd mode defined in eq. (6.3.61) and (6.3.62).

$$Z_{c,e} = \varepsilon_0 \sqrt{\frac{\mu_0}{\varepsilon_0}} \frac{1}{\sqrt{C_e C_{0e}}} \qquad Z_{c,o} = \varepsilon_0 \sqrt{\frac{\mu_0}{\varepsilon_0}} \frac{1}{\sqrt{C_o C_{0o}}} \tag{6.3.71}$$

[26] The factor $\sqrt{\varepsilon_{re}/\varepsilon_r}$ in eq. (6.3.66) is written ambiguously in [108], probably for typographic problems, and it is not immediately clear if the factor is at the numerator or denominator. The reference expressions is the one appearing in [23].

$$\varepsilon_{\text{eff},e} = \frac{C_e}{C_{0e}} \qquad \varepsilon_{\text{eff},o} = \frac{C_o}{C_{0o}} \tag{6.3.72}$$

The terms bearing the "0" are those for dielectric material replaced by air.

To take into account trace thickness t, trace width w is corrected deriving an equivalent quantity, as done for the microstrip, but in this case for the even and odd modes separately:

$$w_{e,e} = w + \delta w \left[1 - 0.5 \exp\left(-0.69 \frac{\delta w}{\delta t}\right)\right] \qquad w_{e,o} = w_{e,e} + \delta t \tag{6.3.73}$$

$$\delta t = \frac{h\,t}{s\,\varepsilon_r} \tag{6.3.74}$$

and δw is the same given in eq. (6.3.3).

6.3.3 Coplanar waveguide

This is a particular type of microstrip, where the central hot conductor is sided by two metallic areas at the same potential of the return plane beneath (the same potential is guaranteed by an adequate number and distribution of shorting vias). Separation between conductors at the upper layer is s, the width of the hot conductor is again w and the width of the two lateral conductors is not relevant, provided it is much larger than w and of signal wavelength. The thickness of all metal layers is t and that of the dielectric substrate is h.

The fundamental mode is not exactly TEM, for which the expression "quasi-TEM" is preferred: as shown in Figure 6.3.7, the lines are not only lying in the paper plane, but are also orthogonal to it.

This is not an unusual transmission line structure: it is encountered all the times that a microstrip line is routed to a connector that requires some pins surrounding the hot pin to be used for grounding and shielding (e.g. reducing crosstalk with adjacent transmission lines; see sec. 7.1.3.2 and 7.1.3.3). This is a common practice for high-frequency connectors, where the pattern of the grounding pins is also selected and optimized in order to preserve characteristic impedance when going through the connector itself.

Also when considering microwave probes of the GSG type (see sec. 6.2), the land pattern for probe tips is that of a coplanar waveguide.

Characteristic impedance and attenuation may be calculated with a conformal mapping approach, by adding to the basic conductor structure of the upper layer all the other elements one by one: substrate thickness, finite thickness of conductors, presence of the ground return plane beneath.

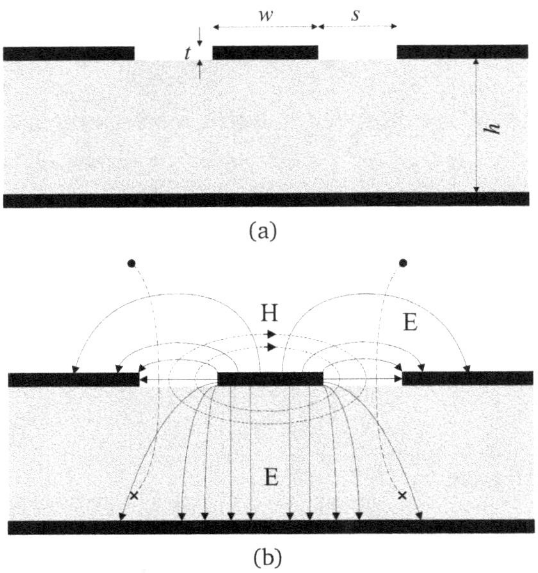

Figure 6.3.7 – Coplanar waveguide geometry and electric and magnetic field distribution: (a) geometry, (b) field distribution.

6.3.3.1 Characteristic impedance

All details on solution methods that are omitted here may be found e.g. in [108], chap. 10. As done for other structures additional elements are taken into account with modified or "effective" expressions of basic parameters.

Characteristic impedance is given by

$$Z_0 = 30\pi \frac{\sqrt{\mu_{re}}}{\sqrt{\varepsilon_{re}}} \frac{K(p_t')}{K(p)} \tag{6.3.75}$$

where ε_{re} is the effective relative permittivity already encountered for the microstrip line. The basic expression is

$$\varepsilon_{re} = \frac{\varepsilon_r + 1}{2} (\varepsilon_1 + \varepsilon_2) \tag{6.3.76}$$

where the factors ε_1 and ε_2 are determined as proposed in [142]:

$$\varepsilon_1 = \tanh(1.785 \ln(h/s) + 1.75) \tag{6.3.77}$$

$$\varepsilon_2 = \frac{ps}{h} [0.04 - 0.7p + 0.01(1 - 0.1\varepsilon_r)(0.25 + p)] \tag{6.3.78}$$

These expressions have an accuracy better than 2% when compared to experimental results, provided that $h/s \geq 1$.

The effect of conductor thickness t on ε_{re} (from which the notation of p' directly with subscript t in the formula above) was evaluated in the literature and proposed in curve fitted form by Gupta, Garg and Bahl [142]:

$$\varepsilon_{re,t} = \varepsilon_{re} - \frac{0.7(\varepsilon_{re}-1)\,t/s}{K(p)/K(p')+0.7\,t/s} \tag{6.3.79}$$

where the quantity $K(\cdot)$ is the complete elliptical integral of the first kind.

The effect of conductor thickness t is included in p by means of an extra width

$$\delta w = \frac{1.25t}{\pi}\left[1+\ln\left(\frac{4\pi w}{t}\right)\right] \tag{6.3.80}$$

and $w_t = w + \delta w$ and $s_t = s - \delta w$ will be used in place of the original quantities.

$$p_t = \frac{w_t}{w_t+2s_t} \qquad p'_t = \sqrt{(1-p_t^2)} \tag{6.3.81}$$

6.3.3.2 Attenuation

Similarly to sec. 6.3.1.2, the attenuation for such a structure may be divided into three terms, due to conduction losses, dielectric losses and radiation.

The expression for conduction attenuation in [142] is

$$\alpha_c = 4.88\,10^{-4} R_s\,\varepsilon_{re}\,Z_0 \frac{K^2(p)\,Q}{K^2(p')\,\pi s}\left(1+\frac{w}{s}\right)\frac{N}{D} \tag{6.3.82}$$

with coefficients given by the following expressions:

$$Q = \begin{cases} \dfrac{p}{\left(1-\sqrt{1-p^2}\right)(1-p^2)^{0.75}} & \text{for } 0 \le p \le 1/\sqrt{2} \\[2ex] \dfrac{K^2(p')}{K^2(p)}\,\dfrac{1}{(1-p)\sqrt{p}} & \text{for } 1/\sqrt{2} \le p \le 1 \end{cases} \tag{6.3.83}$$

$$N = \frac{1.25}{\pi}\ln\left(\frac{4\pi w}{t}\right) + 1 + \frac{1.25t}{\pi w} \tag{6.3.84}$$

$$D = \left\{2+\frac{w}{s}-\frac{1.25t}{\pi s}\left[1+\ln\left(\frac{4\pi w}{t}\right)\right]\right\}^2 \tag{6.3.85}$$

Dielectric losses are accounted for by an expression that was already used for microstrip (see eq. (6.3.46) in sec. 6.3.1.2):

$$\alpha_d = \frac{20\pi}{\ln 10}\,\frac{1/\varepsilon_{re}-1}{1/\varepsilon_r-1}\,\frac{\tan\delta}{\lambda_0}\sqrt{\varepsilon_{re}} \tag{6.3.86}$$

that gives the result in dB p.u.l..

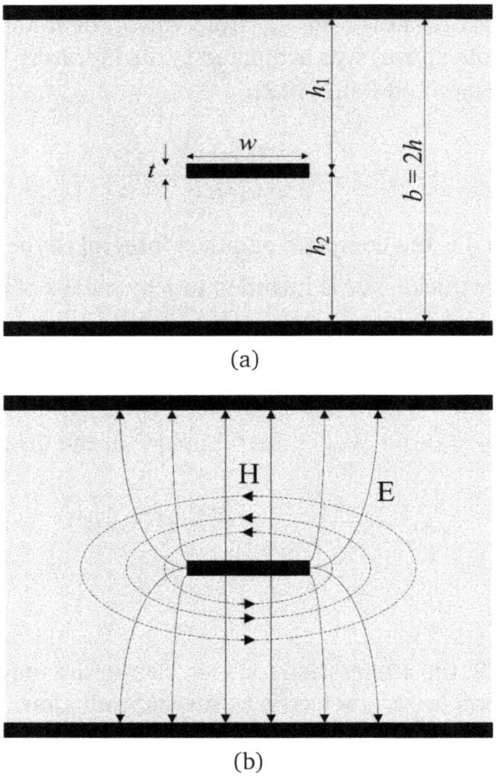

Figure 6.3.8 – (a) Stripline geometry and (b) electric and magnetic field distribution.

6.3.4 Stripline

A stripline structure requires three conductive layers and two dielectric layers: the hot conductor is sandwiched between two dielectric layers and externally by two return planes, or, in other words, it is surrounded by a dielectric lined-up with two conductor planes. The properties of the dielectric layer are: ε_r relative permittivity and μ_r relative magnetic permeability (with again $\mu_r = 1$ as for the microstrip); conductors are of alike thickness t. The medium around the hot conductor is homogeneous and there will be no need to introduce the effective relative permittivity. The separation between upper and lower return planes is named b. The thickness of dielectric layers above and below the central hot conductors may be equal ($h_1 = h_2 = b/2$) or different ($h_1 + h_2 = b$): in the first case we say that the stripline is symmetrical.

The dielectric is homogeneous because the hot trace is surrounded by dielectric and two ground planes, and the fundamental mode is TEM (Figure 6.3.8), with field lines lying in the paper plane.

6.3.4.1 Characteristic impedance

The use of elliptic integrals for the solution of field equations introduces the quantity $K(\cdot)$, called "complete elliptical integral of the first kind":

$$K(m) = \int_0^1 \frac{1}{\sqrt{(1-t^2)(1-mt^2)}}\, dt \tag{6.3.87}$$

It may be calculated using for example the Matlab function `ellipke(m)`.

The argument m of these elliptic integrals is assigned a pair of variables, $p = 1/\cosh(\pi \mathrm{w}/4\mathrm{h})$ and $q = \sqrt{1-p^2} = \tanh(\pi w/4h)$[27].

The approximation of the ratio of $K(p)/K(q)$ depends on the w/h ratio, so

$$\frac{K(p)}{K(q)} = \begin{cases} \dfrac{1}{\pi} \ln\left[2\dfrac{1+\sqrt{p}}{1-\sqrt{p}}\right] & \text{for } w/h \le 1 \\[2ex] \pi/\ln\left[2\dfrac{1+\sqrt{p}}{1-\sqrt{p}}\right] & \text{for } w/h \ge 1 \end{cases} \tag{6.3.88}$$

Thus the expressions of characteristic impedance depend similarly on w/b (or w/h) ratio:

$$Z_c = \begin{cases} \dfrac{30\sqrt{\mu_r}}{\sqrt{\varepsilon_r}} \ln\left[2\dfrac{1+\sqrt{p}}{1-\sqrt{p}}\right] & \text{for } w/h \le 1 \\[2ex] \dfrac{30\pi^2\sqrt{\mu_r}}{\sqrt{\varepsilon_r}} \ln\left[2\dfrac{1+\sqrt{q}}{1-\sqrt{q}}\right] & \text{for } w/h \ge 1 \end{cases} \tag{6.3.89}$$

Bahl and Bhartia [23], Table 2.5, included also hot conductor thickness t, re-writing characteristic impedance as

$$Z_c = \frac{30}{\sqrt{\varepsilon_r}} \ln\left\{1 + \frac{4}{\pi}\frac{2h-t}{w_e}\left[\frac{8}{\pi}\frac{2h-t}{w_e} + \sqrt{\left(\frac{8}{\pi}\frac{2h-t}{w_e}\right)^2 + 6.27}\right]\right\} \quad \text{for } \frac{w_e}{2h-t} \le 10 \tag{6.3.90}$$

where the corrective term $\delta w = w_e - w$ is given by

$$\delta w = \frac{t}{\pi}\left\{1 - \frac{1}{2}\ln\left[\left(\frac{t}{4h-t}\right)^2 + \left(\frac{0.0796\,t}{w+1.1t}\right)^m\right]\right\} \tag{6.3.91}$$

and

$$m = \frac{2}{1 + \dfrac{2}{3}\dfrac{t}{2h-t}} \tag{6.3.92}$$

---

[27] A different notation may be encountered using b [108] or $2h$ [23] for the overall height of the substrate.

The declared range of validity of eq. (6.3.90) refers to an accuracy of 0.5%, so that the proposed expression is useful even well beyond the value of 10.

Including conductor thickness by stray capacitance term A different approach was used by Cohn [85], that included also conductor thickness t, introducing corrective terms related to stray-capacitance effect, obtaining

$$Z_c = \begin{cases} \dfrac{60}{\sqrt{\varepsilon_r}} \ln\left[\dfrac{4h}{\pi\phi}\right] & \text{for } w/(2h-t) \leq 0.35 \\ \dfrac{94.15}{\sqrt{\varepsilon_r}\,[w/(2h-t) + C_f/0.0885\varepsilon_r]} & \text{for } w/(2h-t) \geq 0.35 \end{cases} \tag{6.3.93}$$

where

$$\phi = \frac{w}{2}\left\{1 + \frac{t}{w}\left[1 + \ln\left(\frac{4\pi w}{t}\right) + 0.51\left(\frac{t}{w}\right)^2\right]\right\} \tag{6.3.94}$$

$$\begin{aligned} C_f &= \frac{0.0885\varepsilon_r}{\pi}\left[\frac{4h}{2h-t}\ln\left(\frac{2h}{2h-t}+1\right) + \right. \\ &\left. -\left(\frac{2h}{2h-t}-1\right)\ln\left(\frac{1}{(1-t/2h)^2}-1\right)\right] \end{aligned} \tag{6.3.95}$$

with C_f in the original formulation written to be expressed in [pF/cm].

Of course it is possible to re-write C_f in $\mathrm{pF/m}$ by removing the two leading 0s and get back to 8.85, that is the permittivity of free space. The same may be done in the expression above (eq. (6.3.93)), where C_f is again divided by "$0.0885\varepsilon_r$" that becomes "$8.85\varepsilon_r$". In a simpler way the ε_0 factor expressed in pF/m ($8.85\varepsilon_r$) may be deleted from both equations (6.3.93) and (6.3.95) with a simplified notation, but the correcting factor C_f has no longer the unit of a p.u.l. capacitance.

Looking at the expressions that may be found in Collin [88], eq. (6.3.95) is re-written to demonstrate the equivalence. By hand calculation and re-arranging the fractions at the denominator all terms inside the natural log and multiplying them match except the first one: Collin reports "2" instead of "$4h/(2h-t)$", that is close to two and matches if the thickness t is neglected. Then the multiplying coefficient $0.0885\varepsilon_r$ has no reason to exist as long as the stray capacitance C_f is used to correct the so-calculated characteristic impedance, being no need that it has the unit of capacitance p.u.l.. Collin leaves the multiplying $1/\pi$ and the expression has a simpler structure.

IPC formulation The expression of characteristic impedance of a stripline reported by IPC [176] has a form that is quite similar to that of microstrip, but its origin is not explicitly declared.

$$Z_c = \frac{60}{\sqrt{\varepsilon_r}} \ln\left(\frac{1.9(2h+t)}{0.8w+t}\right) \tag{6.3.96}$$

The validity of this expression is declared for $w/h < 2$, with thickness that preserves the following conditions: $w/(h-t) < 0.35$ and $t/h < 0.25$.

Comparison of expressions The formulations above have a different approach and one may wonder about which one is best, depending for example on the frequency range or on geometry, or if they in general agree and under which conditions. To this aim a sensitivity analysis has been performed with respect to relative permittivity of the substrate (see Figure 6.3.9(a)), thickness of substrate (see Figure 6.3.9(b)), thickness of copper trace (see Figure 6.3.10(a)) and width of copper trace (see Figure 6.3.10(b)). IPC expressions have been discarded because the results were inconsistent (much different from the other two formulations that substantially agree) and often falling outside their interval of validity.

6.3.4.2 Attenuation

For striplines radiation losses are excluded, being the central hot conductor completely surrounded by return planes. Based on the same basic quantities and approximations used to express characteristic impedance, Bahl and Bhartia [23] provide an expression for the conduction loss coefficient α_c:

$$\alpha_c = \frac{0.0231\, R_s\sqrt{\varepsilon_r}}{Z_c}\frac{\partial Z_c}{\partial w_e}\left[1+\frac{2w_e}{b-t}-\frac{1}{\pi}\left(\frac{3t}{2t-b}+\ln\frac{t}{2t-b}\right)\right] \tag{6.3.97}$$

where the partial derivative is expressed as

$$\frac{\partial Z_c}{\partial w_e} = \frac{30e^{-A}}{w_e\sqrt{\varepsilon_r}}\left[\frac{3.135}{Q}-\left(\frac{8}{\pi}\frac{b-t}{w_e}\right)^2(1+Q)\right] \tag{6.3.98}$$

with

$$A = \frac{Z_c\sqrt{\varepsilon_r}}{30\pi} \qquad Q = \sqrt{1+6.27\left(\frac{8}{\pi}\frac{b-t}{w_e}\right)^2}$$

Other expressions for conduction loss constant α_c were proposed by Cohn [85], using the ratio between trace width w, trace thickness t and substrate thickness b:

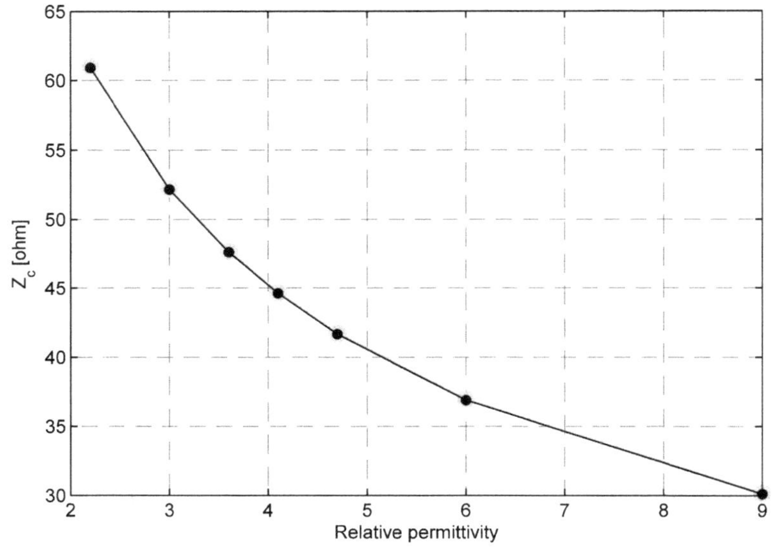

(a)

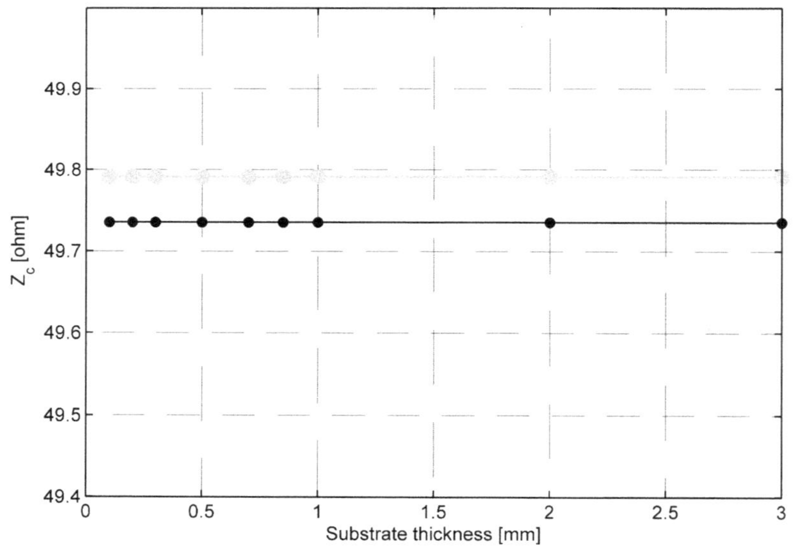

(b)

Figure 6.3.9 – Sensitivity of the expressions for the characteristic impedance of a Stripline versus (a) the substrate permittivity ε_r ($w = 0.25\,\text{mm}$, $b = 2h = 0.4\,\text{mm}$, $t = 0.018\,\text{mm}$) and (b) the substrate thickness b ($\varepsilon_r = 4.7$, $w = 0.25\,\text{mm}$, $t = 0.018\,\text{mm}$): Bahl & Garg (black), Cohn (gray).

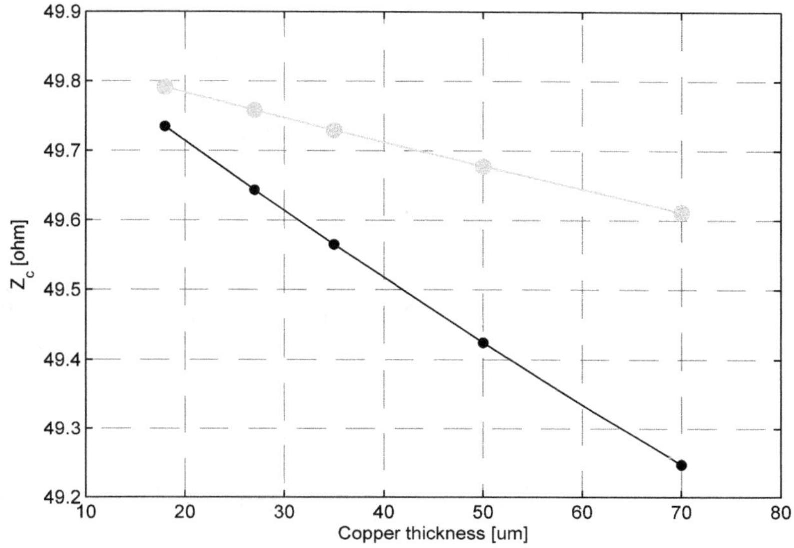

(a)

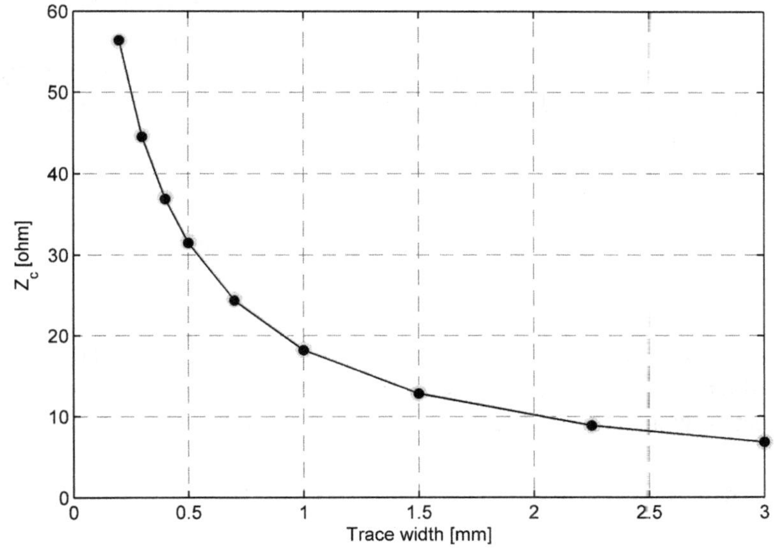

(b)

Figure 6.3.10 – Sensitivity of the expressions for the characteristic impedance of a Stripline versus (a) trace copper thickness t ($\varepsilon_r = 4.7$, $w = 0.25\,\text{mm}$, $b = 2h = 0.4\,\text{mm}$) and (b) trace width w ($\varepsilon_r = 4.7$, $b = 2h = 0.4\,\text{mm}$, $t = 0.018\,\text{mm}$): Bahl & Garg (black), Cohn (gray).

$$\alpha_c = \begin{cases} \left[k+\frac{2w}{b}k^2+\frac{(1+t/b)k^2}{\pi}\ln\left(\frac{k+1}{k-1}\right)\right] & \text{for } w/(b-t)\geq 0.35 \\ \frac{R_s}{2\pi b\zeta}\left\{1+\frac{b}{d}\left[0.5+0.669\left(\frac{t}{w}\right)+\right.\right. & \text{for } w/(b-t)\geq 0.35, \\ \left.\left. -0.225\left(\frac{t}{w}\right)^2+\frac{1}{2\pi}\ln\left(\frac{4\pi w}{t}\right)\right]\right\} & t/b\leq 0.35,\ t/w\leq 0.11 \end{cases} \tag{6.3.99}$$

where

$$P = 4R_s\,\zeta\,\frac{\varepsilon_r\sqrt{\varepsilon_r}}{b(120\pi)^2} \tag{6.3.100}$$

$$k = \frac{1}{1-t/b} \tag{6.3.101}$$

$$d = \frac{w}{2}\left[1+\frac{t}{w}\left(1+\ln\left(\frac{4\pi w}{t}\right)+0.51\left(\frac{t}{w}\right)^2\right)\right] \tag{6.3.102}$$

and d has the meaning of radius of the cylindrical conductor equivalent to the stripline hot conductor.

For the dielectric loss constant α_d there are simpler expressions, provided by Bahl and Bhartia [23], Table 2.5, and Di Paolo [108], sec. 3.6, and identified as $\alpha_{d,BB}$ and $\alpha_{d,DP}$, respectively:

$$\alpha_{d,BB} = 27.3\,\sqrt{\varepsilon_r}\,\frac{\tan\delta}{\lambda_0} \tag{6.3.103}$$

$$\alpha_{d,DP} = \frac{\pi\sqrt{\varepsilon_r \tan\delta}}{\lambda_0} \tag{6.3.104}$$

with "$\tan\delta$" indicating the tangent delta of the substrate material. They are evidently not equal, the most apparent aspect being the $\tan\delta$ factor outside and inside the square root, respectively; also the multiplying factor is quite different.

6.3.4.3 Offset stripline

For practical reasons it is possible that the two dielectric layers have no equal thickness, so $h_1 \neq h_2$; in this case it is still possible to have handy expressions for characteristic impedance and assume that the attenuation is not largely affected.

The reason for offsetting the inner conductor is simply that the fabrication of the PCB does not provide layers of equal thickness in that position of the stack. Simply speaking, we put the three "copper layers" in a four-layer PCB where dielectric layers are build using pre-preg sheets starting from the symmetry plane; the two dielectric layers will be then chosen very close, but will not be equal.

For a not-too-offset stripline, with $0.2 \le h/b \le 0.8$, (20% minimum thickness of one of the two dielectric layers), the resulting characteristic impedance is [108], sec. 3.7:

$$Z_c = \begin{cases} \dfrac{\zeta_0}{2\pi\sqrt{\varepsilon_r}} \operatorname{acosh}(A) & \text{for } w/(b-t) \le 0.35 \\ \\ \dfrac{\zeta_0}{\sqrt{\varepsilon_r}} \dfrac{1}{[\rho/\gamma + \rho/(\beta-\gamma) + 2\phi_2/\varepsilon_0\varepsilon_r]} & \text{for } w/(b-t) \ge 0.35 \end{cases} \tag{6.3.105}$$

where

$$A = \sin\left(\frac{\pi h}{b}\right) \coth\left(\frac{\pi d}{2b}\right) \tag{6.3.106}$$

$$\rho = \frac{w}{b} + (1-\frac{t}{b})^8 \left[\frac{K(p')}{K(p)} - \frac{2}{\pi}\ln 2 - \frac{w}{b}\right] \tag{6.3.107}$$

$$\gamma = \frac{h}{b} - \frac{t}{2b} \qquad \beta = 1 - \frac{t}{b} \tag{6.3.108}$$

$$\phi_2 = \frac{\varepsilon_r \varepsilon_0}{\pi} \left[2\ln\left(\frac{1}{\gamma(\beta-\gamma)}\right) + \frac{1}{\gamma(\beta-\gamma)}\left(F\left(\frac{t}{2b}\right) - F\left(\frac{h}{b}\right)\right)\right] \tag{6.3.109}$$

with

$$F(x) = (1-2x)\left[(1-x)\ln(1-x) - x\ln x\right] \tag{6.3.110}$$

The declared accuracy of these formulas with respect to numeric computer simulations using a finite difference method is better than 2%.

6.3.5 Coupled stripline

A coupled stripline has the structure of a stripline, but the inner hot conductors are two and they may belong to the same layer (in this case we have a "side-coupled stripline") or to adjacent layers (in this case we have a "broadside-coupled stripline"). Its use is to create a differential line buried between two ground planes to drive e.g. a differential pin pair of a connector or device. The electromagnetic field lines for the fundamental TEM mode are shown in Figure 6.3.11.

Symmetrically, two characteristic impedance expressions for the even and the odd mode are written as [85]:

$$Z_{c,e} = \frac{Z_0}{4\sqrt{\varepsilon_r}} \frac{K(q_e)}{K(p_e)} \tag{6.3.111}$$

$$Z_{c,o} = \frac{Z_0}{4\sqrt{\varepsilon_r}} \frac{K(q_o)}{K(p_o)} \tag{6.3.112}$$

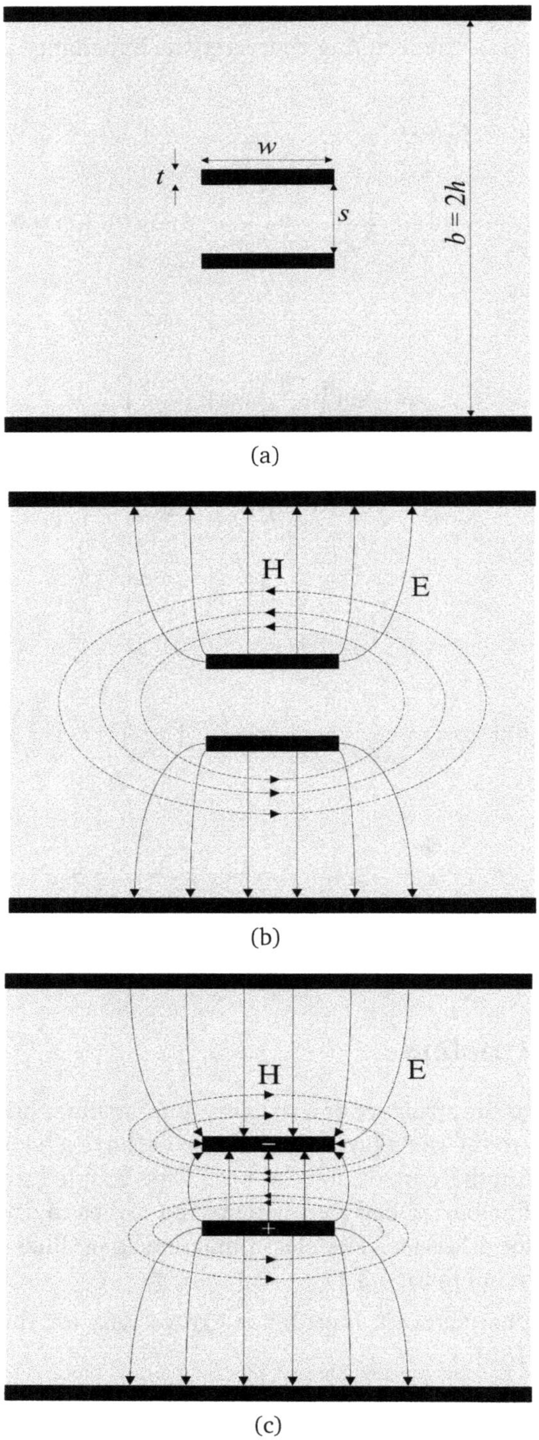

Figure 6.3.11 – Broadside-coupled stripline geometry and electric and magnetic field distribution: (a) geometry, (b) field distribution with even mode excitation, (c) field distribution with odd mode excitation.

where

$$p_e = \tanh\left(\frac{\pi w}{4h}\right) \tanh\left(\frac{\pi}{4}\frac{w+s}{h}\right) \tag{6.3.113}$$

$$p_o = \tanh\left(\frac{\pi w}{4h}\right) \coth\left(\frac{\pi}{4}\frac{w+s}{h}\right) \tag{6.3.114}$$

and, as before, the q quantities are: $q_e = \sqrt{1-p_e^2}$ and $q_o = \sqrt{1-p_o^2}$.

6.3.5.1 Side-coupled stripline

For the side-coupled stripline, neglecting the effect of thickness t of the hot trace, even- and odd-mode characteristic impedances are

$$Z_{c,e} = \frac{Z_0}{4\sqrt{\varepsilon_r}} \frac{1}{\frac{w}{b} + \frac{1}{\pi}\ln 2 + \frac{1}{\pi}\ln\left(1+\tanh\left(\frac{\pi s}{2b}\right)\right)} \tag{6.3.115}$$

$$Z_{c,o} = \frac{Z_0}{4\sqrt{\varepsilon_r}} \frac{1}{\frac{w}{b} + \frac{1}{\pi}\ln 2 + \frac{1}{\pi}\ln\left(1+\coth\left(\frac{\pi s}{2b}\right)\right)} \tag{6.3.116}$$

Including conductor thickness by stray capacitance term (Collin [88], sec. 3.15) As it was done for the coupled microstrip, the finite thickness t of conductors is taken into account by reverting to the stray capacitance term already shown in eq. (6.3.95). For the sake of simplicity it is rewritten in the simpler form presented by Collin [88] and separated by the expressions reported in the paragraph below, both attributed to Cohn, but with much different notations and ranges of validity.

$$Z_{c,m} = \frac{Z_0(2h-t)}{4\sqrt{\varepsilon_r}\left[w + \frac{h}{\pi}C'_f A_m\right]} \qquad \text{for } w/h \geq 0.7,\ m = e,\ o \tag{6.3.117}$$

$$A_e = 1 + \frac{1}{\ln 2}\ln\left(1+\tanh\left(\frac{\pi s}{4h}\right)\right) \tag{6.3.118}$$

$$A_o = 1 + \frac{1}{\ln 2}\ln\left(1+\cosh\left(\frac{\pi s}{4h}\right)\right) \tag{6.3.119}$$

$$C'_f = 2\ln\left(\frac{4h-t}{2h-t}\right) - \frac{t}{2h}\ln\left(\frac{4ht-t^2}{(2h-t)^2}\right) \tag{6.3.120}$$

Including conductor thickness by stray capacitance term (Cohn [86]) As said above, this is the other formulation attributed to Cohn and reported in [108]. Modal characteristic impedances are determined from terms calculated for various combinations of conductor width w, thickness t and substrate thickness $2h$. The term Z_c indicates the stripline characteristic impedance given in eq. (6.3.93).

Under the condition that $w/b \geq 0.35$ and $t/b \leq 0.1$:

$$Z_{c,e} = \frac{1}{\dfrac{1}{Z_c} - \dfrac{C_f}{C_f(t=0)}\left[\dfrac{1}{Z_c(t=0)} - \dfrac{1}{Z_{c,e}(t=0)}\right]} \tag{6.3.121}$$

$$Z_{c,o} = \begin{cases} \left\{\dfrac{1}{Z_c} + \dfrac{C_f}{C_f(t=0)}\left[\dfrac{1}{Z_{c,o}(t=0)} - \dfrac{1}{Z_c(t=0)}\right]\right\}^{-1} & \text{for } s \geq 5t \\ \left\{\dfrac{1}{Z_{c,o}(t=0)} + \dfrac{1}{Z_c} - \dfrac{1}{Z_c(t=0)} - \dfrac{2\,\delta C_f}{377} + \dfrac{2t}{377s}\right\}^{-1} & \text{for } s \leq 5t \end{cases} \tag{6.3.122}$$

where C_f is the fringe capacitance term for a simple stripline as in eq. (6.3.95); when evaluated for zero thickness ($t=0$) the result is $0.03985\varepsilon_r$ pF/cm.

6.3.5.2 Broadside-coupled stripline

For the broadside-coupled stripline the even- and odd-mode characteristic impedance are given by [108], sec. 6.4.2:

$$Z_{c,e} = \frac{Z_0}{2\sqrt{\varepsilon_r}} \frac{1}{\dfrac{w}{2h-s} + 0.4413 + C} \tag{6.3.123}$$

$$Z_{c,o} = \frac{Z_0}{2\sqrt{\varepsilon_r}} \frac{1}{\dfrac{w}{2h-s} + \dfrac{w}{s} + \dfrac{h}{s}C} \tag{6.3.124}$$

where

$$C = \frac{1}{\pi}\left[-\frac{s}{2h-s}\ln\left(\frac{s}{2h}\right) - \ln\left(1 - \frac{s}{2h}\right)\right] \tag{6.3.125}$$

For completeness, eq. (6.3.125) is reported in [23] as [33] and indicated as C_{f0}. The term "0.4413" appearing above, and also below (originally written in [108], eq. (6.4.19), and [88], sec. 3.15) appears on the contrary as "0.443" in [23]; a straight 2-out-of-3 logic would indicate that 0.443 is the result of a typing mistake (by the way the error would be very small in any case, around 0.5% of difference). The correct expression[28] is $2/\pi\ \ln(2) = 0.441271$, inaccurately calculated by Cohn as 0.4407.

For completeness, the two slightly different formulations reported by Collin, and then by Bahl and Barthia, are shown below: the difference stands mainly in the corrective rightmost terms at the denominator, including the correction for conductor thickness.

Collin formulation [88], sec. 3.15, eq. (3.188a) and (3.188b):

[28] Eq. (13) in Cohn [86].

$$Z_{c,e} = \frac{Z_0}{2\sqrt{\varepsilon_r}} \left\{ \frac{w}{2h-s-2t} + 0.4413 + \frac{1}{\pi}\left[\frac{s+2t}{2h-s-2t}\ln\left(\frac{2h}{s+2t}\right) + \ln\left(\frac{2h}{2h-s-2t}\right)\right]\right\}^{-1} \tag{6.3.126}$$

$$Z_{c,o} = \frac{Z_0}{2\sqrt{\varepsilon_r}} \left\{ \frac{w}{2h-s-2t} + \frac{w}{s} + \frac{2}{\pi}\left[\frac{h-t}{2h-s-2t}\ln\left(\frac{2h-2t}{s}\right) + \frac{h-t}{s}\ln\left(\frac{2h-2t}{2h-s-2t}\right) + \left(1+\frac{t}{s}\right)\ln\left(1+\frac{t}{s}\right) - \frac{t}{s}\ln\left(\frac{t}{s}\right)\right]\right\}^{-1} \tag{6.3.127}$$

The arguments of logarithms clearly indicate that the interval of validity is at least restricted to values of h, s and t that give positive values: the separation s between the internal edges of the two traces cannot be larger than the overall thickness of the substrate $2h$. Collin states the following conditions for validity of expressions above: $w \geq 0.35s$, $w \geq 0.7h(1-s/2h)$, and $s < 2h$ is implicit in the necessity of a positive lower limit for w.

Bahl and Bhartia formulation [23], eq. [31] and [32]:

$Z_{c,o}$ and $Z_{c,e}$, have the same structure with one difference:

$$Z_{c,e} = \frac{Z_0}{2\sqrt{\varepsilon_r}} \left\{ \frac{w}{2h-s} + 0.443 + \frac{1}{\pi}\left[\frac{s+2t}{2h-s}\ln\left(\frac{2h+2t}{s+2t}\right) + \ln\left(\frac{2h+2t}{2h-s}\right)\right]\right\}^{-1} \tag{6.3.128}$$

$$Z_{c,o} = \frac{Z_0}{2\sqrt{\varepsilon_r}} \left\{ \frac{w}{2h-s} + \frac{w}{s} + \frac{2}{\pi}\left[\frac{h-t}{2h-s-2t}\ln\left(\frac{2h-2t}{s}\right) + \frac{h-t}{s}\ln\left(\frac{2h-2t}{2h-s-2t}\right) + \left(1+\frac{t}{s}\right)\ln\left(1+\frac{t}{s}\right) - \frac{t}{s}\ln\left(\frac{t}{s}\right)\right]\right\}^{-1} \tag{6.3.129}$$

The difference consists of where trace thickness t appears, numerator or denominator; however, t is always in combination with s or $2h$, so that appreciable difference between the two formulations should appear only for extremely thick traces and/or tiny dielectric substrates.

Comparison of expressions Formulations above have a different approach and one may wonder about which one is best, depending for example on the frequency range or on the geometry, or if they agree and the results match, and if so, under which conditions. To this aim a sensitivity analysis has been performed with respect to relative permittivity of substrate (see Figure 6.3.12), thickness of substrate (see Figure 6.3.13), thickness of copper trace (see Figure 6.3.15), width of hot traces (see Figure 6.3.14) and their separation (see Figure 6.3.16). This sensitivity analysis is complicated by the many conditions on the geometric parameters imposed by the range of validity of expressions of odd- and even-mode characteristic impedance.

There is a substantial agreement between some expressions, even if superposition of curves is alternated: two formulations agree for the odd-mode characteristic impedance and they are not the same two agreeing for the even-mode one. Differences in Figure 6.3.15, where trace thickness is considered, were expected. In some cases some curves are interrupted prematurely simply because the combination of geometric parameters violates validity conditions.

6.3.6 Coplanar strip

It is a planar transmission line made of two traces on the same layer and with no return plane beneath, so that it looks like it is obtained from a coupled microstrip removing the return plane. For calculations the two conductors are supposed to be supported by a nearly infinite dielectric, even if grounded conductors may be present in any case, such as other lines nearby, metallic parts, etc. The assumption of infinite dielectric is valid if its thickness is five times the overall width of the coplanar strip: $h > 5(s+2w)$.

Because the number of electric and magnetic field lines in the air is higher than the number of the same lines in the microstrip case, the effective dielectric constant ε_{re} of coplanar strip is typically 20% lower than that of a microstrip of the same size; consequently, characteristic impedance values are higher. In addition, to avoid field radiation in the air, it is very important to use substrates with high permittivity (e.g. 10, as for Alumina), so the electromagnetic field lines are mostly confined inside substrate.

6.3.6.1 Characteristic impedance

As for coplanar waveguide (see sec. 6.3.3), characteristic impedance is given by

$$Z_0 = \frac{\zeta_0}{4} \frac{\sqrt{\mu_{re}}}{\sqrt{\varepsilon_{re}}} \frac{K(p_t')}{K(p)} \tag{6.3.130}$$

where ε_{re} is the effective relative permittivity. The basic expression is

$$\varepsilon_{re} = \frac{\varepsilon_r + 1}{2} (\varepsilon_1 + \varepsilon_2) \tag{6.3.131}$$

and the factors ε_1 and ε_2 are determined as:

$$\varepsilon_1 = \tanh\left(1.785 \ln\left(\frac{h}{s}\right) + 1.75\right) \tag{6.3.132}$$

$$\varepsilon_2 = \frac{pw}{h} \left[0.04 - 0.7p + 0.01(1 - 0.1\varepsilon_r)(0.25 + p)\right] \tag{6.3.133}$$

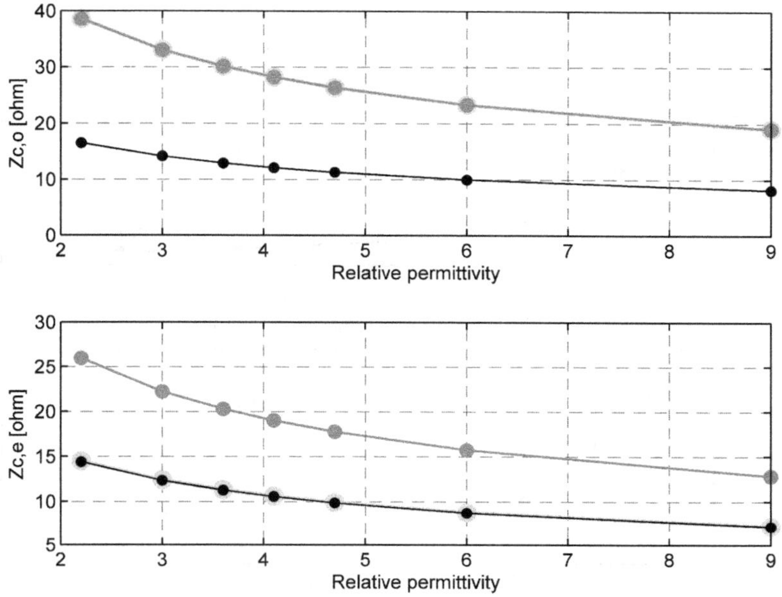

Figure 6.3.12 – Sensitivity of the expressions for the characteristic impedance of a Broadside Coupled Stripline for odd and even mode versus substrate relative permittivity ε_r ($w = 1$ mm, $2h = 0.4$ mm, $t = 0.035$ mm, $s = 2$ mm): Di Paolo (black), Collin (dark gray), Bahl-Bhartia (light gray).

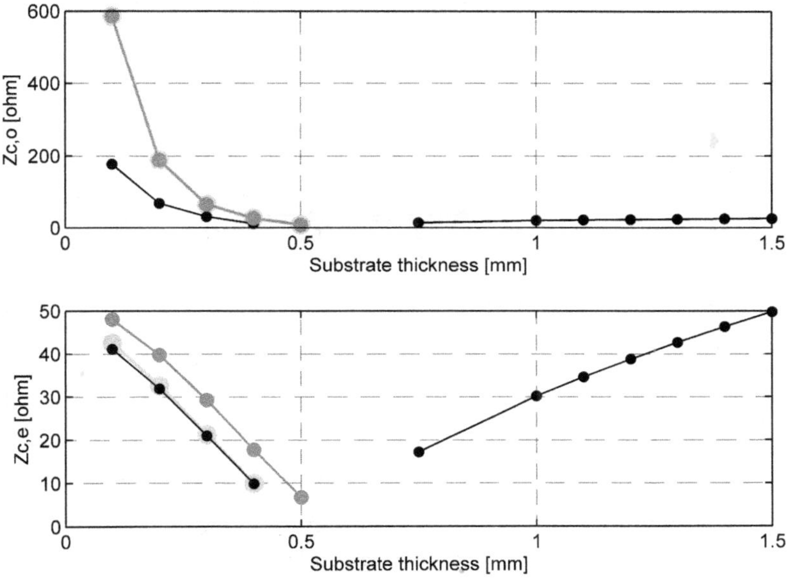

Figure 6.3.13 – Sensitivity of the expressions for the characteristic impedance of a Broadside Coupled Stripline for odd and even mode versus substrate thickness $2h$ ($\varepsilon_r = 4.7$, $w = 1$ mm, $t = 0.035$ mm, $s = 2$ mm): Di Paolo (black), Collin (dark gray), Bahl-Bhartia (light gray).

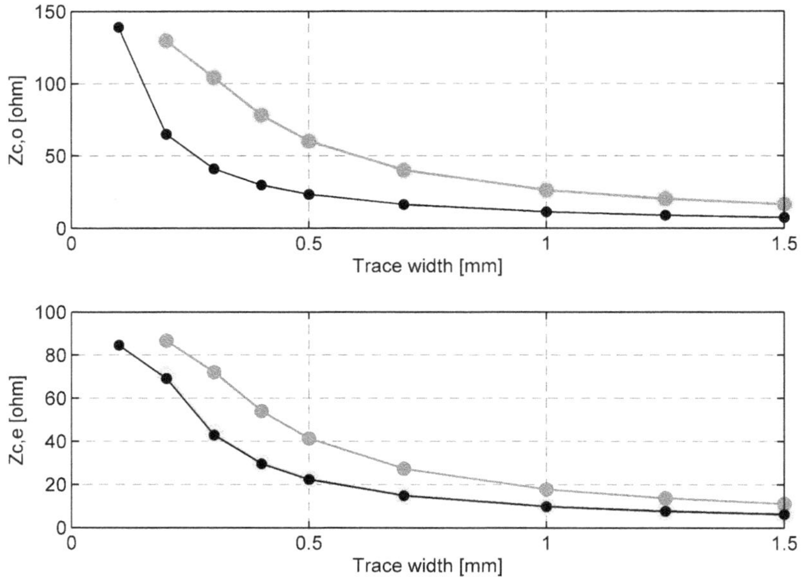

Figure 6.3.14 – Sensitivity of the expressions for the characteristic impedance of a Broadside Coupled Stripline for odd and even mode versus trace width w ($\varepsilon_r = 4.7$, $2h = 0.4\,\text{mm}$, $t = 0.035\,\text{mm}$, $s = 2\,\text{mm}$): Di Paolo (black), Collin (dark gray), Bahl-Bhartia (light gray).

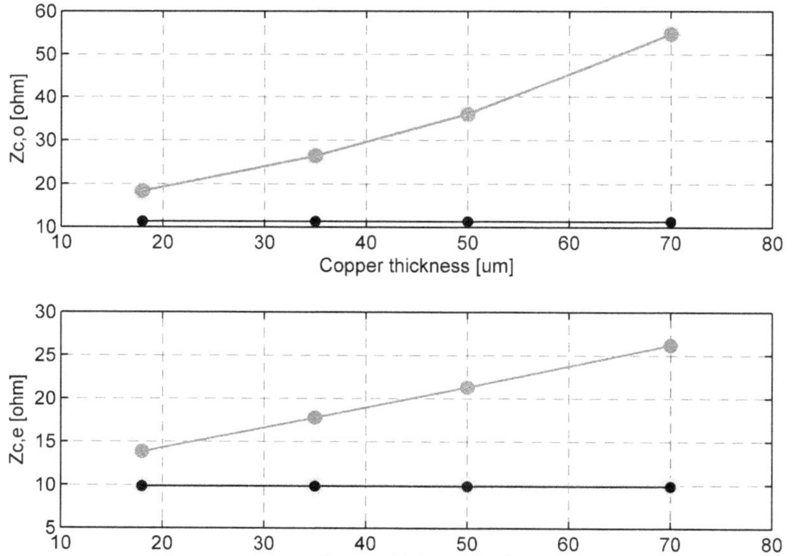

Figure 6.3.15 – Sensitivity of the expressions for the characteristic impedance of a Broadside Coupled Stripline for odd and even mode versus trace thickness t ($\varepsilon_r = 4.7$, $w = 1\,\text{mm}$, $2h = 0.4\,\text{mm}$, $s = 2\,\text{mm}$): Di Paolo (black), Collin (dark gray), Bahl-Bhartia (light gray).

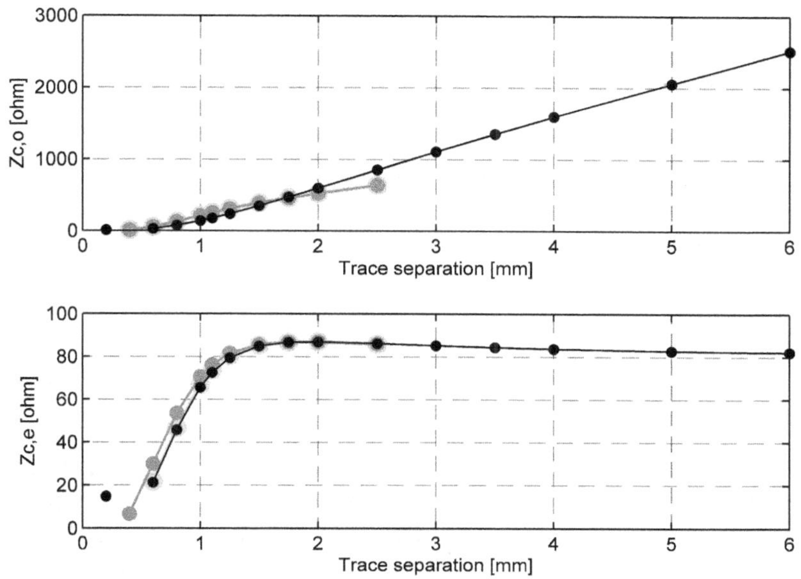

Figure 6.3.16 – Sensitivity of the expressions for the characteristic impedance of a Broadside Coupled Stripline for odd and even mode versus trace separation s ($\varepsilon_r = 4.7$, $w = 1$ mm, $2h = 0.4$ mm, $t = 0.035$ mm): Di Paolo (black), Collin (dark gray), Bahl-Bhartia (light gray).

Pay attention that equations are in all identical to those of coplanar waveguide except for the use of width w in place of separation s at numerator of ε_2.

To include the effect of trace thickness t, we may replace trace width w and separation s, as already done for coplanar waveguide, with equivalent values $w_t = w + \delta w$ and $s_t = s - \delta w$, where

$$\delta w = \frac{1.25t}{\pi}\left[1 + \ln\left(\frac{4\pi w}{t}\right)\right] \tag{6.3.134}$$

Conversely p and p' variables become

$$p_t = \frac{w_t}{w_t + 2s_t} \qquad p'_t = \sqrt{(1 - p_t^2)} \tag{6.3.135}$$

and the new corrected effective relative permittivity $\varepsilon_{re,t}$ is:

$$\varepsilon_{re,t} = \varepsilon_{re} - \frac{1.4(\varepsilon_{re} - 1)\dfrac{t}{s}}{K(p)/K(p') + 1.4\dfrac{t}{s}} \tag{6.3.136}$$

The factor 1.4 that replaces 0.7 accounts for the smaller resulting dielectric permittivity mentioned at the beginning.

Another formulation reported in [23], Table 2.7, and attributed to Ghione and Naldi [135] starts from an equation identical to eq. (6.3.75), but with a different expression of the two arguments p and p_t, called in that case k and k_1, respectively.

$$Z_c = \frac{\zeta_0}{4\sqrt{\varepsilon_{re}}} \frac{K(p)}{K(p')} \tag{6.3.137}$$

$$p = \frac{a}{b} \qquad a = \frac{s}{2} \qquad b = \frac{s}{2} + w \qquad p_t = \frac{\sinh(\pi a/2h)}{\sinh(\pi b/2h)} \tag{6.3.138}$$

$$\varepsilon_{re} = 1 + \frac{\varepsilon_r - 1}{2} \frac{K(p')K(p_t)}{K(p)K(p_t')} \tag{6.3.139}$$

The same expression for ε_{re}, again attributed to Ghione and Naldi, when reported by [88], sec. 3.16, is slightly different and divided into three groups of calls to the $K(\cdot)$ function (we use u instead of the original "k" for the argument of $K(\cdot)$ to avoid confusion and we keep the original "q" in the dielectric permittivity formula):

$$\varepsilon_{re} = 1 + q(\varepsilon_r - 1) \tag{6.3.140}$$

$$q = \frac{K(u_1)/K(u_1')}{K(u_2)/K(u_2') + K(u)/K(u')} \tag{6.3.141}$$

with

$$u = \frac{s}{s+2w} \qquad u_1 = \frac{\tanh(\pi s/4h)}{\tanh(\pi(s+2w)/h)} \qquad u_2 = \frac{\tanh(\pi s/4h_1)}{\tanh(\pi(s+2w)/h_1)} \tag{6.3.142}$$

As usual terms with the prime indicate $\sqrt{1-(\cdot)^2}$ operation. The term u_2 takes care of dielectric above coplanar waveguide of thickness h_1, that in [23, 108] was not included.

6.3.6.2 Attenuation

In a way perfectly identical to coplanar waveguide, we may write conduction and dielectric loss terms.

The expression for the conduction attenuation factor in [142] is

$$\alpha_c = 17.34 \frac{R_s}{Z_0} \frac{K^2(p)\,Q}{K^2(p')\,\pi s} \left(1 + \frac{w}{s}\right) \frac{N}{D} \tag{6.3.143}$$

with the coefficients given by the following expressions:

$$Q = \begin{cases} \dfrac{p}{\left(1-\sqrt{1-p^2}\right)\left(1-p^2\right)^{0.75}} & \text{for } 0 \leq p \leq 1/\sqrt{2} \\ \\ \dfrac{K^2(p')}{K^2(p)} \dfrac{1}{(1-p)\sqrt{p}} & \text{for } 1/\sqrt{2} \leq p \leq 1 \end{cases} \tag{6.3.144}$$

$$N = \frac{1.25}{\pi} \ln\left(\frac{4\pi w}{t}\right) + 1 + \frac{1.25t}{\pi w} \tag{6.3.145}$$

$$D = \left\{1 + \frac{2w}{s} - \frac{1.25t}{\pi s}\left[1 + \ln\left(\frac{4\pi w}{t}\right)\right]\right\}^2 \tag{6.3.146}$$

Modifications with respect to the expressions of coplanar waveguide are only the first part of α_c eq. (6.3.143) and of D eq. (6.3.146).

Dielectric losses are accounted for by an expression that was already used for the microstrip (see eq. (6.3.46)) and for the coplanar waveguide (see eq. (6.3.86)):

$$\alpha_d = \frac{20\pi}{\ln 10} \frac{1/\varepsilon_{re} - 1}{1/\varepsilon_r - 1} \frac{\tan\delta}{\lambda_0} \sqrt{\varepsilon_{re}} \tag{6.3.147}$$

that gives the result in dB p.u.l..

6.4 Signal routing and transmission

A few concepts are reviewed regarding the behavior of PCB elements and how to model it, as for signal transmission and signal integrity. The effect of vias, layers, clearance between traces, pads is considered by means of equivalent inductance and capacitance, and crosstalk.

6.4.1 Transmission of signals and discontinuities

If the previous section was dedicated to review the implementation of transmission lines and approximations between theoretical design methods, here the focus is on those elements that are connected along transmission lines and cause discontinuities affecting signal transmission. Some were already reviewed for their influence on line properties and are again considered for their influence on signal propagation.

After estimating the bandwidth related to a digital pulse as a function of its rise time (see sec. 3.2), in the IPC-D-317A [176] it is proposed a corresponding trace length s covered by the pulse rise time while propagating and half of that value is indicated as the threshold to establish if a line is electrically long and must be analyzed by means of transmission line theory. A justification based on practical reasoning about the effect of reflections is that for path lengths longer than $0.5s$, reflections from a mismatched load are received back at the source after the pulse has reached its maximum plateau value, and overlapping that occurs under these circumstances may lead to false triggering of a device. Conversely, for path lengths shorter than $0.5s$ reflections received back at the source arrive while the pulse is still in its transition time interval, that is not used for any logic function.

While in principle all changes of cross section of a planar transmission line are discontinuities, with this criterion many of them may be ruled out as negligible: all the time the hot trace is driven to a connector or device pin, there will be a change of cross section due to space constraints; in some cases grounded pins are located next to the target signal pin and a coplanar waveguide is normally used. However, these are rarely considered as relevant discontinuities, depending on speed and bandwidth.

Whenever a low characteristic impedance trace with a significant width is brought to the device pin/pad there will be a significant reduction to its size to fit pin size and pin spacing: there might be size constraints not only when joining the trace to the pad, but also to preserve clearance prescriptions from adjacent pins; device packages are getting smaller, going from the SO (Plastic Small Outline) SMD package ensuring 50 mils between the axis of adjacent pads with a pad size of about 20 mils, to the tiny TSSOP (Tiny Small SO) or QFP (Quad Flat Pack) with only 20 mils of distance, and the pad size is shrinking correspondingly, dropping to about 9 mils. It is evident that a $50\,\Omega$ line built on a two-layer board with 1.5 mm of substrate thickness is unsuitable with its 2 mm or so width of the hot trace, and going to a four-layer board with 0.4 mm thickness for each substrate layer is a must. Keeping the substrate dielectric permittivity large, i.e. around 4 to 6, helps in reducing trace width, besides confining a great deal of electromagnetic field in the substrate beneath the trace, rather than around and above it.

Routing traces to connectors, whether of SMD or through-hole type has similar problems, with step change of characteristic impedance when approaching the first external row of contacts, in case the target pin is on an internal row: when the application is a high-speed high-frequency one, signal pins are surrounded by grounded pins, so that a coplanar waveguide scheme may be used; changing separation s between hot trace and lateral traces, it is possible to correct the impedance value that increases while the width of the center trace is getting smaller (see sec. 6.3.3 on coplanar waveguides).

Other criteria may be set depending on the transmission line use. In [45], p. 301, setting a 10% threshold for noise and amplitude of reflections may be suitable for not too fast digital signals; for very fast ones, margins are tight and the threshold might be slightly lower (in sec. 6.4.3 a few % limit is set for crosstalk-induced disturbance); when the transmission line is used in a measurement setup that shall be accurate, by comparison to other sources of inaccuracy such as the VNA and calibration standard uncertainty, 1% is probably a good estimate for a challenging, but achievable, limit. Lossy lines will of course reduce the problem of reflections, exalting however those related to dispersion (see sec. 2.7.3) and introducing unacceptable signal attenuation.

Even in the presence of a carefully designed transmission line, relevant discontinuities that cause significant reflections are possible if the line is branched or capacitively loaded. Long branched lines, even if correctly terminated onto their own characteristic impedance, represent an unacceptable discontinuity at the point of branching (see sec. 2.7.2). If unavoidable, there are various techniques to cope with this and to reduce or totally eliminate the problem: series resistors are added improving matching to the downstream lines (what is done by a resistive power divider); if the branch line is quite short (e.g. the line is derived to connect to a nearby device), then it is a lumped load; any load that involves pads and pins shall be optimized balancing stray inductive and capacitive terms, trying to compensate by design between the different elements (e.g. vias inductance and capacitance, trace extra capacitance, pad capacitance, pin inductance and device input capacitance); with a large impedance termination (high input impedance device) the effect of stray capacitance will always be slightly exalted; any deviation from a straight line (e.g. bending and corners of transmission line routing) introduces parasitics, as seen in sec. 6.3.1.3).

Capacitive loading not all the times worsens line response: in some lucky (or carefully designed) cases some loading slows down line response, reducing steep edges and reflections. For a well-distributed capacitive load (e.g. when memory banks are placed side by side or in a traveling wave amplifiers tubes or transistors, and their capacitance, are distributed along the line) the resulting transmission line will behave like it has a lower characteristic impedance, summing the additional capacitance to its own distributed capacitance.

Inductive discontinuities may also be created by peculiar arrangements of unsuspected elements, such as a ground plane: cuts and gaps in ground planes caused e.g. by added traces, large vias for power components, etc. behave like inductive discontinuities, with a complex influence to nearby transmission lines; if unavoidable, small capacitors may be located at the two sides of the gap trying to compensate the inductive part, by diverting the reclosing currents otherwise taking a long path around the aperture [45], Fig. 8-36. Similarly, board areas where connectors are located may undergo discontinuities of ground planes.

Any short stub that is far from its first resonance frequency and thus is not changing dramatically the overall line response, represents in any case an inductive load that modifies slightly line impedance, the longer the stub the larger the inductance: examples are connector pins, component leads, vias, and even passive devices. It is known that connectors and packages are getting smaller and shorter as the frequency range is broadening, and that connectors specific for very high-frequency high-density boards are often of the "interposer" type; it is also advisable to use blind vias, shorter than through vias, cutting their length by a factor of three already in a four-layer board; of course, useless to say, smaller packages have less inductance, as for 0805, 0603 down to 0402 components, with a corresponding saving in board space, but some issues for soldering, mechanical robustness and interaction with coating and potting agents (0603 is a good compromise for most not-too-demanding designs). The effect of such discontinuities in terms of inductive loading may be estimated as follows: they may be in in parallel or in series to the line. Any inductive component will result in an increasing reactance for increasing frequency: for parallel loading, when the reactance is large enough it is self-excluding in the parallel combination with the line; for series connection, it will cause jumps in line voltage, in correspondence of the steepest fronts (some illustrative examples are in [45], sec. 8-18). The other adverse effect of added inductance in the signal path is additional delay, that might affect synchronization.

The small capacitance C_c needed to compensate inductive effects, placed before and/or after the inductive discontinuity (the total compensating capacitance being however C_c), may be determined so to create a π-cell, featuring the same impedance of the line characteristic impedance:

$$C_c = \frac{L_s}{Z_c^2} \tag{6.4.1}$$

This technique might work well for large inductive terms; when going to very small inductive terms for already optimized geometries and devices, the compensating capacitance value is so small that shall take into account the already existing stray capacitance terms due to the devices themselves and their pads. Only if needed, the compensation may be equally applied after a trial-and-error procedure.

6.4.2 Crosstalk

Crosstalk for PCB transmission lines may be treated with a similar theoretical approach as presented in chap. 8 and applied to cables (see sec. 5.1.6), that will not be repeated here. The attention is on available expressions that support the evaluation of this phenomenon specifically for planar transmission line, namely microstrips and striplines. Of course, a complete circuit formulation is always possible, including all self and mutual inductive and capacitive terms and putting it to work in a circuit simulator or in a custom multiconductor transmission line tool. It is also possible to revert to a full finite element method modeling, where geometries and material properties are all fed to the simulator. However, closed-form expressions have the added value of simplicity and immediacy.

6.4.2.1 IPC expressions

IPC, after describing the phenomenon of line reflections and focusing the attention on the highest frequency components related to edge transitions, deals with crosstalk for microstrips and stripline line structures. They propose a set of equations for the evaluation of what they call "forward" and "backward" crosstalk voltage, that in a more modern terminology is the near-end and far-end crosstalk (see sec. 8.3, and in particular sec. 8.3.4, for their definition and properties).

Microstrip Two forward and backward crosstalk coefficients, k_B and k_F, are considered, applied to the source voltage V:

$$k_B = \frac{V_B}{V} \qquad k_F = \frac{V_F t_r}{V\,L} \tag{6.4.2}$$

where

$$k_B = CZ\frac{k_L + k_C}{4\tau} \tag{6.4.3}$$

$$k_F = -0.5CZ(k_L - k_C) \tag{6.4.4}$$

$$k_L = 0.55\exp\left(-A_2\frac{s}{h} - B_2\frac{w}{h}\right) \qquad k_C = 0.55\exp\left(-A_1\frac{s}{h} - B_1\frac{w}{h}\right) \tag{6.4.5}$$

$$A_1 = 1 + 0.25\,\ln\frac{1+\varepsilon_r}{2} \qquad A_2 = 1 + 0.25\,\ln\frac{1+\mu_r}{2} \tag{6.4.6}$$

$$B_1 = 0.1\sqrt{1+\varepsilon_r} \qquad B_2 = 0.1\sqrt{1+\mu_r} \tag{6.4.7}$$

with the usual meaning of symbols related to the geometry: w is the hot-trace width, h is the substrate thickness and s is the separation of the two hot traces, measured between trace centers (so, differing from the one used to derive coupled microstrips expressions, that was taken between the internal edges of the two traces). Moreover, the two hot traces are supposed equal.

IPC does not say anything on the effect of terminations; the two microstrip lines are assumed correctly terminated onto their characteristic impedance, but it is not clarified how characteristic impedance is considered (each microstrip alone, most likely, or taking into account the effect of the other line).

Stripline When considering the same problem for a pair of striplines, IPC immediately says that the forward crosstalk coefficient (the far-end term) is ideally zero because $k_L = k_C$, while the backward crosstalk is twice that for microstrip. The assumed geometry is symmetric with same distance h of hot traces from the ground planes. Thus measuring far-end crosstalk for stripline geometries is a direct measure of asymmetry and non-ideality, as performed in [299]: FEXT curves are shown commenting the effects of fabrication tolerances, such as misregistration, detailed in the next section 6.4.3.

6.4.3 Practical realization and fabrication tolerances

In many cases transmission lines designed with the previously shown equations shall compromise with space constraints, presence of other interfering elements (such as discrete components of considerable size, connectors, screws) and in general economize with available layers. Design and routing of planar transmission lines is thus complicated by the concomitant exigencies of trace separation, connection to devices, avoid discontinuities that are relevant to signal integrity and ensure a solid reference ground, especially for high-speed digital signals.

We will see that shorting vias at tiny dimensions close to fabrication tolerances may be a disadvantage and they should be used sparingly. To avoid voltage bounces between planes and traces that are supposed to be at the same potential, shorting vias are needed in the most critical areas; voltage bounce is due to the coupling with a hot trace carrying a transitioning signal (e.g. a step-like signal) and inadequately large return impedance. The number of vias to be added and their distribution along the circuit cannot be accurately defined with formulas suited for all cases and geometries, so that only some general criteria can be given, and these criteria are those already known in the high-speed digital world: distance shall be much smaller than the shortest wavelength we expect, that for the steepest wavefront is given by the equivalent bandwidth; the advisable distance between the edge of the return lane or trace is not a unique exact number and in general an even distribution criterion is followed, with the annular ring of vias separated from the edge by at least two drill diameters; there is a small influence of the drill size on the via inductance, so that the latter should be chosen large, but not too much with respect to the other signal vias for practical reasons (let's say that anything between 20 and 30 mils should work).

At a coarse scale our routing will be such that i) the impedance seen by the signal is constant, so that there are no reflections and ii) the distance of the hot trace from nearby lines to be protected is maximized, unless the signal is distributed with a scheme like coplanar waveguide, that ensures the confinement of a large share of electromagnetic field. The first requisite translates in many requirements at a finer lower level, regarding the absence of mechanical discontinuities or soldering areas for devices extraneous to the signal net and that shall be skipped around: it is evident that breaking the return plane beneath the hot trace of a microstrip is not acceptable, and sometimes this happens because return planes are added at the end of the signal routing with the "copper pour" technique; this part of PCB design, thus, deserves some care in restoring plane portions large and continuous enough to be satisfactory for our purpose. This kind of things does not occur for PCBs designed from the start for signal integrity at the largest frequency, where this exigency is well known. Protecting weak signal lines is also necessary and some transmission line schemes may work better than others: a stripline or a coupled stripline is confined in all directions by ground planes and guard traces that may be added at a convenient distance not to alter its response, but this comes at the expense of some board space (three/four layers depending on the choice between side or broadside stripline). Microstrips are the preferred choice for board space and connection to top-layer devices with no additional vias. Microstrips are dispersive and tend to have their characteristic impedance changing when the frequency is increasing above some GHz; probably an optimized scheme with well controlled materials is able to work well up to a tens of GHz. Beyond these values and for long traces stripline should always be preferred because it is better controlled. A trace is to be judged long compared to a fraction of wavelength around one tenth or so (please see sec. 6.4.1 above for more detailed considerations); also discontinuities along the trace play a role, so please have a look at sec. 6.4.1 below (for further reading on typical PCB discontinuities and their effect see [45]).

There are differences between a PCB designed and built as a test setup, e.g. hosting one or few carefully arranged devices, and a populated board for a real application, containing several active and passive devices and a wide range of signals with complex routing and harmonization of extraneous elements, such as fixings and board-edge connectors. First of all space requirements are much more relaxed, except maybe for routing towards and inside the pin field of dense connectors; routing for the former has more freedom and may be optimized avoiding as much as possible sharp bends, closeness to other lines, discontinuity and floating elements, that may behave like resonators and in any case capacitively load the line. In particular vias and pads are responsible for worsening of crosstalk and common-to-differential mode conversion in dense boards at Gbit/s speed: pads and buried vias used in adjacent layers break the continuity of the separating ground planes and coupling symmetry, increasing crosstalk and C-D mode conversion (such elements are usually called "anti-pads"). For high-density high-speed applications, as noticed before, striplines are preferred to microstrips, and they are buried in internal layers, alternating one signal and one grounded layer, forming many coupled striplines (see sec. 6.3.5). Crosstalk occurs in both directions, i.e. horizontally, for adjacent striplines lying in the same layer, and vertically, for striplines located on stacked layers, separated by one ground layer. When approaching connectors or components, several buried and blind vias are used to make the striplines emerge to the top (or bottom) layer where this connector or component is located; the creation of so many anti-pads breaks the continuity of the ground planes, besides loading capacitively the adjacent striplines. In [81] the influence of fabrication tolerance relative to the position and unwanted shift of anti-pad position and consequential break in ground continuity was analyzed[29], resulting in 1.2-1.5% peak near-end crosstalk measured in time domain for the vertically coupled crosstalk, that is normally ruled out as not relevant, because of the good efficiency of the interposed ground layer. The authors underline that for a 1 V swing, this amounts to 12-15 mV, that is in the order of the voltage margin ("few tens of millivolt") that ensures the required very low bit error rate (BER) of 10^{-12}. This is quite commonplace with high-speed data bus operating at few or many Gbit/s rate: the required BER is quite low because of the huge amount of data exchanged over the link and to keep data-link availability high; the voltage swing is again low, guaranteeing the required speed and driving bus and data link capacitance with moderate current pulses; thus, the margins are tiny and the crosstalk is always a concern for the design, first, and whenever the implementation deviates from the design because of the already commented fabrication tolerances.

6.4.3.1 Sensitivity to discontinuities

The effect of discontinuities is simply that of making the surrounding ground plane discontinuous. For a microstrip compromising the continuity and homogeneity of the return plane affects seriously the distribution of current and of field between hot trace and return plane. This example was analyzed in [40], where two phenomena are considered: the simple effect of a discontinuity of the return conductor (circular offset hole) and the arrangement of such discontinuities in an array of evenly spaced holes on both sides of the hot trace. The latter is likely to cause heavy resonance effects, when tuning to a specific frequency. The holes are not conductive, but they have impact in creating a lack of copper in the return plane, besides removing dielectric substrate around the hot trace, thus modulating also the dielectric permittivity (this phenomenon was never considered and is quite difficult, or at least cumbersome, to treat analytically). The authors in [40] show that small amounts of discontinuity (i.e. small diameter holes, smaller than or similar to the hot trace width of 2.5 mm) create appreciable fluctuations and secondary reflections for increasing diameter, but do

[29] The authors assume a 6 mils worst-case misregistration, that confirms the values of 2 mils and 5 mils tested in [299].

not modify appreciably the average reflection coefficient of the microstrip that is the correct one for the given characteristic impedance (the nominal value is $52.5\,\Omega$ with a reflection with respect to $50\,\Omega$ of 2.44%). On the contrary, large holes around the trace (from 1.5 to 2.5 times the trace width) exacerbate fluctuations and increase the reflection coefficient of the microstrip (with the reflection coefficient going to about 5%, indicating an increased characteristic impedance to about $55.5\,\Omega$). Moreover, the periodicity is clearly visible in the oscillations, that in the S_{21} curve in the frequency domain create dips at about $6.9\,\mathrm{GHz}$ and $13.4\,\mathrm{GHz}$ of about 10 dB and more! Simulations have confirmed the experimental results quite accurately. The use of simulation tools allowed also to test configurations with hole on one side only or symmetric on both sides, and to change parameters such as trace-hole distance, including a hole underneath the trace; the results may be summarized as:

- reducing trace-hole separation increases losses (i.e. dips in S_{21} curve), more remarkably with the order of harmonic resonance, that is for values above going from $6.9\,\mathrm{GHz}$, to $13.4\,\mathrm{GHz}$ and to $20.2\,\mathrm{GHz}$; the increase is appreciable when separation is less than about 50% of trace width; the increase is steeper for higher harmonics;
- differences between one side and both side configurations appear but are slight variations of the amount of losses at the resonance frequencies;
- when holes occur also underneath the trace area, losses increase significantly, from about 4 dB for partial 50% superposition to 10 dB for full superposition.

6.4.3.2 Sensitivity to fabrication tolerances

The influence of tolerances and misalignment introduced during fabrication and assembly of PCB is quite complex and partly studied in some works. The considered planar line performances are those affecting signal transmission and that can be measured with a VNA: discontinuities that cause reflections and worsening of return loss, crosstalk with nearby lines (both near-end, NEXT, and far-end, FEXT, see sec. 8.3.4 and 10.4.3), common-to-differential (C-D) mode conversion (see sec. 10.4.2).

Vias on either side of the line shall be balanced, in terms of number and distance from the line, otherwise the line is not perfectly symmetric any longer and so the speed of propagation of odd and even modes will be different, and significant C-D conversion will take place. Analogously, misregistration may alter line symmetry, for example shifting two otherwise vertically aligned coupled striplines, resulting again among others in crosstalk and C-D conversion increase.

For striplines the far-end crosstalk is ideally zero, so that it's evaluation is a good indicator of any deviation and non-ideality. Tests performed by IBM [299] confirm that only 2 mils of misregistration may cause 8-12 dB increase of FEXT (due to C-D conversion) over a 100 mm line length up to 14 GHz, while above it the increase is even more remarkable; increasing misregistration to 5 mils brings in another 8-10 dB, reaching values of FEXT of about -20 dB @ $3\,\mathrm{GHz}$ and -14 dB above about 10 GHz. The amount of grounded vias of course has its own influence, so for the same configuration using two different percentages of grounded vias (S/G of 2:1 and

1:1, the latter bringing the number of grounded vias equal to the number of signal vias), when using S/G of 1:1 the previous curves are improved by nearly 3-4 dB. The similarity of the two phenomena (variation and dependency of FEXT and C-D conversion) suggests a common cause, that is however not considered in more detail in the paper. They discover also a minor dependency (2-3 dB) of crosstalk (while C-D conversion is almost unaffected) on whether the lines are straight or diagonal, worse for the latter, but without a clear explanation (they underline that diagonal lines are slightly longer, 71 against 68 mm, but the reason is to be sought in glass-weave effect, see sec. 6.1.2), for which the difference might depend on the way the problem is posed, e.g. spacing between vias using the diagonal ($\sqrt{2}$ times longer than the side).

6.5 Performance and test methods

Planar transmission lines have been reviewed for design and calculation and for the effects of tolerances and approximations; correct signal transmission and performance have been defined in terms of return and insertion loss, including reflections, crosstalk, common-to-differential mode conversion and eye diagrams. Material variability, fabrication tolerances and consequential geometry changes may be evaluated by simulation approaches, but the complexity of the problem and the many unknowns, possibly suffering incomplete information and characterization of input parameters, require experimental confirmation. To this aim in the last twenty years methods have been developed and improved that allow to assess performance of planar transmission lines, and indirectly to estimate properties of PCB materials; test coupons, test patterns, reference planar lines and connector arrangements are all conceived for reproducibility by users aiming at verifying their own designs, a fabrication process, a material, or all of them together.

Standards for determination of single properties and performance have already been reviewed: IPC-TM-650-2.5.5.1 [178], -2.5.5.10 [186] and -2.5.5.13 [189] for example deal with determination of permittivity and loss tangent of PCB laminates, proposing different methods to be mainly adopted by manufacturers of laminates and PCBs; the counterpart of IPC standards are IEC standards of the group 61188 and 61189, that are maybe at a less mature stage of development. With increasing bandwidth occupation and data rates, test protocols have been improved, reaching an agreement on geometry and probing points, and improving methods and techniques for new materials, extending frequency range and improving accuracy and consistency. The IPC-TM-650-2.5.5.12 [188] addresses signal loss of planar lines, proposing four methods, both time and frequency domain, supported by a considerable number of publications and that are reviewed in the following.

All methods that follow require test coupons and standard arrangements to hold probes and probing connectors and to verify accurately and completely PCB characteristics and planar transmission lines behavior. Besides the already reviewed land patterns for test probes (see sec. 6.2), boards for testing (often called "test vehicles" or "test cards") are manufactured with a certain number of standard transmission lines, with horizontal, vertical and diagonal arrangements (glass-weave effect is not

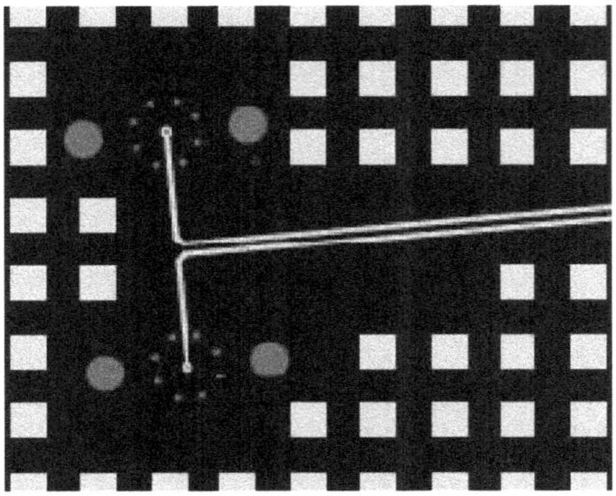

Figure 6.5.1 – Details of test boards: orthogonal short launching strips from coax connectors to planar transmission lines; lines are often diagonal and in zig zag fashion to avoid strong glass weave effect; the exagonal pattern of return pads for the coaxial launchers is clearly visible (two bolts, one on each side, are for Molex-like SMA, 3.5 mm, 2.92 mm or 2.4 mm coax PCB snap-in, or "compression", connectors [113]).

isotropic, see sec. 6.1.2), using arrangements and precautions to minimize unwanted effects. Examples are given in [105]:

- M.R. Harper, N.M. Ridler and M.J. Salter, "A Proposed Standard for Loss Measurements on Printed Circuit Boards," 3rd Annual Seminar on Passive RF and Microwave Components, 2012;
- S. Hinaga, M.Y. Koledintseva, P.K.R. Anmula and J.L. Drewniak, "Effect of Conductor Surface Roughness upon Measured Loss and Extracted Values of PCB Laminate Material Dissipation Factor," IPC APEX Expo 2009 Conference, Las Vegas, NV, USA, March 2009.

Some details are shown in Figure 6.5.1.

6.5.1 Signal loss and propagation (IPC-TM-650-2.5.5.12)

Publications have considered the five test methods in IPC-TM-650-2.5.5.12 [188], discussing merits and demerits and attempting improvements, and in some cases the publications themselves were the starting point for the development of the method, with the authors and their companies participating to IPC works. The case of the Short Pulse Propagation method reviewed below in sec. 6.5.1.3 is emblematic, since it has grown and gone farther than the others, by embedding redundant determination of dielectric properties, mathematical post-processing and support by simulation tools, etc. SET2DIL also is gaining a lot of interest for its simplicity, even if the literature on it is much less abundant.

6.5.1.1 Effective bandwidth (EBW) method

This method is a natural extension of a time domain transmission (TDT) technique. It simply applies a step with a specified rise time to an unterminated planar transmission line (the IPC standard says "conductor"); the degradation of rise time at the other end of the line is a measurement of several effects altogether: vias influence, skin effect, dielectric losses, etc. The method in reality uses the maximum slope of the rising edge as definition of rise time (see sec. 3.2.1 and 3.2.3), rather than other maybe more comfortable definitions, such as 10-90% or 20-80%. However, when specifying the minimum rise time, the standard makes equivalent a specification of maximum slope or rise time; in any case no prescriptions for limits are given, left to be "agreed upon between customer and vendor."

One point is underlined that is quite common to all TDR/TDT measurement methods: due to different locations of measuring points for TDR and TDT and possibly different probes, the rise time of the signal inevitably degrades and shall be compensated for. Rise-time compensation for probe-limited bandwidth may be achieved either by direct correction (see sec. 3.2.4 and 3.2.2), or by calibration and de-embedding (in the VNA sense as in sec. 11.7, or by means of a reference short line, such as a Through). An example of rise-time degradation is given in [107], Fig. 30 and 31.

The method (as declared in the standard) is not an absolute method; rather it determines a factor called EBW (equivalent bandwidth) to be used for comparison between similar or identical items or lots.

The length of the test line shall be greater than $2\,\mathrm{in}$; results for shorter lines are too exposed to disuniformity due to via coupling.

The selection/adjustment of the rise time determines the tested frequency interval; results are susceptible to noise and jitter, but this is in common to almost all oscilloscope-based measurements. Time jitter (expressed as rms) shall be less than $40\,\mathrm{ps}$, time base accuracy better than $2\,\mathrm{ps}$. Step aberration shall be better than 1% of step amplitude.

Again, agreement between customer and vendor is called, but a reference value of $250\,\mathrm{ps}$ is reported for 10-90% rise time. Correspondingly, a horizontal scale setting of the oscilloscope of $200\,\mathrm{ps/div}$ is recommended with 4000-sample recordings.

6.5.1.2 Root Impulse Energy (RIE) method

Also this method is an extension of TDT techniques: it applies a step to two separate unterminated open-ended planar transmission lines of different lengths, measuring a calculated impulse response, based on the gated derivative of the edges of the reflection response. The impulse responses are then integrated, obtaining an estimation of the energy of the two pulses, and then ratioed; the ratio is the RIE parameter.

The RIE gives as the EBW above an aggregate estimate of line losses. The two lines shall have quite a different length in order to correctly estimate line attenuation parameter α; the specified length values are $1\,\mathrm{in}$ for the shortest line (called reference line) and between $6\,\mathrm{in}$ and $12\,\mathrm{in}$ for the other line. If the board size does not allow the full prescribed length, foldback is accepted for striplines, provided that a distance of $1\,\mathrm{in}$ minimum is maintained; foldback is not acceptable for microstrip lines (the reason staying in the wider reach of the fringe field in the air region). The characteristic impedance of the two lines shall match within 5%.

The fundamental limitation of the method is that the derivative is heavily based on the high frequency content of the captured reflected signals, that in turn depends on the bandwidth and slew rate of the used equipment, namely the oscilloscope and voltage probes. Comparison between different systems and different test setups depends on the assumption of same bandwidth.

Moreover, since the derivative operation is particularly susceptible to noise, it is advisable to use a filter algorithm on the captured data before differentiation, that shall however not impact on the genuine high frequency content of the signal.

Time jitter (expressed as rms) shall be less than 40 ps, time-base accuracy better than 1% of the full range. Sampling time shall be ≤ 25 ps, that corresponds to 40 GS/s sampling rate, that is quite a demanding requisite, suitable for high-end oscilloscopes.

The requirement on step aberration is relaxed with respect to the EBW method: 5% of the step amplitude in the first 400 ps and 2% afterward.

6.5.1.3 Short Pulse Propagation (SPP) method

The SPP method requires that measurements are taken on two planar transmission lines of different length and as identical for the remaining characteristics as possible (e.g. cross section, dielectric homogeneity, etc.). The recommended combinations of length values are: a) 3.0 cm and 10.0 cm, preferred, but may allow some inhomogeneity of physical structures; b) shorter lines of 2.0 cm and 8.0 cm, suggested for thin boards thinner than 1 mm; c) longer lines, 5.0 cm and 15.24 cm, that allow easier measurements but require larger boards. We will assume that the longest line is called line 2, the other being line 1 [105].

The method recommends that for thick boards above 0.254 cm the length of vias and their effect as resonating stubs and lumped loading elements are reduced by back-drilling, use of blind and buried vias, micro-vias or top milling. The method requires also that the dielectric constant of the material is determined separately; to this aim three techniques are suggested: disc capacitance, short line and delay.

Dielectric permittivity: disk A disk of the recommended 12.7 mm diameter is specified; the diameter is required to be larger than 100 times the dielectric layer thickness to neglect fringe field effects. The disc will be measured at a conventionally low frequency (1 MHz is mandatory by standard) with a LCR meter; the electrical connection of the disk is by means of plated through hole (PTH) vias, and for the correction of the additional capacitance of the short trace length connecting them to the disk and to the reference plane above or below it, an additional dummy PTH/trace structure of the same design and size is required for a separate measurement that determines the correction capacitance value to be subtracted. The dielectric permittivity is determined by means of the parallel planes capacitor.

Dielectric permittivity: short line A short line is used as an equivalent capacitor: the measurement is made at low frequency using the same LCR meter and the line inductive behavior may be neglected, especially if the line is short; of course too a short line will result in a too small capacitance and correspondingly increasing uncertainty. The measurement is supported by modeling with a 2-D field solver: the dielectric permittivity value is adjusted until the calculated and measured capacitance values match. The advantage of this technique is not only that requires less board space, but that it is more accurate in measuring board inhomogeneity in relationship to glass weave, its more or less even distribution, and resin rich regions of the board.

Dielectric permittivity: line delay As a byproduct of the measurement of the line propagation delay τ, determined without the effect of launchers, pads, etc. because it is obtained by subtraction of two delay values for lines of different lengths, the relative permittivity may be determined by simply estimating the propagation speed. The standard reports a quantity, that is the per-unit-length delay that we indicate with τ': $\sqrt{\varepsilon_r} = c\tau'$, where c is the speed of light in vacuum.

Determination of line delay Two sets of elements affect line delay: board elements such as launchers, vias, pads, etc. may modify line propagation and increase line delay; the measurement system and its limited bandwidth increase the rise time and also the resulting time delay by widening the test pulse. The former effect is the most relevant and is ruled out by determining line delay as subtraction of two identical lines of different length: $2.0\,\mathrm{cm}$ and $8.0\,\mathrm{cm}$ lines are suggested; the normalized per-unit-length delay is $\tau' = (\tau_2 - \tau_1)/(l_2 - l_1)$ with obvious meaning of symbols. Regarding the determination of the influence of system rise time, the influence of the measurement channel and probe is that of adding their own rise time degradation, that at a first approximation is responsible of an increase of the measured delay. However, being the two delays τ_1 and τ_2 measured with the same probe and channel combination, and thus disappearing in the subtraction, it is not clear why the IPC standard indicates the rise time of the measuring system as necessary for line delay evaluation. A sensible hypothesis is just to check that the measurement system contribution is not excessive.

Determination of measurement system rise time The addition of a third line with negligible length (or "zero length line") allows to evaluate the rise time of the measurement system, including the probe and the receiving channel, but also the effect of one launcher. The standard suggests $0.25 - 0.45\,\mathrm{mm}$ length. This is clearly quite similar to a reference line for calibration, as the one used in SET2DIL (see sec. 6.5.1.4).

Determination of signal loss The measurement of two pulses propagating through a line pair is made and the resulting waveforms are time windowed, adjusting the overall time duration of the window, zero padded and Fourier transformed. Time windowing is a simple multiplication by a rectangular window whose time extension is long enough to contain both pulse waveforms, each one with its own delay, assuming that the time instant at which the pulse at the end of line 1 detaches from the zero axis is the initial reference time $t_0 = 0$. Then the tail of the second pulse at the end of the longest line 2 is considered, making sure that its front tail is well included and it is not cut too early: this defines the final time T. If the time recording of each channel for the two signals is not long enough to supply the required points for the time interval $[0, T]$, the two signals are zero padded. The standard recommends great caution in keeping the relative time distance between pulses unchanged while zero padding. The two signals are then Fourier transformed and the standard specifies a power of 2 number of samples, namely 8192 or 16384; the possibility of re-sampling to meet this requirement is also mentioned. In reality the standard specified that the signals are recorded with 512 or 1024 points, but modern systems have the possibility of extending memory depth from several tens or hundreds of thousands samples to millions of samples at the maximum sampling rate.

The two voltage $v_1(t)$ and $v_2(t)$ are thus transformed into $V_1(f)$ and $V_2(f)$, that the standard expresses as $V_1(f)=A_1(f)e^{-j\phi_1(f)}$ and $V_2(f)=A_2(f)e^{-j\phi_2(f)}$. The line propagation constant is then defined readily as

$$\gamma(f)=\alpha(f)+j\beta(f)=-\frac{1}{l_1-l_2}\ln\left(\frac{A_1(f)}{A_2(f)}\right)+j\frac{\phi_1(f)-\phi_2(f)}{l_1-l_2} \tag{6.5.1}$$

Frequency-dependent line parameters Regarding the identification by fitting of line parameters, including thus their frequency dependence, there is no definition of suitable code, only a few references to IBM tools and in general the sensation that any well-designed routine should in principle work, except for a few details that are quite relevant whenever fitting experimental curves:

- input data cleaning for small aberrations and outliers, e.g. using averaging or, better, improved smoothing routines or median filter;
- fitting to rationale polynomial functions to preserve physical meaning, including Debye model for capacitive terms;
- fitting criterion based for example on least mean square, maximum absolute deviation, or a combination of various criteria;
- wise use of weighting of portions of data samples, extremely useful to tailor fitting to curve characteristics, such as ensuring better accuracy around major resonances, balancing poorly represented parts of the frequency range, etc..

The Matlab® environment offers functions for smoothing (`smooth()` using weighted linear least squares and various polynomial degrees, also in robust version), fitting (`fit()` function that includes various models, using linear and quadratic polynomials, piecewise linear and cubic, spline curves) and linear and non-linear optimization methods to be used on user's supplied model functions (e.g. `fminsearch()`, `lscurvefit()`, `lsqnonlin()`).

Pros and cons of the method Limitations are first of all those of any TDR/TDT method: because of time-domain measurement, the signal-to-noise ratio is smaller than for frequency-domain equipment and increases slightly with frequency; conversely, the impact on accuracy for frequency-domain equipment is represented by calibration and de-embedding. Then, when measuring line with significant attenuation and dispersion, high frequency components may be too small for a satisfactory and accurate reconstruction of $\tan\delta$; however, a few data points are required to reconstruct the behavior at high frequency, when good materials have smooth variations of complex permittivity, that meets a Debye model. In general changes smaller than 5% can be detected for typical materials [105]. Accuracy in frequency domain is in any case improved by additional measurements to determine low-frequency resistance, dielectric capacitance and insertion loss, performed with frequency-domain equipment.

The strong side of SPP stays in the additional fitting of line parameters and model extraction, able to use mixed-mode time and frequency domain data, starting from initial values calculated using a 2-D field solver with built-in Debye function and causal models for the parameters (the overall model and some experimental results for new dielectric materials are shown and commented in [106]). This is also a weak side of the procedure, because it is not explicitly reported in the standard and steps and possible choices of parameters and convergence criteria are many.

In general, when compared to pure frequency-domain methods, the advantage of such time-domain method is that it rules out many non-idealities due to the launching and measurement system and that windowing can effectively remove the artifacts caused by discontinuities if not too strong, much better than calibration and de-embedding (see sec.) can do in frequency domain. By the way time gating (see sec. 11.9) was conceived for frequency domain equipment just to allow for discrimination of discontinuities (either selecting them or notching them out).

In [105] a general $\pm 6\%$ reproducibility was observed in measurements of the same sample at different times (including repeatability) and using different probes. For SPP however a more demanding specification of rise time is indicated: “11 to 35 ps or less”.

6.5.1.4 Single-Ended TDR to Differential Insertion Loss (SET2DIL) method

This method was proposed by Loyer and Kunze [226] in 2010: the measurement of the transmission mixed-mode differential S parameter S_{dd21} (see sec. 10.4) requires in general four probing points, two at the beginning of the planar transmission line and two at the end; reasoning with single-ended VNA measurements a four-port VNA is necessary. The SET2DIL method provides for measurements on differential structures using two-port single-ended measurements and enables the use of TDR instruments and oscilloscopes. As clarified in sec. 10.4, the relationship between differential and single-ended S parameters is $S_{dd21} = (S_{21} - S_{23} - S_{41} + S_{43})/2$. If the transmission line is symmetric ($S_{21} \simeq S_{43}$, $S_{41} \simeq S_{23}$), then only two parameters are necessary: $S_{dd21} \simeq S_{21} - S_{41}$. In reality, many details and practical issues limit this simplification and S parameters are only approximately equal: fabrication tolerances, solder mask imperfections and uneven loading by other PCB elements cause asymmetry. Loyer presented bidirectional SET2DIL three years later to cope with possible relevant asymmetries [227].

The relationship between single-ended and differential parameters is valid both in frequency and time domain, so that TDT techniques may be used as well: a step generator launches at the near end and two oscilloscope probes measure the single-ended voltages at the far end, minimizing the number of measuring points and using commonly available instrumentation.

The further simplification carried out by SET2DIL method is that launching and measurement are performed at the same terminals: the transmission line has no longer

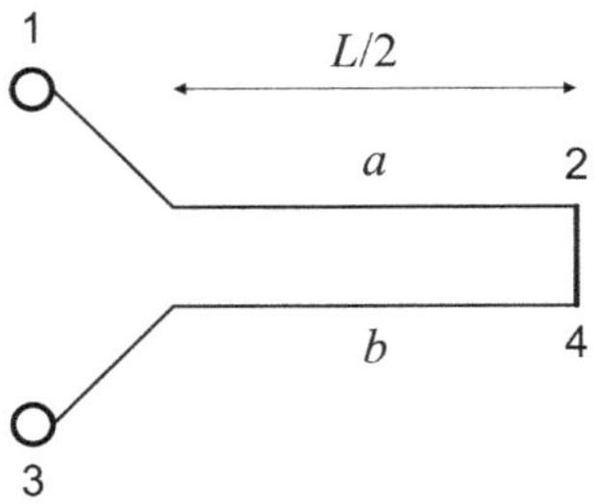

Figure 6.5.2 – SET2DIL setup: the short circuit at the far end is indicated as externally applied but in reality line a is bend to continue into line b.

physically separated near-end and far-end terminals; only near-end side is accessible and at the far end line conductors are shorted together (see Figure 6.5.2).

Being a coupled loop the SET2DIL test structure and related waveforms are not immediately clear: launching from terminal 1 NEXT and FEXT are immediately induced onto the adjacent line, each propagating in its own direction; however, also the launched signal propagates to the end and recloses onto the adjacent line thanks to the short circuit at the far end. For this other terms (reflections and NEXT) overlap to the desired waveforms, i.e. transmitted signal and FEXT. The method thus aims to correct for this before using the measured waveforms.

The procedure that manipulates the measured waveforms to progressively subtract the unwanted terms and responses is long, but uses quantities and parameters that may be easily determined by inspecting waveforms on oscilloscope. Measurements provide a preliminary q_0 waveform on the short "through" line (a sort of calibration that will be used at the end of the procedure). The waveforms measured at the two near-end terminals are called q_1 and q_3, with a one-to-one correspondence with port numbering (port 2 and 4 are the far-end ports that are shorted to create the loopback).

Applying the step voltage V between conductors a and b corresponds in reality to apply a differential signal $\pm V/2$, supported by a common mode voltage $+V/2$.

1. Raw measurements and initial situation:

    (a) $V_1(t)$ has an initial rising edge ($t = t_0$), reaches an upper level that is more or less constant and remains until $t = t_1$; $t_1 = T_d + t_0$, where T_d is the full propagation time along conductor a to the far end and back along conductor b to the other port, and t_0 is the initial delay that corresponds to the short time length of traces going from the connectors to the beginning of the planar transmission line and may be determined by measuring the through, that has same shape and same length;

    (b) at port 3 on conductor b the near-end crosstalk from conductor a, $V_{ne,b}^{(a)}$, is immediately present and is constant until time $t = t_1$ is reached; the rising edge of $V_3(t)$ at t_1 is much slower than the initial rising edge of $V_1(t)$, because there is the combination of concomitant signals;

(c) at $t=t_1$, $V_1(t)$ features the sharp drop due to a combination of far-end crosstalk induced terms: far-end crosstalk on conductor b, $V_{fe,b}^{(a)}$, progressively builds up while the signal is traveling towards the far end along conductor a; then it travels through the loopback and propagates back to port 1 along conductor a; the polarity of $V_{fe,b}^{(a)}$ is negative and it will subtract from $V_1(t)$ voltage applied by the source;

(d) at $t=t_1$, we have observed an increase of $V_3(t)$ because the step $V_1(t)$ has traveled along conductor a and back along conductor b reaching port 3;

(e) beyond t_1 there is another T_d time interval in which reflected portions of the signals that have reached port 1 and port 3 may be reflected on their way back;

2. Identification of time instants:

(a) the time instants used intuitively during the description of the initial situation are measured by locating specific points on the $V_1(t)$ waveform:

i. t_0 at the rising edge leaving the small launching initial section that corresponds to the Through;

ii. t_1 in relationship to the rising edge of $V_3(t)$, rather than the drop in $V_1(t)$;

iii. $t_2=2t_1-t_0=2T_d+t_0$ might not correspond to clearly identifiable characteristics of the waveforms;

3. Manipulation of $V_1(t)$:

(a) remove dc offset of $V_1(t)$ not zeroing it by subtracting the average over an undefined time interval, but subtracting the value in t_1, $V_1(t_1)$;

(b) replace with zeros the samples of $V_1(t)$ before t_1; transition should be smooth enough not to introduce erroneous high-frequency components;

(c) taking the value in t_2 $V_1(t_2)$ draw a straight line joining to a zero point at the end of the time axis;

(d) now $V_1(t)$ contains only the FEXT term;

4. Manipulation of $V_3(t)$:

(a) this waveform contains the propagated signal and thus information on the transmission; again remove dc offset by subtracting the value in t_1 $V_3(t_1)$;

(b) replace with zeros the sample of $V_3(t)$ before t_1; transition should be smooth enough not to introduce erroneous high-frequency components;

(c) taking the value in t_2 $V_3(t_2)$ draw a straight line joining to a point at the end of the time axis equal to the "initial value" (this expression is not further explained by it makes sense that it corresponds to the voltage value coming out of the launching short section, in all similar to the Through;

5. Final calculation of transmission:

(a) two manipulated waveforms, $V_1'(t)$ and $V_3'(t)$, are available that contain reflections to remove and transmission at the far end of the line, respectively;

(b) the desired time-domain transmission for the folded planar line is given by the difference $T_{dd21}(t) = V_3'(t) - V_1'(t)$;

(c) it is possible to obtain the equivalent frequency-domain transmission S_{dd21} by Fourier transformation, after conversion of step response into impulse response by differentiation.

The mentioned Bidirectional SET2DIL repeats the experiment by exchanging the polarity of the step voltage, so that we come up with two waveforms one for transmission from port 1 to port 3 along conductor a and then b (the one just calculated) and one for transmission from port 3 to port 1 along conductor b and then a. In this way it is possible to correctly measure and weight the two terms S_{23} and S_{41} that are indicated by Loyer as those not always equal in case of asymmetry.

The SET2DIL is probably quite diffused and used extensively because of the cheap setup and instrumentation: a published contribution that describes results regarding PCB insertion losses made at the same company, Intel, is Chu-tien Chia, R. Kunze, D. Boggs and M. Cromley, "A study of PCB insertion loss variation in manufacturing using a new low cost metrology," IPC APEX Expo Conference, San Diego, CA, USA, Feb. 28–March 1, 2012.

6.5.1.5 Frequency Domain (FD) method

Frequency Domain method is a catch-all definition for the typical VNA-based measurements, that are often criticized in this context for using the VNA itself (seen as a complex and expensive instrument) and the need of undeniably complex calibration procedures. It was however proposed [105] to adopt time gating (see sec. 11.9) to avoid calibration: this is not completely true, in the sense that the meaning of "calibration" in this paper covers the compensation of launchers and short launching line (as already mentioned above for SET2DIL), while VNA calibration is still needed. The second suggestion is to stop the processing of VNA frequency-domain data when they are anti-transformed to time domain, and then take benefit of all time-gating functions, smoothing and interpolation, line parameters fitting, etc. reviewed for SPP method (see sec. 6.5.1.3); in this case the overall method would benefit of the extremely large signal-to-noise ratio and dynamic range of native frequency domain measurement.

A pure frequency-domain method was proposed by Marks and Williams[30] to measure the propagation constant, that in turn allows the determination of the characteristic impedance, probably always at the expense of the limiting assumption of quasi-TEM propagation; the method was tested on a lossy coaxial cable and a coplanar waveguide, with much different size and behavior.

[30] R.B. Marks and D.F. Williams, "Characteristic impedance determination using propagation constant measurements," *IEEE Microwave Guided Wave Letters*, vol. 1, pp. 141–143, 1991.

An interesting approach for the determination of S parameters applied to the SET2DIL setup originally conceived for time domain was proposed by Pupalaikis and Doshi[31]. They started from the observed weaknesses of the original SET2DIL (the proposed processing of recorded waveforms is not strongly theoretically founded, and sensitivity and uncertainty are difficult to determine) and based their method on S-parameter modeling of the SET2DIL setup. The method claims that only a two-port VNA measurement is necessary on a transmission line structure like that of SET2DIL: the reason is that perfect symmetry and no mode conversion are assumed on the grounds that such line is purposely made for testing. Then a clever technique based on a signal flow graph representation in terms of reflection ρ and attenuation L (this is the notation of the original paper) is presented, that solves for two S parameters, S_{11} and S_{21}, and this is valid for even and odd modes, independently: $S_{11} = \frac{\rho(1-L^2)}{1-\rho^2 L^2}$ and $S_{21} = \frac{L(1-\rho^2)}{1-\rho^2 L^2}$. When transformed into a single-ended representation the total number of S parameters to measure is four. Fitting of solution is proposed in the least mean square sense, in order to reduce the effects of noise and increase accuracy at higher frequency. For determination of ρ, however, the proposed solution is the expected one: use of inverse DFT of recorded S_{11}, excluding initial transition and reflection from the far end, i.e. using time gating, during an initial calibration (that assumes that the far end is accessible and can be terminated as desired). Additionally, the usual amount of de-embedding for probes and setup elements is necessary, unless time gating or additional calibrations are used.

When the authors in [105] indicate the "possibility of introducing additional errors in the translations between time domain and FD" they probably overestimate the negative aspects, because frequency-to-time and time-to-frequency transformations are well documented, their non-ideality is well known (think of leakage in both domains, equivalence between time and frequency resolution, etc.). Additionally, the Debye model for capacitive terms and line-parameter fitting are applied in frequency domain, so, first, even the SPP method necessitates of a time-to-frequency transformations, and, second, such phenomena are better known and modeled in frequency domain, so that using directly frequency domain does not represent a disadvantage. It is true, however, that the VNA (see Chapter 11) is an additional expensive equipment to procure and that its calibration (in the proper meaning) requires great care in designing, selecting and using calibration standards and procedures (see sec. 11.5). Conversely, scientific literature is quite advanced and many contributions exist dealing with accuracy and uncertainty, optimized calibration procedures and verification methods.

6.5.2 Requirements from controlled impedance standards (IPC-2141 [190, 191], EN 61188-1-2 [76])

The often used "controlled impedance" expression has no real meaning unless the tolerance is explicitly stated; it is only a remark that the process is aware (or shall be

[31] P.J. Pupalaikis and K. Doshi, "A Fast and Inexpensive Method for PCB Trace Characterization in Production Environments," DesignCon 2010, Santa Clara, CA, USA, Jan. 28-31, 2013.

aware) of the problem of ensuring that the fabrication matches impedance requirements of the designer, but the tolerance and frequency range might be left unspecified. It is in general designer's responsibility to verify that correct requirements are communicated and that the adopted fabrication methods, tolerances, etc. are suitable. However, for not too demanding designs the routine checks during "controlled impedance fabrication" are more than enough.

The IPC 2141 [190, 191] (the IEC/CENELEC 61118-1-2 standard being identical to parts of the IPC-2141) report many considerations that regard good design, but we focus our attention on the few requirements regarding testing and predisposition to test printed boards. Generic recommendations such as not to reverse signal and ground connections, to avoid influence of the operator's fingers touching the coupon or to ensure good connections of probe tips are not dealt with and further considered.

What is important is that both standards leave undetermined the method to estimate the impedance from TDR time domain signals: the concept of impedance is typical of frequency domain, the methods of calculation require the use of time-to-frequency transforms, including tapering windows and possibly other techniques (such as interpolation and zero padding), fitting to rational functions is also a valid technique to reduce inconsistencies and ensure causality (as seen in sec. 6.5.1). So, leaving this part undefined ("should be agreed upon") brings to a wide spread of results and possible mistakes or inconsistencies, where the competence and experience of the parties play a fundamental role, unless they decide tout court to rely on the test methods reviewed in sec. 6.5.1 and well defined in other IPC standards.

Section 9.2 of IPC-2141 describes equipment characteristics necessary for effective measurements: TDR rise time and its ability to resolve lines of some length (e.g. $200\,\mathrm{ps}$, corresponding to $1.75\,\mathrm{GHz}$ of bandwidth, for $\geq 3\,\mathrm{in}$ length; $35\,\mathrm{ps}$, corresponding to $10\,\mathrm{GHz}$ of bandwidth, for $\geq 0.25\,\mathrm{in}$ length). Deviations from a perfect step signal shall be identified, such as flatness of the upper state and any aberration due to internal reflections, slew rate limitations and non-linearity; a maximum deviation of $\pm 3\%$ in amplitude is acceptable in a critical portion of the signal at about $1\,\mathrm{ns}$; fast aberrations are normally filtered out by cables and some averaging implicit in calculations.

Regarding calibration, it is recommended to have a reference coupon available with a known impedance; airlines are also mentioned and they are a recognized standard also for frequency domain equipment for both calibration and verification (see sec. 11.12). Airlines would be used for verification of mismatch and reflection terms of instrumentation and setup. By the way, there is no need of accurate calibration (see sec. 6.5.1), because reflection terms of interest are spotted out by windowing; the important thing being a good setup close enough to $50\,\Omega$ impedance[32].

Then, when measuring a coupon and the voltage at its terminals V, its impedance Z is determined as difference with respect to the measured voltage for the reference impedance Z_{ref} used for calibration:

[32] In sec. 5.1.1. of IPC-650-2.5.5.7 [184] the $50\,\Omega$ impedance is recommended for values of the characteristic impedance of the line under test between $30\,\Omega$ and $85\,\Omega$; the airline should have the same impedance of the line under test for even farther apart values.

$$Z = Z_{ref} \frac{V}{2V_{ref} - V} \tag{6.5.2}$$

The IPC-2141 (1996) says that differential lines are not so common, it mentions the possibility of testing each line of the pair separately, instead of a full differential test at once, but does not give any detail nor mention how to quantify performance, how even and odd modes may be considered and how common-to-differential and differential-to-common transformations take place.

Calibration is detailed in another IPC standard, IPC-650-2.5.5.7 [184], where they mention a "stored reference method" and an "*in situ* reference method". With the stored reference method the cable that is going to be connected to the test coupon is connected to the reference airline, that is in turn terminated on another good cable open circuited at the far end, so that the airline is terminated on the cable characteristic impedance (not said, but it shall be the same of the airline), with reflections from far end only after a significant time interval has elapsed, so not disturbing the readings for the calibration: this is step 1, "zero reflection reference voltage". Then, the ending cable is replaced by the probe that will be used for measurements and the voltage wave, called "incident wave voltage", is measured in step 2. The former represents the base of voltages (being in matched conditions), the latter the top voltage reached in open-circuit conditions through the probe used in the measurements; their difference is the total height spanned by the waveform and is called "incident voltage" V_{inc}.

The IPC-2141 was updated in 2004 [191] reporting a complete procedure for the characterization, or calibration, of the TDR setup (see Figure 6.5.3): as with the stored reference method, the measurement of a reference line is performed with the aim of characterizing first the line that connects the TDR to the PCB transmission line. The equations governing the determination of the desired transmission line characteristic impedance Z_{TL} based on the estimated characteristic impedance of the connecting line Z_L, based on the reference line impedance Z_{REF} are:

$$Z_L = \frac{V_{2,r} - V_{0,r}}{2V_{1,r} - V_{0,r} - V_{2,r}} Z_{REF} \qquad Z_{TL} = \frac{V_{2,tl} - V_{0,tl}}{2V_{1,tl} - V_{0,tl} - V_{2,tl}} Z_L \tag{6.5.3}$$

The voltage quantities are shown in Figure 6.5.3 with an idealized TDR waveform: the three voltage levels indicate the base voltage (V_0), the first step reflection from the line connecting the TDR to the PCB transmission line going up to voltage V_1 (the difference $V_{\text{inc}} = V_1 - V_0$ is the effective incident voltage), the second reflection is the desired one, occurring in the PCB transmission line and reaching the voltage level V_2 (this time the voltage step $V_{\text{refl}} = V_2 - V_1$ amounts to the reflected voltage by the PCB planar transmission line). The standard recommends to measure the desired Z_{TL}, using the central portion of the reflected voltage step, once the ringing following the rising edge has vanished and the trace is approximately flat. Regarding what is meant with "central portion" in a more quantitative way, the IPC-650-2.5.5.7 standard [184] indicates guard times of 15-25% with respect to the time instants where waveform

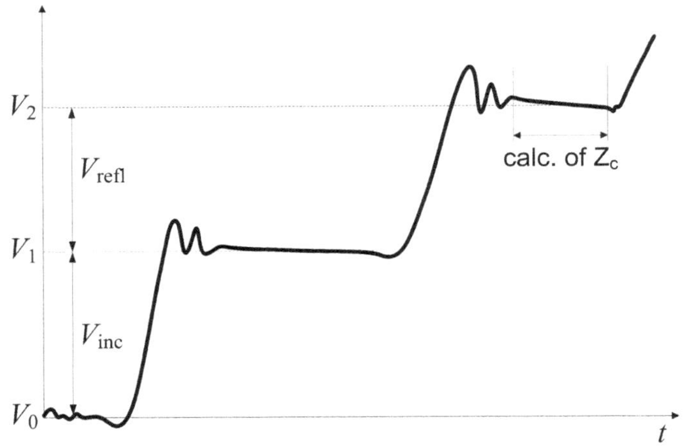

Figure 6.5.3 – Example of TDR waveshape and reference points for the determination of voltages and characteristic impedance (it is imagined that the slope is larger in the planar transmission line section due to larger attenuation and that the rise time increases as the step signals propagates).

edges are located; in the figure the time interval for determination of the characteristic impedance is identified as "calc. of Z_c".

There are a few issues with respect to the ideal case that deserve a comment:

- a voltage drop will occur along transmission lines due to their own attenuation (conduction loss), in particular for planar transmission lines; this will be seen as a downward slope, or droop, of the otherwise flat portion of the voltage step; for this reason it is advisable to use the very first portion of the waveform, soon after the initial ringing has finished or is tolerably low;

- other effects of losses, taking into account frequency dependency, may appear during rise time, increasing it by reducing the slope and making more confused waveform portions at the detachment around 10%; when around 90% ringing begins to appear; for this reason it is often prescribed to make estimates of time and bandwidth using the central portion of the rising edge (see sec. 3.2);

- moderate oscillations or "wiggle" may be seen in traces when excitation is a fast rising edge and PCB laminate suffers heavily of unevenness of dielectric constant, e.g. due to glass weave (see sec. 6.1.2 and 6.5.1.3);

- whereas time-domain techniques (e.g. TDR) are always preferable with respect to the identification and exclusion of unwanted reflections and influence of line discontinuities (e.g. launchers), it is in principle possible to transform to frequency domain (e.g. S parameters), in order to apply de-embedding (see sec. 11.7) of discontinuity, provided that its matrix representation may be somehow estimated separately (see sec. 11.8).

7

Connectors

If cables come in a wide range and variety for characteristics and performance, for connectors maybe the situation is even worse: the impact on signal integrity is smaller because of their reduced dimensions, mostly appreciable at higher frequency and thus there might be a general lack of complete characterization by some manufacturers for most low and medium frequency connectors, leaving such information only for high-performance high-frequency connectors, and often upon customer's request or in the presence of mandatory normative requirements.

To introduce the subject, just simplifying a little the range of cases and configurations, connectors represent the interface between cables and boards, where there might be connectors between cables, between boards located on panels, at board edge, or simply at the end of a cable splice. At some extent they influence the electromagnetic termination condition, notwithstanding that the overall behavior is mainly determined by the circuit that is connected at the other side of the interface. Connector influence may be broadly classified as:

- lumped capacitive or inductive loading, because of change of conductor separation and thickness;
- creation of short transmission lines with different characteristic impedance, due to different geometry and electrical characteristics;

- change of characteristic impedance due to line-to-line coupling occurring inside the connector and influence for the terms of odd- and even-mode characteristic impedance (see Chapter 8);
- delay skew, for multi-conductor lines, when signal routes have different lengths or undergo different inductive/capacitive loading inside the connector.

Several characteristics and problems may be extended to adapters and launchers.

7.1 Connector performance and characteristics

7.1.1 Coaxial connectors

The US Army-Navy RF Cable Coordinating Committee (ANRFCCC) was established in the early 1940s to develop electrical and mechanical standards for coaxial cables, connectors, and rigid high-frequency transmission lines for use with communications radios and radar systems.

The ANRFCCC introduced the Type N connector with its threaded coupling nut and air interface in 1942 to replace the old and no longer satisfactory UHF connector. The new connector was named after Paul Neill of Bell Laboratories, also a member of the ANRFCCC. A high-voltage version of the Type-N connector, called HN connector, was then released. It was followed by the Type C connector with a twist-lock coupling mechanism for rapid connection and disconnection; it was named after its inventor, Carl Concelman of Amphenol Corporation. The MIL-C-39012, was issued in 1964 to control the specifications of Type N connectors for military applications. Special N connectors for test and measurement are able to operate up to 18 GHz, but the standard N connectors as covered by MIL-C-39012 operate up to 11 GHz (in many situations, when calibration uncertainty and its traceability are of concern, the frequency range is reduced even more, to about 8-9 GHz, yielding in favor of smaller connectors).

Smaller coaxial connectors would have followed, including the bayonet-type BNC and threaded TNC connectors developed by Neill and Concelman.

At present, one of the more popular coaxial RF/microwave connectors is the subminiature Type A or SMA connector. It began its life as the Bendix real miniature (BRM) connector, designed by James Cheal of Bendix Research Laboratories in 1958 and eventually becoming the SMA connector, as a result of work by Omni-Spectra (now M/A-COM Technology Solutions), to convert it into their Omni-Spectra miniature (or OSM) connector. The SMA connector was originally intended for use with 0.141 in. diameter semi-rigid coaxial cable, with the center conductor of the cable serving as the center pin for the connector. It would later be modified for use with flexible cables, using center pins soldered to the cable center conductor. In 1968, the SMA connector was incorporated into the MIL-C-39012 specification, where it was given its current designation as a sub-miniature A, or SMA, connector. Its specifications may be found in the MIL-STD-348. Standard versions of SMA connectors can readily operate from DC to 18 GHz, with precision versions available up to 26 GHz.

The SMA was not conceived for frequent connect/disconnect, with a target number of about only 500-1000 operations; a joint venture between Hewlett-Packard and Amphenol in the early '70s came out with an APC (Amphenol Precision Connector), the APC 3.5 mm with air dielectric: it is compatible with the SMA connector, slightly more extended operating frequency and is able to withstand several thousands connect/disconnect cycles. Accurate specifications of all precision connectors of this kind can be seen in the IEEE Std. 287 [172]. The 3.5 mm was designed as a laboratory grade connector for SMA-equipped devices, so that the SMA-3.5 mm combination is not only allowed, but produces better results than a SMA-SMA connection.

Further, to extend the frequency range, the 2.92 mm connector, still mateable to the SMA, began to appear in the late '70s, first as a Maury project, then used by Weinschel and by Wiltron (now Anritsu), which was producing instruments up to 40 GHz, naming the connector as "K connector" (synonymous of 2.92 mm connector), indicating a K band of operation. In the end the SMA proved to be a valid workhorse for the mostly used types of precision connectors until nowadays.

Most connectors employ threaded coupling, including Type N and SMA connectors. Some connectors, such as SMB connectors, use a snap-on mating technique. One of the early, unique mating approaches is used in BNC connector, which has a typical frequency range of DC to a few GHz (depending on quality of surfaces, mating and mechanical precision) and can be manufactured for 50 or 75 Ω applications. Commonly found in low-power signal generators, oscilloscopes, and other test equipment, the BNC employs a bayonet-type retention collar that enables simple mating, but also provides a sound and repeatable electrical connection and helps prevent accidental disconnections, especially in high-vibration environments. The BNC features two bayonet lugs on the female connector, and quickly connects and disconnects with only a quarter turn of the coupling nut required for positive mating. Unfortunately, this coupling approach has its shortcomings at higher frequencies. The TNC connector is a threaded version of BNC for higher frequency use, beyond 4 GHz, and shares all interface dimensions with the exception of their coupling nuts and mating surfaces. For precision applications, such as the use in high frequency, high sampling rate, oscilloscopes, there exist versions that adopt a bayonet/threaded version of the locking mechanism, guaranteeing a better and firmer coupling.

Numerous stories claim to tell the origins of the three-letter abbreviation for this connector, including British Navy Connector and the Bayonet Node Connector. But, as yet another creation of Neill and Concelman, the most likely full name is the Bayonet Neil-Concelman connector. BNC connectors are suitable for use with cables ranging in size from RG-174/U to RG-213/U, including RG59/U coaxial cables common in old closed-circuit-television (CCTV) systems. Their specifications are covered by International Electrotechnical Commission (IEC) standard IEC 60169-8 and MIL-C-39012.

Antennas and replacements that must be tightly controlled according to Federal Communications Commission (FCC) requirements not to exceed a specified power level and/or operating bandwidth are subject to FCC Part 15 unique connector requirements. By incorporating reverse-polarity interfaces, where gendered center conductors are reversed, or by the use of reverse threading on the connectors, it is possible to ensure that components using Part 15 compliant interfaces or coupling will not mate with standard connectors and their components. The reverse-polarity/reverse-threading approaches are used with various coaxial connector types, including Type N, BNC, TNC, MMCX, SMB, and SMA connectors. We know that nowadays reversed adapters are easily available, making almost all connector types compatible with each other (and only making life more complicated for technicians who shall carry a huge range of adapters, ... just in case).

Other connectors that are increasingly popular especially for high-density applications are SMP and FMC[1], the former able of amazing performance with nominal frequency for some types up to 40 GHz, while the latter is used for high-density data transmission up to some GHz (nominal frequency 6 GHz). Both can be installed in tiny board space and have the so-called microstrip mounting, that is SMD technology with one flat pin on one of the four sides and four thicker pins at the four corners for mechanical stability and grounding.

High-frequency connector impedances are typically 50 or 75 Ω, in relationship to the characteristic impedance of connected cables; they were however used also for RG 62 cables in network applications (ARCnet, or ARCNET, operating at 2.5 Mbps), thus with a 93 Ω characteristic impedance, but at a moderate data rate. For the choice of the reference impedance values long time ago, please see at the end of sec. 2.2.

Gendered connectors can be coplanar (the mating reference plane is the same for the inner and the outer conductors) or non-coplanar, and are usually gender-matched in both inner and outer conductors, like Type N and SMA connectors. However, there are examples of sexless high performance connectors, such as APC-7 and some high-frequency board-to-board interposers. For a coplanar connector, the center and outer conductors mate in the same plane (such as SMA connectors). For non-coplanar connectors, the center conductors do not mate in the same plane as the outer conductors (such as Type N, BNC, and TNC connectors). Coaxial connectors can be designed with air or solid dielectrics, with e.g. Polytetrafluoroethylene (PTFE or Teflon®), Delrin®, Polyethylene (PE), Fluorinated Ethylene Propylene (FEP), and even ceramic materials, serving as common solid dielectric materials. Also for conductive elements different materials are available, such as brass or stainless-steel base material, with a variety of finishes, including gold, silver, nickel, white bronze and surface treatments.

7.1.1.1 Connector materials and plating

Materials for connectors can be evaluated in terms of their mechanical, electrical, and environmental properties, as well as how well the materials allow attachment to other materials, via soldering, crimping, or other process. Chosen materials should have good electrical conductivity, minimal resistance, good machinability, good stability over time, and good hardness to withstand repeated coupling cycles with minimal wear and performance deterioration. Many metals can suffer from surface corrosion, which can degrade electrical performance over time, so only appropriate metals are used. The choice of materials for a coaxial connector can greatly determine both reliability and electrical performance. Stainless steel, for example, is a steel alloy with

[1] A good range of SMP connectors can be found in the Rosemberg catalog "Communication products". For FMC connectors e.g. Hirose has a wide range of affordable models.

a small amount of chromium content. It is formulated not to rust or corrode when exposed to moisture. It is a hard material that is extremely durable and often used for connector housings, although not for contact parts because of its hardness and relatively low electrical conductivity. It is higher in cost than bronze or brass, but features high stability, high durability, and outstanding corrosion resistance for high reliability in a variety of operating environments.

Less durable, but lower in cost, brass is also used for connector housings, as well as for connector contacts. It is essentially a copper-zinc alloy that is easily machined, being considerably softer than stainless steel. It is an excellent conductor of both heat and electricity, and provides good durability in most industrial and marine environments because it is not subject to corrosion from contaminants found in those environments.

Connector parts are plated with different metals for various reasons, including: to improve electrical and thermal conductivity, to improve contact between conductors, and even to improve easiness to solder or weld a part. Noble metals, among them gold and silver, are excellent conductors, and resistant to corrosion, but they are expensive materials, so that thin layers are used on top of other metals when fabricating connectors. This makes it possible to take advantage of the electrical and thermal properties of the plated metal, while using as little of the material as possible[2].

Gold, for example, is an excellent conductor and it features very good oxidation resistance. It greatly enhances the electrical conductivity of connector parts made from copper or brass. But because of its high cost, gold is plated in thin layers, which can sometimes suffer from diffusion or wearing away of the gold finish. To minimize gold diffusion, nickel is used as an underplating beneath the gold layer. In many connectors gold and nickel substrate have normal thickness in the order of few to ten μm for gold plating and several tens to a hundred of μm for the underlying nickel support. Nickel and gold may be applied with different processes, that is chemically or mechanically, and using also electro-deposition. While the main problem with tiny gold is that it may wear out with use and thus has a limited number of guaranteed operations, the nickel substrate has the drawback of even modest outcrop appearing at the surface: nickel is much harder than gold and wears by friction the surrounding surfaces much faster, it has a poorer conductivity and is not as resistant to oxidation; all this results in shorter life for the connector and worse contact resistance. The commercial names for the various "plating" operations are "electrogold", "flash gold", "hard gold", "electroless nickel immersion gold" (or ENIG), etc., differentiating for the method of deposition of gold, but for most of them only for the thickness of nickel and gold layers, usually ranging between 30 and 100 μm the former, few to 15 μm the latter. Special production may increase gold thickness up to 30 − 35 μm (not thicker, except maybe few special works where heavy gold plating is specified). At tiny gold thickness the preparation of the substrate is very important to avoid porosity, with which the protection offered by gold is compromised. Rhodium was used as an alternative to gold, but rhodium is much harder and creates problems of damaging of the underlying base under heavy pressure and wearing out by friction the other metals with which it comes into contact. When using silver for underplating some problems of diffusion and contamination

[2] A.M. Fowler, "Radio Frequency Performance of Electroplated Finishes," *Proceedings of IREE Australia*, May 1970, pp. 148-164.

were found during the lifetime of contacts used for telecommunication applications, probably promoted by the electric field and current in the contact zone, but beginning to appear also in the early days of the connector, at its first installation[3]

Silver is also a fine electrical and thermal conductor, less expensive than gold; silver has a very attractive feature: its oxide is conductive and thus durability is prolonged also in heavy duty circumstances, such as high ampacity contacts. It also can be plated on materials like copper and brass to improve their electrical performance, can carry high current loads with very low loss, and is particularly good at minimizing PIM at high power levels (see sec. 7.4.3). But, as with gold, silver has its drawbacks, and its main disadvantage is its tendency to tarnish when exposed to some contaminants, including sulphur-based materials and ozone. Fortunately, the effects of tarnishing can be minimized by passivation: passivation can mean different things, but usually refers to a process that restores the protective oxide layer to the surface of a metal, making it more resistant to rust and corrosion. In general silver is used for high quality cable shields, for relay and switch contacts, but less frequently for connectors.

White bronze is an alloy composed of copper, tin and zinc (about 55-60% copper, 20-25% tin and 15-20% zinc) that has been proposed as a cheap alternative to plating: when exposed to harsh environments, white bronze does not discolor or tarnish; no oxidation builds up and contact resistance keeps quite constant, even at high temperatures or varying humidity levels. Plating is always exposed to the possibility of cracking and blistering if heavy temperature cycles are applied, due to unavoidable different thermal expansion coefficients[4]; white bronze has nearly the same thermal coefficient of brass, with which it is a good combination for the construction of non-expensive heavy duty radiofrequency connectors.

So far, plating has been considered an ideal process that deposits a smooth uniform film of "precious" metal, without focusing on the non-ideality of such surface. Depending on the type of process, on the base metal that receives the plating and on thickness, the plated surface may result in excess porosity, that worsens electrical conduction and makes it more exposed to environmental agents; different finishing are possible to smooth the surface, in particular by redistributing the material, such as buffing, burnishing, lapping. When the surface is cold-worked it is possible that some residual strain remains, depending on the type of plating, and successive heat treating may be necessary.

Designers should be aware that suppliers offer connectors for measurement and laboratory applications in three different grades: commercial grade (for production and general-purpose use on standard components), general precision grade (GPC, as used on instrumentation and test equipment), and laboratory precision grade (LPC, for calibration and measurement standards). For general applications we know that the shades are many and the offered range is very wide; when using connectors not purposely designed for measurement and laboratory use, a general assessment of metro-

[3] G. Kovacs, "Failure Analysis of Contaminated Gold-Plated Connector Contacts from Operating Communication Equipment," *IEEE Transactions on Components, Hybrids and Manufacturing Technology*, Vol. 5, no. 1, March 1982, pp. 95-101.

[4] Good plating with correct thickness and deposition technique is however quite robust and in general tested for extremely wide temperature changes.

logical performance is advisable, at least by evaluating performance (return and insertion loss, for instance), repeatability, susceptibility to vibrations and movement, and so on. However, there are always imponderable factors and phenomena that cannot be fully characterized on the spot, such as non-linearity and intermodulation effects (e.g. see sec. 7.4.3 at the end of this Chapter).

Any passive component that exhibits a non-linear behavior, including a simple coaxial transmission line, can contribute to distortion. For connectors and cable-connector pairs we must look to skin effect: we know that it depends on frequency, and also on electrical conductivity and magnetic permeability, so that a material susceptible to magnetism (or ferromagnetic material) can have a more pronounced skin effect. Connectors may contain metals that are not used for cables: if plating and finishing is made with nickel, that is durable, resists to wear, but is ferromagnetic, intermodulation distortion when mixing signals of relevant amplitude at different frequencies may occur; for this reason nickel-plated connectors are not accepted any longer by the administration and they were removed from the MIL-C-39012 standard. Since nickel is also used for PCB plating and finishing and in many commercial connectors, one may wonder what to do with this issue; a reasonable trade-off is that the use of nickel is limited in thickness to the amount necessary to support the final gold plating, accepting a minimum extra skin effect and non-linearity that for small PCB connectors is really a minor problem, while for bulky RF connectors used in power applications may be not. As observed later, adding a small percentage of Phosphorus to Nickel prevents the formation of magnetic domains and alleviates the skin effect problem.

7.1.1.2 Surface plating and electric behavior

Plating was already addressed above with some considerations on the electrochemical process and mechanical properties. Here the attention is focused on the electrical characteristics of plating and its performance versus frequency.

When plating some contamination may occur by diffusion of the metals at the interface: regions of metal alloys or mixture thus form and the resulting electrical conductivity is lower than the original metals; in case of copper adding small percentages of another metal may reduce remarkably electrical conductivity. For example adding 0.1% of Phosphorus a 50% reduction of conductivity is observed, while the same reduction is observed for nearly 0.2% of Iron; about 0.3% of Tin or Nickel reduces the conductivity of copper to 80%, while with the same amount of Aluminum about 70% is observed. In the following pure metals are assumed without contamination.

Let's consider two metal layers, one of considerable thickness with electrical conductivity σ_b and the other in the form of a deposited film of thickness t and electrical conductivity σ_p. Two considerations may be done regarding the effectiveness of the added plating layer: depending on its electrical conductivity and its thickness it will be effective in parallel to the metallic base starting from some high frequency because of skin effect; at high frequency conduction takes place in the thin layer whose electrical properties become relevant; at low frequency, on the contrary, bulk conduction occurs and the two metal regions work in parallel, with the largest current transported through the bulk base.

We recall that the skin depth δ depends on electrical conductivity $\sigma = (\sigma_b, \sigma_p)$, on magnetic permeability μ (assumed equal to vacuum permeability μ_0, since ferromagnetic materials are normally avoided) and on frequency f:

$$\delta = \frac{1}{\sqrt{\pi f \mu \sigma}} \tag{7.1.1}$$

Skin depth may be plotted versus frequency for some metals that are normally used in connector construction, retrieving the information on their conductivity from Table 1.1.1 in Chapter 1.

Under these assumptions, at sufficiently large frequency, it may be stated that almost all the current is confined in the plating layer and thus the location of the current elements for calculations is quite accurately known. For example conductor inductance may be accurately determined without uncertainty on the distribution of current at the frequencies where such inductance is relevant, e.g. for rise-time determination, interaction with stray capacitance and in general connector response.

However, with a more complex structure of deposited layers, with different thickness and electrical properties, the overall resistance for increasing frequency may be more complex. In a real situation, besides the effects of contamination reviewed above, also surface roughness and porosity should be considered, that at very high frequency create a sinuous path for the current, thus increasing the equivalent resistance and also inductance. If the metals have been hard-worked, then the deformed outer layer will exhibit a better conductivity than the bulk material.

An effect similar at a first glance to contamination was recognized in silver-plating treatment for cosmetic effect, adding a brightning agent: two to three times larger resistivity was observed. A similar problem of resistivity increase was reported in conjunction with different techniques of electro-deposition (with and without current reversal) and then solved by cold working, so that surface porosity in the end was identified as the real problem for the second issue, while for the use of brightning agents contamination seems most likely, since cold working is ineffective in reducing surface porosity [230].

In his work[5] Graham tests repeated connections and disconnections thus addressing fretting and wearing altogether. He does not point the finger at porosity, but demonstrates that too hard alloys with electrodeposited gold on cobalt is extremely subject to wearing and chemically active atmosphere, while soft gold overflash-deposited keeps almost intact after 25000 cycles (in addition, the gold layer was really thin, $0.08\,\mu\mathrm{m}$, deposited onto a supporting Palladium-Nickel of the same thickness, $0.8\,\mu\mathrm{m}$); contact resistance remained nearly unchanged around $4\,\mathrm{m}\Omega$. No information, however, is given for ac and radiofrequency behavior.

Finishing is another aspect of the same problem: in [230] a variability of 0.1% of characteristic impedance of an airline is reported while changing the surface finishing from a $2\,\mu\mathrm{m}$ finished surface to a 10 times better finishing. The authors underline the

[5] A.H. Graham, "Wear Resistance Characterization for Plated Connectors," *IEEE Transactions on Components, Hybrids and Manufacturing Technology*, Vol. 8, no. 1, March 1985, pp. 142-147.

relevance with respect to an equivalent change of diameters and they admit that no theoretical explanation is available.

Moreover, plating thickness is a badly controlled quantity, depending in particular to the used plating method. In general, in precision connectors, where dimensional accuracy is quite high, this is the first source of uncertainty due to construction.

When plating is strongly necessary for protection, e.g. for mechanical hardness or environmental conditions, or simply to improve the appearance of the device, then a viable solution is choosing a low-conductivity plating that does not alter the designed performance at radiofrequency and microwaves (this is of course possible if the plated surface is not part of an electrical contact): Tin-Nickel, Copper-Tin-Zinc, Chromium are all possible choices. Nickel, when Phosphorus is added, increases its resistivity remarkably, and above about 8% of Phosphorus loses its magnetic properties, with benefit for skin effect and passive intermodulation effect (see sec. 7.4.3).

7.1.1.3 Connector geometry optimization

Connectors may have quite different applications: power, radiofrequency and microwaves, fast digital signals, weak signals, or a combination thereof, with conflicting constraints and exigencies. Thus, when designing a new connector, its optimization is a lengthy iterative process that copes with different requirements: connector density in terms of pin displacement and arrangement, remaining thickness of insulating material and electrical insulation, mating and electrical contact resistance, current carrying capability, mechanical properties, insertion and retention forces, fastening system and required space, etc..

It is not possible to include all these elements thoroughly and comprehensively, being the subject vast and articulated, and in many cases knowledge is spread among few companies manufacturing high performance connectors and even optimal choices may be subject to designer's experience and preferences, as well as on company's philosophy and design margins, harmonization with other products, customers, etc.. Methods, guidelines and supporting tools are various, as various are the involved disciplines and the required background.

General considerations are equally applicable to multi-pin connectors, while the following is focused on coaxial connectors. Step changes of coaxial structures are the fundamental reason of connector mismatching and are subject to optimization, in the attempt of improving the useful frequency range and return loss.

Whinnery, Jamieson and Robbins (1944) A pioneering work that is quite complete and of broad application to planar and coaxial structures is that written by Whinnery, Jamieson and Robbins [337]: although original formulations are given for parallel-plane transmission lines, instructions are given also for application to coaxial lines, whose analysis was completed in a later paper in the same year [338]; the experimental verification of theoretical expressions is performed on coaxial lines. The authors recognize a previous work by Hahn[6] acknowledged for the inspiration and the original ideas. Reference cases are considered of step discontinuities and the corresponding equivalent circuit showing the lumped capacitance term. They can be combined for more complex geometries without worrying too much for the interaction between individual discontinuities and related field distortion and errors; limits of validity are however given, including also the effect of too a wide spacing between conductors and the distance of the line termination.

Corrections are needed when spacing between conductors (i.e. distance between parallel planes or between the inner and outer cylinder of a coaxial line) is approaching half wavelength, at which resonance occurs; a frequency-dependent factor $F = Y_d/\omega C_d \geq 1$ is given with C_d indicating the extra capacitance for the discontinuity calculated under low-frequency assumption. Up to a spacing of 0.1λ the value of F is nearly unity, going to about 1.08-1.10 at 0.2λ, 1.2-1.28 at 0.3λ and 1.5-1.7 at 0.4λ (see [338], Fig. 13).

Regarding the relevance and mutual influence of discontinuities depending on their type and separation, the authors divide them into two classes, as shown in Figure 7.1.1: for those on the left, field lines in the "constricted portion" are straight enough so that the two sides are "shielded" one from the other and there is no significant mutual influence; for those on the right the admittance equivalent to one discontinuity is decreased by the presence of the other, and remarkably, if the spacing between discontinuities is much less than twice the diameter.

Ragan (1948) Ragan [278], page 186, presents a comprehensive state of the art of coaxial discontinuities and related expressions, yet, dating back to 60 years ago. Figure 7.1.2 reports the four most common discontinuities together with expressions that allow quite an accurate estimate of capacitance, based on the factor $C'_d(\alpha)$ corresponding to the curve shown on the right as a function of α.

Reported expressions for extra capacitance and underlying assumptions are valid when measurements are taken at a distance from the discontinuity which is large enough to ensure that the higher order fields have been attenuated to negligible values. The discontinuities shown in Figure 7.1.2 are all modeled as a shunt admittance, directly related to the calculated capacitance as $j\omega C_n$, $n = 1 \ldots 4$.

The shunt capacity created in the Type-N connector by the step change of diameter of inner and outer conductors was originally reduced (at the time of its design) by replacing polystyrene with PTFE, so passing from a 2.5 to 2.15 relative permittivity. Moreover, the inserted PTFE bead that keeps the inner pin centered has a length that is about a quarter wavelength at 3 GHz (about 8 mm), thus improving connector behavior in that frequency range by compensating the mentioned shunt capacitance. Such improvement allowed to introduce again some mechanical tolerance in the center pin and plug-and-jack mating mechanism, preventing the spreading of jack fingers. The introduction of the center pin in the hosting jack is always the major source of variability with wearing and aging of such connectors, as commented in sec. 7.4.2 later

[6] W.C. Hahn, "A new method for the calculation of cavity resonators," *Journal of Applied Physics*, vol. 12, pp. 62-68, Jan. 1941.

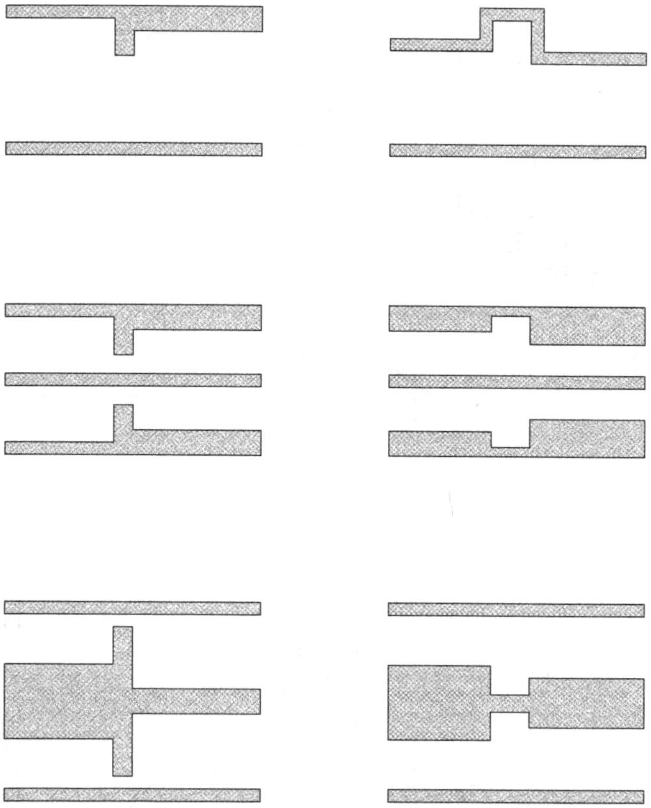

Figure 7.1.1 – Double discontinuities in which (left) there is no relevant reciprocal influence and (right) the mutual influence decreases each individual capacitive admittance (see [338], Fig. 4 and 5).

on: observed abrasion of slotted fingers was prevented by undercutting jack fingers, giving some mechanical tolerance when the plug is introducing.

MacKenzie and Sanderson (1966) MacKenzie and Sanderson [230] present nearly twenty years later a thorough revision of design principles for precision coaxial connectors: the focus is on more modern issues such as: i) the gap at the mating plane of slightly recessed inner conductors, necessary to ensure that mechanical stress or damage is not applied while moving and rotating connector body; ii) the rules to size the dielectric support — if any — and to keep its size and cross section homogeneous, minimizing impedance mismatch along connector; iii) contact mating and allowance for axial and radial tolerances, the former included in the just mentioned gap, the latter optimized and limited at best.

Among various characteristics at the interface, they describe well the advantages of sexless connectors, graphically showing a butt-joint example with optimized pattern of current flow and self-cleaning action through wiping of contact areas, that however has no impact such as wearing on critical geometry for microwave performance. They identify the tiny gap between the two facing inner contacts as the only impedance dis-

(a) $C_1 = 2\pi r_1 k_{eB} C_d'(\alpha); \alpha = a_1/b_1$

(b) $C_2 = 2\pi r_2 k_{eB} C_d'(\alpha); \alpha = a_2/b_2$

(c) $C_3 = 2\pi r_1 k_{eA} C_d'(\alpha); \alpha = a_1/b_1$
 Total capacity is $C_c = C_2 + C_3$

(d) $C_d = \dfrac{C_1 C_2}{C_1 + C_2}$ (See text)

Discontinuity capacity, $C_d'(\alpha)$ $\mu\mu fd$/cm

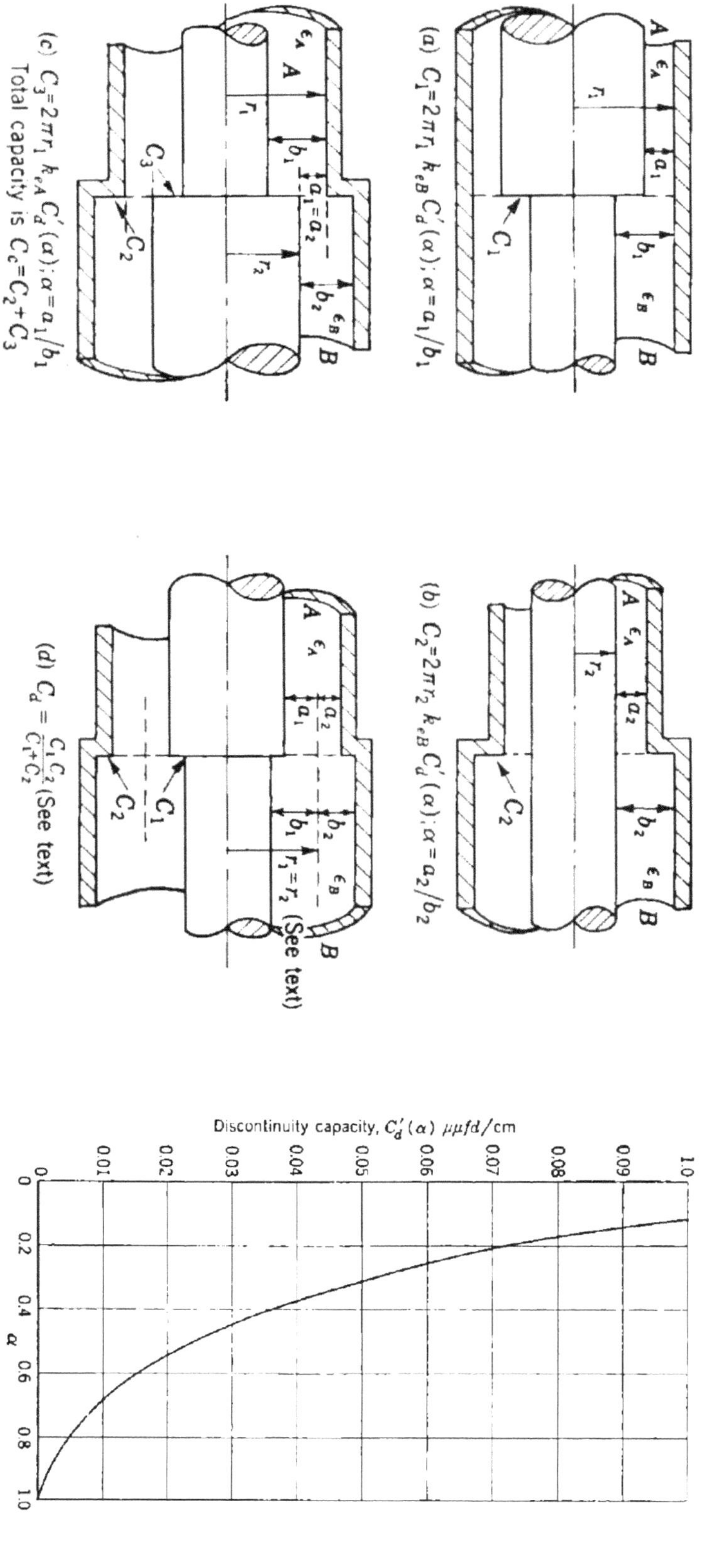

Figure 7.1.2 – Step discontinuities of coaxial structures and expressions for the estimate of the capacitance [278].

continuity, ruling out significance for its reduced size (1-3 mils of thickness, 10 mils of width). We might go ahead, remembering that a slight gap by recessing inner contact plane is indeed a desirable feature to "linearize" connector response and compensate for unavoidable minimal fabrication tolerances, relevant in the tens of GHz range, as pointed out by Wong and Hoffman [341] in sec. 7.1.1.7.

In their work they treat also the effect of inner conductor slots and then their effect of interface gap introduced to correct the discontinuity at the inner conductor junction.

The change of characteristic impedance due to inner slots effect is

$$\delta Z_c = 12.5 N_s \left(\frac{w}{d}\right)^2 \tag{7.1.2}$$

where N_s is the number of slots, w is the slot width and d is the inner conductor diameter. The same is proposed also for slots in the outer conductor.

This formula was tested by the authors using a modified airline (four slots of relevant size purposely created) with an impressive accuracy, down to the measurement of accuracy of 0.05%.

A change of inner or outer conductor diameter is proposed to compensate for slots effect:

$$\Delta d = +104 N_s \frac{w^2}{d} \qquad \Delta D = -104 N_s \frac{W^2}{D} \tag{7.1.3}$$

where the quantities are in inches and the result is in mils.

The effect of the gap introduced in the vast majority of connectors to compensate for inner junction inaccuracies (and slots influence) is also expressed in terms of SWR:

$$\mathrm{SWR}_{\mathrm{gap,i}} = 0.064 f g \ln\left(\frac{\pi d - N_s w}{\pi d_g - N_s w}\right) \tag{7.1.4}$$

where g is the gap width and d_g is the inner conductor diameter in the gap region, and the $\mathrm{SWR}_{\mathrm{gap}}$ is expressed in %. Similarly, for a gap in the outer conductor we have

$$\mathrm{SWR}_{\mathrm{gap,o}} = 0.064 f g \ln\left(\frac{\pi D_g - N_s W}{\pi D - N_s W}\right) \tag{7.1.5}$$

with obvious meaning of quantities.

Lewandowski reports eq. (7.1.4) by calculating the equivalent impedance Z_g and adds a surface-impedance component due to skin effect in the gap walls and connector pin:

$$Z_g = \left[j\omega \frac{\mu_0}{2\pi} g + \frac{1+j}{\pi} \sqrt{\frac{\omega \mu_0}{2\sigma}} \right] \ln\left(\frac{\pi d - N_s w}{\pi d_g - N_s w}\right) \tag{7.1.6}$$

As a quick reference the axial width of the gap is about $10-15\,\mu\mathrm{m}$ for all precision coaxial connectors (N, 3.5 mm, 2.4 mm, 1.85 mm); their diameters are however different: $d = 3040$, 1520, 1042, $804\,\mu\mathrm{m}$ and $d_p = 1651$, 927, 511, $511\,\mu\mathrm{m}$, respectively.

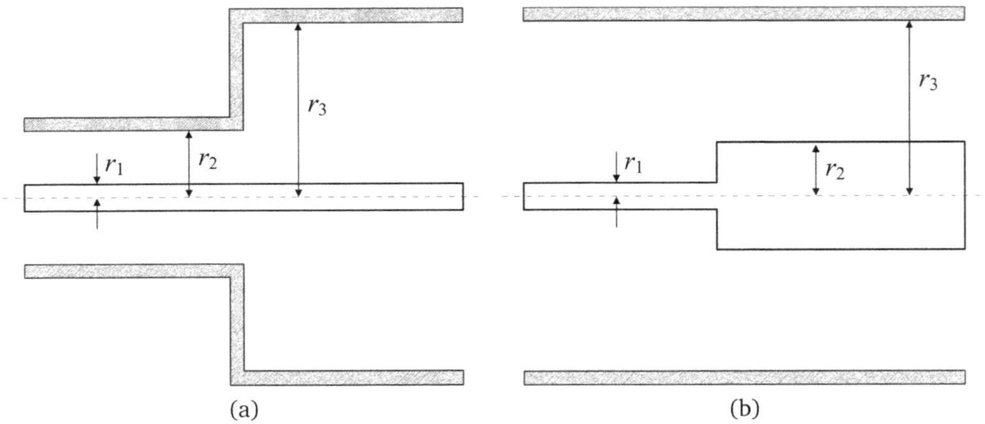

Figure 7.1.3 – Step discontinuities of coaxial structure: (a) step on inner, $\alpha=(r_3-r_2)/(r_3-r_1)$, $\tau=r_3/r_1$ and (b) step on outer conductors, $\alpha=(r_2-r_1)/(r_3-r_1)$, $\tau=r_3/r_1$.

Somlo (1969) In his work [306] Somlo reviews the two papers written by Whinnery, Jamieson and Robbins [337, 338], identifying generically the inaccuracy of the charts as about 5% (supported by his two references [4] and [5]), but recognizing the relevance of those works in 1944, remaining the only accurate reference for step discontinuities for more than twenty years. Somlo adds also a series of typing mistakes that if not corrected would make the expressions appearing in [338] useless.

Two simple and straightforward expressions are given then for two most common step discontinuities of a coaxial structure, the "step on inner" conductor and "step on outer" conductor, as shown in Figure 7.1.3, where the two normally used parameters α and τ are also defined.

$$C_{d,i} = \frac{\varepsilon}{100\pi}\left[\frac{\alpha^2+1}{\alpha}\ln\frac{1+\alpha}{1-\alpha}-2\ln\frac{4\alpha}{1-\alpha^2}\right]+1.11\,10^{-15}(1-\alpha)(\tau-1)\ \text{[F/cm]} \tag{7.1.7}$$

that produces a maximum error of $\pm0.3\,\text{fF/cm}$ for $0.01\le\alpha\le1$ and $1\le\tau\le6$.

$$C_{d,o} = \frac{\varepsilon}{100\pi}\left[\frac{\alpha^2+1}{\alpha}\ln\frac{1+\alpha}{1-\alpha}-2\ln\frac{4\alpha}{1-\alpha^2}\right]+4.12\,10^{-15}(0.8-\alpha)(\tau-1.4)\ \text{[F/cm]} \tag{7.1.8}$$

that produces a maximum error of $\pm0.6\,\text{fF/cm}$ for $0.01\le\alpha\le0.7$ and $1.5\le\tau\le6$. The first part of the two expressions was obtained by approximation of the values reported in [337], while the second term represents the perturbation determined under the assumption that the applied step, inner or outer, is small.

The author underlines that other methods exist for the determination of low frequency, or static, capacitance of step discontinuities, such as electrostatic problem formulation or interpolation of pre-calculated values for some parameter combinations[7]. However, the effect of frequency shall be accounted for separately by a correction factor K, that in turn depends on the frequency itself, α and τ, and on the type of step, inner or outer. The details of formulations do not appear in the paper and Somlo invites to request the computer program written in Fortran; only the curves appearing in his Fig. 3 remain, but Somlo gives indications of the steps in realizing the program, including details of mistakes identified in the reference work of Whinnery, Jamieson and Robbins, so that in principle it is possible to rewrite the software[8].

7.1.1.4 Sensitivity to geometry variations

Considerations and expressions derived for cables in sec. 5.3.2 are equally applicable to connectors with cylindrical symmetry. Regarding the effective impact on connector performance of fabrication tolerances we shall distinguish between normal connectors and connectors undergoing a tight quality control, up to the highest performance versions for high-end metrology.

The IEEE Std. 287 [172] reports tolerance of the inside diameter for the outer conductor for Laboratory Precision Connector, LPC, class and General Precision Connector, GPC, usually with two to four times larger tolerances than the LPC. Due to unavoidable limitations of tooling and finishing processes and related controls, tolerance is increasing for smaller connectors, so it is 0.1% and 0.2% for 7 mm connectors for LPC and GPC, respectively, 0.2% and 0.4% for 3.5 mm connectors, up to 0.6% and 1.2% for 1.00 mm connectors.

Wong[9] describes the dimensional verification that precision shorts, precision opens and airlines for primary metrology undergo at Agilent, ensuring uncertainties of 1.71 µm for center conductor diameter, 3.33 µm for outer conductor internal diameter, 1.73 µm for length, 1.00 µm for pin depth, 0.025 µm for surface finish, and 1.6 µm for roundness, using a variety of techniques, such as laser scanning, air gauging, etc.. In many cases, such as for shorts and opens, dimensional verification is also accompanied by finite element modeling, by which corrective coefficients for stray parameters may be estimated (see Figure 11.5.6) and then matched with measured scattering parameters (the sample results at the end of the paper show that agreement between model and measurement is in general within 1-2 dB up to 50 GHz).

What is not considered for high-end applications is the mechanical stress exerted by improper usage or due to cable bending , as seen in sec. 5.2.4, 11.3 and 11.12.

---

[7] Some interesting considerations on the electric field distribution at step discontinuities are given in: P. Silvester and I.A. Cermak, "Analysis of Coaxial Line Discontinuities by Boundary Relaxation," *IEEE Transactions on Microwave Theory and Techniques*, Vol. 17, no. 8, pp. 489-495, Aug. 1969.

[8] Routines like these are available in Agilent, formerly Eagleware, Genesys.

[9] K.H. Wong, "Characterization of Calibration Standards by Physical Measurements," Proc. of the 39th ARTFG Conference, vol. 21, June 1992, pp. 53-62.

	Type N	3.5 mm	2.4 mm	1.85 mm
D [mm]	7	3.5	2.4	1.85
d [mm]	3.04	1.52	1.042	0.804
ΔD [µm]	5	2.5	2.5	2.5
Δd [µm]	2.6	2	2	2
$\max\left(\frac{\Delta Z_c}{Z_c}\right)$ %	0.19	0.24	0.35	0.46
ΔS_{11} %	0.095	0.12	0.175	0.23
$\max\left(\frac{\Delta R_c}{R_c}\right)$ %	0.08	0.11	0.17	0.21
ΔS_{21} %/1 cm	0.0016	0.0018	0.0025	0.0035

Table 7.1.1 – Variation of return loss S_{11} and insertion loss S_{21} of coaxial connectors for common geometry and conductivity tolerance as per IEEE Std. 287 [172]. For connectors with an overall length of few cm maximum, the impact on S_{21} is negligible (S_{21} is expressed per 1 cm length); the main factor is conversely S_{11} that for the % values given above translates into −60.4 dB, −58.4 dB, −55.1 dB and −52.8 dB for the four connectors.

		Type N	3.5 mm	2.92 mm	2.4 mm	1.85 mm	1.0 mm
HP	Δd [µm]	±25.4/±50.8	±7.6/±12.7	±5/±12.7	±5/±12.7	±5/±12.7	±5/±12.7
	ΔS_{11} [dB]	−45/−39	−50/−46	−52/−44	−50/−42	−47/−40	−43/−34
NIST	Δd [µm]	±12.7	±7.6	±5	—	±5	—
	ΔS_{11} [dB]	−50	−49	−51	—	−46	—

Table 7.1.2 – Variation of return loss S_{11} of coaxial connectors for common geometry and conductivity tolerance as per IEEE Std. 287 [172].

7.1.1.5 Effect of geometry on connector reflection and transmission

Analogously to what was made for cables (see sec. 5.3.2.4), with the same model used for cylindrical structures and the same expressions for return loss and insertion loss variations, it is possible to estimate the effect on reflection and transmission of ordinary geometry and conductivity variations (see eq. (5.3.24) and (5.3.25)).

Assuming typical connector fabrication tolerances, as set forth by IEEE Std. 287 [172] for precision connectors, it is possible to estimate the effect on the final ΔS_{11} and ΔS_{21}, as shown in Table 7.1.1.

In the IEEE Std. 287 there are reported simulation results obtained at HP and NIST assuming oversized pin conditions at various extents and using two different simulation programs: the results are not all comparable because of different pin oversizing, different frequencies at which return loss is calculated and absence of explicit comparison and equivalence statement between the two sets of results. The standard reports results both in tabular and graphical form; the synthesis is shown in Table 7.1.2.

7.1.1.6 Connector usage

Regarding mechanical robustness and the influence of bending, movement and vibration, it is not possible to draw general rules and quantify the influence of each factor on cable electrical and transmissive properties. US military standards, for example, recommend not to bend a cable at a radius that is smaller than ten times its diameter (a 10 mm cable shall not be bended below a 10 cm radius). Results on induced amplitude-modulation and phase-modulation noise are presented later in sec. 7.4.3.

Often it is said that connectors shall be tightened to the prescribed torque; but what is this torque? Connector guides written by Skinner [304], Table D1, and Bailey [24], Table 1, report a general $1.1-1.4\,\mathrm{Nm}$ torque value for N connectors and $0.9-1.1\,\mathrm{Nm}$ for the air-filled smaller connectors, a lower torque of $0.5-0.6\,\mathrm{Nm}$ (officially $0.56\,\mathrm{Nm}$) for SMA (dielectric filled), down to about $0.3-0.35\,\mathrm{Nm}$ for 1 mm connectors. Values may change slightly between prescribed and recommended maximum torque.

In many manuals it is prescribed that connectors shall be clean and void of particles: the cleaning agent should be a low-residue one, such as isopropyl alcohol (aggressive solvents shall not be used in general not to damage dielectric); cloth or paper that release no fluff or lint is necessary. Compressed air is again very effective, especially initially to remove the easiest and largest particles; a low-pressure one should be used. When cleaning male connectors, the narrow portion of space around the pin is accessible e.g. using tooth-pickers or other small wooden sticks, wrapped with some cloth and used gently. About the use of lubricants to reduce wearing on connectors threads, there are opposite opinions: it is clear that lubricant residue attracts dust and may worsen slightly the contact resistance, but promote smooth turning and less troublesome insertion. Obvious, but it takes a short time to say, when mating two connectors the surfaces that are allowed to slide and fret are those of the external retaining thread, while the inner surfaces of male pin and female socket are those to preserve to minimize wearing: to this aim, please, avoid to turn both connectors; one should be kept in place, the other turned gently at the beginning when it is engaging onto the other, avoiding misalignment, shaking and bending movements. The torque wrench may be used at the end, beginning only using fingers; when applying torque to the moving connector just turned, the other shall be kept in place with a tool (spanner) to avoid friction when the inner surfaces are mating for the last μm. The reader is invited to look at the connector guide prepared at NPL [304], where a lot of high-quality pictures, advices on handling and references to previous publications may be found; the guide is quite all-comprehensive for the category of precision coaxial connectors and there are two appendices where repeatability is also considered (see sec. 7.4.2).

7.1.1.7 Evaluation/test of connector defects

Coaxial connector defects were already addressed when talking about geometry optimization and considering those details adversely affecting connector performance; so "defective" and "under-performing" are synonymous. Rather than reviewing once

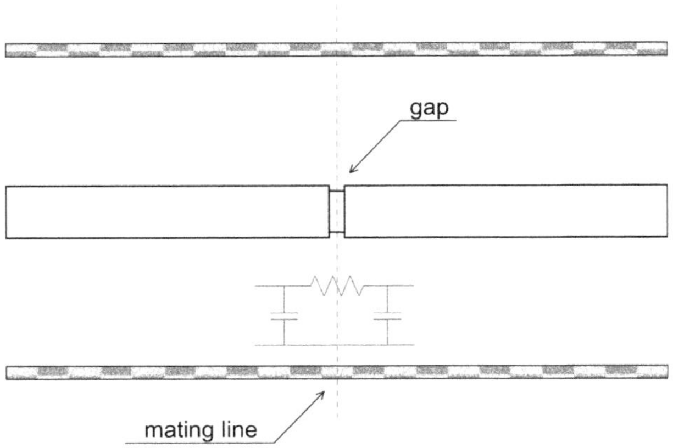

Figure 7.1.4 – Scheme of connector "defect" and its equivalent circuit.

more possible geometry and material defects, the attention is on how practically handle and verify connector defects. As already done in [249], a connector may be modeled as a lumped-parameter cell, usually with resistance and capacitance as preferred parameters, corresponding to the most likely effects of geometry deviations. In [95] a cable terminating in a possibly defective connector was tested simply and effectively with a short open-circuited line section.

The connector is modeled as a π cell with a longitudinal resistance R and shunt capacitance terms on both sides for a total capacitance $2C$: the normalized resistance r and admittance y are readily obtained by normalizing with respect to the line characteristic impedance Z_c.

$$r = \frac{R}{Z_c} \qquad y = j\omega C Z_c \tag{7.1.9}$$

It is possible to include series inductance $x = j\dfrac{2\pi f L}{Z_c}$ as proposed by Juroschek [202]. The S-parameter matrix representation of the π cell is symmetrical:

$$S_{11} = S_{22} = \frac{r}{2} + \frac{x}{2} - y \qquad S_{21} = S_{12} = 1 - \frac{r}{2} - \frac{x}{2} - y \tag{7.1.10}$$

The reflection coefficient at the point of insertion of the connector, where the defect is located, is given by

$$\Gamma = S_{11} + \frac{S_{12} S_{21} e^{-2\gamma l + j\phi}}{1 - S_{22} e^{-2\gamma l + j\phi}} \tag{7.1.11}$$

where γ is the propagation constant of the line between connector and open-circuit termination, l the length of the section, and ϕ characterizes the open circuit, that is assumed to have a unity reflection coefficient with phase ϕ ($e^{j\phi}$).

By replacing the expressions above into eq. (7.1.11), it is possible to obtain two expressions that were used by Juroschek and Daywitt: Juroschek was interested in evaluating the maximum impact of each connector parameter, together with the frequency and phase condition of reflection coefficient Γ, at which it occurs; Daywitt wanted to analyze the behavior of measured traces and to be able to evaluate connector parameters by looking at the ripple of magnitude and phase curves (to this aim his model neglects connector series inductance).

Juroschek obtained the following expression that includes the reflection coefficient of the airline $\Gamma_l = e^{-2\gamma l + j\phi}$,

$$\Gamma \simeq \Gamma_l + (1-\Gamma_l)^2 \frac{r}{2} + (1-\Gamma_l)^2 \frac{x}{2} - (1+\Gamma_l)^2 y \tag{7.1.12}$$

Daywitt separated the propagation constant γ in real and imaginary parts, α and β,

$$\Gamma = e^{-j2\beta l + j\phi} \{1 - 2\alpha l - r\,[1-\cos(2\beta l - \phi)] - 2y\,[1+\cos(2\beta l - \phi)]\} \tag{7.1.13}$$

$$|\Gamma| = 1 - 2\alpha l - r\,[1-\cos(2\beta l - \phi)] \tag{7.1.14}$$

$$\angle\Gamma = -2\beta l + \phi - 2\,|y|\,[1+\cos(2\beta l - \phi)] \tag{7.1.15}$$

showing that the magnitude depends on r and the phase on y. This eases the interpretation of results, allowing one to evaluate magnitude and phase curves separately.

Juroschek calculated derivatives of reflection coefficient with respect to r, x and y:

$$\frac{\partial\Gamma}{\partial r} = \frac{(1-\Gamma_l)^2}{2} \qquad \frac{\partial\Gamma}{\partial x} = \frac{(1-\Gamma_l)^2}{2} \qquad \frac{\partial\Gamma}{\partial y} = -(1+\Gamma_l)^2 \tag{7.1.16}$$

and neglecting airline attenuation ($\Gamma_l \simeq e^{-j\beta f}$), assumed terminated on a reflective load (e.g. open), he soon indicated that at $\beta f = \pi,\ 3\pi,\ \ldots$ current maxima occur

$$\frac{\partial\Gamma}{\partial r} = \frac{\partial\Gamma}{\partial x} = 2 \qquad \frac{\partial\Gamma}{\partial y} = 0 \tag{7.1.17}$$

while at $\beta f = 2\pi,\ 4\pi,\ \ldots$ current minima

$$\frac{\partial\Gamma}{\partial r} = \frac{\partial\Gamma}{\partial x} = 0 \qquad \frac{\partial\Gamma}{\partial y} = -4 \tag{7.1.18}$$

If the airline is on the contrary terminated on a matched load, then $\Gamma_l = 0$ and

$$\frac{\partial\Gamma}{\partial r} = \frac{\partial\Gamma}{\partial x} = \frac{1}{2} \qquad \frac{\partial\Gamma}{\partial y} = -1 \tag{7.1.19}$$

Juroschek's results for reflective load are in line with the conclusion in Daywitt's work, where one has to inspect simply the peak-to-peak ripple of the measured curve:

- for the magnitude curve, the peak-to-peak ripple gives $2\,|r|$, because the term $[1-\cos(2\beta l - \phi)]$ oscillates between -2 and $+2$;
- for the phase curve, the peak-to-peak ripple gives $4\,|y|$, again for the same reason.

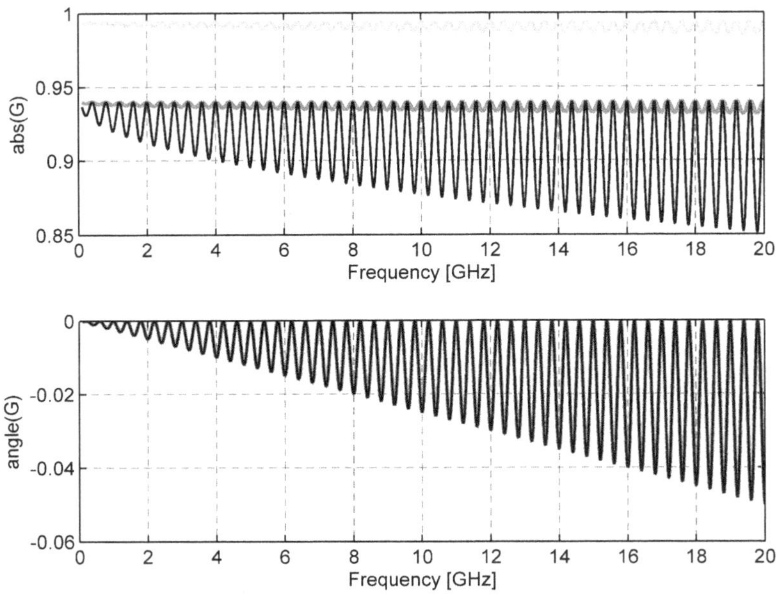

Figure 7.1.5 – Ripple curves of return loss Γ amplitude (above) and de-trended phase in radians (below): the upper light gray trace is obtained with a low-loss airline and low-resistance defect ($r = 0.1\%$); adding losses to the airline shifts downward the trace, but keeps the peak-to-peak ripple constant (dark gray trace); increasing the normalized resistance to 1% produces the more common black trace; the phase is unaltered because the normalized admittance (the defect capacitance) was unchanged and the three traces are overlapped. For the black curve the amplitude envelope shows the assumed $\sqrt{f}$ increase of defect resistance. The defect capacitance was set to $0.002\,\text{pF}$ as in [95].

In theory the peak-to-peak ripple of the return loss curve increases with frequency for both magnitude and phase. The magnitude curve has a downward slope typical of the return loss of a reflect termination with increasing losses and the ripple spread with frequency, being the resistance approximately proportional to the square root of frequency due to skin effect (it is not excluded that other phenomena take place, but this is the most significant one). For Γ phase the expression contains the term $-2\beta l + \phi$ that is not only useless for the determination of y, but also "counterproductive", giving a linear trend that impedes to appreciate small variations; the phase curve thus needs de-trending, as already commented for measurement of phase delay in sec. 3.3.1. If the line section used for the test is an assurance line, or any other kind of well-known well-characterized line, then the de-trending may be done simply by introducing an artificial delay in calculations to subtract the known phase constant term.

What is interesting in Juroschek's work [202] is that he presents several measured curves and calculates the resulting values of the joint parameters, giving the reader the possibility of cross-checking. Observing his curves, he underlines that the largest variability between repeated measurements occurs at current maxima, that is to say that resistive and inductive joint parameters are subject to the largest variability, than capacitance terms. Moreover, the height of these peaks at current maxima increases with frequency, due to skin effect for resistance r and/or to the linearly increasing inductive reactance x. A typical example that exaggerates cable bending and connector stress, where this behavior is easily visible, is shown in Figure 5.7.1.

Pucic and Daywitt[10] write down some additional considerations. They suggest a double reflect measurement (short and open), both featuring the usual ripple in a complementary, or specular, fashion, so that the sought curve is exactly in the middle and may be defined graphically or numerically. Second, they identify the necessity of correcting for losses in the used reflect terminations, although they underline that for good calibration-grade reflects this is hardly necessary for a wide frequency range; correction is however applied by shifting upward the estimated connector or adapter reflection curve by a fixed amount, that is the estimated reflect loss. The same, of course, is applicable to correct for airline.

Regarding the meaning and the extent of the defect shown above in Figure 7.1.4, it should be emphasized that in different contexts this term "defect" (or analogously "poor performance") may assume different meaning, from a coarse defect that might impair the significance of results to a quite "subtle" influence on connector repeatability or VNA performance, that might go unnoticed in many cases, but that Wong and Hoffman [341] point out when talking about the "gap" effect at connector interface: they observe that reducing the gap to "close to zero" has the negative effect of being more sensitive to slight unavoidable variations due to even tight fabrication tolerances (Fig. 7 in [341]). They apply thus a thin dielectric spacer to ensure a minimum pin gap whose capacitance takes rid of the smaller fluctuating term causing repeatability issues; the slight increase of total capacitance at the gap is of no real concern because it is absorbed and compensated during VNA calibration. The spread of successive calibrations measuring the same matched load has improved from about 10 dB to just 1; the reader might wonder how there are 10 dB of spread of assumed correct and "top of class" calibrations, but it is underlined that such spread regards a matched load return loss measurement in any case steadily below -40 dB around 10 GHz!

To underline how difficult is to find agreement on such minor discrepancies that are indeed quite difficult to track consistently and require always expensive equipment and consolidated knowledge and expertise, Lewandowski [222], page 99, citing other references[11] indicates as a plain fact that the gap at the pin interface is modeled as a series inductance, while we have seen proposed also a series capacitance model. Frequency values at which such evaluations take place are around 50-70 GHz, so even socket fingers and their shape have relevance; on the other hand any model may be in principle corrected for, provided that it has no singularities or range issues.

---

[10] S.P. Pucic and W.C. Daywitt, "Single-port technique for adaptor efficiency evaluation," 45th ARFTG Conference, Orlando, FL, USA, May 19, 1995, pp. 113-118. doi: 10.1109/ARFTG.1995.327114

[11] B.B. Szendrenyi, "Effects of pin depth in LPC 3.5 mm, 2.4 mm, and 1.0 mm connectors," Proc. IEEE MTT-S International Microwave Symposium, Vol. 3, June 11-16, 2000, pp. 1859-1862.
J.P. Hoffmann, P. Leuchtmann, and R. Vahldieck, "Pin gap investigations for the 1.85 mm coaxial connector," Proc. European Microwave Conference, Oct. 9-12, 2007, pp. 388-391.

7.1.2 Multi-pin connectors

Multi-pin connectors exist in much more different types, sizes, versions than coaxial connectors; broadly speaking, the considerations reported in previous sections for coaxial connectors are equally applicable to many, not to say all, multi-pin connectors:

- materials and their properties, especially with respect to surface and bulk conductivity and variability with frequency;
- surface treatment and finish, such as plating, passivation, etc. and again their impact on electrical performance.

Whilst they were once limited to low- and medium-frequency applications, e.g. slow digital applications in the '80s and early '90s, they are nowadays used as a valid replacement for coaxial connectors in many applications up to several GHz. The advantages with respect to coaxial connectors are much higher density, weight, easier PCB installation: higher board densities and lower profiles are thus possible, together with the possibility of building board stacks with interposer connectors; the advantage is not only in weight and size reduction, but also in better mechanical characteristics, as board assemblies are much more robust and rigid, and may be used in demanding applications, such as automotive, aerospace and military ones.

The old multi-pin connectors which we are familiar with are those for data connection outside a computerized equipment (e.g. sub-D, high-density SCSI connectors, etc.); they have progressively moved to higher density, supporting higher speed and throughput, but keeping them at a handy size and cost efficient (e.g. FireWire following the IEEE 1394 standard, USB connector moving from 1.1, 2.0 versions to 3.0, HDMI). Even higher densities and performance are attained with inter-board connectors, either soldered or plunger-equipped interposers: transmission of fast data rates of internal buses, as well as of radiofrequency signals, are possible.

While it is generally true that measuring instruments are interfaced using coaxial connectors, multi-pin connectors may be suitable e.g. when the number of tested ports is very large, when they equip an existing board, such as an evaluation board, or in general when required by the setup and DUTs.

Multi-pin connector performance may be assessed by specifying requirements and evaluating (similarly to coaxial connectors) input impedance mismatch and attenuation (by means of return and insertion loss), with several other additional parameters that are characteristic of differential pairs (mixed-mode scattering parameters, mode conversion, etc.) and signal transmission quality (crosstalk and eye-diagram degradation). In this respect multi-pin connectors are similar to planar transmission lines, when considering coupling and symmetry. Despite the length is much less, multi-pin connectors feature at least one of the following characteristics:

- connector pins are much closely packed than the traces of connected planar transmission lines and even more than multi-pair cables;
- all connectors cause some impedance discontinuity due to several reasons: bare difference of characteristic impedance, propagation of different higher modes, creation of reactive field at geometry discontinuities (e.g. edges and corners);

- there are more impedance changes due to changes of geometry and electrical parameters (pin-to-pin distance, pin diameter, characteristics and thickness of the supporting dielectric material, mating zone) than in homogeneous planar transmission lines or cables.

While attenuation is in any case limited by the short length, impedance discontinuity occurs at some longitudinal position, but necessitates a sufficiently high-frequency content of the applied signal to become relevant. This is of course related to the concept of impedance in time domain that is quite misleading and that to be expressed needs the definition of an equivalent frequency, or frequency interval. The same occurs with coaxial connectors, but multi-pin connectors have two distinct features: they do not have circular symmetry and allow coupling between lines and pins; as a consequence, mode conversion may occur and more complex representations are necessary (e.g. mixed-mode S parameters, generalized modal S parameters, coupling and crosstalk, deviation from TEM or quasi-TEM propagation and higher modes).

Pin-to-pin coupling depends not only on the local geometry and parameters, but extends also to nearby pins and changes also with location (connector center, at its periphery near the edge, etc.) and with the way pins are driven and the assigned potential (e.g. grounded pins, low or high impedance sources, etc.).

Because of the larger pin density, also board pad patterns have become more complex, inaccessible to through-hole technology and requiring multi-layer boards for the complete routing. Many patterns for hot and return pins may be conceived: the degrees of freedom in pin positioning depend on pin size and grid pattern, that may be quite different for power connectors, signal connectors and hybrid connectors (mixing power and signal in the same connector). Circular and flat pins, different socket geometry and mating mechanisms, the presence of springs, plungers, slotted contacts, etc., characterize each pin and its geometry. Pin patterns define how pins are assigned and deployed inside connector area (e.g. aligned rows, staggered rows, grouped, etc.). For high-frequency, large data rate and high signal integrity applications, coaxial-like structures are repeated inside multi-pin connectors in different ways: several grounded pins surround the signal pins mimicking a circular coaxial shield, or the connector is subdivided into "islands" that are shielded by additional grounded metallic elements (sometimes called "ground blades"), or custom coaxial lines are embedded within the connector. Shielding may be implemented with individual shields, pin arrays or the external connector enclosure; much depends on the application and connector installation, e.g. between sandwiched boards, between distant boards, inside the equipment, bolted to its metallic chassis and enclosure. Different solutions exist to ensure electrical continuity for signal integrity and electromagnetic interference (EMI), all addressing the problem of the integrity of the overall electrical shielding in terms of shielding effectiveness and transfer impedance. Ensuring a low-impedance circumferential electrical bond between cable shield, connector shell and equipment enclosure is definitely a complex problem where several factors come into play:

- for cables using screens of metalized plastic (e.g. Aluminum on Mylar) the circumferential electrical contact onto the cable gland and connector shell is troublesome, being impossible to lean solely on the drain wire;

- for many commercial products the design faces the exigency of a light connector, robust and able to withstand several mating cycles and years of service, and yet satisfactorily fulfilling EMI requirements;
- mechanical contact between plug and socket transmitting the electrical continuity of bonded shields, screens and grounded parts cannot rely on complicated solutions: at the separable interface, detents are used to guarantee contact at several points around the periphery, while for connector body in contact with the equipment enclosure washers and gripping profiles are quite satisfactory.

The assessment of EMC performance of connectors and cable-connector assemblies is covered by standards prescribing testing of shielding effectiveness and transfer impedance, using test methods as in sec. 5.1.7 for cables. A thorough discussion of test methods for board, backplane and cable connectors can be found in [241].

Since density has grown and multiple rows of SMD pins shall be accommodated for, also assembly and production testing are more complex: while with through-hole technology optical inspection of the board was possible and gave satisfactory results, multilayer boards using surface mount technology require a new set of inspection methods. X-Ray inspection is often used, trimmed to scan pads and solder, but also automated electrical test, both of the mated and non-contact type is widely used. Mating probing connectors give in principle a better picture of signals and behavior of the connector under test, but there is the risk of bad mating, due to small misalignment, seating problems, etc., while contactless sensing is faster and of no impact in any sense.

Contactless testing is broadly based on sensing near-field emissions from the pins of the connector under test, once these pins are suitably driven with ac test signals: sensing is much more effective for signal pins, than for supply and ground pins, usually parallel connected, so that the area of emissions is wider and the emission patterns are less sharp and localized. Near-field sensing of some pin emissions is influenced by the coupling of this pin to surrounding pins: in high-performance controlled-impedance connectors, where care is taken to stabilize and keep constant the local characteristic impedance, electromagnetic coupling to surrounding pins is quite stable and predictable, easing the task of contactless testing and the interpretation of results. High-performance sensing is based on something more than detecting "higher-than-usual" electromagnetic emissions, but on a complete verification of the multi-pin connector as a multiconductor transmission line, sensing defects as e.g. lack of symmetry and deviation from normal behavior and performance. To this aim, not only sophisticated models based on time and frequency domain measurements may be used, but also pattern recognition and classification techniques, using a set of "normal" readings as training set. It may be finally underlined that such a diagnostic test system, not only detects real defects (such as open pins), but is extremely helpful also to detect abnormal situations featuring unusual parameter deviations; the drawback is a high "false fail" rate in some cases, that however is difficult to exactly estimate and point at, considering the various shades of defects from open pins to badly terminated pins, to just unusually loaded pins and violation of controlled-impedance requirements.

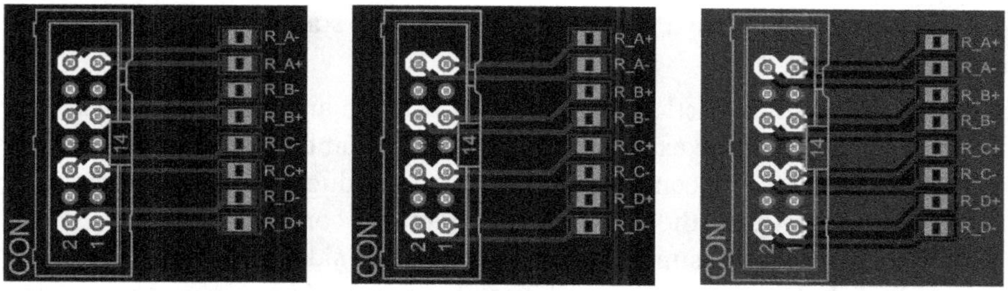

(a) (b) (c)

Figure 7.1.6 – Routing and grounding of four differential pairs on a 100 mils spacing connector: (a) ground plane on bottom connected to pin rows 2, 4 and 6 from top to bottom, separating each differential pair; differential signals have different routing length; (b) routing length is balanced by swapping the trace origin and displacing it with respect to pins; (c) ground plane is added on top layer reducing crosstalk between PCB traces.

7.1.3 Signal integrity performance and geometry optimization

7.1.3.1 Routing inductance and grounding

Through-hole connectors require more PCB space for pin vias and clearance around them, so that for the smallest pitch values it is quite difficult to drive traces and ground plane through the pin field. In this case the ground plane would pour only around the connector area with comprehensible poorer performance regarding signal integrity: ringing due to stray inductance to ground, crosstalk, EMI. A first countermeasure is to use some of the pins to break the whole connector into smaller areas, each one separated from the adjacent ones by grounded traces and pins (Figure 7.1.6(a)). Those pins may or may not be connected to corresponding grounded conductors of the connector cable: either for single-ended or differential signal lines there will be return conductors, e.g. simple conductors, individual shields for small coaxial lines (in use for some video standards and for high-integrity sub-ns applications) or twisted pairs (as used for SSTP Cat. 6), overall cable shield. For high-frequency applications it is quite a natural choice, since impedance matching is pushed from the connecting cable down through the connector body and it is normally reached by optimization of materials and pin patterns (see sec. 7.1.3.3 below).

Joining traces to connector pins is a compromise of routing inductance, routing balancing for differential signals, coupling optimization and ground distribution. For a dual-in-line connector with 100 mils pin spacing routing is rather easy, even if there is always room for improvement, as shown in Figure 7.1.6(b).

The reduction of the routing inductance and crosstalk is possible by reducing trace-to-trace separation and in general going for smaller isolation clearance values, provided it is compatible with the withstanding voltage requirements, pollution degrees, etc.. Alternatively, traces may be brought very close to each other, being placed overlapped on adjacent layers: the separation is the layer thickness, that for multi-layer boards is of same order of the minimum clearance required for PCB production (usually around 8-12 mils, that is 0.2-0.3 mm), but ensures a much higher voltage insulation (when using the 0.4 mm thickness of standard four-layer boards voltage insulation is ensured at very high levels and compatible with electric safety standards at rated voltages as high as 160 V).

Stacking of traces and reduction of routing inductance around pins is quite effective with SMD connectors, at the expense of an increased number of vias: as known, traces may be routed beneath the connector pins reducing inductance to minimum, but vias are needed to jump up to the outer layer where the connector is soldered. In the example above, the resistors might be put on opposite sides of the board, even if this might have impact on the mounting and assembly process (e.g. higher costs); given the size of SMD components, this is rarely needed, maybe only at the largest frequencies well in the GHz range.

As said, when moving to surface mounted connectors (SMT technology), the routing inductance is reduced to a minimum with a ground layer supporting all PCB lines connected to connector pins. This approach is exactly the same used for high-density SMT integrated circuits with SO, TS-SO, BGA layout (to cite the most commonly used) [177]. Bonding to the ground potential traces on the top layer is done through micro-vias and at high frequency is also backed up by the stray capacitance between layers: the already low inductance of vias and micro-vias used to pass between layers can be compensated, if necessary, by the stray capacitance of the dielectric substrate; since this compensation is valid only if tuned to the characteristic impedance and frequency range, the size of vias and the optional extra amount of pad copper shall be accurately trimmed accordingly.

7.1.3.2 Electrical bonding of ground pins

Connectors bring not only signals, but also the return signal and/or ground reference: return signal may be considered as part of the signal path, as for what concerns line impedance, crosstalk, etc.; however, in general, ground potential is brought through connectors as well, by means of pins and using connector shell and guard rings.

Grounded pins are normally used to reach a better electromagnetic separation between connector pins and to ensure a smoother transition from PCB (or cable) to connector in terms of impedance, preserving geometrical symmetry and uniformity of line cross-section (see sec. 7.1.3.3 below).

Separate grounded pins are also used to transmit the supply ground reference, so that such pins are usually mixed with those of the supply rails (for some standards, such as HDMI, they are in a separate part of the connector, using larger pins).

Finally, pins at ground potential are also used to terminate shields, the connector shell or frame and cable screen. In many EMC standards and books ensuring "360° electrical continuity" between cable and metallic connector shell is required, as confirmed in sec. 5.1.7, where we see that any even small defect in the cable shield translates in a worse transfer impedance. For multi-conductor / multi-pair cables it is possible to distinguish between the outer overall shield and individual shields of each pair that may be terminated differently with their own connector pin (e.g. with a small series resistance avoiding that large interference currents flow into the board). This is particularly relevant when connecting a low-level sensor: a double-shield cable is advisable and current diversion/filtering at board edge may be applied. The same is applicable when considering an entire equipment and the metallic connector shell electrically bonded to enclosure: chassis currents and common mode currents are the real threat, and they may be kept under control only by electrically separating connector body and selectively grounding it; additionally internal reference conductors are kept separate from the grounded circuit bonded to the outer shield.

7.1.3.3 Pin grid optimization

Signal integrity requires the minimization of stray inductance in ground distribution and trace routing, balance of impedance and electrical length of differential lines and impedance matching. Of course, the final result stems from the harmonized design of PCB stack-up, PCB traces, connector pins and the way the former are routed to the latter. Differential line balance is ensured when the two hot traces have the same length, but are also routed uniformly, paying attention also to uneven coupling to elements nearby (see sec. 8.3.1, where it is observed that the input impedance of a line depends on the signals applied to an adjacent parallel line). Impedance matching and preservation of the line impedance when entering the connector and through it is obtained by guiding signals on connector pins amid a pattern of grounded pins, that creates both a coax-like structure and reduces crosstalk between lines. Normalization of the coupling terms occurs, as it was briefly commented at the end of sec. 8.3.3.

Pin-grid patterns leave several degrees of freedom : selection of connector and distribution of return and grounded pins for single-ended and differential lines is often based on designer's sensibility and his/her knowledge of the problem, besides other constraints such as connector complexity, size and costs. Extensive simulation at design time confirms that even small changes exert a significant influence on performance, less when the connector is available as a definitive product and only routing and connections shall be optimized. All this is dismissed hastily in the reference standard for connector testing, EN 60512-5-7 [75]: "a 2:1 signal to ground ratio shall be used (one signal pair for each ground return)", without specifying how to distribute grounds and suggesting that the use of more grounded pins is not recommended.

Depending on frequency range and impedance matching, crosstalk and signal integrity requirements, there are various possibilities, more or less "expensive" in terms of the required number of pins and space needs for each signal line.

Starting from low-frequency applications (e.g. up to some hundreds of MHz), return conductors may be simply placed next to the hot pins, maybe distributing grounded pins to reduce crosstalk, not only as signal return (see Figure 7.1.7(a)). A more performing solution is obtained when the hot pins are completely surrounded by the return/ground pins (see Figure 7.1.7(b)), emulating a coaxial structure but with a square pin grid: the number of used pins per line increases and such solution is advisable only when the frequency range increases and isolation between lines shall be high. The solution may be further refined by observing that for square patterns of pin distribution, the distance of the ground pins from the center hot pin is not uniform and this may cause problems with field uniformity and characteristic impedance at the highest frequencies. The viable solutions are to remove some grounded pins from the pattern, letting them float[12] (they are grayed in Figure 7.1.7(b2)), or to select a pin-grid scheme where pin rows are alternated, so that a coax structure may be implemented with a symmetric hexagonal pattern (Figure 7.1.7(c)). For differential lines this pin-grid arrangement is not much more advantageous, even if pin pairs may be oriented every 120°: some attempts are shown in Figure 7.1.7(d), but the number of required grounded pins between pairs to effectively decouple them is still quite large. The hexagonal arrangement of return pins mimics a coaxial structure and it is quite effective up to very large frequency to improve shielding; there are high-density commercial connectors that allow such arrangement such as Amphenol HDM and VHDM, Samtec FOURRAY® quad row staggered or HDLP and HPH series by Smiths Connectors, reaching an impressive pin number and pin density. When the characteristic impedance and in general signal integrity are of concern, dielectric and pin separation and arrangement shall be optimized to the specific needs and off-the-shelf connectors might be not sufficient any longer. There are situations, especially for differential signal lines, in which uneven pin distribution is followed, increasing slightly the pin pitch and thus the distance between adjacent lines while preserving a tighter coupling for the pins of the same line; this allows to reach a satisfactorily low crosstalk, with very few grounded pins: the space wasted increasing pin distance is possibly less than that regained by reducing the number of grounded pins. In some cases grounded pins may have a special design, called e.g. "ground blade", run transversely in the connector and separating a whole row of signal pins.

A simple but effective pin-grid arrangement is shown in Figure 7.1.7(a): shielding of the hot pins is arranged north-south east-west and hot pins are aligned on diagonals; despite the hot pins are not fully surrounded by grounded pins (and this might imply low-frequency applicability), this pattern was used for tests up to 20 GHz [109]. The measured crosstalk curves are smooth without hints of internal resonances and critical frequencies. The distance is a good criterion to predict crosstalk between pins: as commonly done in such geometry, under the assumption of a large enough connector so that discontinuity at the edges is neglected at the moment, there are a few characteristic mutual positions, the other obtainable with rigid transformations (rotation, tilting, and so on). Such positions are shown in Figure 7.1.8 and corresponds to: 1-5, adjacent diagonal pins with no grounded pins in between, expecting thus the largest coupling and crosstalk; 1-3, adjacent vertically aligned pins, separated by one grounded pin (this position is equivalent along the other orthogonal orientations such as 1-12 and below); 1-10, analogous to 1-3, but for pins aligned along a diagonal, for which a slightly lower crosstalk than 1-3 is expected; 1-9 is an order-two distance, with two signal pins and two grounded pins nearly in between, and thus characterized by an even lower crosstalk.

Crosstalk curves are indicated by the corresponding pin numbers: crosstalk between two adjacent diagonal pins indicated as S_{15} is steadily between $-20\,\mathrm{dB}$ and $-30\,\mathrm{dB}$,

[12] It is extremely questionable to leave pins floating, without giving them a reference potential. In [109] a 5 to 10 dB peak-to-peak ripple appears in the near-end and far-end measured crosstalk due to the influence of floating dummy vias resonating like stubs (see sec. 6.1.4 on PCB vias); for traces near the dummy vias (not distributed evenly over the entire PCB) cause a remarkable resonance peak besides the already mentioned ripple, reaching a nearly 20 dB variation; usual frequency range for such resonances is between some GHz and some tens of GHz.

while that of two vertically placed hot pins separated by one grounded pin (S_{13}) is much lower, around $-40/45$ dB; increasing further the distance by jumping to farther pins like 9, crosstalk is even lower, with more effective shielding between pins. With lower values of crosstalk some resonances become visible: a 2.5 GHz ripple and a major resonance around 18/19 GHz. It is not clear if the ripple is due to VNA performance or some residual deficiency of calibration, but the main resonance is clearly physical.

Let's reconsider now the assumption of disregarding edge effects and discontinuity for the pins under test: in all cases, a lot of pins will be near connector edges, not immersed in and surrounded by a repetitive uniform pattern of pins; this breaks the symmetry and creates two main issues, increased crosstalk and worsening of transmission and impedance matching performance. For signal integrity, lines shall be assigned depending on their susceptibility, emissions and specifications of characteristic impedance, higher modes and phase delay. Pins at the periphery will be normally assigned to slow signals and in particular power supply, besides grounded pins.

Connectors in many cases have a grounded metallic band that is not only useful for mechanical reasons, but also to normalize the fringe field at the edges. Distinguishing between single-ended and differential pairs:

- single-ended lines, when operating at a sufficiently high frequency, are arranged in a S/G (signal/ground) pattern with 1:1 proportionality; in reality, due to the edge influence, the pattern of signal pin cannot extend to the edge and the "efficiency" of the pattern will be lower, with slightly less than 50% of signal pins out of the total number of pins;

- lower crosstalk is achieved if the ground pins surround completely the hot pins, but connector usage drops to nearly 25%;

- alternatively, a coaxial arrangement like the one shown in Figure 7.1.7(c) and (d) is used, for high-frequency high-shielding low-crosstalk applications, preserving accurately characteristic impedance and propagation (this is achieved by adjusting the distance of the surrounding ground pins with the help of an electromagnetic simulator); for this configuration the connector usage is even lower, with six return pins for one signal pin, plus extra grounded pins in between, so around 12%;

- several arrangements are possible for differential pairs where the lines are positioned mutually orthogonally to reduce crosstalk and the number of the necessary ground pins is lower; when crosstalk is much more important than preservation of characteristic impedance, anti-diagonal orientation is able to reduced differential crosstalk at high frequency[13];

- when shielding and signal integrity are to be maximized, ground pins are added that surround each differential pair; the wasted space for the large number of grounded pins is partially compensated in modern connectors by reducing the pin pitch (separation) and the pin diameter;

---

[13] W.P. Siebert, "High Frequency Cable Connector for Twisted Pair Cables," *IEEE Transactions on Components and Packaging Technologies*, Vol. 26, no. 3, Sept. 2003, pp. 642-650.

- alternatively, high-performance, high-density connectors for backplane applications (e.g. Teradyne Gbx, Tyco-AMP Z-Pack HM-Zd, etc.) feature grounded plates (or "blades") between rows of pins that electromagnetically separate them quite efficiently, freeing the user from excessive grounding at the board level.

Other pin-grid arrangements and related performances appear in [241].

With multi-pin connectors for high-speed, high-signal-quality applications, the designer/user shall cope with the necessity of reducing drastically the number of signal lines or in other words that the connector efficiency in terms of S/G ratio will be low. If the voltage insulation and the current rating of the lines are not particularly demanding, small pins may be used reaching much higher densities (a distance between pins of 1 mm is attainable, even for spring contacts, so that one square centimeter holds a hundred contacts).

7.2 Standardized tests and performance

We shall focus on standards for commercial grade connectors (such as CENELEC and IEC standards, e.g. the 60512 family) and for precision, either general or laboratory type[14] (such as the IEEE Std. 287 [172]). Other standards such as for railway, avionics, space applications, may complement this basic information with prescriptions, details or clarification for specific applications. In general these standards address a wide range of specifications for mechanical, environmental, electrical, electromagnetic characteristics and for signal perfromance they reveal quite generic with respect to the state of the art proposed by instrument and connectors manufacturers; usually details and procedures appear in standards and codes released by interest groups and category alliances, working on a specific product (e.g. a new connector for FPGA mezzanine cards, for USB 3.0, etc.). The vast majority of connectors comply to industry standards and they are anyway the first source for the occasional experimenter; conversely, metrological laboratories and research centers have familiarity with precision connectors, their characteristics and performance.

Focusing on CENELEC/IEC standards, the 60512 family is dedicated to connector characteristics and related test methods; the reason for briefly considering CENELEC/IEC standards is that they are probably the most widespread normative reference, quite known in the industrial and technical world. For CENELEC standards connectors are at a less mature stage than their counterpart, cables, as regards test methods, approaches to verify requirement fulfillment and scientific substantiation.

The measurement methods proposed in the IEEE Std. 287 are similar to those that will be reviewed in Chapter 10 and 11.

7.2.1 Contact resistance

The methods for the measurement of contact resistance are specified in the EN 60512-2-X standards, distinguishing between the "millivolt level method" (EN 60512-2-1),

[14] They are named GPC (general precision connectors) and LPC (laboratory precision connectors).

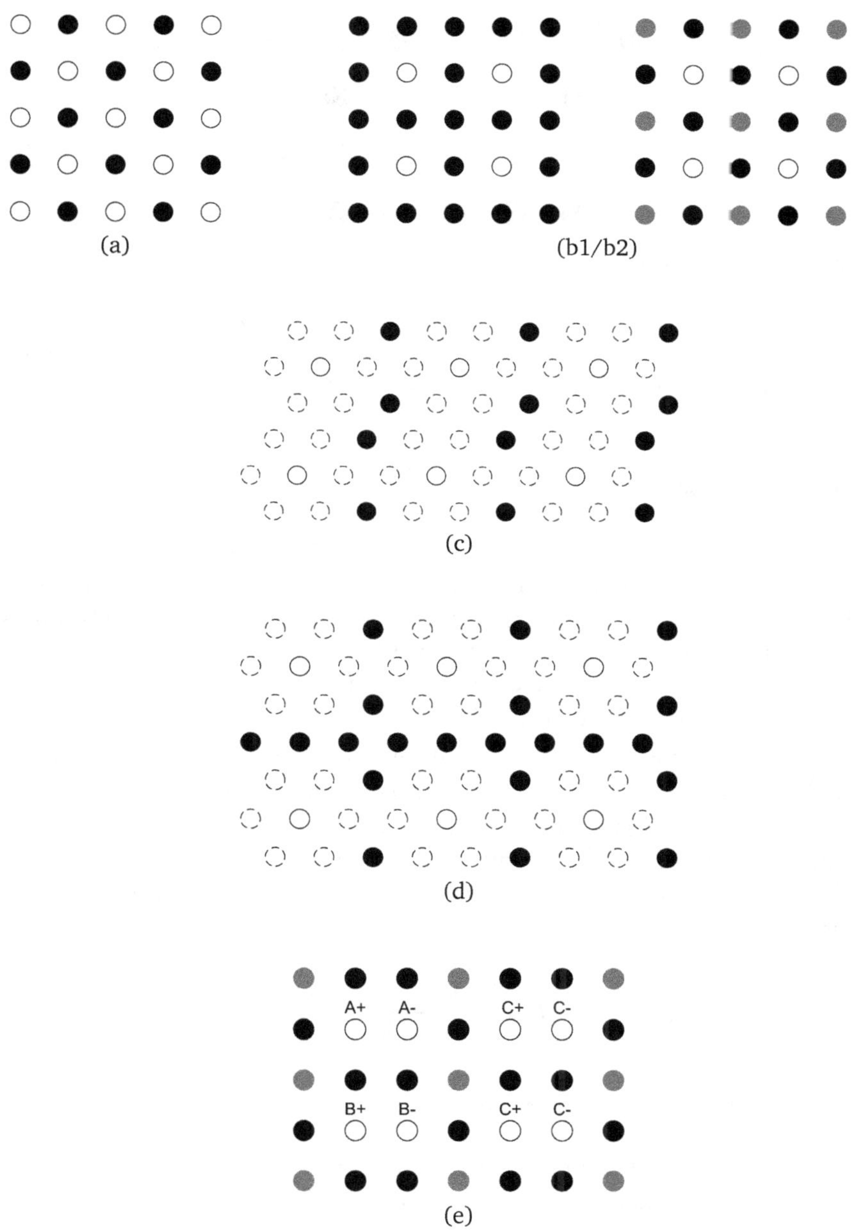

Figure 7.1.7 – Pin-grid array patterns: (a) simple single-ended diagonal arrangement, (b1) single-ended arrangement with signal pins completely surrounded, (b2) same as (b1) but with farther pins left floating (in gray), (c) single-ended coaxial arrangement with minimum number of grounded pins (the dashed pins are the coax return pins, that may be grounded or kept at the potential of each return circuit), (d) in case of independent floating coax lines, a row of grounded pins separate them more effectively, (e) differential configuration with pairs surrounded by grounded pins with farther pins possibly left floating (in gray).

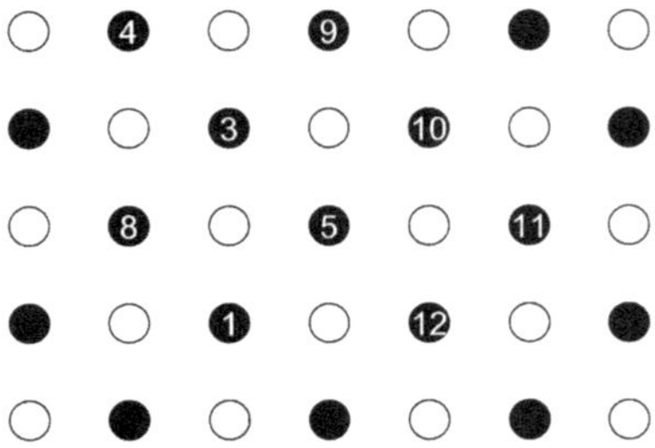

Figure 7.1.8 – Pin-grid array pattern used in [109] for the determination of crosstalk.

the "specified test current method" (EN 60512-2-2), "contact resistance variation" (EN 60512-2-3), "contact disturbance" (EN 60512-2-5) and "housing electrical continuity" (EN 60512-2-6).

7.2.1.1 Millivolt level method

The EN 60512-2-1 regards contact resistance in dc, R_{dc}, determined very straightforwardly by applying a direct current (limited to 100 mA) and reading the voltage drop with a "millivoltmeter" (a multimeter with adequate sensitivity), with full-scale voltage limited to 20 mV not to damage any insulating film inside the connector. The test may be performed at dc or ac, with a maximum frequency of 2000 Hz; in case of mismatch or disagreement, the dc test rules.

When the test is carried out at dc, the standard requires two readings with opposite current flow and voltage polarity, in order to compensate for offsets, any galvanic contact and thermoelectric effect due to dissimilar metals. In that case the estimation of dc contact resistance is obtained with the difference of the two voltage readings, noted as "f" for forward and "r" for reverse, taken with their sign divided by the total applied current:

$$R_{dc} = \frac{|V_f - V_r|}{|I_f| + |I_r|} \tag{7.2.1}$$

7.2.1.2 Specified test current method

The test may be executed at dc or ac, but with a larger voltage, never below 1 V dc or peak value for ac. No limitation is indicated for the test current. All other details are the same as the previous test method.

7.2.1.3 Contact resistance variation

Again, test setup and method identical to those described in sec. 7.2.1.1, but with a maximum test current of 50 mA. The test is indicated as "dynamic", but any detail is to be specified in a dedicated product or application standard. By direct experience external conditions may be applied, such as temperature, vibrations, shocks, while contact resistance is kept being monitored.

7.2.1.4 Contact disturbance

The test is performed at dc with test current limited to 150 mA and test voltage to 10 V. The time intervals to be measured to quantify the contact disturbance range from about $1\,\mu s$ to abut $10\,ms$. A closed contact is considered "disturbed" when the voltage across it exceeds 50% of the voltage source level. An open contact is considered "disturbed" when the measured voltage across it drops below the 50% of the voltage source level. Again the test is suitable to test the influence of externally applied conditions.

7.2.1.5 Electrical continuity of housing

The EN 60512-2-6 standard is not so informative, limiting itself to a verification of the electrical connection, while focus should have been at least on the measurement of the bonding resistance and inductance of connector shell, as well as the overall shielding effectiveness, to support signal integrity and EMC.

7.2.2 Insulation

Part 3 of the EN 60512 standards is dedicated to the test and measurement of insulation resistance; it will be no further considered.

7.2.3 Voltage stress test

The voltage withstanding capability of the connector is considered in Part 4 of the EN 60512, distinguishing between applied voltage test and partial discharge test.

The test is performed between each pin and the other pins connected together with the housing, or alternatively the pins may be arranged in groups and then the test voltage applied between groups, to reduce test duration. The applied voltage is almost steady, dc or ac (in the latter case the values are referred to the peak voltage) with the specification that the rate of application shall not exceed 500 V/s; the final value of the test voltage is not specified and depends on the connector characteristics and the product or application standard. The test voltage shall be applied for 1 minute and there shall be no evident breakdown or flash-over and the maximum permissible leakage current is 2 mA. It is evident that this is only a minimum requirement and that better performance may be required in the specific product or application standard.

7.2.4 Current rating

Part 5 of the EN 60512 standards regards current-carrying capacity (or current rating). Current rating of connectors is to be assessed by measuring any excess of temperature and local overheating, with the determination of connector derating as a function of external environmental temperature. The temperature sensor shall be located as close as possible to the points where the largest overheating is expected. Since the length of the cable that wires the connector has influence on its thermal response, the minimum conductors length is specified. Normally product standards by VDE, IEC and UL specify quite well all test conditions.

7.2.5 Transfer Impedance

Part 23 of the EN 60512 standards regards the measurement of transfer impedance. The method is quite similar to the triaxial method for cables (see sec. 5.6.1), but care is taken in removing any influence of connecting cables inside the test cell (e.g. excessive length of connecting cables and contact resistances, because they add to the test result), so that only the lower transfer impedance of the connector under test is measurable. The simple method to improve shielding of cables and to lower their transfer impedance is to cover them with a concentric solid tube that is bonded to the connector at one side and to the cylindrical test cell end at the other side. This method is friendly called "tube-in-tube" and it is described in a recent work where testing methods for both cables and connectors as appearing in the IEC 62153-4-4 are collected and discussed [149].

As already done for cables, the required quantity is the ratio of the power captured by the outer cylinder and the applied test power, and is called *screening attenuation*. Care shall be taken to collect the maxima of the screening attenuation curve, as already commented for cables, so that the measurement can be performed also for partially mismatched setups; to this aim frequency values that can be tested are limited by the cut-off frequency, that is proportional to the length of the so-formed setup coaxial structure. Be it the minimum length to hold the connector under test, the usable frequency range would start in the GHz range, because connectors have lengths of some tens of mm; to this aim, the setup is prolonged using a feeding section of cable inside the external cylinder that shall have no influence on the results, so that it is made of a solid tube; from this the name "tube-in-tube".

The setup is connected to the generator and to the receiver (or to the VNA ports) by standard RF coaxial connectors, while the connector under test need to be paired by a mating connector to close the circuit inside the test fixture; so connectors can be tested as mating pairs.

7.2.6 Signal integrity characteristics

With signal integrity characteristics crosstalk, reflections, attenuation, rise time degradation and propagation delay, and their variability, are meant. They are addressed in Part 25 of the EN 60512, where tests are specified to quantify the "crosstalk ratio" (-25-1), the "insertion loss" (-25-2), the "rise time degradation" (-25-3), the "propagation delay" (-25-4), the "return loss" (-25-5), "eye pattern" and "jitter" (-25-6), "impedance", "reflection coefficient" and "VSWR" (-25-7), "alien crosstalk" (-25-9). Many elements are in common with cables and many considerations of Chapter 5 are equally applicable.

7.2.6.1 Crosstalk

As for cables, Near-End Crosstalk (NEXT) and Far-End Crosstalk (FEXT) are considered and two test methods are specified, naming them "time domain method" and "frequency domain method".

Time-domain method The EN 60512-25-1 standard [69] states a relationship between crosstalk and rise time, without specifying which equation or expression is assumed; crosstalk expressions for time and frequency domain are presented in sec. 8.3.4 later on. The method proceeds with first the setup calibration, that is called "measurement of the fixture":

- the rise time is measured first and shall fulfill the following: 100 ps for an expected rise time of the c.u.t. between 100 and 500 ps, 500 ps for c.u.t. rise time between 500 and 1000 ps, 1000 ps when rise time is larger than 1000 ps;
- then the intrinsic crosstalk of the fixture (i.e. the pre-existing crosstalk) is measured: the fraction of voltage appearing on the induced line is measured and recorded (amplitude and sign), to be used later.

Two methods are in reality proposed for the calibration of the setup fixture, that is zeroing the fixture with the c.u.t. removed (and connecting the near-end and the far-end sides together), or using a reference fixture (that is going to be used whenever the connection of the two sides influences the result): this is clearly done for generality, but without a scheme or a practical example, the methods are not so clear and the reader might get confused.

The only quantitative specification is that the rise time shall be intended as 10-90% rise time (see sec. 3.2.1 and 3.2.3).

The test fixture compensation is based upon the possibility of measuring the test fixture behavior with the c.u.t. removed. For PCB-based test fixtures the c.u.t. is often soldered to PCB pads and removing it there is no connection left between the opposite sides, so that a control trace of the same dimensions and construction is often added for calibration; alternatively, a connection of the same physical extension of the c.u.t. but of known characteristics may be used for the calibration. This is what the standard names "reference fixture", whose implementation highly depends on the physical characteristics of the c.u.t. and its final application.

When the c.u.t. is of the coaxial type, its removal is much easier and when measuring in time domain there is no need to compensate for the shortening of the physical length between the step generator and the oscilloscope; in case it is needed for compatibility with the frequency-domain method an equal phase insertable or a known connector may be used to keep the physical length constant.

Frequency-domain method The procedure is very similar to the time-domain one and does not add any insight. The standard distinguishes between measurements performed with a VNA and with a spectrum analyzer, but does not give any useful detail.

A "through" calibration is required by connecting the VNA to the "appropriate locations"; for the spectrum analyzer the standard mentions a "reference measurement". However, details of VNA calibration are not given.

The only quantitative specification is that the number of frequency points shall be larger than 200, but without specifying the frequency range.

7.2.6.2 Insertion loss

Insertion loss measurement procedure is not much different from the crosstalk measurement performed with the frequency domain method. The EN 60512-25-2 standard [70] clearly specifies when using a VNA a 12-term calibration of the "through" type (also named SOLT, referring to the most common calibration, see sec. 11.5); when using a spectrum analyzer the details are not given and one simply records the measured attenuation to subtract it later from the final measurement. However, when indicating the reference planes for the VNA calibration, the standard prescribes to include the feeding cables, but not the test fixture itself, thus stopping at the feeding connectors of the test fixture. If this is done to have the possibility of later including the test fixture either by a separate measurement or by calculation is not clear.

One further method is reported, based on the use of an impedance analyzer; however, after requiring the measurement of the impedance of the test fixture with the c.u.t. with the other end either open circuited or short circuited, it is not clear how the square root of their product should give the desired attenuation, while we know that when applied to a transmission line it gives the characteristic impedance (see sec. 2.4).

The use of a TDT method is also proposed, but there are no operational details and the simple FFT transformation of time-domain data is indicated as the means to get a frequency-domain response.

7.2.6.3 Return loss

Also the return loss measurement indicated in EN 60512-5-5 [73] adopts a standard method, as described in sec. 10.2. The standard cites time domain and frequency domain methods without giving any details or further reference. On the contrary, the standard at sec. 5.4 "Time domain" states that the line loss constant α and the phase constant β may be "calculated over a very wide frequency range": this is stated generically for a return loss measurement, without making reference more correctly to the insertion loss measurement (the EN 60512-5-2).

7.2.6.4 Impedance, reflection coefficient and VSWR

These three quantities are addressed in a separate standard, the EN 60512-25-7 [75], that reports clearly the fundamental relationships, useful also if one ever needs a normative reference to cite for such matter. Details are given for example regarding the termination of adjacent lines, the measurement on "single-ended" and "differential" lines (the terms "unbalanced" and "balanced" would have been maybe preferable), the exact meaning of "specimen measurement" with direct method and open/short method, etc.

Impedance, as well as return loss, may be measured using VNA (S_{11}), once the c.u.t. (i.e. the specimen) is inserted and correctly terminated: when the c.u.t. is terminated at the far end with the characteristic impedance (what the standard calls "specimen environment impedance"), then the impedance is determined by a direct measurement; alternatively, open circuit and short circuit may be applied in place of the matched termination and two readings may be made, so that the impedance is determined as the square root of their product (see sec. 2.4).

However, the standard does not specify what is the resulting impedance value: it would be the characteristic impedance of the line between the measurement point (input port 1) and the termination point (output port 2) if the line is homogeneous. The line on the contrary is made of the feeding line of the test fixture, the c.u.t. and the terminating line going to the output connector where the termination is applied (port 2); so what is measured is the overall impedance of such a transmission line, assumed homogeneous. To separate the c.u.t. contribution, a calibration of the test fixture would be needed and this is only addressed in the Annex C of the standard, even if it is the most relevant and critical part of the whole procedure. Annex C for the first time addresses the problem of using PCB traces as the lines connecting the c.u.t. to the external connectors; this part is considered separately below (see sec. 7.3), since it is common to many measurements considered here.

7.2.6.5 Rise time degradation and propagation delay

These two measurements (rise-time degradation as in EN 60512-5-3 [71] and propagation delay as in EN 60512-5-4 [72]) require a full-time domain method, using a step generator and an oscilloscope.

The rise time is evaluated as the time interval between the 10% and 90% points of the rising edge under the assumption of Gaussian profile, in order to fully apply the theory that we have seen at sec. 3.2.2 and 3.2.4.

Propagation delay is evaluated as the time difference between corresponding points of the considered waveforms both at 10% and at 50% (half-value) of the amplitude. However, it is not said what to do with the two values, if they shall be shown both, averaged, the worst one retained, or else.

The "considered waveforms" term indicates waveforms that depend on the type of measurement: they may be i) the input and output waveforms at the test fixture connectors with the c.u.t. inserted, or ii) the two output voltage waveforms measured at the same point with and without the c.u.t., or iii) the output voltage waveforms measured at the same point of the test fixture with the c.u.t. and of the reference fixture (in all equal to the test fixture but without the c.u.t.). Of course for the methods ii) and iii) we need to store one of the waveforms and to recall it for comparison with the last acquired waveform; to this aim a common timebase reference point shall be selected and this is usually triggered with the step voltage rise time at the source; if

Connector type	10-90% rise time [ns]
RG-58 BNC crimp	0.011
RG-174 BNC crimp	0.011
RG-8 N-type	0.004

Table 7.2.1 – Experimental values of the 10-90% rise time for commercial coax male connectors [199]; rise time is proportional not only to the connector length, but it is also influenced by the geometry and coupling between parts (see text).

the connecting cables are not too long, there is no need for a deep channel memory to store all the sample data points.

Of course, when measuring fast rise times, the rise time of the oscilloscope and any used voltage probe shall be accounted for, using eq. (3.2.15) in sec. 3.2.4. We know that when the rise time of the measurement system is half of the measured rise time, the effect is already only 11.8%, and when it drops to one third the influence is 5.3%. The standard requires that the rise time of the measurement system be less than 70% of the rise time value with the c.u.t. inserted.

A few experimental data appear in [199] and are reported here in Table 7.2.1 for completeness. In the original table also connector inductance was reported and the rise time is exactly related to the time constant of the LR circuit formed by inductance and the terminating $50\,\Omega$ at each sides; the used proportionality between the rise time and the time constant is 2.2. These values of rise time were determined with measured VSWR values, as they appear in the connector datasheet; then the assumption of purely inductive mismatching brings to the estimated rise time. The reflection coefficient is due to the reactive inductance X_L added to the otherwise perfectly matched connector impedance, thus

$$\Gamma = \frac{X_L}{2Z_c} \qquad L = \frac{2\Gamma Z_c}{\omega} \tag{7.2.2}$$

The problem is to fix which value assign to ω: the assumption that the impedance mismatch is purely inductive leads to the necessity of accepting that the mismatch increases linearly with frequency, that does not occur in practice. This is the weak point of the rationale; at high frequency inductive and capacitive terms compensate and connector impedance begins fluctuating between inductive and capacitive behavior. We may assume that the best value to use for the reflection coefficient is the one measured at the lowest frequency that excludes the relevance of the capacitive reactance.

Once the inductance is determined, then the rise time is given as the resulting time constant, that is

$$t_r = 2.2\,\frac{L}{Z_c} \tag{7.2.3}$$

For the values shown in Table 7.2.1, the estimated inductance was $0.25\,\text{nH}$ for the first two rows and $0.09\,\text{nH}$ for the third one.

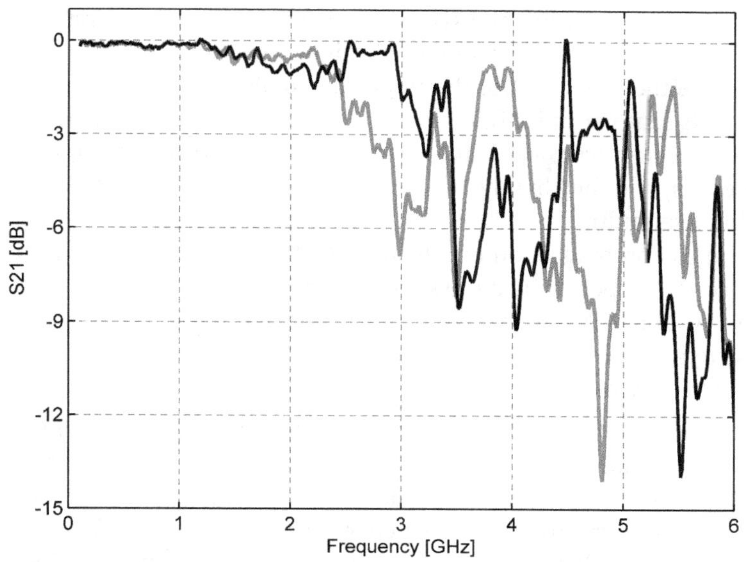

Figure 7.2.1 – Example of major resonance of insertion loss curve with smaller repetitive peaks and valleys at the electrical length resonance.

7.2.7 Interpretation of measured characteristics

The insertion loss curve is probably the mostly widespread adopted performance to decide on the suitability of a connector to transmit a signal or class of signals. In most cases the $-3\,\mathrm{dB}$ intercept is used to determine the useful bandwidth and compared to the rise-time bandwidth of the signal (see sec. 3.2.1 and 3.2.2). However, for any transmission system the insertion loss measured in the frequency domain exhibits peaks and valleys due to resonances and unavoidable mismatches between measuring ports; for connectors, in particular, when losses are moderate or negligible, such phenomenon may be quite relevant and the exact $-3\,\mathrm{dB}$ intercept point may be anticipated or retarded, and more than one crossing of the ideal $-3\,\mathrm{dB}$ horizontal line observed. In this case, depending on the spacing between peaks and valleys, significant under- and over-estimates of the "right" intercept may occur. It is also remembered that even the $-3\,\mathrm{dB}$ intercept rule is approximate, being derived by simplifying assumptions on signal rise time and a low-pass 1-pole model for the connector (see sec. 3.2.2). The use of a smoothed version of the curve is advisable, as well as the identification of any significant resonance of the transmission line, besides the expected resonances across the electrical length: such resonance can be recognized as a hump in the curve contoured by the always-present smaller peaks and valleys, as it is shown in Figure 7.2.1.

Deciding on connector "bandwidth" and data throughput is even more complicated, because not only rise time and transmission rate matter, but also the statistical properties of the transmitted sequence of symbols, that is in the end the power spectral density of the signal. This is a typical problem of determining spectral properties of digital modulations and is well covered, among others, in [274].

A simple approach is to double the symbol rate and that gives a worst-case estimation of band occupancy. However, "symbol rate" might be either the payload data rate (the "clean" data rate the user sees) or the physical exploitation of the channel in terms of channel symbols sent per unit of time. In the first case we are underestimating slightly the real rate, because overheads due to addressing, control and data protection are not accounted for, and they highly depend on the protocol. When considering complex protocols that use multiple levels (and not a simple base-band binary encoding) and add channel noise probing or other advanced techniques, a more accurate evaluation based on spectral properties is necessary. In the second case, depending on channel coding (e.g. Alternate Mark Inversion, AMI) and correlation between symbols, the spectrum occupancy may be variable, but resulting in a bandwidth always less than the raw estimate based on symbol rate.

Also the input impedance of the connector is important, because it causes reflections on the feeding line and, as a consequence, not only power loss (the resulting signal amplitude is lower), but also crosstalk and mode-conversion (see sec. 5.1.6, 8.3.4 and 10.4.1). Connectors have their own characteristic impedance when seen as short transmission lines and this impedance may vary along the connector itself as the geometry changes (cable or conductor insertion, then the center pin, the mating point to the other sex, or sexless, connector, and so on). On the other hand, we know that such impedance changes are appreciable when the signal wavelength is approaching the size of the geometrical detail. So, there will be a varying impedance offered by the connector when the input signal goes through a transient, such as a rise time, due to the instantaneous frequency; in reality this is better seen with a time domain reflectometry measurement.

The insertion loss measurement in frequency domain S_{21} gives also the return loss curve S_{11}, because of conservation of power (see sec. 10.1.3). However, this is of limited application if we are interested at what happens during the development of the rise-time front, for which the concept of instantaneous frequency is hardly applicable in practice due to limitations in the number of samples and constraints of time-frequency transforms (resolution in time and frequency is basically constrained by the Gabor's inequality, see sec. 3.1.3.2 at page 104): the VNA addresses this problem when operating with time-domain option (see sec. 11.8 and 11.9). Practically speaking, depending on the instantaneous frequency of the signal portion (e.g. during a transition), the connector will show a different input impedance and time- and frequency-domain measurements will inevitably differ somewhat.

When rise-time degradation is measured, this is a clear indication of the impact of connector on signal line performance, because the resulting rise time is in relationship with signal edges and eye pattern (see sec. 3.3.3). Signal integrity and receiver performance specifications for high-speed protocols are often specified in terms of "eye mask", that is the area inside the eye pattern profiles that shall be clear of signal traces, to ensure that the decision criteria implemented at the receiver work. This is a direct visual interpretation of the impact of connector, obtained by measurement or simulation of the response to synthetic data streams. Other quantities are the bit or frame error rates (BER and FER), that establish the resulting effect of noise, bandwidth and signal distortion, fully taking into account the characteristics and spectral properties of the protocol, thus weighting symbol occurrence and correlation in the proper way. Eye masks can be applied to a full signal connection path, from transmitter to receiver, but hardly can be used to quantitatively pinpoint the impact of the connector and assuring that a positive result with respect to the eye mask will remain such when the whole transmission system, including e.g. cables, is assembled and tested.

7.3 Test fixture design and calibration

When it comes to measure connector performance, especially for new connectors, there are a few aspects that deserve some attention: connectors cannot be tested alone but need a test fixture for installation, connection and signal feeding; they are almost always tested in mating pairs, each held in place by the test fixture; often the connector under test has performances comparable or better than the test fixture, so that calibration plays a paramount role. Interposers stay between two sandwiched PCB boards and may be in principle tested alone, even if there is some convenience in testing two of them to bring the signal back onto the same board, as we will see below. For high-performance connectors the design and construction of planar transmission lines included in the test fixture requires the selection of stable and performing materials, as well as an accurate manufacturing process; tolerances and variability shall be accounted for and line dispersion and attenuation modeled and evaluated at best. The particularly large dispersion of microstrips, as well as the influence of vias and parasitic stubs shall be considered.

7.3.1 Coaxial fixture

Coaxial test fixtures are used either to test new coaxial connectors (not so frequent, they are quite well standardized) or to evaluate the transfer impedance and shielding effectiveness of connectors and cable-connectors assemblies. To this aim various test setups are used quite similar to the triaxial method for cable transfer impedance (see sec. 5.6.1 and 7.2.5), where additional shielding enclosures are added to include connectors of various sizes[15].

7.3.2 Planar PCB fixture

The EN 60512-25-7 standard [75] in Annex C gives some advices regarding the PCB test fixture and the desirable characteristics for calibration, using a SOLT method (see sec. 11.5). The c.u.t. is located on the PCB at a distance l_1 from port 1 (input port) and l_2 from port 2 (output port); the c.u.t. itself has its physical length l. All these length values may be translated into equivalent electrical lengths or propagation times

[15] See for example:
B. Mund and T. Schmid, "Measuring EMC of HV Cables & Components with the Triaxial Cell - Part 1, - Part 2," *Wire & Cable Technology International*, Jan. 2012, pp. 88-90 and 116-118;
L. Halme, R. Kytönen, V. Nässi, M. Nupponen, M. Wollitzer, T. Schmid, E. Rodig and B. Mund, "Measurement of the Shielding or Screening Effectiveness of Feed-throughs and Electromagnetic Gaskets up to 4 GHz and above," 56th Intern. Wire & Cable Symposium, Orlando, FL, USA, Nov. 11-14, 2007, pp. 303-311.

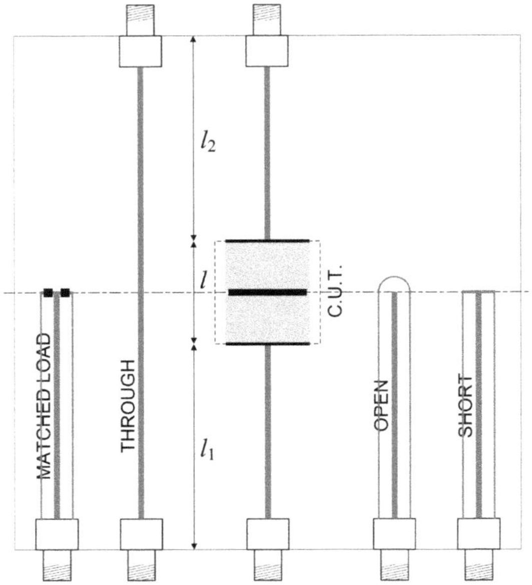

Figure 7.3.1 – Example of 2-D test fixture in the form of a PCB with the c.u.t. in the center and SOLT calibration standards.

when propagation velocity is known, and of course they will be different for PCB planar lines and c.u.t.. Considering a c.u.t. made of two mating halves, the board may correspondingly be realized separable and mirrored around the mating plane; otherwise, the mated c.u.t. may be installed without separating the board as a single unmodifiable component (see Figure 7.3.1).

The SOLT calibration is made presenting to the VNA ports a Short-circuit, an Open-circuit, a matched Load and a Through condition: this translates into as many lines to be carefully built on the PCB, in addition to the line that is effectively used to connect the c.u.t.. Depending on the need of including the c.u.t. length into the calibration (to avoid phase errors), the equivalent length of the PCB lines shall be selected to include not only l_1 or l_2 depending on the port, but also $0.5l$, so that the terminating condition is located ideally at the center of the c.u.t.. It will be thus necessary to build two shorted lines, two open-circuited lines and two matched lines of length $l_1+0.5l$ and $l_2+0.5l$ each, respectively; then a full length l_1+l_2+l line is added that represent the Through term. All these lines (a total of seven) shall be accessible by means of suitable connectors located on the PCB edge[16], matching the standard that is used for the VNA measurement (let's assume SMA). Setting $l_1=l_2$ is of course a great simplification because the number of different calibration lines is reduced to four, one for each type of calibration standard.

It is also advisable, if the physical size has no major constraints, to allow an electrical length of the lines long enough so that transitions at the launchers and at the c.u.t. are separable when applying time gating (see sec. 11.9). If the test fixture calibration is successful, time gating is normally not necessary, but it might be useful anyway, for

[16] Coax edge connectors may be replaced by other types of PCB coax connectors, see sec. 6.2.2.1.

example, to isolate the performance of launchers and further improve the design. As pointed out in sec. 11.8 and 11.10, the effective spatial resolution is limited by the maximum frequency; using a 6 GHz VNA 5 cm are an adequate choice for l_1 and l_2.

One major source of uncertainty of test-fixture geometry is that pads or pins have non-zero dimension. For example, when setting the Through line to have the same electrical length of the line holding the c.u.t., the length of the latter is determined and subtracted from the edge-to-edge overall physical length. Its length is subject to unavoidable approximations depending on where the reference points are, inaccuracies in positioning and soldering, etc.; taking mid-points between edges is a good solution on average, but it is likely that some tenths of mm of uncertainty remain.

Additionally, pads and soldering contribute parasitics that if not properly and reliably modeled to be subtracted by de-embedding (see sec. 11.7) should be included in the calibration, that is added to the planar lines that represents the calibration standards.

7.3.3 3-D PCB fixture

If the the mating and soldering planes of the c.u.t. are parallel, or the c.u.t. is an interposer[17] that may be housed between two sandwiched PCBs, the fixture geometry develops in 3-D and different arrangements are possible (see Figure 7.3.2).

With just one interposer available the configuration shown in Figure 7.3.2(a) is used, with launchers on two different PCBs on opposite sides of the intermediate c.u.t. and a land pattern of pads on each board position the interposer; the interposer shall be removed to perform calibration. The problem of calibration is awkward because Open and Short application can be realized on the boards and even accommodated with good accuracy on duplicated land patterns thus including their parasitics, but Matched load and Through standards are difficult to realize. A dummy interposer may be realized with a slice of known coaxial element and used for the calibration, but it is extremely uncertain. Otherwise a set of homogeneous calibration standards may be realized separately on one of the boards similar to Figure 7.3.1, including the length of the interposer (that is the physical separation of the two boards). The c.u.t. would be thus projected onto the horizontal plane of the board: the length of Matched load and Through lines would be as before, using $l = h$, with h the height of the c.u.t..

The question remains if unfolding the interposer to an equivalent planar transmission line is a correct approach, provided that c.u.t. pads shall be included in the Short and Open, and observing that in this way the 90° coupling from the PCB to the interposer becomes part of the measured interposer response; the use of a fake c.u.t. with known characteristic impedance has the advantage of calibrating also for pads and 90° rotation (that disappear from the measured c.u.t. response), despite inaccuracy related to the electrical contact and positioning.

When two identical interposers are available, then the signal may be brought back to the original PCB where both launchers are located, simplifying thus calibration (see Figure 7.3.2(b)). Both Matched load and Through are realized on the launching board

---

[17] Examples of coaxial interposers are Cinch SMPM series or Custom Mated Pair Spring Probe Connectors by Hypertac.

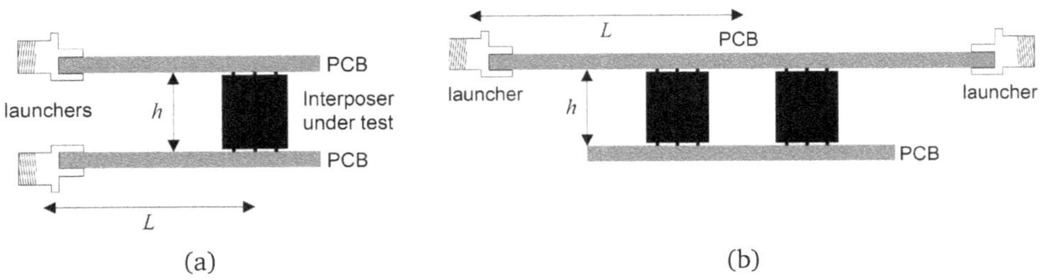

Figure 7.3.2 – Sketch of 3-D PCB fixtures to test interposers sandwiched between two PCBs: (a) launchers on both PCBs with test signals traveling through only one c.u.t., (b) launchers on the same unique PCB (shown for clarity on opposite ends) and signals traveling up and down through two c.u.t. (or two parts of the same c.u.t., if its physical size, the number of pins and the level of coupling between allow).

(top board): the reference axis for the load is half-way between the two c.u.t. and the Through simply runs between two similar launchers parallel to the line that connects to the c.u.t.. Such a parallel line is sometimes named “control signal path”.

The test will return the measured insertions loss of two c.u.t. and necessitates to be halved, while the return loss is simply related to the input impedance mismatch.

If working with multi-pin c.u.t. and testing several lines, the PCB planar lines used to connect to the launchers shall have the same length and be balanced, making them sufficiently long to mask very small differences and also to perform a satisfactory gating as commented in sec. 11.9. A normally adopted solution is that of spanning them radially on the board. As already pointed out vias and glass-weave effect shall be duly considered. With a sufficiently accurate design and paying attention to manufacturing tolerances and typical defects or non-ideality reviewed in Chapter 6, the same calibration may be uniformly adopted for all lines with a consistent distribution of errors.

7.4 Results on repeatability and non-linearity

Attention is on less known characteristics regarding repeatability of connector characteristics, variability of connector performance between manufacturers, influence of connector use non-linear behavior, e.g. giving rise to distortion and intermodulation effects.

There are small connector types, such as SMA (and other 3.5 mm categories), or smaller 2.92 mm, 2.4 mm, etc., suitable for almost all signal applications over an extended frequency range, that help in reducing the number of necessary calibration kits, cables and adapters. For some telecommunication applications the choice falls onto e.g. SMB and MCX, that conversely suffer the lack of calibration standards. Generally speaking, under the same name there is a wide range of connectors for performance, price and ease of procurement: going for very expensive items made by renowned manufacturers probably one never goes wrong with performance, but over a more than 5:1 price range, the price-performance relationship is not so obvious. So, allowing budget issues to condition a little our choices, as well as being aware of which are the true performances of off-the-shelf items, is a normal and understandable approach, considering that in most of our experiments the frequency interval is limited to some GHz and that the overall uncertainty is on one side not too demanding, as for primary calibration, and, on the other side, it is mainly influenced by cable and PCB performance. That does not mean that connectors may be neglected when repeatability and performance reduction with wearing and aging are considered.

In general a wise approach is to consider the ensemble cable-connector or PCB-planar transmission line-connector, operating the necessary tradeoffs, but most of all being aware of inconsistencies and weak points: cable flexure and induced stress (reviewed in sec. 5.7.1), inhomogeneity of PCB and variability of planar transmission line response (considered in Chapter 6), pin grid optimization in multi-pin connector applications (see sec. 7.1.3.3 above).

7.4.1 Intermating of SMA, 3.5 mm and 2.92 mm connectors

The SMA, 3.5 mm and 2.92 mm connectors are mechanically compatible, but their internal geometry (and performance) differ, so one may be interested in understanding the variability of performance depending on mutual tolerances and other geometry characteristics while mating. Pin gap was already identified as the reason for bad response and/or repeatability issues; in sec. 7.1.1.7 with reference to Daywitt work pin gap was considered a connector defect.

Variable recess (so negative pin height, or in other words increasing pin gap) was tested in [270], where the pin was recessed progressively from 0 to -5 mil, 1 mil at a time, starting from the nominal pre-existing recess value (-0.12 mil for 3.5 mm type, -0.65 mil for 2.92 mm type and -1.5 mil for SMA type). The return loss was measured each time for the following combinations of connectors: 3.5 mm-3.5 mm, 2.92 mm-3.5 mm and SMA-2.92 mm; ranking of resulting performance taking a snapshot of all values is in this order. Changes of impedance and return loss were increasing with pin recess, but not all best performance occurring for the same amount of recess: the 3.5 mm-3.5 mm pair has a minimum reflection better than -50 dB for the initial existing recess of 0.12 mil and keeps around -50 dB for recess up to 1.12 mil; when mating 3.5 mm and 2.92 mm some recess is beneficial and the minimum of reflection at -50 dB is obtained for 3.65 mil (3 mil above the existing 0.65 mil recess); for the SMA-2.92 mm pair the initial return loss is already worse than the others (around $-40/45$ dB) and increases both with frequency and with increasing recess (between 18 and 22 GHz $4-6$ mil of recess cause a return loss of $-32/34$ dB).

7.4.2 Repeatability of Return Loss and Insertion Loss

Pioneering work in the experimental determination of insertion loss characteristics of RF coax connectors of the same standards used today dates to the early '70s [34, 260]. In these works SMA and N connectors were considered and the test setup was enough sophisticated and performing to represent a valid and significant reference even today.

7.4.2.1 Direct testing of connector repeatability

References for quantitative results of direct evaluation of connector or adapter repeatability are only a few, with connectors undergoing many cycles of simulated use and measured by instrumentation and setups with very low uncertainty. Collected references are quite old even if applied already to "modern" connectors, that is N and SMA type connectors; it is possible that experiments of this kind never stopped and have been performed more recently by the many manufacturers and high-end users, but so complete and useful results were published any longer to author's knowledge.

Bergfried and Fischer (1970) Simulated "real use" conditions were adopted [34]: repeatability of insertion loss values was verified after 10000 connect-disconnect cycles. The considered connectors were of the precision and semi-precision type, the latter affordable for many users. A ratio measurement was used to avoid the effect of slight changes of the RF source level; care was taken to assure thermal stability, avoiding heat transfer between parts and from the operator's hands, and to control mechanical conditions, such as positioning the connectors vertically to reduce transverse forces and torque, due to their own weight and the weight of connected devices. The largest uncertainty source for these measurements was the stability of the ratio meter, which, when stabilized to within 1°C, had a drift of less than 0.0006 dB in 10 minutes. The sample size for each connector type was three or four. After the initial repeatability evaluation for the whole sample, the connectors underwent an overall 10000-operation cycling, subdivided in four sets of 2500 operations. The number of repeated measurements for each connector is not specified, but when talking about the stability of the ratio meter, the authors mention 30 measurements in 10 minutes; it is a sensible value to adopt as the number of measurements per connector, before a new one is put in place, very similar to the one used in [260].

The better performance of SMA connectors with respect to wearing is mainly ascribed by the authors to the observed abrasion of slotted fingers of the female center conductor of the N connector, causing loose contact and adhesion problems. They observed that undercutting fingers on the internal side, as done in the precision N connector, greatly reduces friction without sacrificing electrical performance.

Pakay and Torok (1977) Similar measurements were performed on a wider range of connectors [260]: three or four samples for each type, twenty to thirty repeated measurements for each condition. Even if non explicitly said, it looks like precision connectors were considered (the authors mention some semi-precision connectors as unsuitable for testing above 1 GHz, that, if it is for sure too a drastic statement, confirms the adoption of high-performance connectors of the precision type). The overall drift of the measurement system was quite the same of the previous work, so less than 0.0006 dB over 10 minutes. The authors analyze also the effect of the applied torque on measured insertion loss and for SMA a minimum 0.2 Nm value was recommended. Focusing on the comparison between SMA and N connectors, the former behave much better with an average dispersion of insertion loss readings that is 50-60% smaller. The results are reported in Table 7.4.2; also in this case measurements were performed with connectors held vertical, except for one set of readings at 10 GHz, marked with

Connector	4 GHz	8 GHz	12.4 GHz	18 GHz
Type N, silver-plated brass, general purpose [UG-29B/U]	0.006 0.006	0.006 0.0082	0.006 0.0082	—
Type N, stainless steel, semi-precision [Weinschel mod. M1510N, F1510N]	0.0025 0.0046	0.003 0.003	0.003 0.007	0.005 0.0105
SMA, gold-plated, semi-precision [OSM mod. 217, 218]	—	0.0055 0.0082	0.0055 0.0082	0.0055 0.016
SMA, gold-plated, stainless steel, precision [WPM Weinschel miniature]	—	0.0028 0.0038	0.0028 0.0038	0.005 0.007

Table 7.4.1 – Repeatability of SMA and N connectors measured by Bergfried and Fischer [34].

Connector	0.1 GHz	1 GHz	6 GHz	10 GHz	10 GHz^{h}	12 GHz
Type N, stainless steel, semi-precision	0.0002	0.0006	0.0009	0.0010	0.0014	0.0014
SMA, gold-plated, semi-precision [Amphenol]	0.0003	0.0004	0.0004	0.0006	0.0008	0.0010

Table 7.4.2 – Repeatability of SMA and N connectors measured by Pakay and Torok [260].

"h" (for "horizontal"). Even if the work by Bergfried and Fischer [34] is cited as the first bibliographic reference, there is no explicit comparison of the obtained results: looking at the two tables, it is evident that these values of repeatability presented in [260] are four to six times better than the previous ones.

Jesch (1976) In Jesch [197] a very comprehensive set of results regarding SMA connectors is reported, for both return and insertion loss repeatability. The measurement system is quite performing despite the measurements were made nearly 40 years ago! The system has 7 mm connectors (N type) and goes to the SMA under test with N-SMA adapters.

Thus, when evaluating return loss repeatability, the N-SMA adapters and the 50 Ω SMA terminations were tested: ten repetitions are made for each connector and at each measurement the connector is rotated by 30°. Results on return loss dispersion are expressed as fraction of the absolute value of the reflection coefficient at the various tested frequencies; it is observed that there are some discrepancies between samples in terms of values and dependency on frequency. Moreover, when testing a third set of SMA connectors connected to the cables used for the successive insertion loss tests (so with a more realistic configuration), the resulting dispersion of the return loss values is nearly ten times larger.

Tests of repeatability for insertion loss measurements are performed on the N-SMA adapters and on the cable samples (of course, not on the SMA matched loads) and the resulting dispersion is expressed in dB. The number of repeated measurements is twenty for each sample. The behavior with frequency is more regular, with the dispersion increasing smoothly with increasing frequency and the curves are better grouped together. The numeric values are perfectly comparable with the previous ones obtained by Bergfried and Fischer, and by Pakay and Torok.

When calculating the dispersion associated to a limited number of different samples (less than the ten or twenty repeated measurements on the same sample), the author indicates a reference for the statistical processing and warns that the confidence intervals are not symmetric: it is evident that a Student t distribution (see sec. 4.1.3.8) was used with a not completely clear number of degrees of freedom; we may summarize the result saying that the overall span of the 99% probability confidence interval is approximately equal to the mean value and it is located approximately 1/3 below the mean value and 2/3 above it.

The results of return loss and insertion loss repeatability are reported in Table 7.4.3 and 7.4.4, respectively.

Juroschek (1987) Juroschek's work [202] was already considered when talking of connector defects and their modeling using an airline measurement (see sec. 7.1.1.7). In the cited paper the results of repeated measurements of a 14 mm connector pair are compatible with the results reported in this section: the largest peak value is 0.0025 at 8 GHz, quite close to the figures obtained by Bergfried in Table 7.4.1, the smallest values at 1 GHz are around 0.0003, so close to the other values appearing in the other tables. However, the reason for those measurements was to demonstrate the

Connector	2 GHz	4 GHz	6 GHz	8 GHz	10 GHz	12 GHz	14 GHz	16 GHz	18 GHz
N-SMA adapters (three samples)	0.0001 0.0001	0.00013 0.00017	0.0001 0.00016	0.0001 0.00025	0.00012 0.0003	0.00012 0.00036	0.00012 0.00046	0.00029 0.00057	0.00040 0.00063
SMA matched load terminations (three samples)	0.00009 0.00016	0.00008 0.00021	0.00014 0.00043	0.00015 0.00064	0.0002 0.00061	0.0003 0.00052	0.00045 0.00062	0.00055 0.00078	0.00040 0.00076
SMA terminated semi-rigid cables 0.141" (three samples)	0.0002 0.00035	0.0002 0.0008	0.0002 0.0012	0.00022 0.0016	0.00025 0.002	0.00035 0.0027	0.0004 0.0042	0.00045 0.00435	0.00065 0.0029

Table 7.4.3 – Repeatability of the insertion loss of SMA connectors measured by Jesch for adapters, matched loads and cable terminations; average values fraction of the reflection coefficient out of the three samples [197].

Connector	2 GHz	8 GHz	12 GHz	18 GHz
N-SMA adapters (three samples)	0.0025	0.0045	0.0050	0.0068
SMA terminated semi-rigid cables 0.141" (three samples)	0.0035	0.0065	0.0095	0.0125

Table 7.4.4 – Repeatability of the insertion loss of SMA connectors measured by Jesch for adapters and cable terminations; average values in dB out of the three samples [197].

model of connector defect, not to evaluate connector repeatability; the author himself underlines that the measurements were done with several VNA calibrations and thus other factors might exert their influence.

Lewandowski (2004) Lewandowski [222] presents several results while discussing the main topic of his Ph.D. thesis, that is modeling of VNA uncertainty. Fig. 6.6 in [222] reports the repeatability of 16 measurements performed on 1.85 mm connectors, where an offset short, an offset open and a matched load where tested (see Table 7.4.5). The real and imaginary parts of the error of reflection coefficient[18] measurements for the three samples are plotted with the 16 traces overlapped followed by the 95% probability confidence intervals predicted by his stochastic model. The frequency range is quite extended, up to the maximum frequency available in the used VNA, namely 67 GHz.

IEEE Std. 287 (2007) The standard reports in Figure 6, 7 and 8 overlapped curves of S_{11} measurements taken up to 40 GHz for a pair of mated 2.92 mm connectors GPC grade and up to 26.5 GHz for 3.5 mm connectors, distinguishing between slotted and slotless versions. Besides the general very low return loss of the tested connectors (around $-70\,\mathrm{dB}$ or better for the first half of the respective frequency ranges, up to $-60\,\mathrm{dB}$ occasionally in the remaining part of the frequency range), the slotless connector show impressively better dispersion, with the curves all with a similar shape and grouped in a band of less than $10\,\mathrm{dB}$; for the other connectors some spread and marks of impedance mismatch are visible, accounting for more than double dispersion.

Regarding insertion loss, there are reported results of S_{21} measurements up to 67 GHz for a pair of mated 1.85 mm connectors GPC grade, showing better than $0.001\,\mathrm{dB}$ spread (not dispersion) up to half the frequency range, slightly increasing by a factor of two for the remaining frequency range. For the phase there are linear changes in the positive and negative half plane for about $0.1°$/10 GHz on average, up to $0.23°$ in a few cases.

The table appearing at page 31 in IEEE Std. 287 is an amazing concentrate of information for all the standardized coax precision connectors if considered together with the various Annexes. Summarizing the prescriptions for repeatability under the maximum allowed tolerances and variability of each connector type, the values appearing in Table 7.4.6 are obtained.

Stewart (1945) Finally, it is important to recall a very old work on a completely different class of connectors[19], where however there is the important information of the influence of the soldering of connectors to cables and the influence of the person who performed the soldering: for reflection coefficients ranging between 0.09 and 0.2, variations of 0.02-0.04 are quite common between operators and there are larger sporadic values of 0.05-0.1, indicating a defect or unsuitable technique. In general, after

---

18 For the reflection coefficient results are expressed in linear quantities and not in dB.

19 C. Stewart, "Electrical Testing of Coaxial Radio-Frequency Cable Connectors," *Proceedings of the IRE*, Sept. 1945, pp. 609-619.

Connector	5 GHz	15 GHz	25 GHz	35 GHz	45 GHz	55 GHz
6.3 mm female offset-short	$\pm0.013+j\pm0.05$	$\pm0.013+j\pm0.08$	$\pm0.018+j\pm0.12$	$\pm0.02+j\pm0.13$	$\pm0.033+j\pm0.22$	$\pm0.033+j\pm0.24$
5.4 mm female offset-open	$\pm0.02+j\pm0.04$	$\pm0.02+j\pm0.1$	$\pm0.024+j\pm0.12$	$\pm0.033+j\pm0.18$	$\pm0.04+j\pm0.27$	$\pm0.03+j\pm0.22$
matched load	$\pm0.015+j\pm0.018$	$\pm0.03+j\pm0.03$	$\pm0.04+j\pm0.035$	$\pm0.04+j\pm0.05$	$\pm0.035+j\pm0.06$	$\pm0.08+j\pm0.08$

Table 7.4.5 – Repeatability of the reflection coefficient of 1.85 mm connectors (6.3 mm female offset-short, 5.4 mm female offset-open and matched load) reported by Lewandowski [222] (16 repeated measurements): the displayed values are in % for the real and imaginary part (the latter always the largest) and correspond to the maximum span observed in the original figures, being impossible to calculate the dispersion (values are derived from the plotted curves by visually weighting the data samples around the frequency values shown in the table header).

Connector	Nominal $\|S_{11}\|$ [dB]	Repeatability $\|S_{11}\|$ [dB]	Nominal $\|S_{21}\|$ [dB]	Repeatability $\|S_{21}\|$ [dB]
Type-N	34 / —	55 / 60	0.04 / 0.035	0.01 / 0.01
3.5 mm	32 / 36	55 / 60	0.3 / 0.3	0.008 / 0.008
2.92 mm	30 / 34	41 / 41	0.3 / 0.3	0.01 / 0.01
2.4 mm	24 / 28	55 / 60	0.15 / 0.135	$0.0115+0.0025\sqrt{f}$ / $0.0115+0.0025\sqrt{f}$

Table 7.4.6 – Repeatability of the return loss S_{11} and insertion loss S_{21} under maximum tolerance and variability of connector characteristics indicated in IEEE Std. 287 [172], Table E.1 (type N), F.1 (3.5 mm), G.1 (2.92 mm), H.1 (2.4 mm).

soldering a bulk-head connector to a cable, a check of its performance is absolutely necessary: of course, for the expected and required performance in terms of return and insertion loss, the uncertainty of calibration and setup shall be correspondingly low.

7.4.2.2 Extrapolation of connector performance and repeatability

A modern work [261] addresses the estimation of SMA characteristics when installed on a PCB setup: the mathematical manipulation of the estimated effect of PCB microstrip line allows the subtraction from the raw measurement, leaving only the input and output SMA connectors. The tested connectors are of the PCB edge type; the installation by soldering to the landing patterns on PCB is quite repetitive, pushing the connector against the PCB edge while it is soldered and centering to the microstrip hot trace, and no significant variability or operator's influence is expected, provided that solder flux and clean surfaces are used. The method is based on the assumption that the characteristics of the microstrip are quite constant and that the determination of its matrix representation based on measurements on a long section may be extrapolated to a much shorter section, that is used to keep the input and output SMA together, ruling out its influence and giving the transfer function of the two edge SMA alone. Three transmission matrices T_{C1}, T_M and T_{C2} are used for input connector 1, microstrip and output connector 2 (transmission matrices are described in sec. 1.3.1.3). T_{C1} and T_{C1} are built around an assumed lumped parameter model for SMA connectors: the authors propose an asymmetric L-C circuit where the total stray inductance and capacitance are assigned to each of the two terms, respectively, but a T or π cell would work as well and would probably be a more elegant solution. A further assumption, necessary to exchange matrix multiplication order and the measurements between setups, is that the SMA connectors are identical. The authors propose the use of a genetic algorithm to fit experimental data and to calculate the parameters of the equivalent circuits: of course, any interpolation, fitting and adjustment prohibits the appreciation of small variations due to repeatability and tends to return an average behavior. However, the method is straightforward and promising: the transmission matrix T_M may be determined by the measurement of the scattering parameters on the longer section by using T-to-S and S-to-T transformations (see sec. 1.3.1.6 and the first equations); then, $T_{C1} = T_{C2}$ may be determined by matrix inversion and simplifications. Once a model reference with two unknowns (stray inductance and capacitance of SMA connectors) is postulated, then it is also possible to determine them from T or S matrix representation, this time only to extract the equivalent circuit of the SMA connector, without introducing those parameters in the main calculations.

The same approach may also be applied when the c.u.t. (always a pair of them, assumed identical) are connected through a cable, rather than a PCB transmission line: extrapolation to nearly zero-length (the minimum for interconnecting the c.u.t.) by e.g. linear regression and matrix manipulation follow. When measuring the long cable section to characterize and compensate for the cable effect, cable screening and its transfer impedance (see sec. 5.1.7) may be improved not to influence appreciably the results, by e.g. wrapping the cable with shielding tape or a metallic pipe.

Of course, the presented method is first useful to determine connector characteristics, rather than its repeatability; however, this configuration is extremely useful to determine the variability of assembled connectors, where the setup may be reproducing the real use, only modifying the overall attenuation and contribution of the connecting line, in favor of connectors under test, to spot out and characterize them at best.

7.4.2.3 Practical evaluation of connector repeatability

In this section we briefly consider an experiment where the effect of connector repeatability was evaluated by simulating various connections/disconnections operations and measuring the response every some cycles.

The VNA was calibrated with a full 2-port SOLT standard, using a low-loss flexible precision Sucoflex® cable as Through. The same Through cable was then connected and disconnected many times, under different conditions. The verification of repeatability of standard SMA and N connectors was made during the measurement of the attenuation of 6 dB pads (see Figure 11.7.2 for results regarding 6 dB attenuators). Making measurements with some dB of attenuation in the Through connection has the advantage of reducing greatly reflections coming from mismatch errors at the far VNA port; this possibility is foreseen in the VNA calibration methods (in Eul and Schiek's classification the "A" stands for attenuation, see sec. 11.5.3.8).

The 6 dB pads were measured by inserting them one by one between one end of the Through cable and port 2 of the VNA (results are shown in Figure 7.4.1). After the entire set of measurements (18 in all), the setup was verified and the Through cable measured again (light gray curve); in this case the N-SMA adapters were still in place and never removed. After this, other measurements were performed and then the repeatability was tested by first unscrewing and screwing again both N-SMA adapters, with care to the tightening torque and a correct vertical insertion of the N connectors onto the VNA ports (dark gray curve); the N-SMA adapters have been operated the least number of times. Then the SMA connectors of the cable were operated by different technicians with no specific knowledge on microwaves, repeatability and uncertainty (thin black curves represent some of the measurements), just instructed and conscious that the SMA connectors shall be firmly fastened with general care not to damage the thread and mating surfaces. Finally the cable was connected again using the usual care for the tightening torque and cleanliness of surfaces and threads (the thick black curve).

The three reference tests (thick lines) are quite repeatable: the other two curves show a maximum 0.015 dB deviation from the light gray curve; the black curve of the last test closing the series is only slightly worse (in the S_{11} curve this is more evident and differences are "amplified"), demonstrating that the system has no memory of the past measurements, all performed within 1 hour, with only a slight change of the internal temperature of the VNA. Of course, the observed deviations are larger than those of the former experiments of Bergfried-Fischer and Pakay-Torok, but reflect the typical real use of connectors. Small fluctuations of internal instrumentation temperature were unavoidable; once the warm-up time has elapsed, internal temperature regulation operates autonomously and there is no external conditioning possible that avoids the

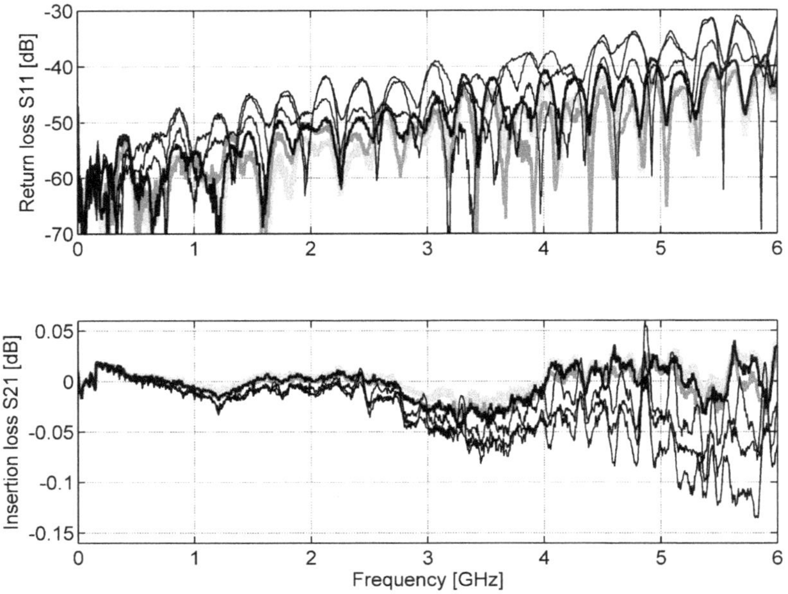

Figure 7.4.1 – Practical repeatability of SMA and N connectors (commercial grade): practical verification during a test by repeatedly measuring the Through cable of the calibration: initial measurement of the Through (thick light gray); repeated measurement after 18 operations (thick dark gray); several measurements with no specific attention to mating and tightening torque (thin black); final measurement with specific care to mating and tightening torque (thick black).

small fluctuations of about 1-2 °C; on the contrary, avoiding the influence of the operator's hands temperature is of no relevance in this case, with the VNA port connector and the connected adapter at about 40 °C of internal temperature.

7.4.3 Passive intermodulation (PIM)

As already introduced at the beginning of this Chapter, connectors may be a source of intermodulation products. The mechanism is quite complex and depends on many factors (such as finishing, type of surface, used metals for body and plating), but there is general agreement that it occurs at surface contact: the identified mechanisms are dirt and surface roughness, tunneling effect at oxide barriers, self heating, ferromagnetic materials. A general treatment of PIM sources is available in N. B. Carvalho and J. C. Pedro, *Intermodulation Distortion in Microwave and Wireless Circuits*, Norwood, MA, Artech House, 2003.

An experimental evaluation is well described in [154], where different connectors were tested at a relatively large power level (43 dBm) and 3rd order PIM products were observed in the range -60 to $-140\,\mathrm{dBm}$. The considered connectors are Pasternack and Spinner connectors, of the BNC, SMA, N and 7/16 type and for some connectors two versions were available, gold plated (only the center pin, not the outer contact that was made of nickel or stainless steel) and silver plated. The authors report a wide range of values, but the many measurements performed on each connector type are all grouped together, indicating a very good consistency and repeatability:

- Pasternack Enterprises SMA connector, center pin gold plated, outer contact nickel (part # PE 9103, PE 9081, and PE 9082) or stainless steel (part # PE 9433 and PE 9506): PIM between -60 and $-70\,\mathrm{dBm}$ with a measurement error $< 0.5\,\mathrm{dB}$;
- Pasternack Enterprises BNC connector, center pin gold plated, outer contact nickel (part # PE 9002 and PE 9127): PIM between -70 and -70 dBm with a measurement error $< 2\,\mathrm{dB}$;
- Pasternack Enterprises N connector, center pin gold plated, outer contact nickel (PE 9311) or stainless steel (PE 9426): PIM between -75 and -85 dBm with a measurement error $< 0.5\,\mathrm{dB}$;
- Spinner GmbH N connector, center pin and outer contact silver plated (part # BN 950890 and BN 203834): PIM between -115 and -125 dBm with a measurement error between $-8\,\mathrm{dB}$ and $+4\,\mathrm{dB}$;
- Spinner GmbH 7/16 connector, center pin and outer contact silver plated (part # BN 756404 and BN 203391): PIM $< -125\,\mathrm{dBm}$, lower than the residual intermodulation distortion of the measurement system and thus with a not quantifiable error.

A similar evaluation is synthetically reported in a slide collection[20], where Pasternak connectors were tested for two-tone PIM at variable frequency separation of the two sinusoidal tone signals with the following results:

- Pasternack Enterprises PE 6154 shows a PIM of about -70 to $-90\,\mathrm{dB}$ for tone separation larger than 100 Hz (fed at 2 W);
- Pasternack Enterprises PE 6152 shows a PIM of about -100 to $-115\,\mathrm{dB}$ for tone separation larger than 100 Hz (fed at 2 W);
- Pasternack Enterprises PE 6097 termination shows a PIM of about -75 to $-105\,\mathrm{dB}$ for tone separation larger than 100 Hz (fed at 5 W);
- Pasternack Enterprises PE 6035 termination shows a PIM of about -85 to $-105\,\mathrm{dB}$ for tone separation larger than 100 Hz (fed at 10 W).

In these slides there are also reported further references to PIM subject:

- J. R. Wilkerson, K. G. Gard, A. G. Schuchinsky, and M.B. Steer, "Electro-thermal theory of intermodulation distortion in lossy microwave components," *IEEE Transactions on Microwave Theory and Techniques*, vol. 56, no. 12, pp. 2717-2725, Dec. 2008.

---

[20] M. Steer, G. Mazzaro and J. Wetherington, "Passive Intermodulation Distortion, Part 1", "Passive Intermodulation Distortion, Part 2", North Carolina University, 2011

- J.R. Wilkerson, P.G. Lam, K.G. Gard, and M.B. Steer, “Distributed passive intermodulation distortion on transmission lines,” *IEEE Transactions on Microwave Theory and Techniques*, vol. 59, no. 5, pp. 1190-1205, May. 2011.
- J.R. Wilkerson, K.G. Gard, and M.B. Steer, “Electro-thermal passive intermodulation distortion in microwave attenuators,” Proc. 36th Eur. Microwave Conf., 10-15 Sept. 2006, pp. 157-160.
- G.J. Mazzaro, M.B. Steer, and K.G. Gard, “Intermodulation distortion in narrowband amplifier circuits,” *IET Microwaves, Antennas, & Propagation*, April 2010, pp. 1149–115.
- A. Christianson, W.J. Chappell, “Measurement of ultra low passive intermodulation with ability to separate current/voltage induced nonlinearities,” 2009 IEEE MTT-S International Microwave Symposium Digest, June 2009, pp. 1301–1304.
- A. Christianson, J. Henrie, and W.J. Chappell, “ Higher order intermodulation product measurement of passive components,” *IEEE Transactions on Microwave Theory and Techniques*, July 2008, pp. 1729–1736.
- J. Henrie, A. Christianson, W.J. Chappell, “ Linear–nonlinear interaction and passive intermodulation distortion,” *IEEE Transactions on Microwave Theory and Techniques*, May 2010, pp. 1230–1237.

... and a short presentation of time-frequency effects noticed in microwave components in the last 70 years.

To scientific literature the IEC 62037 family of standards may be added: the general title of standards from Part 1 to 6 is “Passive RF and microwave devices, intermodulation level measurement”, detailing then for coaxial cable assemblies, coaxial connectors, coaxial cables and filters and antennas. Publications dates are between 2012 and 2013.

Of course not tightening connectors has a remarkable effect on distortion: the authors show a loose N connector with variations in few seconds of 40 dBm of PIM intensity, stabilizing to the lowest values when tightened with the prescribed torque.

Regarding, on the contrary, the dependence of PIM on the number of connection and disconnection cycles or in general the variability from connection to connection, the made experiments confirm that a SMA connector after 50 cycles has shown no trend in the PIM level and a very small dispersion of only 0.24 dB.

When higher test power levels are not available, to extrapolate the measured PIM, assuming the PIM is due to 3rd order intermodulation distortion, there will be an increase by 3 dB for each dB of power increase at the fundamental signals.

PIM is not strongly frequency dependent, but being in part influenced by skin effect, a slight increase with frequency may be expected, even if until now there is no experimental evidence.

Finally, the PIM is influenced by temperature, that affects the quality of the electrical contact at the mechanical interfaces, but there is no clear evidence of the effect and if the temperature dependency is positive or negative.

8

Coupled transmission lines and crosstalk

It is effective to begin with the definition given by Wadell [332], who says that "when two transmission lines are in close enough proximity for their field patterns to be disrupted (relative to their isolated patterns) it is discovered that a portion of the signal in one line is also present in the second line. This property can be desired (...) or undesired (...). In the former case we use term the second signal *coupling*, in the latter we refer to *crosstalk*."

Two approaches may be used to model the behavior of multiple transmission lines and their mutual effect, including that of terminations: a circuit approach, where the complete set of equations for conductor voltages and currents is laid down and then solved, and a modal approach, where propagation modes are distinguished and a more synthetic and engineering oriented approach is applied. The first approach is suitable for circuit simulation (e.g. using Spice); the second approach formulates the problem in the engineer's way, using parameters related more to practice and measured values, and aims at reaching sound, even if approximate, results to put in relationship with experimental evidence.

8.1 Coupled transmission line equations

Two lines are said to be coupled when the electromagnetic field supported by a transmission line can induce an electromagnetic field on another transmission line with a non-negligible intensity. What "non-negligible" means, of course, highly depends on the application, on susceptibility of connected devices, on signal amplitude, etc..

It is preferred to derive first a set of equations that are applicable to the cases that are considered in the rest of the chapter. The reader may skip this section and the next one if a more practical approach is desired.

We choose the most general circuit scheme, consisting of two transmission line pairs (or simply "pairs"), that might or may not share the return conductor: pair 1 is the culprit (as it is usually called in the "EMC jargon"), featuring a signal source at one side (the near end) and a load at the other side (the far end); the other line is the victim line and has its two ends terminated.

The telegrapher's equations valid for a single transmission line are integrated by mutual terms, the mutual impedance Z_{ms} and admittance Y_{mp}, representative of magnetic and electric coupling terms, respectively.

$$\frac{\mathrm{d}v_1(x)}{\mathrm{d}x} = -Z_{s1}i_1(x) - Z_{sm}i_2(x) \tag{8.1.1}$$

$$\frac{\mathrm{d}v_2(x)}{\mathrm{d}x} = -Z_{s2}i_2(x) - Z_{sm}i_1(x) \tag{8.1.2}$$

$$\frac{\mathrm{d}i_1(x)}{\mathrm{d}x} = -Y_{p1}v_1(x) - Y_{pm}v_2(x) \tag{8.1.3}$$

$$\frac{\mathrm{d}i_2(x)}{\mathrm{d}x} = -Y_{p2}v_2(x) - Y_{pm}v_1(x) \tag{8.1.4}$$

The Z and Y terms are expressed in p.u.l., even if not strictly necessary.

Let's assume constant impedance and admittance terms with respect to position, so the lines are uniform. It is possible to derive once eq. (8.1.1) and (8.1.2), and

$$\frac{\mathrm{d}^2v_1(x)}{\mathrm{d}x^2} = -Z_{s1}\frac{\mathrm{d}i_1(x)}{\mathrm{d}x} - Z_{sm}\frac{\mathrm{d}i_2(x)}{\mathrm{d}x} \tag{8.1.5}$$

$$\frac{\mathrm{d}^2v_2(x)}{\mathrm{d}x^2} = -Z_{s2}\frac{\mathrm{d}i_2(x)}{\mathrm{d}x} - Z_{sm}\frac{\mathrm{d}i_1(x)}{\mathrm{d}x} \tag{8.1.6}$$

Then, substituting the remaining eq. (8.1.3) and (8.1.4), we finally obtain a harmonic equation in the voltages for each transmission line

$$\frac{\mathrm{d}^2v_1(x)}{\mathrm{d}x^2} = -A_1v_1(x) - B_1v_2(x) \tag{8.1.7}$$

$$\frac{\mathrm{d}^2v_2(x)}{\mathrm{d}x^2} = -A_2v_2(x) - B_2v_1(x) \tag{8.1.8}$$

The notation for A and B coefficients have been purposely chosen in capital letters not to confuse them with the coefficients used to indicate the amplitude of the incident and reflected waves used elsewhere (see Chapter 10). They are related to Y and Z quantities as follows:

$$A_1 = -Z_{s1}Y_{p1} - Z_{s1}Y_{pm} \quad A_2 = -Z_{s2}Y_{p2} - Z_{s2}Y_{pm} \tag{8.1.9}$$

$$B_1 = -Z_{s1}Y_{pm} - Z_{sm}Y_{p2} \quad B_2 = -Z_{s2}Y_{pm} - Z_{sm}Y_{p1} \tag{8.1.10}$$

The two equations are clearly coupled by the presence of quantities belonging to both transmission lines in each equation. By further substitution, it is possible to uncouple the two equations, obtaining two fourth-order linear differential equations in $v_1(x)$ or $v_2(x)$ only.

$$\frac{\mathrm{d}^4 v_1(x)}{\mathrm{d}x^4} - (A_1 + A_2)\frac{\mathrm{d}^2 v_1(x)}{\mathrm{d}x^2} - (A_1A_2 - B_1B_2)\, v_1(x) = 0 \tag{8.1.11}$$

$$\frac{\mathrm{d}^4 v_2(x)}{\mathrm{d}x^4} - (A_1 + A_2)\frac{\mathrm{d}^2 v_2(x)}{\mathrm{d}x^2} - (A_1A_2 - B_1B_2)\, v_2(x) = 0 \tag{8.1.12}$$

It is known that the transmission line propagation problem admits an exponential wave as solution, here of the form $v_1(x) = V_1 e^{\gamma x}$, but in this case the exponent has four solutions: $\gamma_1 = \gamma_I$, $\gamma_2 = -\gamma_I$, $\gamma_3 = \gamma_{II}$ and $\gamma_4 = -\gamma_{II}$. The two quantities γ_I and γ_{II}[1], i.e. the wave velocity factors (or propagation constants, see sec. 2.1.1) for the two line modes, are:

$$\gamma_I = \sqrt{\frac{1}{2}(A_1 + A_2) + \frac{1}{2}\sqrt{(A_1 - A_2)^2 + 4B_1B_2}} \tag{8.1.13}$$

$$\gamma_{II} = \sqrt{\frac{1}{2}(A_1 + A_2) - \frac{1}{2}\sqrt{(A_1 - A_2)^2 + 4B_1B_2}} \tag{8.1.14}$$

and they correspond to two modes propagating along the line with forward and backward terms each, corresponding in turn to the minus and plus sign of the solution. Thus the general solution for $v_1(x)$ is a linear combination of four exponential waves, and the same may be said for the other quantities, $v_2(x)$, $i_1(x)$ and $i_2(x)$.

$$v_1(x) = V_{1I}e^{-\gamma_I x} + V_{1I}e^{+\gamma_I x} + V_{1II}e^{-\gamma_{II} x} + V_{1II}e^{+\gamma_{II} x} \tag{8.1.15}$$

Proceeding in the analysis of coefficients that characterize propagation and coupling between lines, if we take the ratio $R(x)$ of voltages for the two solution pairs $\pm\gamma_I$ and $\pm\gamma_{II}$, we have:

[1] Indicated as γ_c and γ_π in the original work by Tipathi: V.K. Tripathi, "Asymmetric Coupled Transmission Lines in an Inhomogeneous Medium," *IEEE Transactions on Microwave Theory and Techniques*, Vol. 23, no. 9, Sept. 1975, pp. 734-739. doi: 10.1109/TMTT.1975.1128665

$$R(x)=\frac{v_2(x)}{v_1(x)}=\frac{B_2}{k^2-A_2}=$$

$$=\begin{cases}\dfrac{\left[A_2-A_1+\sqrt{(A_1-A_2)^2+4B_1B_2}\right]}{2B_1}=R_c \quad \text{for } \gamma=\pm\gamma_I \\[2ex] \dfrac{\left[A_2-A_1-\sqrt{(A_1-A_2)^2+4B_1B_2}\right]}{2B_1}=R_p \quad \text{for } \gamma=\pm\gamma_{II}\end{cases} \tag{8.1.16}$$

The voltage $v_2(x)$ appearing on the other line, by invoking the ratio we have just defined, is

$$\begin{aligned} v_2(x) &= R_I V_{1I} e^{-\gamma_I x} + R_I V_{1I} e^{+\gamma_I x} + \\ &+ R_{II} V_{1II} e^{-\gamma_{II} x} + R_{II} V_{1II} e^{+\gamma_{II} x} \end{aligned} \tag{8.1.17}$$

The solution for the current $i_1(x)$ is derived again from eq. (8.1.15):

$$\begin{aligned} i_1(x) &= Y_{I1} V_{1I} e^{-\gamma_I x} - Y_{I1} V_{1I} e^{+\gamma_I x} + \\ &+ Y_{II1} V_{1II} e^{-\gamma_{II} x} - Y_{II1} V_{1II} e^{+\gamma_{II} x} \end{aligned} \tag{8.1.18}$$

$$\begin{aligned} i_2(x) &= Y_{I2} R_I V_{1I} e^{-\gamma_I x} - Y_{I2} R_I V_{1I} e^{+\gamma_I x} + \\ &+ Y_{II2} R_{II} V_{1II} e^{-\gamma_{II} x} - Y_{II2} R_{II} V_{1II} e^{+\gamma_{II} x} \end{aligned} \tag{8.1.19}$$

where the Y terms have been calculated from self and mutual impedance terms, including the respective propagation constants γ_I and γ_{II} :

$$Y_{I1}=\gamma_I \frac{Z_{s2}-R_I Z_{sm}}{Z_{s1}Z_{s2}-Z_{sm}^2} \qquad Y_{I2}=\frac{\gamma_I}{R_I}\frac{R_I Z_{s1}-Z_{sm}}{Z_{s1}Z_{s2}-Z_{sm}^2} \tag{8.1.20}$$

$$Y_{II1}=\gamma_{II} \frac{Z_{s2}-R_{II} Z_{sm}}{Z_{s1}Z_{s2}-Z_{sm}^2} \qquad Y_{II2}=\frac{\gamma_{II}}{R_{II}}\frac{R_{II} Z_{s1}-Z_{sm}}{Z_{s1}Z_{s2}-Z_{sm}^2} \tag{8.1.21}$$

Now a few considerations follow on some peculiar conditions.

If the transmission lines are identical (i.e. "symmetrical"), then $Z_{s1}=Z_{s2}=Z_s$ and $Y_{II1}=Y_{II2}=Y_{II}$, so that the A and B coefficients are also equal ($A_1=A_2$ and $B_1=B_2$), and thus $R_I=1$ and $R_{II}=-1$.

Well, the condition $R_I=1$ brings to a corresponding mode that is called *even*, where the voltages on the two lines compatibly with the propagation constraints are identical and with the same phase; conversely, for $R_{II}=-1$ the mode is said *odd* and the voltages are opposite in phase (180° out of phase), but always of the same amplitude. The resulting Y terms are then named $Y_e=Y_{c1}=Y_{c2}$ and $Y_o=Y_{p1}=Y_{p2}$ for the even- and odd-mode, respectively.

This condition has resulted into a useful decomposition of modes in even and odd ones, that in reality may be used as a decomposition basis in general, also when the transmission line is not symmetrical, by simply invoking the superposition of effects. There will be the need of defining two different characteristic impedances for the two modes: $Z_{c,e}$ is the *even-mode characteristic impedance*, when even excitation is applied and the other line is present (i.e. its influence is taken into account, see sec. 8.2.2 below); analogously, $Z_{c,o}$ is the *odd-mode characteristic impedance,* when odd excitation is applied and the other line is present.

Another condition that is worth a comment is the one that brings to $\gamma_I = \gamma_{II}$: it occurs if both conditions

$$Y_{II1} Z_{s1} = Y_{II2} Z_{s2} \,, \qquad Y_{IIm} / Z_{sm} = -(Y_{II1} Y_{II2}) / (Z_{s1} Z_{s2})$$

are satisfied. In this case the transmission lines are said to be in "homogenous medium".

8.2 Mode decomposition

Signals at the ports propagating through a line pair may be decomposed into modes to ease interpretation and analysis. From previous section it is evident that modes where the electrical quantities are subtracted and summed (odd and even, we might say) are the preferred ones: odd and even modes will be attributed to the propagation inside the transmission line, while differential and common mode are used for the signals at the ports, either e.g. applied by a source or received by a connected load.

We now consider further the case of two coupled transmission lines described by their even- and odd-mode representations, that allow to represent any applied signal, while keeping the two lines separate.

Let's take two transmission lines (as shown in Figure 8.2.1), defined by four single-ended ports named 1, 2, 3 and 4, with an excitation source E at port 1 and terminating impedances Z_1, Z_2, Z_3 and Z_4 (Z_1 is e.g. the internal impedance of the voltage source or a combination of the internal source impedance and added impedance. The reason for invoking this port numbering is to establish a notation that will be used also in the next chapters for S parameters and VNA.

For the values of terminating impedances three cases are normally considered in the literature, depending on the use and aim of the analysis:

- if it deals with RF devices and technology [108], the assumption of equal terminating impedances matched to the reference system impedance (e.g. $50\,\Omega$) is almost always applied: results are based on this case, followed by considerations on input impedance matching, coupling factor between lines, etc.;
- if focusing on RF EMC, such as for signal integrity [45] and noise coupled on high data-rate lines, conclusions cover the problem of common-mode signals rejection, mode transformation, including slight deviations from the matched terminating impedance conditions because of non-ideality or imperfections;

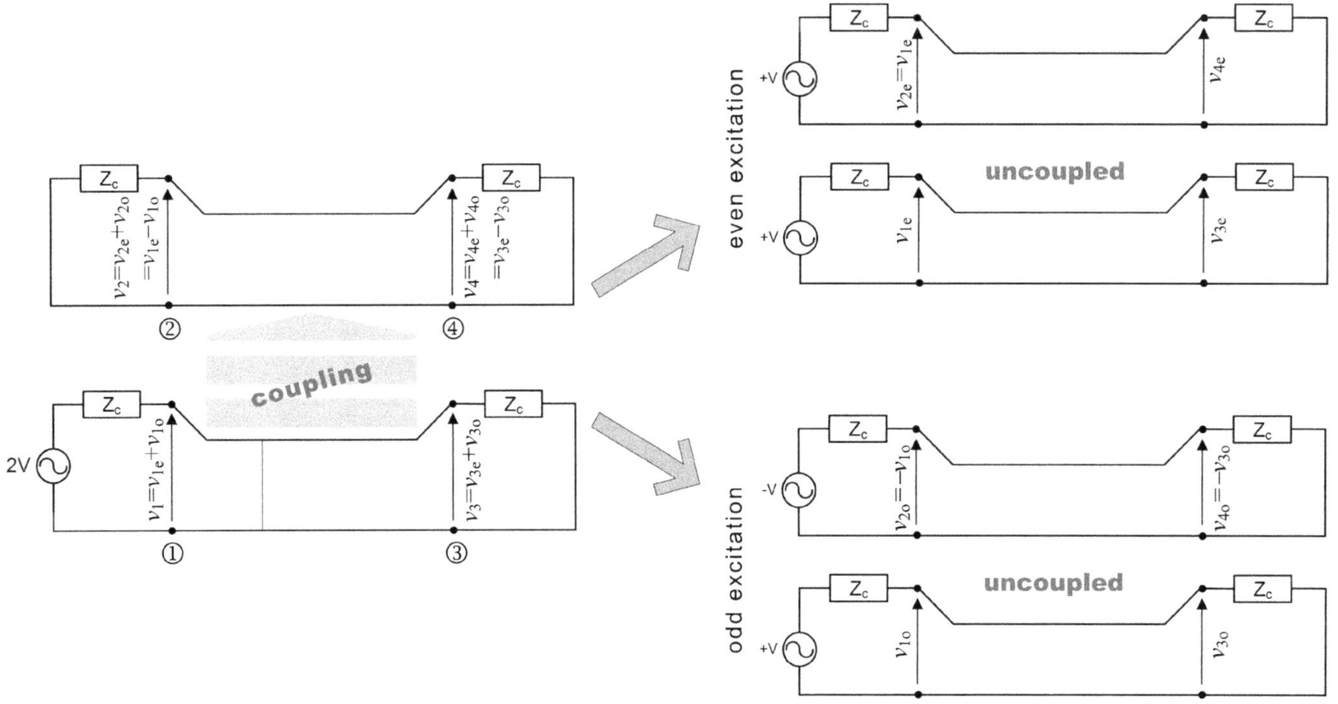

Figure 8.2.1 – Two transmission lines with excitation and terminating impedances: (left) complete circuit; (right top) even-mode circuit; (right bottom) odd-mode circuit

- if on the contrary the focus is on a wide range of coupling phenomena occurring at a moderate high frequency on fairly long lines, as in the case of large installations and their cable harness, then terminating impedance values may vary largely, from a nearly short-circuit condition to very large values, assimilated to an open-circuit condition; Paul presents a theoretical frame [264] to model crosstalk between an arbitrary number of conductors and the so-formed transmission lines, called Multiconductor Transmission Line (MTL) modeling, that is however beyond the scope of this book.

It is remembered that when talking about characteristic impedance, it shall be interpreted as the characteristic impedance of one line with the other present, i.e. including the effects of mutual coupling and applied signals, as in sec. 8.3.1.

Let's take two single-ended transmission lines that share the same return path (see Figure 8.2.2(a)); the structure can be visualized as two PCB traces running parallel on one layer and sharing a common return plane underneath. The two traces might even represent the two conductors of a differential pair, if properly driven; but they share a common plane that is normally used for voltage reference. This is the simplest structure to study crosstalk [45, 264]. When considering PCB implementation, each conductor forms a microstrip (see sec. 6.3.1) with the return plane; in the more general case we have two transmission lines that are sharing the return conductor at some extent (that is the spread of the image current of one trace in the return plane might overlap with the other, depending on frequency, material properties and geometry, as it will be seen in sec. 8.3.3).

It is extremely important to distinguish two aspects of mutual influence between lines:

- crosstalk occurs when the signal propagating on one line (line 1) causes a signal induced on the other line (line 2) by mutual coupling (reactive coupling is due to mutual inductive or capacitive terms, conductive coupling is due to common impedance, e.g. through the return plane); this signal may than superimpose and being received by loads of line 2;

- the characteristic impedance of line 1 and line 2 taken alone is correctly defined in terms of self terms, being the mutual terms zero; when the two lines are considered together, the mutual coupling terms modify the so obtained characteristic impedance; being the mutual coupling terms driven by the signals on line 1 and line 2, the effective value (or degree) of mutual coupling is a result of all the signals applied in a given instant of time; this means that the characteristic impedance of one line (line 1) is subject to change with the signals applied on the other line (line 2), and the characteristic impedance has influence on the amount of reflections and signal integrity of the same line (line 1).

A more complex configuration is when two transmission line pairs form two distinct transmission lines, with or without a common plane (see Figure 8.2.2(b)). Odd and even modes are still used and they are applied selecting one mode on each line and defining the relationship between them; this representation is that of mixed-mode scattering parameters, reviewed in sec. 10.4.2.

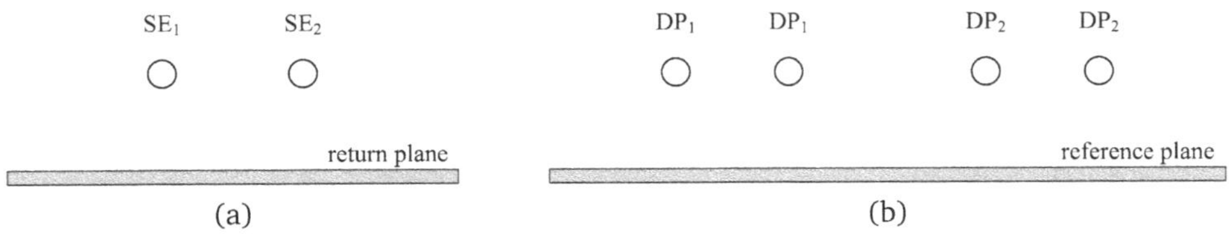

Figure 8.2.2 – Line configurations used to evaluate coupling: (a) single-ended lines with a common return path, (b) two distinct differential pairs with common reference plane (if correctly used differential pairs do not need the return plane to correctly propagate signals and thus the name "reference plane" was preferred).

To analyze coupling we must introduce the concept of common mode (c.m.) and differential mode (d.m.) for externally applied signals[2]; even mode and odd mode are characteristic of the line structure, and not of the external excitation.

8.2.1 Differential and common mode decomposition

Considering the example in Figure 8.2.1, differential and common modes at the ports are simply defined by considering the terminal voltages as:

$$V_d = V_1 - V_2 \qquad\qquad V_c = \frac{V_1 + V_2}{2} \tag{8.2.1}$$

We might guess that for currents (assumed entering the line section for a matter of clarity) we will have a similar notation; yes, but with a slight asymmetry, because signal power in the two notations (that of terminal quantities, P_1 and P_2, and that of differential and common-mode quantities, P_d and P_c) shall preserve.

$$I_d = \frac{I_1 - I_2}{2} \qquad\qquad I_c = I_1 + I_2 \tag{8.2.2}$$

In this way the power terms are:

$$\begin{aligned} P_{tot} &= P_d + P_c = V_d I_d + V_c I_c = (V_1 - V_2)\frac{I_1 - I_2}{2} + \frac{V_1 + V_2}{2}(I_1 + I_2) = \\ &= \cdots = V_1 I_1 + V_2 I_2 = P_1 + P_2 \end{aligned} \tag{8.2.3}$$

8.2.2 Odd- and even-mode decomposition

Besides being supported by demonstration in sec. 8.1, odd- and even-mode decomposition is pragmatically based on the straight choice of difference and sum of terminal components, so that they differ from the previously defined differential- and common-mode quantities only for a factor $\frac{1}{2}$ in either of the two relationships.

---

[2] It is commented briefly in the next section on the use of the expressions "common and differential signals", rather than "common mode and differential mode signals".

Besides the purely electric-geometrical interpretation of driving the two conductors either at the same or at opposite potentials, the two modes in reality are the eigenvectors of the circuit for a symmetric configuration; for the general case of asymmetric lines, two eigenvectors result from a linear combination (whatever the representation of the line, in terms of impedance matrix, admittance matrix, etc.).

8.2.3 Derivation of modes in the general case

In the general case of a line for which a complete impedance representation at the two ports is available,

$$\begin{bmatrix} V_1 \\ V_2 \end{bmatrix} = \begin{bmatrix} Z_{11} & Z_{12} \\ Z_{21} & Z_{22} \end{bmatrix} \begin{bmatrix} I_1 \\ I_2 \end{bmatrix} \tag{8.2.4}$$

odd- and even-mode impedance may be determined by applying an odd- or even-mode current, respectively, and measuring the resulting voltage.
For odd-mode current, with $I_1 = -I_2 = I$ and $I_d = (I_1 - I_2)/2 = I$

$$V_1 = (Z_{11} - Z_{12})I \qquad V_2 = (Z_{21} - Z_{22})I \tag{8.2.5}$$

$$V_d = V_1 - V_2 = [Z_{11} - Z_{12} - (Z_{21} - Z_{22})]\, I \tag{8.2.6}$$

and thus

$$Z_{dm} = (Z_{11} + Z_{22}) - (Z_{12} + Z_{21}) = * = 2(Z_{11} - Z_{12}) \tag{8.2.7}$$

where * indicates simplification by invoking line symmetry.
Analogously, for even mode, $I_1 = I_2 = I$ and $I_c = I_1 + I_2 = 2I$

$$V_1 = (Z_{11} + Z_{12})I \qquad V_2 = (Z_{21} + Z_{22})I \tag{8.2.8}$$

$$V_c = \frac{V_1 + V_2}{2} = \frac{Z_{11} + Z_{12} + Z_{21} + Z_{22}}{2} I \tag{8.2.9}$$

and thus

$$Z_{cm} = \frac{Z_{11} + Z_{12} + Z_{21} + Z_{22}}{4} = * = \frac{Z_{11} + Z_{12}}{2} \tag{8.2.10}$$

Recalling that odd- and even-mode characteristic impedances are defined as the ratio of the respective voltages $Z_{c,o} = V_o/I_o$ and $Z_{c,e} = V_e/I_e$, and that odd- and even-mode voltages and currents are difference and sum of single-ended quantities

$$Z_{c,o} = \frac{V_1 - V_2}{I_1 - I_2} = \frac{1}{2} Z_{dm} \qquad Z_{c,e} = \frac{V_1 + V_2}{I_1 + I_2} = 2Z_{cm} \tag{8.2.11}$$

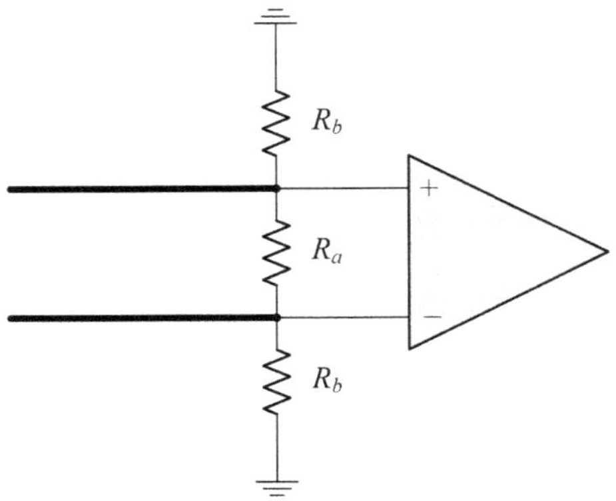

Figure 8.2.3 – Example of line pair driving a receiver and its optimal impedance-matching termination.

Checking on a simplified case of two 50 Ω coaxial lines running parallel and absolutely decoupled, $Z_{11} = Z_{22} = 50\,\Omega$ and $Z_{12} = Z_{21} = 0\,\Omega$, and then $Z_{dm} = 100\,\Omega$ and $Z_{cm} = 25\,\Omega$; as a consequence we may calculate $Z_{c,o} = 50\,\Omega$ and $Z_{c,e} = 50\,\Omega$.

Example 8.1: Impedance-matching termination of line pair

The general case of a line pair carrying a digital signal in differential format (e.g. LVDS, see sec. 8.3) connected to a receiver at the far end is considered: the receiver features ideally an infinite input impedance (no load voltage pick-up) and we are interested in matching both modes of propagation by including a passive resistive network as shown in Figure 8.2.3. The resistor between the two line conductors when a common-mode signal is applied with the same potential for both conductor is not "seen" by the signal because its voltage drop is zero; it is thus identified as the resistor for differential-mode components. Conversely, we add two more resistors between each conductor and ground: they are surely effective for common-mode components propagating in even mode, but exert some effect also for the differential part. We shall thus select the values so that

$$R_b = Z_{c,e} \qquad R_b // \frac{R_a}{2} = Z_{c,o} \tag{8.2.12}$$

$$R_a = \frac{2 Z_{c,e} Z_{c,o}}{Z_{c,e} - Z_{c,o}} \tag{8.2.13}$$

In the simplified configuration of two decoupled coaxial cables it is evident that they do not need differential mode termination (the denominator goes to zero and R_a is infinite) and that R_b is exactly the single-ended characteristic impedance, each coax terminated independently.

8.2.4 Mode conversion

Mode conversion occurs when feeding the line in one configuration that is propagating a pure mode (e.g. even mode, with equal currents applied) and then the other mode component happens not to be zero, in this case the differential-mode component.

Based on an impedance perspective, a straightforward rationale is the following.

For odd-mode current, with $I_1 = -I_2 = I$

$$V_1 = (Z_{11} - Z_{12})I \qquad V_2 = (Z_{21} - Z_{22})I \tag{8.2.14}$$

and the common-mode conversion is given by

$$V_{c,\text{cnv}} = \frac{V_1 + V_2}{2} = \frac{Z_{11} - Z_{12} + Z_{21} - Z_{22}}{2} I \tag{8.2.15}$$

For even-mode current, $I_1 = I_2 = I$

$$V_1 = (Z_{11} + Z_{12})I \qquad V_2 = (Z_{21} + Z_{22})I \tag{8.2.16}$$

and the differential-mode conversion is given by

$$V_d = V_1 - V_2 = [Z_{11} + Z_{12} - Z_{21} - Z_{22}]\, I \tag{8.2.17}$$

No mode conversion occurs if the expressions in square brackets are zero and this corresponds to the condition that the line is symmetric: $Z_{12} = Z_{21}$ and $Z_{11} = Z_{22}$.

8.2.5 Delay skew and differential mode attenuation

Time skew may occur because of both different propagation velocities for the two modes (and this will undergo the same requirement of symmetric line) and different lengths for either of the two conductors of the line pair: using complex quantities, impedance values Z_{ij} will result rotated and thus the previously stated equality for symmetry will not hold any longer if phase or complex numbers are considered. More practically, time delay t_d or phase rotation $2\pi f t_d$ may be introduced and then voltages at the other end of the line calculated, using e.g. sinusoidal signals as reference: for both V_d and V_c propagated at the end of the line we may calculate the corresponding power terms that will result in a reduction and increase for increasing time-delay skew, respectively.

We have already used interchangeably delay and phase rotation to mean that either a time-domain or a frequency-domain interpretation may be used.

$$V_1 = \text{V}\cos(2\pi f_0 t) \qquad V_2 = -\text{V}\cos(2\pi f_0 t) \qquad V_{2,\text{sk}} = -\text{V}\cos(2\pi f_0 (t + t_d)) \tag{8.2.18}$$

Decomposing arguments always to have $\pi f_0 t_d$ terms isolated,

$$V_{d,\text{sk}} = V_1 - V_{2,\text{sk}} = 2\cos\left(2\pi f_0\left(t + \frac{t_d}{2}\right)\right)\cos(\pi f_0 t_d) \tag{8.2.19}$$

$$V_{c,\text{sk}} = \frac{V_1 + V_{2,\text{sk}}}{2} = \frac{1}{2}\sin\left(2\pi f_0\left(t + \frac{t_d}{2}\right)\right)\sin(\pi f_0 t_d) \tag{8.2.20}$$

The respective magnitudes may be easily calculated, as well as power terms, once the two components are squared and divided by the respective impedances, Z_{dm} and Z_{cm}.

The rightmost term in eq. (8.2.19) reduces progressively as long as the argument increases, that is either the delay skew t_d or the signal frequency f_0 (and by extension its bandwidth) increase: thus, having assumed lossless lines, the differential component transmitted in the presence of delay skew is attenuated as if the line pair has attenuation (simply it is transformed into useless common-mode signal); delay skew shall be limited not to reduce too much the effective channel bandwidth, that is the maximum transmittable frequency for an acceptable signal attenuation, e.g. $-3\,\text{dB}$. Correspondingly, the $\sin(\pi f_0 t_d)$ term in the common-mode voltage increases with t_d and f_0, thus supporting the statement that differential-to-common mode conversion occurs.

8.3 Two coupled single-ended lines and a common return plane

The combination of two conductors and a common return plane to form a pair is the first example of coupled transmission line: two voltages may be applied independently at either conductor of the pair, both referenced to the underlying return plane. Let's call V_1 and V_2 the voltage applied to conductor 1 and 2, with a subscript i for input side and o for output side. In the normal and most intuitive use of the pair, a "differential signal" is applied at the input side and is transmitted along the pair to the other end, where a differential receiver measures it. Yet, single voltages applied to conductors may be characterized also by a non-null common mode component.

This is more clear if a typical example of a differential signal is considered. LVDS drivers apply two strictly positive node voltages swinging between 1.125 V and 1.375 V with a signal difference of 250 mV, which is our differential signal. Common-mode and differential-mode components are thus 1.25 V and 125 mV, respectively, as shown in Figure 8.3.1. It is evident that a useful signal of 125 mV is transmitted at the expense of a ten times larger underlying c.m. signal. The driver operates similarly to the Emitter Coupled Logic (ECL) driver: collectors of the two transistors of a differential amplifier are quickly swinging up and down symmetrically around the collector bias point, following the base input signals; the average c.m. voltage ensures transistor biasing and the lower voltage level is related to transistors saturation voltage.

To simplify, unless explicitly said, lines are assumed symmetric with expressions as in sec. 8.2.3; an example of asymmetric line is considered in sec. 8.3.2.3.

8.3.1 Coupling between lines

Transmission lines that are so close to have mutual influence of their respective fringe electric and magnetic fields are "coupled". Thus we will have a mutual capacitance and mutual inductance term, scalar for the case of two transmission lines sharing the return plane. As the two conductors 1 and 2 get closer, self-capacitance and self-inductance terms get smaller and the mutual capacitance and inductance terms get larger. However, there is no direct correspondence between the decrease of the former

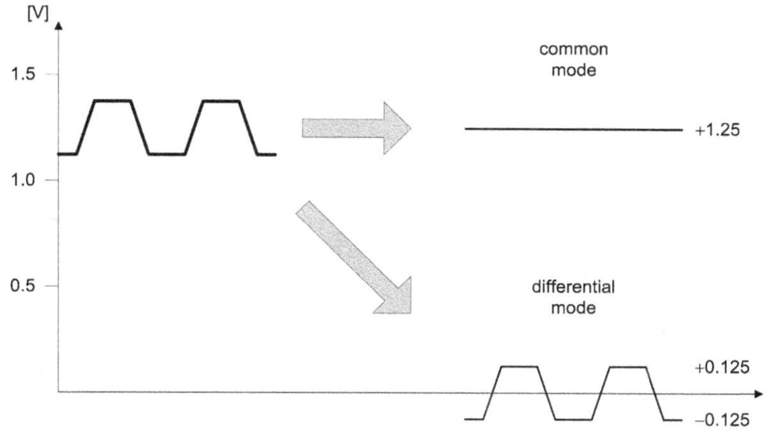

Figure 8.3.1 – Example of decomposition of a signal (LVDS format) into common-mode and differential-mode components.

and the increase of the latter, since the expected reduction of self terms is in the order of few % up to about ten %, while the change in the mutual terms might be more than a order of magnitude. The self terms keep larger than the mutual terms and the resulting characteristic impedance is more influenced by the former than by the latter, so that it is not expected to vary dramatically. We will review soon the odd- and even-mode characteristic impedance for variable coupling between lines. For the moment we state that the input impedance of each transmission line is not changing so much, as the two lines are moving closer; so, the input current sourced by the generator feeding one of the two lines (let's say line 1) may be considered almost unchanged.

While the signal on line 1 propagates, the mutual coupling terms (let's focus on mutual capacitance, for example) are subject also to the voltage on line 2 at the other end. This means that if the two voltages V_1 and V_2 on the two lines have the same polarity, a bootstrap effect on the mutual capacitance is expected and little displacement current will flow through it; on the contrary, if V_2 has opposite polarity, then this magnifies the current through the mutual capacitance. This is exactly what happens when considering the stray capacitance terms of transistors amplified by the voltage swing at the collector in a common emitter architecture, or, conversely, reduced by the bootstrap of the emitter voltage in a voltage follower architecture. This effect is normally called "Miller effect".

If we interpret the impedance of line 1 always as driving voltage over input current, then the current on line 1 changes depending on the signal on the other line and so the input impedance; this means that line 2 has influence on line 1 and its input impedance.

8.3.1.1 Mutual coupling terms and line impedance

For two single-ended lines 1 and 2 sharing a common return path we may define the inductance and capacitance matrices containing self terms on the diagonal and mutual terms off diagonal (symmetric). Let's assume the two lines identical, so that

	Separation s [mm]							
	1	**2**	**4**	**6**	**10**	**15**	**23**	**35**
C_{11}[pF/m]	2.57	2.69	2.73 (2.78)	2.77 (2.84)	2.85 (2.91)	2.92 (2.98)	2.98 (3.05)	3.05 (3.12)
C_{12}[pF/m]	38.7	15.8	10.8 (13.6)	8.9 (10.84)	7.22 (8.52)	6.24 (7.22)	5.39 (6.16)	4.72 (5.33)

Table 8.3.1 – Example of variation of inductance and capacitance matrices for two straight parallel wires at variable separation s.

the self terms are also equal: the exact values of self and mutual inductance $L_s = L_{11}$ and $L_m = L_{12}$ and of self and mutual capacitance $C_s = C_{11}$ and $C_m = C_{12}$ may be determined by various techniques, e.g. using theoretical formulations and closed form expressions as seen in Chapter 6 for PCB transmission lines or relying on finite element modeling and field solvers[3]. In general, we will observe an increase of mutual terms as the separation distance between lines gets smaller (and a slight reduction of self terms at the same time). Let's review this behavior for some common lines and problem geometries: in the following two examples are considered, two conductors in air and two PCB traces on the same layer.

Two conductors in air Two cylindrical conductors of $d = 0.8\,\mathrm{mm}$ diameter are held straight parallel and their separation s measured between conductor centers is made variable between two diameters (i.e. there is one diameter of air between the two conductors) and $3.5\,\mathrm{cm}$ (i.e. the two conductors are much far away, separated by more than 40 diameters), see Table 8.3.1.

Self and mutual capacitance terms have opposite behavior versus separation. The slight increase of C_{11} is a consequence of two facts: by separating the wires they get closer to the walls of enclosing box used as reference at infinite for the simulation, that is however $50\,\mathrm{cm}$ away and, second, some variation is a consequence of the more relevant change occurring to the mutual capacitance term. As expected, C_{12} varies significantly in the first mm in a nearly linear fashion (inversely proportional to separation s) and then as approximately a square root function, both approximations of the known logarithmic behavior.

Two PCB traces A simple case of two PCB traces 1 and 2 running straight parallel over a dielectric substrate and an underlying return plane is considered: the width of the PCB traces is $40\,\mathrm{mils} = 1\,\mathrm{mm}$, trace separation measured between internal trace edges is $48\,\mathrm{mils} = 1.2\,\mathrm{mm}$ or $80\,\mathrm{mils} = 2\,\mathrm{mm}$, the thickness of the PCB board is $60\,\mathrm{mils}$ (the usual $1.5\,\mathrm{mm}$ two-layer board). Two guard traces of the same width are added and a victim trace 3 (the distance of the victim trace is measured from the nearest trace between internal edges). The results of the resulting inductance and capacitance per-unit-length are shown in Table 8.3.2.

[3] A 2-D field solver for this kind of calculations is Ansoft Maxwell 2D (ver. 9 is used for these calculations), once freely available for downloading as a stand-alone tool in its student version.

	$s = 1.2\,\mathrm{mm}$ no guard no victim	$s = 2\,\mathrm{mm}$ no guard no victim	$s = 1.2\,\mathrm{mm}$ $s_g = 1\,\mathrm{mm}$ no victim	$s = 2\,\mathrm{mm}$ $s_g = 1\,\mathrm{mm}$ $s_v = 5\,\mathrm{mm}$	$s = 2\,\mathrm{mm}$ $s_g = 3\,\mathrm{mm}$ $s_v = 5\,\mathrm{mm}$	$s = 2\,\mathrm{mm}$ no guard $s_v = 5\,\mathrm{mm}$
$C_{11}\,[\mathrm{pF/m}]$	27	39.6	45.4	59.7	41.6	46.8
$C_{22}\,[\mathrm{pF/m}]$	27	39.6	45.4	64.6	40.1	44.0
$C_{33}\,[\mathrm{pF/m}]$	—	—	—	52.4	64.7	45.5
$C_{12}\,[\mathrm{pF/m}]$	20.9	14.1	18.9	18.4	13.4	10.7
$C_{13}\,[\mathrm{pF/m}]$	—	—	—	0.30	0.58	0.65
$C_{23}\,[\mathrm{pF/m}]$	—	—	—	1.32	2.0	4.68
$L_{11}\,[\mathrm{\mu H/m}]$	1.91	1.91	1.91	1.91	1.91	1.91
$L_{22}\,[\mathrm{\mu H/m}]$	1.91	1.91	1.91	1.91	1.91	1.91
$L_{33}\,[\mathrm{\mu H/m}]$	—	—	—	1.86	1.82	1.84
$L_{12}\,[\mathrm{\mu H/m}]$	1.49	1.43	1.49	1.41	1.40	1.41
$L_{13}\,[\mathrm{\mu H/m}]$	—	—	—	1.18	1.18	1.19
$L_{23}\,[\mathrm{\mu H/m}]$	—	—	—	1.27	1.26	1.27

Table 8.3.2 – Example of variation of inductance and capacitance matrices for two straight parallel wires at variable separation s.

As it appears, inductance is fairly insensitive to the position of guard traces as long as the two traces are driven differentially; the same may be said for the mutual inductance with the external victim trace.

Capacitance terms are largely influenced by the guard trace, as it will be considered later in this Chapter. The effect on the source line capacitance is to increase self-capacitance with the electric field lines directed not only towards the return plane beneath, but also to the guard traces siding the source line (when also the victim line is present, a slight further increase may be observed). Mutual capacitance between the two source traces is inversely affected and significantly smaller than self capacitance. The mutual capacitance values between the two traces 1 and 2 and the victim line (C_{13} and C_{23}) are not equal as expected and largely influenced by the presence of the guard trace in between: there is a difference of a factor of 2 for C_{13} (coupling between farthest elements) and more than 3 for C_{23}; for the position of the guard trace the influence is not so remarkable and limited to some %. The self capacitance of the victim line C_{33} is of course influenced by the position of the guard trace because the line is assumed single ended and returning through ground.

8.3.1.2 Effect of coupling and potentials on line impedance

We start from what was observed above regarding the effect on one line of signal (or potential) applied to the other line in terms of "perceived" impedance: the mutual terms are "modulated" by the potential, that may be of the same sign or opposite sign with respect to that applied to the line of interest, thus reducing or magnifying the effect of mutual terms. The problem is now considered in more quantitative terms, following two directions: the expressions for characteristic impedance are reported and commented, and the loading effect of the other line is considered and evaluated by observing the change to the input current under ideal driving conditions.

Characteristic impedance formulation The effect of potentials for mutual terms is implicitly accounted when writing the expressions for characteristic impedance of odd and even mode, because the relationship with the potentials appearing across the other conductor is clear and uniquely defined: odd mode indicates that the other conductor is driven opposite with respect to the return plane, even mode on the contrary clarifies that both conductors have the same polarity.

Distinguishing self and mutual parameters as $L_{11} = L_s$ and $L_{21} = L_m$ for the self and mutual inductance and $C_{11} = C_s$ and $C_{21} = C_m$ for the self and mutual capacitance (with $C_t = C_s + 2C_m$ the total capacitance "seen" by a single conductor), and omitting resistance and conductance terms for simplicity, the odd-mode and even-mode characteristic impedance may be written as:

$$Z_{c,o} = \sqrt{\frac{L_s - L_m}{C_s + 2C_m}} \qquad Z_{c,e} = \sqrt{\frac{L_s + L_m}{C_s}} \tag{8.3.1}$$

Loading effect of the other line For the same reason (i.e. amplification of mutual capacitive effect because of driving potentials) each line sees itself loaded by the other line as a function of the respective potentials. Focusing only on the input current drawn by capacitive terms, in odd mode the total capacitance seen by line 1 input is $(C_s + C_m)$, that is larger than that seen when both lines are driven by the same potential in even mode and mutual capacitance C_m is bootstrapped and vanishes. This observation leads to conclude that the absorbed current will be larger in odd mode, thus exposing more to slew-rate limitations and peaky current. This is also in agreement with the fact that the impedance for the even mode excited by common mode signals is larger than the odd mode impedance (see sec. 8.3.2).

Assuming an equal voltage V that is applied with either of the two polarities to the two conductors, thus resulting in V of common-mode voltage or $2V$ of differential-mode voltage, we will see the doubling of the mutual capacitance, and between the odd and the even mode there will be a "delta current" of capacitive nature with respect to the "rest position", where each line sees exactly $(C_s + C_m)$ of total input capacitance: in odd mode this delta current I'_C is positive and adds to the normal capacitive current I^0_C, while it changes sign in even mode and subtracts from the same current.

$$I^0_C = j\omega(C_s + C_m)V \qquad I'_C = j\omega C_m V \qquad \begin{cases} I^{odd}_C = I^0_C + I'_C \\ I^{even}_C = I^0_C - I'_C \end{cases} \tag{8.3.2}$$

Meaning of "characteristic impedance" So far we have seen that the concept of impedance, both as characteristic impedance and as input impedance seen from the input terminal, is not unique when mutual coupling with adjacent lines comes into play, which is relevant for moderate separation in the order of one or few trace widths.

The need of specifying if the other line (line 2) is driven with the same polarity or opposite polarity or held to ground justifies thus the introduction of the odd mode and even mode representation: when we consider the effect of the other line, we are simply considering the set of two single-ended lines as one system, made of a pair of conductors that are both driven by a potential, and this is a differential line that may be described by decomposing it in modes, odd and even (see sec. 8.3.2). This representation keeps valid also for asymmetric lines.

8.3.2 Odd and even modes and related impedance

8.3.2.1 Odd- and even-mode characteristic impedance

So far common-mode impedance Z_{cm}, differential-mode impedance Z_{dm}, even-mode impedance Z_e and odd-mode impedance Z_o have been considered. Some authors prefer to use the terms "common impedance" and "differential impedance", omitting the word "mode": the reason is that "common" and "differential" terms are refer to signals, while the word "mode" is suited for propagation modes inside the transmission line and belongs thus to the "infrastructure". We keep the "m" of mode to avoid confusion with the "c" alone used for "characteristic" and because the use of the word "mode" is quite diffused; so, for example, the use of the terms "differential impedance" and "differential mode impedance" will be interchangeable.

For symmetric lines an externally applied differential signal propagates in odd mode and a common-mode signal propagates in even mode.

The differential impedance is the impedance seen when a differential signal is applied; the odd-mode impedance is the impedance when the pair is driven in odd mode. Thus, when applying a differential signal the impedance seen across the two conductors of the pair is twice the impedance seen by one of them with respect to the underlying return plane.

$$Z_{dm} = 2Z_o \tag{8.3.3}$$

The common-mode impedance is the impedance seen when a common-mode signal is applied; the even-mode impedance is the impedance of one of the two lines when the pair is driven by even-mode signals. Since the common-mode signal sees the two lines in parallel, the resulting common-mode impedance is half the even-mode impedance.

$$Z_{cm} = \frac{1}{2}Z_e \tag{8.3.4}$$

These expressions shall not be misused: when thinking of the correct termination of a differential pair carrying a high frequency digital signal on a PCB and the necessity of setting it to $100\,\Omega$ differential impedance, this doesn't come out automatically by taking two single lines with a $50\,\Omega$ characteristic impedance with respect to the return plane; we have already seen that the two lines are mutually coupled and what we are interested in are the odd- and even-mode characteristic impedances, that take into account the presence of the other line (see Example 8.1).

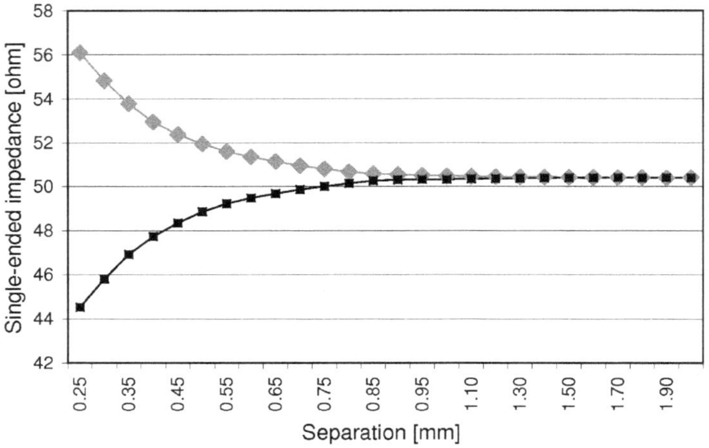

Figure 8.3.2 – Example of common (gray) and differential (black) impedance of two planar single-ended lines designed for nearly $50\,\Omega$ characteristic impedance alone for increasing separation between the two (the two impedance values are measured as single ended to be displayed together).

Changing the physical separation between lines affects directly the amount of physical coupling, whatever the potentials applied to the lines. Odd- and even-mode characteristic impedances are thus different for tightly coupled and loosely coupled lines: in the first case, with self and mutual terms nearly equal, $Z_{c,o}$ is quite small and $Z_{c,e}$ assumes a larger value than that at very large distance (see Figure 8.3.2); for the second case, when mutual terms are negligible, the two characteristic impedances are equal to the characteristic impedance of the single line.

Additionally, differences between modal characteristic impedances are not only related to the feeding scheme and how line series and shunt parameters are driven, but to an intrinsic difference of the parameters.

The physical explanation for the difference is that the effective dielectric constant might be different because the electric field distribution changes for the two modes, and the same is in principle valid also for the magnetic field distribution, even if less common in normally used lines.

Cables are more homogeneous in cross-section and materials and the terms μ and ε are directly related to the per-unit-length parameters of inductance l and capacitance c, and so are the mode propagation velocities. A line on a PCB is more complex because there are geometries that are not homogeneous, such as for microstrips, where traces are immersed in a medium made of substrate dielectric beneath and air above, with different permittivities and dispersion (please see Chapter 6). For an edge-coupled microstrip the electric field lines for odd mode are distributed below and above the ideal horizontal plane containing the traces and the dielectric permittivity is the result of the combination of substrate dielectric and air; conversely, when considering even mode most of the electric field lines are in the substrate towards the underlying reference plane and, for this reason, the equivalent dielectric permittivity is smaller and the propagation velocity is correspondingly larger.

Propagation time For the propagation time two expressions for odd and even modes may be derived, still using the simplified inductance/capacitance representation of the two lines:

$$\tau_o = \sqrt{(L_s - L_m)(C_s + 2C_m)} \qquad \tau_e = \sqrt{(L_s + L_m)C_s} \tag{8.3.5}$$

When the two lines are tightly coupled, the odd mode is much faster than the even mode, because the difference of self and mutual inductance terms goes to zero (not quite zero, but really small!); for loose coupling the two propagation times are equal. So, odd- and even-mode components propagating along the system will reach the other end at different times in all cases, except for uncoupled lines. This effect, besides any asymmetry of parameters due to inhomogeneity, defects and any other practical issue, warns us about the dispersion of time of arrival and signal edge, causing degradation of signal integrity and eye pattern (see sec. 3.3.3).

It is also interesting to put together eq. (8.3.5) and (8.3.1) and derive inductance and capacitance quantities assuming that propagation time and impedance can be determined from measurements (TDR/TDT measurements):

$$\begin{aligned} L_s &= \frac{1}{2}(Z_e\tau_e + Z_o\tau_o) \qquad & L_m &= \frac{1}{2}(Z_e\tau_e - Z_o\tau_o) \\ C_{tot} = C_s + 2C_m &= \frac{1}{2}\left(\frac{\tau_o}{Z_o} + \frac{\tau_e}{Z_e}\right) \qquad & C_m &= \frac{1}{2}\left(\frac{\tau_o}{Z_o} - \frac{\tau_e}{Z_e}\right) \end{aligned} \tag{8.3.6}$$

Complete formulation with reactive and resistive terms Finally, to include resistance and conductance it suffices to express for each configuration the series impedance Z_{ser} and the shunt admittance Y_{sh}: it is known that the characteristic impedance is the square root of the ratio of the two (see sec. 2.2), and that the propagation time is the reciprocal of propagation velocity, that is the square root of the product of the two.

When the pair is driven in even mode: the mutual admittance terms are bootstrapped to zero, because they see no voltage difference; the mutual impedance terms add to the self terms because the current in the two lines is of same polarity.

$$\begin{aligned} Z_{\text{ser},even} &= r_{11} + r_{12} + j\omega(l_{11} + l_{12}) \\ Y_{\text{sh},even} &= g_{11} + j\omega c_{11} \end{aligned} \tag{8.3.7}$$

The resulting even-mode impedance and velocity are thus:

$$Z_{c,e} = \sqrt{\frac{r_{11} + r_{12} + j\omega(l_{11} + l_{12})}{g_{11} + j\omega c_{11}}} \tag{8.3.8}$$

$$v_e = \frac{1}{\sqrt{(r_{11} + r_{12} + j\omega(l_{11} + l_{12}))(g_{11} + j\omega c_{11})}} \tag{8.3.9}$$

When the pair is driven in the odd mode; the resulting current is flowing into one line (line 1) and back through the return plane, while the same current is flowing out of the other line (line 2) and back through its portion of the return plane. The total admittance is made of the self term of one line (line 1, g_{11} and c_{11}) in series through the return plane with the self term of the other line (line 2, g_{22} and c_{22}), that should lead to half the value (under the assumption that the two lines are identical and "11" and "22" terms are equal), but under twice the voltage; and the mutual term of the line admittance also is undergoing twice the voltage. The impedance mutual terms subtract from the self terms because the current in the two lines is of opposite polarity. Thus the total resistance and inductance are given by the self and mutual terms subtracted; the conductance and capacitance terms are given by the self terms summed to twice the mutual terms that see twice the voltage.

$$\begin{aligned} Z_{\text{ser},odd} &= r_{11} - r_{12} + j\omega(l_{11} - l_{12}) \\ Y_{\text{sh},odd} &= g_{11} + j\omega c_{11} + g_{12} + j\omega c_{12} \end{aligned} \tag{8.3.10}$$

The resulting odd-mode impedance and velocity are thus:

$$Z_{c,o} = \sqrt{\frac{r_{11} - r_{12} + j\omega(l_{11} - l_{12})}{g_{11} + j\omega c_{11} + 2(g_{12} + j\omega c_{12})}} \tag{8.3.11}$$

$$v_o = \frac{1}{\sqrt{(r_{11} - r_{12} + j\omega(l_{11} - l_{12}))(g_{11} + j\omega c_{11} + 2(g_{12} + j\omega c_{12}))}} \tag{8.3.12}$$

8.3.2.2 Mode propagation

We have just seen that coupling between lines affects symmetry of mode impedances and mode propagation, extended in general to transmission line response and propagation and casting doubts on the possibility of using effectively multi-conductor transmission lines. Let's consider again our coupled transmission lines, sharing a common return path; we summarize the influence of the signal polarity on signal propagation:

- by applying a step signal on line 1, while keeping at zero the signal in the other line (line 2), it is possible to observe that there is crosstalk between the two and distortion on both lines;
- if both lines have the same signal applied (the step signal) with the same polarity, then there is no instantaneous voltage difference between the two (ruling out the capacitive coupling term) and the current flowing is the same in both lines (ruling out the inductive coupling term); it may be concluded that in this case the two signals propagate undistorted;
- if now we apply the same step signal to line 1 and 2, but with opposite polarity, it is evident that there is an instantaneous voltage difference between the mutual coupling terms and that both signals generate far-end noise (see sec. 8.3.4): line 1 causes a negative pulse on line 2, larger as the signal progresses to the far end,

and at the same time there is a reduction of the voltage on line 1; similarly, line 2 causes a positive pulse on line 1 of the same magnitude and its voltage amplitude reduces of the same amount; for this reason the mutual effects compensate and the two signals propagate undistorted.

So, when a pair, made of two conductors and a reference plane, is excited in one of these two modes, the applied signals propagate undistorted. For the special case of a symmetric pair, like one made of two edge-coupled PCB traces lying on the same layer with a reference return plane underneath (it's a coupled microstrip, see sec. 6.3.2), then these two modes correspond to a common-mode and a differential-mode excitation. In reality, this example is tailored on a symmetric pair, where the common and differential modes of the applied signals correspond to the even and odd modes of the coupled transmission line. In general, the even mode and odd mode pertain to the propagation inside the transmission line system, the common mode and differential mode are characteristic of the externally applied signals.

8.3.2.3 General model

Let's consider now a counter example with a transmission line where the even and odd modes are not so intuitive: if a coupled planar line on PCB is made of two edge-coupled microstrips with quite different width of hot traces and sharing the same return plane, then, being the self terms of each microstrip different, it is expected that the line is more difficult to represent and is not characterized by symmetrical signals.

In order to correctly represent asymmetric lines in general, a more general modal approach based on matrix representation is briefly considered.

When inductance and capacitance matrices $\mathbf{L}$ and $\mathbf{C}$ are formulated (e.g. calculated using a wave solver), then the determination of eigenvalues and eigenvectors for their left and right products determine the propagation modes along the structure:

$$\mathbf{M}_v = \text{eigenvector}(\mathbf{LC}) \qquad \mathbf{M}_i = \text{eigenvector}(\mathbf{CL}) \tag{8.3.13}$$

$$\mathbf{\Lambda} = \sqrt{\text{eigenvalue}(\mathbf{LC})} = \sqrt{\text{eigenvalue}(\mathbf{CL})} \tag{8.3.14}$$

The so obtained eigenvectors are not normalized and the most straightforward solution is to divide evenly each term for the square root of the determinant. Then the characteristic impedance may be obtained by two equivalent formulations

$$\mathbf{Z}_c = \mathbf{L}\mathbf{M}_i\mathbf{\Lambda}^{-1}\mathbf{M}_i^{-1} \qquad \mathbf{Z}_c = \mathbf{M}_v\mathbf{\Lambda}\mathbf{M}_v^{-1}\mathbf{C}^{-1} \tag{8.3.15}$$

Gentili and Salazar-Palma [133] then comment on the three possible diagonalizations of matrix $\mathbf{Z}_c$ that lead to three different formulations of the modal characteristic impedance, that they determined correspondingly in the same paper with three different approaches to the definition: using voltage and current ratio (the one usually used), the ratio of power and squared current or the ratio of squared voltage and power[4]. They show that the three definitions give slightly different values of the modal characteristic impedance matrix, with the V-I definition equal to the geometric mean of the other two and lying thus in the middle.

$$\mathbf{Z}_m^{\mathrm{VI}} = \mathbf{M}_v^{-1}\mathbf{Z}_c\mathbf{M}_i \qquad \mathbf{Z}_m^{\mathrm{PI}} = \mathbf{M}_i^t\mathbf{Z}_c\mathbf{M}_i \qquad \mathbf{Z}_m^{\mathrm{PV}} = \mathbf{M}_v^{-1}\mathbf{Z}_c(\mathbf{M}_v^t)^{-1} \tag{8.3.16}$$

The example below (example 1 appearing in [133]) considers the PCB geometry shown in Figure 8.3.3 and is used for the discussion of behavior and modal decomposition of such asymmetric line.

The calculated inductance and capacitance matrices are:

$$\mathbf{L}/\mu_0 = \begin{bmatrix} 0.13059 & 0.04963 \\ 0.04963 & 0.2846 \end{bmatrix} \qquad \mathbf{C}/\varepsilon_0 = \begin{bmatrix} 54.604 & -6.634 \\ -6.634 & 22.704 \end{bmatrix} \tag{8.3.17}$$

The eigenvalues correctly are

$$\mathbf{\Lambda} = \begin{bmatrix} 8.847 & 0 \\ 0 & 8.098 \end{bmatrix} 10^9 \tag{8.3.18}$$

and the characteristic impedance matrix is

$$\mathbf{Z}_c[\Omega] = \begin{bmatrix} 18.76 & 6.33 \\ 6.33 & 42.92 \end{bmatrix} \tag{8.3.19}$$

The modal characteristic impedance matrix is diagonal and the calculated values as anticipated are slightly different for the adopted method: $Z_{c,o} = 23.95,\ 26.69,\ 29.74\,\Omega$ and $Z_{c,e} = 25.70,\ 28.64,\ 31.91\,\Omega$ for the PI, VI and PV criterion, respectively. The values obtained with the three formulations of the modal characteristic impedance matrix span by nearly 22%, the usual definition based on voltage and current ratio staying in the middle, so with a maximum difference of about 10%. One may argue that provided that all we use the same definition, there will be no inconsistency related to this aspect; yet, values to compare may originate from measurements (for which probably we control the post processing algorithm), but also from simulations performed with various circuit simulators, field solvers, etc. (probably using one of the two definitions based on power, the latter often used as the criterion to verify convergence).

This example looked weird: who on the planet is going to build a line pair with so different conductors for the two polarities? a more realistic example is when the two traces are as equal as possible by design, but one of the two is inductively and capacitively loaded through unwanted coupling with an extraneous parallel trace nearby; this is a far more common case over a densely populated PCB. The victim trace considered in Figure 8.3.2 was loading the rightmost signal trace causing a loss of symmetry of the capacitance matrix. Similarly a parallel supply or ground trace.

As said, both odd-mode impedance and differential impedance decrease when the separation between the two traces is reduced, and correspondingly the even-mode

---

4 The same problem of equivalence between formulations was commented in [46] when evaluating simulated and measured values of characteristic impedance for microstrips; please see sec. 6.3.1.1.

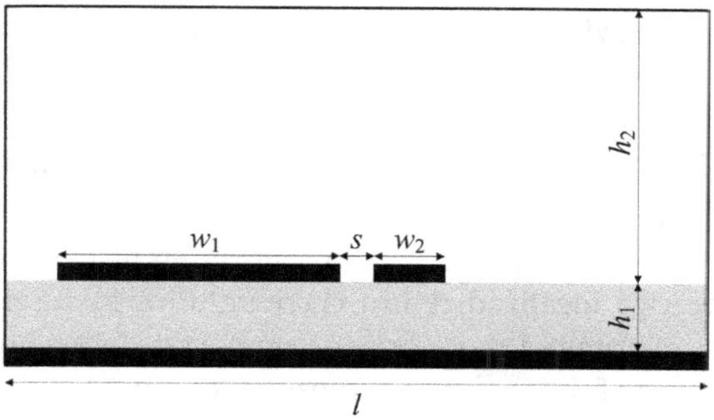

Figure 8.3.3 – Asymmetric coupled planar transmission line: $w_1 = 0.4l$, $w_2 = 0.1l$, $h_1 = 0.1l$, $h_2 = 0.4l$, $s = 0.02l$, $\varepsilon_{r1} = 10$, $\varepsilon_{r2} = 1$ (the microstrips are surrounded by an equipotential surface with size $l \times h_2$).

impedance and the common impedance both increase. On PCBs space is at premium and it is quite common that different lines run along with a separation distance that is minimal, dictated by the fabrication process and required tolerance (usually between 8 and 10 mils for commercial PCBs, about one half for high-quality PCBs). This aspect is quite relevant for both the final value of the modal characteristic impedance and for the amount of crosstalk between the lines (see sec. 8.3.4).

8.3.2.4 Guard traces

In many real world configurations hot signal traces are separated from nearby transmission lines by guard traces brought at the ground potential, the same potential of the common return plane beneath. This configuration is adopted for many reasons: the most obvious is to shield the stray electromagnetic field that might couple to nearby lines, reducing coupling and disturbance and preserving signal integrity; another reason is to offer a more stable normalized environment, so that the line impedances do not depend on the adjacent nearby lines; last, in many case a guard trace creating a coplanar waveguide structure (see sec. 6.3.3) is effective in reducing the total line characteristic impedance.

An estimate of the low frequency loading effect (capacitive loading) was done in sec. 8.3.1.1 and the results shown in Table 8.3.2: self and mutual capacitance terms of a pair of PCB traces are shown for two different positions of guard traces on both sides.

8.3.3 Return current distribution for coupled lines

If we apply a differential signal to a pair made of line 1, line 2 and the common return plane, we may see a positive signal between line 1 and the return plane, and a negative signal between line 2 and the same return plane. This means that the current enters line 1 returning through the return plane and at the same time there is a negative current that leaves line 2 and flows in opposite direction in the return plane. Focusing on the two lines and their hot conductors, we see always two opposite currents, i.e.

one entering and one leaving, as it should be for a conductor pair, where we apply the driving signal to a port. In this interpretation we do not include any detail of the return plane and we have a pure differential pair.

As long as the two transmission lines are far away, the two return currents flow in separate regions of the return plane and do not influence each other. We know that the total current in the return plane is zero, but different current distributions may occur: there will be two localized return currents of opposite polarity beneath the two hot conductors, distributed more or less compactly, depending on geometry and electrical characteristics of the so-formed transmission lines.

As the two lines are brought closer, the two return currents will more and more merge and overlap and there will be a direct influence on the single-ended impedance of the line (that we call even-mode impedance, Z_e) and on the differential impedance of the line (that we call odd-mode impedance, Z_o): the former will increase, while the latter will decrease (this justification is complementary to what was said about mutual coupling terms in sec. 8.3.2). Looking at the current distribution in the return plane, if the return plane is moved farther away from the two lines (increasing the thickness of the PCB substrate), there will be a higher degree of overlap between the two return currents with a larger deal of cancellation, until the effect of the return plane is negligible and the two lines will look like a differential pair. At large distance the return plane does not influence the differential impedance any longer; the single-ended impedance continues to increase with the distance of the return plane, because it is always defined between one conductor and the return plane as reference conductor.

If the return plane is replaced by a shield, such as in a shielded twisted pair cable, then the distance is normalized to a value that is approximately the same for all the conductors in the cable, that on average are located in the center of this cylindrical shield. In this case the distribution of the return current in the shield due to conductor 1 and conductor 2 is the same, but with opposite sign, and the total net effect is null, like if the shield does not exist for the differential mode. If the pair conductors have an offset with respect to the shield longitudinal axis, then the return current distribution is not perfectly uniform along the circumference and there is a slight influence on differential impedance. Ideally, there is no difference for the differential impedance of an unshielded pair and a pair with a shield of large diameter. On the contrary, the shield is mostly useful in closing the return path of common-mode signals, thus reducing the common-mode electromagnetic emissions. Keeping the return plane at a distance larger than the separation of the pair conductors (let's say twice or more) has slight benefit on differential impedance, but increases common-mode impedance.

8.3.4 Crosstalk model, near-end and far-end crosstalk

A simplified model that assumes an electrically short transmission line is considered in order to determine near-end and far-end crosstalk terms. More complete models are possible, especially with simulation tools (Multiconductor Transmission Lines methods [264], method of characteristics, in general circuit simulators, both in time and frequency domain). An accurate solution deserves an accurate model of losses and attenuation and they are frequency dependent, a concept that is not easily transferred into time domain and that needs some compromise, such as interpolation, fitting to a reference model (as done in sec. 6.5.1.3), or using Laplace transform.

Recalling Wadell's definition, we may simply say that crosstalk quantifies the amount of signal coupled onto the victim line sourced by the culprit line, and, by the linearity of the phenomenon, it represents a transfer function. The quantities on either side of the transfer function may be differential- or common-mode signals, taken at either end of the two lines, so a total of four different transfer functions are possible (see sec. 5.1.6 for definitions of terms and measurements for cables).

8.3.4.1 Crosstalk model for lines with common return plane

From a quantitative viewpoint, the simplest configuration of two hot conductors (one for the source and one for the victim) and a common return conductor are considered.

The used notation assigns "1" to the source conductor, "2" to the victim conductor and "0" to the return conductor. The two transmission lines 1-0 and 2-0 have the usual p.u.l. parameters representation; the two lines are terminated at the two ends with arbitrary impedances, that to simplify are assumed pure resistors.

Writing telegrapher's equations for a line section of length Δz and separating currents and voltages, the following equations result.

$$\boldsymbol{v}(z,t) = \begin{bmatrix} v_1(z,t) \\ v_2(z,t) \end{bmatrix} \qquad \boldsymbol{i}(z,t) = \begin{bmatrix} i_1(z,t) \\ i_2(z,t) \end{bmatrix} \tag{8.3.20}$$

$$\begin{aligned} \frac{\partial \boldsymbol{v}(z,t)}{\partial z} &= -\boldsymbol{r}\,\boldsymbol{i}(z,t) - \boldsymbol{l}\,\frac{\partial \boldsymbol{i}(z,t)}{\partial z} \\ \frac{\partial \boldsymbol{i}(z,t)}{\partial z} &= -\boldsymbol{g}\,\boldsymbol{v}(z,t) - \boldsymbol{c}\,\frac{\partial \boldsymbol{v}(z,t)}{\partial z} \end{aligned} \tag{8.3.21}$$

where the four matrices of the so-formed three-conductor transmission line contain self and mutual parameters:

$$\boldsymbol{r} = \begin{bmatrix} r_{10}+r_0 & r_0 \\ r_0 & r_{20}+r_0 \end{bmatrix} \qquad \boldsymbol{l} = \begin{bmatrix} l_{11} & l_{12} \\ l_{21} & l_{22} \end{bmatrix} \tag{8.3.22}$$

$$\boldsymbol{c} = \begin{bmatrix} c_{10}+c_{12} & -c_{12} \\ c_{12} & c_{20}+c_{12} \end{bmatrix} \qquad \boldsymbol{g} = \begin{bmatrix} g_{10}+g_{12} & -g_{12} \\ -g_{12} & g_{20}+g_{12} \end{bmatrix} \tag{8.3.23}$$

If now the system is solved in the frequency domain using phasor approach, the expressions appearing in [265], chap. 9, may be obtained,

$$\frac{\mathrm{d}\boldsymbol{V}(z)}{\mathrm{d}z} = -(\boldsymbol{r}+j\omega\boldsymbol{l})\,\boldsymbol{I}(z) = -\boldsymbol{Z}\,\boldsymbol{I}(z) \tag{8.3.24}$$

$$\frac{\mathrm{d}\boldsymbol{I}(z)}{\mathrm{d}z} = -(\boldsymbol{g}+j\omega\boldsymbol{c})\,\boldsymbol{V}(z) = -\boldsymbol{Y}\,\boldsymbol{V}(z) \tag{8.3.25}$$

that can be solved with a second differentiation of the voltage or current vector:

$$\frac{\mathrm{d}^2 \boldsymbol{V}(z)}{\mathrm{d}z^2} = \boldsymbol{Z}\boldsymbol{Y}\,\boldsymbol{V}(z) \qquad \frac{\mathrm{d}^2 \boldsymbol{I}(z)}{\mathrm{d}z^2} = \boldsymbol{Y}\boldsymbol{Z}\,\boldsymbol{I}(z) \tag{8.3.26}$$

paying attention to the order of multiplication of matrices.

If one is interested in decoupling the equations for source and victim lines, further manipulation with a linear transformation allows to operate in the modal domain, that we have already introduced: odd and even mode components. The resulting modal variables are indicated by the subscript $m = o,\, e$ and they are two for each mode, one component for the propagation along the source line and one along the victim line. The transformation matrix $\boldsymbol{T}$ is defined as a left-multiplicative matrix on modal variables (two columns, one for each mode, and two rows, for source and victim lines).

$$\boldsymbol{V}(z) = \boldsymbol{T}\,\boldsymbol{V_m}(z) \qquad \boldsymbol{I}(z) = \boldsymbol{T}\,\boldsymbol{I_m}(z) \tag{8.3.27}$$

It is observed that the transformation matrix is the same for voltage and current, because the transformation is a topological transformation. The resulting modal system of equations is

$$\frac{\mathrm{d}^2 \boldsymbol{V_m}(z)}{\mathrm{d}z^2} = \boldsymbol{T}^{-1}\boldsymbol{Z}\boldsymbol{Y}\boldsymbol{T}\,\boldsymbol{V_m}(z) \qquad \frac{\mathrm{d}^2 \boldsymbol{I_m}(z)}{\mathrm{d}z^2} = \boldsymbol{T}^{-1}\boldsymbol{Y}\boldsymbol{Z}\boldsymbol{T}\,\boldsymbol{I_m}(z) \tag{8.3.28}$$

As expected, the equations take a diagonal form and the source and victim circuits are decoupled: this is demonstrated by the characteristics of the matrix product $\boldsymbol{T}^{-1}\boldsymbol{Z}\boldsymbol{Y}\boldsymbol{T}$.

$$\boldsymbol{T}^{-1}\boldsymbol{Z}\boldsymbol{Y}\boldsymbol{T} = \gamma_{\boldsymbol{m}}^2 = \begin{bmatrix} \gamma_o^2 & 0 \\ 0 & \gamma_e^2 \end{bmatrix} \tag{8.3.29}$$

The two quantities γ_o^2 and γ_e^2 are the eigenvalues of the product of admittance and impedance matrices and represents the propagation constants for individual modes.

We have a decoupled propagation of modes, that e.g. for currents takes the form:

$$\boldsymbol{I_m}(z) = \boldsymbol{I_m^+} e^{-\gamma_m z} - \boldsymbol{I_m^-} e^{+\gamma_m z} \tag{8.3.30}$$

The original voltages and currents in the two physical lines can be recovered by applying the initial transformation given in eq. (8.3.27). As expected, the two equations for voltages and currents are related by the characteristic impedance, that for our modal system may be derived with some manipulation as

$$\boldsymbol{Z_c} = \boldsymbol{Z}\boldsymbol{T}\gamma^{-1}\boldsymbol{T}^{-1} \tag{8.3.31}$$

It is possible to apply the modal representation of the three-conductor source-victim transmission line example to the problem of crosstalk for this configuration, that represents the simplest crosstalk case.

Source and victim lines are terminated onto impedances with an obvious notation: $Z_{1,ne}$, $Z_{1,fe}$, $Z_{2,ne}$ and $Z_{2,fe}$. The former two are what Paul [265], Chap. 9, calls R_S and R_L, because they correspond to generator and load impedances of the source line; the four impedances connected respectively at the near- and far-end may be in short indicated as vectors $\boldsymbol{Z_{ne}}$ and $\boldsymbol{Z_{fe}}$. A voltage generator V_S is located at the near end of the source line 1; correspondingly the victim line 2 is terminated on its own impedance, but has no voltage source, so that the source vector is $\boldsymbol{V_S} = \begin{bmatrix} V_S \\ 0 \end{bmatrix}$.

The resulting terminal equations for voltages and currents at the near-end ($z=0$) and far-end ($z=L$) ports are thus:

$$\begin{aligned} \boldsymbol{V}(0) &= \boldsymbol{V_S} - \boldsymbol{Z_{ne}}\boldsymbol{I}(0) \\ \boldsymbol{V}(L) &= \boldsymbol{Z_{fe}}\boldsymbol{I}(L) \end{aligned} \tag{8.3.32}$$

The mathematical details are kept limited and we proceed directly to the solution. The modal voltages and currents at the two ends of the line are:

$$\boldsymbol{V}(0) = \boldsymbol{Z_c}\boldsymbol{T}\,(\boldsymbol{I_m^+} + \boldsymbol{I_m^-}) \qquad \boldsymbol{I}(0) = \boldsymbol{T}\,(\boldsymbol{I_m^+} - \boldsymbol{I_m^-}) \tag{8.3.33}$$

$$\boldsymbol{V}(L) = \boldsymbol{Z_c}\boldsymbol{T}\,(\boldsymbol{I_m^+}e^{-\gamma L} + \boldsymbol{I_m^-}e^{+\gamma L}) \qquad \boldsymbol{I}(L) = \boldsymbol{T}\,(\boldsymbol{I_m^+}e^{-\gamma L} - \boldsymbol{I_m^-}e^{+\gamma L}) \tag{8.3.34}$$

Then the line currents may be solved by combining the equations above

$$\begin{bmatrix} (\boldsymbol{Z_c} + \boldsymbol{Z_{ne}})\boldsymbol{T} & (\boldsymbol{Z_c} - \boldsymbol{Z_{ne}})\boldsymbol{T} \\ (\boldsymbol{Z_c} - \boldsymbol{Z_{fe}})\boldsymbol{T}e^{-\gamma L} & (\boldsymbol{Z_c} - \boldsymbol{Z_{fe}})\boldsymbol{T}e^{+\gamma L} \end{bmatrix} \begin{bmatrix} \boldsymbol{I_m^+} \\ \boldsymbol{I_m^-} \end{bmatrix} = \begin{bmatrix} \boldsymbol{V_S} \\ \boldsymbol{0} \end{bmatrix} \tag{8.3.35}$$

If the transmission matrix representation of the line is considered (see sec. 1.3.1.3), relating voltage and current appearing at the input port at $z=0$ with voltage and current at the other end $z=L$, we obtain a compact form that transfers the solution of eq. (8.3.35) for $z=0$ (input current vector) to the other side of the line. Numbers "1" and "2" were already used to identify the hot conductors of the two lines, so the near-end and far-end ports have been identified using the letters "n" and "f".

$$\begin{bmatrix} \boldsymbol{Z_{fe}}\boldsymbol{I}(L) \\ \boldsymbol{I}(L) \end{bmatrix} = \begin{bmatrix} \varphi_{nn} & \varphi_{nf} \\ \varphi_{fn} & \varphi_{ff} \end{bmatrix} \begin{bmatrix} \boldsymbol{V_S} - \boldsymbol{Z_{ne}}\boldsymbol{I}(0) \\ \boldsymbol{I}(0) \end{bmatrix} \tag{8.3.36}$$

where

$$\begin{bmatrix} \boldsymbol{V}(L) \\ \boldsymbol{I}(L) \end{bmatrix} = \begin{bmatrix} \varphi_{nn} & \varphi_{nf} \\ \varphi_{fn} & \varphi_{ff} \end{bmatrix} \begin{bmatrix} \boldsymbol{V}(0) \\ \boldsymbol{I}(0) \end{bmatrix} \tag{8.3.37}$$

$$\varphi_{nn} = \frac{1}{2}\mathbf{Y}^{-1}\boldsymbol{T}\,(e^{+\gamma L} + e^{-\gamma L})\boldsymbol{T}^{-1}\mathbf{Y} \tag{8.3.38}$$

$$\varphi_{nf} = -\frac{1}{2}\mathbf{Y}^{-1}\boldsymbol{T}\gamma\,(e^{+\gamma L} - e^{-\gamma L})\boldsymbol{T}^{-1} \tag{8.3.39}$$

$$\varphi_{fn} = -\frac{1}{2}\boldsymbol{T}\,(e^{+\gamma L} - e^{-\gamma L})\gamma^{-1}\boldsymbol{T}^{-1}\mathbf{Y} \tag{8.3.40}$$

$$\varphi_{ff} = \frac{1}{2}\boldsymbol{T}\,(e^{+\gamma L} + e^{-\gamma L})\boldsymbol{T}^{-1} \tag{8.3.41}$$

8.3.4.2 Crosstalk for the lossless line case

It is now possible to simplify, considering the case of lossless lines. This approximation holds for good quality lines and also up to moderately high frequency: thin conductors have no dramatic skin effect and good dielectrics have negligible loss.

The near-end and far-end crosstalk terms across the victim line 2 correspond to the second component of each vectors appearing in eq. (8.3.36): the near-end voltage $V_{2,ne}$ and far-end voltage $V_{2,fe}$ across victim line 2 are:

$$\begin{aligned} V_{2,ne} &= \frac{S}{D}\left[\frac{R_{2,ne}}{R_{2,ne}+R_{2,fe}}\, j\omega l_{12} L\left(C + \frac{j2\pi L/\lambda}{\sqrt{1-k^2}} Q_{ne} S\right) I_1 + \right. \\ &+ \left. \frac{R_{2,ne}R_{2,fe}}{R_{2,ne}+R_{2,fe}}\, j\omega c_{12} L\left(C + \frac{j2\pi L/\lambda}{\sqrt{1-k^2}} Q_{ne} S\right) V_1\right] \end{aligned} \tag{8.3.42}$$

$$V_{2,fe} = \frac{S}{D}\left[-\frac{R_{2,fe}}{R_{2,ne}+R_{2,fe}}\, j\omega l_{12} L\, I_1 + \frac{R_{2,ne}R_{2,fe}}{R_{2,ne}+R_{2,fe}}\, j\omega c_{12} L\, V_1\right] \tag{8.3.43}$$

where:

$$C = \cos(\beta L) \text{ and } S = \frac{\sin(\beta L)}{\beta L}$$

$$D = C^2 - S^2\omega^2\tau_1\tau_2\left[1 - k^2\frac{(1-\alpha_{1,ne}\alpha_{2,fe})(1-\alpha_{1,fe}\alpha_{2,ne})}{(1+\alpha_{1,ne}\alpha_{1,fe})(1+\alpha_{2,ne}\alpha_{2,fe})}\right] + j\omega CS(\tau_1+\tau_2)$$

$$\tau_1 = \frac{l_{11}L}{R_{1,ne}+R_{1,fe}} + (c_{11}+c_{12})L\frac{R_{1,ne}R_{1,fe}}{R_{1,ne}+R_{1,fe}}$$

$$\tau_2 = \frac{l_{22}L}{R_{2,ne}+R_{2,fe}} + (c_{22}+c_{12})L\frac{R_{2,ne}R_{2,fe}}{R_{2,ne}+R_{2,fe}}$$

$$k = \frac{l_{12}}{\sqrt{l_{11}l_{22}}} = \frac{c_{12}}{\sqrt{(c_{11}+c_{12})(c_{22}+c_{12})}}$$

$$V_1 = \frac{R_{1,fe}}{R_{1,ne}+R_{1,fe}}V_S \qquad I_1 = \frac{1}{R_{1,ne}+R_{1,fe}}V_S$$

$$\alpha_{1,ne} = \frac{R_{1,ne}}{Z_{c,1}} \qquad \alpha_{1,fe} = \frac{R_{1,fe}}{Z_{c,1}} \qquad \alpha_{2,ne} = \frac{R_{2,ne}}{Z_{c,2}} \qquad \alpha_{2,fe} = \frac{R_{2,fe}}{Z_{c,2}}$$

where $Z_{c,1}$ and $Z_{c,2}$ are the characteristic impedances of each line without the effect of the adjacent one:

$$Z_{c,1} = \sqrt{\frac{l_{11}}{c_{11}+c_{12}}} \qquad Z_{c,2} = \sqrt{\frac{l_{22}}{c_{22}+c_{12}}} \qquad Q_{ne} = \frac{\alpha_{1fe}+\alpha_{2fe}}{1-\alpha_{1fe}\alpha_{2fe}}$$

8.3.4.3 Simplified crosstalk model

For an electrically short line (physical length shorter than wavelength) there is a further simplification, that is applied gradually, first with the *low-frequency assumption*, and then with *weak coupling*.

With the low-frequency assumption C and S terms become unitary and simplify, so that the expression of D becomes

$$D_{lf} = 1-\omega^2\tau_1\tau_2\left[1-k^2\frac{(1-\alpha_{1,ne}\alpha_{2,fe})(1-\alpha_{1,fe}\alpha_{2,ne})}{(1-\alpha_{2,ne}\alpha_{2,fe})(1-\alpha_{1,ne}\alpha_{1,fe})}\right] + j\omega(\tau_1+\tau_2) \qquad (8.3.44)$$

Another simplifying assumption is applicable to the vast majority of cases: the two lines are assumed to be weakly coupled, so that coupling occurs only in one direction (from the source to the victim) and there is no significant feedback coupling to the source. Then D may be further simplified, dropping the term in square brackets:

$$D_{lf,wk} = 1-\omega^2\tau_1\tau_2 + j\omega(\tau_1+\tau_2) = (1+j\omega\tau_1)(1+j\omega\tau_2) \qquad (8.3.45)$$

and the two parameters τ_1 and τ_2 show clearly as time constants.

The resulting near-end and far-end crosstalk voltages are

$$V_{ne(lf,wk)} = \frac{1}{D}\left[\frac{Z_{2,ne}}{Z_{2,ne}+Z_{2,fe}}j\omega l_{12}L\,I_1 + \frac{Z_{2,ne}Z_{2,fe}}{Z_{2,ne}+Z_{2,fe}}j\omega c_{12}L\,V_1\right] \qquad (8.3.46)$$

$$V_{fe(lf,wk)} = \frac{1}{D}\left[-\frac{Z_{2,fe}}{Z_{2,ne}+Z_{2,fe}}j\omega l_{12}L\,I_1 + \frac{Z_{2,ne}Z_{2,fe}}{Z_{2,ne}+Z_{2,fe}}j\omega c_{12}L\,V_1\right] \qquad (8.3.47)$$

with the leading fraction $1/D$ disappearing if the frequency is low enough, that is below the two critical frequencies defined by time constants $f_1 = 1/(2\pi\tau_1)$ and $f_2 = 1/(2\pi\tau_2)$.

The terms summed or subtracted between square brackets are indicated by Paul [265] as M_{ne}^{ind} and M_{ne}^{cap} for the first near-end equation and M_{fe}^{ind} and $M_{fe}^{cap} = M_{ne}^{cap}$ for the second far-end equation. The change of sign for the inductive term between the near- and the far-end coupling equations may be intuitively interpreted as two inductive terms propagating on victim line 2 in the same and opposite direction of the source current.

The terms increase linearly with frequency and they will be complemented by a common impedance coupling term (due to the common return conductor and consisting of its longitudinal resistance), that completes the low-frequency simplified crosstalk model. The coupled voltages at near and far end due to the voltage drop V_0 on the return conductor is

$$V_0 = r_0 L\, I_1' = V_S \frac{r_0 L}{r_0 L + R_{1,ne} + R_{1,fe}} \cong V_S \frac{r_0 L}{R_{1,ne} + R_{1,fe}} \tag{8.3.48}$$

$$V_{ne}^{cond} = V_S \frac{r_0 L}{R_{1,ne} + R_{1,fe}} \frac{R_{2,ne}}{R_{2,ne} + R_{2,fe}} \tag{8.3.49}$$

$$V_{fe}^{cond} = -V_S \frac{r_0 L}{R_{1,ne} + R_{1,fe}} \frac{R_{2,fe}}{R_{2,ne} + R_{2,fe}} \tag{8.3.50}$$

They represent conductive coupling terms and add to the equations above (8.3.46) and (8.3.47) for inductive and capacitive coupling terms.

8.3.4.4 Descriptive crosstalk formulation

Rather than using the complete set of equations, in the PCB world IPC proposes two coefficients, k_F and k_B, called *forward crosstalk coefficient* and *backward crosstalk coefficient*, respectively, corresponding to near-end and far-end crosstalk, and analyzes how the coupling from the source to the victim line occurs while signals are propagating. Such a representation is well suited thus to depict logic signals switching from one state to the other, with inductive and capacitive coupling occurring only during the transition in the form of coupled pulses onto the victim line.

Recalling the scheme of a source line 1 and a victim line 2, let's consider a voltage step as source that begins propagating along line 1: arriving at a first position $z = z_1$ we may observe a certain amount of inductive and capacitive coupling that causes two signals to appear on the victim line. Focusing on the inductive coupling terms, it is observed that each small section dz of inducing line couples onto the victim line and causes a voltage pulse to appear at the two ends of this small mutually-coupled circuit with a negative pulse at the side towards the end of the victim line and a positive pulse at the side towards the beginning of the victim line. The voltage step continues traveling along the source line and arrives at a second position $z = z_2$ after a time interval $\Delta t = t_2 - t_1$ and the coupling to the victim line is similar: capacitive coupling will cause two symmetrical currents to go back to the near end and the far end, while the inductive coupling will create an opposite current. However, at position z_2 we are ahead of the previous position and the elapsed time is Δt: this means that the voltage pulse caused at position z_1 and going to the far end has traveled side by side with the voltage step in line 1 and will sum with the new coupled voltage proceeding towards the far end; the longer the lines the larger the number of coupled voltage terms that sum all along towards the far end. Conversely, the voltage pulse created at z_1 and going towards the near end is much farther and while the voltage step has traveled in the positive direction for Δt, the backward pulse progressed in the opposite direction towards the near-end for another Δt, so that any coupled voltage at position z_2 will be $2\Delta t$ far in time. So, there will be a series of pulses created on the victim line traveling back to the near end, separated by $2\Delta t$ and that do not influence each other, so that the line length has no influence on their amplitude; for $\Delta t \to 0$ this means that the near-end crosstalk voltage will be fairly constant and will last for twice the propagation time of the line.

To conclude, while the step signal on the inducing line propagates along line 1, several negative pulses are inductively generated on the forward sides of each elementary dz transformer and propagate towards the far end, arriving there at the same time, so that a reinforcement is observed with a final negative pulse that is the sum of all the elementary pulses. Conversely, the positive pulses travel back towards the near end and those generated when the step signal had just left the voltage source arrive at the near end almost immediately; as long as the voltage step propagates along line 1, the generated backward pulses take more and more time to reach back the near end, so that the result is a flat pulse lasting for twice the propagation time τ.

Considering the capacitively coupled current terms, since the transverse mutual capacitance sees no difference in either direction of propagation on the victim line, it divides into two currents flowing in opposite directions towards the near-end and the far-end. Both are positive pulses reaching the far end all together and the near end distributed on a 2τ time interval.

While the capacitively coupled pulses have the same polarity, the inductively coupled pulses have opposite polarity: the inductive current will sum to the capacitive current when going to the near-end side and subtracts when going to the far-end side. This explains the change of sign evident in the two equations above (8.3.46) and (8.3.47). Since it was already commented that the resulting terms coupled into the victim line are of derivative nature (both have a multiplying $j\omega$ factor), the waveshape of crosstalk terms is that of a short pulse, with a time duration that is much similar to the rise time t_r of the source voltage step.

The result is that at the far end the two coupling terms partially or totally compensate, but the resulting pulse for each of them is the sum of all the elementary pulse, and thus the intensity is proportional to the length of the line; at the near end the two terms sum up, but the individual intensity is much lower because the individual pulses are spread over twice the propagation time. For lines with balanced inductive and capacitive coupling terms, the resulting far-end crosstalk is quite under control. Yet, the backward term propagating towards the near-end, once there, is very likely reflected back, because in many digital lines the driving circuit may have a low impedance and it is thus mismatched with a negative reflection coefficient. This means that after one propagation time τ, a negative flat pulse of time duration 2τ will begin to appear at the far end too. Of course, terminating drivers and receivers correctly, impedance matching removes crosstalk terms reflection.

This explanation has assumed that the inducing and victim lines have the same propagation velocity, so that the voltage step and the far-end induced terms in z_1 have traveled side by side. Removing this assumption makes the description much more complex. The general case that includes also mismatched terminations and line attenuation is to be analyzed by simulation methods and loses its simplicity.

Another deviation from the assumed ideal is that the step edge smears while traveling along the source line for many reasons: line losses increasing with frequency, variable delay (or dispersion) and the coupling itself. The consequence is that the coupled voltage pulses will be smaller and less sharp, the near-end crosstalk term will be no longer constant and the far-end crosstalk will not be as large as it was announced.

Finally, the assumption was implicitly made that the propagation time τ of the source line is much longer than the rise time of the voltage step; this is quite relevant, since the induced pulses are all due to a derivative effect, that originates at the signal edges and the pulses have a time duration more or less equal to the signal rise or fall time. What happens if the lines are shorter? If the propagation time $\tau < t_r$, then the near-end crosstalk terms will be smaller, by the ratio $2\tau/t_r$ and the far-end crosstalk terms too, for the simple fact that the high frequency content is smaller, again by $1/t_r$.

The expressions for the backward and forward crosstalk coefficients used frequently in the analysis of PCB transmission lines are:

$$k_B = \frac{1}{4\tau}\left(\frac{L_{12}}{Z_c} + C_{12}Z_c\right) \qquad k_F = -\frac{1}{2}\left(\frac{L_{12}}{Z_c} - C_{12}Z_c\right) \tag{8.3.51}$$

where the meaning of symbols is as before in this chapter: the mutual inductance L_{12} and capacitance C_{12} are the total values for the whole line and the characteristic impedance Z_c is the same for the source and victim line.

Interpreting the two crosstalk coefficients with respect to odd- and even-mode characteristics of the line [45], we may write

$$k_B = \frac{1}{2}\frac{Z_{c,e} - Z_{c,o}}{Z_{c,e} + Z_{c,o}} \qquad k_F = \frac{1}{2}\left(\frac{1}{v_o} - \frac{1}{v_e}\right) \tag{8.3.52}$$

The term $\dfrac{Z_{c,e} - Z_{c,o}}{Z_{c,e} + Z_{c,o}}$ is sometimes called "coupling". Two traces farther apart with thus nearly equal odd- and even-mode characteristic impedance will have negligible crosstalk.

For transmission lines of the microstrip type Johnson and Graham [199] propose a simple rule for the far-end crosstalk:

$$\hat{k}_F = \frac{1}{1 + \left(\frac{d}{h}\right)^2} \tag{8.3.53}$$

where d is the separation between traces and h is the height of the substrate separating the traces from the common return plane beneath. It is easy to see that for $d = 2h$ the crosstalk is 20% and for $d = 3h$ it drops to 10%. Adding a grounded guard trace simply reduces further the crosstalk to about one half, as demonstrated by their experiments. When tested against experiments, this estimate looks quite accurate and only slightly overestimating.

For the near-end crosstalk, as a first approximation, it is possible to rule out time dependency by taking the maximum value, called *saturated crosstalk*.

$$V_{ne,sat} = k_B V_S \tag{8.3.54}$$

For slow rise times (longer than twice the propagation time) the saturated crosstalk voltage is never achieved.

Another factor that differentiate real situations from an ideal coupling configuration is that the coupling length is always assumed equal to the physical length of the lines, whilst it might be the case that the source and the victim line couple significantly only for a fraction of their physical length, when e.g. the separation between them is forced to reduce (for example when passing through a multi-pin connector or because of board space constraints). This of course complicates the analysis, but at a first approximation the calculation of the crosstalk impact may be done using the coupling length as line length, while propagation time is that for the whole length.

In real situations more than one aggressor may be active at a time, as it happens when a parallel digital bus has many lines switching synchronously from one state to the other, with same or opposite polarity: it is thus evident that there is superposition of the various terms and many combinations are possible; however, the worst-case situations may be isolated (e.g. all aggressors switching with the same polarity, equal or opposite to that of the victim line) and calculations carried out under this assumption, thus using again eq. (8.3.51) with varying values of inductive and capacitive terms.

8.3.4.5 Samtec reference board example

Samtec designed and manufactured a board for internal use, containing a coupled microstrip with guard traces at the ground potential and a ground plane underneath. The geometry of the board is shown in Figure 8.3.4 (microstrip width, gaps between traces, trace thickness and substrate thickness are drawn to scale) and the detailed information on geometry and materials is the following: $w = 75\,\text{mils}$, $s = 25\,\text{mils}$, $w_0 = 1000\,\text{mils}$, $h = 59\,\text{mils}$, $t = 14\,\text{mils}$, generic FR4 dielectric material with dielectric permittivity around $\varepsilon_r = 4.3$. No information appears on Samtec's website or any of their publications (as per web search), but various characteristics can be retrieved from presentations and leaflets prepared by board and instrument manufacturers.

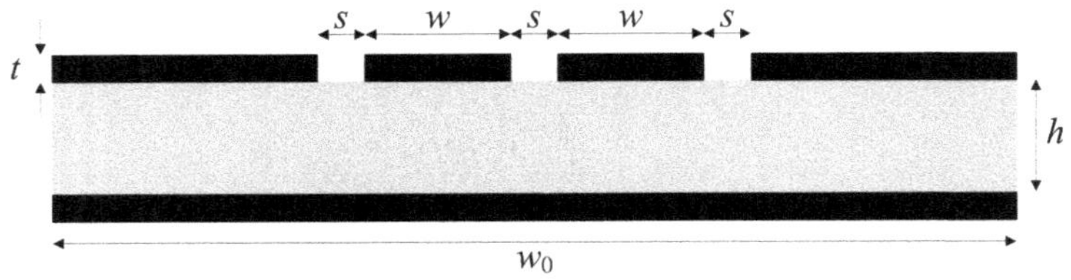

Figure 8.3.4 – Samtec Golden Board used as reference for mixed-mode S parameters measurement and other benchmarking parameters.

Freq. [GHz]	S_{11} [dB]	S_{21} [dB]	S_{31} [dB]	S_{41} [dB]
1.0	−23	−0.5	−14	−16
2.0	−21	−1.3	−15.5	−10
3.0	−20	−2.3	−17	−7
4.0	−18	−4.5	−18.2	−5
5.0	−18	−7.0	−17.8	−3.1
6.0	−19	−12	−14.9	−2.8
7.0	−17	−25*	−15	−2.8
8.0	−12	−15*	−16	−4
9.0	−10	−10	−12.5	−8
10.0	−7	−11	−15	−18**

Table 8.3.3 – Measured S parameters (magnitude) of Samtec Golden Board: * resonance of S_{21} at 7.2 GHz, ** resonance of S_{41} at 9.8 GHz.

Characteristic impedances resulting from simulations and time-domain measurements are: $Z_{c1} = Z_{c2} = 54\,\Omega$ (characteristic impedance of each microstrip alone), $Z_{c,e} = 67\,\Omega$ (even-mode characteristic impedance), $Z_{c,o} = 43\,\Omega$ (odd-mode characteristic impedance), $Z_{c,diff} = 2Z_{c,o} = 86\,\Omega$ (impedance seen by a differential signal), $Z_{c,com} = Z_{c,e}/2 = 33\,\Omega$ (impedance seen by a differential signal) all in agreement with simulated values. The two microstrip traces run parallel to each other for $5.15''$ and then diverge to connect to the respective launchers. Measured results obtained with a VNA are reported in Table 8.3.3: assuming, as demonstrated by measurements, that the two microstrips are symmetrical, only S_{11} is shown for the return loss; then S_{21} is the insertion loss, S_{31} is the near-end crosstalk and S_{41} is the far-end crosstalk.

As seen, the resonance in FEXT S_{41} curve is experimentally determined at 9.8 GHz; theoretical calculations show that this resonance is variable and depend on line losses, thus mainly on dielectric losses of substrate: in lossless condition it should be 13.6 GHz, whereas for typical FR4 losses it drops down to 8 GHz; any FR4-like material with lower losses increases it around the measured value.

Analogously, the insertion loss S_{21} curve has a resonance at 7.2 GHz due not to a defect, discontinuity or mismatch, but simply to the coupling with the adjacent line, confirmed by the maximum in the FEXT curve.

9

Spectrum Analyzer

This chapter doesn't aim to cover the whole architectural details of a Spectrum Analyzer (SA) and all features, settings and options that might be available and in many cases distinguish equipment following manufacturers' choices and views. Rather, details and characteristics that are quite relevant for accurate measurements and to determine SA performance are considered. Good descriptions of SA operation and architecture may be found in manuals of major manufacturers [6, 280], together with pictures, diagrams and examples of advanced measurements, based often on options and commands of their own instruments: EMC measurements, channel power (in particular for modern digital modulation schemes), phase noise, distortion and intermodulation, to cite a few. Often these examples are tailored to a specific architecture, so their usefulness and applicability might be limited.

It is preferred to focus on the basic settings that equip the vast majority of spectrum analyzers, on how to check the correct operation of the instrument and that results are in-line with expectations and are technically correct and consistent.

9.1 Architecture and block diagram

A general architecture of a super-heterodyne spectrum analyzer is considered, that, as the time goes by and numeric techniques and digital sampling are used more extensively, should be updated to resemble the so-called real-time spectrum analyzers; these analyzers are in principle based on a pure time-domain capture and digital processing, but necessitate of a first mixer and down-sampling stage to extend their applicability to the highest frequency part of the spectrum. A pure Fourier transform approach is limited by the maximum ADC sampling frequency (even with architectures with high interleaving factors) and by the huge number of samples needed; under-sampling and band-pass processing are used nowadays to push the limit to very high frequencies.

The vast majority of spectrum analyzers are still based on super-heterodyne architecture, that is more appropriate to illustrate the most common functionality and behavior of spectrum analyzers in practical situations; it's quite common that super-heterodyne architecture sets the reference to evaluate performance of other architectures and solutions. Any super-heterodyne architecture has some digital processing (including Fourier transform) applied to signals, once they have been demodulated and brought to a conveniently low intermediate frequency (IF) interval. When relevant, distinction will be made between analog and digital processing.

The typical block diagram of a Spectrum Analyzer is shown in Figure 9.1.1.

9.1.1 Input attenuator

Attenuators were introduced in Chapter 1, describing their architecture and defining the behavior in terms of input impedance and attenuation (see sec. 1.2), as well as matching capability for mismatched loads (see sec. 1.2.5).

The input attenuator is responsible for the adjustment of input signal power, in order to feed successive blocks with the correct signal amplitude; the correct operating point of the mixer, for example, is extremely important for its operation with respect to linearity, distortion, compression point, dynamic range.

Moreover, since it is a passive resistive network positioned at the beginning of the SA block chain, it ensures protection against moderate external overvoltage. Using a large input attenuation is always advisable when a measurement is performed for the first time, involving a new setup and the possibility of external disturbance; in case of more severe hazards, an external attenuator (or a filter, a transient protector, or a combination) shall be added.

Another useful feature of the input attenuator is that it ensures a better impedance matching at the input port, improving the accuracy related to reflection (VSWR) on the input line. In Figure 1.2.5 performance of attenuators featuring varying attenuation was shown, when the connected load has a VSWR variation as large as 5:1; the results have confirmed that 10 dB of attenuation ensure a VSWR lower than 1.1 for a 3:1 variation of load resistance. It is not uncommon that a calibration is performed with 10 dB of attenuation to ensure a better uncertainty (that shall be reproduced as is for certificate to be valid!); in general 5 or 10 dB of input attenuation are advised by all manufacturers somewhere in their manuals and application notes, even if for some particular measurements the full input dynamic range is needed, such as when measuring distortion and phase noise (see sec. 9.4 and 9.5).

Last, when connecting the spectrum analyzer to a device with a widely changing impedance (such as an antenna) and with far from ideal conditions (such as on site

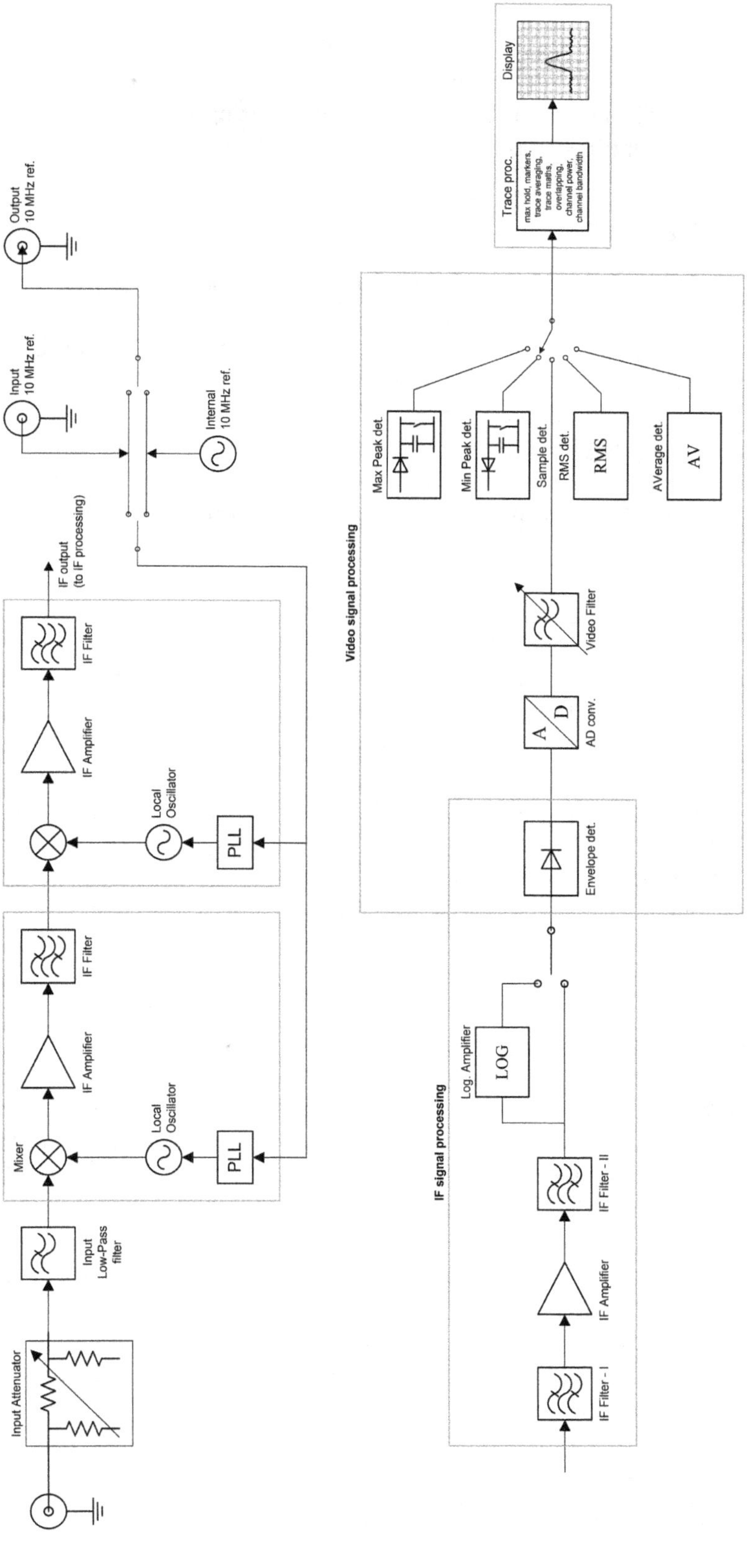

Figure 9.1.1 – Spectrum Analyzer architecture and block diagram

and using quite a long connection cable, exposed even to common-mode resonances with ground), the use of attenuators at both cable ends is also advisable.

As customary, we will indicate the input attenuator setting by using the word "Att." followed by the amount in dB, as usually appears on SA display.

9.1.2 Pre-amplifier

The task of the pre-amplifier is to reduce the noise level by amplifying the input signal at the beginning of the SA block chain: based on the theory of noise through cascaded linear time-invariant systems (see sec. 4.3.1.6), the signal-to-noise ratio is increased if the amount of amplification before the internal noise sources is increased. Also in the case of two internal noise sources, increasing the amplification of the system blocks between the two improves the signal-to-noise ratio at the output by substantially ruling out the second noise source from the result. This is what is done by the pre-amplifier: some residual noise, however, will be still present due to the pre-amplifier itself and the input attenuator.

The usual gain of the pre-amplifier is in the range of 15-25 dB, 20 dB the most common, making the rapid interpretation of results easier, when not compensated automatically in old spectrum analyzers. Due to hardware limitations, and in particular when using different modules to cover the entire frequency range, pre-amplifier gain may change[1]: in this case a measurement that spans across two frequency ranges is more complicated and exposed to errors when correction is applied.

Internal noise reduction can be measured when turning on the pre-amplifier, quantifying it in terms of equivalent noise resistance. In Figure 9.1.2 there are the noise displayed levels for RBW = 10 kHz and with input attenuator changing from 20 dB to 0 dB, with and without the pre-amplifier; it is possible to estimate the noise contribution of each element separately and derive an estimation of the equivalent noise resistance (see sec. 4.3.1.8 for the mathematical aspects of equivalent noise resistance). The noise level for Att = 10 dB is $-116.03\,\mathrm{dBm}$; stepping down to Att = 0 dB brings the noise level to $-125.64\,\mathrm{dBm}$. The noise contribution of the attenuator should then be estimated as the square root of the difference of the squares of the two noise levels, assuming – as it is in reality – that the measurements are uncorrelated. However, being the levels so different, attenuator noise can be readily estimated as the entire measured noise level, while the exact calculation gives $-116.05\,\mathrm{dBm}$. The displayed noise level is thus entirely caused by the input attenuator.

The use of pre-amplifier lowers the noise level to $-135.91\,\mathrm{dBm}$ starting from a 0 dB attenuation configuration: the effect of the pre-amplifier gain is clearly visible in the 10 dB step; assuming a perfect 10 dB power gain, the slight difference once the gain is compensated for may be used to have a rough estimate of the pre-amplifier noise term. The square root of the difference of the squares of the corrected 0 dB att. noise level and the noise level measured with the pre-amplifier on gives an estimate of $-140\,\mathrm{dBm}$ as pre-amplifier noise contribution.

[1] For example it occurred in one of the first battery supplied models, where the pre-amplifier gain had a ugly step change around 2 GHz.

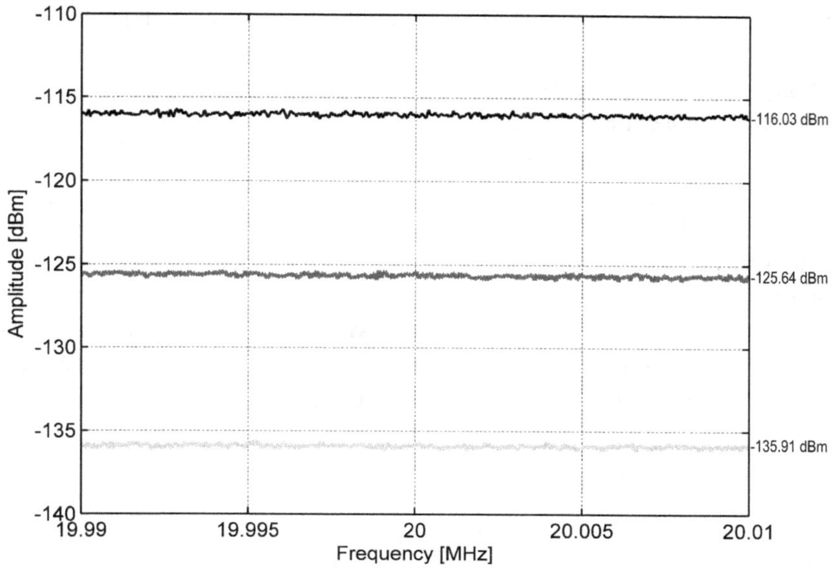

Figure 9.1.2 – Display noise level caused by input attenuator and pre-amplifier: Att = 10 dB and pre-amp off (black), Att = 0 dB and pre-amp off (dark gray), Att = 0 dB and pre-amp on (light gray); the RBW was set to 10 kHz.

The equivalent noise resistance (ENR) values may be determined using eq. (4.3.11). When the input attenuator is set to 10 dB, the calculated ENR is around $140\,\Omega$, not so far from the values of the internal resistors of the attenuator and masking any other internal noise source; when the attenuator is removed (0 dB attenuation) resistance drops to about $15\,\Omega$ and then to $1.5\,\Omega$, when the pre-amplifier is switched on, much lower than the noise level of a $50\,\Omega$ resistor.

9.1.3 Local Oscillator

In modern spectrum analyzers the Local Oscillator is implemented with digital synthesis starting from a PLL (Phase Locked Loop) fed by a high-stability crystal oscillator (XO), generally of the temperature-compensated (sometimes named improperly temperature-controlled) type, TCXO, or in high-end models with an oven-controlled crystal oscillator (OCXO), where a heating element really controls the temperature around the crystal oscillator and keeps it in the most stable environmental conditions possible. Any kind of digital oscillator or voltage controlled oscillator (VCO) is initially fed by a crystal; fully electronic oscillators (either digital or analog) are of course possible [165], but their stability is absolutely worse than crystal-based oscillators. YIG (Yttrium Iron Garnet) technology may be also adopted because of the low phase-noise that characterizes it, and it was certainly widely used in the past generation of spectrum analyzers, as well as RF generators.

The information on the stability is given in instrument characteristics and can be measured directly from the 10 MHz output connector, provided that an order of magnitude better stability is available in the external instrument (e.g. frequency meter). It is observed that for an OCXO LO the short-term stability is already of the same kind of the best GPS frequency references: where GPS information comes into play is for long-term stability. For the evaluation of the stability of a frequency reference, please see sec. 9.5 on phase-noise measurements.

The Local Oscillator is adjusted in steps during the sweep, whose amplitude depends on the number of points (or pixels, see sec. 9.1.9 and 9.2.7 on detectors) and the selected resolution bandwidth (see sec. 9.1.5 and 9.2.2). The number of points may be variable, so the user has some control on frequency resolution. The frequency step is always related to the adopted resolution bandwidth for a very practical exigency, that is not to miss parts of the spectrum while sweeping; when performing emissions measurements, for instance, it is always required that the frequency step is smaller than the adopted resolution bandwidth with a margin.

9.1.4 Demodulation and Mixer

The mixer ideally "mixes" two input signals at its ports, namely the input signal, fed by a preceding stage, and Local Oscillator signal for down-conversion. The working principle is that of the product of the two signals and the resulting output has a spectrum that contains components at the difference and the sum of the individual frequencies: the principle can be explained easily by visualizing the product of two sinusoidal tones and using the prosthaphaeresis algorithm. The reader is invited to have a look at some references for details on how mixers are built and the principle used to perform the multiplication of signals (Horowitz [165]; S.A. Mass, *The RF and Microwave Circuit Design Cookbook*, Artech House, Boston, 1998, entirely dealing with mixers; Collin [88], sec. 12.7 to 12.11).

In modern spectrum analyzers, featuring an extended frequency range, there might be more than one mixer, progressively demodulating the incoming signal to lower frequency until the IF (Intermediate Frequency) stage before log-amplifier and detectors.

Heterodyne demodulation by means of a local oscillator at frequency f_{LO}, converts an incoming mixer tone at frequency f_{in} to an intermediate frequency f_{IF}:

$$|m\, f_{LO} \pm n\, f_{in}| = f_{IF} \tag{9.1.1}$$

that for $m = n = 1$ simplifies to $|\,f_{LO} \pm f_{in}| = f_{IF}$. This relationship is obtained by trigonometric manipulation of mixer product (see Figure 9.1.3).

Conversely, $f_{in} = |\,f_{LO} \pm f_{IF}|$ and this clearly indicates that for each pair of local oscillator frequency f_{LO} and intermediate frequency f_{IF}, there exist two input frequency values that satisfy the relationship, and that are thus captured and demodulated: the desired one towards dc and a second one, at higher frequency, called "image frequency". Extending this reasoning to the interval of frequencies that "nearly" satisfy the relationship, or in other words, that are close to f_{in} within a given interval, we talk of "desired frequency band" and "image frequency band". Of course at the output of the mixer also f_{LO} and f_{in} components are visible: for a sufficiently low amplitude of the input signal the mixer operates in the linear region and these four components are the only ones that populate the spectrum. This is a simplified reasoning, because for diode mixers the repeated switch on and off of the bridge diodes creates many inter-harmonics of input and LO frequencies (the explanation in sec. 10.4 of A.W. Scott, *Understanding Microwaves*, John Wiley & Sons, 1993 is very clear).

Two inconveniences may affect mixer operation:

- the signal tone that is being read is accompanied by another component that is at the same frequency distance from f_{LO} but at the opposite side of the frequency axis, i.e. this component is the image component; there is no way to separate the two resulting frequency differences if considering amplitude only, because both fall exactly at the same frequency, but have an opposite phase; if this additional information can be used, then it is possible to separate the two components (the "real" one and that caused by the "image tone", that is however a perfectly real and legitimate component of the input signal); this is accomplished by the so-called "image rejection" mixer that uses two balanced bridges at 90°;
- the amplitude of the input signal increases and the mixer non-linearity creates inter-modulation products, that spread over the frequency axis: with one single input tone, the sum/difference combination of harmonics of f_{LO} and f_{in} produces frequencies that are integer multiples of the basic difference $|f_{\mathrm{LO}} - f_{\mathrm{in}}|$ that are called *single-tone intermodulation* products, or spurious terms; if more than one tone is present at the input (e.g. f_{in1} and f_{in2}), then the number of combinations increases, and sums and differences occur of integer multiples of f_{LO}, f_{in1} and f_{in2} thus producing terms that may fall next to the original components around the IF band; this is called *two-tone intermodulation* and with distortion products is reviewed in sec. 3.4; harmonics and intermodulation may be kept well under control when using the so-called double-balanced bridge.

The mixer is the core of the spectrum analyzer and is largely responsible for its performance in terms of distortion and spectral accuracy. As just considered, key performance indexes of a mixer are the presence of harmonics and inter-modulation products, identified as *harmonic suppression* and *spurious suppression*, respectively.

Another performance index that describes the spectral properties and quantifies the amount of LO signal that leaks into the output is the *LO-to-RF isolation*. It is the ability to keep separate the components fed at each port, conveying the input uniquely towards the mixing process and the output; this is normally addressed by evaluating the leakage of the LO frequency back to the input and into the output; the overall effect is named "Local Oscillator feed-through"and is considered later in sec. 9.2.6.

Last, passives and semiconductors inside the mixer act as noise sources and add to the noise coupled through the signal and LO ports: as described in sec. 4.3.1.6 and 4.3.1.8, the mixer is characterized by its noise figure, that shall be determined taking into account that only a fraction of the incoming signal (and noise) reaches the output (as indicated by the conversion loss) and that some signal regeneration is operated by the internal IF amplifier, that is in turn characterized by its own noise figure.

The amount of intermodulation and distortion is primarily a mixer characteristic and depends on its operating point; this is often addressed by specifying the range of acceptable input power level for linear operation and compressions points, when the mixer deviates from its ideally linear input-output relationship (see sec. 3.4 and 9.4).

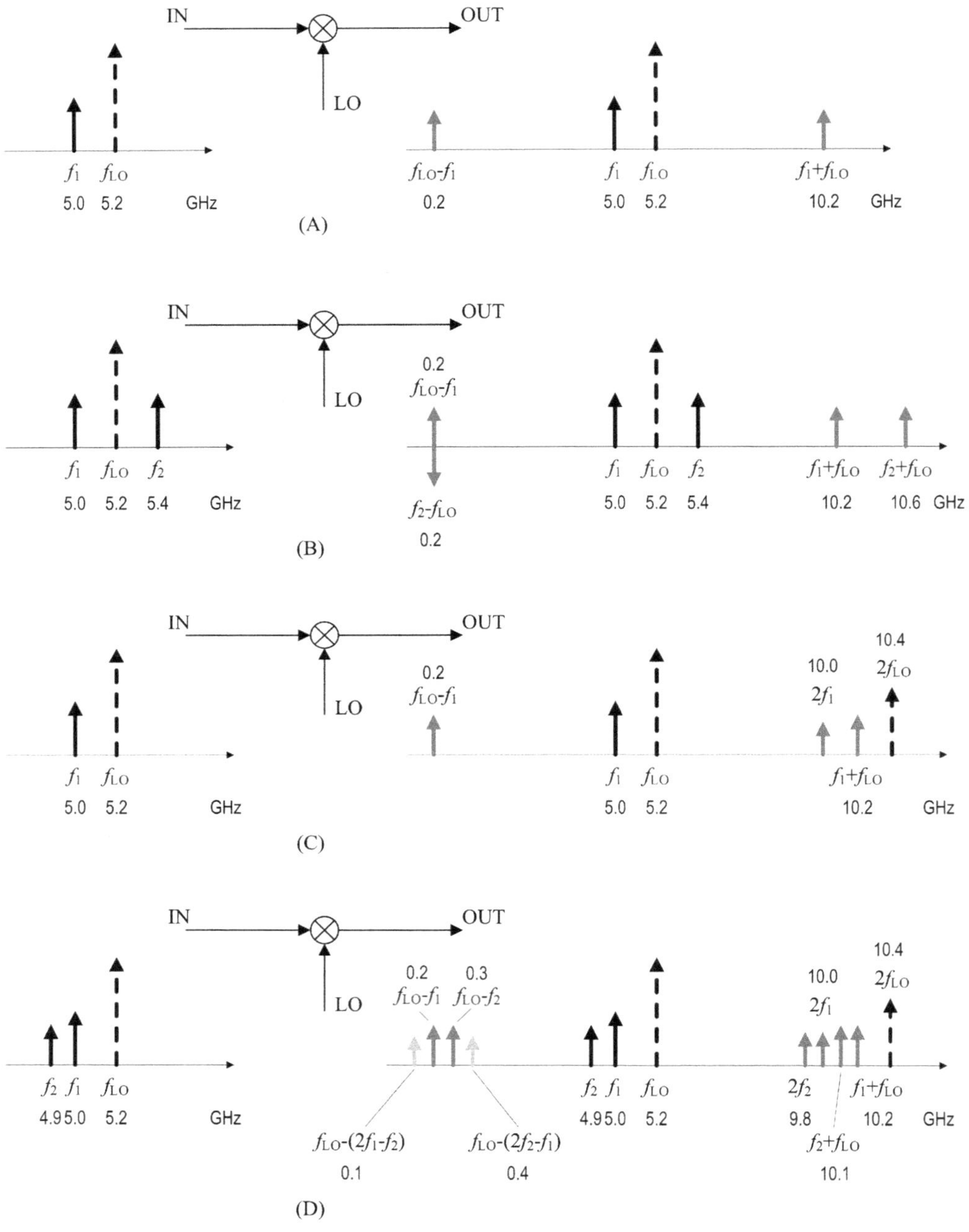

Figure 9.1.3 – Mixer operation with ideal line spectrum of input signal (solid black, made of either one or two tones at f_1 and f_2), local oscillator tone (dashed black at f_{LO}) and mixer output (including original components, in black, and mixing products, in gray): (A) simple mixing under linearity assumption and one input tone f_1; (B) two input tones mirrored around f_{LO} (f_2 is the image frequency) create concomitant IF components with same frequency and different phase; (C) first-order intermodulation products with one input tone at f_1; (D) second-order intermodulation products with two input tones at f_1 and f_2 (only the main second-order intermodulation terms in the IF band are shown in light gray).

To avoid that strong out-of-band high-frequency components are captured and demodulated, overlapping to the correct desired components, a pre-selector filter may be added after the input attenuator: it is a tunable filter or filter bank that is optional in spectrum analyzers and mandatory in EMI receivers, and that rejects extraneous components that would otherwise be indistinguishable from the desired ones after demodulation. The rejection of the unwanted image frequency may be accomplished differently: an image-rejection mixer may be used distinguishing the different phase relationships; a tunable low-pass filter might be able to reject the image frequency, but it would be of difficult implementation over a much extended frequency range; for this reason it is preferred to have a high enough f_{IF} value that pushes the image-frequency interval well above the desired input frequency range, so that a fixed low-pass filter can be used instead. The low-pass filter helps also in decoupling the RF input from the internal RF sources of disturbance, that is IF and LO.

For a 9 kHz - 3 GHz spectrum analyzer such IF is in the range of about 3.1-4 GHz and the LO 3 GHz above it. The resolution bandwidth (RBW) filter that accurately weights the various portions of the input signal (see sec. 9.2.2) can be hardly applied directly to such a high IF; the same may be said for the algorithms used for detection and signal processing (e.g. video filtering and trace processing in sec. 9.1.10 and 9.2.8). Thus, conversion to a lower IF is necessary and this conversion is done in steps passing normally through an intermediate IF and a final IF value, the latter then fed to the Analog-to-Digital Converter (ADC). To exemplify some values, the intermediate IF might be in the range of some hundreds MHz and the final IF around some tens of MHz, compatible with the RBW values and not so different from the values of some old single-IF spectrum analyzers [280]. Usually the higher the maximum input frequency, the higher the first IF frequency and the down-converting IF frequency; when the difference is too large, an IF stage is added.

For some time when requiring too a large first IF frequency for very wide frequency ranges (tens of GHz), due to merely technological problems the IF method was replaced by direct conversion with a tracking band-pass filter (e.g. using the YIG principle). Today, even very large bandwidth instruments generally use the IF method and increase the number of intermediate IF stages.

The mixer, as any other RF or microwave device, is characterized by reflection and transmission:

- reflection at the ports (quantified as return loss) impedes the complete coupling of the input and LO signals due to impedance mismatch: the mixer is a complex device whose input impedance may vary with frequency and operating point; lacking a pre-selector filter, this impedance and its variation are visible from the input connector through the input attenuator and for this reason it is always advisable to ensure a minimum of attenuation (see sec. 9.1.1);
- the specification of its attenuation is more complex: the mixer is a non-linear device that redistributes spectrum components, so that the ability to transmit the useful signal is measured by the *conversion loss*; different mechanisms like internal mismatches and simply voltage drops contribute to conversion loss, but

the manufacturer may set corrective coefficients in the subsequent processing blocks, if such losses are not detrimentally large[2].

With respect to the input level, the power reaching the mixer depends on the setting of the input attenuator G_{att} and the inclusion/exclusion of the pre-amplifier (G_{pre}): $P_{mix} = P_{inp} - G_{att} + G_{pre}$. The optimization of the range of power levels fed to the mixer for signal-to-noise ratio and internal distortion products will be considered in sec. 9.2.11 and 9.4.

It is worth underlying that when measuring wideband noise-like signals, the perceived power for the chosen RBW value may be much lower than the total power applied to the mixer by e.g. one or some orders of magnitude (it suffices to remember that RBW is a fraction of the frequency range and that the crest factor may be large for noise-like signals). It is usually advisable to set the mixer input power level around -10 dBm, so that displaying a -20 dBm or -30 dBm power level with a narrow RBW ensures that the ideal limit of +10 dBm is never overcome. Too a high level before damaging the mixer will drive it into its 1 dB compression point with signal compression and distortion by-products (harmonics of the input signal) as a consequence (see sec. 9.4.3). Modern spectrum analyzers, not only warn the user for exceeding the advised maximum operating point, but often have hardware or software protections.

Similarly, it is not advisable to feed the mixer with too a weak signal to avoid that internal noise becomes relevant, unnecessarily reducing the signal-to-noise ratio; it is however a good policy to begin with a larger attenuation to protect the mixer and then reducing it by observing the effect.

9.1.5 IF filter

IF signal processing is performed at the last intermediate frequency (at some tens of MHz, it was said); here the signal is amplified again and resolution bandwidth and envelope detector are applied. The gain G_{IF} at this last IF can be adjusted in small steps (often 0.1 dB steps), so the maximum signal level can be kept constant when passed to the subsequent signal processing, regardless of attenuator setting and the mixer level. Of course, wrong settings compromise the signal-to-noise ratio, the dynamic range and may introduce some distortion; this operation is generally performed automatically within the spectrum analyzer and the user has no control on it. In case of more than one mixer, and correspondingly more than one IF processing block, there will be several IF gain terms $G_{\mathrm{IF}1}$, $G_{\mathrm{IF}2}$, etc. as well as different mixer levels.

The IF filter is used to define that portion of the IF-converted input signal that is further analyzed and then displayed; the IF filter is thus responsible for the accurate definition of the Resolution Bandwidth (RBW), that weights many spectral characteristics and features a very wide range, normally from few Hz to a tens of MHz.

Implementation choices are obviously those of analog and digital filtering, but also the use of the Fourier transform is an option. There is a major tradeoff in IF filter construction, combining the opposite requirements of selectivity and fast transient response (so

[2] Post-processing correction is not effective if the signal-to-noise ratio worsens too much, because correction is blindly applied to signal and noise at mixer output.

to sweep the frequency axis by changing the LO frequency at a convenient pace). The tradeoff has converged onto the Gaussian architecture for the analog/digital implementation. For details regarding filter performance and for Fourier transform characteristics please see sec. 9.2.2 and 9.2.9.

Analog filters are often used for the largest RBW, let's say above some hundreds kHz for the difficulty to implement digital filters with the necessary sampling rate[3]; analog implementation is conceived to have similar performance in terms of selectivity and shape factor at least down to 20 dB with respect to center band. Since analog filters are in any case built assembling different stages to obtain the desired roll-off, it is customary to split the filter into two parts, one soon after the IF mixer output, followed by the *IF amplifier* and then by the second part of the filter.

Digital filters have a much better selectivity than analog ones, that comes at hand when selecting the narrowest RBW values. Digital filters are also more stable with respect to temperature, drift and aging, and thus more reliable and preferable as for uncertainty and the need for periodic calibration. Moreover, since they exhibit faster transient response than their analog equivalent, they allow slightly faster sweep times. However, they shall be preceded by the ADC stage, in turn preceded by an analog anti-aliasing pre-filter, and this slowed down their adoption at the beginning.

Last, when reaching low RBW values and/or requiring fast sweep times, it is much more convenient to rely on Fourier transform (FFT) and direct frequency filtering. The use of windows for frequency leakage suppression affects the effective frequency resolution and requires a longer *transformation window* (named also *observation time*). However selectivity is much better than analog and digital implementations: e.g. with a flat-top window a shape factor of 2.6 is attainable. The *shape factor* or *skirt selectivity* is defined as the ratio between two bandwidths, one at high attenuation, e.g. down by 20, 40 or 60 dB, and the other taken as reference, usually the -3 dB bandwidth (normally used to define the RBW value itself); further details and values for the most common implementations can be found in sec. 9.2.2.

9.1.6 Log-amplifier

The log-amplifier implements an ideal logarithmic function (linear on a log-scale) of the incoming power between the two extreme points of a range that may span several decades. When years ago the log-amplifier could be built starting from a good diode (or a transistor in diode connection), attention was focused on the excellent logarithmic law needed for the base-emitter junction and datasheets of the most suitable transistors were expressively reporting the deviation from the pure log law (the 2N1711 was an example of a suitable device with nearly 5 decades of excellent logarithmic behavior). For a spectrum analyzer this performance hides behind the figures of "scale fidelity" or "linearity error" appearing in the datasheet.

[3] With the increasing computational power available first in DSPs and then in FPGAs, the only bottleneck remaining is the ADC sampling rate, that requires a tradeoff between resolution, linearity and the sampling rate. In the last years ADCs offering 20-60 MS/s sampling rate at 12-14 bit resolution are available, allowing the full digital implementation up to some MHz, even recurring to an up-sampling/down-sampling block in the FPGA.

The log amplification acts as a compressor for large noise (or signal) peaks; a peak of ten times the average level is only 10 dB higher on a power log-scale. Instantaneous near-zero envelopes, on the other hand, contain no power, but are expanded toward negative infinity decibels, and this might bias averaging (see sec. 1.1.5.2), while for display purpose they can be removed quite easily. The combination of these two aspects of the logarithmic curve affects the noise power statistical distribution and leads to read values lower than the true noise power for some settings (see sec. 9.1.9.2 and 9.3) [4]. By the way, when spotting sinusoidal tones out from noise, log compression ensures a reduced noise floor and thus an increased display dynamic range, in particular if used in conjunction with Min Peak detector (see sec. 9.2.7).

9.1.7 IF envelope detector

The IF signal after band-pass filtering is passed to the envelope detector block. This block removes the IF frequency and takes the lower frequency modulating signal, that is the base-band image of the incoming signal. This is like when recovering the modulating signal from the envelope of the incoming signal in AM modulation; the basic scheme for AM demodulation is that of a diode followed by a RC circuit that establishes the time constant of demodulation. With envelope detection only the magnitude of the envelope is used, while the phase information is lost; on the contrary, when processing is entirely performed by Fourier transform, as for real-time spectrum analyzers, the full information is retained.

The power content of the incoming signal is centered around the IF frequency like if it were the carrier, it is split towards dc and higher frequency (at twice the IF frequency), so that the desired band may be isolated and further processed. The Video Filter (see sec. 9.1.8 and 9.2.8) low-pass filters the detected IF output signal and eliminates the high frequency sub-band. When processing is performed digitally for the last IF block, the envelope is determined from the samples of the IF signal and a numeric implementation of the video filter is applied. For the first "digital" spectrum analyzers appearing in the '80s and for some models still in use today, the digital sampling was not operated on the IF signal directly, but on the signal at the output of the envelope detector fed to the video filter. This of course largely reduces the requirement on ADC sampling frequency, but limits the availability and flexibility of post-processing functions, especially needed today with sophisticated modulations featuring a wide dynamic range and demanding time-frequency performance.

The dynamic range of the envelope detector is very important because it largely affects the overall dynamic range of spectrum analyzer; modern spectrum analyzers feature normally 100-120 dB of dynamic range of the IF signal (see sec. 9.2.11). The user modifies the dynamic range by changing the reference level: when using this option, it is possible to widen the display dynamic range, and correspondingly to change the IF gain (but it is SA-model dependent); mixer operating point is not influenced, because down-conversion occurs before the IF envelope detector block. So, at this stage we can amplify the signal stretching it onto the displayed range, but we have no influence on signal distortion and compression, for which the input attenuation and the mixer operating point are the key settings (see sec. 9.2.11 and 9.4).

The IF signal before being passed to the envelope detector may undergo amplification and logarithmic compression by means of the log-amplifier described in sec. 9.1.6.

9.1.8 Video filter

The video filter in old analog spectrum analyzers was a simple RC filter following directly the output of the envelope detector (as for the basic AM demodulation circuit mentioned before) and drove directly the signal for the cathode ray tube. The name has been preserved, but all modern spectrum analyzers use analog-to-digital conversion (ADC) to implement digital processing functions, even if the video filter is still a low-pass filter with the aim of smoothing the trace, removing the jagged noise profile and increasing somewhat the displayed signal-to-noise ratio. The ADC may be located inside or after the IF envelope detector[4]: in all cases the digital implementation of Detectors and Video Filter is assured. The combined operation of detectors and video filter is quite articulated for the many options of modern spectrum analyzers, so that a clear all-comprehensive detailed connection scheme is not easily attainable.

The ADC also has its own recommended input power level specified by the manufacturer, that corresponds to its full scale range probably with some margin to avoid signal clipping at the two extremes; the margin necessary to avoid clipping must again take into account the crest factor of superimposed noise: considering Gaussian noise a crest factor of 2.6 is met 99% of the time (see sec. 4.1.1), resulting in 3 to 5 dB of margin for the equivalent power.

Details of Video Filter operation and performance with respect to the selected detector and linear or logarithmic scale are reported in other sections of this chapter: sec. 9.2.8 on video filtering and sec. 9.3.2 on channel power measurements.

If the Video Bandwidth VBW is smaller than the Resolution Bandwidth RBW, then the sweep time is further increased with respect to the bare requisite regarding the transient response of the RBW filter (see sec. 9.2.4).

9.1.9 Detectors

Detectors available in a modern spectrum analyzer are Peak (as Max Peak and Min Peak), Average, RMS, Sample. Their description and the explanation of the basic operation follow; for a more complete analysis of their performance, also in combination to other settings, please see sec. 9.2.7.

The explanation of their operation is based on the correspondence between a portion of the spectrum and its representation on the display that is made of a certain number of pixel columns (that we call for brevity "pixel"): a single pixel may contain the spectral information of a relatively large frequency interval; in this case the information contained in the interval, after being processed with the detector, is synthesized in a single value that is then displayed.

[4] In other words, the complete IF output signal that is fed to the envelope detector or only the envelope detector output may be digitized.

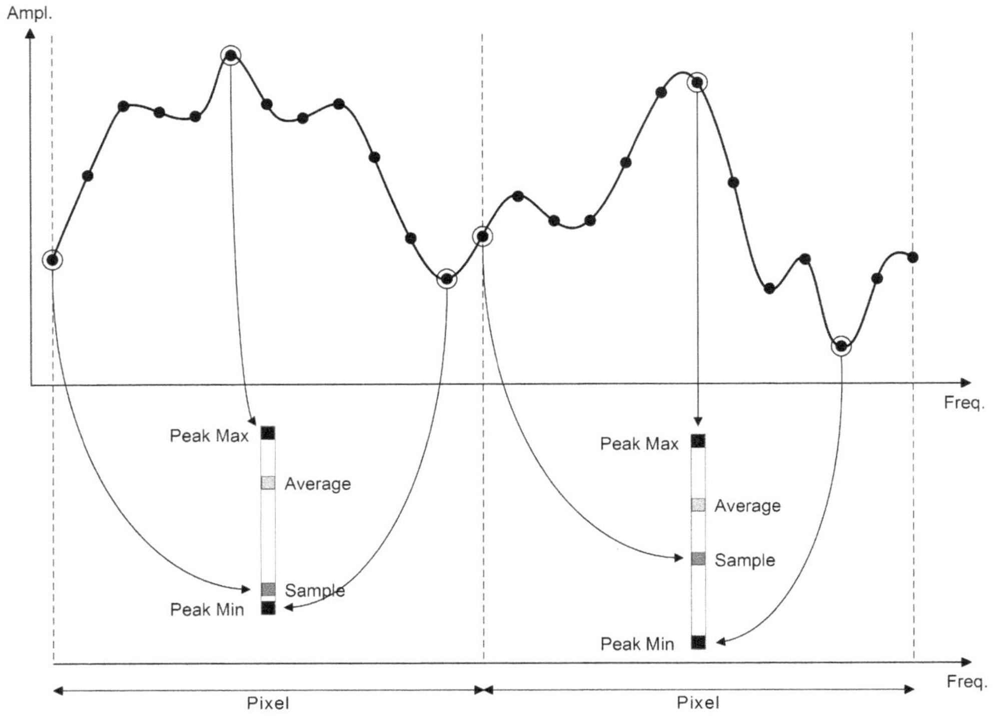

Figure 9.1.4 – Operation of detectors exemplified.

9.1.9.1 Definition of detectors

The operation of detectors is graphically described in Figure 9.1.4.

When considering random signals, such as noise, detector performance may be quite different and may depend also on the number of samples fed to the detector at each step: each detector operates on a pixel basis and collects the samples pertaining to that pixel in the chosen frequency span and resolution; when the number of samples per pixel is limited, this might bias the final spectral estimate and influence detector performance (see sec. 9.2.7 and 9.2.8).

Max Peak detector The Max Peak detector displays the maximum value. From the samples allocated to a pixel the one with the highest level is selected and displayed. Even if wide spans are displayed with very poor resolution bandwidth (this occurs when span/RBW is greater than the number of pixels on the frequency axis), no input signals are lost. Therefore this type of detector is particularly useful for EMC measurements and to capture at best transient signals with a wide frequency occupation.

Min Peak detector The Min Peak detector selects from the samples allocated to a pixel the one with the minimum value for display. Its usefulness alone is limited, but together with the Max Peak detector it is able to implement a maximum span detector.

Auto Peak detector It is the combination of the Max Peak and the Min Peak detectors, displayed simultaneously.

Sample detector The Sample detector samples the IF envelope for each pixel of the trace to be displayed only once. That is, it selects only one value from the samples allocated to a pixel, e.g. the first one, to be displayed. If the span to be displayed is much greater than the resolution bandwidth (again for a span/RBW greater than the number of pixels on the frequency axis), input signals are no longer reliably detected.

RMS detector The RMS (root mean square) detector calculates for each pixel the power of the displayed trace from the samples allocated to that pixel. The result corresponds to signal power within the span represented by the pixel. For RMS calculation samples of the envelope are required on a linear level scale. The reference resistance value (50 Ω) may be used to calculate power from voltage readings.

AVerage detector The AV detector calculates the linear average for each pixel of the displayed trace from samples allocated to a pixel. For this calculation samples of the envelope are required on a linear level scale. Averaging on a log scale wouldn't be useful because the largest values that are relevant to the result would be log compressed; it is however possible to enable this function by selecting a narrow Video Filter bandwidth (VBW) after processing the IF envelope on a log scale. In all cases averaging is also obtained by using the Video Filter after the Sample detector.

Needless to say, averaging of AV detector is performed on subsequent samples of the IF envelope signal, while *trace averaging* option (see sec. 9.1.10) performs averaging of trace pixels, far apart in time (each separated by a sweep time ST interval).

Quasi-Peak detector The Quasi-Peak (QP) detector "yields a measure roughly correlated to the subjective annoyance effect on AM broadcast services", and "has also found restricted use for measuring interference in television" [20]. So, its intended use is for interference measurement applications, adopting defined charge and discharge times that weight severity in terms of amplitude and rate of occurrence for periodic and intermittent disturbance. These times are established in Table 2 of the cited ANSI/IEEE C63.2 standard and in Table 2 of CISPR 16-2-1 [83] (see Table 9.2.3 later at the end of sec. 9.2.4).

9.1.9.2 Detector output for different signals

Depending on the type of input signal, different detectors may provide different measurement results. Choosing the wrong or inappropriate detector leads thus to a measurement error larger than the high-performance uncertainty characteristics of our spectrum analyzer we paid for.

Sinusoidal signal Assuming that the spectrum analyzer is tuned to the frequency of the input signal (span = 0 Hz), the envelope of the IF signal and thus the video voltage of a sinusoidal input tone with sufficiently high signal-to-noise ratio are constant.

Therefore, the level of the displayed signal is independent of the selected detector, since all samples exhibit the same level and since the derived average value (AV detector) and RMS value (RMS detector) correspond to the level of the individual samples.

Being different the amount of resulting noise, if the original signal-to-noise ratio is not sufficiently large, this contribution will be relevant and there will be differences in the displayed amplitude. It is easy to calculate the expected deviation by adding the difference in detector readout D (see the next paragraph) to the signal-to-noise ratio SNR, both expressed in dB, and reversing the log-relationship to extract the deviation around the tone amplitude (see sec. 9.2.12).

Noise The behavior is much different, however, taking random signals, such as noise or noise-like signals, in which the instantaneous power varies with time. If we consider the crest factor of the noise signal, then it is clear that the Max Peak and Min Peak detectors will report values as much as different from the RMS values as 10-12 dB if a Gaussian noise is observed for a sufficiently long time (please see sec. 4.1.1 for crest factor and related probability for a Gaussian distribution, and sec. 9.3.2 for a quantitative estimate), with the Max Peak and Min Peak detectors respectively over-weighting and under-weighting the noise level.

When measuring noise with low span/RBW ratios, the number of samples per pixel will be low (down to only one for extreme settings) and the Sample detector will give the same result of the Max Peak detector. In reality, using Sample detector, the spectrum analyzer may have lower effective video bandwidth than in Max Peak detection mode, because of limitations of the sample-and-hold circuit that precedes the A/D converter; examples include the Agilent 8560E-Series spectrum analyzer family with 450 kHz effective sample-mode video bandwidth, and a substantially wider bandwidth (over 2 MHz) in the Agilent ESA-E Series spectrum analyzer family.

The RMS detector calculates for each pixel the total rms of data samples pertaining to that pixel. The use of additional video filtering is not necessary and not advisable, since it would lower the displayed values down to a theoretical maximum error of 2.51 dB for Gaussian noise input.

The Average detector performs the average of samples pertaining to a pixel. If a linear envelope detector is used (no log compression), then applying video filtering by narrowing the VBW value brings a similar result and the two settings are compatible. If the log-amplifier is used, on the contrary, video averaging leads to a lower value than the true average value. In the case of Gaussian noise this difference is 1.45 dB.

We may say that performing an RMS detection of signal envelope is substantially equivalent to evaluate the average of signal power; however, there is a difference for the power calculated respectively as RMS average and the square of AV detector output.

The demonstration of the difference between RMS and AV detector output can be achieved by considering the Rayleigh distribution f_{Rayleigh} of the amplitude of noise at the output of the IF filter (considering the distribution of the magnitude, with in-phase and in-quadrature components, both Gaussian): the AV envelope detector gives the average envelope voltage v_{IF}, while the RMS gives the average power of the envelope voltage with respect to a reference resistor value R.

$$\begin{aligned} \bar{V} &= \int_0^\infty v_{IF}\, f_{\text{Rayleigh}} dv = \sigma\sqrt{\frac{\pi}{2}} \\ \bar{P} &= \int_0^\infty \frac{v_{IF}^2}{R}\, f_{\text{Rayleigh}} dv = \frac{2\sigma^2}{R} \end{aligned} \tag{9.1.2}$$

Thus the ratio of the power estimated as square of the average voltage $\bar{V}$ over a reference resistor R and the power indicated by the RMS value is

$$\Delta P = 10\ \log_{10}\left(\frac{\bar{V}^2/R}{\bar{P}}\right) = 10\ \log_{10}\left(\frac{\pi}{4}\right) = -1.05\,\text{dB} \tag{9.1.3}$$

The different performances of detectors for input noise signals are also considered in sec. 9.2.7 and 9.3.2 from a more operative viewpoint.

Repetitive pulses A pulse train is the reference signal used to characterize the behavior of detectors as per CISPR 16-2-1 [83]: it is a reproducible and repeatable transient by which the overall impulse response may be evaluated, including not only the specific detector response (in particular Quasi-Peak detector), but also the IF filter response at various RBW values, accounting thus for not only the $-3\,\text{dB}$ or $-6\,\text{dB}$ bandwidth specifications (possibly integrated by a shape factor specification, see sec. 9.1.5), but also the entire frequency response, in amplitude and phase (see sec. 4.3.1.5 and 9.2.2.1).

A pulse train is also extremely useful as example of discrete spectrum signal, to evaluate the effect of RBW and VBW settings: this is considered below in sec. 9.2.3 and 9.2.13.

9.1.10 Trace processing

As already pointed out, modern spectrum analyzers allow some post-processing of acquired traces, such as averaging, almost always implemented as a running average of new samples since the last reset (see par. "Averaging operator" at page 214): for each pixel new traces are accumulated and averaged arithmetically or using an aging factor (see sec. 4.3.2.3). The averaging operation is performed taking into account display settings, so linearly or logarithmically, depending on the chosen display units and scale. Averaging in post-processing faces thus the same problems already considered for video filtering (a form of averaging by low-pass filtering) and for AV detector itself: the use of Sample detector and linear-scale setting yields correct results, while the log scale setting underestimates by 1.45 dB, as already said; with Min Peak, Max Peak or Sample detectors using Video Filter (with low enough VBW/RBW ratio) helps the running average to converge.

The *max hold* option is again applied in trace post-processing and consists in keeping the envelope of the maximum value for each pixel; the result represents thus a worst-case estimate of the measured amplitude and may be useful e.g. to obtain an upper bound of the background noise, as it is done during measurements of electromagnetic emissions. Additionally, when sweeping over the variable-frequency carrier signals of digital modulations, such as OFDM, the overall spectrum and the graphical representation of the band occupation, it is useful to capture signal portions on the fly and accumulate them on the display trace.

Channel power measurement is another example of trace processing that requires the definition of the channel bandwidth (see sec. 9.3.2.1). Many spectrum analyzers have built in parameters for many standard formats, such as cellular and wireless network transmission protocols. The possibility of covering channel bandwidth with several adjacent samples, rather than reading it altogether with a single measurement at the largest RBW, allows to measure even modern wideband spread-spectrum modulations without spending in expensive equipment: depending on modulation characteristics and statistics, it may be necessary to accumulate several readings.

Trace saving and recall allows to superpose past and new traces for visual comparison or to apply markers.

Trace markers are quite useful in spotting out numeric values accurately, more than the display and resolution limitation. They allow also to calculate the difference in frequency and/or amplitude between two points in the same trace or with respect to a saved trace. Of course, with modern spectrum analyzers that allow to save and download to an external computer the raw numeric values, all kinds of post-processing are possible, using general purpose math packages such as Matlab, Octave, Mathcad.

Enhanced marker functions implement more complex operations and calculations (as in the case of phase noise in sec. 9.5), saving time and the risk of mistakes.

9.2 Performance and Operations

This section considers the operation of SA elements presented in the previous section, with emphasis on practical methods to verify and quantify performance and how manufacturers present them. Statistical analysis is heavily used squeezing information from repeated tests and the analysis of distributions and expectations.

9.2.1 Input VSWR and reference impedance value

The input Voltage Standing Wave Ratio is often specified in terms of “less than” (so an upper bound over a frequency interval is often given) and the upper bound of the inequality ranges usually from about 1.2 to 1.8. It is first to note that 1.2 is in any case a pretty large VSWR, resulting in a 20% difference between the maximum and the minimum of the voltage wave at the RF input connector, or in other terms $\pm 10\,\%$ (or ± 1 dB) around an average point. Most important, the VSWR specification is always accompanied by attenuator setting, that is almost always set to 10 dB. Higher attenuation values may further improve the VSWR but are less desirable for noise floor and dynamic range; smaller attenuation in turn has a bad impact on input VSWR.

The reference impedance is 50 Ω for almost all spectrum analyzers, some of them with the option for a 75 Ω input setting. In any case measurements of a 75 Ω (or any other reference impedance) system are still possible with a 50 Ω spectrum analyzer: the possible solutions are i) adding a 25 Ω resistor in series (that matches approximately the measured line), ii) using a coupling transformer or iii) inserting an attenuator.

The coupling transformer changes the impedance of the system under measurement Z_{sys} connected to its primary winding by the square of the turns ratio n when looking into the secondary side, where the 50 Ω spectrum analyzer is connected. Thus the ruling equation for the turns ratio is $n^2 = Z_{sys}/50$. In reality, the correct definition of n, despite its name, is based on the winding inductance, so $n^2 = L_1/L_2$.

Attenuators are commonly used to correct impedance mismatch: a typical example is an antenna with highly variable input impedance connected to the spectrum analyzer, where the attenuator normalizes the VSWR for changing frequency. An attenuator is usually built on a π or T resistor network, with resistor values set to have the desired attenuation and to match the input and output reference impedance (see sec. 1.2).

9.2.2 Resolution bandwidth

The setting of the Resolution Bandwidth (RBW) may span over about six or seven decades, from one or few Hz to tens of MHz; in old spectrum analyzers the possible choices where just two or three per decades (distributed in 1-3-10 or 1-2-5-10 scales), while in modern ones digital processing in principle allows almost any value, limited by practical reasons (e.g. pre-calculated values of digital filter coefficients or a limited number of installed analog filters).

9.2.2.1 Resolution bandwidth, impulse bandwidth and skirt selectivity

Let's focus on the definition of RBW as IF filter bandwidth: for example, the -3 dB bandwidth is the frequency spacing between two points of the transfer function at which the insertion loss of the filter has increased by 3 dB relative to the center frequency; analogously a very similar -6 dB bandwidth definition may be adopted. Why so much interest in the definition of 3 dB and 6 dB bandwidths? Since the definition is based on the frequency response we are talking of "amplitude dB", so with a factor of 20 in front of the log function: the 3 dB and 6 dB insertion loss correspond to 0.707 and 0.5 the center band gain; it is clear also that the -6 dB definition corresponds to half the amplitude gain and the -3 dB definition to half the power gain. From a different viewpoint the -6 dB bandwidth for Gaussian filters can be made correspond roughly to the *impulse bandwidth*.

The impulse bandwidth is defined as

$$B_i = \frac{1}{H_0} \int_0^{+\infty} H(f)\,\mathrm{d}f \qquad (9.2.1)$$

Multiples of $B_{-3\,\mathrm{dB}}$	Analog 4-pole	Analog 5-pole	Gaussian
$-6\,\mathrm{dB}$ bandwidth	1.480	1.464	1.415
ENBW	1.129	1.114	1.065
Pulse bandwidth	1.806	1.727	1.506

Table 9.2.1 – Relationships between -3 dB, -6 dB, ENBW and pulse bandwidths (Rohde & Schwarz [280]).

where $H(f)$ is the amplitude (or voltage) frequency response. The Equivalent Noise Bandwidth of a filter was defined based on the square of the frequency response, because this definition is based on power (see sec. 4.3.1.5).

For EMC and EMI measurements, where the spectrum of pulses, clicks and transients is often measured, 6 dB bandwidths are exclusively specified and the CISPR 16 standards make reference to 6 dB bandwidth for the standardization of IF filters of EMI receivers.

The definition of the -20 dB bandwidth is sometimes added (that corresponds to a decade in amplitude, or an insertion loss of 10) as a first characterization of the roll-off (or skirt) of the filter. The *shape factor* (or *skirt selectivity*) is specified down to the -60 dB bandwidth as $SF_{60/3} = B_{-60\,\mathrm{dB}}/B_{-3\,\mathrm{dB}}$.

Depending on the analog or digital implementation, the effective bandwidth measured at -3 dB and -6 dB points may vary and, as a consequence, also the ENBW is subject to change. Furthermore, even when specifying the IF filter as a four-pole, five-pole or Gaussian, different manufacturers and implementations have slightly different values.

In [280] Rohde & Schwarz gives a complete set of values for the relationship between the -3 dB, -6 dB, ENBW and pulse bandwidth for the IF filters used in their spectrum analyzers; they are reported in Table 9.2.1.

Agilent [4] gives very similar figures for the 4-pole and 5-pole synchronous filter implementation (less clearly called "4 filter" and "5 filter" by Rohde & Schwarz in [280]): ENBW = $1.128\,B_{-3\,\mathrm{dB}}$ and $1.111\,B_{-3\,\mathrm{dB}}$, respectively. For FFT-based processing Agilent gives ENBW = $1.056\,B_{-3\,\mathrm{dB}}$, not so different from the Gaussian filter specified in Table 9.2.1 (1.065). In the same Agilent application note the impulse bandwidth of a 4-pole analog and a Gaussian filter are declared $1.62\,B_{-3\,\mathrm{dB}}$ and $1.499\,B_{-3\,\mathrm{dB}}$ respectively: while the latter matches the 1.506 value shown in Table 9.2.1, the 1.62 value is much different from the 1.806 one. It is thus underlined that while 4-pole and 5-pole implementations may differ remarkably between manufacturers (as we have seen), there is a general agreement on the Gaussian response filter performance.

Such calculations are possible if the IF filter is fully characterized and this information is often known only to the manufacturer itself, unless the filter model was extensively documented, maybe because it was used for tutorials and application notes.

9.2.2.2 IF filter imaging

There is however the possibility of obtaining the frequency response of the IF filter directly from the spectrum analyzer by applying a pure sinusoidal tone, that, once

RBW	$B_{-3\,\mathrm{dB}}$	$B_{-6\,\mathrm{dB}}$	$B_{-20\,\mathrm{dB}}$	$B_{-60\,\mathrm{dB}}$
1 MHz	1.00 / 1.00	1.382 / 1.401	2.5509 / 2.473	4.073 / 4.273
100 kHz	101.82 / 103.64	145.46 / 143.64	258.18 / 258.18	425.46 / 429.09
10 kHz	10.182 / 10.181	14.546 / 14.545	26.182 / 26.181	42.909 / 42.181
1 kHz	1.018 / 1.019	1.454 / 1.418	2.546 / 2.545	4.072 / 4.109
100 Hz	102.0 / 101.0	146.0 / 142.0	254.0 / 255.0	408.0 / 407.0

Table 9.2.2 – Measured bandwidths for two SA operating modes: (left) FFT, (right) analog/digital filter implementation. Measuring units in each cell are those of the corresponding RBW value.

the span and the RBW are set, is used to "scan" the IF filter: while the LO frequency is swept, the IF filter bell-shape begins to touch the rightmost tail of the tone vertical line, proceeding with the center of its frequency response and then with the leftmost tail; the applied sinusoidal tone may be an external test tone coming from a RF generator or the 10 MHz reference sinusoidal signal as well (as used in the following examples). For a satisfactory representation what is needed is a small span/RBW ratio, such as 10 or 20, and a very narrow VBW, to clean the curve profile; the detector may be any, but AV or Min Peak are preferable to lower the noise floor (see sec. 9.1.9.2). Regarding the operating point of the mixer and the corresponding choices of the input attenuation and reference level, what is needed is a displayed dynamic range of at least 80 dB, in order to include the -60 dB points that we have used to define the shape factor of the IF filter (see sec. 9.2.2).

The mechanism of scanning and imaging the IF filter response curve by means of a probing sinusoidal tone is shown in an ideal case in Figure 9.2.1: a complete evaluation is worked out, where the IF filter response is centered around 10 MHz, taken from the local reference oscillator output; it is a frequency value large enough to measure satisfactorily even the 1 MHz RBW. Using the 10 MHz local reference signal has also the advantage of removing the phase noise resulting from the local oscillator feed-through (see sec. 9.2.6). The results for FFT implementation and for a mixed analog/digital implementation of the resolution bandwidth filter[5] are shown in Figure 9.2.2 and 9.2.3, respectively, and summarized in Table 9.2.2, where the values of the -3 dB, -6 dB, -20 dB and -60 dB bandwidth resulting from measurements are reported.

A final check is reported in Figure 9.2.4, where a few curves extracted from FFT mode and Analog/Digital filter mode figures, are overlapped and compared; the 10 Hz RBW case is considered with the RBW filter implemented in digital form and the results are perfectly equivalent to those obtained with FFT mode.

---

5 The SA manufacturer does not clearly indicate if analog implementation is used and for which RBW setting; it is absolutely possible that the implementation is fully digital.

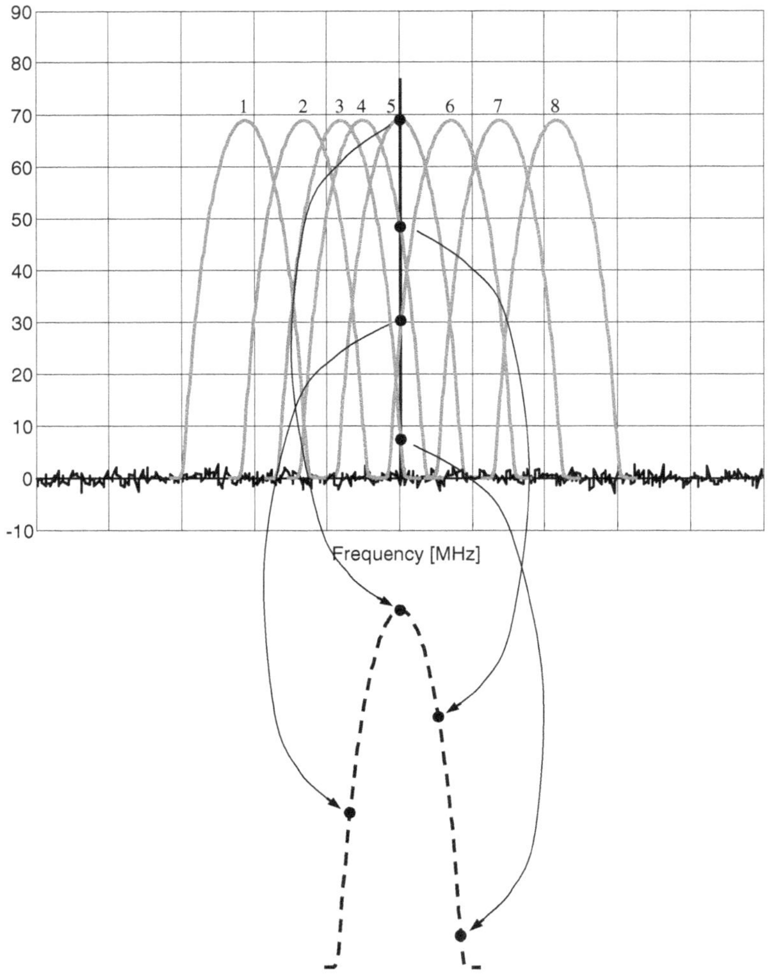

Figure 9.2.1 – IF filter imaging using a sinusoidal tone and small span/RBW ratio.

9.2.2.3 Input attenuation correction

The input attenuator is one of the most relevant sources of error, and, if corrected, of uncertainty, in absolute amplitude measurements: the error of combined attenuator and IF filter gain has systematic components, so it is not always true that the same attenuator setting gives the same amplitude error whatever the chosen IF RBW.

For example, in the tested unit there is a certain amount of over compensation of input attenuator, so that the 0 dB attenuation (exclusion of input attenuator) gave an amplitude reading of 189.27 – 189.33 dBμV that resulted always, or almost always, in the lowest reading. At RBW = 1 MHz the error for the 10 dB and 20 dB attenuation settings is +1.8 dB with respect to 0 dB attenuation; at RBW = 100 kHz the spread that contains all the tested attenuator settings is only 0.25 dB and it is 0.18-0.22 dB at the remaining RBW values from 10 kHz down to 10 Hz. There is no unique correction for the attenuator setting alone, but the error is related to the input attenuation-RBW

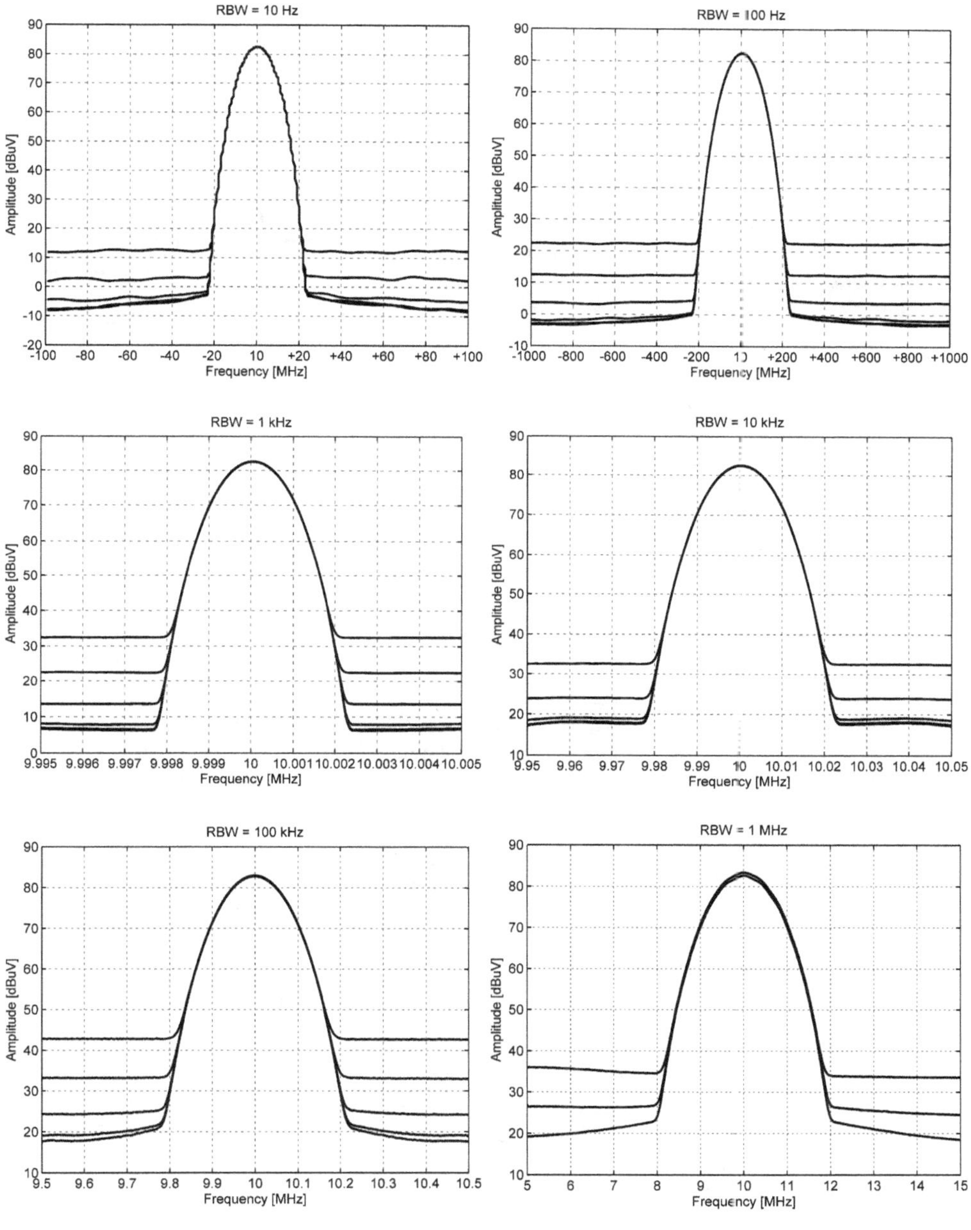

Figure 9.2.2 – IF filter frequency response "imaged" with the help of a sinusoidal input tone (different curves for increasing RF attenuation): FFT mode. Please note that the displacement with respect to the central frequency is indicated in Hz for the lowest RBW values.

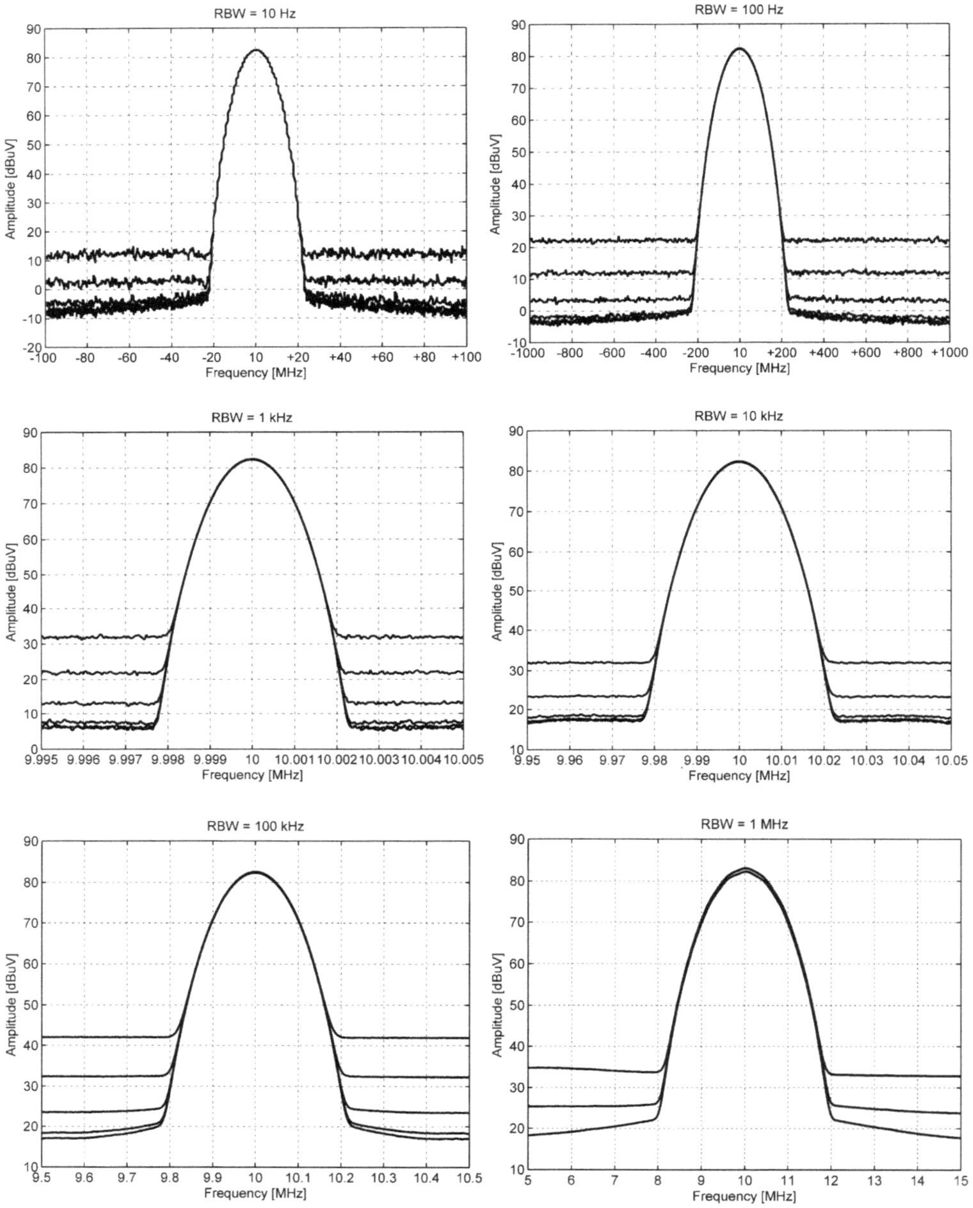

Figure 9.2.3 – IF filter frequency response "imaged" with the help of a sinusoidal input tone (different curves for increasing RF attenuation): Analog/Digital IF filter for Swept mode. Please note that the displacement with respect to the central frequency is indicated in Hz for the lowest RBW values.

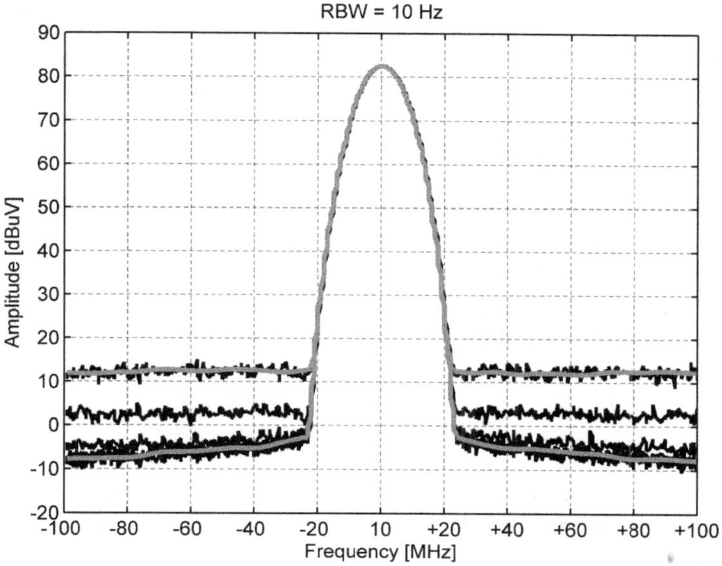

Figure 9.2.4 – IF filter frequency response compared for FFT mode (gray) and swept mode (black) at RBW=10 Hz.

setting pair. Such an error, if not compensated, affects the absolute amplitude accuracy by about ±0.1 dB; uniform distribution might be assumed, having found no specific correlation with the RBW setting except at 1 MHz. Moreover, the FFT operating mode gives consistently slightly larger amplitude estimates of about 0.15 dB on average.

Except extreme cases, such values are within the declared absolute accuracy. The best use of Spectrum Analyzer is for relative amplitude measurements (e.g. attenuation and frequency response), while a power meter is always a much better choice as reference for absolute readings.

It is thus advisable to verify always SA response by using a reference source that for the case of the internal IF filter response may be the 10 MHz reference itself; as a matter of fact we are looking for inconsistencies in the correction of attenuator and RBW combined gain settings, so trying to compensate for a systematic error. On the contrary, when stability and phase noise are questioned, an external reference shall be used (please see also sec. 9.5).

9.2.3 Frequency resolution and effective resolution

The displayed trace is stored digitally as a vector of points that can be downloaded to a USB memory, a local hard disk or through a cable connection. The points are separated on the frequency axis by the so-called *step* (or *frequency step*), that results simply from the selected span divided by the number of points the SA uses to store the trace; the latter may be varied going to system settings and normally ranges between 400 and 2000 (several thousands in EMI receivers, that cover wide frequency ranges in one single sweep). The step represents the frequency resolution of the internal sample representation, but it is not the effective resolution, so does not indicate the ability to discriminate between spectral elements really present in the input signal.

The IEEE Std. 748 [173] defines the *effective resolution* as "the ability to display adjacent responses discretely; the measure of resolution is the frequency separation of two responses which merge with a 3 dB notch."

The effective resolution is to be evaluated on the final result of the measured spectrum, so including RBW and the IF filter selectivity — as it might be expected —, but also the sweep time and the influence of the video filter and its transient response. For every combination of frequency span and sweep time there exists a minimum achievable resolution value, called optimum (or minimum) resolution, defined as

$$R_{opt} = K\sqrt{\frac{\mathrm{SPAN}}{\mathrm{ST}}} \tag{9.2.2}$$

where the frequency span, or simply the span, is the whole frequency interval swept by the SA, and the sweep time ST (that is considered below in sec. 9.2.4) is the time needed to perform the measurement from the start to the stop frequency. This expression is derived from the eq. (9.2.5) given to determine the minimum sweep time; the parameter K here is to be distinguished from k there: $k \propto K^2$ and they refer to resolution bandwidth and frequency resolution, respectively. The resolution in this context has an operative definition, as soon as we go to the next paragraph, where a measurement method is proposed; in eq. (9.2.5) the resolution is the Resolution Bandwidth and it is considered a fixed characteristic, not a measurable performance.

Based on the test method indicated by IEEE Std. 748, the effective resolution may be verified and quantified in the following way (quoted from the standard and only slightly modified):

- Equal Amplitude Signals. Apply two variable frequency, equal amplitude continuous wave (CW) signals through a combiner to the input of the spectrum analyzer. Adjust the frequency of the two signals, so their responses appear on the screen, and adjust the separation between signals, so their responses merge with a 3 dB notch between them. (The amplitude of the two responses must be equal). The frequency difference between the signals is the resolution. The use of the video filter will aid in observing the 3 dB notch exactly, by rejecting most of the short-term noise.

- Unequal Amplitude Signals. Apply two variable frequency and variable amplitude CW signals through a combiner to the input of the spectrum analyzer. Adjust the frequency of the two signals so their responses appear on screen. Adjust the amplitude difference between the two responses to specifications (that is, 20 or 60 dB). Adjust the frequency separation between the signals, so their responses merge with a notch that is 3 dB down from the peak of the lower amplitude response. The signal separation is the skirt resolution. The use of the video filter again will aid in observing the 3 dB notch.

The impact of effective resolution on spectrum interpretation may be tested in various situations:

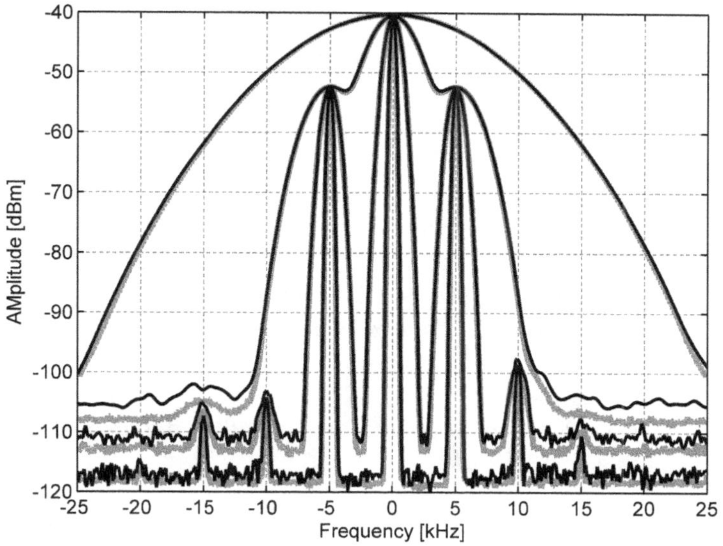

Figure 9.2.5 – Amplitude modulation (20 MHz carrier at $-40\,\mathrm{dBm}$, 50% modulation at 5 kHz) as appearing for different RBW values (10 kHz, 3 kHz, 1 kHz, 300 Hz, RBW/VBW=10) and using the RMS (gray curve) and Peak (black curve) detectors. The sweep time was always adjusted at about five times the minimum value (particularly relevant for the RMS detector, when measuring modulation peaks, not the noise floor).

- if amplitude modulation (AM) is considered, with carrier at f_1 and an applied modulation tone of frequency f_m, then sidebands and carrier are separated on the display as soon as the RBW gets down to the modulation frequency value and below it; an example is shown in Figure 9.2.5, where the same carrier with 50% AM modulation at 5 kHz is measured with both RMS and Peak detector, varying the resolution bandwidth around the modulation frequency (300 Hz, 1 kHz, 3 kHz, 10 kHz): the spectra are perfectly equivalent for the evaluation of the amplitude of the main components, that appear clearly separated when the RBW is 1 kHz or less; the noise floor is as expected larger for the Peak detector and affects the accuracy with which the smaller harmonic byproducts of the generator are measured;
- a pulse train, that is a discrete spectrum signal (or line spectrum signal), may result in a continuous or discrete spectrum depending on the RBW value with respect to its fundamental f_1 (corresponding to the repetition interval T), that is also the frequency separation of harmonic components (see sec. 9.2.13 for a more complete analysis of the spectrum analyzer behavior with discrete spectrum signals).

9.2.4 Sweep Time

The Sweep Time (ST) defines the time it takes to sweep the frequency range for the N points set from the start to the stop frequency (the span); each point (or pixel) has thus a fraction of the total ST, inversely proportional to N. Being the chosen number of points in close relationship with the span/RBW ratio, it is said that the time T' apportioned to measuring at one frequency point with a bandwidth RBW is directly proportional to RBW[6]:

$$T' = \frac{\mathrm{ST}}{\mathrm{SPAN}}\,\mathrm{RBW} \tag{9.2.3}$$

The main constraint for the ST setting is the transient response of the IF filter, that depends not only on the chosen RBW value, but also on the type of filter (analog, digital, FFT) and its order (see sec. 9.2.2). For a correctly damped filter, with no remarkable overshoot, taking a readout of its output before the transient response has vanished results in under-estimating the real output value and it is necessary to wait for the rise time to pass (see sec. 3.2.1). The rise time of the filter is thus inversely proportional to the RBW value, provided that all other details and corrections are conveyed in a suitable coefficient k:

$$t_r = \frac{k}{\mathrm{RBW}} \tag{9.2.4}$$

Combining the above quantities and making $T' = t_r$, a compact expression for the minimum sweep time $\mathrm{ST}_{\mathrm{min}}$ for the used RBW is obtained:

$$\mathrm{ST}_{\mathrm{min}} = k\,\frac{\mathrm{SPAN}}{\mathrm{RBW}^2} \tag{9.2.5}$$

where k hides all details on filter structure and tradeoff for transient response error[7].

There is some disagreement on the value of k, stemming out from different statements and conclusions belonging to different sources of information, even if there is no explicit comparison or discussion: Agilent in [6] indicates a range of 2 to 3 for Gaussian (or nearly Gaussian) filters, while Rohde & Schwarz [280] for the same class of filters indicates $k = 1$ and assigns $k = 2.5$ to analog filters (without details on their exact implementation). Let's say that the definition of k derives primarily from the difference in the adopted rise-time definition with respect to a classical $1/(\pi \mathrm{bandwidth})$ definition and that RBW corresponds to "bandwidth". With the information contained in sec. 3.2.1 and 3.2.3 we may evaluate the expected behavior of more classical two-pole and Bessel analog filters with respect to a Gaussian response: in Table 3.2.2 the $-3\,\mathrm{dB}$ bandwidth is somewhat smaller for Gaussian implementation and the rise-time values for the 20-80% definition, the inverse of $B_{-3\,\mathrm{dB}}$ and t_σ are all longer; it is thus sensible to rely on larger k values for Gaussian filters, about 20% larger than for other analog filters. The exact value to assign depends of course on the time instant at which we stop following the rise time and we assume the transient response has vanished: again

5 The use of RBW rather the number of points N aims at deriving a compact expression for the minimum sweep time that involves IF filter quantities and not the display size, that is N. It is understood that N depends on the frequency step and not on the IF resolution filter RBW.

7 It is always possible for manufacturers to anticipate the rise time and correct, at least partially, for the transient response that has not yet reached its steady state value: the response time is evidently improved at the cost of a slightly increased uncertainty, however kept under control by the fact that all the boundary conditions and possible settings are well known.

observing Table 3.2.2 and assuming a residual error less than 10% of the 10-90% rise time, it is sensible to require a much longer time than the $t_{-3\,\mathrm{dB}}$ definition, e.g. by a factor of 2. Accumulating the "20% more" and the "factor of 2" we reach a probably conservative $k = 2.5$ for Gaussian filters and $k = 2$ for the other analog filters (removing the 20% margin).

For digital filters it is quite difficult to indicate a straight neat figure, even if it is known that many constraints are removed and that the frequency selectivity vs. time response tradeoff is more favorable. A $k = 1$ may be guessed as the other extreme of the interval of values applicable to eq. (9.2.5).

Finally, the transient response may be corrected, if the decision is taken to shorten the sweep time beneath the limit established by eq. (9.2.5): of course this should be implemented automatically in the spectrum analyzer by the manufacturer, who knows well the details of the implemented filters and their complete time response. A 10% or less correction in amplitude is not much relevant for the residual uncertainty and allows the reduction of the IF filter time response by about a factor of two (please see sec. 9.2.13 and desensitization factor, compulsory when measuring fast signals).

If the FFT option is used instead, ST is dramatically reduced: in theory for smaller RBW values the advantage is orders of magnitude, but in practice, due to the computational overload, the ST reduction may be no more than about two orders of magnitude. The transient response time is replaced by the *observation time* (or *time window length*), that is the duration of the transformation window (see sec. 3.1.1 and 3.1.3), not directly related to span, but to frequency resolution. Keeping the latter constant ensures that the required "sweep time" is not changing; larger spans, beyond the maximum number of points of the buffer, are subdivided in sub-spans keeping the frequency resolution constant, increasing the required computational effort.

Reducing the sweep time below the minimum value (if allowed by the instrument software) causes unacceptable errors in the measured spectrum: the amplitude is underestimated (due to non-vanished transient response) and the frequency location of spectrum peaks is shifted. When accurate measurements are required, even meeting the minimum sweep time value (as established by the manufacturer) doesn't ensure that the required accuracy is attained and three-five times longer sweep times are normally advisable as a safety margin.

9.2.4.1 Comparison with experimental results

In Figure 9.2.6 a comparison for sweep time is reported of theoretical expressions and real evaluations for different span/RBW values. It is evident that the FFT mode is about two orders of magnitude faster than the swept mode, as expected. Moreover, it may be observed that the sweep time in Swept Mode set by the manufacturer as the minimum value is always much longer than the minimum value established by eq. (9.2.5), thus ensuring very accurate readings. The lowest RBW values (for the spectrum analyzer used in these examples it is 1 Hz) are omitted, because internal compensation of the transient response is surely adopted and the resulting sweep time value is much shorter than expected.

The analysis is repeated fixing the span to 10 MHz and changing RBW only; the results are shown in Figure 9.2.7.

9.2.4.2 Influence of Video Bandwidth

Until now Video Filter was excluded by setting the VBW well above the RBW value (a factor of 3 is the minimum ratio to keep its influence smaller than 0.15 dB, see sec. 9.3.2; in the present analysis the experimental data of Figure 9.2.6 and 9.2.7 had VBW/RBW=10). When using the Video Filter, its influence shall be taken into account: roughly speaking, when using analog filters, a VBW/RBW = 1 or larger allows to keep the minimum sweep time nearly unchanged; analogously, for digital filters, when a correction of the transient response has been implemented leading to $k=1$ as above, the ratio VBW/RBW shall be set to 3 to ensure that the Video Filter is not exerting any influence on the required minimum sweep time value $\mathrm{ST}_{\mathrm{min}}$. If the SA does not adjust the sweep time automatically to account of the reduced VBW (quite uncommon, modern SA always do), with an insufficient sweep time the trace will be distorted in amplitude and shape, resulting also in a wrong location of peaks on the frequency axis.

However, even at very large resolution and video bandwidths, the sweep time cannot be reduced without limitation. To set the local oscillator to each new frequency point and to collect measured values, some minimum time is required. There is always an intrinsic lower limit that cannot be reduced further, even under the most favorable circumstances: for example the permissible tuning speed of the local oscillator [280], that for old YIG, due to the inertia of magnetic tuning, was in the order of tens of ms.

If the spectrum analyzer remains tuned to a fixed frequency during the measurements, which is referred to as *zero span* (see sec. 9.2.5 below), the minimum measurement time only depends on the ADC sampling frequency. The minimum measurement times achievable nowadays in this mode are very short, in the μs range[8].

9.2.4.3 CISPR 16-2-1 specifications

Last, the requirements in the CISPR 16-2-1 standard [83] are considered, i.e. the minimum required sweep time (they call it "scan time") to perform EMC measurements with the various detectors already examined in sec. 9.2.7: Table 1 and Annex D of the standard report the scan times required for the four frequency bands A, B, C and D and for the Peak, Quasi Peak and Average detectors, that are the detectors accepted for EMC measurements and for which limits are specified by the various EMC standards.

9.2.5 Zero span

The zero-span mode consists of a direct reading of IF filter output after the envelope detector, having the LO frequency held to a fixed value defined by the center frequency setting. In all modern digital spectrum analyzer the signal is digital and taken at the

[8] Taking a look at the datasheet of my old Tektronix 2754P, I see that going to the *new* "digitized display" mode only 10 ms/div are possible, while, remaining on the *reliable* "real-time" analog mode, the sweep time could be reduced down to 10 μs!

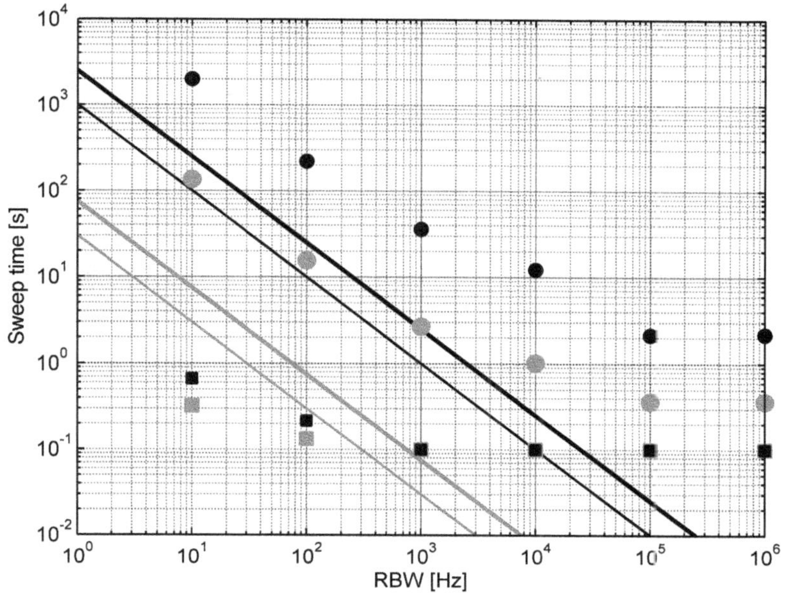

Figure 9.2.6 – Required Sweep Time as a function of the Resolution Bandwidth for a given span/RBW excluding Video Bandwidth (VBW/RBW set to 10). The sloped lines correspond to estimations: eq. (9.2.5) for a span/RWB=1000 with $k = 2.5$ (black thick) and $k = 1$ (black thin) and for a span/RBW=30 with $k = 2.5$ (gray thick) and $k = 1$ (gray thin). Both Swept mode (circle) and FFT mode (square) were tested in the two span/RBW ratios (black for span/RWB=1000, gray for span/RWB=30).

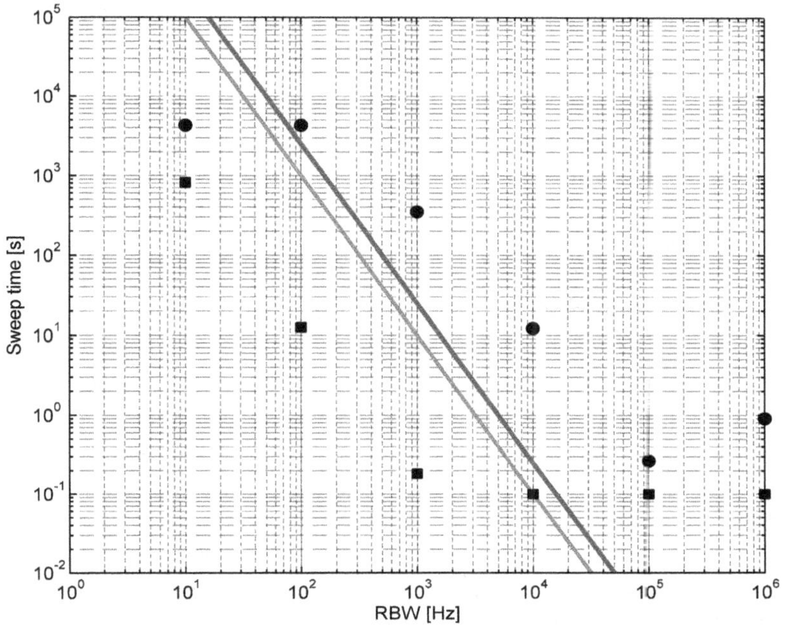

Figure 9.2.7 – Required Sweep Time as a function of the Resolution Bandwidth for a span=10 MHz, excluding Video Bandwidth (VBW/RBW set to 10). The sloped lines correspond to estimations: eq. (9.2.5) with $k = 2.5$ (gray thick) and $k = 1$ (gray thin). Both Swept mode (circle) and FFT mode (square) were tested.

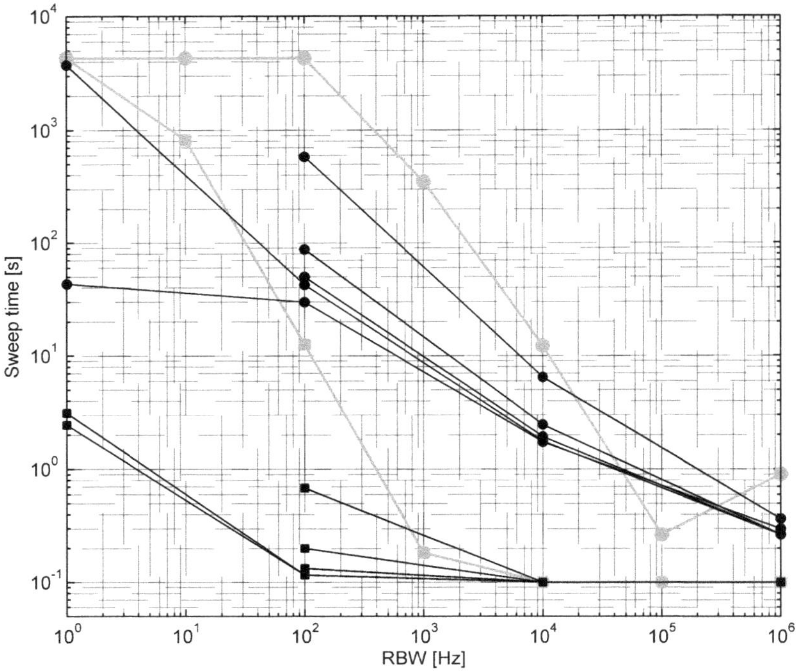

Figure 9.2.8 – Required Sweep Time as a function of the Resolution Bandwidth for a span=1 MHz, including the effect of the Video Filter Bandwidth VBW. Results for VBW/RBW = 0.01, 0.1, 0.3, 1, 3 are shown as black circles (Swept mode) and black squares (FFT mode) joined by a solid line; for too small RBW values the lowest VBW/RBW ratios were not attainable and there are only two data points for RBW=1 Hz. The 10 MHz span data already appearing in Figure 9.2.7 are shown as gray circles for swept mode and gray squares for FFT mode, joined by straight lines to ease recognition.

ADC output, so that it is no longer possible to tap the IF output and send it to external equipment[9]. In old spectrum analyzers reading the IF output is a safe and easy way to have a tunable band-pass filter: the signal may be sent to a time-domain equipment, such as an oscilloscope, a multimeter, a data acquisition system, and band limited measurements become possible, to record transients in noisy environments, to perform crossing rate distribution measurements etc.. Nowadays, signal processing functions are replacing this approach, but it comes at a price, so that a modern digital sampling oscilloscope with a moderate analog bandwidth of let's say 60 to 100 MHz and an old cheap spectrum analyzer may be a good alternative.

Zero span is quite useful in tracking time behavior of modulations (e.g. setting the center frequency to one of the lateral bands carrying the modulation information, or measuring the instantaneous power over a channel bandwidth), in measuring weak signals of known spectral distribution, but hidden by noise in time domain, and in measuring the time response of the IF filter itself. Regarding the former, channel power measurements may be performed in swept or zero-span modes (see details in sec. 9.3 and in particular 9.3.2). Regarding the latter, if the spectrum analyzer is fed

[9] Reconstruction by a dedicated DAC reconversion of the digital ADC output would be needed.

Frequency band		Peak det.	Quasi-Peak det.	Average det.
A	9 - 150 kHz	14.1 s	2820 s = 47 min	1255 s = 20.9 min
B	0.15 - 30 MHz	2.985 s	5970 s = 99.5 min	5134 s = 85.6 min
C-D	30 - 1000 MHz	0.97 s	19400 s = 323.3 min	8051 s = 134.2 min

Table 9.2.3 – Minimum scan times as specified by CISPR 16-2-1 [83]

with a pulse RF signal and the zero-span is centered at the carrier frequency, then the rise time of the IF filter may be measured, provided that the input pulse rise time is fast enough (see sec. 9.6): in case the input pulse rise time is not much shorter than the desired RBW filter rise time, correction may be applied as shown in sec. 3.2.4.

When operating in zero span on very long time intervals, observing phenomena that are narrowband (such as a lateral band), the local oscillator stability and drift may have impact: small fluctuations of the measured amplitude may appear caused by the change of the portion of signal spectrum intercepted by the IF filter centered at the supposed fixed center frequency, in reality probing around with erratic behavior. Also warmup and temperature fluctuations may be of concern (see sec. 9.2.10.2), when measuring weak signals and the spectrum analyzer noise floor is fluctuating.

9.2.6 Local Oscillator feed-through

The local oscillator leakage was anticipated when talking of mixer and its ability to control LO signal leakage (see sec. 9.1.4). The phenomenon of feed-through from the local oscillator (LO) has consequences that influence somewhat measurement results and might be interpreted as a limitation of the spectrum analyzer performance, or even worse as a true genuine measured signal indicating e.g. an excess of disturbance or band occupation at the lowest frequencies. It is also a not-so-often documented phenomenon and shall be tested by the user.

When performing swept measurements over an extended frequency band while starting from low frequency (as in the case of conducted or radiated emission measurements starting at 9 kHz), there is an evident increase of the noise floor towards the lowest frequency values. At the input mixer, operating at the largest local oscillator frequency, there might be a certain degree of coupling and as a consequence some leakage of LO signal in the IF path. So, if a low enough start frequency is selected, such a low frequency input signal is converted to an IF frequency that is practically the same of the LO, that feeds through and is only slightly or at all attenuated by the IF filter. The resulting noise floor has a peculiar shape, that is larger towards dc, with a profile like flicker noise or what in the FFT calculation is called "snow-drift profile". The consequences are reduced dynamic range and sensitivity, a wrongly suggested presence of excess noise, and low-order harmonics leaking in the incoming signal.

The phenomenon is well described in Figure 9.2.9, where several sweeps between 5 and 145 kHz are shown in both swept and FFT mode: the curves start from below for RBW=100 Hz and move up as RBW is set to 300 Hz, 1 kHz, 3 kHz, 10 kHz and 30 kHz; the video bandwidth VBW is always one tenth of RBW to smooth the curves just for graphical exigency. It is apparent that for the smaller RBW values the curves are spaced by about 5 dB each, meeting the assumption of broadband incoherent noise, but for the largest RBW there is no further increase because of the coherent LO leakage, that adds to the underlying noise. All the curves are not horizontal except for the lowest RBW values and in the last data points at the right end of the interval: this is again a symptom of the LO leakage and of a typical phase noise profile (see sec. 9.5). The behavior is roughly the same for swept mode and FFT mode (slightly worse for the latter), to proof that the phenomenon is not related to the processing method, but occurs before the IF filter. One brief comment on the small spikes appearing in the lowest curves in FFT mode: it is typical of FFT mode to create artifacts, often of small amplitude, that appear when using the smallest RBW values and, as a consequence, the lowest noise floor (this is considered further in sec. 9.2.9 below).

The importance of LO leakage and the significance of this small portion of the frequency range is better understood if EMC measurements are considered. Typically there are many applications that require EMC measurements (both conducted and radiated) starting at 9 kHz: band A indicated by CISPR [83] extends right from 9 to 150 kHz; other standards requiring such measurements are MIL-STD-461 [246] for military applications, and EN 50121-2 [56] and EN 50212-3-1 [57] for railway applications, to cite a few. The adopted RBW values are those prescribed or tolerated by the EMC standards and range between 100 and 1000 Hz: 1 kHz is sometimes allowed to reach faster sweep times when measuring non-steady phenomena, as it is for rolling stock in the EN 50121-3-1; the MIL-STD 461 also allows the use of a larger bandwidth, but application of corrective factors is forbidden. In addition, as commonly specified in EMC standards, no Video filter will be used, setting the RBW/VBW ratio to 10, but it is of no relevance to evaluate the impact of LO feed-through.

It is quite evident, then, that LO feed-through phenomenon is able to impair a test campaign if not properly considered: at RBW = 1 kHz the slope is still evident, with an increase of 4-5 dB with respect to the noise floor at the other end of the frequency interval; the use of a larger RBW, at 3 kHz, is unacceptable.

9.2.7 Detectors

Detectors performance have already been considered in other sections of this Chapter (see sec. 9.1.9 and 9.1.9.2) and will only be briefly considered and evaluated here, showing the results of a number of tests evaluating their response to Gaussian noise for two different span/RBW ratios (20 and 500) and for two different sweep time values (10 seconds and 2 minutes). A large span/RBW results in a reduced number of samples per pixel (or even one only), so that all detectors will operate in an unfavorable condition and their response will be not so different. The larger the number of accumulated data points per pixel in each scan, the better the convergence to the expected output for the known statistics of the input signal (the Gaussian white noise). Similarly, fast sweep times (however above the minimum sweep time compatible with the chosen RBW and VBW, see sec. 9.2.4) tend to underestimate for most detectors, because again the number of collected data points per pixel is small.

Results appearing in Table 9.2.4 may be used to substantiate previous considerations:

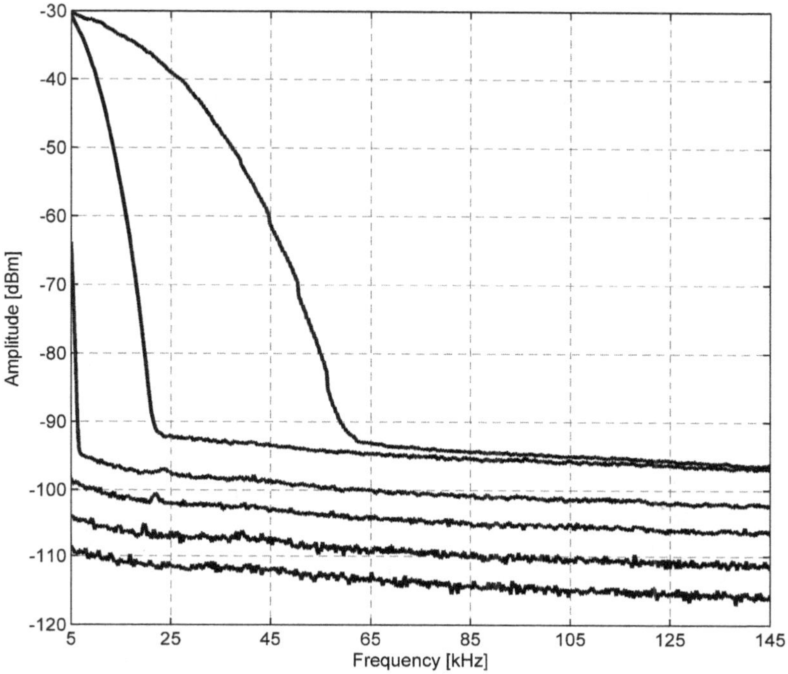

(a)

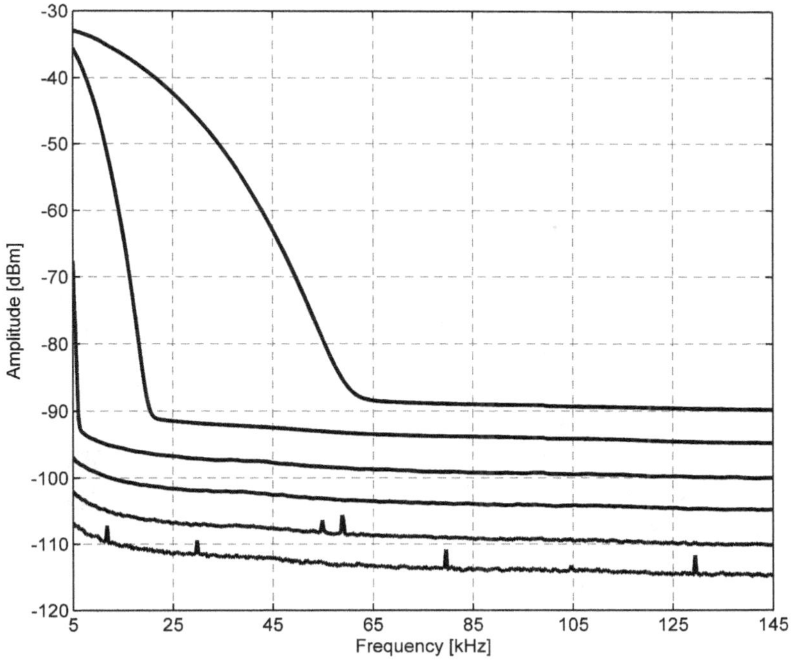

(b)

Figure 9.2.9 – Example of Local Oscillator feed-through for different RBW choices (100 Hz, 300 Hz, 1 kHz, 3 kHz, 10 kHz and 30 kHz): (a) swept mode, (b) FFT mode.

	Large span: span/RBW = 500			Small span: span/RBW = 20		
	RBW/VBW=0.1	RBW/VBW=1	RBW/VBW=10	RBW/VBW=0.1	RBW/VBW=1	RBW/VBW=10
RMS Det. (lin.)	-94.7254 -94.3398	-94.7104 -94.2103	-94.7079 -94.1777	-94.579 -94.501	-94.542 -94.498	-94.556 -94.479
RMS Det. (log)	-97.200 -96.685	-97.198 -96.686	-97.168 -96.676	-97.083 -97.016	-97.064 -96.986	-97.057 -96.978
Peak Det. (lin.)	-85.090 -83.628	-85.972 -84.249	-90.639 -89.353	-85.062 -83.913	-85.786 -84.525	-90.496 -89.597
Peak Det. (log)	-85.010 -83.575	-85.972 -84.337	-92.161 -90.726	-85.022 -83.967	-85.791 -84.597	-91.977 -90.933
Sample Det. (lin.)	-97.151 -96.250	-95.985 -96.010	-94.924 -94.247	-97.081 -96.700	-96.536 -95.974	-94.852 -94.531
Sample Det. (log)	-97.273 -96.818	-97.405 -96.495	-97.148 -96.544	-96.827 -96.742	-97.210 -96.992	-97.062 -96.883

Table 9.2.4 – Detectors performance with respect to sweep time (for each cell, the value above is for 10 s, the one below for 2 min.), RBW/VBW (=0.1, 1 and 10) and span/RBW (20 and 500); both linear and log compression cases are considered for each detector.

- for RMS detector the difference between the row for values using "linear" option and the row below applying "log compression" should lead to the already announced 2.51 dB difference: it is easy to see that for large span/RBW ratios (left side of the table) the difference is lower, with an average of 2.457 and a minimum of 2.34; for small span/RBW, so with many data samples per pixel, the difference on average is 2.505 (the confirmation we were looking for), with a negligible standard deviation of only 0.01 dB;
- the output of Peak detector is larger than that of RMS detector and is in general not much consistent, so that large variations from one sweep to the next may be expected; if a long sweep time is selected, then readings are more consistent and results for small and large span are comparable; the averaging operated by Video Filter counteracts Peak detection and at RBW/VBW = 10 this smoothing effect reduces the output by 5 dB, 1 dB more if log compression is used; it is evident what already said, not to apply video filtering choosing a unitary, or smaller, RBW/VBW, when the Peak detector is selected (as advised by EMC standards); the log compression reduces the output, as already explained in sec. 9.1.9.2;
- the output of Sample detector is smaller than that of RMS detector, as expected; however, a large RBW/VBW ratio applies a good deal of averaging and behavior of the two detectors is similar, both with linear and log compression option: with linear option RMS output is slightly larger, and in log compression smaller, than Sample output; the average difference in the two cases is 0.12 and 0.03 dB, with a standard deviation of 0.16 and 0.10 dB, respectively.

9.2.8 Video filter and Video Bandwidth VBW

In its original analog implementation integrated in the detection block, the Video Filter was a low-pass filter of the first order. Aboard modern spectrum analyzers its implementation is digital, after the ADC stage; in this case any suitable transfer function might be implemented, but the required and most useful behavior is always that of the analog low-pass filter.

A few tests are performed in order to complement results of other sections as far as the effects of video bandwidth VBW for spectra with large dynamics.

9.2.8.1 Influence of VBW for AM modulated tone

The first test consists of measuring a repetitive pulsed signal with a sufficiently large instantaneous peak value variation and with a spectrum characterized by several peaks and a variable noise floor. The spectrum resulting from four different VBW settings (VBW/RBW ranging between 10 and 0.01) is shown in Figure 9.2.11: what is evident is that — as expected — smaller VBW values reduce trace noise. However, the effect on the two tested detectors is different; while the main peak values keep constant, the smaller ones and the noise floor have significant variations for two different reasons: the Peak detector overestimates always noise-like signals and intermittent

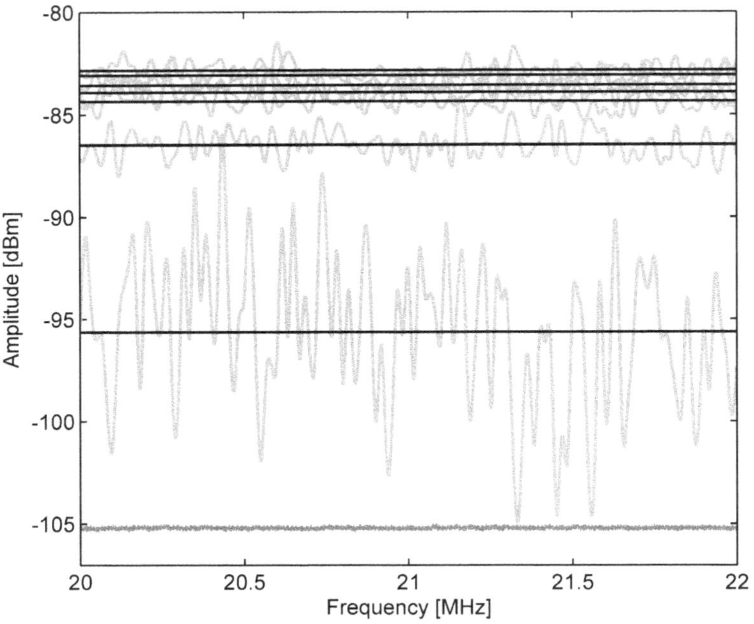

Figure 9.2.10 – Peak detector behavior for increasing sweep time: RMS detector reference curve (thick gray), Peak detector curves from bottom to top for increasing sweep time (ST = 0.1, 0.3, 1, 3, 10, 30, 100, 300, 1000 s); the horizontal black lines are the mathematical average of each curve.

components, with smaller VBW values introducing a significant averaging that contrast this behavior, with the black trace converging to the one of the Sample detector; the Sample detector is far more noisy and this is the reason of the inaccuracy when estimating the smaller peaks, but it is underlined that no significant biasing occurs with all noise-like trace portions centered around the same noise floor level.

So, what is the best choice in this specific case? the spectra are nearly identical as for largest peaks (see sec. 9.2.12 on the accurate estimate of signal amplitude in the presence of noise); lower peaks are overestimated more frequently by Peak detector (by its own nature) than by Sample detector, while at the smallest VBW value the two nearly coincide. The peak detector is strongly advisable for a quick measurement if a very small VBW value cannot be afforded; the use of Sample detector avoids biasing for non-sinusoidal, intermittent or noise-like signals.

The subject is further developed in sec. 9.2.13 when considering discrete spectra for pulse-modulated repetitive signals.

9.2.8.2 Influence of VBW for noise-like signals

The behavior with respect to noise and noise-like signals is considered adopting the two applicable detectors, RMS and Sample detector. Video Bandwidth VBW is changed while the trace is sweeping and measuring internal noise, with settings such to have a purely Gaussian noise source; with RBW = 1 kHz, the VBW settings are 10 Hz, 100 Hz,

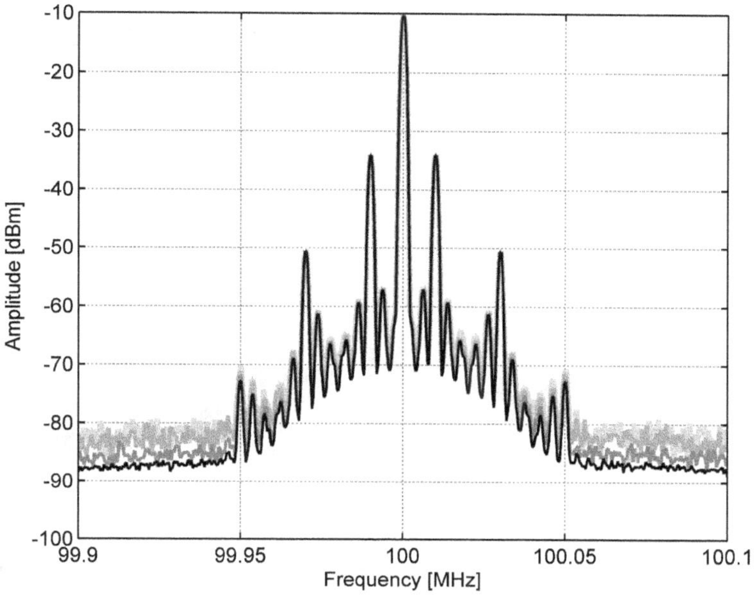

(a)

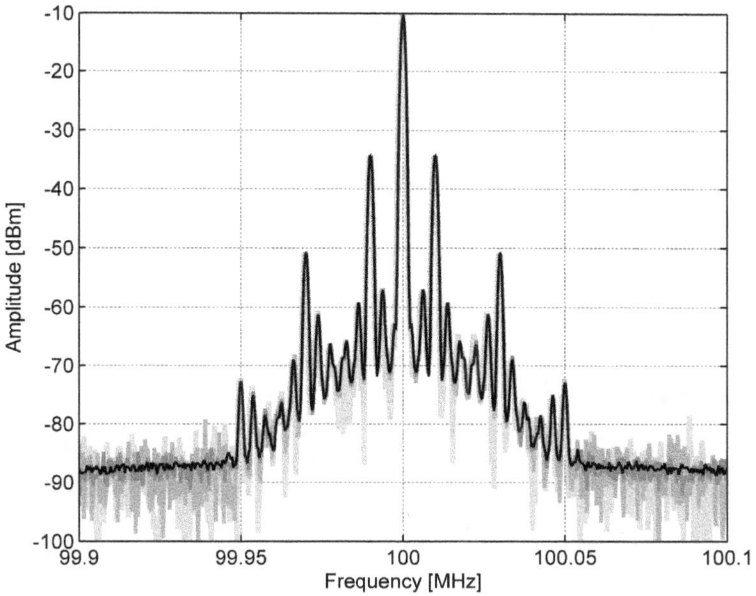

(b)

Figure 9.2.11 – VBW effect on pulsed-modulated signals (100 MHz sinusoidal signal modulated by a 10 kHz square wave with 10% modulation depth); RBW=1 kHz and four VBW values (10 kHz, 1 kHz, 100 Hz and 10 Hz from light gray to black): (a) peak detector, (b) sample detector.

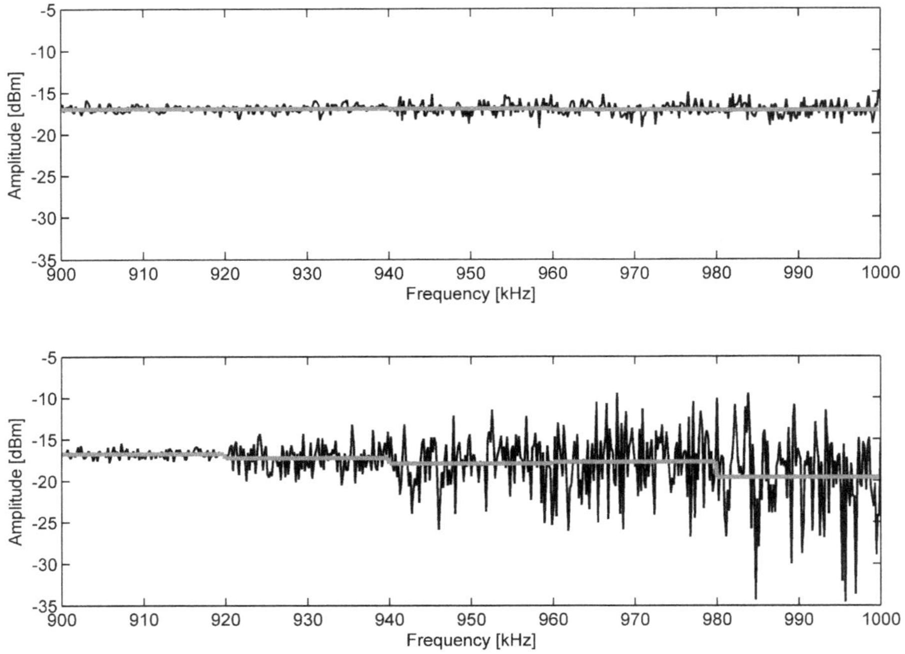

Figure 9.2.12 – VBW effect on the internal Gaussian noise for (above) RMS Detector, (below) Sample Detector. The VBW is step changed as 10 Hz, 100 Hz, 300 Hz, 1 kHz and 10 kHz while sweeping every two horizontal divisions.

300 Hz, 1 kHz and 10 kHz, the change applied every two divisions (i.e. after the sweeping point has moved by two divisions). Results are shown in Figure 9.2.12.

The results shown in Figure 9.2.12 were also numerically evaluated, calculating sample mean and standard deviation for each sub-record, made of the samples collected in each 2-division sub-interval, accounting for 110 samples each (550 samples in total). The results are shown in Table 9.2.5.

As it may be observed in Table 9.2.5 the σ/μ ratio is reduced only when VBW is smaller than RBW, with a ratio above 1:1 for Sample detector and with the already said 1:3 for RMS detector. It is evident that the RMS estimate is quite consistent, with a stable average value and a reducing dispersion, while the average value obtained from Sample detector values looks much less stable; the two are much closer for the lowest VBW value, but the difference is always quite large. A graphical synthesis of the table is shown in Figure 9.2.13.

9.2.9 Swept versus Fourier analysis

Modern spectrum analyzers that apply analog-to-digital conversion at the output of the IF block allow the user to choose between a traditional "swept analysis" and a frequency-domain image, obtained by Fourier transform of digitized IF data.

The Fast Fourier Transform (FFT) allows to reduce the minimum sweep time and to keep it almost unaltered while changing settings; sweep times of only hundreds of ms

	Video Bandwidth VBW				
	10 Hz	**100 Hz**	**300 Hz**	**1 kHz**	**10 kHz**
μ_{RMS}	-16.9288	-16.9252	-16.9314	-17.0799	-17.1464
σ_{RMS}	0.3953	0.4805	0.7594	0.7131	0.8383
$\sigma_{\text{RMS}}/\mu_{\text{RMS}}$	0.0233	0.0284	0.0449	0.0418	0.0489
μ_{SMP}	-16.7386	-17.2229	-17.8786	-17.7008	-19.5667
σ_{SMP}	0.4697	1.5529	2.8339	3.8120	5.0923
$\sigma_{\text{SMP}}/\mu_{\text{SMP}}$	0.0281	0.0902	0.1585	0.2154	0.2603

Table 9.2.5 – Performance of the Video Filter using as input the internal Gaussian noise and RBW = 1 kHz: "RMS" and "SMP" stand for the RMS and Sample detectors.

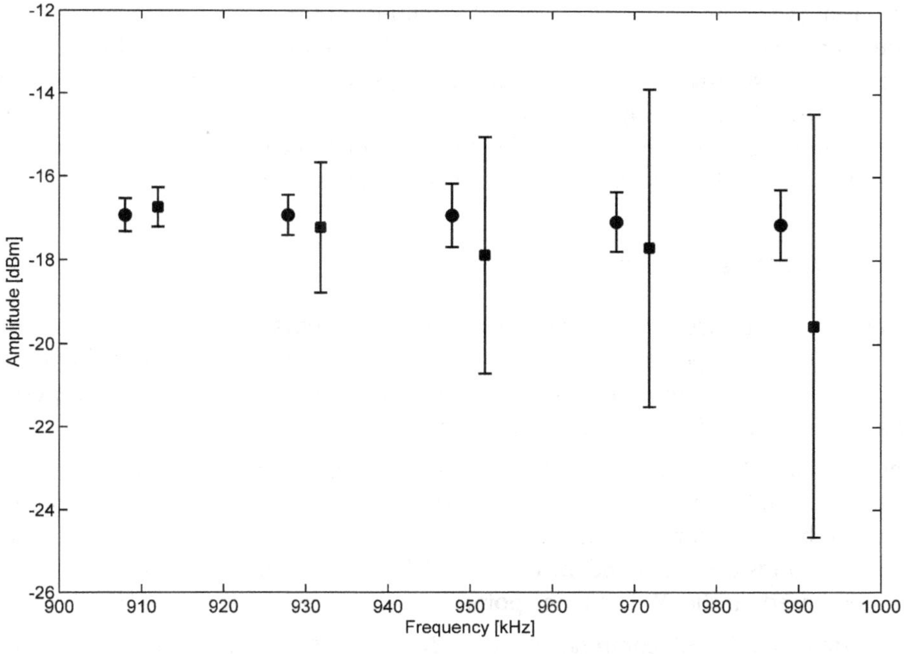

Figure 9.2.13 – Mean and dispersion of the noise floor using the RMS and Sample Detector. The VBW is step changed to 10 Hz, 100 Hz, 300 Hz, 1 kHz and 10 kHz, while sweeping over every 200 kHz on the frequency axis.

may be achieved for many combinations of settings. FFT however is not universally applicable to all measurement conditions and some extreme settings that in Swept Mode analysis lead only to some inaccuracy and biasing of the trace, in FFT mode may turn out into incomprehensible traces and weird behavior: steps and discontinuities, asymmetric traces, fluctuations.

Since FFT analysis is applied to the digitized IF envelope signal (as RBW filter would do), the FFT resolution is adjusted for the same effective resolution of RBW filter setting in Swept Mode; the number of spectra and any internal averaging operation are adjusted to match VBW setting and the desired sweep time value (a sweep time input value is still indicated as reference). The combined effect of smoothing windows for frequency leakage suppression and possible internal averaging over successive transforms displays smoother spectra with almost no apparent noise-jagged profile.

When discussing variance reduction by averaging N points of the same trace, the independence of the data points was assumed (see sec. 9.1.10 and 9.2.10.1). This independence is ensured in swept mode, due to the time required to sweep from one measurement point to the next one; in FFT mode the degrees of freedom may be less, due to the correlation resulting from the shorter time record and the influence of windowing, thus impacting slightly on the expected variance reduction. To author's knowledge there is no literature reference that gives quantitative results, so this problem is addressed experimentally in sec. 9.3.1.

When measuring transients, it is necessary to select the right method for the fastest sweep attainable: normally, at lower RBW values and moderate spans, FFT mode is the fastest; however, for large RBW and spans the answer is not straightforward and needs some reasoning. Transients, such as steps and pulses, have a wide spectrum with a narrow time location and they cannot thus be considered narrowband; from this observation the choice of a large frequency span and a correspondingly large RBW to keep the sweep time as fast as possible (please see sec. 9.2.4).

9.2.10 Internal noise

When preparing to use the SA for spectrum measurements, even for the measurement of EUT noise, it is important to evaluate accurately the noise behavior of the SA itself. There are several assumptions and properties that deserve to be verified experimentally and that are reviewed here before starting the measurements.

Flatness of noise PSD appears when incoherent averaging of acquired traces is applied and the expected reduction of noise floor variance is proportional to the square root of the number of records J (see sec. 9.3.1). Assuming the underlying flatness and that the PSD bins are independent (see sec. 4.3.2.3), the variance of each spectrum is estimated based on the N frequency points of a single record representing the sample.

Noise sources inside SA combine so that EIN is an AWGN (Additive White Gaussian Noise, see sec. 4.2.3.3) process: this statement is based on the fact that we may assume that noise sources are many and the Central Limit Theorem holds (see sec. 4.1.3). It is not true, however, when considering an extended frequency interval altogether, where different noise mechanisms or statistics of internal sources may take place, as for

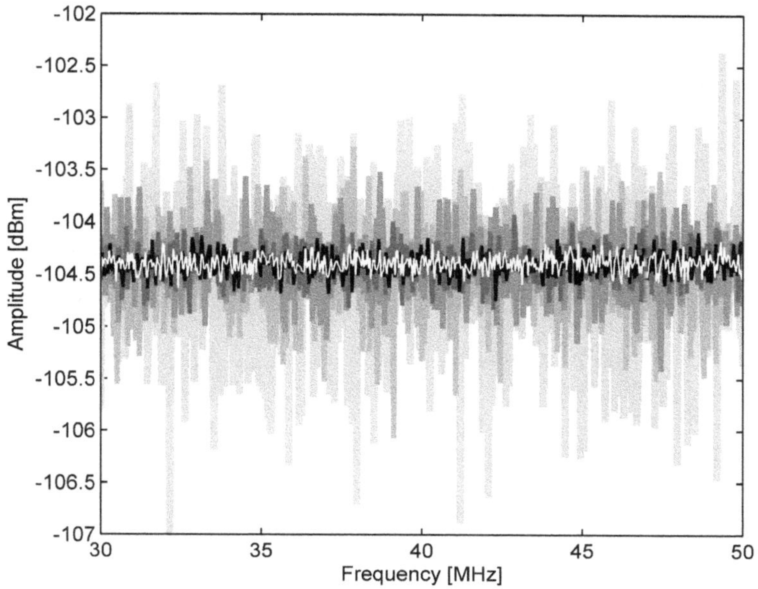

Figure 9.2.14 – Example of noise floor averaging: peak-to-peak excursion reduces progressively while increasing the number of averages (1, 4, 9, 25, 64, 100).

flicker noise (see sec. 4.3.1.3) in the lower portion of the frequency interval or, more important, for phase noise due to Local Oscillator feed-through (see sec. (9.2.6)).

9.2.10.1 Reduction of noise floor variance with incoherent averaging

A spectrum analyzer, when averaging acquired traces, can implement only incoherent averaging (see sec. 4.3.2.3), operating on absolute value of data points, without any phase information. Even thinking of using the FFT mode, since averaging is an operation performed after detection (see sec. 9.1.10) and the time-domain IF signal is not available, coherent averaging cannot be implemented. For real-time spectrum analyzers operating directly on a wideband IF signal more processing options are available, including joint time-frequency transforms.

Averaging an increasing number J of traces reduces their variance, not the mean value, so the noise floor is not reduced, but its variance is (see sec. 4.3.2.3 and Figure 4.3.6). If the J traces are uncorrelated, then the variance of the noise floor reduces with $\sqrt{J}$ as $\sigma^2_{nf} = \frac{\sigma^2_{tr}}{\sqrt{J}}$, where σ^2_{tr} is the variance of the single trace (the reader is warned to pay attention if the results are displayed in e.g. dBm or $dB\mu V$, for the observed variations will be scaled to 10 or 20 dB respectively). The concept is graphically demonstrated in Figure (9.2.14). For extreme settings, such as very large span and very fast sweep time, the data points are not completely uncorrelated and a deviation from the expected reduction should be observed; this is considered further in sec. 9.3.1.

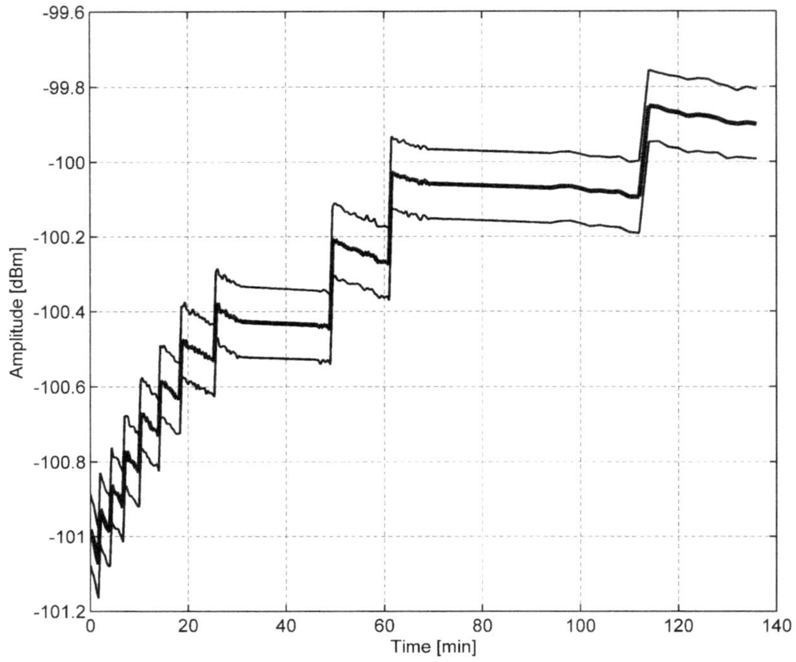

Figure 9.2.15 – Mean value of the noise floor and $\pm 1\sigma$ boundaries of the spectrum analyzer noise floor during warmup: RMS detector, sweep time of 20 s, RBW = 30 kHz, VBW = 1 kHz, Att = 10 dB, no pre-amplifier.

9.2.10.2 Effect of warmup and thermal stability

A warmup time of more or less 30 minutes is always advised in instrument manuals, in order to bring the spectrum analyzer into nominal conditions, in which the declared performances are valid. What may be observed with a simple experiment is that during warm-up and lasting for more than an hour there is a complex process of thermal stabilization, identified by means of a slight increase of the noise floor: the various parts of the instrument warm up with their own dynamics, that may differ from the natural self heating because of variable power consumption; when temperature control is operating, it is possible that the warm-up process proceeds in steps, with a correspondingly step-like noise floor. Each instrument has its own behavior and it is always worth spending a couple of hours doing a test. Warm-up profiles may be quite peculiar, as shown in Figure 9.2.15, where several successive scans were performed over about two hours after the switch-on instant (and boot of the instrument that requires approximately 2 minutes). SA settings are such to obtain a clear curve (VBW/RBW =0.030) and the sweep time is fast enough (20 seconds) to track significant drifts due to temperature variations, while keeping the RMS detector readouts stable and accurate for statistical consistency.

It may be immediately observed that the noise floor increases with the increasing internal temperature and that the slope is steeper in the first 20 minutes, corresponding to the known "warmup period". It would be expected that thermal equilibrium is reached and that the curve gets horizontal and flat, but temperature gradients and

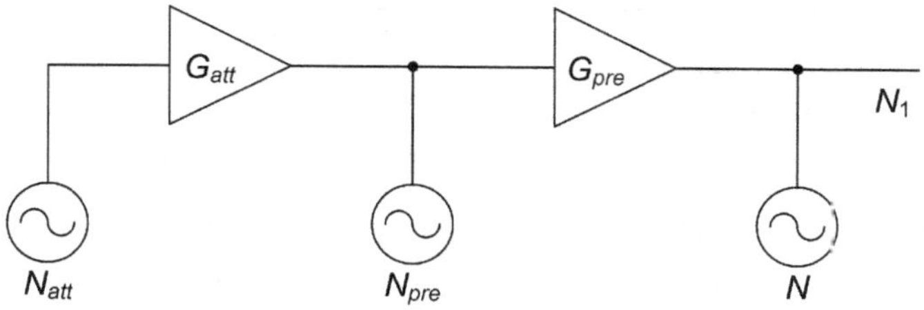

Figure 9.2.16 – Block diagram of attenuator and pre-amplifier for calculation of noise terms.

temperature control are responsible for small fluctuations and step-like behavior; such changes are in any case very small and could be appreciated only using averaging and narrow VBW.

9.2.10.3 Effect of attenuator and pre-amplifier

The composition of noise sources connected by cascaded blocks was considered in sec. 4.3.1.4, 4.3.1.5 and 4.3.1.6. Now we apply those concepts to quantify the noise originating from internal sources, distinguishing the effect of input attenuator (see sec. 9.1.1) and pre-amplifier (see sec. 9.1.2). A block diagram is presented in Figure 9.2.16, where G_{pre} and N_{pre} indicate the gain and the equivalent noise of pre-amplifier, G_{att} and N_{att} those of attenuator, and N is the noise added to the pre-amplifier output by e.g. the mixer and other SA blocks. The total noise *measured* by the SA is given by

$$N_1 = N + G_{pre}(N_{pre} + G_{att}N_{att}) \tag{9.2.6}$$

whilst the noise *displayed* by the SA is corrected for the applied pre-amplifier gain, so

$$N_1' = N_1/G_{pre} = \frac{N}{G_{pre}} + N_{pre} + G_{att}N_{att} \tag{9.2.7}$$

When not using the pre-amplifier the resulting noise term is

$$N_2 = N + G_{att}N_{att} \tag{9.2.8}$$

It is thus evident the reason why pre-amplifier improves the amount of displayed noise and the signal-to-noise ratio: all noise terms after the pre-amplifier block are reduced by its gain, with its equivalent input noise N_{pre} substantially determining the overall trace noise. This was already commented by observing the noise floor level of three comparative measurements in sec. 9.1.2.

While with a few measurements it is possible to verify that pre-amplifier gain and attenuator effect on noise correspond to nominal behavior, it is also possible to perform a more complex analysis for a more accurate determination, without using any external signal. By inspection of Table 9.2.6 it appears that pre-amplifier gain is about 10 dB

	Pre-amp ON	Pre-amp OFF
Att = 0 dB	-125.0465	-114.7400
Att = 5 dB	-119.9095	-109.4665
Att = 10 dB	-115.0610	-104.5720
Att = 20 dB	—	-94.4201

Table 9.2.6 – Different internal noise levels (DANL) in dBm (obtained averaging the 550 points of each trace) for RBW=10 kHz and VBW=30 Hz in a small frequency range around 10 MHz (not to have biasing fluctuations of the noise profile), 106 s sweep time, 60 minutes warm-up. The input is terminated on 50 Ω load: readings with the input connector left open show a slightly smaller noise only for Att = 0 dB setting, by about 0.11-0.16 dBm.

power and that the effect of the attenuator is to increase the noise level of 10 dB at each 10 dB step, as expected. As an exercise and sometimes to better determine the values of the above quantities (if they don't appear already in the instrument datasheet or or manual), it is possible to perform measurements and solve a set of linear equations.

The internal noise appearing on the display as DANL for a given resolution bandwidth (here RBW=10 kHz and VBW=300 Hz) is reported in Table 9.2.6 for different settings: the attenuator is set to 0 dB, 10 dB and 20 dB, the latter not available when the pre-amplifier is on. We will see that for the solution of the equation system we need additional readings, so that a further 5 dB attenuation measurement is added.

In the rightmost column the values appearing for 10 dB step change of the attenuator are effectively changing by 10 dB with an error not larger than 0.1 dB. Based on the readouts and the equations above, it is possible to write:

$$
\begin{aligned}
&N_2 + 1\,N_{att,0} &&= -114.7400\,\text{dBm} && \text{(I)}\\
&N_2 + 0.316\,N_{att,5} &&= -109.4665\,\text{dBm} && \text{(II)}\\
&N_2 + 0.1\,N_{att,10} &&= -104.5720\,\text{dBm} && \text{(III)}\\
&N_2 + 0.01\,N_{att,20} &&= -94.4201\,\text{dBm} && \text{(IV)}\\
&(N_2 + G_{pre}(N_{pre} + 1\,N_{att,0}))/G_{pre} &&= -125.0465\,\text{dBm} && \text{(V)}\\
&(N_2 + G_{pre}(N_{pre} + 0.316\,N_{att,5}))/G_{pre} &&= -119.9095\,\text{dBm} && \text{(VI)}\\
&(N_2 + G_{pre}(N_{pre} + 0.1\,N_{att,10}))/G_{pre} &&= -115.0610\,\text{dBm} && \text{(VII)}
\end{aligned}
\tag{9.2.9}
$$

By replacing the first and second equation (9.2.9).I and (9.2.9).II into the fourth and the fifth (9.2.9).IV and (9.2.9).V, two equations in the three unknowns N_2, N_{pre} and G_{pre} are obtained. G_{pre} may be determined by making an additional measurement at a different attenuator setting, with the problem that too a close attenuation value of e.g. 5 dB will lead possibly to a ill-conditioned solution of the system of equations, so that the solution might be approximate. In any case, such a method may be used with various levels of accuracy depending on the number of pre-amplifier/attenuator combinations and SA settings (sweep time and VBW).

9.2.11 Dynamic range and mixer operating point

The dynamic range should not be confused with the display range, in relationship to the *reference level* and the selected scale. The dynamic range provides information about SA capability to simultaneously process signals with very different levels, including e.g. a signal and its harmonics; of course, for the evaluation of harmonics and in general harmonic distortion, the dynamic range is not the only limiting factor: also mixer linearity plays a role, followed by other less relevant non-linearity issues.

The limits of dynamic range depend on the measurement to be performed. The lower limit is determined by thermal noise and phase noise, both contributing at a different extent to the noise floor: while the former is more or less flat, the latter decreases when the carrier offset increases (see sec. 9.2.6 and 9.5). The upper limit is set either by the 1 dB compression point or by distortion products occurring in case of over-driving. When the input signal at the first mixer approaches its 1 dB compression point, the mixer will operate non-linearly and there will be very high levels of distortion terms. Operating at small resolution bandwidth and reducing the noise floor, distortion products may become visible and the user should undertake a number of checks: their amplitude is plausible? does it reduce drastically reducing the input signal amplitude? or, similarly, do they reduce by increasing input attenuation? When the spectrum analyzer is operated automatically, the reference level and the combination of attenuator and IF amplifier are such that the mixer operating point is monitored and the 1 dB compression point is never reached.

Trying a raw estimate of the dynamic range in modern spectrum analyzers, simply comparing the minimum noise and the maximum input levels, we may conclude that amplitudes that differ by as much as 160 or 180 dB may be displayed. However these range limits cannot be reached at the same time for the same SA settings: internal blocks such as mixer, log-amplifier and ADC have a much smaller dynamic range and their operating points shall be carefully decided, in order to maximize the available dynamic range, that normally result in a smaller range up to a hundred of dB[10].

Considering first the maximum input level, it is to be underlined that values around 25-35 dBm are normally encountered, that shall be reduced to 10 dBm or less, not to damage or over-drive, the first mixer. Let's consider more in detail the influence of the mixer and its correct operation on the resulting dynamic range.

A compromise is to be found in the selection of mixer level. If RF attenuation is high and mixer level is low, the levels of distortion and intermodulation products are also low, but at the same time the signal-to-noise ratio is small. In this case the dynamic range is limited at the lower end by the inherent noise. If, on the other hand, the mixer level is high, then distortion and intermodulation products are generated, whose levels exceed the inherent noise level and therefore become visible.

[10] This is why for phase noise measurements instruments such as phase noise analyzer can reach better performance than a general purpose spectrum analyzer.

Example 9.1: Optimizing the mixer operating range

Considering a 13 dBm input tone and a 1 dB compression point located at +10 dBm for the mixer, selecting a +5 dBm, or even 0 dBm, operating point for the mixer is advisable. Since the input attenuator may be often set only in 10 dB steps, then the choice is 10 dB in the first case, and 20 dB in the second one. In both cases the obtained signal level at the mixer input does not allow to fully exploit the SA dynamic range, losing 2 dBm and 7 dBm, respectively. There is no fine adjustment possible, such as acting on the IF-amplifier, because this block comes after the mixer and has no influence in maximizing the mixer operating range. The only solution is to add an external attenuator, that reduces the input tone amplitude to a multiple of 10 dB with respect to the desired mixer operating level: an 8 dB attenuator would be what we are looking for in the first case, but forcing the input attenuator to be set to 0 dB, that might be questionable in some cases (please see sec. 9.1.1); in the second case a 3 dB external attenuator with an internal 10 dB input attenuator settings brings the signal level exactly to the desired mixer 0 dBm reference level.

At the lower end of the dynamic range, the relevant elements are the harmonic components originating from internal non-linearity (e.g. mixer, log-amplifier and ADC distortion), phase noise (to evaluate for a carrier offset that is in the order of the signal harmonic components we desire to measure) and noise floor (resulting from the resolution bandwidth value, and video filter and detector settings). Intermodulation and harmonic distortion have already been considered for mixer operation (see sec. 9.1.4). Phase noise and its measurement are considered in sec. 9.5. Some processing of the trace can reduce the noise variability, or "noise variance" (see sec. 9.3 on averaging for noise reduction), as well as its average value (see sec. 9.2.7 on detector operation and performance)

Depending on whether intermodulation products or harmonics limit this range, one speaks of an "intermodulation-free range" or "maximum harmonic suppression", or "harmonic free range". Both parameters depend on mixer level and selected resolution bandwidth. A maximum is attained if the levels of the intermodulation products or harmonics of higher order are equal to the noise level; of course, beneath this point nothing can be said on those components. The ideal mixer level required for this purpose can either be calculated or graphically determined, as it is shown in sec. 9.4.

9.2.12 Accuracy of amplitude measurements

9.2.12.1 Scalloping loss

A first source of error in the measurement of amplitude of a peaky spectrum, such as a sinusoidal tone, is due to the scalloping loss for a spectrum analyzer, that is the fact that for too a large span/RBW ratio, there is no overlap between the IF filter frequency response ideally located on each trace pixel and during the sweep some parts of the real spectrum underneath will be unavoidably lost. Reducing the span/RBW ratio and thus the step size to less than RBW greatly enhances the amplitude estimate, leaving the possibility that a tone located midway between two trace pixels will undergo the maximum error (exactly for the same reason that was presented for the FFT scalloping loss, see sec. 3.1.3.1). Thus the scalloping error is

$$\varepsilon_{sclp,\max}[\mathrm{dB}] = 3.01 \left(\frac{span}{\mathrm{RBW}} \frac{1}{N-1} \right)^2 \tag{9.2.10}$$

Setting the step to half the RBW as required by some EMC standards, the maximum scalloping error is 0.75 dB. Using a step of 1/3 brings the error down to 0.33 dB. Negligible scalloping errors (below 0.03 dB) are ensured for a 10:1 ratio between the step and the RBW.

9.2.12.2 Internal noise

When taking into account the effect of internal noise (be it phase noise or thermal noise), amplitude readings of externally fed input signals may always be corrected, provided that a reliable estimate of internal noise is available. It is intuitive that the "true" value of the input signal level P_s corresponds to the measured one P_m minus the effect of noise. Since the internal noise (its power being P_n) is assumed uncorrelated from the input signal (identified by P_s) with a random phase relationship, subtraction occurs in the sense of root mean square, or incoherent subtraction:

$$P_s = P_m - P_n \tag{9.2.11}$$

Subtraction shall be performed going from dB to linear quantities, and once the subtraction of power terms has been done, going back to the original dB quantities. The procedure starts from the original P_m and P_n values and to estimate the correction that is multiplicative in linear scale (dB to subtract), it normalizes the difference $P_m - P_n$ by P_m itself, thus leading to

$$\Delta P_s[\mathrm{dB}] = 10 \log_{10} \left(1 - \frac{P_n}{P_m} \right) \tag{9.2.12}$$

that depends only on the ratio of P_n and P_m, that is their distance in dB, not on their absolute values; the resulting correction values ΔP_s to subtract to the measured value for the corrected estimate $\hat{P}_s$ of input signal power are reported in Table 9.2.7. It is extremely important that SA settings are such to smooth the noise floor and to reduce its variance (see sec. 9.2.7, 9.2.10.1 and 9.3.1), for a consistent correction and to reduce uncertainty. At a first approximation the confidence interval that brackets the noise floor distribution reflects directly in the dB correction and is transferred as uncertainty onto the estimated signal power $\hat{P}_s$.

9.2.12.3 Absolute vs. relative reading and related errors

Amplitude measurements may be absolute or relative. It is intuitive that most error terms related to amplitude quantification are ruled out when performing a relative amplitude readout, because many internal blocks hold the operating point with comparable errors, when two measurements are taken one after the other, at a short distance in time (this is what in many instruments is called "short-term uncertainty" or "short-term variability").

$P_{meas}[\text{dB}] - P_n[\text{dB}]$	$\Delta P_s[\text{dB}]$ **(additive)**	$\Delta P_s\,\%$ **(multiplicative)**
1	-6.87	20.6
2	-4.33	36.9
3	-3.02	49.9
4	-2.20	60.2
5	-1.65	68.4
6	-1.26	74.9
7	-0.97	80.1
8	-0.75	84.2
9	-0.58	87.4
10	-0.46	90.0
15	-0.14	96.8
20	-0.044	99.0
25	-0.014	99.7
30	-0.0043	99.9

Table 9.2.7 – Correction values in dB for the difference-of-power estimation of measured signal power.

Regarding absolute reading, i.e. the measurement of an amplitude and its expression as a voltage or power, it is instructive to review the most relevant error terms and the way they are expressed and quantified in manufacturer's datasheets. First of all it is recalled that any quantification of an error (called *calibration*, but not be confused with the Vector Network Analyzer calibration) is subject to change and drift with temperature and time; from this the need of warm-up and a specified operating temperature interval, the increase in the uncertainty for the time past the last calibration and the need for periodic calibrations. Calibrations may be performed as a general periodic check against the internal reference signal generator, that is usually temperature-compensated or oven-controlled for better stability, or against factory or laboratory standards. In any case calibration can reduce the error term by quantifying corrections that may be stored in the spectrum analyzer for later use, but is deemed by an intrinsic residual uncertainty, due to used references, test setup and adopted procedures. What is normally stated in manufacturer's datasheet is the residual uncertainty due to all these factors and declared by the laboratory for the specific calibration, after all systematic errors have been removed; so the declared uncertainty is the least attainable uncertainty in ideal conditions. Declared uncertainty is deemed to increase with the time past the last calibration, until a new calibration is necessary. It is remembered that the internal reference signal source usually operates at a single or few frequency values and signal levels; while lower levels may be obtained by inserting a high accuracy attenuator (in case the calibration signal is available on an external connector and is not routed internally), a direct check at other frequencies is not possible. An external generator and a reference amplitude measurement are then needed for a more complete characterization when periodically checking SA performance, rather than sending it *tout court* to an accredited laboratory for calibration.

On the contrary, relative reading is quite immune to many of the uncertainty issues considered for absolute reading: the difference between two points of the measured spectrum (or between two successive spectra) may be accomplished by using delta markers or other external trace processing. While long-term variability and drifts, as well as any offset or deviation, vanish in the difference, residual uncertainty and short-term variability are the only factors that matter: what is really "short term" depends not only on operating characteristics, but also on our knowledge of such characteristics and the choice of best settings. In an attempt to simplify, we may say that short-term variability is made mainly of noise terms, internal temperature changes, linearity error (that may affect two points of the spectrum at different power levels), and short-term drift, usually expressed as %, ppm or dB over square root of unit of time (e.g. hours, days, months ...). Once the time interval over which measurements are taken is established, some of these terms may be estimated with the information appearing in the instrument manual (such as extrapolating the drift); others may be more vague or rarely characterized, such as internal temperature changes (for which coarse assumptions may be made). Putting altogether, uncertainty budget may be estimated using a Type B approach [39][11].

9.2.12.4 Uncertainty budget

Uncertainty estimation is considered with a closer look to the terms that influence the accuracy of amplitude measurement. The aim is identifying the sources of information for the factors that we have reviewed in the previous section and how to characterize their uncertainty. Depending on manufacturer's approach and quality system, uncertainty may be declared with a more or less explicit statement of distribution (e.g. uniform, Gaussian, U-shaped) and coverage factor (possibly only standard uncertainty is given, i.e. $k = 1$, but most often $k = 1.96$, or 2, will be given, not to say $k = 4$, as Agilent did for many instruments). Even if not clearly stated nor commented in the literature, the impression is that the declared uncertainty is usually quite overestimating: while this sentence wouldn't make sense if the uncertainty terms were coming from a Type A estimate, if they are in turn the result of a Type B method applied to single components or blocks, it is possible that margins are introduced. This is the impression whenever there is the possibility of confirming the Type B statements with repeated measurements and a Type A approach.

Frequency response is relevant if samples to compare are distributed over a large interval: while the frequency response is nominally flat (zero error on average), depending on frequency and used RBW the manufacturer will declare a varying uncertainty; frequency response tends to worsen in the very low frequency range and gives its best between a few MHz and a few, or several, GHz; uncertainty may vary between a fraction of dB to even 1 dB for extreme settings and conditions.

---

[11] Some examples of uncertainty budget for VNA measurements are given in the Appendix of EURAMET guide [123].

Linearity is always quite good and for the advised mixer operating interval (e.g. below $-10\,\mathrm{dBm}$ down to about $-70/-90\,\mathrm{dBm}$ depending on instruments) it is usually limited to $0.1\,\mathrm{dB}$.

The input attenuator might be a source of uncertainty when switching between steps: some manufacturers declare $0.1\,\mathrm{dB}$ or even $0.2\,\mathrm{dB}$ of uncertainty, and we have seen when comparing averaged noise floors that the correspondence is in fact quite good and that practically uncertainty is maybe lower than $0.1\,\mathrm{dB}$.

Amplitude accuracy rather than being estimated based on single uncertainty terms (as discussed before) is often presented with a catch-all statement based on some assumptions, e.g. a value or range for RBW, some frequency points, a range for the input signal and a minimum input attenuation (e.g. $10\,\mathrm{dB}$). It is evident that in this case the declared uncertainty shall be quite large and it is, being normally above $1-2\,\mathrm{dB}$, and in the end not so useful. By the way we have seen that changing detector or varying Video Bandwidth and sweep time have impact on amplitude reading, but rarely these details are considered.

Uncertainty related to input VSWR is always dramatic: VSWR values are always declared "less than" with quite a large upper limit; while a minimum of input attenuation is always advisable (except when maximizing the dynamic range, e.g. for distortion and phase noise measurements), when connecting troublesome sources such as long cables and antennas, additional external attenuation may be needed, and this adds to the overall uncertainty budget.

Temperature effects and necessary warm-up are not always completely characterized and either a general indication of uncertainty is given for a temperature range or the necessity of minimum warm-up time is indicated to reach declared performance and stability. It is just recalled that a higher temperature might increase slightly internal noise that follows also internal temperature regulation and gradients (see Figure 9.2.15), where after 1 hour it is evident that the noise generation process hasn't reached yet a stable state. Considering two sample readings at about 80 minutes, the difference in the noise power level is about $0.4\,\mathrm{dBm}$, reducing steadily in the successive hour. This difference affects compensation of measured amplitudes taken at different times from switch-on (see sec. 9.2.12.2) when using a unique noise floor value for compensation, but are not relevant if each amplitude measurement is accompanied by a noise floor measurement immediately before or after.

When performing power measurements (as will be considered in sec. 9.3 below) RBW accuracy is also relevant: ENBW directly weights noise-like signal power; apart from any RBW accuracy declaration, we saw that IF filter imaging allows an accurate quantification of the overall IF filter response, thus leading to bandwidth determination, including ENBW (see sec. 9.2.2.2).

9.2.13 Repetitive pulse signals, discrete and envelope spectrum

First, we must stress the difference between pulse base-band signals and pulse modulated RF signals: the former belong to signal formats used in data networks and digital links (such as Ethernet, USB, SATA to cite a few known technologies), the latter characterize digital modulations, such as modern cellular communication and wireless network formats (GSM, UMTS, WiFi) and radar signals. Pulse-modulated signals are those usually assumed in spectrum analyzers manuals for examples and they are considered in the following; an example of a pulsed base-band signal will be also considered.

The energy of periodic pulse signals is concentrated at discrete frequencies $n \cdot f_1$, where f_1 is the fundamental given by the inverse of the repetition period T. The envelope $\mathrm{sinc}(\cdot)$ function has nulls at integer multiples depending on the mark-to-space ratio (duty cycle) τ/T. If the pulse signal is used for modulation of a carrier, the spectrum will be symmetrically distributed above and below the carrier frequency. Depending on the adopted resolution bandwidth, three cases are possible:

- if resolution bandwidth RBW_1 is small relative to spacing of spectrum lines $\Delta f = 1/T$, then individual spectral lines can be resolved and a *line spectrum* is obtained; if resolution bandwidth is further reduced to RBW_2, amplitude and number of spectral lines do not change (i.e. they are narrowband), but the noise level reduces and thus the signal-to-noise ratio improves as $10\ \log_{10}(\mathrm{RBW}_1/\mathrm{RBW}_2)$;
- if the resolution bandwidth RBW_1 is larger than the spacing of spectrum lines Δf, but smaller than the first null of the $\mathrm{sinc}(\cdot)$ envelope function at $1/\tau$ from the carrier frequency, spectral lines cannot be resolved and the amplitude height of the envelope depends on bandwidth; the amplitude within the selected RBW_1 depends on the number of underlying spectral lines collected within RBW_1. In this case we speak of an envelope display; the envelope amplitude decreases with decreasing bandwidth as $20\ \log_{10}(\mathrm{RBW}_1/\mathrm{RBW}_2)$.
- if the resolution bandwidth RBW is larger than the envelope null spacing $1/\tau$, selectivity is no longer effective and the amplitude distribution in the spectrum cannot be recognized any more; increasing RBW the IF filter impulse response approaches the time function of the pulse-modulated carrier.

The change of measured spectrum shape is shown in Figure 9.2.17 to give a first graphical description of the phenomenon: RMS detector is used and we will see that for correct quantification of the pulsed waveform a Peak detector should be used instead.

A $\tau = 6.5\ \mu\mathrm{s}$ rectangular pulse with $f_0 = 10\,\mathrm{kHz}$ repetition frequency, modulating a 200 MHz carrier of $E = -40\,\mathrm{dBm}$ (2 mV peak) amplitude[12] is considered, using RBW values ranging from 1 kHz to 100 kHz, and setting VBW=RBW. The results are plotted in Figure 9.2.17. The effect of larger RBW values is an increase of the spectrum, a smoother curve and a loss of frequency resolution. The IEEE Std. 376 [18] confirms that the amplitude in linear units of spectrum components around the sine wave frequency is $f_0 \tau E$; in the same standard we read that the first null of the spectrum (provided that the frequency resolution is sufficiently narrow to identify it) occurs at $2\pi/\tau$ from the center frequency, and this is wrong, unless we revert to use pulsation instead of frequency: the correct expression for the first null is $1/\tau$ and indeed it gives 156 kHz in the present case. The increase of the spectrum measured at the center frequency of the 200 MHz carrier gives a first indication of the desensitization phenomenon: at 100

[12] The used symbols for the rectangular pulse and the sinusoidal carrier are those of the IEEE Std. 376.

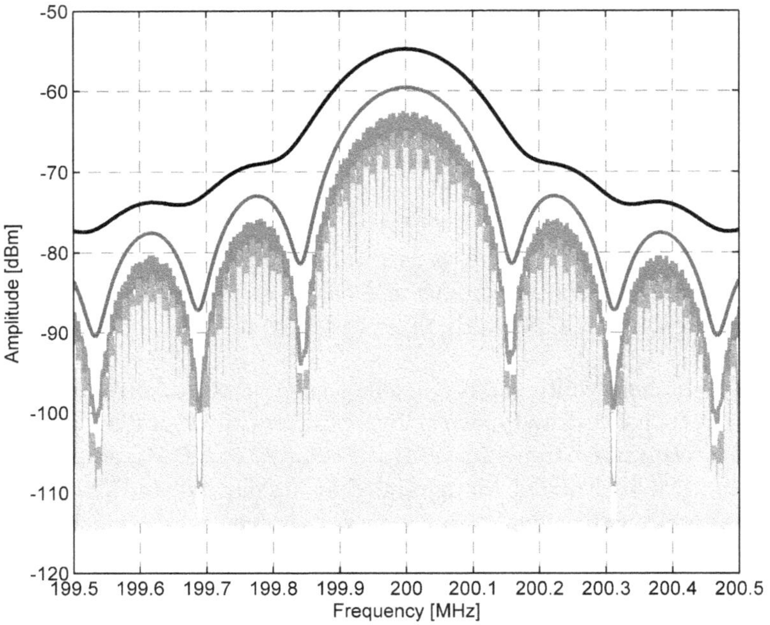

Figure 9.2.17 – Example of desensitization for a 200 MHz carrier at -40 dBm, modulated by a square pulse of 6.5 μs time duration and 10 kHz repetition frequency; measurements are done with RMS detector and RBW=1, 3, 10, 30 and 100 kHz from light gray to black.

kHz the marker at the center is $-54.77\,\mathrm{dBm}$, at 30 kHz it is $-59.56\,\mathrm{dBm}$, then -62.47 at 10 kHz, -64.93 at 3 kHz and -67.02 at 1 kHz. These measurements and the readings at the center peak were made using the RMS detector with 1 minute sweep time: the "true" value by analysis of the signal spectrum is $20\ \log_{10}(\tau/T)$, that is $-11.87\,\mathrm{dB}$ below the $-40.0\,\mathrm{dBm}$ value of the amplitude of the non-modulated 200 MHz sinusoidal tone (so $-51.87\,\mathrm{dBm}$) for the 6.5% duty cycle used in the tests, neglecting cable losses and amplitude errors.

The reduction of the envelope amplitude with decreasing bandwidth is called *pulse desensitization*. We have said that the amplitude of spectrum lines does not change with bandwidth, provided that it is small enough with respect to line spacing Δf. Pulse desensitization, often indicated by parameter α, refers to the envelope spectrum and to the coarse line spectrum that is obtained when resolution bandwidth is of the same order of spectrum line spacing. The usual application of pulse desensitization is when an accurate estimate of carrier amplitude is necessary, but the carrier is pulse modulated and "available" thus for a short time. In order to correctly return the amplitude of the underlying sinusoidal carrier, a Peak detector shall be applied; RMS detector will always give a lower output, proportional to the duty cycle.

The pulse desensitization factor

$$\alpha = 20\ \log_{10}(\tau k B) \tag{9.2.13}$$

RBW [MHz]	Desens. factor α [dB]	Measured [dBm]	Estimated [dBm]
10	0	-40.00	-40.00
3	0	-43.15	-43.15
1	-9.12	-50.40	-41.28
0.3	-19.58	-61.72	-42.14
0.1	-29.12	-69.95	-40.83

Table 9.2.8 – Estimated values of the desensitization factor α and corrected amplitude when measured with Peak detector for a $0.4\,\mu s$ pulse modulating a $-40\,dBm$ 200 MHz RF carrier: calculated α values for RBW=100 kHz, 300 kHz and 1 MHz are bracketed by an uncertainty interval of about $\pm0.25\,dB$ due to the uncertainty of the pulse generator thresholds for logic 0 and 1 when driven by the pulse digital control signal (about 15 ns on the time axis); the effect of desensitization is visible already for 3 MHz RBW, that is however not included in the correction of eq. (9.2.13) and results in a substantial error. The applied k value is 1, as for Gaussian filters.

depends on the duration of pulse τ, on bandwidth B and on shape factor k, that weights the effect of the type of RBW filter: $k=1$ for Gaussian filters and $k=1.5$ for a rectangular filter.

As easily seen in the expression above, pulse desensitization is a matter of IF resolution bandwidth and IF filter response to pulsed RF waveforms: a correct application and accurate evaluation of theory against experimental evidence require that the effect of the duty cycle is ruled out, leaving the pulse width alone. The desensitization factor allows the correction of measured pulse intensity, provided that its value is not too large, otherwise the measured amplitude on which the correction is applied is too small and significantly affected by noise, with a significant increase of uncertainty.

To this aim a Peak detector is best suited, provided that the sweep time is long enough for a stable maximum reading: in the example that follows the pulse repetition rate was fast enough (10 kHz) to hit the IF filter more than once at each sweep step (pixel); a sweep time ST = 1 min ensures that the scan of one pixel point lasts for about 50-150 ms for the most common choices regarding display resolution and that a minimum of 500 hits is ensured for each display point. The use of Peak detector gives a higher trace than the one obtained with RMS detector, well above the noise floor even for the largest RBW values. Reading the peak value at each pixel with enough hits per pixel, the trace is nearly flat, losing resemblance with the expected spectrum (the one in Figure 9.2.17, obtained with RMS detector, was better for illustration); its plotting does not bring any additional information or further insight, and for this reason is omitted. The result of the experiment with a $0.4\,\mu s$ pulse modulating a $-40\,dBm$ 200 MHz RF carrier are reported in Table 9.2.8.

The application of the desensitization factor is quite common when measuring fast pulses as for radar signals: modern SA with RBW values as large as 10 or 30 MHz are barely sufficient to follow a 100 ns rise-time RF pulse, still with a significant amplitude error, that necessitates compensation for desensitization; when measuring ultra-wideband (UWB) signals, rise-time values are even shorter and desensitization compensation is absolutely necessary.

A compromise shall be found since small RBW values reduce the amplitude too much, whereas large RBW values degrade resolution and ability to identify spectral lines. The optimal relationship between τ and B was found empirically as

$$\tau B = 0.1 \tag{9.2.14}$$

Spectrum analyzers operating in FFT mode are not suitable for pulse measurements.

Analogous behavior with respect to the quantities characterizing the modulating pulse (repetition frequency and pulse width) is expected when considering a base-band signal: two square waves of 100 mV amplitude and 10 kHz repetition frequency with small 6.4% duty cycle (thus resembling a rectangular pulse of $6.4\,\mu\mathrm{s}$ width) and a balanced 50% duty cycle (thus giving a line spectrum of odd harmonics), respectively, are considered in the example shown in Figure 9.2.18. Both cases demonstrate what we know already about the correctness of amplitude measurement with respect to resolution bandwidth: a larger RBW (upper half of figure for 3 kHz, lower half of figure for 1 kHz) gives a higher reading of peak amplitude.

Moreover, a narrower resolution bandwidth lowers the noise floor, as it is apparent by comparing the depth of notches: not only the noise floor is lower, but spectrum lines are well separated and there is a negligible contribution from the adjacent ones (stopped by skirt selectivity), so that the base floor appears with a noisy look.

For the short $6.4\,\mu\mathrm{s}$ rectangular pulse, the spectrum is characterized by a first notch of the envelope correctly located at the inverse of its time duration, that is $156\,\mathrm{kHz}$. For the 50% duty cycle the envelope goes as the expected $1/n$, with n the harmonic order, if amplitude is considered; the spectrum was measured in dBm, so a $1/n^2$ envelope reduction is correct for power (passing from the first to the second peak, that is from the fundamental to the third harmonic, the amplitude reduction is nearly 10 dB, that is an order of magnitude).

9.3 Power measurement

Methods and settings suitable for measurement and quantification of the power of the incoming signal are many, depending on the signal itself, its characteristics and the corresponding limitations of spectrum analyzer and its settings. We may distinguish between noise signals (true noise or signals with noise-like statistics) and sinusoidal or narrowband signals (as for many modulation schemes), steadily occupying a frequency interval (channel bandwidth) or occurring in bursts.

9.3.1 Averaging and noise power

In this Section we add some technical details and reconsider the "averaging" operation, that we have already analyzed in Chap. 4 on an exclusively statistical basis. A spectrum

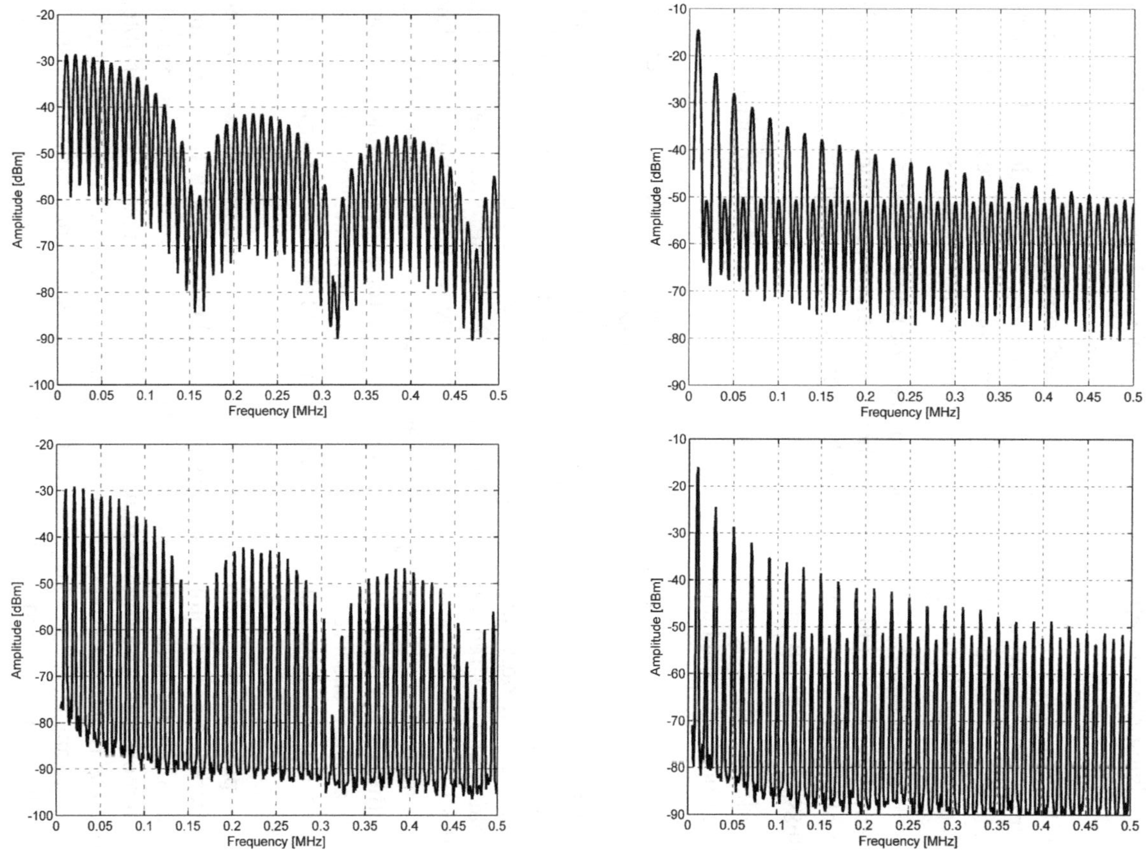

Figure 9.2.18 – Line spectra of square waves for (a) 6.4% and (b) 49.6% duty cycle; the effect of reduced resolution bandwidth (RBW = 3 kHz above, = 1 kHz below) is reflected in the lower noise floor between spectrum lines and a slightly lower (and more correct) height of spectrum lines.

analyzer has many options that implement some sort of averaging or smoothing, as we have already seen in the previous sections:

- IF filter RBW implements merging and integration of signal components over RBW itself, with larger RBW values giving smoother spectra; using zero-span combined with a large enough RBW allows measurement of power spectral density, by applying the concept of analog filtering followed by rms detection (see sec. 4.3.2.4);
- a slower ST (Sweep Time) is able to "accumulate" on the same frequency for a longer time, thus allowing potentially a stronger integrating action; the words "accumulate" and "potentially" aim at indicating that the final result (estimate error) depends heavily on the chosen detector and signal statistics;
- AVerage and RMS detectors are the best choices for noise power measurements: the expectation may be implemented explicitly as average or implicitly in the calculation of the rms itself; the problem is the choice of the correct representation, i.e. voltage or power, linear or log scale;
- setting Video Bandwidth VBW to a fraction of RBW reduces noise variance, but keeps the noise floor unchanged.

Considerations behind true power measurements and reduction of variance of the estimate for noise and noise-like signals are reviewed once more.

Regarding the use of average and rms, it is useful to consider more deeply on which quantities the average is operated. Averaging log data (that is data expressed in log scale, namely in dB) not only causes the aforementioned 2.51 dB under-response, it has also a higher than desired variance. Negative spikes of the envelope resulting from taking the log of small power levels add significantly to the variance of the log average, even though they bring very little power. Conversely, large power spikes are compressed by the log operation, and two orders of magnitude for example translate into 40 dB. We already observed that the result of a power measurement made by averaging power is lower than that made by averaging the log of power by a factor of 1.64 (2.51 dB).

Average may be performed on a number of channel power measurements x_i corresponding to different traces acquired successively: it is known that if the traces are independent and their number is large enough, then the Central Limit Theorem may be invoked to demonstrate the Gaussian distribution of the average random variable, for which we know that the variance is inversely proportional to the number of collected samples N, that is σ_x^2/N. Assuming noise-like statistics, including a Gaussian distribution for the measured signal, ensures that the distribution of samples is already Gaussian. This is strictly true if they are represented on a linear scale, otherwise log-scaled power readings have not a Gaussian distribution any longer. However, being each sample (display pixel) the result of many underlying real measurements summed together and averaged, the power level has no longer a distribution like the "logged Rayleigh" (see sec. 4.3.2.3), but rather look Gaussian, and there is no need to convert dB-format values back to absolute power. Also, their distribution is sufficiently narrow that the log (dB) scale is linear enough to be a good approximation of the power scale.

# avgs J	span/RBW=200 Swept mode	span/RBW=20 Swept mode
1	-104.4872 (0.7781)	-104.4977 (0.7179)
4	-104.4577 (0.4138)	-104.3919 (0.4110)
9	-104.3931 (0.2922)	-104.3865 (0.2405)
25	-104.4060 (0.1607)	-104.3838 (0.1420)
64	-104.4008 (0.1043)	-104.3762 (0.0870)
100	-104.3944 (0.0817)	-104.3984 (0.0715)

Table 9.3.1 – Noise averaging results for large and small span/RBW ratios: sample mean (and sample standard deviation between round brackets) of the noise floor (each trace is more than 500 points).

The just mentioned averaging operation that ensures that the Central Limit Theorem holds for trace samples refers to two underlying averaging operations: the average of all measured values belonging to the same pixel performed by the selected Average/RMS detector and the average performed by the resolution bandwidth filter, when collecting the signal power within its measurement bandwidth RBW. Moreover, for the Central Limit Theorem to strictly hold, samples shall be independent.

If the number of samples is low (a few samples, or just one, are contributing to each pixel) and the independence assumption is weak (as when the measurement time is short using FFT mode instead of swept mode), the variance reduction is less pronounced, reducing slightly the theoretical speed advantage of Fourier transform mode to attain the same signal-to-noise ratio.

The noise mean value is substantially unmodified as the number of averaged traces is increased (a small reduction is observed only for the lowest number of averaged traces, however equal to about 20% of the standard deviation). The reduction is more evident, as expected, when the span/RBW ratio is small, so that more samples fall into the same pixel increasing the significance of the average operation; this means also that at least one of the two cases (large or small span/RBW ratio) is not following the expected rule of reduction of variance with the number of averages. By calculating the ratio of each pair of standard deviation values, it is possible to verify that

$$\frac{\sigma_{J_k}}{\sigma_{J_{k+1}}} = \sqrt{\frac{J_k}{J_{k+1}}} \tag{9.3.1}$$

and this expression meets accurately the measurements for $J_k \geq 9$.

Reduction of variance is plotted in Figure 9.3.1 for the two cases, graphically compared to the expected reduction given by the reciprocal of the number of averages.

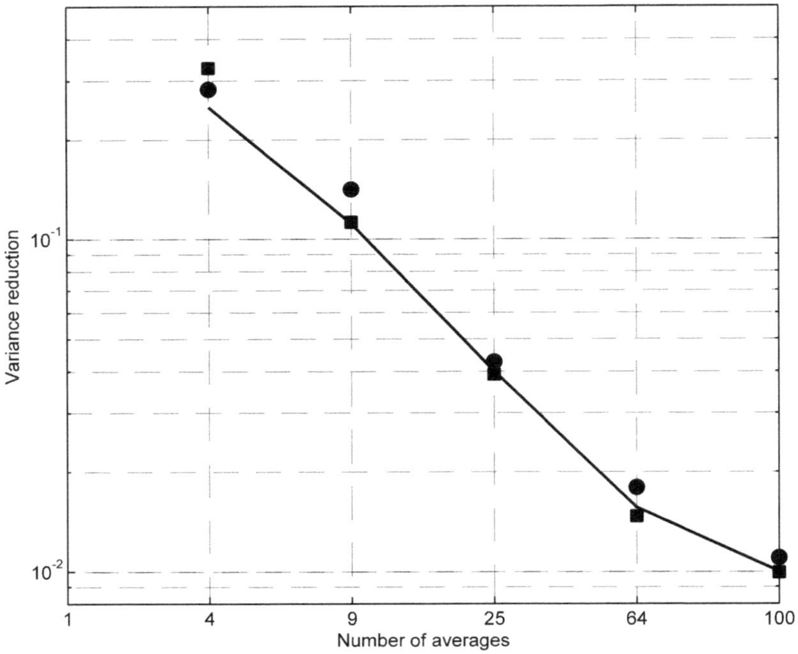

Figure 9.3.1 – Reduction of variance for trace averaging in the cases of large span/RBW (circle) and small span/RBW (square).

9.3.2 Channel power measurement

Resolution bandwidths offered by spectrum analyzers are limited to some MHz or a ten of MHz, not suitable to cover the entire channel bandwidth of modern modulations; moreover, the IF filter is not selective enough to reject power terms in adjacent channels and to give an accurate measurement. For these reasons the measurement of power across the entire channel bandwidth is mostly carried on by a suitable narrower RBW and successive integration. Selecting a narrow RBW around a few % of the channel bandwidth ensures that the noise floor of the spectrum analyzer (the DANL) is kept conveniently low and that, when measuring at the edge of the channel bandwidth, we are not letting in too much of the adjacent channels. The total channel power P_{ch} is obtained by integration of readouts corresponding to trace pixels P_i [dBm] falling inside the channel bandwidth (that we assume go from index N_{lo} to N_{up}); power values are obtained with a much smaller bandwidth (RBW) with respect to channel bandwidth B_{ch} and thus shall be normalized to the latter; finally, the average of the true power levels of each trace pixel on a linear scale gives the estimate of channel power.

$$P_{ch}\,[\mathrm{dBm}] = 10\log_{10}\left[\frac{B_{ch}}{B_n}\,\frac{1}{N_{up}-N_{lo}}\sum_{i=N_1}^{N_2} 10^{P_i/10}\right] \tag{9.3.2}$$

where B_n is the ENBW of the used RBW, B_{ch} is the nominal channel bandwidth, $N_{up}-N_{lo}$ is the number of points that make the trace that falls in the channel bandwidth and effectively used for the average. The transformation from dB to the linear power scale avoids the problem related to the log influence on the average result.

A similar operation may be accomplished using the Video Filter to perform the average. If operating on dB power values distributed logarithmically, there will be the known underestimation: in the case of a noise signal with Gaussian distribution this causes the already said 2.51 dB underestimation. If the average is performed with the average option of the acquired traces represented again in dB on a log scale, then the result is the same. On the contrary, using the AV detector, it is possible to operate directly on the sample points, represented in power linear scale rather than dB.

Since the 2.51 dB correction holds only for Gaussian distribution of noise and it is thus not universally applicable, it is advisable not to perform any averaging before the traces are fed to the channel power algorithm and that VBW is set to minimum 3 times RBW, so to exclude it (with a maximum error of 0.05 dB on channel power).

Now the question is "how the power sample points P_i shall be taken in terms of correct detector to use?" As said, the AVerage detector applied to the true power samples is a correct choice to obtain the average power reading. If using eq. (9.3.2) to estimate the average power, the use of the Peak Max detector might slightly bias the final result; the Sample detector gives unbiased results.

Regarding the biasing of Peak Max detector, we observe that the peak of noise will exceed its power average by an amount, that increases (on average) with the length of time over which the peak is observed (see sec. 9.1.9.2 and 9.2.7 that introduce this problem). A quite accurate formula that has been checked against some experimental observations in the literature [4] puts in relationship the observation time T (that is the sweep time divided by the number of points) and the impulse bandwidth of the chosen RBW setting (the exact relationship between the two depends on the IF filter type, see sec. 9.2.2[13]) with the observed ratio between the average power obtained with peak detector $P_{av,pk}$ and with sample detector $P_{av,smp}$.

$$\frac{P_{av,pk}}{P_{av,smp}} = 10\ \log_{10}\left[\ln\left(2\pi T B_n + e\right)\right] \tag{9.3.3}$$

Agilent in [4] indicates a better fitting of measured values to the theoretical curve of eq. (9.3.3) if the multiplying 10 coefficient is replaced by 10.7, even if they underline that there is no theoretical support for this change.

Regarding the use of FFT mode or Swept Mode, the time behavior of the measured signal and its modulation shall be considered.

Let's consider a constant-envelope modulation, such as FM (analog) or Frequency Shift Keying (digital), used for GSM cellular phones, satellite communications and noise-protected transmissions. The constant-envelope makes the measured power constant when that power is measured over a bandwidth wide enough to include all the relevant spectral terms, that is the channel bandwidth. Using the FFT mode over a wide enough frequency interval to cover the channel width will lead to channel power mea-

[13] The impulse bandwidth is 1.6-1.8 the −3 dB bandwidth for older four-pole synchronously tuned filters, but more commonly for modern digital spectrum analyzers it is 1.5 the −3 dB bandwidth for a digitally implemented Gaussian filter; see sec. 9.2.2 and Table 9.2.1.

surements with very low variance. On the contrary, swept mode, that is typically performed with a RBW value that is a fraction of the span and narrower than the symbol rate, will show large fluctuations and the channel power measurement will thus have a larger variance.

9.3.2.1 Carrier power

The measurement of amplitude (or power) of a sinusoidal tone with steady characteristics was already considered, when talking about the compensation of the internal noise contribution by subtraction (see sec. 1.1.5 and 4.3.1.7) and when observing that with sinusoidal tones several detectors give the same result (see sec. 9.1.9.2).

When the sinusoidal tone, or the entire modulated carrier, is not steady, but occurs in bursts, then a zero-span mode measurement is advisable: the largest RBW compatible with the modulation band is applied to the center frequency, which the LO is locked onto; then the time domain dynamics of the signal are tracked on the display. The measurement of the power is done by averaging samples, provided that each point corresponds to a true power measurement by selecting the RMS detector. Since there are intervals of time when the carrier is on and others when the carrier is off, the time intervals for the computation are selected manually, by applying a criterion of minimal amplitude of the measured rms, e.g. at least -20 dB (10%) of the full carrier rms (that if unknown may correspond to the largest recorded value). With modern spectrum analyzers automatic triggering is possible by setting such an amplitude value.

Regarding the use of Video Filter, it is underlined that the characteristics of the transient carrier signal are not those of a pure sinusoidal tone and the effects of the modulation may make it look like a noise signal, but probably with much different statistics: it is thus advisable to exclude the Video Filter by setting the VBW above the chosen RBW (e.g. with the already suggested minimum ratio of 3:1). This might turn into a problem, especially when spectrum analyzer software locks the minimum VBW/RBW ratio to 1, for high data rate protocols (e.g. hundreds kbps or some Mbps), where bursts occur in μs or hundreds of ns.

9.3.2.2 Channel power statistics

Besides the measurement of carrier power alone, it might be necessary to characterize the behavior of the modulated signal, e.g. in terms of peak power, peak-to-average power ratio, etc.

The peak power may be interpreted as the maximum of either the total channel power or the power over a portion of the frequency interval; in any case it is a measurement that can be made using the same settings for carrier power measurement, provided that the spectrum analyzer is able to track signal dynamics.

Analogously, when estimating peak-to-average power ratio, the value may be variable depending on bandwidth and sweep time used to catch the peak envelope power. Normally, a zero-span measurement setup with a large resolution bandwidth is the only possibility; it is advisable to verify result sensitivity to the used RBW value.

Given the probabilistic nature of the measured signals, a curve of probability for the different peak-to-average levels is quite common to adequately describe the statistical behavior of a modulation: the required curve is the cumulative distribution function that gives the probability that some peak-to-average level is overcome during measurements; of course the curve shall be accompanied by the number of acquired samples to give an indication of the statistical significance. Such characterization is not only relevant to possible interference to adjacent channels and to third-parties, but also to correctly sizing electronic circuits (most of all amplifiers) in transmitter and receiver paths, quantifying the headroom with respect to thermal power sizing and the number of times the signal amplitude is near the maximum.

9.3.2.3 Adjacent-Channel Power (ACP)

The measurement consists in evaluating the ratio of the carrier power in a given channel and in the adjacent one (or the second adjacent one, called "alternate" channel), where channels are used by a transmission protocol and carry some kind of modulation. For noise-like modulations, since the interest is in the ratio of the two, either the true RMS detector or the log-averaged power (by means of the Video Filter) may be used. On the contrary, if the modulation is not noise-like or if two separate channel power values are required, then the same caution for the use of video filter reviewed before shall be considered (i.e. video filter exclusion) and the measurement is to be performed with RMS detector.

When performing this kind of measurements the power in adjacent channels is estimated as for a channel power measurement, specifying the bandwidth and also the relative distance between the carrier frequency of the potentially interfering channel and the adjacent channels. Corrections might be necessary to account for channel filters and symbol shaping filters of the specific protocol.

The terms that leak into and potentially interfere with adjacent channels have their origin in non-linear distortion of the original signal (in this case we speak of harmonics and inter-modulation products that are narrowband) and in frequency instability of the original signal (phase noise with a broadband behavior). These spectrum components are normally identified with the overall definition of "spectral regrowth".

When measuring such terms it is important that the internally generated components of the spectrum analyzer may be distinguished and that they are much lower and under control: intermodulation products would originate from non-linear mixer operation (see sec. 9.1.4 and 9.4), while broadband phase noise products might be caused by frequency instability (spectrum analyzers are quite stable on the short term and may be hooked to GPS reference for long-term stability) and local oscillator feed-through (see sec. 9.2.6). Unless there is evidence of significant narrowband components in adjacent channels, adjacent channel disturbance will be largely of the noise-like type [4], and measurement is less troublesome than channel power measurement.

9.4 Dynamic range and distortion measurement

The dynamic range quantifies the ability to measure the larger fundamental tone and the smaller distortion products. The widest range that a spectrum analyzer can distinguish between, respectively, signal and distortion, signal and noise, or signal and phase noise may be all named as the spectrum analyzer's dynamic range, but in reality with respect to different phenomena. However, since a unique setting for the dynamic range is used, then a single dynamic range is considered, that is the result of the tradeoff operated on the three distinct characteristics.

The mixer level used to optimize dynamic range can be determined from the *second-harmonic distortion*, *third-order intermodulation distortion*, and *displayed average noise level* (DANL) specifications of the spectrum analyzer (distortion and inter-modulation terms are discussed in sec. 3.4). From these specifications, a graph of the internally generated distortion and noise versus mixer level can be made, that helps in determining the best mixer level, and thus in turn the input attenuator setting and the inclusion/exclusion of pre-amplifier.

9.4.1 Increasing the dynamic range

When performing distortion measurements we strive for the largest dynamic range attainable: the correction of the measured amplitude of a sinusoidal tone for the noise floor (see sec. 9.2.12) works better for larger signal-to-noise ratio, so it is instructive to review quickly the settings that improve it.

Figure 9.4.1 shows the effect of three actions that progressively reduce the DANL, thus increasing the dynamic range. The top trace (trace 1) is the reference trace obtained for RBW = 3 kHz, Att. = 10 dB; for all traces the RBW/VBW ratio is 10. The second trace from top shows the result of reducing the resolution bandwidth with power averaging set on RMS detector (RBW = 300 Hz); then, the third trace is obtained by reducing the input attenuation (passing from Att. = 10 dB used for the previous traces to Att. = 0 dB). Finally, the lowest trace (trace 4) is obtained using logarithmic power averaging, that we know lowers the noise floor by another 2.51 dB; however, it is important to stress that it does not work with all signals and their modulations, because in the case of on/off keying (with a square wave AM modulation), time intervals with low or no signal power have the same effect of lowest noise values in lowering the average after log operation.

Let's get a closer look at the dynamic range improvement after the three operations we have just described: such improvement is measured by the increased signal-to-noise ratio and by the approaching of the peak marker value to the expected "exact" value of the signal. The signal-to-noise ratio in reality does not give the real complete picture, because in the end it is the confidence interval that matters, and the assigned confidence level (let's assume 95% probability and thus a coverage factor $k = 1.96$ under Gaussian assumption). Yet, noise floor statistics are not constant during the experiment, because it is known that larger RBW values yield less dispersed noise floors, and the log operation at last increases dispersion even more, besides modifying

Curve	Peak marker [dBm]	Noise floor (mean) [dBm]	Noise floor (std.dev.) [dBm]	Estim. signal power [dBm]
1	-104.46	-107.37	0.2735	-107.583 (-0.645/+0.500)
2	-106.74	-117.51	0.5560	-107.119 (-0.115/+0.087)
3	-107.50	-127.34	0.5542	-107.549 (-0.013/+0.01)
4	-108.18	-130.27	0.6770	-108.203 (-0.0096/+0.007)

Table 9.4.1 – Statistics and values of the example of Figure 9.4.1 on improved dynamic range for the measurement of signals out of noise clutter. Noise statistics were calculated on 400 points far from signal peak at center and with a span/RBW=33.

the distribution slightly, for the negative tail (see sec. 1.1.5.2 and Figure 4.3.8). The increased dispersion is shown in Table 9.4.1, fourth column.

The "exact" value was estimated by averaging 30 traces of a narrow RBW measurement (10 Hz) with a conveniently narrow span (span/RBW=50) and long sweep time (2 minutes), resulting in a reading of $-107.47\,\mathrm{dBm}$.

It is important to observe that the log option flats out the noise floor with a narrower effective confidence interval suggesting a consistent estimate, whilst the operation reduces correspondingly the signal amplitude, significantly biasing the estimate. As expected, reasoning on the left tail of the logged distribution extending to negative values, the resulting value will be negatively biased.

Using pre-amplifier reduces DANL even more: roughly speaking, for a good pre-amplifier with a conveniently large gain, the total equivalent input noise becomes that of pre-amplifier alone (see sec. 9.1.2), besides the noise of the input attenuator (see sec. 9.1.1).

9.4.2 Optimizing mixer operating point

By the way, when measuring distortion, reducing DANL is only one side of the problem; a distortion measurement is a ratio of amplitude of a spurious harmonic component to the amplitude of the fundamental (or carrier), so that the full dynamic range shall be displayed. This exigency is common with other measurement tasks, such as phase noise measurement (see sec. 9.5). It is opportune to stress that the settings for this kind of measurement, when achieving very high linearity and minimizing the influence of internal noise and distortion, are established in an open-loop fashion, without the possibility of seeing a feedback on the display: if you notice something that is apparently wrong in amplitude, then you are already too far away. The correctness shall be ensured *a priori*, by meeting requisites and optimizing dynamic range.

In this case the quantity that rules is the mixer input level: the SA settings shall be selected in order to reduce the DANL, yes, but in reality to maximize the dynamic range,

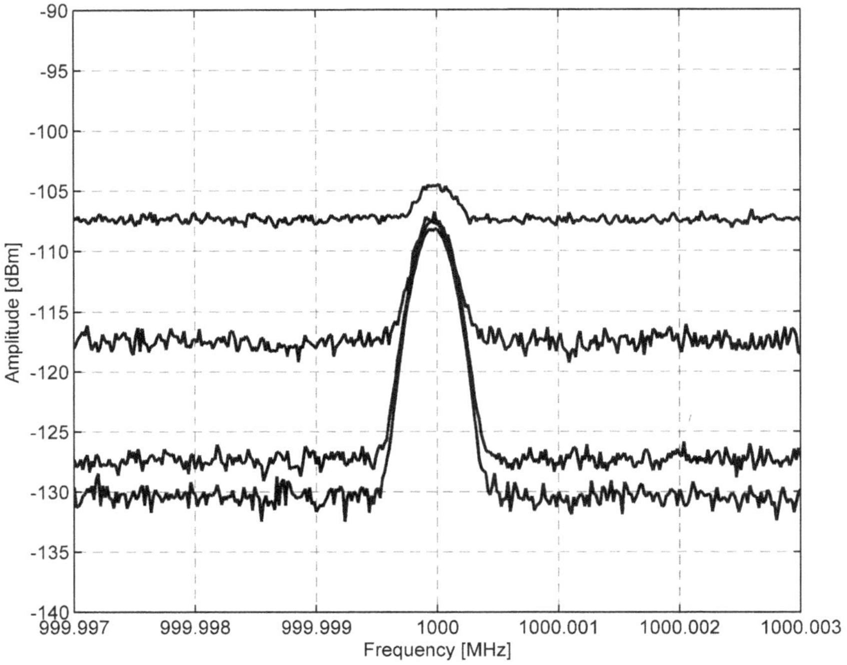

Figure 9.4.1 – Increasing the Spectrum Analyzer dynamic range: (from top to bottom) 1) basic curve (RBW = 3 kHz, Att. = 10 dB), 2) selecting a lower RBW (RBW = 300 Hz), 3) reducing input attenuation (Att. = 0 dB), 4) using logarithmic power average. The example is built around the measurement of the 5th harmonic of a supposed pure sinusoidal signal at 200 MHz, -40 dBm amplitude. RBW/VBW = 10, span/RBW = 33 and sweep time of 100 seconds for each trace.

between the minimum and maximum points, that we have discussed in sec. 9.2.11. We want that the mixer input level (i.e. its operating point) lies in the range of levels for which its operation is linear and does not cause further distortion: mixer operation can be characterized by the 2nd harmonic distortion curve and its intercept, by the 3rd harmonic distortion curve and its intercept, and by DANL. The intercepts are taken with respect to a reference level that for convenience is an ideal carrier/fundamental level, all other quantities measured in dBc (dB below carrier amplitude). This is shown in Figure 9.4.2, where three noise floor profiles are shown at the bottom left for as many RBW values and the assumed phase noise level is $-70\,\mathrm{dBc}$.

As explained in [5], the 2nd harmonic line has a slope of 1 (one order of magnitude on each axis) because what matters is the added distortion and, despite the 2nd harmonic amplitude increases with an order 2 with respect to the mixer input level (2 dB every 1 dB of increase), the difference with the fundamental increases by an order 1 (1 dB every 1 dB of increase); the same may be said for the slope of 2 assigned to the 3rd harmonic distortion line.

The dynamic range allowed for mixer input level has the upper bound established by the amount of maximum distortion that we accept for the fundamental, that is assumed to be positioned around the largest mixer level, let's say $-7\,\mathrm{dBm}$ for $-40\,\mathrm{dBc}$ distortion (as assumed and shown in Figure 9.4.2). Looking at the lower bound of the

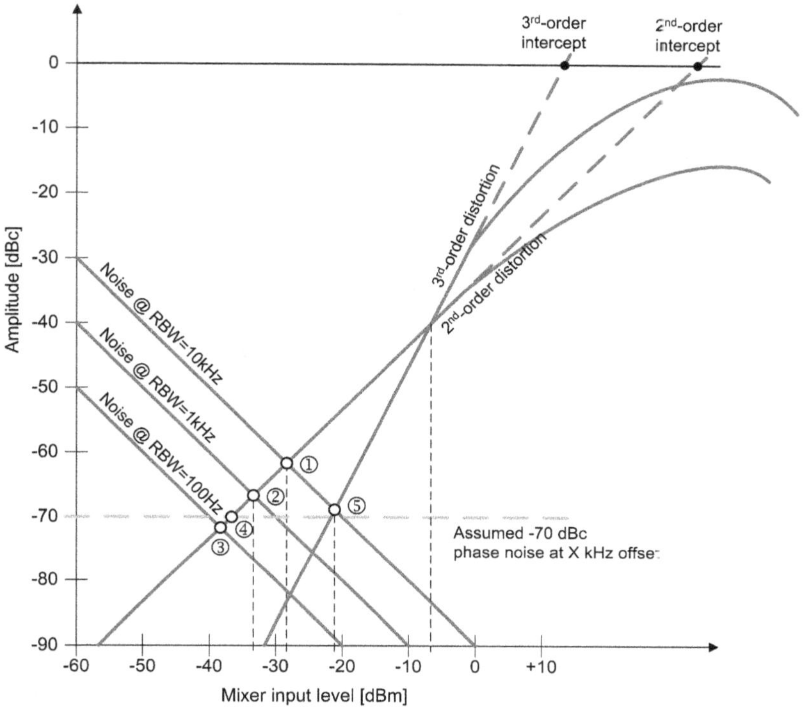

Figure 9.4.2 – Dynamic range as a function of mixer distortion loci and noise.

mixer level range, it is apparent that selecting the largest RBW (and the largest noise floor), the second-order harmonic distortion is crossed at point ①, corresponding to about $-28\,\mathrm{dBm}$ of mixer input level. If a wider mixer input range is needed, then the RBW shall be reduced to 1 kHz (moving to point ②) and then to 100 Hz (point ③), improving by an order of magnitude on the vertical axis each time: for the 100 Hz RBW however, point ③ is in conflict with the coherent distortion caused by the internal phase noise (assumed quite pessimistically at $-70\,\mathrm{dBc}$) and the minimum second-order distortion achievable is limited to point ④. For the third-order harmonic distortion, lower than the second-order one for all the admissible mixer input levels, the minimum is nearly immediately reached at point ⑤ crossing the assumed phase noise level. For a given frequency assumed for a measurement, the phase noise is determined and fixed and represents the ultimate limit that sets the hard lower bound for mixer level.

The elements that characterize internal distortion and phase noise are reviewed in the next sections.

9.4.3 Internal distortion terms

The effect of SA internal distortion terms can be seen when measuring distortion of an external signal and can be quantified with some calculations.

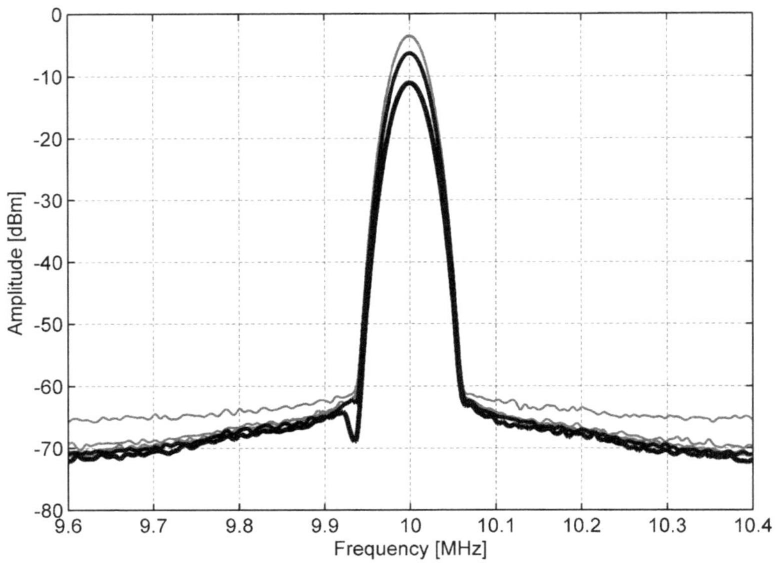

Figure 9.4.3 – Check of internally generated distortion effect: compression of external $-3.5\,\mathrm{dBm}$ sinusoidal tone (gray curves for Att = 40 to 10 dB, first thick black curve Att = 5 dB, second lower thick black curve Att = 0 dB).

9.4.3.1 Effect on a distortion measurement

Taking a sinusoidal signal source we want to quantify the amplitude of its 2nd harmonic (or any other harmonic); the SA center frequency is put onto the 2nd harmonic frequency and a measurements is taken for a given RBW, VBW, input attenuation, and using the RMS detector. After storing the first measurement, the input attenuation is reduced and a second reading is taken and compared to the first one: any increase of the measured harmonic amplitude at the lower input attenuation setting is due to distortion introduced by the spectrum analyzer.

Now, let's focus on the values of settings in order to obtain the results shown in Figure 9.4.3. A good signal source is taken, so that the expected small distortion products of the spectrum analyzer may appear without not too extreme settings; the signal source is set to 0 dBm. The SA was configured for a power measurement with RBW = 30 kHz, VBW = 0.1 RBW, Peak detector, and variable attenuation from Att. = 40 dB down to 0 dB, so that the mixer input level crosses the -10 dBm reference level and is then driven with the full input signal. Such a configuration of course is the result of some trial and error, in order to produce the right figure; the input signal level may be adjusted either by acting on the signal source or by introducing some external attenuation on the input line. The reduction of input attenuation for the last 10 dB is responsible for the displayed erroneous reading indicating signal compression: at Att = 5 dB with $-8.5\,\mathrm{dBm}$ into the mixer compression is already visible ($2.8\,\mathrm{dB}$) and when the input attenuation is removed (Att = 0 dB) it increases to $7.5\,\mathrm{dB}$[14].

[14] In many spectrum analyzers the user is warned when overdriving the mixer; this configuration was obtained by suppressing the warning and forcing the desired level.

9.4.3.2 Estimate of internal distortion

An experiment is made aiming at determining internal distortion terms of the spectrum analyzer, expressed as coefficients of non-linear input-output relationship $a_1 = 1$, a_2 and a_3 (as defined in sec. 3.4), rather than compression point, intercept point, etc.

The spectrum analyzer was fed with a low distortion sinusoidal tone $A_1 \cos(\omega t)$, that yet has its own distortion terms, assumed corresponding to the second and third harmonic $A_2 \cos(2\omega t)$ and $A_3 \cos(3\omega t)$. These components may be determined by measurement at mixer input levels that are far from its 1 dB compression point and do not generate additional harmonic distortion. Then the coefficients of the third-order non-linear input-output relationship are determined with some math, together with some of the typical performance indexes we have reviewed in sec. 3.4, namely the 1 dB compression point and the IP_2 and IP_3 intercepts, extrapolated rather than directly measured, as suggested in sec. 3.4.4. The signal level feeding the mixer has been made varying in steps of 1 dB by adjusting the external RF source level; the internal attenuator can be changed only in steps of 5 dB and so it doesn't lend itself to fine tuning of the mixer input level. The source signal is a 200 MHz sinusoidal tone made varying between 0 and +5 dBm. The results are shown in Table 9.4.2.

The 1 dB compression point may be readily identified observing that at +2 dBm of source level, moving the attenuator from 30 dB to 10 dB, the drop in the measured amplitude of the fundamental is $+0.78 + 0.39 = 1.17\,\text{dB}$ and the mixer input level is $+2\,\text{dBm} - 10\,\text{dB} = -8\,\text{dBm}$, that may be taken as an approximation of the 1 dB compression point, that probably remains around $-7.8\,\text{dBm}$.

Source distortion is estimated by taking the measured values at the largest attenuation, so for mixer levels far from its non-linear region (as done also in Example 3.8):
$A_2 = -40.15 + 1.12\,\text{dBm} = -39.03\,\text{dBm}$ and $A_3 = -48.74 + 1.12\,\text{dBm} = -47.62\,\text{dBm}$.

The slope of the 2nd-order and 3rd-order distortion curves, ending in the intercept points IP_2 and IP_3, may be determined by linear regression of measured points, selecting the correct interval of mixer levels, where the internal distortion prevails on source distortion and before the line becomes strongly curved at higher mixer input levels: it is the interval of values where the linear approximation holds as shown in Figure 9.4.2.

Of course making a two-tone measurement is easier because the amplitude of intermodulation products is larger than that of harmonics; however, to prepare such test two signal generators are needed and a power combiner.

9.5 Phase noise measurement

The phase noise of an oscillator or signal source is a measure of its short-term stability (please, see sec. 3.5 for the theory and some mathematical insight). The spectrum analyzer may be used if the dynamic range requirements are not demanding and the sweep time shall be set accordingly, taking into account the expected frequency drift versus time of the oscillator under test. If the frequency drift is too large (so for very low quality oscillators), the results might be arguable.

Source	Internal attenuator [dB]					
level [dBm]	30	20	15	10	5	0
	-1.12	-1.08	-1.09	-1.39	-5.28	-10.22
0	-40.15	-40.20	-40.19	-40.23	-40.39	-40.66
	-48.74	-48.52	-48.42	-48.02	-47.16	-44.66
	-0.20	-0.16	-0.17	-0.84	-4.81	—
+1	-38.74	-38.76	-38.77	-38.82	-38.98	-39.31
	-47.60	-47.50	-47.37	-46.90	-45.84	-42.69
	+0.78	+0.85	+0.82	-0.39	-4.58	—
+2	-37.18	-37.17	-37.22	-37.26	-37.43	-37.82
	-46.48	-46.38	-46.25	-45.67	-44.40	-40.28
	—	—	—	—	—	—
+3	-35.61	-35.60	-35.64	-35.71	-35.88	-36.32
	—	—	—	—	—	—
	—	—	—	—	—	—
+4	-34.01	-34.03	-34.10	-34.16	-34.34	-34.88
	—	—	—	—	—	—
	—	—	—	—	—	—
+5	-32.42	-32.42	-32.48	-32.55	-32.76	-33.49
	—	—	—	—	—	—

Table 9.4.2 – Measurement of distortion terms with a 200 MHz sinusoidal tone with intensity variable between 0 and +5 dBm, using RBW=10 kHz, VBW=1 kHz, RMS detector, 1 minute sweep time and a reference level of +10 dBm. For each cell the values of the fundamental (200 MHz), the second harmonic (400 MHz) and the third harmonic (600 MHz) are reported. The existing source distortion may be assumed to correspond to the measured values at the largest attenuation of 30 dB.

As seen in sec. 3.5, phase noise is a concept related to instantaneous frequency instability: any change of frequency of a simple sinusoidal signal will result in a modulation of the phase angle and frequency instability (or jitter) is seen as a phase instability, or better phase noise (the phase is the integral of the instantaneous frequency).

When using a spectrum analyzer to measure phase noise, we should remember that the measured phase noise is always the sum of desired phase noise under test, spectrum analyzer phase noise and spectrum analyzer thermal noise (or noise floor). In general, phase noise measurements made with spectrum analyzers have the following limitations: i) the usual IF bandwidth values are too large and choice is restricted to the lowest values (1 Hz most often), mainly to bring down the noise floor and to spot out small phase instabilities; ii) the resulting dynamic range is somehow limited, because of internal noise and also due to internal distortion, and care shall be taken in optimizing the dynamic-range settings (namely input power level, attenuation, reference level, pre-amplifier); iii) LO instability may be relevant and affect results, if one is measuring a high quality oscillator; iv) the spectrum analyzer measures AM and PM noise altogether and cannot separate them.

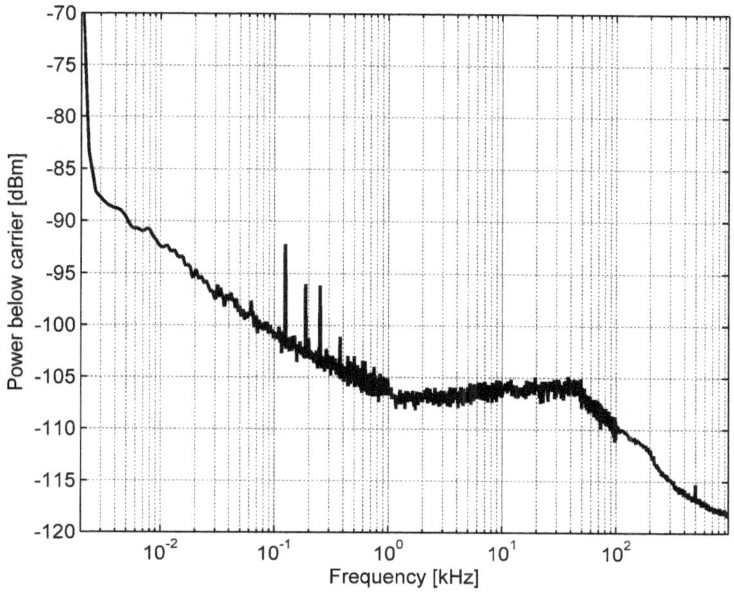

Figure 9.5.1 – Internal phase noise measured with respect to the −33.76 dBm amplitude of the 10 MHz reference signal and normalized to 1 Hz, displayed as offset from carrier frequency: FFT mode and RBW=VBW=1 Hz up to 10 kHz, then RBW=3 Hz up to 100 kHz and 10 Hz up to 1 MHz. The different RBW/VBW values and trace averaging options make the curve look more noisy below 100 kHz.

Regarding the fact that both AM and PM noise are included in a spectrum analyzer measurement, there is nothing to do: it may be said that AM is usually closer to the carrier, while the effect of phase instability reaches frequency intervals far from it. When saturated mixers are used (external to the spectrum analyzer), the AM is rejected; this is one of the most used phase detectors, but it doesn't come for free with the spectrum analyzer!

So, for a direct phase-noise measurement using our spectrum analyzer, very low RBW values shall be used, minimizing the noise floor and maximizing the dynamic range, limited only by distortion products. On the contrary, LO instability and internal phase noise of the spectrum analyzer shall be known and evaluated beforehand: in Figure 9.5.1 the spectrum analyzer is fed with its own 10 MHz reference signal and the resulting phase noise is measured, thus accounting for all the successive blocks after the internal reference oscillator. The intention is thus to evaluate as much as possible the influence of internal blocks when measuring an external generator, fed by the internal SA 10 MHz reference.

Measuring phase noise is often a race to squeeze the last dB of dynamic range by setting careful combinations of the following: reference level, internal attenuator, resolution bandwidth, use of pre-amplifier and offset from the carrier. When optimizing spectrum analyzer settings for the widest dynamic range and the lowest noise floor, first we must accept that measurements are taken at some offset from the carrier (the minimum offset is determined by the used RBW and the skirt factor of the IF filter); second, the lowest noise floor is attained with the narrowest RBW value, the use of the pre-amplifier and the least internal attenuation (departing from the exigency of protecting the mixer); finally, the sweep time, if long enough can get that last dB more out of the measurement, as we have already commented regarding the dwell (or observation) time for each pixel (knowing that the span/RBW needs to be in general quite large[15]). Trace averaging may be used to smooth the curve profile, not so much for a cosmetic effect, as for spotting out small narrowband components, if we are interested in them. In any case, the use of too a large span/RBW ratio, larger than the number of display points, has the drawback of giving a coarse grain pattern, missing narrow-band components; adjusting RBW to larger values is detrimental for the noise floor. These two limitations are removed with specific instruments such as signal analyzers and phase noise analyzers, featuring lower internal noise, narrower RBW values available and a much larger number of trace sample points, in the order of several thousands or some tens of thousands.

Practically speaking, when measuring phase noise with a spectrum analyzer, an initial good setup is needed that gives plausible results and then further small improvements are possible, tuning some parameters, such as increasing the amount of signal, provided that the mixer input signal is still in the linear range. While the attenuator can be adjusted in 5 or 10 dB steps, the input signal level may be adjusted usually down to 1 dB or a fraction of dB, so that fine tuning is more conveniently performed at the input side. Finally, it is always a good practice to switch off input RF power and perform a noise floor assessment between measurements or every some measurements, to keep track of the adjustments done, of the applied changes and their effect. We should also remember that the noise floor may increase slightly during the measurements because of self heating and internal thermal control (see sec. 9.2.10.2).

We have seen that the typical profile of a phase noise spectrum is a downward curve with mostly $1/f$ noise near the carrier (but not only, see sec. 3.5), corresponding to a -20 dB/decade slope in amplitude, and with nearly horizontal parts at higher frequency, corresponding to uncorrelated noise. In Figure 9.5.2 the gray curve has such a profile close to the carrier, while the other two have a steeper slope; the flat part of the spectrum begins around 500 Hz for two curves out of three. Many variations of this general structure are possible: in particular narrowband components and humps may appear both at small and large frequency offsets from the carrier, that increase locally the spectrum by some dB. In Figure 9.5.2 the black curve has such a hump between 0.9 and 6 kHz, while the other two curves have a nearly flat profile; all the curves have narrowband components between 100 and 360 Hz. Getting to about 40 kHz all the curves begin going down again with a nearly $1/f$ profile reaching a second pedestal located between $-114\,\mathrm{dBm}$ and $-109\,\mathrm{dBm}$.

Interpreting a phase noise spectrum is not an easy task because each oscillator or generator has its own characteristics of stability, internal sources, warm up and temperature drifts. Generally, starting from the carrier and getting farther away with increasing offset, it may be said that low-frequency drift and wander is normally due to the clock reference, be it a quartz oscillator or an OCXO, possibly including a first digital stage for base clock generation. When the curve reaches its first flat part it is

[15] Phase noise measurements with narrow RBW and an extended frequency range of tens of kHz or even some MHz are incompatible with the need for low span/RBW values.

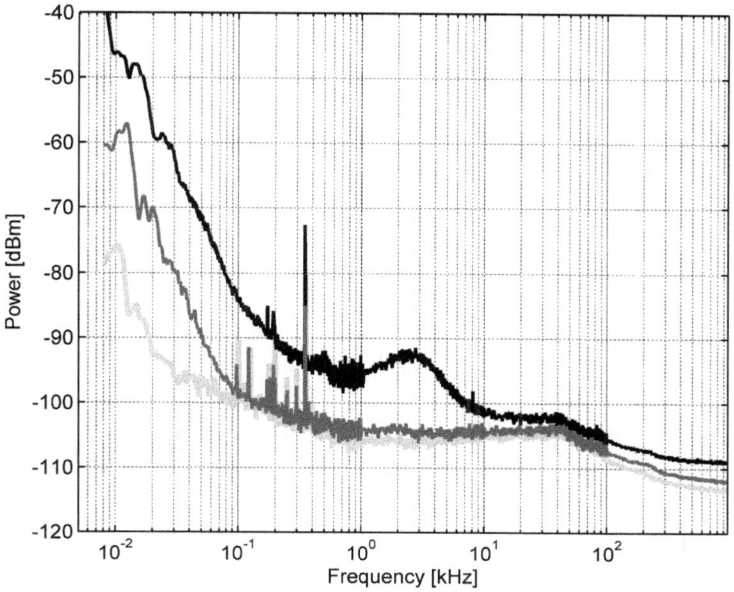

Figure 9.5.2 – Phase noise measurements with respect to the −40 dBm amplitude of a 200 MHz signal and normalized to 1 Hz, displayed as offset from carrier frequency, ATT=0 dB, minimum carrier offset 8 Hz with curves made by separate measurements extending to 200 Hz, 1 kHz, 10 kHz, 100 kHz and 1 MHz; trace averaging used for smoothing; carrier frequency of 20 MHz (light gray), 200 MHz (dark gray) and 2 GHz (black).

customary to assign it to the complex frequency synthesis, that in digital generators is composed of several Phase Locked Loop (PLL) circuits; the flat horizontal curve portion indicates a broadband white noise behavior, even if digital PLL-based circuits are also a potential source of narrowband components, highly depending on the combination of input and output frequency of each stage, even for fractional PLLs. The remaining part of the curve may or may not be revealed, depending on measurement conditions: too a high SA noise floor has impact, not to say masks, the real phase noise far from the carrier, that gets to look as a flat broadband noise. If the noise floor is low enough it may be observed a second slope more or less pronounced, whose origin may be the output oscillator and that highly depends on the circuit type (YIG oscillator, digital-to-analog converter, completely analog). The three curves in Figure 9.5.2 show such profile, nearly the same for all, indicating a common reason, with the three chosen carrier frequencies spanning two decades.

Considering again the phenomenon of narrow-band components appearing in the phase-noise spectrum, they may be caused by many non-idealities of the internal mechanism of clock and waveform generation, in particular when digital synthesis is used. Periodicity is thus created by many beating frequencies and unavoidable errors due to finite precision of integer factors used in the synthesis process: this is quite a known phenomenon also in other fields, such as power converter modulation, when the cycle time of the processor is not perfectly aligned with the carrier and modulation frequencies. The results shown in Figure 9.5.3 refer to slightly different frequencies generated around the desired nominal 200 MHz tone: the shift is between 0 and +7 Hz which

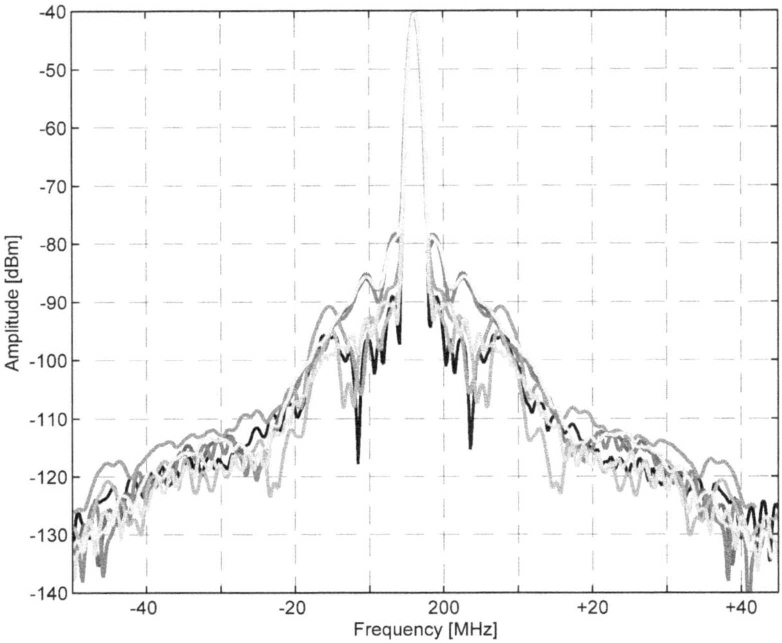

Figure 9.5.3 – Phase noise for a RF generator using digital synthesis measured with RBW=1 Hz: sinusoidal tone at 200 MHz and -40 dBm (a constant frequency error of -4 Hz is visible). The curves refer to different values of the nominal tone frequency offset from the nominal 200 MHz to explore periodicity and beating tones in the digital synthesis; the offset ranges from 0 Hz to +7 Hz (corresponding to curves from black to light gray) and it is manually adjusted to have the measured traces overlap.

the various phase noise profiles correspond to; the carrier is located in the original position having the generator an initial offset of −7 Hz, the other curves were moved manually leftward and overlapped during post-processing.

For some values of the generated sinusoidal tone, evident lateral bands of the carrier appear, caused by periodicity in the generation process. This periodicity is about 0.33 Hz and it is clearly visible when a zero-span reading is made at some offset from the carrier, as shown in Figure 9.5.4.

Using long sweep times has the drawback of possible long-term drift influence coming into play, due to the sweep mechanism. Long-term drift of the generator is excluded if the 10 MHz reference input is tied to the 10 MHz reference output of the spectrum analyzer: the internal OCXO of the spectrum analyzer features a good, but not excellent, stability, but we are not looking for an absolute reference, while simply trying to avoid that one of the two instruments is drifting away.

9.6 Time gating and pulse measurement

A transient signal with repetitive nature has a discrete spectrum (or "line spectrum") that results from the measurement, only if the used resolution bandwidth is low

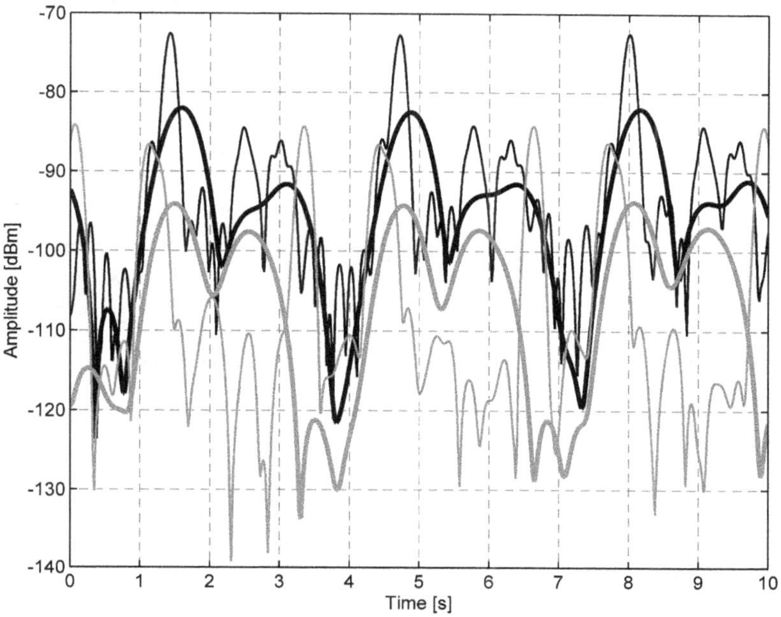

Figure 9.5.4 – Zero-span reading at offset frequency from carrier of +7 Hz (black curves) and +15 Hz (gray curves); RBW = 1 Hz (thick line), RBW = 3 Hz (thin line).

enough, otherwise the envelope spectrum is displayed with possible amplitude underestimation (see sec. 9.2.13).

If the spectrum analyzer is triggered by the incoming pulse, so that we are operating in a synchronized mode, the measurement of the pulse spectrum is quicker and more controllable in terms of analysis settings. This kind of synchronization is called *time gating* or *time triggering*, because a time event, such as the arrival of the pulse of the RF burst, triggers and switches one or many internal blocks:

- when the video signal after the envelope detector is enabled (using a time delay, or pre-trigger, for the IF filter to overcome problems due to its transient response) we speak of "video gating" (this is possible in zero span mode, so locking onto a specific frequency point of the interval with a given RBW value);
- in very high-end spectrum analyzers (and signal analyzers) the local oscillator may be enabled following the incoming pulse, so that a short sweep can be made and many frequency points measured at a time (this mode is called "gated LO");
- the same can be done triggering a Fourier processing of the incoming pulse points rather than a sweep, and in this case we talk about "gated FFT".

The latter mode, however, is not so much different from using an oscilloscope triggered on the incoming pulse and acquiring the whole signal for later processing, applying e.g. the Fourier transform on-board the oscilloscope itself or on an external computer. However, as already noted, the use of a spectrum analyzer is beneficial for the signal-to-noise ratio thanks to the intrinsic narrow-band processing: the sampling of

	Rise time definition [μs]			
	t_{10-90}	t_{20-80}	$t_{max.sl.}$	$t_{-3\,\mathrm{dB}}$
1 MHz	0.5901 (0.0443)	0.3977 (0.0523)	0.8853 (0.2418)	0.318
3 MHz	0.277 (0.0105)	0.1808 (0.0120)	0.2680 (0.0132)	0.106
10 MHz	0.0834 (0.0102)	0.0542 (0.0062)	0.0875 (0.0306)	0.032

Table 9.6.1 – Measured values of rise time for the square pulse IF signal shown in Figure 9.6.1; $t_{-3\,\mathrm{dB}}$ results from RBW bandwidth and $-3\,\mathrm{dB}$ rise-time definition as $1/(\pi\mathrm{RBW})$. The displayed values are the mean value of three repeated tests with interpolated base time to improve resolution; the value between round brackets is the spread, obtained as max minus min value.

the IF output, however, might be externally implemented by connecting a sampling oscilloscope to the IF output connector, if present (see sec. 9.2.5). A band-limited triggering might be beneficial in some cases, when the incoming burst is disturbed by other signals and time-domain triggering is not easy, nor possible. Alas, several modern spectrum analyzers haven't this feature any longer.

Video gating is tested now against a purposely generated pulse signal (see Figure 9.6.1): it is evident that larger RBW values are faster in following the pulse edge, but result in a larger noise floor, thus with a negative impact on safe and repeatable triggering (the trigger threshold was set once for all at $-70\,\mathrm{dBm}$ resulting in some little delay between the various curve groups for different RBW values). Three tests were made for each RBW and the resulting curves are shown overlapped using the same color.

The rise time of the demodulated square pulse profiles for RBW = 1, 3 and 10 MHz is reported in Table 9.6.1 using different definitions of rise time (see sec. 3.2.1), including rise time that may be calculated directly from the IF filter $-3\,\mathrm{dB}$ bandwidth. To avoid biasing of rise time estimate coming from the limited time-domain resolution and quantization error of the time step, a 100:1 up-sampling was made, using Matlab interpolation function `interp1()` and the `'pchip'` option (cubic interpolation). The values reported in Table 9.6.1 are the resulting mean value and spread. Both 10-90% and 20-80% rise times suffer from the error in defining the pedestal from which the 10% or 20% shall be gauged: the pedestal is quite noisy and using the average value of the trace before the square pulse is detected is highly inaccurate. Using the mean plus two standard deviations is much better, but a manually set threshold is also a good solution: for fast rise times the introduced error is not relevant, while for the slowest ones there might be a difference as large as 20%.

The rise time that best suits the theoretical definition based on $-3\,\mathrm{dB}$ bandwidth and Gaussian assumption is the one defined with 20%-80% thresholds, used in general for noisy environments, when there is no clear detachment from the baseline (as in the present case). Regarding the rise time based on $-3\,\mathrm{dB}$ bandwidth, values obtained from the straight application of the $1/(\pi\mathrm{RBW})$ expression result in too a fast rise time and $0.35/\mathrm{RBW}$ is probably better.

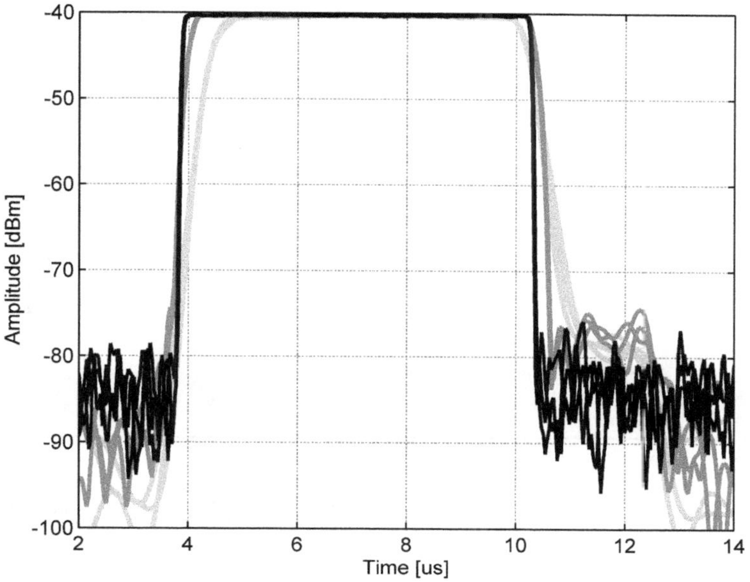

Figure 9.6.1 – Example of video gating and measurement of a pulse modulated signal: RBW = 10 MHz (black), RBW = 3 MHz (dark gray), RBW = 1 MHz (light gray).

It may be interesting to have a brief look at what is the effect of VBW on the demodulated video signal. We have seen that VBW helps in smoothing traces and might flatten the noise floor, thus improving triggering. The effect of RBW/VBW=10 for the maximum RBW of 10 MHz is shown in Figure 9.6.2: the digital video filter (see sec. 9.1.8 and 9.2.8) implementation is quite evident, having samples held constant with a first-order hold, squaring considerably the original IF stage signal. At this stage, before any smoothing operated by subsequent detection, the use of a video filter bandwidth VBW smaller than RBW is of no practical usefulness.

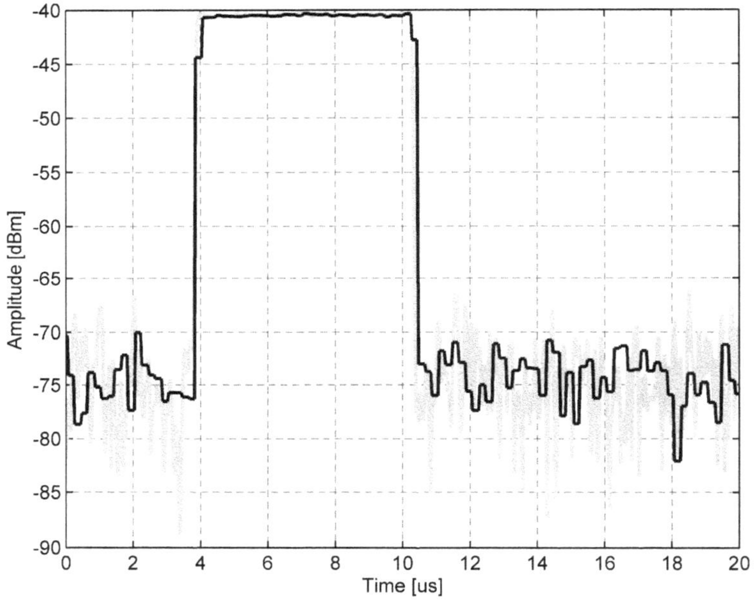

Figure 9.6.2 – Effect of VBW (black trace) on the video signal using a RBW/VBW=10: evident flattening, no clear relationship with the characteristics of the gray curve in the noisy part.

10

Scattering parameters

10.1 Fundamentals

Scattering (or S) parameters are a simple way of expressing typical RF and microwave quantities: they are defined at ports with respect to a reference impedance and can represent forward and backward terms, as well as express relationships such as port impedance and port-to-port transmission. Matrix network representations were already considered at the beginning in sec. 1.3.

The idea of using S parameters to represent networks and devices is nearly a hundred years old: it dates back to 1920[1] as indicated by Carlin [53], but dealing with a different topic. Probably the same Carlin's work may be considered the first complete work on scattering parameters applied to network theory and RF or microwave circuits, developing theory and main properties of S-parameter representation, including one-port and n-port networks. Many of the basic properties are well described in the work by Kurokawa [216], nearly ten years later: relationships with reflection coefficients, reciprocity, conservation of energy, lossless and lossy lines are all there. Then many textbooks may be found dealing with scattering parameters: Mavaddat [243] is very complete and clear.

---

[1] G.A. Campbell and R.M. Foster, "Maximum Output Networks for Telephone Substation and Repeater Circuits," *Transactions of the AIEE*, Vol. 39 (1920), pp. 231-280.

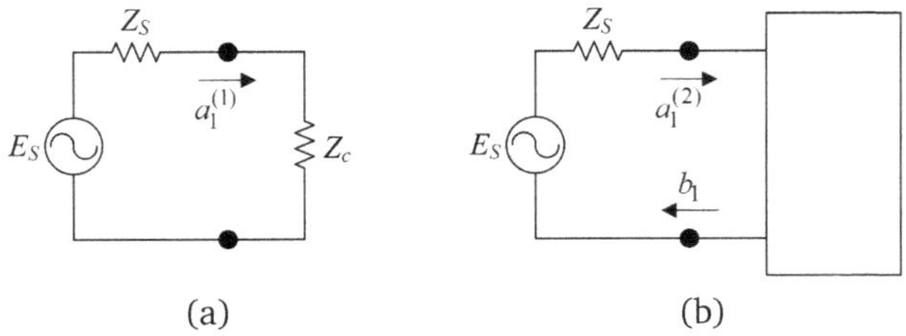

Figure 10.1.1 – Definition of S parameters of 1-port network: (a) defining the incident variables for a reference load, (b) adding the reflected variables by replacing the load with the actual network.

S parameters are common also to time domain, where they take a slightly different identification ("T" parameters, see sec. 10.4.4) and are therefore quite useful to exchange between VNA or TDR measured data, as well as to implement known corrections, such as de-embedding or moving reference planes. As measured or calculated quantities, they need to be characterized also for their uncertainty: this rather new topic has attracted attention only in the last ten years, turning out sometimes in unexpected complexity (see sec. 10.5).

S parameters are defined as the ratio of input and output waves, referred to the same or different ports: when considering the same port a "forward" and "backward" wave are directly related to reflection (or scattering) at the interface; S parameters referring to different ports describe a more complex mechanism of multiple reflections and transmissions at the various interfaces, as well as propagation, and may be seen as "attenuation", or "transmission", as they are usually called, and in general as a transfer function. S parameters are complex numbers, expressing an amplitude ratio and a phase difference versus frequency; for this reason their amplitude varies from 0 to 1 and is more comfortably expressed in dB.

10.1.1 Scattering parameters of a 1-port network

We consider a voltage source E_S with arbitrary internal impedance Z_S feeding the network port; this circuit is split into two circuits whose superposition allows the definition of the forward and reflected wave terms. The voltage source is first loaded by the reference impedance R (see Figure 10.1.1(a)). When the voltage source is connected to a load, the voltage at its terminals and the sourced current are defined as the total voltage V_1 and current I_1 at the port 1.

The incident voltage and current are calculated for the case (a), when the generator is loaded by the reference impedance:

$$I_1^i = \frac{E}{2R} = \frac{V_1 + RI_1}{2R} \qquad V_1^i = \frac{E}{2} = \frac{V_1 + RI_1}{2} \tag{10.1.1}$$

Next we replace the reference load with the actual network connected through its port 1 (see Figure 10.1.1(b)): there will be a reflection and we denote with V_1^r and I_1^r the reflected voltage and current waves. By definition reflected quantities are related to incident and total ones by

$$I_1^r = I_1^i - I_1 \qquad V_1^r = V_1 - V_1^i \tag{10.1.2}$$

and recalling the equations above

$$I_1^r = \frac{V_1 + RI_1}{2R} - I_1 = \frac{V_1 - RI_1}{2R} \qquad V_1^r = V_1 - \frac{V_1 + RI_1}{2} = \frac{V_1 - RI_1}{2} \tag{10.1.3}$$

The scattering parameters are defined as the ratio of the reflected wave to the incident wave, be it a voltage or a current. With the particular choice of the reference impedance equal to the driving impedance of the voltage source, the two voltage and current S parameters happen to be equal.

$$S_{11}^V = S_{11}^I = \frac{V_1 - RI_1}{V_1 + RI_1} \tag{10.1.4}$$

In general the definition of S parameters goes through the definition of two quantities, called "incident wave amplitude" and "reflected wave amplitude", which share the value of the reference impedance by taking each the square root of it:

$$a_1 = \sqrt{R}\, I_1^i = \frac{1}{\sqrt{R}}\, V_1^i = \frac{V_1 + RI_1}{2\sqrt{R}} = \frac{E}{2\sqrt{R}} \tag{10.1.5}$$

$$b_1 = \sqrt{R}\, I_1^r = \frac{1}{\sqrt{R}}\, V_1^r = \frac{V_1 - RI_1}{2\sqrt{R}} \tag{10.1.6}$$

so that the S parameter for port 1 is readily

$$S_{11} = \frac{b_1}{a_1} \tag{10.1.7}$$

When we remove the assumption that the generator impedance is equal to the reference impedance R, the formalism is still valid, provided that we take into account the impedance mismatch and the new reflection at the source; this reflection will be caused by the reflected waves coming back from the network port and going towards the voltage source, that has a mismatched impedance and thus is characterized by a non-null reflection coefficient Γ_S. Thus, we may consider that the incident wave amplitude a_1 is composed of the former main incident term a_1' sourced by the generator and an additional term a_1'', related to a reflected wave back towards the voltage source.

$$a_1' = \sqrt{R}\, \frac{E}{Z_S + R} = \frac{E}{2\sqrt{R}}\,(1 - \Gamma_S) \qquad \Gamma_S = \frac{Z_S - R}{Z_S + R} \tag{10.1.8}$$

Considering the reflected wave b_1 coming back to the voltage source interface, by applying the reflection coefficient above, we obtain readily

$$a_1'' = \Gamma_S\, b_1 \tag{10.1.9}$$

Thus the total incident wave amplitude $a_1$is given by

$$a_1 = a_1' + a_1'' = \frac{E}{2\sqrt{R}}(1-\Gamma_S) + \Gamma_S S_{11} a_1 \tag{10.1.10}$$

where S_{11} appears, replacing b_1/a_1. Solving for a_1 gives

$$a_1 = \frac{E}{2\sqrt{R}} \frac{1-\Gamma_S}{1-\Gamma_S S_{11}} \tag{10.1.11}$$

It is interesting to observe that going back to the situation of a matched voltage source, that is $\Gamma_S = 0$, we obtain again the result of (10.1.5).

The same result might be attained considering the infinite series of incident and reflected waves deriving from the multiple reflection model; summation for an infinite number of terms (when reaching steady state) converges to (10.1.11).

To conclude, if we separate the contribution of the reflection at the source by defining a reflected wave amplitude b_S, we have:

$$b_S = \frac{E}{2\sqrt{R}}(1-\Gamma_S) \tag{10.1.12}$$

$$a_1 = \frac{b_S}{1-\Gamma_S S_{11}} \qquad b_1 = S_{11} a_1 \tag{10.1.13}$$

The concept of passivity normally related to positively valued resistance (i.e. positive real part of impedance) is translated into the requisite that the input reflection coefficient modulus S_{11} be ≤ 1. So, the limit condition for passivity is in the former case the vertical line that divides the left from the right half plane, and for scattering parameters and reflection coefficients it is a circle of unitary radius.

We may consider as well the power transfer from the generator to the network through the connecting port; for the source we define the available power as $P_S = e_S^2$, where

$$e_S = \frac{E}{2\sqrt{R}} \tag{10.1.14}$$

It is possible in general to define incident and reflected power terms as the squared moduli of the a_1 and b_1 terms and the input power at port 1 is then

$$P_{1,in} = a_1 a_1^* - b_1 b_1^* \tag{10.1.15}$$

Assuming that the source impedance is the reference impedance R, we obtain easily

$$P_{1,in} = \frac{1}{2}(I_1 V_1^* + I_1^* V_1) \tag{10.1.16}$$

that is right the power delivered into the network (the relationship is valid for any source impedance).

It is possible also to express the input power in terms of the scattering parameters and reflection coefficients by replacing the "b" terms in eq. (10.1.15) above:

$$P_{1,in} = a_1 a_1^* \left(1-|S_{11}|^2\right) = |b_S|^2 \frac{1-|S_{11}|^2}{|1-\Gamma_S S_{11}|^2} \tag{10.1.17}$$

and referring to the available power from the source e_S, it is possible to obtain

$$P_{1,in} = e_S^2 \frac{\left(1-|S_{11}|^2\right)\left(1-|\Gamma_S|^2\right)}{|1-\Gamma_S S_{11}|^2} \tag{10.1.18}$$

When the input impedance and the source impedance are equal, the delivered power corresponds to the available source power (maximum power transfer).

The generalized S parameters are not treated in detail, defining conversion relationships later on; more detailed presentation and examples may be found in [243].

10.1.2 Scattering parameters of a 2-port network

In this case there are two ports in the network under study and we define V_1, I_1 and V_2, I_2 as the total voltage and current at the two ports; they may be grouped into two column vectors **I** and **V**. Each port is considered connected to a test voltage source with internal impedance equal to the network reference impedance R.

By repeating the previous operations for 1-port network, we connect a reference load R to each voltage source, thus defining the incident wave quantities, that may be assembled in vectors $\mathbf{V}^{\mathrm{i}}$and $\mathbf{I}^{\mathrm{i}}$ of 2 components, for port 1 and 2. The two vectors are related by a diagonal matrix of R values, since the two loads are decoupled and independent.

$$\mathbf{V}^{\mathrm{i}} = \begin{bmatrix} R & 0 \\ 0 & R \end{bmatrix} \mathbf{I}^{\mathrm{i}} \tag{10.1.19}$$

The incident currents and voltages are defined as the current through and the voltage across the two ports when the 2-port network is not connected and each generator is loaded by its reference impedance R. Hence

$$\mathbf{I}^{\mathrm{i}} = \begin{bmatrix} \dfrac{E_1}{2R} \\[2ex] \dfrac{E_2}{2R} \end{bmatrix} = \begin{bmatrix} \dfrac{V_1+RI_1}{2R} \\[2ex] \dfrac{V_2+RI_2}{2R} \end{bmatrix} \qquad \mathbf{V}^{\mathrm{i}} = \begin{bmatrix} \dfrac{E_1}{2} \\[2ex] \dfrac{E_2}{2} \end{bmatrix} = \begin{bmatrix} \dfrac{V_1+RI_1}{2} \\[2ex] \dfrac{V_2+RI_2}{2} \end{bmatrix} \tag{10.1.20}$$

Replacing the two test loads with the 2-port network, we define also the reflected variables as the difference between the defined incident current and voltage and the total current and voltage, respectively:

$$\mathbf{I}^{\mathrm{r}} = \boldsymbol{I}^{i} - \mathbf{I} \qquad \mathbf{V}^{\mathrm{r}} = \boldsymbol{V}^{i} - \mathbf{V} \tag{10.1.21}$$

$$\mathbf{I}^{\mathbf{r}} = \begin{bmatrix} \dfrac{E_1}{2R} - I_1 \\ \\ \dfrac{E_2}{2R} - I_2 \end{bmatrix} = \begin{bmatrix} \dfrac{V_1 - RI_1}{2R} \\ \\ \dfrac{V_2 - RI_2}{2R} \end{bmatrix} \qquad \mathbf{V}^{\mathbf{r}} = \begin{bmatrix} V_1 - \dfrac{E_1}{2} \\ \\ V_2 - \dfrac{E_2}{2} \end{bmatrix} = \begin{bmatrix} \dfrac{V_1 - RI_1}{2} \\ \\ \dfrac{V_2 - RI_2}{2} \end{bmatrix} \tag{10.1.22}$$

The relation between reflected and incident waves may be written as

$$\mathbf{I}^{\mathbf{r}} = \mathbf{S}^{\mathbf{I}}\mathbf{I}^{\mathbf{i}} \qquad \mathbf{S}^{\mathbf{I}} = \begin{bmatrix} S^I_{11} & S^I_{12} \\ S^I_{21} & S^I_{22} \end{bmatrix} \tag{10.1.23}$$

$$\mathbf{V}^{\mathbf{r}} = \mathbf{S}^{\mathbf{V}}\mathbf{V}^{\mathbf{i}} \qquad \mathbf{S}^{\mathbf{V}} = \begin{bmatrix} S^V_{11} & S^V_{12} \\ S^V_{21} & S^V_{22} \end{bmatrix} \tag{10.1.24}$$

The so-defined scattering parameters remain unchanged, provided that each port has its own reference impedance and its variables are defined upon that (the two reference impedances are taken equal).

Going from incident and reflected currents/voltages to a corresponding concept for wave quantities, the notation and the number of variables necessary to describe the network is simplified. Currents and voltages differ for R appearing at the denominator; splitting it into two multiplicative $\sqrt{R}$ terms split among current and voltage, the latter become equal, so that unique wave quantities may be defined: they are a for incident terms and b for reflected terms, and both are vectors, one row for each port:

$$\mathbf{a} = \begin{bmatrix} \dfrac{V_1 + RI_1}{2\sqrt{R}} \\ \\ \dfrac{V_2 + RI_2}{2\sqrt{R}} \end{bmatrix} \qquad \mathbf{b} = \begin{bmatrix} \dfrac{V_1 - RI_1}{2\sqrt{R}} \\ \\ \dfrac{V_2 - RI_2}{2\sqrt{R}} \end{bmatrix} \tag{10.1.25}$$

$$\boldsymbol{a} = \boldsymbol{S}\,\boldsymbol{b} \qquad \mathbf{S} = \begin{bmatrix} S_{11} & S_{12} \\ S_{21} & S_{22} \end{bmatrix} \tag{10.1.26}$$

$$\mathbf{S} = \mathbf{S}^{\mathbf{I}} = \mathbf{S}^{\mathbf{V}} \tag{10.1.27}$$

Taking a simple two-port network example, the most relevant parameters are:

- S_{11} identifies the reflection coefficient at the input port 1, and is commonly referred to as "return loss", for it indicates the amount of voltage that is returned reflected at the incident port and is not transmitted through to the other port;

- S_{21} identifies what is not transmitted to the output port 2, and is commonly referred to as "insertion loss", for it indicates the amount of signal that is lost in traveling from port 1 to port 2 (even if what is really indicated by S_{21} is the amount of signal that is transmitted from port 1 to port 2).

The two parameters, S_{11} and S_{21}, are not independent on each other; by invoking the conservation of energy we may write $S_{11}^2 + S_{21}^2 = 1$. The conservation of energy can be invoked if no significant coupling to adjacent traces and radiation losses occur. A coaxial cable is a type of interconnect that meet these requisites over a wide frequency range, while for PCB traces this is not absolutely true due to radiation and coupling to nearby lines especially for planar transmission lines of the open type (such as microstrips). When radiation losses and coupling to external circuits is relevant deviations from the expected normality may be seen.

As complex numbers, S parameters can be displayed in separate amplitude and phase plots or in polar plot, as shown below in sec. 10.2.

Reflection and transmission coefficients depend on the impedance matching between source (or load) and transmission line. From this the dependency of S parameters from a reference impedance value, that indicates the source/load impedance. Once S parameters are known with a given impedance reference value, any other impedance value may be selected and the S parameters changed accordingly, by using a matrix transformation. Of course if we are considering measured values, measurement errors may slightly vary due to scale change, thus affecting measurement accuracy (see sec. 11.11); in other words measuring with $50\,\Omega$ or $100\,\Omega$ reference impedance will not lead to the same results in terms of uncertainty.

In case of an arbitrary loading impedance Z_L at the output port 2 of the network

$$S'_{11} = S_{11} + \frac{S_{12}S_{21}\Gamma_L}{1 - S_{22}\Gamma_L} \qquad (10.1.28)$$

If the impedance mismatching is at the input port 1 with source impedance Z_S, then

$$S'_{22} = S_{22} + \frac{S_{12}S_{21}\Gamma_S}{1 - S_{11}\Gamma_S} \qquad (10.1.29)$$

S parameters thus need to be expressed with an explicit reference to the reference impedance (see sec. 10.1.5 below).

10.1.3 Lossless and lossy 2-port networks

For the representation of Figure 10.1.1, where the a quantities are directed to the network ports into the network, while the b quantities are aiming outward, it is possible to establish a relationship between these quantities, extended to the S-parameter matrix, that characterizes lossless and lossy networks.

A lossless network does not dissipate any power internally and the power entering the network is equal to the power exiting it, that is different from the concept of power at port 1 and port 2. For the chosen notation the power entering the network is made of a quantities and conversely the power leaving the network is made of b quantities:

$$|a_1|^2 + |a_2|^2 = |b_1|^2 + |b_2|^2 \qquad (10.1.30)$$

and the resulting S matrix is unitary, that means the even if complex, the norm is unity:

$$S^*S - I = 0 \tag{10.1.31}$$

where the asterisk identifies the complex conjugate operator and I is the identity (or unity) matrix. More simply this relationship for two-port networks translates into

$$S_{11}^2 + S_{21}^2 = 1 \tag{10.1.32}$$

that says that when S_{11} increases towards unity, S_{21} shall correspondingly reduce; vice-versa, good transmission properties with a large S_{21} are accompanied by a low reflection, that is a good input matching (low S_{11}). If the network is lossy, then the amount of dissipated power inside it is given by the difference between the power entering it and the power leaving it, so that the sum of squared b quantities shall be less than the sum of squared a quantities and the S matrix will have a smaller norm.

$$|a_1|^2 + |a_2|^2 > |b_1|^2 + |b_2|^2 \tag{10.1.33}$$

The eigenvalues of the S matrix are thus in the left half plane, the network is a passive stable network and its transient response is characterized by decaying exponentials, with or without a superimposed oscillatory behavior.

10.1.4 N-port structures

Port assignment has some degrees of freedom when treating multiple transmission lines. When considering interconnects with a clear division of ports between two sides, it is advisable to use odd numbering for the input side and even numbering for the output side. Given an input port i and an output port j, the scattering S parameter for the wave coming into i and coming out from j is S_{ji}. This notation is probably less intuitive than S_{ij}, but lends itself easily to some manipulations. If we consider a stimulus voltage vector u_i applied as a vector of input waves, the response voltage vector y_j is given by $y_j = S_{ji}u_i$. Likewise a voltage wave entering the port j and exiting from the port i gives rise to the S parameter S_{ij}.

The number of S parameters that compose the S matrix is equal to n^2, with n the number of ports. For a linear interconnect the S matrix is symmetric and the off-diagonal terms in the uppermost and lowermost triangle are equal; together with the diagonal terms we have $n(n+1)/2$ unique terms.

10.1.5 Conversion to another impedance base

It is not so uncommon to wonder if for the measurement of e.g. a $75\,\Omega$ device, the VNA shall have the same port impedance, and, for a $50\,\Omega$ measuring system, how measured values and derived S parameters can be translated to a different base.

Since scattering parameters are defined with reference to a specific impedance base (a resistive base), an absolute system is needed to pass from one basis to the other, and this system is for example that of impedance parameters, which are uniquely defined.

We write the impedance parameters in terms of the scattering parameters (as already seen in sec. 1.3.1.6 and eq. (1.3.29)):

$$Z_{11} = Z_{c1}\frac{1-S_{11}+S_{22}-\Delta_s}{1-S_{11}-S_{22}+\Delta_s} \tag{10.1.34}$$

$$Z_{12} = 2\sqrt{Z_{c1}Z_{c1}}\,\frac{S_{12}}{1-S_{11}-S_{22}+\Delta_s} \tag{10.1.35}$$

$$Z_{12} = 2\sqrt{Z_{c1}Z_{c1}}\,\frac{S_{21}}{1-S_{11}-S_{22}+\Delta_s} \tag{10.1.36}$$

$$Z_{22} = Z_{c2}\frac{1+S_{11}-S_{22}-\Delta_s}{1-S_{11}-S_{22}+\Delta_s} \tag{10.1.37}$$

Having defined two quantities that express the two reference impedances Z'_{c1} and Z'_{c2} with respect to the previous reference impedances Z_{c1} and Z_{c2}

$$\gamma_1 = \frac{Z'_{c1}-Z_{c1}}{Z'_{c1}+Z_{c1}} \qquad \gamma_2 = \frac{Z'_{c2}-Z_{c2}}{Z'_{c2}+Z_{c2}} \tag{10.1.38}$$

we can define also the new scattering parameters in terms of the old ones. We need to express the new scattering parameters in terms of impedances and then substitute for the various terms[2]; finally we obtain:

$$S'_{11} = \frac{\Lambda_1^*}{\Lambda_1}\,\frac{(1-\gamma_2 S_{22})(S_{11}-\gamma_1^*)+\gamma_2 S_{12}S_{21}}{(1-\gamma_2 S_{22})(1-\gamma_1 S_{11})-\gamma_1\gamma_2 S_{11}S_{22}} \tag{10.1.39}$$

$$S'_{12} = \Lambda_1^*\Lambda_2^*\,\frac{S_{12}}{(1-\gamma_2 S_{22})(1-\gamma_1 S_{11})-\gamma_1\gamma_2 S_{11}S_{22}} \tag{10.1.40}$$

$$S'_{21} = \Lambda_1^*\Lambda_2^*\,\frac{S_{21}}{(1-\gamma_2 S_{22})(1-\gamma_1 S_{11})-\gamma_1\gamma_2 S_{11}S_{22}} \tag{10.1.41}$$

$$S'_{22} = \frac{\Lambda_2^*}{\Lambda_2}\,\frac{(1-\gamma_1 S_{11})(S_{22}-\gamma_2^*)+\gamma_1 S_{12}S_{21}}{(1-\gamma_2 S_{22})(1-\gamma_1 S_{11})-\gamma_1\gamma_2 S_{11}S_{22}} \tag{10.1.42}$$

where Λ terms are defined as follows

$$\Lambda_1 = (1-\gamma_1^*)\sqrt{\frac{1-\gamma_1\gamma_1^*}{(1-\gamma_1)(1-\gamma_1^*)}} \qquad \Lambda_2 = (1-\gamma_2^*)\sqrt{\frac{1-\gamma_2\gamma_2^*}{(1-\gamma_2)(1-\gamma_2^*)}} \tag{10.1.43}$$

and the asterisk indicates the complex conjugate.

[2] All substitutions and partial expressions can be found in Mavaddat [243], sec. 6.3.

Example 10.1: Transformation of 50 Ω measurements to 75 Ω

It is in principle possible to perform VNA measurements using its own 50 Ω port reference even when testing a 75 Ω device: we define the necessary transformation.

$$\gamma_1 = \gamma_2 = \frac{75-50}{75+50} = 0.2$$

$$\Lambda_1 = \Lambda_2 = (1-0.2)\sqrt{\frac{1-0.04}{(1-0.2)(1-0.2)}} = 0.9798$$

They are real numbers so equal to their complex conjugates. The resulting S parameters are:

$$S'_{11} = \frac{(1-0.2S_{22})(S_{11}-0.2)+0.2S_{12}S_{21}}{(1-0.2S_{22})(1-0.2S_{11})-0.04S_{11}S_{22}}$$

$$S'_{12} = 0.96\frac{S_{12}}{(1-0.2S_{22})(1-0.2S_{11})-0.04S_{11}S_{22}}$$

$$S'_{21} = 0.96\frac{S_{21}}{(1-0.2S_{22})(1-0.2S_{11})-0.04S_{11}S_{22}}$$

$$S'_{22} = \frac{(1-0.2S_{11})(S_{22}-0.2)+0.2S_{12}S_{21}}{(1-0.2S_{22})(1-0.2S_{11})-0.04S_{11}S_{22}}$$

with the denominator simplifying for all expressions to

$$(1-0.2S_{22})(1-0.2S_{11})-0.04S_{11}S_{22} = 1-0.2S_{11}-0.2S_{22}.$$

10.1.6 Signal flow graphs

When considering how to represent a network for RF and microwave applications, signal flow graph are particularly useful and effective. However, they are simply a way to put in a graphical form a set of linear equations, relating the a and b quantities.

It is possible to lay down some rules related to particular arrangements of nodes and branches, that are the constitutive elements of signal flow graphs:

- each port corresponds to two nodes, one for the wave entering it, and thus associated to a terms, and one for the wave reflected from the port, and thus associated to b terms;
- branches connecting nodes are oriented and have a multiplicative parameter, such as gain or attenuation, including scattering parameters;
- each node results in the sum of the insisting branches, each weighted positively or negatively depending on the direction of the arrow to or from the node;
- when two branches are in parallel, so between the same pair of nodes, the two terms may be summed, but only if the direction of the arrow is the same; if they are not it is not possible to use any rule to change the direction of the arrow and then combine them altogether;

- when two branches are cascaded, so that one enters a node and the other one leaves the same node, then the result is given by a multiplication.

Some examples follow; comprehensive references are Collin [88], Pozar [273] and Kuhn[3] among others. An interesting and very clear reading are the handouts by J. Stiles at the University of Kansas, Dept. EECS, sec. 4-5.

Considering a two-port network (see Figure 10.1.2), the associated flow graph consists of four nodes, two for port 1 and two for port 2, for the incident and reflected waves; the S parameters are distributed among four branches.

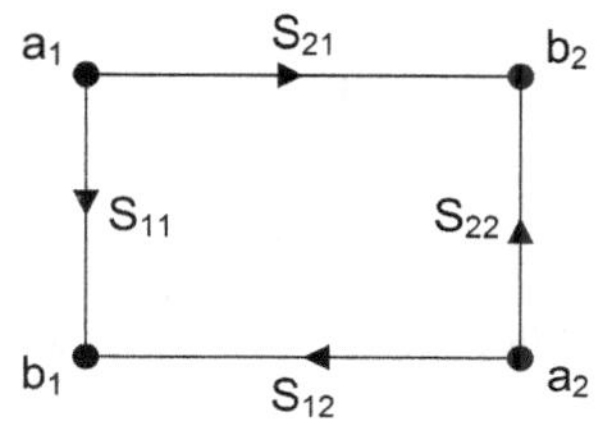

Figure 10.1.2 – Signal flow graph representation of a two-port network.

The graph can be interpreted quickly, observing for example that a signal entering as an incident wave at node a_1 (that gives the amplitude of the signal) is reflected back through node b_1 flowing through the branch identified with parameter S_{11} (reflected term) and is transmitted to node b_2 flowing through the branch identified with parameter S_{21} (transmitted term). Why the signal is transmitted to node b_2 instead of node a_2? It looks like the incident wave once transmitted through the network is transformed into a reflected wave! Yes, it is, if we consider that the positive direction of incidence is from outside the network onto port 2 and thus a wave coming from inside the network has the same direction of a reflected wave at the port interface.

Then a final comment on the direction of the branches: the equations are derived for the b quantities that correspond to nodes with incoming branches (look at the direction of the arrows), so we may write $b_1 = S_{11}a_1 + S_{12}a_2$ and $b_2 = S_{21}a_1 + S_{22}a_2$, but we do not write equations for a quantities that see the back of the arrows leaving their nodes.

A simple load that terminates a transmission line is shown in Figure 10.1.3. The incident wave on node a is reflected back by effect of the reflection coefficient Γ.

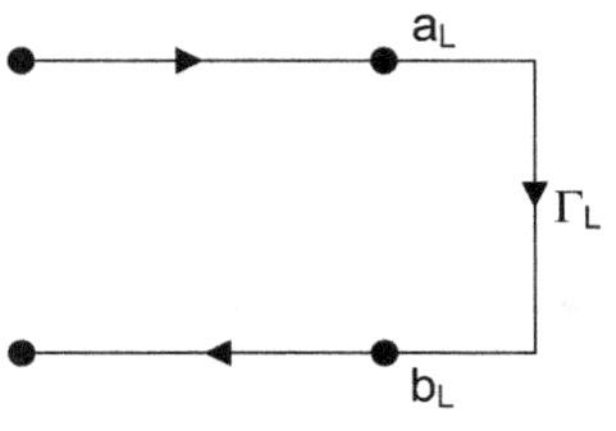

Figure 10.1.3 – Signal flow graph representation of a terminating load.

[3] N. Kuhn, "Simplified signal flow graph analysis," *Microwave Journal*, Vol. 6, no. 11, Nov. 1963, pp. 59.

A voltage source E_S with its own internal impedance Z_S is connected to a transmission line (see Figure 10.1.4): part of the signal is transmitted into the transmission line as forward (or incident) wave and goes thus to node b, the remaining part going to node a is the one reflected back to the source by Γ_S. Again we see that the use of node a and b look like they have been mistakenly reversed: yet, the direction that rules is the one looking from the transmission line into the port, so the wave transmitted to the transmission line is like it is reflected by the network and goes to node b.

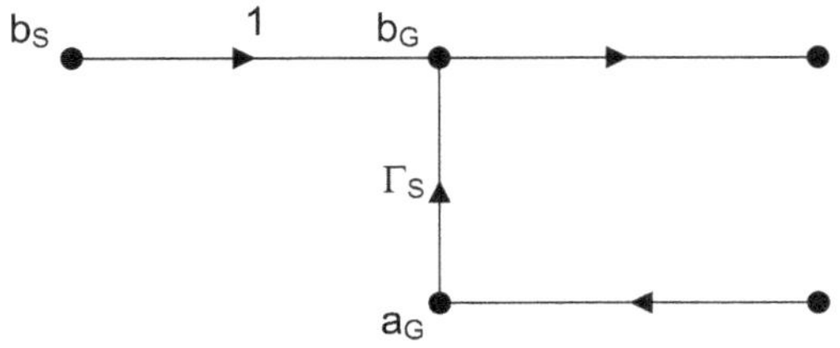

Figure 10.1.4 – Signal flow graph representation of a voltage source.

A self-loop configuration occurs (see Figure 10.1.5) when a branch begins at and ends on the same node: a configuration consisting of single branch in self-loop arrangement may look trivial, but that branch may be the result of a simplification operated for series and parallel connected branches. When a node has such a self-loop with a coefficient γ, then the graph is simplified eliminating the self-loop and multiplying all the other branches insisting on the same node by $1/(1-\gamma)$.

Self-loop is quite a curious arrangement that might seem so uncommon not to be useful, but whenever an equation has the same variable at the left and at the right of equal sign, then a self-loop is there in the logic of signal flow graphs. For example

$$b_1 = \alpha a_1 + \beta a_2 + \gamma b_1$$

after simplification leads to

$$b_1 = \frac{\alpha}{1-\gamma} a_1 + \frac{\beta}{1-\gamma} a_2$$

that confirms the self-loop rule with γ appearing on the self-looped branch.

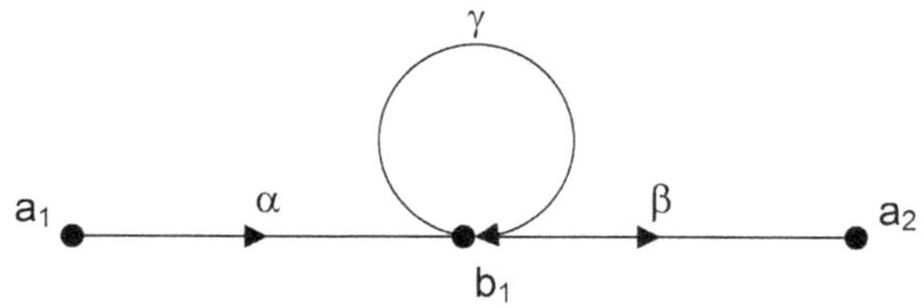

Figure 10.1.5 – Signal flow graph representation of a self-loop.

Finally, a more complete example is that of a two-port network fed at port 1 by a voltage source with its own internal impedance and loaded at port 2 by a load impedance

Z_L. The signal flow graph is shown in Figure 10.1.6. The intention is to derive a known expression for the input reflection coefficient Γ_{in}, by using signal flow graph manipulation rules. The result consists of the S_{11} parameter adding the effect of load mismatch:

$$\Gamma_{in} = S_{11} + \frac{S_{12}S_{21}\Gamma_L}{1 - S_{22}\Gamma_L}$$

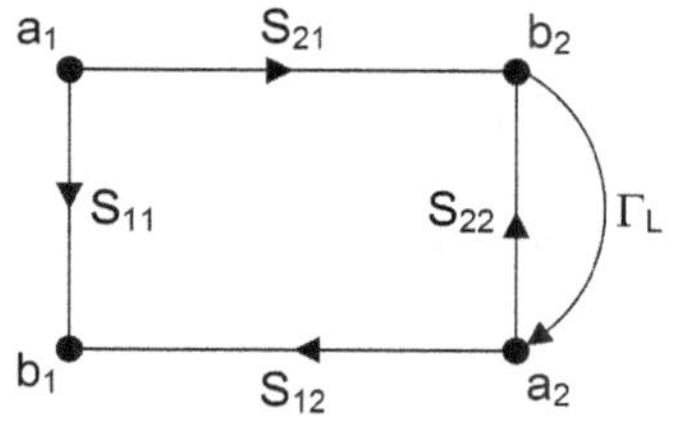

Figure 10.1.6 – Signal flow graph representation of a complete two-port network including a voltage source and a load.

We may want to complicate things a little and examine a real case, where load connection is not immediate and it is through a piece of transmission line: so we have a source with its adapter (with its own scattering matrix S' of subscripts 1 and 2), a cable of length L and propagation constant γ, a second adapter (with a similar matrix S'' of subscripts 3 and 4) and the load. The representation is no more complicated than making a cascade connection of blocks, as shown in Figure 10.1.7.

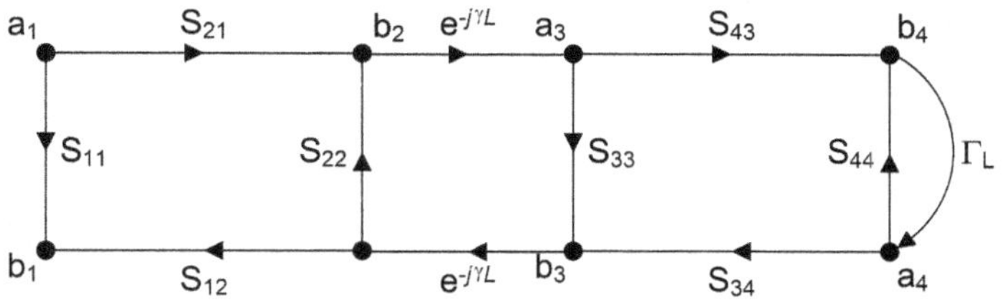

Figure 10.1.7 – Signal flow graph representation of a complete two-port network including a voltage source and a load connected by a cable.

10.1.6.1 Mason's rule (or non-touching loop rule)

A topological procedure to solve a network may be derived, able to find expressions of all nodes, identifying the so-called first-order loops, defined as closed path from a node back to the node itself, moving along branches and other nodes following the direction of the arrows. In our example it is possible to recognize three first order loops: from a_1 through b_1, giving $S_{11}\Gamma_S$, from a_2 through b_2, giving $S_{22}\Gamma_L$, and from a_1 through a_2, giving $S_{21}\Gamma_L S_{12}\Gamma_S$. Based on the definition of the first-order loop, a second-order loop is made of pairs of non-touching first-order loops, that in our example are only the first two, so that $S_{11}\Gamma_S S_{22}\Gamma_L$; if feedback paths are present in the network, then there are many more second-order loops.

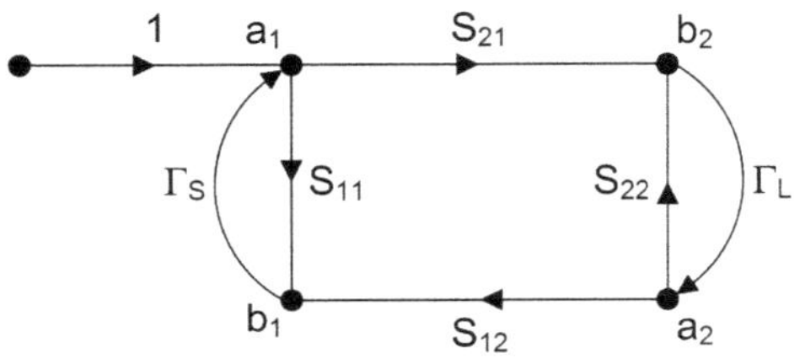

Figure 10.1.8 – Network with nodes and branches for exemplification of Mason's rule.

To determine the ratio between a dependent and an independent variable (the transfer function between the two), it is needed first to determine the paths P_i joining the two variables, and then for each of them to determine the non-touching loops of the various orders k, $L_k^{(i)}$. Then the transfer function is given by a product-sum combination of paths and non-touching loops as

$$T = \frac{\sum_i P_i \left(1 - \sum L_1^{(i)} + \sum L_2^{(i)} - \sum L_3^{(i)} + \ldots\right)}{\left(1 - \sum L_1 + \sum L_2 - \sum L_3 + \ldots\right)} \tag{10.1.44}$$

The summation terms refer for P_i to all the paths joining the two variables of interest with i ranging between 1 and the number of paths. For loops L we must distinguish between non-touching loops at numerator, indicated by (i), for the i-th path the is not to be touched, and generic loops (both touching and non-touching) at denominator.

Example 10.2: Transfer function calculation by Mason's rule (Figure 10.1.8)

The variable b_S is independent on all the others and from it we may determine the variables of the network. If we are interested in b_1 (i.e. the reflected wave when the source is connected), we first have to determine all the paths connecting b_1 to b_S: 1) S_{11}, as expected by definition, and 2) $S_{21}\Gamma_L S_{12}$. Then we determine the non-touching loops for the identified paths: for the path 1 S_{11}, we see that the loop $S_{22}\Gamma_L$ is a non-touching loop because they share no nodes; for the path 2 $S_{21}\Gamma_L S_{12}$, that touches all the network nodes, it is not possible to find any non-touching loop. So, the resulting transfer function is S_{11} multiplied by 1 minus all the first order non-touching loops sum together, in our case only $S_{22}\Gamma_L$, plus the other path multiplied only by 1 since there are no non-touching loops, and at the denominator we put together all the three first order loops and the second order loop, so:

$$\frac{b_1}{b_S} = \frac{S_{11}(1 - S_{22}\Gamma_L) + S_{21}\Gamma_L S_{12}}{1 - (S_{11}\Gamma_S + S_{22}\Gamma_L + S_{21}\Gamma_L S_{12}\Gamma_S) + (S_{11}\Gamma_S S_{22}\Gamma_L)}$$

10.1.7 Relationships with other parameters

It is known that for a linear network several representations of its input-output transfer functions are equivalent. The transformation relationships between the most common two-port representations have already been reviewed in sec. 1.3, where a comprehensive list of matrix relations is given in particular in sec. 1.3.1.6. It is only briefly underlined that the relationships reported in [155] are not complete, while they are reported at a great extent in [99, 115, 131], fully reported and checked in sec. 1.3.1.6. The S parameters are now considered in relationship to a few significant cases of less common parameters, used for amplifier characterization.

10.1.7.1 Transmission line

Most of the elements that are part of RF and microwave setups are transmission lines or may be modeled as such; the relationship is briefly reviewed between transmission line quantities (namely characteristic impedance Z_c and propagation constant γ) and the S parameters measurable at line ports. The transmission line is symmetric and so only S_{11} and S_{22} are calculated:

$$S_{11}=S_{22}=\frac{\left[\left(\frac{Z_c}{Z_r}\right)^2-1\right]\sinh(\gamma L)}{2\left(\frac{Z_c}{Z_r}\right)\cosh(\gamma L)+\left[\left(\frac{Z_c}{Z_r}\right)^2+1\right]\sinh(\gamma L)} \tag{10.1.45}$$

$$S_{21}=S_{12}=\frac{2\left(\frac{Z_c}{Z_r}\right)}{2\left(\frac{Z_c}{Z_r}\right)\cosh(\gamma L)+\left[\left(\frac{Z_c}{Z_r}\right)^2+1\right]\sinh(\gamma L)} \tag{10.1.46}$$

where L is the transmission line length and Z_r is the reference impedance of measured S parameters, so the calibration impedance of VNA (normally assumed equal to the VNA port impedance itself, but not necessarily so).

These relationships can be inverted, so that the transmission line quantities may be determined experimentally using measured S parameters. In general Z_c and γ may be determined by curve fitting of measured S parameters, using e.g. a least mean square approach, provided that the variability with frequency of per-unit-length parameters was adequately modeled: the larger the frequency range the more complex the task (see sec. 6.5.1.3).

10.1.7.2 Amplifier characterization

The forward current gain (hybrid parameter h_{21}) is of course in direct relationship with the insertion loss, using however a minus sign because of the different sign convention for the output current:

$$h_{21}=\frac{-S_{21}}{(1-S_{11})(1+S_{22})+S_{12}S_{21}} \tag{10.1.47}$$

Once the stability factor K is defined and derived from S parameters (under the assumption of a maximum transducer gain with impedance matching at input and output), also the *maximum stable gain* (MSG) and *maximum available gain* (MAG) are obtained:

$$K = \frac{1 + |S_{11}S_{22} - S_{12}S_{21}|^2 - |S_{11}|^2 - |S_{22}|^2}{2\,|S_{12}|\,|S_{21}|} \tag{10.1.48}$$

$$\mathrm{MSG} = \frac{|S_{21}|}{|S_{12}|} \qquad \text{if } K < 1 \tag{10.1.49}$$

$$\mathrm{MAG} = \frac{|S_{21}|}{|S_{12}|}\left(K - \sqrt{K^2 - 1}\right) \qquad \text{if } K > 1 \tag{10.1.50}$$

Three gain terms are useful when considering amplification:

$$G_0 = |S_{21}|^2 \qquad G_1 = \frac{1 - |\Gamma_S|^2}{|1 - S_{11}\Gamma_S|^2} \qquad G_2 = \frac{1 - |\Gamma_L|^2}{|1 - S_{22}\Gamma_L|^2} \tag{10.1.51}$$

The first one, G_0, is also called unmatched power gain and represents the minimum gain of the transistor when unmatched at both input and output. The other two parameters take into account the mismatching at the source and at the load.

So that the *unilateral transducer power gain* G_u is given by

$$G_u = |S_{21}|^2 \frac{1 - |\Gamma_S|^2}{|1 - S_{11}\Gamma_S|^2} \frac{1 - |\Gamma_L|^2}{|1 - S_{22}\Gamma_L|^2} = G_0 G_1 G_2 \tag{10.1.52}$$

Taking the maximum of G_1 and G_2 with respect to the reflection coefficients Γ_S and Γ_L that depend on the arbitrary terminating impedance

$$G_{1,\max} = \frac{1}{1 - |S_{11}|^2} \qquad G_{2,\max} = \frac{1}{1 - |S_{22}|^2} \tag{10.1.53}$$

the *maximum unilateral transducer power gain* $G_{u,\max}$ is obtained

$$G_{u,\max} = \frac{|S_{21}|^2}{\left(1 - |S_{11}|^2\right)\left(1 - |S_{22}|^2\right)} = G_0 G_{1,\max} G_{2,\max} \tag{10.1.54}$$

$G_{u,\max}$ is also called *maximum available gain* disregarding stability issues and represents the maximum gain that can be achieved with the transistor perfectly matched.

However, it might not be possible to achieve the maximum available gain because of unwanted coupling through the feedback parameters of the transistor that causes oscillations and instability; so, there is a gain limit that ensure stability indicated by MSG defined above.

10.2 Interpretation of Return and Insertion Loss

Taking a $50\,\Omega$ transmission line, like most coaxial cables, we go further with the interpretation of the behavior of S_{11} and S_{21} parameters, considering the output of a Vector Network Analyzer. If a sinusoidal wave is applied to port 1 using a $50\,\Omega$ source on a $50\,\Omega$ transmission line, it is known that theoretically the reflection coefficient is 0 and the entire wave is transmitted to the other port (except for line losses and coupling or radiation to the surrounding elements, as already said). The amplitude of S_{11} and S_{21} parameters are thus 0 and 1 respectively. From a practical standpoint, when measured with a VNA, they will look like a large negative value in dB and a nearly 0 dB curve when plotted versus frequency; including losses, the amplitude of S_{21} will begin to decrease with frequency.

10.2.1 Insertion loss S_{21} interpretation

We have seen that the amplitude of S_{21} is expected to depart for increasing frequency from the low frequency value of 0 dB (that implies that the full amplitude of the applied sine-wave reaches the output port 2). The amplitude of S_{21} starts from 0 dB and begins decreasing in the negative dB range (so the figure gets larger, but negative) with a more or less linear shape, which begins to show the predicted periodic oscillations when the line termination begins deviating from the perfect matching or line losses are not negligible. If the oscillations are ruled out by curve smoothing and linear interpolation, the slope indicates line losses, ascribed to conductive losses, dielectric losses and radiation (or coupling) losses (see sec. 2.5.3 and 2.7.3).

The slope of the amplitude of insertion loss curve gives the attenuation constant α: this factor has always been considered in linear units, while an insertion loss measurement will often return a quantity expressed in dB, so that some mathematical manipulation is necessary; eq. (10.2.2) below makes this explicit. The envelope of the real and imaginary part curves appearing in Figure 10.2.1(a) is exactly the magnitude value expressed in linear units. The magnitude is also shown in polar coordinates (Figure 10.2.1(b)), where the spiral is slightly looser at the lower frequency values and then gets denser and more packed.

Conductive losses are more evident at the lowest frequencies, where skin-effect reduction of the useful cross section is more pronounced; dielectric losses begin to be relevant starting from a conveniently high frequency, depending on material properties. Eq. (2.5.19) is useful to determine total cable losses starting from line parameters that are recovered from return loss and insertion loss measurements.

What about the phase of the insertion loss curve? The interpretation of the phase curve requires that the behavior of the line under test is recalled when the sinusoidal test wave is applied (see also sec. 2.1.1): depending on the frequency, less than one period or many periods may be spatially present along the line at a given instant of time, depending on the ratio of wavelength and line length. With impedance mismatch at cable ends, reflections and standing wave will take place. Because perfect matching is never attained, a slight periodicity might be always visible in measured S_{21} curves (and of course in S_{11}). For a line of length L the applied sine-wave of frequency f (and

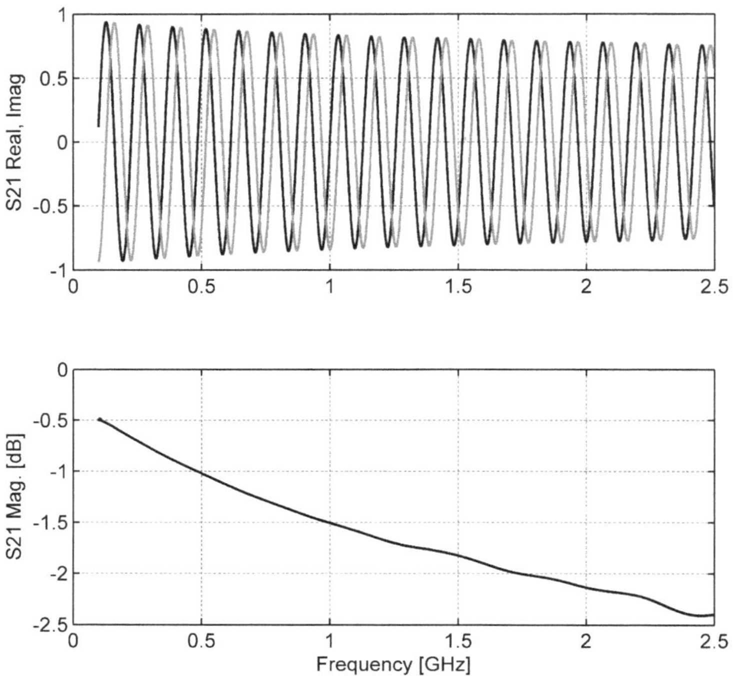

(a)

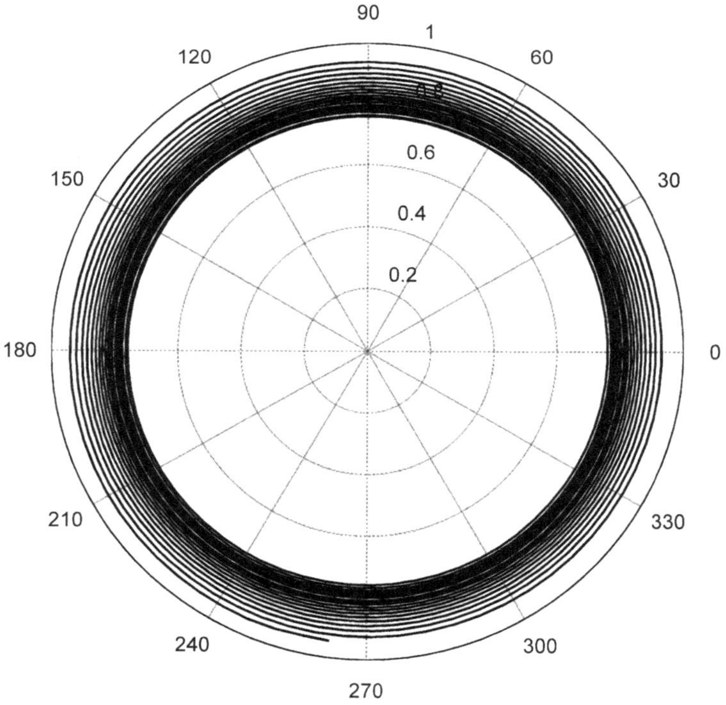

(b)

Figure 10.2.1 – Interpretation of S_{21} for a 10 m low-loss cable: (a) real and imaginary parts (above) and log mag (below); (b) polar plot of magnitude vs. unwrapped phase.

wavelength $\lambda = v/f$, where $v = kc$ is the velocity of light through the line and k is the velocity factor of the cable with respect to the speed of light in vacuum c) will undergo a phase rotation φ of a complete turn when $\lambda = L$. So, at a given instant of time, when the input and the output sine waves are measured by the VNA to calculate S_{21}, their phase displacement will be increasing with frequency passing one period for every wavelength that is contained within line length. With the time delay TD of the line, i.e. $\mathrm{TD} = L/v$, the phase displacement between output and input sine waves will be increasing with frequency. If we observe that the measured phasor of the sine wave at port 2 is an old copy of the phasor applied to port 1 TD seconds ago, that has traveled through the line, and that at the same time instant the new phasor at port 1 that can be measured by the VNA has advanced in time and phase, we understand that the sign of the S_{21} phase is negative. The phase of S_{21} is thus expressed by $\varphi = -f\,\mathrm{TD}$. As long as the line is an ideal lossless line, the phase relationship keeps linear with frequency (see sec. 2.7.3 and 3.3.1).

The starting point of the S_{21} phase curve is theoretically 0, but since a real frequency-domain measurement can start only from the minimum VNA frequency, the first value will be already somewhat negative. This is quite evident in Figure 10.2.1(a), where the frequency axis begins at $100\,\mathrm{MHz}$.

10.2.2 Return loss S_{11} interpretation

The S_{11} interpretation follows that of S_{21}: the phase rotation is explained similarly, by observing that for the round-trip it is twice that evaluated for S_{21} parameter. Assuming a reflective termination, to determination of S_{11} contributes not only the first reflection at the near port 1, but also the signal that propagates to the remote port 2 and that is reflected back to port 1: the phase relationship determines how these two terms compose and if they cancel out partially, totally or if they add in amplitude. At low frequency the reflection at port 1 is always negligible; the wave reflected back at port 2 has an opposite phase with respect to the one entering at port 1, so that they almost cancel out and the S_{11} curve starts from a large negative dB value. As the frequency increases the phase shift for the wave traveling from port 1 along the line increases; at a quarter of the cycle the reflected wave comes back at port 1 with half a period phase rotation: in this case the two signals entering port 1 from the generator and reflected back to port 1 almost cancel out, so that the reflection at port 1 is at its maximum and the S_{11} amplitude is equal to twice the reflection coefficient (assuming equal reflection coefficients for both ports). Correspondingly S_{21} will be at its minimum (by the conservation of energy). With the frequency further increasing S_{11} and S_{21} will alternatively cycle between their minima and maxima.

Observing eq. (2.5.17) one may wonder how effective is a return loss measurement to determine line attenuation, as opposed to the more intuitive use of the insertion loss measurement. With some knowledge of real measurements and the shape of S_{11} curves, we know that a long enough line is needed for the satisfactory determination of α, but at the expense of many phase rotations and very frequent minima and maxima, that clutter the S_{11} curve and require complicated smoothing and interpolation. An example is shown in the next section.

To conclude, S_{11} is conveniently low if both ports are matched terminated (onto the line characteristic impedance) or, when this is not possible, if the electrical length of the line is as short as possible, so that there is no appreciable phase rotation. Connectors are a typical example of this: when they are designed with poor impedance control, the maximum useful frequency is directly determined by their electrical length.

10.2.3 Measuring return and insertion loss

Some examples of measurements made in amplitude only (Spectrum Analyzer) and in amplitude and phase (Vector Network Analyzer) are discussed.

10.2.3.1 Reflection (S_{11}) of a sample coaxial cable

Let's consider a sample of RG8 coax cable of length $L = 3.70\,\mathrm{m}$ and examine the measurement of return loss. The return loss at input port S_{11} with respect to the $Z_0 = 50\,\Omega$ reference impedance may be measured with different cable terminations. An amplitude-only measurement is considered.

$$S_{11} = 10 \log \left[\frac{Z_{in} - 50}{Z_{in} + 50} \right] \tag{10.2.1}$$

By replacing eq. (2.3.12) into eq. (10.2.1) above, a relationship between return loss S_{11}, cable length L and propagation constant γ is obtained:

$$\begin{aligned} S_{11} &= 10 \log \left[\frac{\tanh(\gamma L) - 1}{\tanh(\gamma L) + 1} \right] = 10 \log \left[\exp(-\gamma L) \right] = \\ &= 10 \log \left[\exp(-\alpha L) \right] + 10 \log \left[\cos(-\beta L) + j \sin(-\beta L) \right] \end{aligned} \tag{10.2.2}$$

The first term represents the attenuation and, assuming a constant value of α, it would be a horizontal line, constant with frequency; since α increases with frequency it exhibits a downward slope and $\alpha(f)$ may be estimated by curve fitting. The second term represents a ripple superimposed to the first term, oscillating by the same mechanism shown for eq. (2.5.15), where the distance between two adjacent peaks is such that $2\pi L/\lambda = \pi$ and thus $L = \lambda/2$. By observing the distance on the frequency axis between successive local minima or maxima, the resonant frequency f_0 is derived and wave velocity is $v = 2 f_0 L$.

Example 10.3: Calculation of cable parameters from Return Loss curve

The results of measurements on cable sample are shown in Figure 10.2.2: 17.5 periods are counted between the fourth and the ninth division (a more accurate estimation than counting three periods per division) and each division is equal to $(800-10)/10=79\,\text{MHz}$: $f_0=6\times 79/17.5=27.086\,\text{MHz}$. Cable velocity is $v=2f_0L=200.44\,10^6\,\text{m/s}$ and the velocity factor v/c is exactly 0.667 (as it is known from cable datasheet). For the attenuation constant term eq. (10.2.2) may be simplified as $RL=-\alpha L\,10\log(e)=-\alpha L\,4.3429$.

Let's imagine an ideal line drawn amid the repeated oscillations, the decay is about 0.2 dB @ 10 MHz and increase to 0.83 dB @ 250 MHz, so that $\alpha=0.0461/L=0.0124\,\text{np/m}$ and $\alpha=0.1911/L=0.0517\,\text{np/m}$ at the two frequency values respectively.

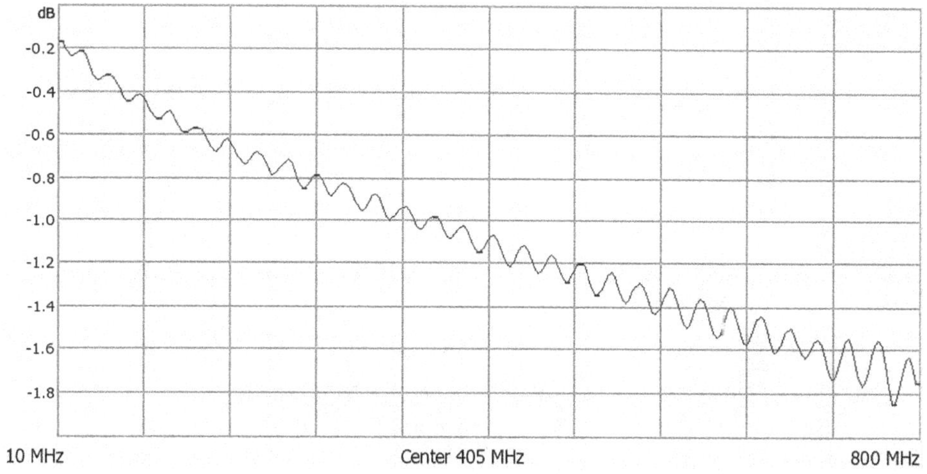

Figure 10.2.2 – Return Loss measurement in open-circuit condition.

10.2.3.2 Reflection of a mismatched cable

Return and insertion loss of cables of different characteristic impedance are preliminarily considered, introducing a topic that will be more extensively treated in sec. 11.10.5.

A RG 62 cable sample of 2 m length is considered as an example of transmission line with mismatched characteristic impedance (for RG 62 $Z_c=93\,\Omega$) measured from a $50\,\Omega$ system, i.e. port 1 of VNA (for details on how the measurement is done and how the VNA is calibrated before each measurement, please see Chapter 11). The test setup is very simple: VNA port 1 is adapted to the cable sample at port 1A and the other end of the cable sample (port 1B) is terminated in different ways (open, short and/or matched load). The expected reflection coefficient measured at port 1 looking into the cable is $\Gamma=\dfrac{Z_1-Z_c}{Z_1+Z_c}=\dfrac{50-93}{50+93}=-0.301$; this value translated in dB is $-10.4\,\text{dB}$. The travel time τ assuming the velocity factor of 0.83 (see sec. 11.10.5) is $7.95\,\text{ns}$. Thus changes of the measured reflection coefficient are expected every 2τ. The first experiment with the cable sample terminated in open condition is shown in Figure 10.2.3 (gray curve). The cable sample has of course also some attenuation, that may

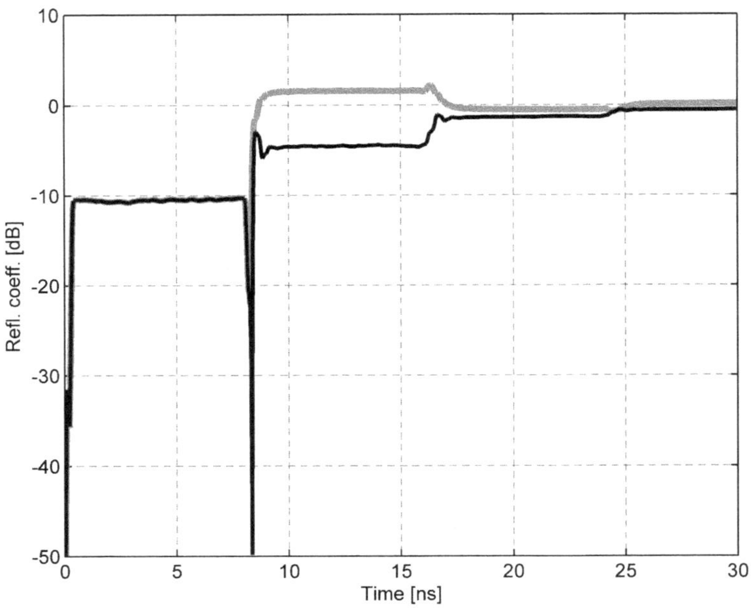

Figure 10.2.3 – Measurement of S_{11} for a 2 m RG 62 cable section terminated in open circuit (gray) and short circuit (black).

be neglected at a first approximation. The shown value is $S_{11} = 20\log_{10}\Gamma$; the profile of the measured S_{11} follows reflections along the cable and properly scaled is a direct measure of voltage at port 1.

The cable sample is now terminated on a short circuit and, as expected, the S_{11} value is not changing in the first signal trip because the termination does not influence the characteristic impedance of the cable. With a short circuit phase rotation is reversed and the reflection coefficient from inside the cable looking into port 1 passes from $+0.301$ to -0.301; the reflected voltage is thus phase reversed and the measured S_{11} curve for multiple trip times now has a different shape (see Figure 10.2.3 (black curve)). The sudden spiky drop of the reflection coefficient to -50 dB at about 8 ns is due to the adapter and its good matching to 50 Ω used to insert the terminations.

If we add another cable section at the end of the first one, again with a different characteristic impedance, we may observe other reflections that superimpose to the previous ones: the second section is made of RG 58, with 50 Ω characteristic impedance, and thus creating an impedance mismatch at port 1B, this time equal to $+0.301$ looking from section I into section II. If the new cable sample is much shorter than the previous one, e.g. 26 cm, there will be small "teeth" in the previous S_{11} curve, due to reflections from the new section reaching port 1 after having traveled 26 cm more back and forth at a velocity factor of 0.66. This situation is shown in Figure 10.2.4 (gray curve), for section II terminated in open circuit. The new section II is responsible for two types of reflections, those from the interface with section I at port 1B and also those coming from the open circuit termination at its far end, that we call port 1C. If we want to remove the latter, we terminate this port with a 50 Ω load and we are able thus to isolate the effect of the impedance mismatch at port 1B between the two cable

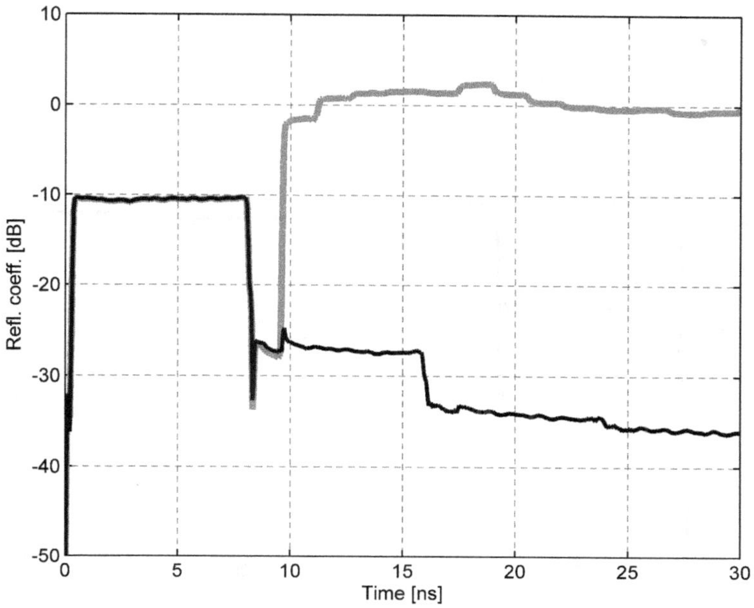

Figure 10.2.4 – Measurement of S_{11} for a 2 m RG 62 cable section followed by a 26 cm RG 58 cable section terminated in open circuit (gray) and matched 50 Ω load (black).

samples, that is shown in Figure 10.2.4 (black curve). The short 1.5 ns portion of the two traces before 10 ns is the matching impedance of the RG 58 section, prolonged in the black trace between 10 and 16 ns due to the matching termination, while the gray trace jumps to a nearly full reflection due to the open circuit termination. The 50 Ω load cancels reflections improving the reflection coefficient from −27 dB to about −35 dB, and improving for subsequent reflections with a slightly downward slope due to cable attenuation.

The effect of losses is to reduce the observed value of reflection coefficient, as a result of smaller amplitude of the reflected voltage terms (this phenomenon is part of "masking" and is considered in sec. 11.10.4). To proof this we arrange the same test setup made of a 2 m section of RG 62 cable introducing a 6 dB attenuator between the feeding port 1 and the cable section. The travel time was shown in the previous graphs in the so called "one way" representation, where the VNA corrects for the double travel for the reflections to reach the measuring port; the VNA presents the results as they were measured in the position along the transmission line where reflection occurs. However, the same cannot be done for amplitude of reflection terms: the signals travel twice through the attenuator before the first reflection reaches port 1 back from the other end of the cable sample and the displayed attenuation will be 12 dB! This is clearly demonstrated in Figure 10.2.5: the difference between the black and the gray curve is 12 dB and the asymptote of the black curve is 12 dB, the ratio between the reflected signal (attenuated twice by the 6 dB pad) and the original source signal. The presence of the attenuator is visible at the beginning of the black trace: it increases line electrical length, so the black curve is slightly shifted rightward.

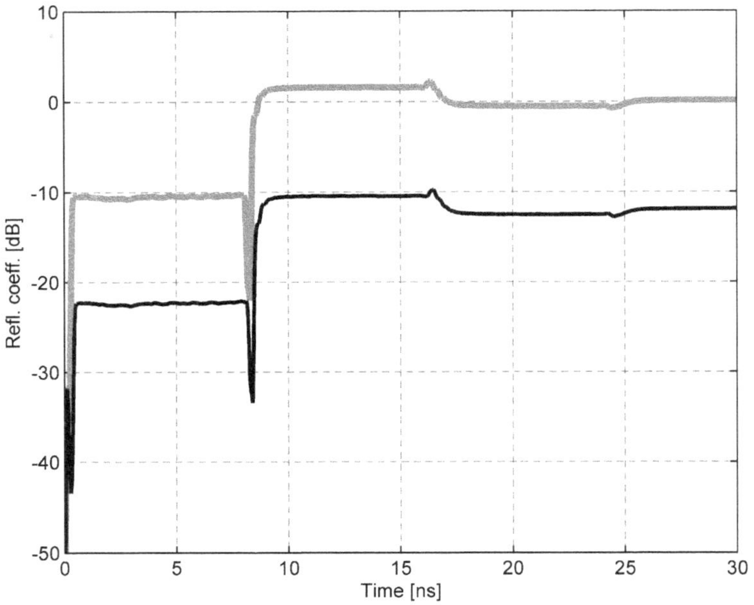

Figure 10.2.5 – Effect of attenuation on the S_{11} measurement: the 2 m RG 62 cable section is preceded by a 6 dB attenuator connected between it and the VNA port. Original RG 62 cable section open circuited (gray) and preceded by the 6 dB attenuator (black).

10.2.3.3 Attenuation of a sample coaxial cable

The direct quantification of attenuation of a cable sample is of course obtained by measuring insertion loss S_{21}: this is a two-port measurement that requires that both ends of the cable sample are accessible and available at the same location to plug into the VNA. The typical S_{21} curve shows the downward slope discussed above in sec. 10.2.1, due to the various loss mechanisms. In Figure 10.2.6 three insertion loss curves are shown for three nearly identical RG 316 cable samples, each approximately 30 m long (curves are normalized to the exact lengths): one is evidently affected by some noise in the S_{11} trace without evident effect in the corresponding S_{21} trace; the S_{21} curve has the expected square-root-of-frequency downward slope and up to about 4 GHz the curves are all nearly perfectly overlapped, indicating a uniform repeatable behavior although this type of cable is never used at such frequencies for the excessive attenuation and rapidly worsening return loss.

One might wonder what measurement can be done when the two cable ends are not available at the same location and a two-port measurement is not possible. Examples are common, whenever testing an already installed cable. There are nowadays very performing solutions using controlled calibration boxes, but even a RF generator connected at the far end is a viable solution if one is interested in attenuation only.

However, there is also a handy solution using the same return loss measurement seen in sec. 10.2.3.1 and Example 10.3: $\alpha(f)$ may be estimated by fitting the S_{11} curve.

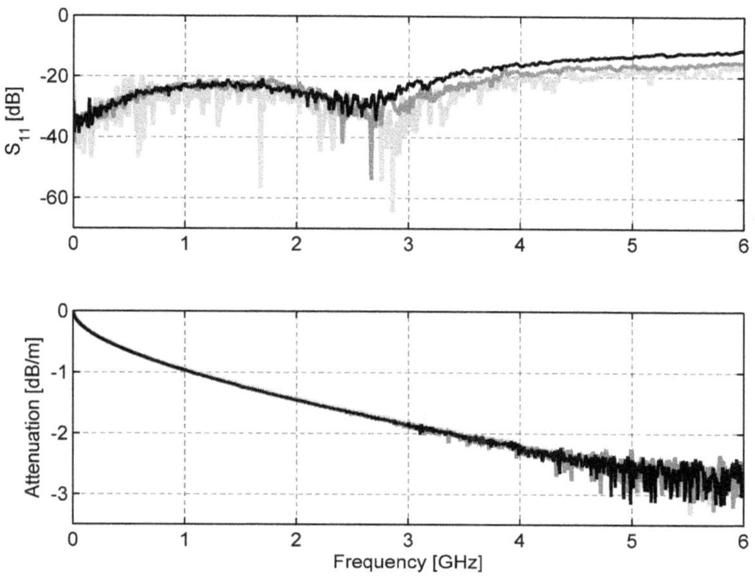

Figure 10.2.6 – Measured attenuation per-unit-length [dB/m] of three different RG 316 coaxial cables of about 30 m length.

10.3 Multi-port S parameters

This section is a continuation of matrix representations considered in sec. 1.3. The number of ports is extended from one (for which the concept of input impedance is relevant), to two (for which the forward and backward transmission adds to the input impedance), to three (or more). In the latter case we still have forward and backward transmission coefficients, acting on more than one port at the same time.

The general representation of an M-port network is given by

$$\begin{bmatrix} b_1 \\ b_2 \\ b_3 \\ \vdots \\ b_M \end{bmatrix} = \begin{bmatrix} S_{11} & S_{12} & S_{13} & \cdots & S_{1M} \\ S_{21} & S_{22} & S_{23} & \cdots & S_{2M} \\ S_{31} & S_{32} & S_{33} & \cdots & S_{3M} \\ \vdots & \vdots & \vdots & \ddots & \vdots \\ S_{M1} & S_{M2} & S_{M3} & \cdots & S_{MM} \end{bmatrix} \begin{bmatrix} a_1 \\ a_2 \\ a_3 \\ \vdots \\ a_M \end{bmatrix} \tag{10.3.1}$$

and for the three-port case the expressions are explicitly

$$\begin{aligned} b_1 &= S_{11}a_1 + S_{12}a_2 + S_{13}a_3 \\ b_2 &= S_{21}a_1 + S_{22}a_2 + S_{23}a_3 \\ b_3 &= S_{31}a_1 + S_{32}a_2 + S_{33}a_3 \end{aligned} \tag{10.3.2}$$

assuming that all three ports have the same reference impedance (e.g. $50\,\Omega$).

Now, by applying a condition to one of the ports, it is possible to simplify the set of equations and derive a closed-form relationship between the other quantities.

10.3.1 2-port to 1-port conversion and component impedance

Particularly interesting is the conversion between a 2-port representation to a 1-port representation: when measuring for example a two-terminal passive device with a full 2-port VNA connection (e.g. with a component test fixture such as those built in microstrip), the characteristics of the component such as its impedance are derived once a 1-port representation is available. The method is based on the representation of the component impedance as a differential impedance between two terminals: $Z_{dd} = Z_{11} + Z_{22} - Z_{12} - Z_{21}$. The 1-port representation is obtaining thus by isolating the S_{11} parameter, but including the effect of the other port 2. Taking eq. (10.3.2) for $M = 2$ and imposing the condition that the other port 2 is connected to ground (so that the reflection coefficient is a full -1 ($b_2 = -a_2$), we can substitute in the first row a_2, having it calculated as a function of a_1 using the second row, and we finally obtain

$$S_{11-1\mathrm{p}} = S_{11} - \frac{S_{12}S_{21}}{1 + S_{22}} \tag{10.3.3}$$

The impedance values may be readily obtained by including the reference characteristic impedance Z_r used during measurements, so that

$$Z_{cmp} = Z_r \frac{1 + S_{11-1\mathrm{p}}}{1 - S_{11-1\mathrm{p}}} \tag{10.3.4}$$

Real and imaginary part of impedance can now be interpreted depending on the type of component: for resistors the imaginary part is related to stray inductance (positive) and capacitance (negative) terms; for inductors the real part represents series inductor losses and the ratio of imaginary and real part gives the factor of merit Q; for capacitors the real part represents capacitor losses, namely dielectric losses given by $\tan\delta$ in a series RC representation.

This subject, as the next one, is marginally treated in some references, such as fragments of the help files of IC-CAP (by Agilent). A few helpful references are:

- M. Danesh and J.R. Long, "Differential driven symmetric microstrip inductors," *IEEE Transactions on Microwave Theory and Techniques*, Vol. 50, no. 1, Jan. 2002, pp. 332-341. doi: 10.1109/22.981285;
- C. Inui and M. Fujishima, "Characterization of T-Shaped Terminal Impedances of Differential Short Stubs in Advanced CMOS Technology," IEEE International Conference on Microelectronic Test Structures ICMTS, Edinburgh, UK, 24-27 March 2008, pp. 183-189. doi: 10.1109/ICMTS.2008.4509336.

10.3.2 Conversion between 2-port to 3-port representations

Examples of three-port networks are less intuitive than the previous case; transistors in some cases (e.g. when measured directly on wafer) are better described by a three-port representation, each terminal identified as a different port. For conversion between three-port and two-port representations [308], it is compulsory to assume the condition of one port of the three-port model: it is quite customary to assume one port grounded, and in this example it may be port 3. Grounding a port is common when considering bipolar transistors in one of the three CB, CC or CE configurations, and similarly for FET and MOSFET transistors. Thus the reflection coefficient Γ_3 becomes -1, and we may replace b_3 with $-a_3$. This allows to express the remaining a_3 in the third line of the equation system (10.3.2) as a function of a_1 and a_2 and replace this expression in the two other lines, obtaining after manipulation two equations between a and b quantities of the first two ports only, so having transformed the original three-port representation to an equivalent two-port one:

$$\begin{bmatrix} b_1 \\ b_2 \end{bmatrix} = \begin{bmatrix} S_{11} - \dfrac{S_{13}S_{31}}{1+S_{33}} & S_{12} - \dfrac{S_{13}S_{32}}{1+S_{33}} \\ S_{21} - \dfrac{S_{23}S_{31}}{1+S_{33}} & S_{22} - \dfrac{S_{23}S_{32}}{1+S_{33}} \end{bmatrix} \begin{bmatrix} a_1 \\ a_2 \end{bmatrix} \tag{10.3.5}$$

It is also possible to augment a two-port representation, reaching a consistent three-port representation. First, it is observed that the sum of the S parameters in a row or a column of the three-port matrix is equal to 1 (we will use this to calculate the remaining S parameters). Then, always referring to the equation system (10.3.2), it is possible to write the first of the new three-port S parameters, S_{33}, as a function of the original two-port S parameters distinguished by a "′" (prime).

$$S_{33} = \frac{S'_{11}+S'_{12}+S'_{21}+S'_{22}}{4-(S'_{11}+S'_{12}+S'_{21}+S'_{22})} \tag{10.3.6}$$

$$S_{23} = S_{32} = \frac{1+S_{33}}{2}\,(1-S'_{12}-S'_{22}) \tag{10.3.7}$$

$$S_{22} = S'_{22} + \frac{S_{23}+S_{32}}{1+S_{33}} \tag{10.3.8}$$

$$S_{31} = 1-S_{33}-S_{32} \qquad S_{13} = 1-S_{33}-S_{23} \tag{10.3.9}$$

$$S_{21} = 1-S_{22}-S_{23} \qquad S_{12} = 1-S_{22}-S_{32} \tag{10.3.10}$$

$$S_{11} = 1-S_{22}-S_{23} \tag{10.3.11}$$

10.4 Single-ended and differential S parameters

A lot have been written on relationships between differential and common-mode (or better, "single-ended") S parameters, because it is the first problem the user faces when performing measurements using either a VNA, a TDR, an oscilloscope, or their combination: first, many instruments of this kind have their ports referred to "common", that for RF measurements we know that almost always shall coincide with "ground";

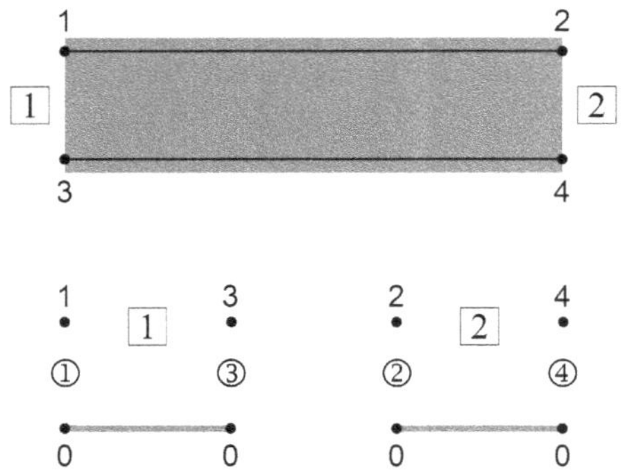

Figure 10.4.1 – Single-ended and differential ports for two single-ended lines and a common return path (top view above, side view below): 1, 2, 3, 4 are the conductor terminal numbers; ①, ②, ③, ④ are the single-ended ports, where each terminal is considered referred to the common return plane 0; the two numbers in square frame 1 and 2 indicate the differential ports, that will be used to identify the mixed-mode S parameters.

second, even with balun decoupled ports, the connectors going to the test fixture (e.g. a PCB fixture) are coax connectors, bonded often to the fixture return plane; only in specific cases coax connectors are used in a floating configuration. However, the topic is rather new and the first complete reference is the paper written by Bockelman and Eisenstadt [43, 45] twenty years ago. Some insight in other subtleties related to mismatched ends, TDR/VNA relationships, and calibration procedures may be found in [155, 156, 329, 330].

Again we consider two straight conductors (such as PCB traces) with a reference plane 0 underneath: they may be seen as two coupled transmission lines with a common return plane or also as a pair featuring a reference plane (see Figure 10.4). In the first case we interpret each port as a pair of terminals, one located on the trace and the other one on the return plane, using ① and ③ for the near end of the two traces, and ② and ④ for the respective far ends. In the second case the concept of port is migrated to the differential pair, so that we have a differential port 1 at the near end and a corresponding differential port 2 at the far end.

For a differential pair featuring a reference plane there is still the possibility of applying differential or common signals (let's call them also differential-mode and common-mode signals, but let's distinguish them from the odd mode and the even mode of the pair). Thus, we end up with the possibility of focusing our attention on all combinations of differential and common signals at port 1 and 2, and we distinguish them by using letters "d" and "c", ordered with respect to port 1 and 2: an example is a differential signal entering at port 1 and a common signal exiting at port 2, for which the insertion loss may be considered, indicated as S_{dc21}; the opposite described by S_{cd21} is also relevant and is normally named common-to-differential conversion, or transformation. S parameters expressed for differential and common signals are called mixed-mode S parameters. Of course, different arrangements are possible of the combinations of "d" and "c" and port 1 and 2; the complete relationship in matrix form for ports and modes is given in Table 10.4.2.

		d.m.		c.m.	
		port 1	port 2	port 1	port 2
d.m.	port 1	S_{dd11}	S_{dd12}	S_{dc11}	S_{dc12}
	port 2	S_{dd21}	S_{dd22}	S_{dc21}	S_{dc22}
c.m.	port 1	S_{cd11}	S_{cd12}	S_{cc11}	S_{cc12}
	port 2	S_{cd21}	S_{dd22}	S_{dc21}	S_{dc22}

(a)

		port 1		port 2	
		d.m.	c.m.	d.m.	c.m.
port 1	d.m.	S_{dd11}	S_{dc11}	S_{dd12}	S_{dc12}
	c.m.	S_{cd21}	S_{cc11}	S_{cd12}	S_{cc12}
port 2	d.m.	S_{dd21}	S_{dc21}	S_{dd22}	S_{dc22}
	c.m.	S_{cd21}	S_{cc21}	S_{cd22}	S_{cc22}

(b)

Figure 10.4.2 – Mixed-mode matrix representations: (a) classification based on modes and (b) classification based on ports. The most commonly adopted is (a).

Let's consider the impedance each port refers to: when the ports are single ended connected between trace and return plane, then the reference impedance is, let's say, 50 Ω; on the contrary, when going to the differential S parameters, the reference impedance is the characteristic impedance seen when applying a differential or a common signal, that is 100 Ω and 25 Ω, respectively. Some degree of coupling is possible so that the resulting odd- and even-mode characteristic impedances may differ, resulting in a slightly lower and slightly larger value, respectively.

For all signal standards using differentially driven pairs to transmit over back-planes, board interconnects or links between equipment (using e.g. LVDS, LVCMOS, Ethernet), the most important parameter is S_{dd21}. This parameter indicates the attenuation of the useful signal: as known, it is mainly determined by conductor losses and dielectric losses. Additionally, it is influenced by resonances with nearby structures that act as resonators and subtract energy to the transmitted signal; the drop in the curve is more evident the higher the Q-factor of the resonating structure. If the wavelength is sufficiently short, even small elements of a PCB can represent a resonator, such as a dead trace going to the pads of an optional component, or a via protruding on the other side of the PCB where it is unterminated and left floating. Phase shift and reflection along these stubs determine the resonance that occurs in quarter-wavelength conditions (see sec. 6.1.4). For vias the countermeasures are to put them only between adjacent layers (i.e. trace jumping is one layer at a time), to use thinner PCBs, or to drive fast signals on internal layers using buried or blind vias (a buried line, such as stripline, has a better frequency behavior than an open line, such as microstrip).

10.4.1 Mode conversion

"Mode conversion" indicates the transformation between differential and common signals when traveling from port 1 to port 2. Of course the appearance of a common signal component even when feeding with a pure differential signal not only reduces the amplitude of the differential signal on its way to port 2, but the common signal may be reflected back transforming into a differential signal component, thus overlapping and affecting the integrity of the transmitted differential signal. More serious, of course, is the specular transformation of common to differential components, directly interfering with the transmitted differential signal. This mechanism of reflection and transformation is dictated by the length of the line and the position of discontinuities causing reflections; it is not related to the rate with which a differential signal is applied at port 1, that depends on the symbol rate of the considered transmission line. This leads to uncorrelated signal deterioration and affects as a consequence the bit error rate in an unpredictable way. In multi-wire buses reflections may originate at different positions along each line, depending on routing rules and coupling to nearby elements, thus creating a multitude of reflections not synchronized between them and the symbol rate; this happens when joining the pins of a multi-pin connector using PCB traces of slightly different lengths because of various constraints (mechanical parts, components, etc.) and when applying fast signals whose spectral content covers the frequencies associated to the time distance between imperfections.

A quick numerical example clarifies the situation: a grid of pins spaced each by 1 mm and located on both sides of the longitudinal axis of a multi-pin connector is fed through PCB traces going parallel to the connector axis and serving the two sides; PCB traces on either side turn to the left or to the right to connect to each pin; let's assume that this creates a discontinuity and a reflection as a consequence: a displacement of one pin for each trace implies a difference of length of 1 mm, that is only about $5-6\,\mathrm{ps}$; however, if pull-up resistors are located next to the connector, they require extra-space, thus creating a small discontinuity every $3\,\mathrm{mm}$, that is $15-18\,\mathrm{ps}$; a $5\,\mathrm{Gbps}$ signal with rise and fall times of $20\,\mathrm{ps}$ will trigger reflections and interference will occur at different positions along the symbol, starting from the rising edge and moving along by $15-18\,\mathrm{ps}$ each.

Common signals radiate more efficiently because they lack the compensation ensured by proximity of positive and negative current flows: this is another reason to evaluate mode conversion. We may briefly consider the main characteristics of common- and differential-mode electromagnetic emissions, to emphasize the relevance of the former. Common-mode emissions increase linearly with frequency, while differential-mode emissions are proportional to the square of frequency: the reason is that the geometry of radiation is a straight wire for the former and a rectangular loop for the latter. However, for the same line length L the area between the single straight wire that carries the common-mode current and its reference (the ground) is much wider than the small area enclosed between the two currents of opposite polarity of the differential-mode configuration: the multiplying coefficient is much smaller for differential-mode emissions, scaling the intensity down and making the square of f behavior irrelevant up to a considerably large frequency [255], sec. 7.5.2.

$$|E_{cm,max}| = 1.257 \times 10^{-6} f L \frac{|I_{cm}|}{r} \quad (10.4.1)$$

$$|E_{dm,max}| = 1.326 \times 10^{-14} f^2 L s \frac{|I_{dm}|}{r} \quad (10.4.2)$$

10.4.2 Conversion of single-ended and differential S parameters

The scheme of single-ended lines (each one connected to a single-ended port) or equivalently differential lines with new ports obtained by pairing conductor terminals that was shown in Figure 10.4.1 is considered again for the equivalence of the two representations: conductors were grouped differently invoking a modal description of the so formed differential pair, that for the original four single-ended ports needs two modes, differential mode and common mode. For this reason the matrix representation of differential S parameters is called *mixed-mode representation*.

In the past, if differential measurements were desired, two baluns (or hybrid coupler) would be needed, feeding the two single-ended lines at one side and the other, and connected to the two single-ended ports of VNA. The problems associated with this method are i) magnitude and phase imbalance of baluns, ii) no way to measure mode conversion (that is, from differential to common-mode), iii) no rigorous definition of mixed-mode S parameters, and iv) calibration of the system with baluns is poorly defined (no differential calibration standards are "standardized"). Today, VNA with balanced ports that can perform directly the measurement on differential ports are available. However, it is interesting (and compulsory, if one lacks such a VNA) to understand how it is mathematically possible to pass from one representation (the single-ended) to the other (the differential) by exploiting the full set of measurements done on the single-ended network.

To derive a matrix representation of the relationship between differential and single-ended S parameters, it is needed first to decide port numbering: there are two possible arrangements, one (arrangement A) with the numbers 1 and 2 on the left side and the 3 and 4 on the right side, both from top to bottom, and the other arrangement, B, in a zig-zag fashion, with 1 and 2 assigned to the first line on the left and the right side respectively, and 3 and 4 assigned to the second line in the same way. Both are acceptable, but give a different transformation matrix with the permutation of the intermediate columns 2 and 3. Back in this chapter the zig-zag numbering was used.

Also for mixed-mode matrix representation it is necessary to decide if modes are joined for differential/common mode or for belonging to a physical port, as already considered in the two representations shown in Table 10.4.2. For a mixed-mode representation like in Table 10.4.2(a), that privileges the grouping by mode "m" (the most commonly adopted), matrix representations are:

$$\mathbf{M_A^m} = \frac{1}{\sqrt{2}} \begin{bmatrix} 1 & 0 & -1 & 0 \\ 0 & 1 & 0 & -1 \\ 1 & 0 & 1 & 0 \\ 0 & 1 & 0 & 1 \end{bmatrix} \qquad \mathbf{M_B^m} = \frac{1}{\sqrt{2}} \begin{bmatrix} 1 & -1 & 0 & 0 \\ 0 & 0 & 1 & -1 \\ 1 & 1 & 0 & 0 \\ 0 & 0 & 1 & 1 \end{bmatrix} \tag{10.4.3}$$

The final matrix relationship that gives the mixed-mode representation $\mathbf{S}_{mm}$ is

$$\mathbf{S}_{mm} = \mathbf{M}\,\mathbf{S}_{se}\,\mathbf{M}^{-1} \tag{10.4.4}$$

with $\mathbf{M} = \mathbf{M_A^m}$ or $\mathbf{M_B^m}$.

After deciding for one of the two single-ended representations (let's say B) and the mixed-mode representation (grouped by mode), it is possible to write down explicit expressions, that clarify how the single-ended parameters are combined [287].

The chosen representation that led to the S_{mm} matrix in eq. (10.4.7) is the one adopted in the `.s4p` file format[4]. Had the other mixed-mode representation been chosen (the one grouping by port and corresponding to that shown in Table 10.4.2(left)), the resulting transformation matrix would have had a permutation of rows (rows 2 and 3 exchanged) and columns (column 2 and 3 exchanged), nothing else. Quite conveniently all transformations from single ended to differential and common mode, and globally to mixed-mode representation, are available in Matlab® RF Toolbox and well described in [276]. Conversely, it is also possible to write reverse relationships that link mixed-mode parameters to single-ended ones, as shown in eq. (10.4.8).

It is underlined that the method described so far for single-ended to mixed-mode conversion assumes that even- and odd-mode characteristic impedances are equal to a common characteristic impedance value, i.e. there is no coupling between the differential signals. This is in general not true and in order to remove this weakness, pointed out by Bockelman and Eisenstadt [43] and by Vaz and Caggiano [329], two parameters are introduced: the coefficients relating the two modal characteristic impedances $Z_{c,e}$ and $Z_{c,o}$ to the single-ended characteristic impedance Z_c [168]:

$$Z_{c,e} = k_e Z_c \qquad Z_{c,o} = k_o Z_c \tag{10.4.5}$$

The mixed-mode matrix transformation becomes

$$\mathbf{S}_{mm} = (\mathbf{M_1 S_{se}} + \mathbf{M_2})(\mathbf{M_1} + \mathbf{M_2 S_{se}})^{-1} \tag{10.4.6}$$

where

[4] Touchstone "SnP" or .snp format is a *de facto* standard used by many simulators and math packages, as well as available on many instruments for data export.

$$S_{mm}=\frac{1}{2}\begin{bmatrix} S_{11}-S_{12}-S_{21}+S_{22} & S_{11}+S_{12}-S_{21}-S_{22} & S_{13}-S_{14}-S_{23}+S_{24} & S_{13}+S_{14}-S_{23}-S_{24} \\ S_{11}-S_{12}+S_{21}-S_{22} & S_{11}+S_{12}+S_{21}+S_{22} & S_{13}-S_{14}+S_{23}-S_{24} & S_{13}+S_{14}+S_{23}+S_{24} \\ S_{31}-S_{32}-S_{41}+S_{42} & S_{31}+S_{32}-S_{41}-S_{42} & S_{33}-S_{34}-S_{43}+S_{44} & S_{33}+S_{34}-S_{43}-S_{44} \\ S_{31}-S_{32}+S_{41}-S_{42} & S_{31}+S_{32}+S_{41}+S_{42} & S_{33}-S_{34}+S_{43}-S_{44} & S_{33}+S_{34}+S_{43}+S_{44} \end{bmatrix} \tag{10.4.7}$$

$$S_{se}=\frac{1}{2}\begin{bmatrix} S_{dd11}+S_{cd11}+S_{dc11}+S_{cc11} & -S_{dd11}-S_{cd11}+S_{dc11}+S_{cc11} & S_{dd12}+S_{cd12}+S_{dc12}+S_{cc12} & -S_{dd12}-S_{cd12}+S_{dc12}+S_{cc12} \\ -S_{dd11}+S_{cd11}-S_{dc11}+S_{cc11} & S_{dd11}-S_{cd11}-S_{dc11}+S_{cc11} & -S_{dd12}+S_{cd12}-S_{dc12}+S_{cc12} & S_{dd12}-S_{cd12}-S_{dc12}+S_{cc12} \\ S_{dd21}+S_{cd21}+S_{dc21}+S_{cc21} & -S_{dd21}-S_{cd21}+S_{dc21}+S_{cc21} & S_{dd22}+S_{cd22}+S_{dc22}+S_{cc22} & -S_{dd22}-S_{cd22}+S_{dc22}+S_{cc22} \\ -S_{dd21}+S_{cd21}-S_{dc21}+S_{cc21} & S_{dd21}-S_{cd21}-S_{dc21}+S_{cc21} & -S_{dd22}+S_{cd22}-S_{dc22}+S_{cc22} & S_{dd22}-S_{cd22}-S_{dc22}+S_{cc22} \end{bmatrix} \tag{10.4.8}$$

$$\mathbf{M_1} = \frac{1}{2\sqrt{2}} \begin{bmatrix} \frac{1+k_o}{\sqrt{k_o}} & -\frac{1+k_o}{\sqrt{k_o}} & 0 & 0 \\ 0 & 1 & \frac{1+k_o}{\sqrt{k_o}} & -\frac{1+k_o}{\sqrt{k_o}} \\ \frac{1+k_e}{\sqrt{k_e}} & \frac{1+k_e}{\sqrt{k_e}} & 0 & 0 \\ 0 & 0 & \frac{1+k_e}{\sqrt{k_e}} & \frac{1+k_e}{\sqrt{k_e}} \end{bmatrix} \tag{10.4.9}$$

$$\mathbf{M_2} = \frac{1}{2\sqrt{2}} \begin{bmatrix} \frac{1-k_o}{\sqrt{k_o}} & -\frac{1-k_o}{\sqrt{k_o}} & 0 & 0 \\ 0 & 1 & \frac{1-k_o}{\sqrt{k_o}} & -\frac{1-k_o}{\sqrt{k_o}} \\ \frac{1-k_e}{\sqrt{k_e}} & \frac{1-k_e}{\sqrt{k_e}} & 0 & 0 \\ 0 & 0 & \frac{1-k_e}{\sqrt{k_e}} & \frac{1-k_e}{\sqrt{k_e}} \end{bmatrix} \tag{10.4.10}$$

VNA connections are now considered to perform two-port measurements on a four-port network, for which we have just defined the mixed-mode representation and transformation matrix (10.4.7). Such a network is made of a differential pair that is accessible for single-ended measurements, so that we can connect each of VNA ports between each conductor and the common reference plane. Such measurements include both differential return/insertion loss and near-end and far-end crosstalk. Cable connections for each insertion scheme are reviewed in the next section.

10.4.3 Mixed-mode parameter measurement

In the following the attention is on the minimum configuration to speak about mixed-mode parameters, that is 2 mode ports at each end of the transmission line (or device) and thus a 4×4 mixed-mode S-parameter matrix. A complete and accurate presentation of generalized mixed-mode S parameters can be found in [125].

10.4.3.1 General scheme

The typical setup for the measurement of not only differential return and insertion loss, but also in particular near-end and far-end crosstalk (see sec. 8.3.4), for two differential pairs is considered. Four conductors are arranged as two coupled differential pairs plus a reference plane: as normally done one line is called "source" (or "aggressor") and the other "victim". Connections of VNA cables are made single-ended (between a single conductor and the common reference plane), so that the reference plane is used to terminate the unused conductors and to define a ground reference (see Figure 10.4.3 and 10.4.4).

When considering Far-End Crosstalk (FEXT), the insertion scheme is much similar (see Figure 10.4.4), but the VNA ports are normally connected at different line ends.

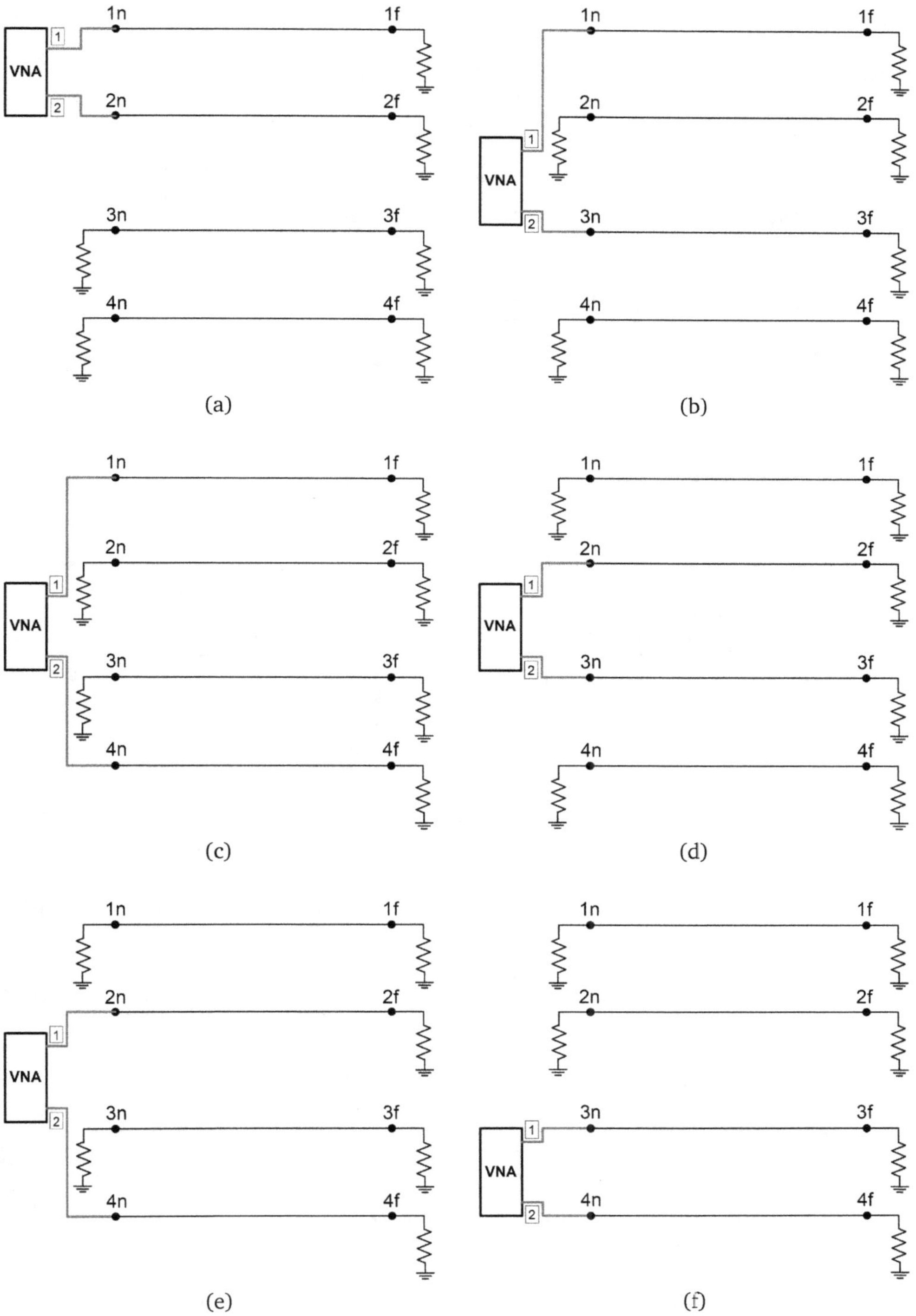

Figure 10.4.3 – Connection scheme for single-ended measurement of Near-End Crosstalk, NEXT, of coupled differential pairs.

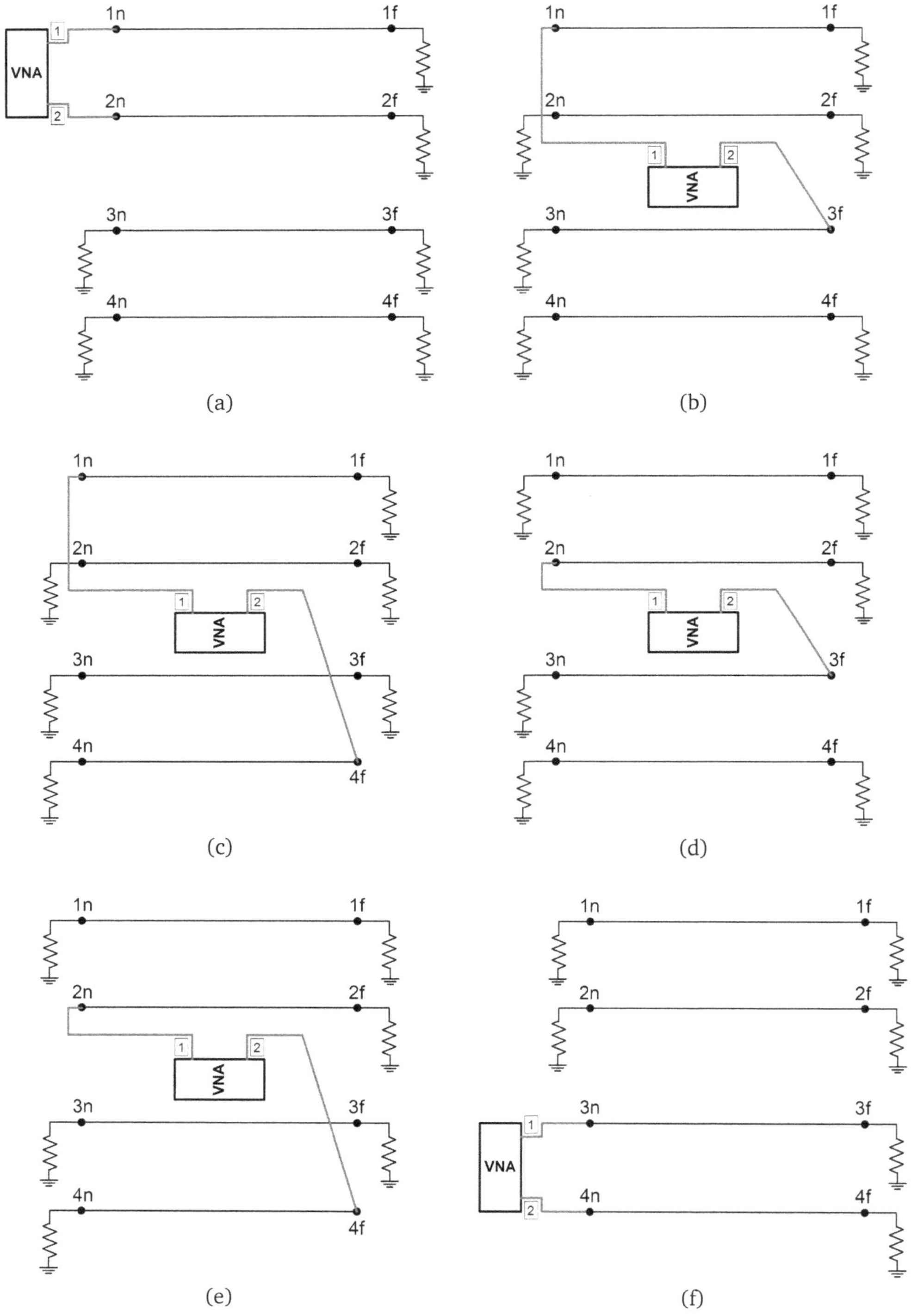

Figure 10.4.4 – Connection scheme for single-ended measurement of Far-End Crosstalk, FEXT, of coupled differential pairs.

Although it is a long procedure that requires connections and disconnections and then some off-line calculations, the measurement of differential pair parameters (and in particular of near-end and far-end crosstalk transfer functions) by means of single-ended measurements is possible and gives accurate results. One important factor is that all unused lines during a measurement shall be terminated and the necessary matched condition for termination is not on the characteristic impedance of the line, but the reference impedance of the measuring system (e.g. as usual $50\,\Omega$).

Fan *et al.*[5] show that the difference between mixed-mode S parameters calculated from single-ended measurements and originally measured with a 4-port VNA with balanced ports is negligible: despite the troublesome quality of plots, observed differences between curves measured in a differential microstrip setup up to 6 GHz are in general a fraction of dB, the largest differences occurring at low frequency, when the measured values are very small for some quantities, i.e. dd or cc; the largest differences are observed for mode-conversion curves, S_{cd21} and S_{dc21} that are not symmetrical any longer above 2 GHz, thus indicating a more significant discrepancy (a few dB on average, due to misplaced small resonances) even focusing attention to one measurement method only, e.g. single-ended S parameters (the cheapest one).

10.4.3.2 Port termination

Depending on the characteristics of the DUT accurate termination of unused ports may be more or less relevant [15]: for "multilateral" devices (featuring a significant transmission between more than two ports, such as couplers, power dividers, etc.) load match error is relevant and for accurate measurements should be corrected; conversely, for devices that are single path (such as cables and planar transmission lines) even approximate termination may work (e.g. by means of attenuators rather than matched loads). Correction is achieved by extending the calibration of the VNA from the usual two port (sufficient for single-path devices) to e.g. a four-port calibration, accounting for the influence of multilateral propagation paths.

One may wonder if it wouldn't be better to terminate the lines on their own characteristic impedance to avoid reflections: this is physically true, but what rules is the reference impedance used during the calibration; if a two-port measurement of ports i and j is turned into a N-port measurement using a N-port VNA, the remaining DUT ports would be connected and terminated onto the other VNA ports; for the two measurements to give (nearly) the same result the termination is to be done on the reference impedance of the calibration. However, it is not forbidden remaining on a two-port calibration scheme and terminating always the unused ports on their characteristic impedance: they will be seen as other lateral paths with a smaller reflection; what is important is then to terminate correctly single-ended or differential ports onto the correct common-mode or differential-mode impedance.

The benefit from a N-port calibration including and correcting for load-match error clearly depends on the characteristics of the DUT: if transmission terms are large, then load-match errors and improper termination will be quite relevant. When single-

[5] W. Fan, A. C. W. Lu, L. L. Wai and B. K. Lok, "Mixed-Mode S-Parameter Characterisation of Differential Structures," Electronics Packaging Technology Conference, 2008, pp. 533-537.

ended S parameters are combined for the calculation of mixed-mode S parameters, a great deal of variability is expected if the single quantities are of the same order of magnitude and cancellation/compensation by subtraction (or phase displacement) is affected by small discrepancies. A typical example is the measurement of a balun (single-ended unbalanced port 1, differential balanced ports 2 and 3): a large differential-mode transmission and a low common-mode transmission are expected; these two quantities are given by the sum and the difference of S_{21} and S_{31}; in particular for the common-mode transmission to be low a nearly perfect 180° phase displacement is expected (as known, S_{cc21} is a measure of the device balance).

10.4.3.3 Connectors and launchers

Depending on frequency range and characteristics of transmission lines and fabrication technology, launching connectors may be a significant element in the tiny return loss budget, and quite often they offer worse matching than the whole planar line under test (see below). Techniques to compensate for their effect will be considered in Chapter 11 and consist of de-embedding, time gating and inclusion in the calibration (when suitable calibration standards for the planar transmission line are available). If the attainable spatial resolution (related to the available bandwidth) is compatible with the setup geometry and separation between elements, time gating is quite effective and may be applied for both time- and frequency-domain measurements.

10.4.3.4 Symmetry exploitation

When a symmetric differential pair is considered, the measuring task is simplified: we don't have to measure every port to determine all the components of the equations for mixed-mode quantities, e.g. $S_{dd21} = (S_{21} - S_{23} - S_{41} + S_{43})/2$. Considering ideal symmetric traces, we may assume $S_{21} \simeq S_{43}$ and $S_{41} \simeq S_{23}$. The equation then simplifies to $S_{dd21} = (S_{21} - S_{41})/2$, that may be carried out by exciting port 1 and measuring port 2 and port 4, to extract the desired S_{dd21}. Performing the complete set of measurements for the remaining ports may be an instructive and useful double-check, that verifies the difference between S_{21} and S_{43}, and S_{41} and S_{23}.

Of course, assuming symmetry without verifying it is hazardous, and we recall that, only in case of close similarity between the individual lines of the differential pair, mode-conversion terms ("cd" and "dc") are small, as observed by Pupailakis [276], page 12, and commented also by Loyer and Kuntze [226], page 5.

10.4.3.5 Differential test signals

Conversely, it is also possible to feed differential signals into the circuit, even if using always the same VNA with single-ended outputs: a differential voltage is available at the output of baluns, 180° hybrids and "rat-race" devices. The use of such devices shall be traded off with their limitations that affect setup performance and uncertainty. Usually VNA balanced-port option is not so expensive and may be worth of, simplifying setup and guaranteeing accuracy and performance over the whole frequency range.

Several low-power baluns in small package and often SMD format are available for telecommunication and radiofrequency applications; many have operating frequency extended in the GHz range, usually never above a few GHz. Baluns are obviously necessary for stand-alone permanent installations, but their use for measurements is still recommended by several standards: an example is shown in Figure 5.1.2 for cable crosstalk measurements discussed in sec. 5.1.6.

However, the use of baluns is exposed to inaccuracy and errors, that can only be partially compensated. The lack of differential calibration standards to connect at the differential side of the balun (facing the DUT), requires that calibration is performed on the input single-ended side where the calibration plane will be located, thus compelling the VNA measurement to include the balun with the DUT. The balun thus shall be "subtracted" *a posteriori*, either by a simple correction for insertion loss, or by a more complete de-embedding: to this aim the balun shall be separately characterized.

An additional issue related to the use of a balun is that the balun does no transmit common mode and thus no mode conversion can be evaluated: pure differential mode responses can be measured. Additionally, focusing on the balun connected at the end of the DUT, the fact that the balun does not "pass through" common mode implies that the reflection for common-mode signals is maximum (either $+1$ for a two-terminal floating balun differential side, -1 for a three-terminal center-tap balun differential side); such reflection goes back through the DUT again partially converted to a differential signal because of mode conversion and partially reflected again. This causes a significant error[6] for all measurements, unless the common-mode signal is offered a way out: this is the reason for match terminating with the correct common-mode impedance put in the middle of baluns.

As a last practical consideration, connection between DUT (and its baluns) and VNA shall be maximally balanced, e.g. using same cables of same length, paying attention to unwanted loading and coupling effects, etc.. A valid countermeasure to any slight asymmetry (as small length differences) is to use de-embedding and port extension, that take into account both electrical length mismatch and losses of connections.

10.4.3.6 Line quantities from single-ended and mixed-mode S parameters

Some quantities that characterize the tested cable or transmission line, already reviewed in Chapter 2 and sec. 5.1.6, can be calculated from measured S parameters.

From the same definition of mixed-mode S parameters there may be defined: return loss (S_{dd11}) and insertion loss (S_{dd21}) for the useful differential signal, common-to-differential mode conversion (S_{cd21}), longitudinal conversion loss, LCL (S_{dc11}) (see sec. 5.1.8). The original single-ended parameters give near-end crosstalk, NEXT (S_{31}) and far-end crosstalk, FEXT (S_{41}), as well as insight on line symmetry at each end of the same conductor or for different conductors (line balance). The determination of mixed-mode parameters from a set of measured single-ended ones has no limitations to the number of ports, for which the number of lines and mixed-mode ports

[6] The equations for reflection-caused errors may be found in: Agilent, "Multiport & Balanced Device Measurement Application Note Series – Concepts in Balanced Device Measurements," *Application Note 1373-2*, 2002.

can be arbitrarily large (see sec. 5.1.6.3 and the measurement scheme shown in Figure 5.1.3): in a multi-pair cable (e.g. a LAN cable with four pairs) there is a number of differential and common mode ports equal to the number of pairs; for completely decoupled pairs the resulting mixed-mode S parameter matrix is block diagonal; the off-diagonal blocks allow to evaluate pair-to-pair coupling for the differential component (e.g. S_{dd31} or S_{dd41}) and common-to-differential transformation during coupling (e.g. S_{cd31} or S_{cd41}); the latter define the amount of crosstalk between pairs (named alien, or exogenous, crosstalk).

Considering again a single differential pair and a the 4×4 mixed-mode S matrix, properties and parameters of the line can be derived as already partially considered: let's focus on propagation delay, propagation constant, its real part the attenuation constant, and characteristic impedance values.

The propagation delay τ is readily available from S_{dd21} by either turning on the corresponding option in the VNA during measurement or by taking the phase curve and de-trending it by adding $f\tau$ and adjusting the amount of delay necessary to have the phase curve nearly flat (see sec. 3.3.1 on group delay and phase response). The difference of propagation delays, or delay skew, is given by the difference of the two group delay estimates based on S_{21} and S_{43}.

The measurement of characteristic impedance of various configurations and modes (each single-ended conductor, odd and even mode, differential and common mode) shall be in principle done in time domain by measuring T parameters with TDR/TDT technique and taking the final values reached by the flat portion of traces once initial reflection at launchers has vanished. Invoking the Final Value Theorem, this may be accomplished considering the value of the low-frequency portion of S parameter curves, and using the following correspondence[7]: S_{11} and S_{33} give the single-ended characteristic impedance $Z_{c,se}$ and shall be equal for balanced symmetrical lines; $S_{11}-S_{13}$ and $S_{33}-S_{31}$ will give the odd-mode characteristic impedance $Z_{c,o}$, $S_{11}+S_{13}$ and $S_{33}+S_{31}$ will give the even-mode characteristic impedance $Z_{c,e}$, the differential characteristic impedance $Z_{c,dm}$ is the half-sum of the two odd-mode characteristic impedances. A more complete expression involving mixed-mode parameters is[8]

$$Z_{c,dm}=Z_{ref}\sqrt{\frac{[(1+S_{dd11})(1+S_{dd22})-S_{dd12}S_{dd21}]\,[(1+S_{dd11})(1-S_{dd22})+S_{dd12}S_{dd21}]}{[(1-S_{dd11})(1-S_{dd22})-S_{dd12}S_{dd21}]\,[(1-S_{dd11})(1+S_{dd22})+S_{dd12}S_{dd21}]}} \qquad (10.4.11)$$

This expression based entirely on differential mixed-mode S quantities is indeed interesting, but exposes to approximations due to the many transformations, if starting from original raw measured single-ended S parameters, and requires that all "dd" S parameters are in a suitable range of validity for the whole frequency range, e.g. not too small to reduce noise effects and VNA inaccuracy, far from resonances, etc. Sim-

---

7 The ideal time-domain waveform is a step signal that cannot be readily transformed into frequency domain, unless correction is applied; with no correction the error is however limited to a fraction of dB or few dB in very extreme cases (see sec. 3.1.4 and 11.10, in particular sec. 11.10.3).

8 This expression appears in Agilent leaflet "Balanced Cable Measurement using the 4-port ENA - FAQs", rev. 031014.

ply speaking, the cable sample shall be not too short and not too long, e.g. between several meters and a hundred meters. Needles to say that when moving to differential components the reference impedance shall be changed accordingly.

We know that the characteristic impedance may come from the line input impedance (S_{dd11}) and the combination of two measurements, in short and open conditions at the far end (see sec. 2.4). Despite the latter is a different additional measurement, but may be worth being done changing the far end termination from matched load to the other two conditions (short and open), it represents a further check of the previous mixed-mode measurements; in case of disagreement, some elements may be cross-checked and verified: at high frequency matched loads may deviate from a clean resistive load; correct reference plane extension behind/beyond the baluns; use of a correct model for baluns; naked unbraided wires used for connection shall be same length and same routing path.

Similarly, when making crosstalk measurements any unbalance of line termination onto its characteristic impedance will raise the measured curve: while the expected behavior in normal conditions is a low value at low frequency increasing almost linearly beyond some critical frequency (e.g. some MHz or tens of MHz), unbalanced terminations will add a nearly constant value.

10.4.3.7 Data formats

This topic is gaining more and more relevance as long as simulation packages and analysis toolboxes are used for a variety of functions: evaluation of circuit modifications and improvements, and in general provisions and countermeasures that shall be simulated before implementation; parameter extraction by curve fitting of measured results; de-embedding support; simple post-processing of results for displaying purposes; analysis of derived quantities, such as eye diagram, bit error rate, etc. using an integrated approach, where the measured characteristics are embedded into a simulated system (e.g. transmitting and receiving devices).

The preferred data format for the widest acceptance and exchange, that constitutes also a *de facto* standard, is the Touchstone "SnP" format, with the most common for the reviewed examples S2P (single-ended S parameters, n=2 the size of matrices), also indicated as `.s2p` format, and M4P (mixed-mode S parameters, n=4 the size of matrices). Alternatively, S parameters may be almost always fed directly using some import/export utility; simulation results also can be displayed in terms of S parameters keeping the representation homogeneous.

We may distinguish between circuit-oriented tools and electromagnetic-field-oriented tools: without any particular endorsement we may cite among the former HSpice Synopsis® and Agilent ADS® (Advanced Design System), while for the latter HFSS® and CST®. Simulation programs tailored to signal integrity analysis, including comparison with measurements, data extraction and model fitting (e.g. for PCB and connector parameters), are less widespread, but well known to specialists (e.g. Simbeor® by Simberian).

Matlab RF Toolbox also provides methods for S-parameter block connection and analysis, with the possibility of loading Touchstone-format descriptions. Scikit-rf[9], an open-source project for a library of mathematical tools for RF and microwaves, uses Touchstone format has the pivot format with which transformations between network matrix representations are performed (see sec. 1.3).

The specifications for the Touchstone format and some examples are shown in Figure 10.4.5 for the most common cases.

10.4.4 Analogy with time-domain TDR/TDT parameters

So far we have considered and used S parameters in frequency domain. What happens if we perform measurements of time-domain reflection (TDR) and time-domain transmission (TDT)? Do we have an equivalent representation of the scattering matrix for time-domain quantities?

The answer is "yes", and the two sets of parameters reviewed so far (single-ended and mixed mode scattering parameters) may be mirrored into time domain keeping the same structure, just changing the "S" in "T"! Again single-ended and mixed-mode T matrices are related by the same type of equations seen above (eq. (10.4.7) and (10.4.8)), but the two domains (time and frequency) are also related, so that with the right transformation all other three matrices can be calculated from any of the four.

When thinking of relationships between time-domain and frequency-domain quantities, it is intuitive that they form a DFT/IDFT pair, either taken singly or under some constraints. The time-domain "T" parameters may be measured with a step or impulse excitation signal: the former bears more energy and reveals details on the type of line discontinuity, while the latter is the one more directly linked to the frequency domain representation under the DFT/IDFT pair assumption (see sec. 3.1.4). As already commented the step is readily available from the calibration connector of oscilloscopes and better reveals line discontinuities, easing the interpretation task (see sec. 11.10.3.1 and 11.10.5). Simply saying that the impulse response may be obtained from the step response by simple differentiation does not take in the due consideration practical problems related to the increased bandwidth and high frequency noise, besides frequency leakage due to truncation. A similar issue was pointed out many years ago, when approaching the problem of getting a frequency-domain representation from step response data (see sec. 3.1.4). However, it is known that the impulse excitation signal benefits from being directly connected to the frequency response of the system and its transformation gives readily the frequency domain quantities, at the expense of poor dc definition and a worsened signal-to-noise ratio.

Starting from waveforms as recorded by a TDR or oscilloscope, when cutting a signal slice over a defined observation window, to ease transformation and interpretation, there might be the need to enforce a few properties: *causality*, that is samples are put to zero before the time instant that is taken as the time origin $t = 0$ and that usually slightly anticipates the most relevant part of the signal, such as the main reflection;

9 scikit-rf, Open-source RF Engineering, [online]: http://scikit-rf-web.readthedocs.org/# (last accessed Aug. 2015)

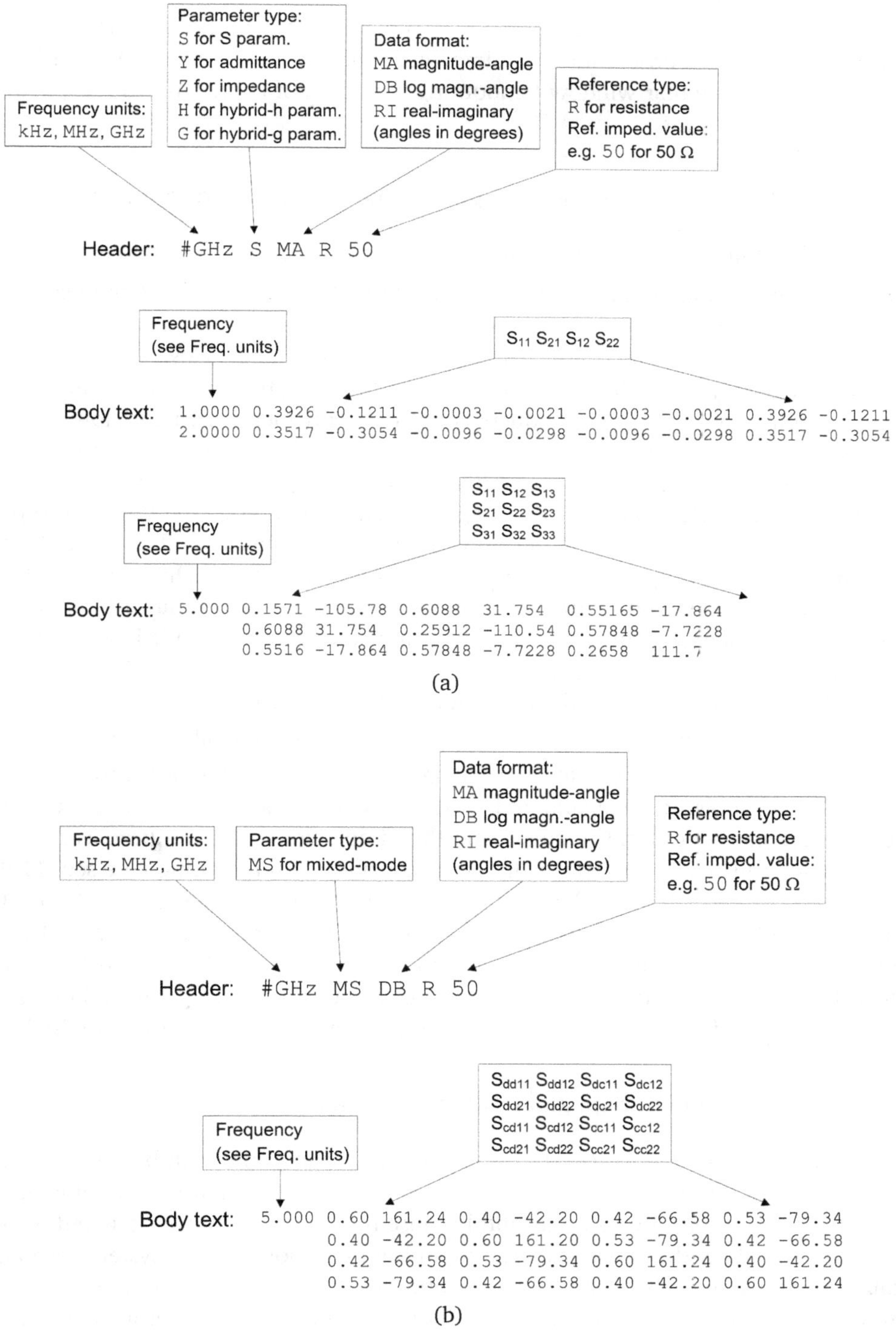

Figure 10.4.5 – Example of Touchstone format for (a) single-ended S parameters `.s2p` and `.s3p`, (b) mixed-mode S parameters `.m4p`. (Source: Touchstone® File Format Specification, rev. 1.1, 2003)

time limitation, putting to zero the samples beyond the end of the established observation window. Both properties are in reality implicit in the definition of windowing of data records over the observation window, that in its simplest implementation is the product for a rectangular window (smoothing windows, of course, may be applied as considered in sec. 3.1.3.2).

10.4.4.1 Advantages and disadvantages of time and frequency domain methods

What are the main differences or the advantages/disadvantages of the two techniques?

For the frequency domain methods considered so far the signal-to-noise ratio is surely the best feature, that brings along protection against external disturbance, while sweeping a specific frequency, and ensures a good dynamic range. However, frequency-domain instruments have the problem of a minimum operating frequency that despite being brought low in the new products is not dc! This means that for slow or long circuits impedance changes may be badly recognized and tracked. Time-domain measurements are more "natural", in that they resemble the classical representation of transmission lines with bounce lattice diagram seen in Chapter 2, and they can be performed by more common and better known equipment, that is oscilloscopes. Moreover, modern oscilloscopes have room in the channel memory for hundreds of thousands samples, contrasted to frequency domain equipment capabilities limited to a thousand samples or so. The oscilloscope is a must for every electronic laboratory and modern oscilloscopes have several good features on-board, including complex mathematical libraries, not to say math environment (e.g. Matlab). The price of course has increased for increasing performance, as for frequency-domain equipment.

The two instruments turn out to be exchangeable in many situations, provided that the right math tools are available and used correctly: basics of time-frequency relationships are considered in Chapter 3 and then implementation and practical details are reviewed in Chapter 11, especially when considering time gating (see sec. 11.9), windowing (see sec. 11.10) and low-pass and band-pass modes (see sec. 11.10.3). The necessary mathematical tools consist of DFT or CZT pairs (see sec. 3.1.1 and 3.1.2), known and reliable procedures for normalization, adjustment of time and frequency intervals, windowing and smoothing (going for a direct straightforward relationship between time and frequency domain quantities as a transform pair is over-simplifying).

10.4.4.2 Calibration in time and frequency domain

When performing native TDR/TDT measurements, calibration shall be done in time domain: details that were relevant in frequency domain (e.g. non-ideality of connectors, cables and so-on) are still relevant, because the frequency range occupied by the excitation signals is the same; calibration still aims at normalizing system response against known standards. Again reflective and matched standards are used, focusing also on time skewness (the role of the calibration plane here is to set a "de-skewed" time reference for all channels, compensating slight delay differences and pulse deformation). When working on differential signals, however, the correct reference value of impedance is twice the single-ended impedance, that is to say $100\,\Omega$; calibration is performed on each single port using 50 Ω standards, that is to say that the connection standard is always a single-ended arrangement (signal on center conductor, ground on external body) combined in pair to get a differential line.

Slightly oversimplifying we may say that calibration in time domain corresponds more or less to adjusting voltage levels (as shown in sec. 6.5.2) estimating longitudinal voltage drops, determining propagation times and round-trip times of major discontinuities, and estimating characteristic impedance with respect to observed step voltage changes. Such operations may be performed on the final setup or on partial setups that allow to isolate elements, such as launchers and transitions, or to terminate lines as desired. Feeding cables and connectors shall be included as for preliminary deskew and as it would be with any frequency-domain calibration. Frequency-domain calibration is treated quite extensively in Chapter 11: a combination of de-embedding, calibration for VNA error model solution and time gating is performed at the desired or accessible reference planes, for which calibration standards are available.

10.4.4.3 Example of combined time- and frequency-domain measurements

Pupalaikis [276] presents a complete set of mixed-mode S-parameter measurements, both in time (indicated as "TDR") and in frequency (indicated as "VNA") domain up to 20 GHz; the setup is a couple of microstrips running in parallel between near- and far-end connectors. Single-ended S parameters were measured and then converted to mixed-mode representation with the expressions in sec. 10.4.2. Mode-conversion terms (where "c" and "d" are paired, evaluating common-to-differential conversion and viceversa) are low, as customary for a well-designed balanced pair of lines; coupling between them is not relevant, they shall be symmetric for these mode-conversion terms to be low (e.g. below $-20\,\mathrm{dB}$). Again confirming line symmetry, matrices are symmetric with $S_{dd11} = S_{dd22}$ and $S_{dd12} = S_{dd21}$ (see sec. 10.4.3.4). It is underlined that line symmetry and small values of mode conversion are not sufficient conditions to establish that the lines have good performance as far as return and insertion loss in differential mode. An example is a well-balanced symmetrical line pair resulting during the development of a test setup for high-density connectors (deviation between single-ended curves lower than $-30\,\mathrm{dB}$ and equal to $-20\,\mathrm{dB}$ at $2\,\mathrm{GHz}$) with mode conversion S_{cd21} lower than $-40\,\mathrm{dB}$ up to $2\,\mathrm{GHz}$, but poor return and insertion loss S_{dd11} and S_{dd21} anyway, larger than $-10\,\mathrm{dB}$ and $-3\,\mathrm{dB}$ at about $1.5\,\mathrm{GHz}$.

Propagation speed for the two line modes is verified during phase unwrapping: the amount of delay used to de-trend the phase curves for differential- and common-mode S parameters (3.7 ns and 3.9 ns, respectively) is made correspond to propagation delays of odd and even modes. When TDR measurements, or transformed VNA frequency-domain measurements, are available a confirmation of propagation delay comes from the measurement of the time that elapses between the two discontinuities at the near-end and far-end launchers; provided that the line is long enough, approximations in spotting out the reference plane and round-offs are absolutely irrelevant.

Two facts are pointed out, illustrating very clearly how the reasoning led to identifying the launchers as the main reason for the observed inconsistency. First, ripple appearing in S parameter curves is unavoidable because of residual mismatches and is in relationship to line length; any longer ripple thus due to a shorter characteristic length is suspicious. Second, the fact that the return response phase may be de-trended with a very short time delay is again suspicious, indicating that the most relevant reflecting mismatch is very close to the input reference plane; probably this might be already spotted out by examination of time-domain reflection response. The conclusion confirmed that the launching connectors are responsible for this, because the center pin was not trimmed and behaves like a resonating stub. Solutions are many: if one can modify the board by trimming or backdrilling, then the problem is solved at the origin. Pupalaikis proposes de-embedding connectors, but does not give details on how the connector matrix representation was obtained: it looks like the recorded frequency responses are time gated to extract the portion related to connectors, that by the way are clearly responsible for the large original return loss. However, being able to perform time gating to extract connector response for de-embedding, the same might have been done to gate out both connectors and recover the correct frequency response. The only reason to use de-embedding is to correct the information obtained with time gating by fitting some equivalent circuit, thus removing largely noise and leakage artifacts. Results appearing in the paper go on showing features of time-domain measurements and the overall evaluation of line behavior with respect to time jitter and eye diagram.

10.5 Uncertainty propagation with S parameters

As anticipated at the beginning of this chapter, the raw measured data returned by the VNA are often used to determine other quantities of setup and of device under test. With a modern measurement approach this operation is accompanied by the evaluation of uncertainty: besides the knowledge of the accuracy of VNA and uncertainty of its calibration, and impact of setup and environment, propagation of uncertainty manipulating S-parameter quantities is a relevant element of the whole picture. Despite being welcome with a shrug, this statement implies that dB quantities, expressed as complex numbers, shall be correctly handled in terms of their probability density function (PDF, see sec. 4.1), degrees of freedom, confidence intervals and so on. With more and more sophisticated and accurate VNAs, the attention to this part of the overall uncertainty budget estimation has recently increased, being possible to appreciate second order details and phenomena, otherwise judged as irrelevant.

Some useful works have appeared starting with the EA (European co-operation for Accreditation) guideline for the evaluation of VNAs around 2000, followed immediately by the papers by Hall [143, 144, 145, 146], Ridler and Salter [282, 283], and Yhland and Stenarson [309, 346], that taken altogether treat in detail all relevant aspects to satisfactorily handle S parameters data and related uncertainty.

The foundations of uncertainty in measurements are contained in the reference BIPM publication, the GUM (Guide to Uncertainty in Measurements), that dates back to 1995, with some modifications in the 2008 revision [39]. While in general the Type A statistical approach based on repeated measurements is applicable to any instrument or setup with a more or less uniform approach, the analytical propagation of uncertainty, or Type B approach, is setup or instrument specific. The latter is the GUM preferred approach for expression of uncertainty, but may be inapplicable when expressions between quantities are complex, contain singular points and in general are not easily differentiable, or are partly based on approximations and assumptions. This is however the approach followed in most of the software for calibration and uncertainty management reviewed in sec. 11.5.3.9, except for MMS4 that gives a real-time estimate of the influence on uncertainty budget of setup quantities.

In the following the attention is focused on propagation of uncertainty for scattering parameters, that is a hot topic that has recently revealed far from trivial.

Ridler and Salter [282] introduce very clearly the problem of applying known methods (such as sample mean, standard deviation, correlation and covariance matrix, confidence intervals) to S parameters, deciding e.g. if polar or Cartesian coordinates should be used, if correlation between quantities exists because of their underlying implicit relationships, which PDF is to be assumed and the extension of the confidence interval.

A complex quantity expressed in real and imaginary parts extends the domain from $-\infty$ to $+\infty$, analogously to real numbers we are used to, but when expressed in magnitude and phase things are changing: the former is positive or zero, while the latter has a cyclical scale and is bounded between $-\pi$ and $+\pi$, or 0 and 2π (with a minor doubt whether the interval is open or closed at the two extremes).

The positiveness of the magnitude scale is the objective of the first example: a return loss measurement of a perfect matched load will give ideally the result $|S_{11}|=0$, but for unavoidable measurement errors and noise the measured values will be somewhat positive, and using averaging to increase the precision has no effect (the result is biased), because there are no negative values to balance the positive ones, while we know that the "right" value is 0, as confirmed by a more complete 2-D plot of magnitude and phase (or real and imaginary parts), where points distribute as a cloud around the origin. From a distribution viewpoint the magnitude-only representation has a Rayleigh PDF, while the 2-D representation will feature a bivariate distribution averaging more or less to zero.

Conversely, operating on phase exposes to the weirdness of the modular behavior of the phase curve, reclosing every turn: two values slightly behind and beyond the $0°$ boundary, such as $+1°$ and $-359°$ if averaged should give 0, but arithmetic on values as they are says "$-179°$"!

It is clear then that the most important advice is to operate on real and imaginary components, always converting back and forth from the original polar values, if necessary.

As observed for sample mean (and any arithmetic operation), the same may be said for sample dispersion, that is the first step to calculate the Type A uncertainty from observations. Applying to the original magnitude and phase data is not possible without getting wrong results. After conversion to real and imaginary values, expressions given in sec. 4.1.1 and 4.1.4 may be used.

Since the attention is focused on uncertainty, and the sample dispersion is the basic uncertainty estimate as a measure of dispersion, the notation $u\{\cdot\}$ will be used in place of the letter s or σ for sample dispersion and standard deviation.

The peculiarity of complex quantities such as S parameters is that in the magnitude and phase representation the probability distributions are and look different, and that often the knowledge of the underlying distribution and related uncertainty privileges the magnitude, rather than the phase. In this case the uncertainty interval is seen as a band around a region with circular symmetry in polar coordinates: a ring shaped distribution of the magnitude centered at the value a indicates that the magnitude average value is a but the phase fluctuates randomly; if the thickness of the ring is not zero, then also the magnitude is affected by uncertainty and the thickness of the band indicates the expected interval where magnitude values may lie. Transformed into real and imaginary parts such distribution takes the form of an $\arcsin(\cdot)$ function in either domain (that of the real or imaginary part) with limits $\pm a$: the resulting standard deviation is $a/\sqrt{2}$.

In general, we might be interested in understanding if the real and imaginary components X and Y of a given complex S quantity are independent or not, and how to estimate the degree of correlation. The reason is that the propagation of uncertainty shall take into account the correlation between factors as second-order mixed terms. Starting from the expression of correlation coefficient, that takes the covariance between two quantities and normalizes dividing by the product of the respective standard deviations, correlation between real and imaginary components (as well as any other two quantities of the problem) is described completely by the second-order central moments.

$$S = X + jY \qquad |S| = \sqrt{X^2 + Y^2} \qquad \phi = \angle S = \arctan(X/Y) \tag{10.5.1}$$

$$\mathbf{V}(S) = \begin{bmatrix} \mathrm{var}(X) & \rho\,\mathrm{std}(X)\mathrm{std}(Y) \\ \rho\,\mathrm{std}(X)\mathrm{std}(Y) & \mathrm{var}(Y) \end{bmatrix} = \begin{bmatrix} u^2(X) & \rho\, u(X)u(Y) \\ \rho\, u(X)u(Y) & u^2(Y) \end{bmatrix} \tag{10.5.2}$$

The notation u indicates the uncertainty and is made corresponding to dispersion (standard uncertainty). The so-defined covariance matrix is usually symmetric. While in general off-diagonal terms are much smaller than diagonal terms for nearly uncorrelated or weakly correlated quantities, in the case of real and imaginary parts of the quantity the correlation is in general quite large and the four elements of the matrix are usually of the same order of magnitude.

Ridler and Salter also ask an interesting question: if the so determined correlation is a reliable estimate of the real correlation between real and imaginary components. The correlation-coefficient estimate ρ for pairs of real and imaginary component values (e.g. a vector of repeated measurements S at a given frequency) may be treated as any other random variable, for the determination of confidence interval and level of confidence (i.e. probability). It is underlined that the number of available measurements is generally small and this implies that the estimate of ρ will be approximate.

The r.v. ρ may be brought back to a normally distributed r.v. using the Fischer's z-transformation[10], $z = \frac{1}{2}\ln\frac{1+\rho}{1-\rho}$ and $r = \tanh(z)$. The resulting standard deviation for z is $1/\sqrt{n-3}$, where n is the number of samples with which we estimate the

[10] R.A. Fischer, "On the probable error of a coefficient of correlation deduced from a small sample," *Metron*, Vol. 1, no. 4, 1921, pp. 3-32.

correlation coefficient ρ. Once an adequate coverage factor is decided (e.g. $k=2$) and the corresponding confidence interval is characterized by a minimum and a maximum value in z, they are converted back to determine the corresponding minimum and maximum values of ρ, that define its confidence interval. This interval will not be symmetric.

In general, the measurand $\mathbf{y}$ depends on m input quantities x_i, each one a complex variable, as complex is the function $f(\cdot)$:

$$\mathbf{y}=f(x_1,x_2,\ldots,x_m)=f(\mathbf{x})=f_{\text{re}}(\mathbf{x})+jf_{\text{im}}(\mathbf{x}) \tag{10.5.3}$$

Under linearity assumption, the covariance matrix $\mathbf{V}(\mathbf{y})$ takes then the dimension $2m\times 2m$, considering the fact that all variables and functions are complex. The propagation from the covariance of $\mathbf{x}$ to that of $\mathbf{y}$ is determined with the Jacobian of f:

$$V(y)=J(y)V(x)J'(y) \tag{10.5.4}$$

Considering again our complex quantity S with respect to its real and imaginary parts X and Y, the Jacobian is calculated by taking the first partial derivatives of $|S|$ and ϕ with respect to X and Y [244]:

$$\mathbf{J}\begin{pmatrix}|S|\\ \phi\end{pmatrix}=\begin{bmatrix}\dfrac{X}{\sqrt{X^2+Y^2}} & \dfrac{Y}{\sqrt{X^2+Y^2}}\\[2ex] -\dfrac{Y}{X^2+Y^2} & \dfrac{X}{X^2+Y^2}\end{bmatrix}=\begin{bmatrix}\dfrac{X}{|S|} & \dfrac{Y}{|S|}\\[2ex] -\dfrac{Y}{|S|^2} & \dfrac{X}{|S|^2}\end{bmatrix} \tag{10.5.5}$$

Propagation from X and Y to $|S|$ and ϕ finally gives:

$$\mathbf{V}\begin{pmatrix}|S|\\ \phi\end{pmatrix}=\begin{bmatrix}\dfrac{X^2\text{var}(X)+Y^2\text{var}(Y)}{(X^2+Y^2)}+2\rho\dfrac{XY\text{std}(X)\text{std}(Y)}{(X^2+Y^2)}\\[2ex] \dfrac{X^2\text{var}(X)+Y^2\text{var}(Y)}{(X^2+Y^2)^2}-2\rho\dfrac{XY\text{std}(X)\text{std}(Y)}{(X^2+Y^2)^2}\end{bmatrix} \tag{10.5.6}$$

The square root of the two lines of the variance gives a direct estimate of the uncertainty for magnitude and phase. For the coverage factor to use for an assigned probability (or level of confidence), the theory involves the F distribution (see sec. 4.1.3.7) [145]. For simpler problems there are calculated values of coverage factor that give the usually adopted 95% level of confidence.

When considering the simple case of equal uncertainty ($\text{std}(X)=\text{std}(Y)=s$) and zero correlation ($\rho=0$), that is a normally used simplifying assumption, $\mathbf{V}(|S|)=s^2$ and $\mathbf{V}(\phi)=s^2/\,|S|$. Knowing that in this case the shape of data in polar plot is circular and we are interested in the magnitude dispersion, the confidence interval for a 95% probability is $u(|S|)=2.45s$.

When the two standard deviations (or standard uncertainties) of real and imaginary parts are not equal any longer ($\text{std}(X)=s_X$, $\text{std}(Y)=s_Y$), but they keep uncorrelated

($\rho = 0$), the obtained region is not circular any longer, but elliptical. In this case with a geometrical interpretation in mind, the question arises of which axis of the ellipse consider for the estimation of magnitude uncertainty (both choices leads unavoidably to an error of under or overestimation with respect to the circular case), or even revert to an rms or average estimation. In reality, using eq. (10.5.6) the dispersion (square root of the two lines) gives the standard uncertainty (simplification occurs using $\rho = 0$). Conversely the max and the rms criterion would correspond to: $u_{\text{max}} = \max\{s_X, s_Y\}$ and $u_{\text{rms}} = \sqrt{s_X^2 + s_Y^2}$. In [244] it is demonstrated that there is a difference with respect to the exact calculation given by eq. (10.5.6), that results in a considerable saving, especially for phase uncertainty (with the phase no longer assumed uniformly distributed over 2π angle).

Extending to n-port devices with n^2 measured S parameters, the coverage factor for a level of confidence p (usually 95%) is obtained with a Chi-square distribution:

$$k = \sqrt{\chi^2_{2n^2,p}} \tag{10.5.7}$$

Table 4 in [282] gives the coverage factor for 95% level of confidence and 1 to 4 ports: 2,45, 3.94, 5.37 and 6.80.

If the number of degrees of freedom ν is finite, then the F distribution is used and

$$k = \sqrt{\frac{2\nu n^2}{\nu + 1 - n^2} F_{2n^2,\nu+1-2n^2,p}} \tag{10.5.8}$$

Table 5 in [282] shows the effect of a limited number of degrees of freedom in the case of a 1-port device: for ν = 2, 3, 4, 5, 6, 8, 10, 100, 1000 the values of k are 28.26, 7.55, 5.05, 4.17, 3.73, 3.29, 3.08, 2.50, 2.45, the latter corresponding to the value reported above neglecting the effect of the number of degrees of freedom.

11

Vector Network Analyzer (VNA)

This chapter initially presents the operating principle of Vector Network Analyzer (VNA), indicating the most relevant elements, characteristics and expected performance. The treatment is brief enough to avoid repeating what is already extensively described in manufacturers' manuals and focuses on more practical details related to its use. The aim is giving enough insight to the reader to manage the procurement of a new VNA and to understand how it works both in frequency and time domain. Nowadays there are VNAs cheap enough for purchase by a small company or even an amateur, possibly looking at reconditioned demos, used units, or the new "virtual" instruments, where saving is achieved by removing panel control and display, and as much as possible of data and trace processing is moved to the on-board DSP/FPGA or to the external computer.

The VNA is a frequency-domain instrument that allows conversion to time domain during post-processing, done in real time by the instrument itself, but a direct time-domain measurement such as TDR is not included. The many settings and options for measurement and post-processing make the VNA a powerful and complex instrument that shall be used appropriately.

11.1 VNA architecture

Architecture may vary between models: cheaper instruments may save on the number of directional couplers, mixers and IF digitizer blocks, and thus on the ability to perform a full 2-port measurement in a single sweep; depending on the maximum span, blocks may be more complex and require more than one conversion; multi-port VNAs of course are proportionally more complex than two-port models and are not much considered here, because they are slightly out of budget and normally used by a restricted group of people who by the way know well the theory behind. Similarly, non-linear VNAs are more expensive and complex equipment to characterize non-linear DUTs with a multi-frequency approach.

A general scheme, accurate enough for our purposes and quite common in for low-medium priced two-port VNAs, is shown in Figure 11.1.1. The most relevant blocks from a measurement viewpoint are:

- RF source that operates when one of the VNA ports feeds signal into the circuit (or device) under measurement; this source is thus cyclically connected to each port as needed for the specific measurement; the other ports are then connected to a matched load by the same switch system;
- IF conversion block that is represented as a mixer (or more than one for decreasing demodulation frequencies, as already discussed for Spectrum Analyzer in Chapter 9) followed by an Analog-to-Digital Converter (ADC); such blocks are many, as needed, e.g. one for each port or one for each direction of waves and each port (that is twice the number of ports); the combined IF conversion block and ADC is often called "IF digitizer";
- second RF source, the Local Oscillator, that feeds mixers for IF conversion; for multi-level cascaded IF conversion more than one LO frequency is needed (see sec. 9.1.3);
- set of directional couplers that connect ports to respective IF digitizer blocks; the overall combination of a directional coupler with the detection part (IF digitizer) is often called "reflectometer".

Then, of course, there are internal modules for data processing, storage, display, etc..

Operation and performance are considered below in sec. 11.2.

The VNA is a formidable machine and it's always amazing that all this hardware is able to deliver accurate readings over quite an extended frequency range (e.g. from kHz to several GHz and tens of GHz) and for variable configurations and signal-to-noise ratios. In reality, due to many limitations of building blocks (e.g. port directivity, port match, etc., all reviewed in the following and in particular in sec. 11.2.3 and 11.11.1), VNA performance is far from ideal, and, while keeping good raw performances, the excellent performances we are now used to are backed up by a lot of software! Code is written to correct non-ideality and unsatisfactory response of some internal blocks against known reference standards (called "calibration", see sec. 11.5), to perform

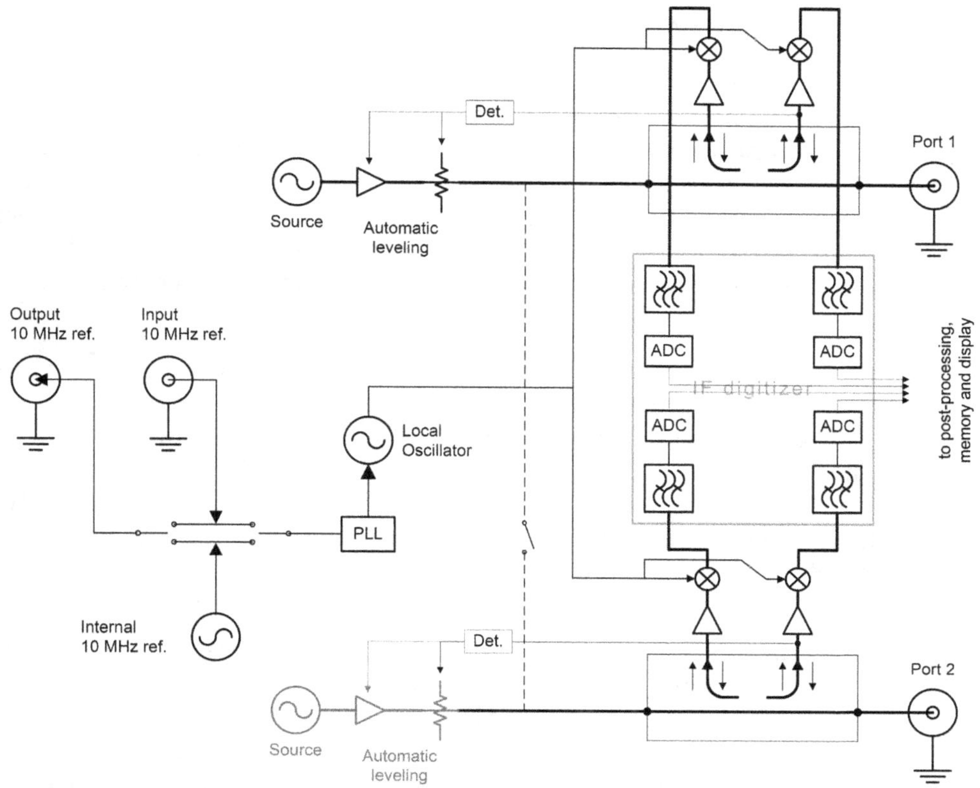

Figure 11.1.1 – Block diagram of a typical VNA architecture.

measurements in frequency, but also indirectly in time domain (see sec. 11.8 and 11.10), implementing useful functions such as exclusion or inclusion of parts of the test setup (in the original frequency domain, by embedding/de-embedding, or in the derived time domain, by time gating; see sec. 11.7 and 11.9, respectively).

11.2 VNA operation and performance

A brief look at building elements, at their operating principle and performance and any practical limitation is the objective of this section.

Frequency sweeping is one of the primary aspects; other relevant elements are how the signal is driven from VNA ports to the IF digitizing block; IF demodulation and digital sampling are performed similarly to Spectrum Analyzer covered in Chapter 9, that is equally applicable for considerations related to local oscillator, IF bandwidth and noise, minimum sweep time, etc. To this aim the VNA has less degrees of freedom related to video bandwidth and detection, the available options and maths focused onto time-frequency transformation, de-embedding, power leveling.

11.2.1 Frequency sweep

Frequency generation and sweeping is a combination of sweeping speed, frequency stability (i.e. phase noise) and spectral purity. Old VNAs were equipped with frequency sweepers that were indeed quite fast (and before about year 2000, they were generally faster than synthesizers), but spectral purity was not as good and there were potentially synchronization issues since LO and IF systems need to be semi-coherent with the source system [317]. As a result, many of the recent VNAs are synthesizer based. When sweeping the frequency range, some points may be considered to better appreciate VNA performance:

- the minimum time needed to move to the next frequency value may be in some cases limited by the rapidity of LO tuning, as peculiar of YIG sources (already considered at the end of sec. 9.2.4 for Spectrum Analyzer), but with modern digital synthesizers this is a minor problem; sweep time is normally influenced more by the transient response of the IF filter and its settings, and using low RBW values (e.g. tens or hundreds of Hz) to lower noise and increase sensitivity slows down the sweep; finally, in general, for small frequency steps the new frequency settles slightly more rapidly;
- the purity (in terms of phase noise, harmonics and spurs) of LO signal is relevant for a spectrum analyzer, but VNAs used in simple sweep to measure S parameters operates at one frequency only, the fundamental synthesized frequency; on the contrary, when considering non-linear and distortion measurements and large power testing, this is a relevant aspect that opens the door to a whole new set of instruments, such as the Microwave Transition Analyzer (MTA) and the Large Signal Network Analyzer (LSNA) [317]); as applications proliferate (IMD, mixer, and non-linear measurements) and need for speed increases, however, one must pay more attention to spectral purity;
- when feeding variable power levels to the DUT (e.g. in situations with a significant noise or when the DUT linearity is measured), also the accuracy with which the power level is controlled is relevant, not to say that calling for fast frequency sweep in this case shall be backed up by a fast leveling of the automatic power control; leveling circuits are used in many applications and are conceptually simple: they use a power detector of some kind and, in the context of a negative feedback loop, compare the detected output to some desired reference voltage (usually from a DAC) and feed the result to a power modulator;
- considering frequency tuning, distortion limitation and noise altogether, it is worth observing that the most modern PLL architectures based on fractional-N structures can offer fast fine-tuning resolution with decent spurious and noise performance (a test of phase noise and near lateral bands while varying the synthesized frequency was done in sec. 9.5.3 for a stand-alone DDS RF generator); wide-bandwidth direct digital synthesizers have recently become more and more common and have had ever-improving spurious performance; the fine-tuning capabilities of such structures are strongly necessary, since VNA tuning resolution must typically be of the order of 1 Hz (or better, if high-order multipliers are part of the tested system).

Regarding the sweeping operation itself, modern digital VNAs operate in stepped sweep, that is the applied frequency is changed at discrete steps, while sweeping the frequency range, and for each step the measurement takes a predetermined dwell time, as required by VNA settings (e.g. IF bandwidth), similarly to what was said for Spectrum Analyzer in sec. 9.2.4. With a progressively changed frequency (the so called ramp sweep), small non-linearities, instabilities and phase noise in general may create noise when performing time domain analysis. The minimum dwell time assigned when using step sweep puts away from problems related to delayed reflection for devices with moderate to long propagation times, as pointed out in sec. 11.10.3.5.

11.2.2 Switches

RF switches are needed in VNAs for a number of reasons, including enabling sources to connect to and drive two or more ports, or to selectively route signals to receivers (i.e. reflectometers). Switches may be a relevant item in the determination of the overall measurement uncertainty depending on VNA architecture (two-port three-reflectometer VNA, two-port four-reflectometer VNA, higher number of ports, e.g. four-port VNAs, etc.); in some cases some are preferable in order to avoid connection and reconnection and thus ruling out switches and their leakage error terms from the error model, calibration and residual uncertainty.

Very often, the demand on switches may be extreme in terms of isolation, insertion loss, bandwidth, and, perhaps, power handling/linearity. Using a 2-port VNA as an example, there is usually a main switch (normally called *transfer switch*, with a single-pole double-throw operation obtained by combinations of semiconductor unipolar elementary switches) allowing the source to drive port 1 or port 2. Its characteristics are quite relevant and influence VNA performance: the isolation of this switch directly translates to the raw isolation of the VNA; the insertion loss and linearity directly affect the maximum available port power; its bandwidth can limit that of the VNA. For a high-performance microwave VNA the switch may be one of the weakest points and the optimization of all the above parameters a challenging combination.

Electromechanical switches used in the past had a good insertion loss / isolation ratio; repeatability of these switches, however, typically no better than a few hundredths to a tenth of dB at microwave frequencies, led to relevant uncertainty. Also, the lifetime was an issue, so that electronic switches are normally used, typically a PIN diode or cold FET circuit, or some combination.

A PIN diode consists of heavily doped P and N layers surrounding a relatively thick intrinsic layer (hence the acronym). Because of this thickness, the reverse biased capacitance of the diode is quite low compared to other diode types. This leads to better isolation when used in series insertion and less insertion loss in a shunt topology. When forward biased, carriers are injected into the intrinsic layer but, do not recombine immediately; this leads to some complications at lower frequencies since the applied RF signal may be on the same scale of the recombination rate and distortion occurs.

An example of cold FET switch is a MESFET, or a similar device, with no drain bias. When the gate is biased strongly negative, no carriers are available in the channel and the device provides reasonable isolation. Like for PIN diode, the off capacitance (drain to source) is quite low thanks to device geometry, so shunt topology insertion losses can be low as well (although typically worse than those of PIN diodes). When the gate is brought near ground potential, carriers are available in the channel and a relatively low series resistance is established. Unlike the PIN diode, the recombination time remains fast, so there are few low-frequency adverse effects. Since one is usually operating against a 0-bias limit, there can be linearity issues, although these have been overcome at least in part with more novel topologies.

Series elements for switches are generally less effective at higher microwave frequencies and shunt elements are preferred, or better, a series-shunt pairs combination built in a single device. At very high microwave frequencies, let's say above 20 GHz, parasitics of filter and bias circuit elements may contribute largely to the total insertion loss; isolation as well may be limited by radiative effects, that depend on layout, device packaging and housing. At millimeter waves also standing waves between elements, and in particular between the various interconnected switches, may become an issue.

11.2.3 Directional devices

Most VNAs use directional couplers or directional bridges to collect incident and reflected waves at their ports. It is intuitive that the most relevant features of the directional coupler are related to the ability of separating the incident from the reflected waves, as well as the separation between ports, besides the usual features of return loss and insertion loss on a single port basis. After getting into the performance details of directional devices (see sec. 11.3.3), it will be possible to accurately evaluate the benefits and strong points of directional coupler and directional bridge architectures.

The first "complication" with directional devices is to understand what is the purpose of the ports and how they are named: the main line through which the signal travels and that needs the largest power handling capability is located between two ports, normally termed "input" and "transmitted", or "through" ("output" would be confusing, since also the tapped forward wave is an output of the directional device); the tapped wave is available at an output port called "coupled", to distinguish it from another port that is isolated for the forward wave and senses the reverse backward wave (this port is normally called "isolated").

The "coupling" (or gain) thus defines the amount of main line signal in the "right" direction that is brought to device "coupled" output; as a consequence the signal traveling through the main line is attenuated at least by the tapped amount directed to the "coupled" output. Conversely, any reflected wave traveling in the opposite direction along the main line shall give no or very little contribution at the output port that we selected for measuring forward waves: this is called "isolation", and highly depends on the internal symmetry of the device. Directivity is then defined as the ratio of the coupling to the isolation (please see sec. 11.3.3 for a more formal treatment). It is underlined that a strong coupling does not imply automatically a proportionally larger directivity, because isolation at large signal power tends to worsen; usual values of coupling are in the range of 10 to 30 dB and directivity between about 20 and 40 dB.

From a more operative viewpoint, limitations in frequency range may come also from the way ports are protected and coupled through their connectors; so, for example, if a port is dc protected (or it is an "ac port"), then a minimum cut-off frequency will be applied, independently from the possibility of squeezing some more MHz downward from the directional device operation.

Finally, as for all RF and microwave devices, return loss characterizes the main line port, so that not the entire signal fed to the port is traveling through the device and is measured, coupling it to output port.

Regarding the operating frequency range of directional devices, there are a minimum and a maximum frequency limit: at the upper limit the signal wavelength is shorter than the physical length of the internal coupled section and stationary waves appear, at the lower limit the fraction of wavelength that is internally coupled becomes smaller and so the coupling, with a typical fall-off of 20 dB/decade or slightly steeper; when also matching problems occur at both ends of the frequency interval, directivity gets even worse. By looking into some product datasheets, it is evident that the frequency behavior depends heavily on coupler architecture and operating principle.

11.2.3.1 Directional coupler

Directional couplers are built around different principles and architectures, the most common based on resistive dividers, Wilkinson and microstrip/stripline, as well as airline, architectures. Briefly we may report the most relevant differences: resistive dividers have no isolation, but can operate from dc, while the other quite popular category, the Wilkinson, is limited to a minimum frequency around tens or a hundred MHz and an operating bandwidth of one or two octaves; both resistive and Wilkinson couplers are in reality power dividers, operating a 50% (-3 dB) power splitting, thus resulting in equal power divided between main line and coupler output, while couplers as such should tap a small fraction of the main line power. Microstrip-based and stripline-based couplers are the most diffused, with a quite extended operating bandwidth even if not reaching dc, not even power splitting (as we know the output is a small fraction of the power in transit on the main line) and a phase shift of the output that is around 90°, slightly fluctuating over frequency. Last, hybrid couplers have a more accurate phase relationship (relevant when treating IQ signals in telecommunications) and split the power in equal parts of -3 dB each. There exist also dividers with 180° phase shift, thus giving a differential signal, that for coupler applications consist of baluns; in modern VNAs some galvanic isolation of this kind is advisable to properly feed the IF digitizing section.

Microstrip-based and stripline-based couplers are in general all sharing the same concept design and scheme: one of the lines is the main line and it is tapped by electromagnetically coupling from a parallel line, that is connected to the "output" terminal [23, 88, 108, 273].

11.2.3.2 Directional bridge

Directional bridges differ from the previous ones in that they are built around a classical measurement bridge architecture, that is a four-arm bridge where measurement is taken in the center (where detector is connected in classical bridges) and the "trans-

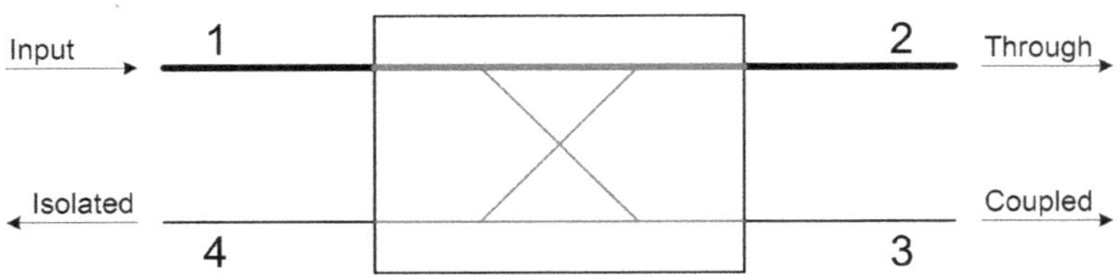

Figure 11.2.1 – Scheme of a directional coupler.

mitted" terminal for the external network/load is one of the bridge arms; the bridge supply is the "input" terminal where the signal source is connected. Good descriptions of directional bridge architectures appear in a tutorial by Agilent and a DIY project by J. Audet[1]. In his tutorial Dunsmore gives an interesting description of the assessment of the various conditions, to estimate coupling and isolation of directional bridge.

The four arms of the bridge are symmetric, one connected – as said – through a cable to the "through" terminal and all featuring the same line characteristic impedance (e.g. $Z_c = 50\,\Omega$); the bridge scheme is shown in Figure 11.2.2. Since one of the corners of the bridge is grounded, either the "input" or the "coupled" lines are coupled through a balun transformer; the turn ratio may be selected to set how the impedance of the external network at the input terminal is transformed when coupled into the bridge. Audet proposes a 1:1 current balun; in Dunsmore's example 5/6 is selected and automatically this defines the amount of signal lost when traveling from input to through port (it is $-1.58\,\mathrm{dB}$). The input impedance is $Z_{in}^2 = R_1 R_4 = R_2 R_3$; since some resistors are in reality the impedances externally connected to ports and they are equal to $Z_c = 50\,\Omega$, their value immediately conditions the other design choices: $R_4 = 50\,\Omega$, then $R_1 = 50\,\Omega$ to match the input, and then R_2 sets the input voltage ratio that we said was decided to be 5/6, so $R_2 = 5R_1 = 250\,\Omega$. Two design parameters characterize the directional bridge, that is the attenuation of the signal through the main line to the "through" port and the coupling amount onto the "coupled" port: we can select one (e.g. the bridge loss in the main line, above $-1.58\,\mathrm{dB}$) or the other (coupling), but not both, and this is accomplished by setting the turns ratio.

11.2.4 System performance and characteristics

To support the on-going discussion we report altogether what are the most relevant performances and characteristics of VNAs:

- *Frequency range*. The frequency range and the number of points may be conditioned by several factors, in particular when non-linear complex measurements are required; for simple measurement in linear conditions the minimum and maximum frequency are set by the required characterization, the desired frequency extension, the size and spatial extension of the test setup, the influence of signal processing settings (such as time gating).

[1] J. Dunsmore, "Network Analyzer basics," 2007, Agilent Technologies, available [online]: `rfic.eecs.berkeley.edu/142/pdf/NABasicsNotes.pdf`;
J. Audet, VE2AZX "Return Loss Bridge Schematic", rev. March 2015, available [online]: `www.ve2azx.net/technical/RLBridges.pdf`.

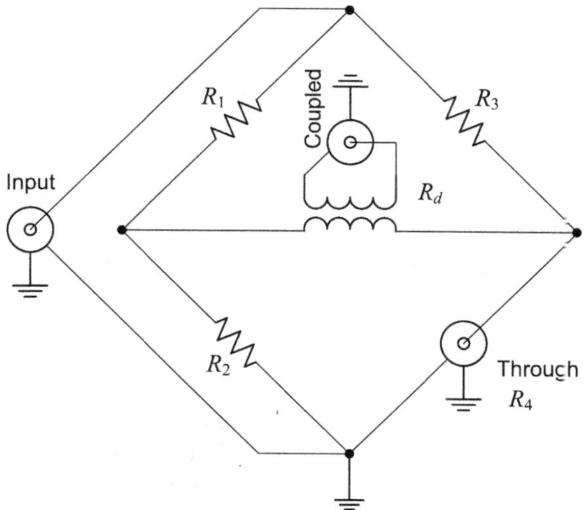

Figure 11.2.2 – Scheme of directional bridge with "input", "through" and "coupled" terminals.

- *Measurement speed.* As for spectrum analyzer, bandwidth and number of points are the most relevant factors; then local oscillator tuning, Analog-to-Digital Converter sampling and hardware processing time exert their influence on this performance index; last, any required additional signal processing function (smoothing, time-gating, etc.) increase the computational burden and depending on the VNA architecture may have a more or less pronounced impact on the overall measurement speed. Even if the declared speed performance of different manufacturers may coarsely be compared for the same main instrument settings, several details and specific settings prevent a direct one-to-one comparison.

- *Dynamic range.* It is quite similar to the long discussion made in Chapter 9 for Spectrum Analyzer, where the two extremes of the dynamic range are influenced by the noise floor and the maximum signal allowed by power handling capability and by distortion and errors caused by signal compression:

  - noise floor is impacted by front-end loss (couplers and attenuators), internal gain and conversion loss; external amplifiers may help and scaling with selected IF bandwidth is linear, as seen for the spectrum analyzer, so minimum IF bandwidth and averaging both reduce apparent trace noise;
  - phase noise from the local oscillator might add to the noise floor for some setting combinations; we shall keep in mind that a VNA measurement is usually a narrowband measurement with a sweeping source, so that the noise contribution of phase noise that falls inside the selected IF bandwidth is relevant and that phase noise may increase at high source-power levels.

- *Trace noise.* As a direct consequence of the noise sources that limit VNA dynamic range at the lowest signal levels, the measured trace will show a noise that is variable as a function of signal intensity; for S-parameter measurement the signal may vary depending if it is a reflection or transmission measurement: for the former a matched load represents the worst case (for this reason the compensation of directivity during calibration may be partly ineffective); manufacturers usually give performance for a fully reflective load, that is a 0 dB reflection; for the latter the worst case is represented by large attenuation measurements (such as filters and attenuators). Thermal and similar noise sources are uncorrelated and will sum in quadrature to the trace (please remember to make this operation with linear units and not dB); phase noise is partially correlated and, as seen, dependent on the applied source level.

- *Port power*. Usually for the port-fed driving source signal VNAs may feature discrete levels or a nearly continuous fine adjustment of the applied power; maximum power normally ranges between +10 dBm and +30 dBm, the latter being quite exceptional; the lowest level available in low- or medium-end VNAs is usually in the range of −20 dBm to −10 dBm, but high-end units have a continuous range of power level down to very small values and allow also for parametric analysis; the power level may be lowered by using internal or external attenuators, e.g. in order to feed active devices appropriately; the maximum power is also limited by non-linearity and distortion of switches, first, and then, mixer, the directional couplers being probably the most robust components. For specific measurements it is also required that the power level may be trimmed accurately, in particular when measuring DUT linearity: gain compression, intermodulation distortion, harmonic content are all quantities that describe non-linearity of DUT versus power level as shown in sec. 3.4; in general, when measuring linear passive devices, being the scattering parameters the result of a ratio between transmitted and received power, small errors in the power level go completely unnoticed.

- *Raw port parameters*. Ports are characterized by quantities that are selected as port performance indexes and are related to the error model used for VNA/setup calibration: directivity, isolation, port match and tracking. These parameters are influenced by several internal components, such as switches and couplers, attenuators, connectors themselves and other construction details, all defined and described by their scattering parameter matrices, if needed, or considered overall in the error-model black box (see sec. 11.3.3 and 11.11.3).

- *Residual port parameters*. We have mentioned the VNA calibration, that, if properly done, removes the systematic error terms, but leaves unaltered the random error terms (for some calibration schemes resulting slightly worse then before the calibration); the distinction between the two is sometimes not so evident and may generate some confusion: errors, or error coefficients, are those parameters of the VNA error model that represent and correct VNA non-ideality (e.g. losses, undue reflections, etc.), transforming the raw port measurements into corrected measurements; such parameters are determined during calibration based on the used calibration standards and method, and of course the underlying VNA error model. What is not corrected, that is by proper terms the "VNA error" in a metrological sense, is named usually "residual error" or "uncertainty"; the reasons for this may be low-quality calibration standards, excessive noise during calibration, drift and variability of some VNA elements.

- *Stability*. Stability of overall VNA performance and error terms, in general and as evolving after the last calibration, is a complex subject, because many factors related to the environment, operator, materials and components come into play: environment temperature, unwanted movements of cables, vibrations, air flow just to cite some, and going inside the VNA, local temperature drift, erratic behavior, local oscillator instability and drift, power level stability, system linearity.

11.3 Elements of the test setup

This section considers the effects of non-ideality of setup elements, other than the VNA itself; the attention is focused on the most relevant elements for the most common measurements, namely cables, connectors and directional devices. Cables and connectors were already discussed in their own chapters 5 and 7; directional devices have been introduced in the previous section.

11.3.1 Cables

Cable selection depends primarily on the frequency range of the measurement and the affordable insertion loss. Other characteristics may be size, fitting of connectors to use, their weight and in particular flexibility/rigidity, that influence the possibility of bending them and adapt their shape to the test fixture. Often the VNA is connected with high quality cables that allow bending and flexure without significant accuracy issues and only the last connections close or within the fixture are prepared with attention to other conditions, environmental or mechanical constraints, etc.

If, for example, the VNA measurement is made at the same time of a vibration test, as part of a more complex experiment, cables shall be selected for their mechanical properties (weight and flexibility) and they shall not transmit vibration along to the connectors, as well as have negligible changes of their electrical properties (please see sec. 5.7 and 7.4 for cables and connectors issues of vibration, flexure and repeatability; for passive intermodulation effects see sec. 5.7.2 and 7.4.3, respectively).

Cable flexure and its effects are in direct connection with connectors used in the setup, also from a mechanical viewpoint: tests shall be made of repeatability of assembled cables in their final installation under credible flexure and bending, in order to derive on the one hand an estimate of expected variability and uncertainty, and on the other hand prescriptions and guidelines to limit such effects. In Figure 5.7.1 extreme flexure was applied with a commercial SMA connector fundamentally weak for this task, resulting in quite remarkable repeatability uncertainty and gross errors.

11.3.2 Connectors

As seen, connectors may also influence results, depending on various characteristics (for a general description of connectors, connector characteristics and performance, please, see Chapter 7).

Several connectors are not accompanied by any statement of their performance versus frequency: at low frequency in general all work more or less well (except some variability of contact resistance and some accelerated wearing), but remarkable differences and even defects may appear at high frequency. In some cases some might even not comply to the average characteristics of their own type; in other words, to exemplify, not all SMA connectors work adequately up to 12 GHz, or even up to 5-6 GHz, although SMA design is rated in general up to 18 GHz (poor geometry, design flaws or fabrication defects may be the reason). This does not imply that more expensive connectors guarantee anything: if a characteristic is not stated clearly, it is not guaranteed and probably will not be there. Conversely, the need to check connector performance before using it, leads to the possibility of saving some money, trying cheaper connectors, that sometimes are well above what the price alone would suggest. When instead going for high-grade precision connectors, one never fails.

The typical design flaws and defects that limit connector performance are approximate internal geometries, bad mating of the cable head and poor materials (such as a dielectric with non-stable dielectric constant or too a large loss, or the use of ferromagnetic materials[2]). Mechanical robustness, such as resistance to externally applied lateral forces, is extremely important for many setups and whenever the environmental stress is not negligible, even at the cost of sacrificing a little electrical and electromagnetic performance. In particular with heavy and semi-rigid or rigid cables[3], that transmit quite efficiently any mechanical stress, tilting the cable causes misalignment of the connector pin as well as poor contact on the external thread, if dirty and loose (jokes of hundredths of mm may be quite relevant in some cases). SMA, 3.5 mm and smaller connectors, compared to the heavier and more robust N connectors, are particularly exposed to this phenomenon; connectors with a long sleeve to hold the cable in place are thus preferable, whether they are laboratory grade special connectors equipping flexible high-performance cables, or home-brew solutions, such as when wrapping around and soldering a metallic tape or fiber tape with resin potting. SMA are quite attractive anyway and often used as the best compromise for weight, performance, size and price, especially in high-density boards and panels. However, for on-site measurements and heavy duty application mechanical stress may be quite relevant and heavier connectors might be preferable.

Low quality connectors come out in the long run: surface wearing and dirt compromise mating and any, even tiny, uncompensated gap between surfaces causes impedance mismatch and unwanted reflections and errors. In addition, imperfections between dielectric bead, inner hole and surrounding wall compromise the accuracy of the reference plane location and have a frequency-dependent impact. A first look at the physical dimension of these elements would suggest that they are relevant only when

2 The use of ferromagnetic materials was prohibited by MIL-STD, but nickel plating is still accepted and used, also because it is a good support for successive gold plating. The use of e.g. phosphorus in the nickel breaks magnetic domains in smaller ones, thus reducing the effect, and leaving the other good properties of nickel (see sec. 7.1.1.2).

3 With rigid cables soldered all around to the connector body the reason for mechanical weakness resides often in the tilting of the center conductor not well supported and locked in place by the soldered connector pin: when the cable is externally bent, the compression of its dielectric applies a tilting force on the inner conductor that is transmitted to the connector pin.

comparable with wavelength, i.e. in the range of tens of GHz and beyond the domain of many applications. However, depending on the required repeatability and the target uncertainty, deterioration may occur already at a few GHz.

Due to many similarities, connectors and adapters share similar characteristics and may be treated with a uniform approach.

11.3.3 Directional coupler

A coupler (or directional coupler) is a device that can separate signals traveling on its main line, dividing forward and backward portions and giving a measurement of their amplitude on assigned ports. In many cases couplers are used to measure only one quantity, i.e. the forward traveling wave; when both directions are measured at the same time we may speak of a bi-directional coupler, but usually only the expression "directional coupler" is used.

11.3.3.1 Scattering parameters and performance

Let's call 1 and 2 the ports of the main line and 3 and 4 the ports at which the measurement of the forward and backward waves is delivered (a sketch was shown in Figure 11.2.1 a few pages earlier). *Coupling* is the ability to couple the desired wave on the assigned port, e.g. the wave from 1 to 2 on port 3; symmetrically, the ability to separate terms and avoid the influence of backward waves on the quantity appearing at port 3 is called *isolation*.

The scattering parameters that define the coupler performance are:

- S_{21} is the insertion loss IL on the main line when passing through the directional coupler, that is the amount of signal loss for using the directional coupler to sense the main line: $IL = -S_{21}$;
- S_{11} is the return loss RL on the main line, that is the amount of signal reflected back and not entering the directional coupler, due to input impedance mismatch;
- S_{31} is the coupling C, that is the amount of the available signal that is transferred to the correct output port, so at a first approximation the ratio of the power at input port P_1 to the power delivered at the coupled port P_3; in reality coupling is a number smaller than one, so that its reciprocal is used: $C = -S_{31}$;
- S_{41} conversely identifies the isolation I, that is the ability to reject and keep small the signal that is transferred to the wrong port (the ratio of the power delivered to the isolated port P_4 to the input power P_1), or, in other words, the amount of leaking of backward waves onto the forward measuring output port: $I = -S_{41}$.
- symmetrically for the other side, all the quantities above may be repeated using the backward wave as the desired one.

Directivity may be thus defined as the amount of useful signal, that is the amount of coupled signal minus the amount of leaking wrong signal, taking into account that the signal through the coupler undergoes some attenuation (insertion loss, negative quantity), but without considering the match of the coupler, that is the fact that not all the signal on the main line is entering the coupler. In other words directivity is given by the difference of coupling and isolation, accounting also for insertion loss:

$$D = I - C + IL = S_{31} - S_{41} + S_{21} \qquad (11.3.1)$$

With low insertion loss this expression is often simplified (optimistically) to [201]

$$D \cong S_{31} - S_{41} \qquad (11.3.2)$$

Measuring directional coupler parameters and performance is quite useful both to verify experimentally that a directional coupler keeps its performance unaltered and to characterize devices for which there is no datasheet (e.g. a coupler found in the laboratory for which it is to decide if it is "good" enough or not).

Insertion loss S_{21} may be measured easily by feeding the directional coupler input port from one VNA port (let's say port 1) and then bringing the signal at the through port to VNA port 2. During this measurement the coupled port shall be terminated on a matched load. If the VNA was calibrated using the same feeding cables they are automatically removed from the measurement, otherwise their attenuation shall be accounted for with an additional measurement shorting them together without the directional coupler in between.

Analogously, coupling S_{31} may be measured by exchanging the roles of coupler ports: VNA port 2 is connected to coupled port, while the through port is terminated on a matched load.

The other way round we are able to measure isolation S_{41}: feeding the coupler from the through to the input port with VNA port 1, what is measured at port 2 connected to the coupled port is the leaking signal due to isolation. Yet, any VNA non ideality will impact on the accuracy of the result. Analogously, problems arise when trying to measure coupler directivity, because of the limited directivity of the VNA itself and the effect of reflections due to source-match error and matched-load residual reflection. In the most favorable case of a reflective termination of the through port, the signal intensity to measure on the coupled port reflected by the load is the directivity; with a matched load termination the signal intensity is low with an almost negligible reflection and the directivity error drops to very low values. For this reason a reverse arrangement is necessary, as shown in sec. 11.3.3.2.

11.3.3.2 Measurement of coupler directivity

Using the VNA in standard two-port connection, it is possible to determine return loss and insertion loss on the coupler main line, terminating correctly at least the coupled port 3: S_{11} and S_{21} may be thus determined. With an identical test it is possible to read the signal on port 3, while feeding the same amount of power at port 1, providing a similar termination for the through port 2: S_{31} is thus determined. This should be sufficient to determine directivity using eq. (11.3.2). However, the limited directivity of the load (either a matched termination or the VNA port termination itself) may lead to a wrong estimation of coupler directivity. What normally happens when measuring matched loads with return loss of the same magnitude of VNA directivity is that the trace has large ripple oscillating between when the two sum in phase (maximum) and when they almost cancel out (minimum), as considered in sec. 11.11.3.1 for VNA directivity itself.

A more accurate technique requires that the directional coupler is reversed, as shown in Figure 11.3.1. Then the measured quantities are arranged to eliminate the influence of other parameters, obtaining a relationship in directivity D alone.

For this reason the technique of the sliding load may be of help: while sliding the load changing its electrical length, a phase shift φ is introduced that makes the measured vector to follow a circle, whose center is the required result; in other words the limited match of the load rules out by itself being able to swing from a maximum to a minimum of mismatch, the correct "zero" value being in the middle.

Let the input voltage $V_2 = V_i$ be applied to port 2 (neglecting return loss that is common to both signal paths seen in the following) and going through the coupler to port 1, where the sliding load is connected: $V_{1,F} = S_{21} V_2 = S_{21} V_i$ is the wave at port 1 "input" insisting onto the sliding load, and a small fraction proportional to the sliding load reflection coefficient Γ_l is reflected back into the coupler, this time with the right direction to be coupled into port 3. The backward voltage onto the coupler input port is $V_{1,B} = \Gamma_l V_{1,F}$, where $\Gamma_l = |\Gamma_l| \, e^{-j\varphi}$ and the coupled voltage is $V_3' = C V_{1,B}$. The reason for using the coupler in reversed orientation was to minimize the amount of signal coupled to the output on the way to the sliding load: the amount of signal going to the coupled port 3 shall be of the same order of magnitude of that just calculated coming back from the sliding load, for the technique to be accurate.

The voltage fraction from port 2 to port 3 is S_{32}, that is the isolation $I = C - D$ in dB, or $I = C/D$ in linear units: directivity D was defined above for port 1, but for port 2 a symmetrical definition holds, that is $S_{42} - S_{32}$ in dB, or S_{42}/S_{32} in linear units. The voltage appearing at port 3 due to the isolation term is thus $V_3'' = I \, V_2 = \dfrac{C}{D} V_2$.

Thus the total voltage at the coupled port 3 is: $V_3 = V_3' + V_3'' = \left[C \, IL \, |\Gamma_l| \, e^{-j\varphi} + \dfrac{C}{D} \right] V_2$.

Moving the sliding load and changing φ a circle is traced in the Re-Im plane, whose center is the required $\dfrac{C}{D}$ value, so accounting for the latter term only; knowing coupling C , directivity may be readily determined.

Otherwise, there are other possible transformations that lead directivity back to ratios of quantities that can be easily evaluated from measurements: calling M and m the ratio of coupled voltage to maximum voltage, and the ratio of maximum voltage to minimum voltage, respectively, then directivity is $D = M \dfrac{2m}{m+1}$ [273].

When a sliding load is unavailable, a few sample measurements may be performed using cable sections of different length, provided that they are short enough so that their attenuation is not relevant.

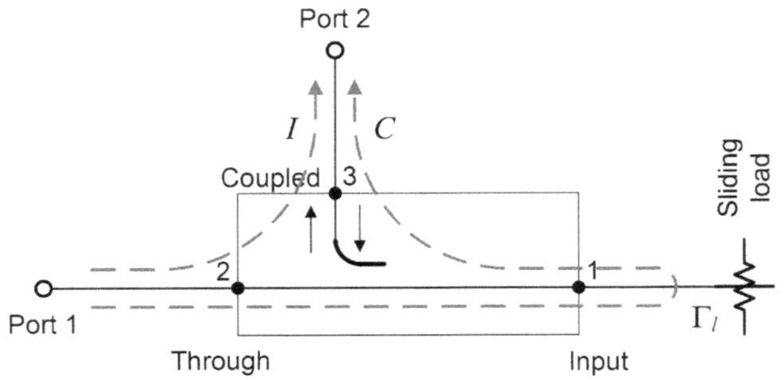

Figure 11.3.1 – Setup for coupler directivity using reversed coupler configuration.

Assuming an ideal coupler that does not alter the main line in terms of impedance mismatch, when measuring the forward wave traveling onto a connected load, any reflection from the load back to the coupler will leak into the measurement through the S_{32} parameter. This is to be evaluated by vector addition of quantities, knowing that the phase relationship with reflected backward wave is arbitrary (e.g. depends on frequency and length of line section). The mathematical steps and reference curves are shown in sec. 11.11.3.1 and 11.11.3.2: based on results shown in those sections, what is underlined here is that 15-20 dB of directivity minimum are needed to ensure forward power measurement errors below about 1 dB. For the reflected power measurement error things are even more complicated, because the coupler shall separate a weak reflected power term from the leakage coming from the strong forward power error terms, so that this time the required 15-20 dB margin on directivity is with respect to DUT return loss; adding thus on average another 10-20 dB for a wide range of DUTs return loss, the required directivity becomes 30-40 dB.

11.4 Reference planes

Reference planes are located at some spatial position along the test setup, where the VNA takes the measurement of forward and backward voltage waves to determine S parameters. Physically, the VNA takes the measurements at its internal reflectometers, that are in turn put in relationship to VNA ports; any reference plane may be referred to the internal reference planes where VNA ports are located, using an offset distance from the zero reference plane of VNA ports. A reference plane may be made corresponding to a specific physical characteristic of the circuit along the setup: cables connecting to the VNA, launchers, other feeding lines connected to the DUT, etc.. The movement of reference planes is achieved by multiplying for suitable matrices that de-embed the portion of transmission line and setup over which the planes are moved (see sec. 11.7). Depending on the adopted technique, reference planes shall or shall not be located at discontinuities: if de-embedding is achieved by additional calibration, then calibration standards will be connected at some internal port, i.e. a discontinuity; conversely, if mathematically de-embedding, a homogeneous section is selected and reference planes should be located within it far from discontinuities. The mating plane of a connector might be chosen as reference plane also for mathematical de-embedding, provided that the two halves of the connector (plug and socket) are separately known and documented for the construction of the de-embedding matrices.

Especially when moving reference planes onto the mating planes of two connector pairs or at a point along a transmission line, the distance to travel shall be determined accurately; to this aim there are various options, such as calculating it from physical dimensions of the elements, measuring it directly with a distance measurement, or having it measured by the VNA itself, spotting it out as a discontinuity. It is evident that the determination of the distance by direct measurement is prone to errors due to unavoidable round-offs in reading and the difficulty of identifying specific points inside connectors and adapters looking from outside (connector datasheets may help finding the mating plane position with respect to some detail accessible from outside). On the other hand, the quantity read on the VNA is time and distance is estimated by assigning a propagation velocity to the possibly heterogeneous setup, in any case leading to some approximation. For quality connectors and adapters used for calibration purposes, the reference plane at the male-female contact point (the "mating plane") is well documented, as for connectors compliant to IEEE Std. 287 [172]. In all other cases the user shall manage the characterization of the connector or adapter (a direct measurement with an accuracy of a fraction of mm and a comparison with mechanical drawings, always drawn to scale even if not explicitly quoted, usually solves the problem up to several GHz). Confirmation of measurements of geometric quantities should be always sought by tracking the discontinuities and their reflections using VNA and TDR S_{11} curve.

It is perfectly understandable that the resolution on the time axis influences directly spatial resolution. Since VNA time-domain measurements are performed originally in frequency domain and then the Inverse Fourier Transform is applied, the larger the bandwidth the finer the time and spatial resolution (please see sec. 11.8 and 11.10).

To reduce the inaccuracy related to the estimate of propagation velocity, measurements may be made for two slightly different cable lengths, L_i: different propagation times $t_{p,i}$ will be summed to the same connector propagation time $t_{p,0}$ and by comparison with the lengths of the connector, L_0, and cable, L_i, it is possible to estimate separately the velocity in the connector and the cable. In general cable datasheets report information on propagation velocity, that is on the contrary much harder to find for connectors; additionally, velocity may be measured more accurately for a long cable than for a short connector.

In general VNAs have the automatic port extension feature that allows the identification of the correct setting of reference planes, provided that additional measurements are performed; this procedure is part of the de-embedding methods and is presented in sec. 11.7.1.2.

11.5 Calibration and Error models

"Calibration" in the VNA jargon indicates the calibration of the test setup, that is its "zeroing", compensating for setup elements and their non-ideality, including the VNA itself, and thus correcting any successive measurements; in many cases without calibration the DUT response may be severely affected and corrupted or even not visible at all. The example of a high-performance connector tested with signals fed from the VNA by two coax cables can help in visualizing the problem (see sec. 7.3): the connector under test has very good impedance matching and low attenuation, better than or comparable to many elements of the setup, and without calibration we wouldn't be able to quantify the performance of connector alone.

This kind of calibration is performed routinely at each new measurement, for example when changing some element of test setup, or when environmental conditions have changed, and even when cables are moved or connectors removed and inserted back into their place. Calibration thus ensures that repeatability is maximized, paying attention, however, that unwanted movements or too tight or too loose connectors are all relevant aspects and have their impact (this is normally recognized as cable and connector repeatability and is considered in sec. 5.7.1 and 7.4.2, respectively).

The term "calibration" might be used also to intend VNA calibration in a metrological sense, so the verification of its operation and accuracy and the quantification of its uncertainty, verified against a set of reference quantities and instruments. This is not possible if the calibration standards are not included: VNA uncertainty, for an instrument that works on corrected measurements, cannot be estimated separated from its calibration standards; the metrological quantities that pertain to the VNA as a stand-alone instrument may be drift and stability, internal noise and other random errors. What the VNA calibration performed by the user does is to compensate for systematic errors that fit the error model and due to non-ideality of test setup and VNA. VNA accuracy in an absolute sense may be then checked against other standards and evaluated as a whole, i.e. VNA repeatability, accuracy and correct definition of calibration standards and correctness of calibration algorithms (normally assumed): this is partly addressed in sec. 11.12.

It is important to stress that calibration of test setup is not an almighty technique, able to compensate equally well for all values of setup parameters; the performance of calibration depends not only on the quality of calibration standards, that will be reviewed later on, and on the type of calibration algorithm (considered in this section), but also on the unavoidable uncertainty and drift characterizing each parameter, relevant in the overall uncertainty budget depending on its magnitude and its influence on results (sensitivity). So, corrections shall be kept as small as possible by selecting a good VNA, good setups and components, and keeping small the influence on the total uncertainty of each parameter.

To understand better, let's take the example of a directional coupler embedded in our setup, with correction by subtraction of its directivity during calibration: any raw directivity value of the directional coupler may be compensated for, but a poor one requires that a large value is introduced in the compensation and any drift or instability is amplified. For example, a measurement performed with a coupler featuring a poor directivity, even if compensated for, will be more sensitive to temperature drift and random instability, requiring also that calibration is repeated more often.

Moreover, the effects of any approximation adopted in the compensation by subtraction are amplified by the amount of compensation made: when de-embedding a cable section feeding the test fixture (see sec. 11.7), the accuracy in measuring the return loss (and thus the impact of directivity) is more relevant, the larger the cable insertion loss (this phenomenon is considered for a reflection measurement in low-pass mode in sec. 11.10.3.1 and when discussing masking in sec. 11.10.4).

11.5.1 General error model

A general error model introduces and identifies the error terms for a full two-port configuration of the VNA, where forward and reverse directions are distinguished. A larger number of ports might be considered as well[4], but for simplicity considerations that follow are limited to the two-port model. The model refers to the two VNA ports (port 1 and port 2) and the DUT connected in between; for each port we have forward and backward terms, a and b, for wave signals insisting at ports, entering and exiting ports, respectively. Such arrangement dates back to the late sixties, when the first automatic calibration algorithms were appearing[5].

Let's consider again the general scheme of the VNA (Figure 11.1.1), where a four-channel (full two-port) VNA is shown and directional couplers are used to separate incident, transmitted and reflected waves at each port, in both forward and reverse direction. The DUT[6] is characterized by four S parameters with usual notation: $S_{11} = b_1/a_1$, $S_{21} = b_2/a_1$, $S_{12} = b_1/a_2$, $S_{22} = b_2/a_2$. The model contains as many measured wave quantities (with "m" subscript) and the related S parameters. In deriving the error model, the effect of elements such as couplers and cables is included in a block, that is modeled as a 4-port network, bringing forward and reverse VNA port quantities (where the DUT is connected) to the internal ports of IF digitizers, with the digitized IF signals called measured waves a_m and b_m.

11.5.1.1 Mathematical development

From a general viewpoint the error model puts in relationship the measured matrix X_m available at the internal VNA IF digitizing block with the device matrix X_d, using two transformation matrices, Φ_1 and Φ_2, that contain the quantification of the error terms of the VNA and are determined during calibration, using known values of X_d (those of calibration standards)[237].

---

[4] See works considered in sec. 11.5.3.9 for an extension to an arbitrary number of ports.

[5] R.A. Hackborn, "An automatic network analyzer system," *Microwave Journal*, Vol. 11, May 1968, pp. 45-52.
J.G. Evans, "Measuring frequency characteristics of linear two-port networks automatically," *Bell Systems Technical Journal*, Vol. 48, no. 5, May-June 1969, pp. 1313-1338. doi: 10.1002/j.1538-7305.1969.tb04270.x
H.V. Shurmer, "New method of calibrating a network analyser," *Electronics Letters*, Vol. 6, 1970, pp. 733-734. doi: 10.1049/el:19700508
W. Kruppa and K.F. Sodomsky, "An Explicit Solution for the Scattering Parameters of a Linear Two-Port Measured with an Imperfect Test Set," (Correspondence), *IEEE Transactions on Microwave Theory and Techniques,* Vol. 19, no. 1, Jan. 1971, pp. 122-123. doi: 10.1109/TMTT.1971.1127466
S. Rehnmark, "On the Calibration Process of Automatic Network Analyzer Systems," (Short Papers), *IEEE Transactions on Microwave Theory and Techniques*, Vol. 22, no. 4, Apr. 1974, pp. 457-458. doi: 10.1109/TMTT.1974.1128250

[6] DUT has the broadest meaning of device when doing measurements and calibration standard when performing calibration, as in this case.

$$X_m = \Phi_1 X_d \Phi_2 \tag{11.5.1}$$

There are many solutions for the Φ matrices, depending on the adopted representation, the type of standards used in the calibration and which reference impedance was used. In many cases the results of different calibrations can be mapped into one another by changes of reference impedance and linear operations, from which we say that they are *consistent*; in other cases, when the result of some combination of calibration readings leads to inconsistent results, such calibration is said *inconsistent*.

The two matrices Φ_1 and Φ_2 are those termed $\mathbf{A}$ and $\mathbf{B}^{-1}$ in [122] or $\mathbf{T}_\mathrm{A}$ and $\mathbf{T}_\mathrm{B}^{-1}$, or even $\mathbf{T}_1$ and $\mathbf{T}_2^{-1}$, and then used by many authors agreeing on "A" to indicate the left and "B" the right parts of the error box cascaded two-port circuit, as in

$$\mathbf{M} = \mathbf{ATB}^{-1} \tag{11.5.2}$$

where the device is described by the matrix T of its scattering parameters

Of course, once Φ_1 and Φ_2 are known, the characteristics of the DUT (X_d) may be determined from the measurements X_m by an inversion operation.

With a closer look at the forward and backward terms involved in eq. (11.5.1), they can be made explicit for a two-port VNA by combining wave terms directed inward and outward the VNA ports and leading to the complete 16-term error model:

$$\begin{bmatrix} b_{1m} \\ b_{2m} \\ a_1 \\ a_2 \end{bmatrix} = \mathbf{S}_e \begin{bmatrix} a_{1m} \\ a_{2m} \\ b_1 \\ b_2 \end{bmatrix} = \begin{bmatrix} \mathbf{S}_{e,11} & \mathbf{S}_{e,12} \\ \mathbf{S}_{e,21} & \mathbf{S}_{e,22} \end{bmatrix} \begin{bmatrix} a_{1m} \\ a_{2m} \\ b_1 \\ b_2 \end{bmatrix} = \begin{bmatrix} e_{11} & e_{12} & e_{13} & e_{14} \\ e_{21} & e_{22} & e_{23} & e_{24} \\ e_{31} & e_{32} & e_{33} & e_{34} \\ e_{41} & e_{42} & e_{43} & e_{44} \end{bmatrix} \begin{bmatrix} a_{1m} \\ a_{2m} \\ b_1 \\ b_2 \end{bmatrix} \tag{11.5.3}$$

The e terms indicate systematic errors introduced by connectors, switches, reflectometers and IF digitizing block. This is a linear model that takes into account "classical" errors such as directivity, port match, tracking, etc., defined below and used in the rest of the Chapter. The off-diagonal terms in $\mathbf{S}_{e,11}$, i.e. e_{12} and e_{21}, quantify the crosstalk occurring internally to the reflectometers, due to poor separation of couplers, e.g. because of switch non-ideality; conversely, terms lying on the anti-diagonal, that link opposite quantities at the input and the output of reflectometers (e.g. e_{32} and e_{41}, and e_{14} and e_{23}), are due to poor isolation between reflectometers, that may in general be neglected for a good construction; last, the off diagonal terms e_{34} and e_{43} in $\mathbf{S}_{e,22}$ indicate crosstalk that occurs externally to, or before, reflectometers and switches, i.e. between port quantities; it may be due to physical coupling between ports, maybe caused by electromagnetic coupling of connecting cables or planar lines. Neglecting all these 8 terms, the 16-term error model reduces to the diffused and well-known 8-term error model.

Alternative formulations To ease the solution of the system there are different arrangements of quantities. Recombining the four-element vectors in two vectors, containing respectively forward and backward quantities, we can make explicit the relationship between homogeneous quantities, that may be made correspond to the previous S-parameter matrix representation:

$$\begin{bmatrix} b_{1m} \\ b_{2m} \end{bmatrix} = \mathbf{S}_m \begin{bmatrix} a_{1m} \\ a_{2m} \end{bmatrix} \qquad \begin{bmatrix} b_1 \\ b_2 \end{bmatrix} = \mathbf{S} \begin{bmatrix} a_1 \\ a_2 \end{bmatrix} \tag{11.5.4}$$

$$\mathbf{S}_m = \mathbf{S}_{e,11} + \mathbf{S}_{e,12}\mathbf{S}\left[\mathbf{I} - \mathbf{S}_{e,22}\mathbf{S}\right]^{-1}\mathbf{S}_{e,21} = \mathbf{S}_{e,11} + \mathbf{S}_{e,12}\left[\mathbf{S}^{-1} - \mathbf{S}_{e,22}\right]^{-1}\mathbf{S}_{e,21} \tag{11.5.5}$$

$$\mathbf{S} = \left[\mathbf{S}_{e,21}(\mathbf{S}_m - \mathbf{S}_{e,11})^{-1}\mathbf{S}_{e,12} + \mathbf{S}_{e,22}\right]^{-1} \tag{11.5.6}$$

The reader may notice that these relationships are non-linear; as a consequence they are more difficult to solve and do not assure that the solution is unique.

It would have been possible to attack the problem in eq. (11.5.3) the other way round, thus swapping two components of the four-element vectors, keeping all the measured quantities on one side and the raw quantities on the other; this is possible if a transmission matrix representation (see sec. 1.3.1.5) is used rather than the scattering parameter matrix[7].

$$\begin{bmatrix} b_{1m} \\ b_{2m} \\ a_{1m} \\ a_{2m} \end{bmatrix} = \mathbf{T}_e \begin{bmatrix} b_1 \\ b_2 \\ a_1 \\ a_2 \end{bmatrix} = \begin{bmatrix} \mathbf{T}_{e,11} & \mathbf{T}_{e,12} \\ \mathbf{T}_{e,21} & \mathbf{T}_{e,22} \end{bmatrix} \begin{bmatrix} b_1 \\ b_2 \\ a_1 \\ a_2 \end{bmatrix} = \begin{bmatrix} t_{11} & t_{12} & t_{13} & t_{14} \\ t_{21} & t_{22} & t_{23} & t_{24} \\ t_{31} & t_{32} & t_{33} & t_{34} \\ t_{41} & t_{42} & t_{43} & t_{44} \end{bmatrix} \begin{bmatrix} b_1 \\ b_2 \\ a_1 \\ a_2 \end{bmatrix} \tag{11.5.7}$$

The relationship with $\mathbf{S}_e$ may be made explicit as:

$$\mathbf{T}_{e,11}\mathbf{S} + \mathbf{T}_{e,12} - \mathbf{S}_m\mathbf{T}_{e,21}\mathbf{S} - \mathbf{S}_m\mathbf{T}_{e,22} = 0 \tag{11.5.8}$$

$$\mathbf{T}_{e,11} = \mathbf{S}_{e,12} - \mathbf{S}_{e,11}(\mathbf{S}_{e,21})^{-1}\mathbf{S}_{e,22} \tag{11.5.9}$$
$$\mathbf{T}_{e,12} = \mathbf{S}_{e,11}(\mathbf{S}_{e,21})^{-1} \tag{11.5.10}$$
$$\mathbf{T}_{e,21} = -(\mathbf{S}_{e,21})^{-1}\mathbf{S}_{e,22} \tag{11.5.11}$$
$$\mathbf{T}_{e,22} = \mathbf{S}_{e,21} \tag{11.5.12}$$

These equations may be reduced for the same reasons above, neglecting the isolation and crosstalk terms, reaching an 8-term error model.

Calibrating for crosstalk always consists in reading the signal at the opposite port while there is no connection between the two, that are terminated on the best matched loads

[7] Just as customary, the terms of the $\mathbf{T}_e$ matrix are named "t", while those of the previous $\mathbf{S}_e$ matrix are named "e".

available. Crosstalk error is relevant when high-isolation DUTs shall be measured: open switches, filters, etc. all feature a large signal attenuation and measurement may be impaired by the small residual crosstalk between VNA ports acting in parallel. Calibrating crosstalk efficiently and correctly is quite difficult, because the measured signal is low and noise can lead to large inaccuracies; trace averaging and narrow IF bandwidth of course both help in reducing noise effects. It is important that the two ports are terminated on the impedance that will be connected during the real measurements. To this aim two DUTs may be connected, one for each port, keeping the signal path between ports absolutely isolated, or if only one, or no, DUT is available, using matched loads; especially when the DUT might have varying input impedance due to tuning or other specific behavior, matched loads to reference impedance are the best solution. In some VNAs it is possible to turn off the reflection receiver (that is, assuming port 1 is launching into the DUT, the reflection receiver and IF digitizer channel on the reverse b_1 quantity) while performing a transmission measurement (that is, again assuming port 1 is launching, while measuring a_2).

Having neglected crosstalk and coupling terms (off-diagonal terms), now the networks pertaining to ports 1 and 2 are separate and may be expressed by two uncoupled systems of equations, described by two separate matrices, $\mathbf{T}_1$ and $\mathbf{T}_2$ (or, as said, in other notations $\mathbf{T}_\mathrm{A}$ and $\mathbf{T}_\mathrm{B}$).

$$\begin{bmatrix} b_{1m} \\ a_{1m} \end{bmatrix} = \mathbf{T}_1 \begin{bmatrix} b_1 \\ a_1 \end{bmatrix} \qquad \begin{bmatrix} a_2 \\ b_2 \end{bmatrix} = \mathbf{T}_2 \begin{bmatrix} a_{2m} \\ b_{2m} \end{bmatrix} \tag{11.5.13}$$

The measured and actual wave quantities at the digitizer interface and DUT interface are again related by

$$\begin{bmatrix} b_1 \\ a_1 \end{bmatrix} = \mathbf{T} \begin{bmatrix} a_2 \\ b_2 \end{bmatrix} \qquad \begin{bmatrix} a_{1m} \\ b_{1m} \end{bmatrix} = \mathbf{T}_m \begin{bmatrix} a_{2m} \\ b_{2m} \end{bmatrix} \tag{11.5.14}$$

and they may be combined as

$$\mathbf{T}_m = \mathbf{T}_1 \mathbf{T} \mathbf{T}_2 \qquad \mathbf{T} = (\mathbf{T}_1)^{-1} \mathbf{T}_m (\mathbf{T}_2)^{-1} \tag{11.5.15}$$

Main error terms The system is simple enough to be useful and is able to describe the most relevant error terms that constitute systematic VNA errors due to non-ideality of internal blocks. For this reason the solution of the system may be achieved in terms of T matrix parameters or of derived parameters, more directly linked to VNA performance[8]:

- E_{DF} and E_{DR} are the directivity errors of reflectometers for forward and reverse direction; by analogy with the operation of the directional coupler (see sec. 11.3.3), signal may couple to the coupled port from the wrong direction due to leakage and also may be reflected back to the coupled port before having reached the DUT; looking into the VNA block diagram of Figure 11.1.1 we may

[8] Available definitions and descriptions of these terms look like all are derived from and are rarely more exhaustive than one-two lines appearing in the App. Note 1287-3 by Agilent [2].

recognize some sources of error and their paths: crosstalk and leakage between ports that shunt the signal directed to DUT e.g. through port 1, insertion loss of reflectometers occurring between the signal measured at the coupler output and the effectively applied signal at the coupler transmitted port (see sec. 11.3.3); "directivity error" indicates the amount of signal that never reaches the reflectometer, mainly due to the input reflection coefficient of the directional coupler;

- E_{SF} and E_{SR} are the source match errors of reflectometers for forward and reverse direction; they indicate the error due to mismatch between VNA ports down to the reflectometer input ports, covering also the effect of multiple reflections: for source match, the reflected signal going to the coupled port of the reflectometer may be reflected back as a further incident signal by the internal source port reflection and perform again a complete travel through to the DUT and back again, overlapping to the first original signal; symmetrically, for load match (with terms E_{LF} and E_{LR}) the far VNA port at the other side of the DUT is responsible for a further reflection that travels through the DUT and overlaps to the reflected signal from the DUT near port;
- E_{RF} and E_{RR} indicate the reflection tracking of the reflectometer, that is the frequency response that includes losses for the elements in the path, i.e. converters, couplers, line sections, switches, connectors; for the more complete 12-term model for two-port measurements they are complemented by transmission tracking errors (indicated by E_{TF} and E_{TR}), with the same meaning, but to be used for transmission measurement; these second pair of terms is in reality not completely independent on the first one, and this is sometimes taken into consideration by some calibration algorithms; reflection and transmission tracking errors are seen as an offset in the measured return- or insertion-loss curve, due to erroneous "gain" or "attenuation" at some frequency intervals;
- E_{XF} and E_{XR} crosstalk terms represent the signal reaching the other VNA port without effectively traveling through the DUT; they may be generally neglected, because coupling between reflectometers and internal connections is absolutely not relevant; crosstalk is quite small and its effect normally covered by VNA noise, so that when measuring devices with large attenuation the calibration should be performed adopting noise reduction techniques, such as averaging and small IF bandwidth; it is underlined that if the noise cannot be effectively reduced by the mentioned techniques, it will lead to erroneous calibration and in this case better results are obtained if crosstalk terms are excluded and considered ideally zero.

Eight-term error model The 8-term error model reduces the number of error parameters to directivity, source match and reflection tracking, that is keeping the two VNA ports decoupled and accounting for six error terms. In reality, one shall take into account that VNA ports and reflectometers are not perfectly symmetric: to this aim there are two additional terms α and β, that are indeed redundant, because the system is solved for any amplitude of the test signals (and of the measured signals, as a consequence), so that any linearly dependent solution will fit and what is relevant is the α/β ratio.

Now, before proceeding, a brief synthesis is proposed:

- there are internal and external quantities, the former intended as "measured" at the IF digitizing stage, the latter being those that characterize the externally connected devices (DUT), whether it is a DUT or a calibration standard;
- the objective is a relationship between these two sets of quantities, such that can be inverted (for mutual calculation in either of the two cases, measurement or calibration); the relationship shall be as most complete as possible, taking into account a satisfactory model of VNA from ports to IF digitizing stage, so that model elements can be determined when performing the calibration;
- calibration from a model viewpoint is nothing more than collecting measured quantities (IF digitizer), corresponding to known external quantities: the larger the number of parameters in the model (i.e. the more complete the model), the larger the number of different measurements to be performed on calibration standards; simple models should be always preferred when possible, to avoid also problems related to the inaccuracy of standards and propagation of uncertainty; on the other hand, simplified models make assumptions and identify insignificant terms (dropped from the model) and equal terms (indicated then by the same quantity);
- model parameters are called often error terms (!) and indicated by "E" or "e"; the reason is that they are the systematic errors that the VNA makes on raw quantities if not properly calibrated; such errors may be compensated for as determined during calibration, resulting in error-free DUT measurement; the residual error that affects "error quantities" will be indicated in the following by putting δ in front of the quantity and is due to several issues related to the quality of calibration standards, noise, numerical approximations, drift and stability, etc. and is in principle "random"; it represents the uncertainty (or inaccuracy) of the measurement;
- another random factor is represented by noise, that is always present in VNA and setup, with a more or less relevant effect depending on settings.

11.5.1.2 Signal flow representation and solution of error models

Two error models (also called "error box models") are proposed using signal flow graph notation, first, for forward and backward directions, following the accepted theory on error modeling [115, 292, 317] and detailing the error terms of eq. (11.5.3) and (11.5.7) by as many branches. Then, the complexity of the graphs is reduced by recurring to the same assumptions we have just discussed. The white paper written by Rytting [292] is a very comprehensive reference for all these models. We will review some calibration methods with insight into error terms and degrees of approximation.

When switched and configured for a forward measurement, the VNA scheme is as shown in Figure 11.5.7(a), a few pages ahead. Conversely, switching to the configuration for reverse direction, the scheme is the one appearing in Figure 11.5.7(b).

The associated error model reported by Agilent in [292] is shown in Figure 11.5.1.

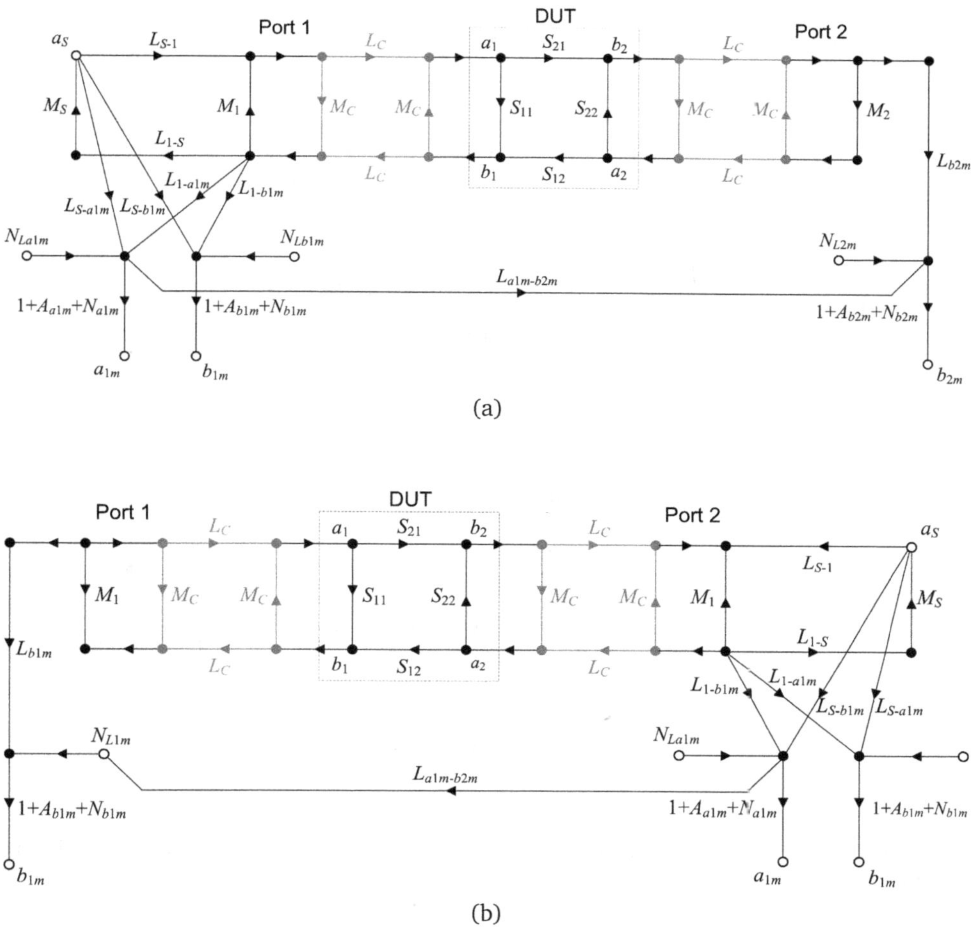

Figure 11.5.1 – Flow-graph of the VNA error models for (a) forward and (b) reverse direction: letters indicate the type of errors and "M" stands for match error, "L" for loss error, "A" for dynamic accuracy error (linearity), "N" for noise. Please note the L_{a0-b3} terms representing the leakage between VNA ports.

Such an error model is quite articulated and complete, but it may be simplified in order to proceed with error correction by calibration, as shown in the following. For example, ruling out the effect of noise and noise terms in Figure 11.5.1 removes the sloped branches in the left part of the graphs. Additionally, being of no interest keeping distinct L and M terms for cables and connectors, they may be merged together and hidden in the other e terms that appear in Figure 11.5.2. Distinction is again made for the forward and reverse error models; the shown model is the 12-term error model, divided into six terms for the forward and six terms for the reverse direction model (thus including crosstalk terms).

The way error terms are indicated is quite convenient, that is with E_{xy}, x and y chosen in the following way: $x = D$ for directivity, $x = R$ for reflection tracking, $x = T$ for trans-

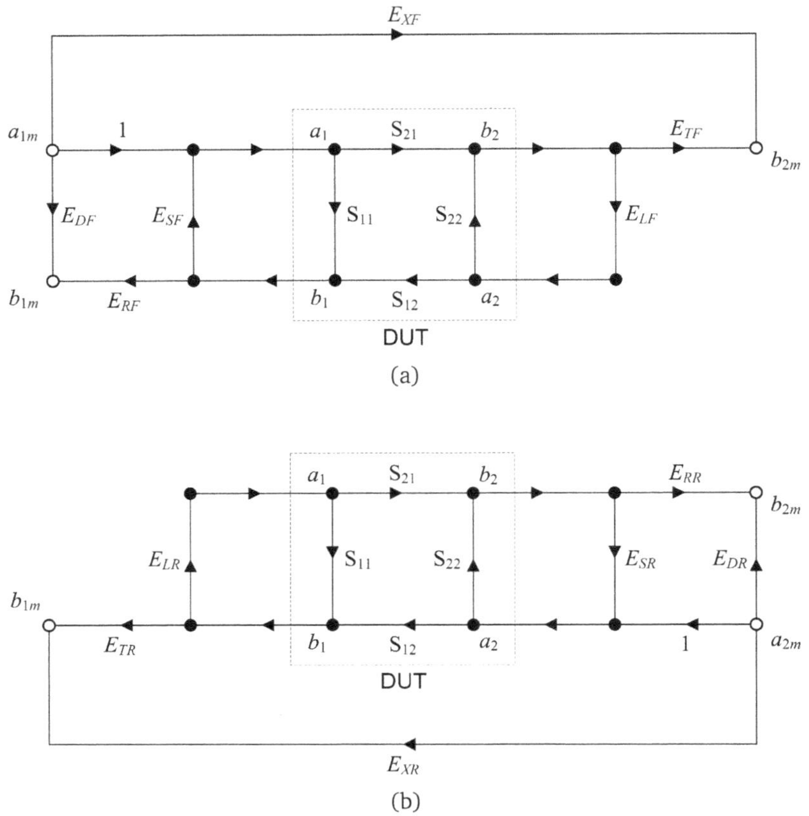

Figure 11.5.2 – Flow-graph of the simplified VNA error models for (a) forward and (b) reverse direction.

mission tracking, $x = S$ for source match, $x = L$ for load match, $x = X$ for crosstalk; $y = F$ for forward and $y = R$ for reverse (or backward).

For generality, there follows the correspondence between the error terms coded with letters for directivity, reflection, source match, forward and reverse and the error terms using only internal port numbers (0, 1, 2 and 3):

- $e_{00} = E_{DF}$, $e_{01} = E_{RF}$, $e_{10} = 1$, $e_{11} = E_{SF}$, $e_{22} = E_{LF}$, $e_{30} = E_{TF}$ for forward model;
- $e_{00} = E_{DR}$, $e_{01} = E_{RR}$, $e_{10} = 1$, $e_{11} = E_{SR}$, $e_{22} = E_{LR}$, $e_{30} = E_{TR}$ for reverse model.

In Marks' paper [236] not only the relationships between the various error models are given in a very clear and comprehensive way, but they are accompanied also by several useful considerations on their validity, together with the explanation behind the fact that 12-term and 8-term error models have indeed only eleven and seven degrees of freedom (or unknowns, or free parameters).

The relationships between the S parameters of the DUT and the measured S parameters inside the VNA for the twelve-error-term model follow.

$$S_{11}^m \equiv \frac{b_{1m}}{a_{1m}} = E_{DF} + E_{RF}\frac{S_{11} - E_{LF}\Delta_S}{1 - E_{SF}S_{11} - E_{LF}S_{22} + E_{SF}E_{LF}\Delta_S} \tag{11.5.16}$$

$$S_{21}^m \equiv \frac{b_{2m}}{a_{1m}} = (E_{XF}) + E_{TF}\frac{S_{21}}{1 - E_{SF}S_{11} - E_{LF}S_{22} + E_{SF}E_{LF}\Delta_S} \tag{11.5.17}$$

$$S_{22}^m \equiv \frac{b_{2m}}{a_{1m}} = E_{DR} + E_{RR}\frac{S_{22} - E_{LR}\Delta_S}{1 - E_{LR}S_{11} - E_{SR}S_{22} + E_{SR}E_{LR}\Delta_S} \tag{11.5.18}$$

$$S_{12}^m \equiv \frac{b_{1m}}{a_{2m}} = (E_{XR}) + E_{TR}\frac{S_{12}}{1 - E_{LR}S_{11} - E_{SR}S_{22} + E_{SR}E_{LR}\Delta_S} \tag{11.5.19}$$

with $\Delta_S = S_{11}S_{22} - S_{12}S_{21}$. The crosstalk terms between round brackets were not in Marks' original expressions.

They are expressed to be used during calibration starting from known S_{ij} parameters (those of calibration standards). Conversely, when in normal measurement mode, the opposite is needed, that is four expressions that give the DUT S_{ij}, starting from the already determined error parameters and the readings at the IF digitizer S_{ij}^m [143].

$$S_{11} = \frac{S_{11}^m - E_{DF}}{QE_{RF}}\left[1 + (S_{22}^m - E_{DR})\frac{E_{SR}}{E_{RR}}\right] - \frac{(S_{12}^m - E_{XR})(S_{21}^m - E_{XF})E_{LF}}{QE_{TF}E_{TR}} \tag{11.5.20}$$

$$S_{21} = \frac{S_{21}^m - E_{XF}}{QE_{TF}}\left[1 + (S_{22}^m - E_{DR})\frac{E_{SR} - E_{LF}}{E_{RR}}\right] \tag{11.5.21}$$

$$S_{12} = \frac{S_{12}^m - E_{XR}}{QE_{TR}}\left[1 + (S_{11}^m - E_{DF})\frac{E_{SF} - E_{LR}}{E_{RF}}\right] \tag{11.5.22}$$

$$S_{22} = \frac{S_{22}^m - E_{DR}}{QE_{RR}}\left[1 + (S_{11}^m - E_{DF})\frac{E_{SF}}{E_{RF}}\right] - \frac{(S_{12}^m - E_{XR})(S_{21}^m - E_{XF})E_{LF}E_{LR}}{E_{TF}E_{TR}} \tag{11.5.23}$$

where

$$\begin{aligned} Q &= \left[1 + (S_{11}^m - E_{DF})\frac{E_{SF}}{E_{RF}}\right]\left[1 + (S_{22}^m - E_{DR})\frac{E_{SR}}{E_{RR}}\right] + \\ &- \frac{(S_{12}^m - E_{XR})(S_{21}^m - E_{XF})E_{LF}E_{LR}}{E_{TF}E_{TR}} \end{aligned} \tag{11.5.24}$$

The expressions above include crosstalk terms E_{XF} and E_{XR} that Marks takes into account in his Appendix B, underlying that correction is straightforward because crosstalk modeling applies outside the error box model. In Figure 11.5.2 the crosstalk terms are shown bridging several branches of the error box model and the reader might wonder why they are simply subtracted in the expressions above as if they were simply connected in parallel: they are, because the quantities from which they are directly subtracted are the S parameters with the m, i.e. those measured between a_{1m} and b_{2m}, or a_{2m} and b_{1m}, and ideally connected between those nodes, that is in parallel to the said crosstalk terms.

The relationship between the 12-term error model and the 8-term error model is quite straightforward for six quantities (E_{DF}, E_{RF}, E_{SF}, E_{DR}, E_{RR}, E_{SR}): they were selected so that they are the same. The influence of parameters α and β is irrelevant and in Marks' model the coefficient on the branch of the S parameter flow graph was set to 1. For the remaining quantities (E_{TF}, E_{LF}, E_{TR}, E_{LR}) of the 12-term model to link with the other remaining quantities of the 8-term model (α and β) we have:

$$\frac{\alpha}{\beta} = \frac{E_{TF}}{E_{RR} + E_{DR}(E_{LF} - E_{SR})} \tag{11.5.25}$$

$$\frac{\beta}{\alpha} = \frac{E_{TR}}{E_{RF} + E_{DF}(E_{LR} - E_{SF})} \tag{11.5.26}$$

The curious fact is that these two expressions are consistent only if a further condition on the 12-term model quantities holds:

$$E_{TF}E_{TR} = [E_{RR} + E_{DR}(E_{LF} - E_{SR})]\,[E_{RF} + E_{DF}(E_{LR} - E_{SF})] \tag{11.5.27}$$

and this explains why the 12-term model has only eleven free parameters (or degrees of freedom). Additionally, this equation is also useful to check the consistency and accuracy of calibration.

Observing then that the two equations above fix only the ratio of α and β, but none of them alone, one of them can be assigned arbitrarily without altering the relationship between the DUT S parameters and the measured S parameters.

Going a little further, once the set of relevant parameters is decided and they are organized in a vector, various techniques for the solution of equation systems may be adopted, rather than a direct solution e.g. by substitution: the obtained solution will be e.g. the best in the least mean square sense (as proposed in [236], App. C, for the determination of α and β), and, provided the criterion satisfies the problem requirement, this is a method that can be implemented on-board the VNA. Some methods as available in stand-alone calibration and uncertainty evaluation tools will be reviewed in sec. 11.5.3.9.

A linear calibration procedure is applied to remove as many of the error terms as possible. The loss and match errors can be greatly reduced depending on the accuracy of the used calibration standards. However, noise and linearity errors cannot be reduced using a simple linear calibration procedure; in fact, they do not appear in the simplified model of Figure 11.5.2 and, in reality, after this calibration, noise and linearity errors might increase by a small amount.

It is recalled, finally, that once the VNA has been calibrated, drift, stability, and repeatability errors will begin degrading system performance; this means that the system will need to be re-calibrated at some time interval, depending on its frequency of use, environmental conditions and required system accuracy.

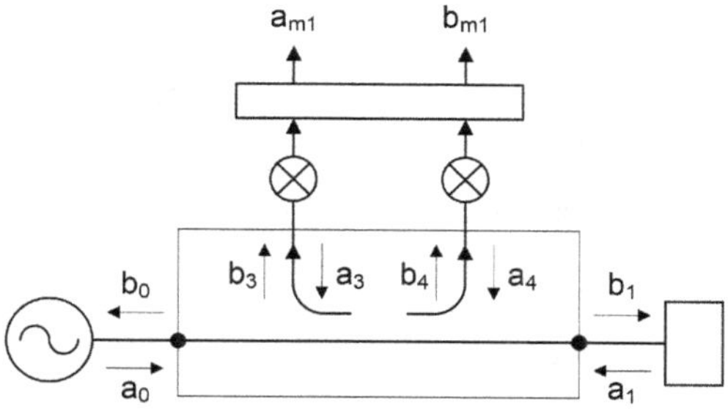

Figure 11.5.3 – VNA block diagram for one-port calibration (signal source is internal).

The flow graph above is quite general and now for better understanding one-port and two-port calibrations are considered more in detail.

11.5.2 One-port calibration

A simplified configuration is considered, the one-port reflection measurement and calibration, that consists of a sinusoidal signal source (identified as port 0), the DUT connected at the VNA port 1, and the resulting forward and backward measured terms indicated by m, derived from the internal couplers that pick up forward and backward terms; VNA ports and internal ports are indicated with numbers 0 and 1, and 3 and 4, respectively (complying to the 8-term error model notation).

Such configuration is quite well described by the selected error terms and has the advantage that neglecting e.g. switch error terms does not bring any inaccuracy.

The usefulness of this configuration is of course limited to single-port return-loss measurements, covering all impedance measurements, including those performed on materials to determine their electrical properties, in particular dielectric permittivity and losses; in general, for many planar transmission lines a one-port measurement is able to asses characteristics and homogeneity of the material (see sec. 6.5).

11.5.2.1 Error model

As said, the basic assumption to develop the error model is the overall linearity for all components, namely cables, connectors, couplers, mixer, ADC etc.. Thus, equations are linear, linking quantities at the IF digitizer with a and b waves at the DUT reference plane. In order for this to hold, VNA components shall be used in its linearity region, that at a first approximation corresponds to the selection of the correct operating point for internal mixers and ADCs.

As anticipated when considering the general model (see sec. 11.5.1), measured wave quantities and VNA port quantities are related compactly as

$$\begin{bmatrix} b_{m1} \\ a_1 \end{bmatrix} = \begin{bmatrix} e_{00} & e_{01} \\ e_{10} & e_{11} \end{bmatrix} \cdot \begin{bmatrix} a_{m1} \\ b_1 \end{bmatrix} \tag{11.5.28}$$

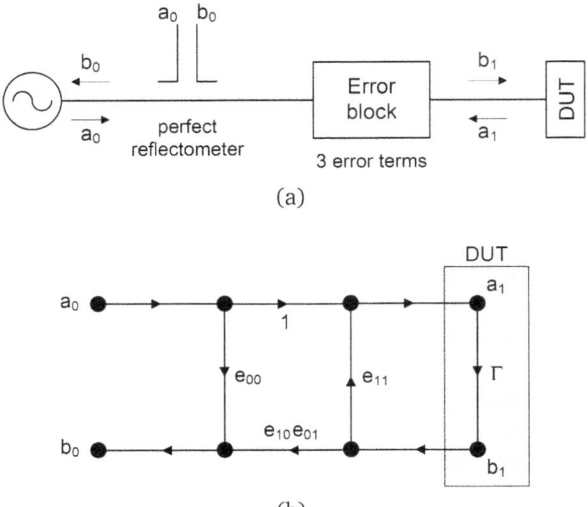

Figure 11.5.4 – 1-port error model: (a) block diagram, (b) flow diagram.

and the e coefficients are related to the previously used E coefficients as: $e_{00} = E_{DF}$, $e_{01} = E_{RF}$, $e_{10} = 1$, $e_{11} = E_{SF}$.

It is recalled that the error network is a fictitious network, where the e terms are complex frequency-dependent quantities, arranged in a scattering-like matrix. This is quite well underlined by Eul in his comments to Soares et al.[9]: it is not possible in general to assume that error boxes are reciprocal networks, because they are "not related to an existing network; they only describe a mathematical relationship between the complex waves at the measurement ports and the output of the measurement channels." For a simple de-embedding task the reciprocity assumption can often be made. Because of one degree of freedom in the error box model (related to the α/β ratio), one can impose a sort of reciprocity by setting the determinant to unity, and this can only be made on one of the two error boxes; doing the same on the other one leads to significant errors. Similarly, it is not possible to invoke passivity.

Speciale, well before, concluded his work on TSD calibration and generalization of a wide range of calibration algorithms [307] (so generally applicable to two-port calibration) commenting on the impossibility to separate the off-diagonal terms on physical grounds, but that their ratio may be set to the square root of the same S parameters of the Through standard, forcing the "error two-ports A and B to share in equal proportions the apparent non-reciprocity due to the system errors."

Let's focus now on the measurement of a DUT with its own reflection coefficient Γ, that is the objective of one-port measurement; it may be put in relationship with the ratio of the measured forward and backward waves, Γ_m, as measured by the VNA. Such measurement is what occurs also during calibration, starting from known standards with known reflection coefficients. When performing calibration it is necessary to use

[9] H.-J. Eul, "Comments on "A Unified Mathematical Approach to Two-Port Calibration Techniques and Some Applications," *IEEE Transactions on Microwave Theory and Techniques*, Vol. 38, no. 8, Aug. 1990, page 1144.

three known Γ values, read the measured reflection coefficient Γ_m for each of them and then solve the resulting system for the error quantities e_{00}, e_{11} and Δ_e:

$$\Gamma_m = \frac{e_{00} - \Delta_e \Gamma}{1 - e_{11}\Gamma} \tag{11.5.29}$$

with

$$\Delta_e = e_{00}e_{11} - e_{10}e_{01} \tag{11.5.30}$$

The reason for using Δ_e instead of explicitly e_{01} is that we thus linearize the equation, avoiding products between e variables (Δ_e is in reality given by the product of the four e quantities, so that the original system of equations is non-linear). With three different known Γ values (as it is in the special case of the Short, Open and matched Load calibration standards) subject to independent measurements, three equations are obtained:

$$\begin{aligned} e_{00} + \Gamma_1\Gamma_{m1}e_{11} - \Gamma_1\Delta_e &= \Gamma_{m1} \\ e_{00} + \Gamma_2\Gamma_{m2}e_{11} - \Gamma_2\Delta_e &= \Gamma_{m2} \\ e_{00} + \Gamma_3\Gamma_{m3}e_{11} - \Gamma_3\Delta_e &= \Gamma_{m3} \end{aligned} \tag{11.5.31}$$

Quite interestingly we may observe that this model works for any reference impedance value, that is not the physical one set by the hardware inside the VNA, but the one established by the adopted calibration set. This is the case when a Load standard is available for calibration, but it is not at the characteristic impedance value of the feeding line from the VNA: this is quite common in reflection measurements made to characterize the dielectric properties of materials, where traceable and nearly-ideal materials are used as calibration standards (e.g. PTFE, pure water, ethanol), but the resulting characteristic impedance is not $50\,\Omega$. The same may be said considering as an example a PCB test fixture or a video or network system, that do not match the $50\,\Omega$ characteristic impedance for connecting cables. For better understanding, we proceed with the solution of the error model for one-port calibration.

11.5.2.2 Solution and calibration using Short, Open and matched Load

The equation system (11.5.31) shall be solved for three known calibration standards (Short, Open and Load in this order), whose reflection coefficients Γ_1, Γ_2 and Γ_3 are known with enough accuracy across the assigned frequency range. The three calibration standards may be chosen to suite different exigencies: ease of construction, stability, confirmation by modeling, uncertainty, etc.

The three measurements shall refer to three reference gamma values, as accurate and easy to build as possible. A common choice is the SOL set of standards: S for Short-circuit, O for Open-circuit and L for matched Load, giving -1, $+1$ and 0 reflection coefficient values, respectively.

Solution for SOL calibration If we assume for the moment that the three standards are ideal and do not need any correction, it is easy to observe that when connecting the Load, and thus $\Gamma_L = 0$, eq. (11.5.29) simplifies to $e_{00} = \Gamma_{m3}$, so that with the measurement of the matched Load the e_{00} directivity parameter is determined.

Then for the remaining values $+1$ and -1 two symmetric relationships are obtained:

$$\begin{aligned} e_{00} + \Gamma_{m1} e_{11} - \Delta_e &= \Gamma_{m1} \\ e_{00} - \Gamma_{m2} e_{11} + \Delta_e &= \Gamma_{m2} \end{aligned} \tag{11.5.32}$$

that can be solved by replacing the already calculated e_{00} and then by direct summation removing Δ_e and determining e_{11}.

$$e_{11} = \frac{\Gamma_{m1} + \Gamma_{m1} - 2\Gamma_{m3}}{\Gamma_{m1} - \Gamma_{m2}} \tag{11.5.33}$$

The last quantity Δ_e is determined by substitution and then it is solved for one of the off-diagonal error terms, under the assumption that they are equal ($e_{10} = e_{01}$).

In the ideal case of perfect directivity (that is infinite isolation between ports and no insertion losses) the solution is $e_{00} = 0$, $e_{11} = 0$ and $\Delta_e = 1$. Directivity is however limited directly by the quality of the matched load standard: the direct measurement of the Load standard gives the directivity, that is in practice unfortunately limited to $-30/-40\,\text{dB}$ because of manufacturing tolerances and matching limitations, quality of materials, etc.; high-performance Loads may be manufactured for even $-50\,\text{dB}$, but they will not preserve such performance over a wide frequency range.

Sliding load This was the reason to develop nearly 40 years ago the so-called *sliding load calibration* [206], using a transmission line with a load that can slide along it, with displacement adjustable by a slider or rotating knob and then reading a gauge. Several measurements are performed for different positions of the slider/knob, that corresponds to a rotation of phase and of the measured point on a Smith chart; the center of the nearly circular set of solution points is the sought e_{00}. Agilent [209] for the VNA internal least-square-circle-fitting method to work properly warns "to move the load element in the same direction, do not move it back and forth. Also, try to slide in non-uniform, not equally spaced, increments."

Errors and uncertainty What are the errors affecting the determination of VNA error model parameters? As shown in [38, 271], using δ before each of the quantities above to indicate uncertainty, directivity error δe_{00} (or δE_{DF}), tracking error δe_{01} (δE_{RF}) and port-match error δe_{11} (δE_{SF}) may be calculated as a function of the three measured standards and their errors, again indicated by a δ preceding the quantity:

$$\delta e_{00} = -\delta \Gamma_{m3} \tag{11.5.34}$$

$$\delta e_{01} = 1 + \frac{1}{\Gamma_{m1}\Gamma_{m2}} \left[\delta e_{00}(\Gamma_{m1} + \Gamma_{m2}) + \frac{\delta\Gamma_{m1}\Gamma_{m2}^2 - \delta\Gamma_{m2}\Gamma_{m1}^2}{\Gamma_{m1} - \Gamma_{m2}} \right] \tag{11.5.35}$$

$$\delta e_{11} = -\frac{1}{\Gamma_{m1}\Gamma_{m2}\delta e_{01}} \left[\delta\Gamma_{m3} + \frac{\delta\Gamma_{m1}\Gamma_{m2} - \delta\Gamma_{m2}\Gamma_{m1}}{\Gamma_{m1} - \Gamma_{m2}} \right] \tag{11.5.36}$$

It is observed that in the weird case $\Gamma_{m1} = \Gamma_{m2}$ the denominators are zero and the error goes to infinite: the reason is simply that we have two identical calibration standards and that the resulting equations are identical and the system is under-determined. For the same reason, this explains why "different" calibration standards should be used and, when more than one of the same type is used (e.g. a short), they shall be offset (e.g. offset short).

Choosing $\Gamma_{m1} = -\Gamma_{m2}$ we can minimize δe_{01}, despite the fact that this might not be easily accomplished at higher frequencies, because Short and Open standards behave differently if their stray parameters are considered.

This type of analysis is expanded in sec. 11.11.4.1, considering in general the uncertainty of one-port calibration methods.

11.5.2.3 Alternative calibration methods and sets of standards

We have seen that the requirements for the solution of the equation system (11.5.31) are the availability of three known calibration standards with reflection coefficients Γ_1, Γ_2 and Γ_3 that shall be "enough" different to allow the solution of the equation system itself. SOL is one possibility, where the distance between the magnitude of the refection coefficients is maximized. However, it was also said that the matched Load characteristics are the limiting factor in determining the VNA directivity and in general Loads are the most difficult to build among the three, if good performance over an extended frequency range is required. Many efforts have been done to shape the field of the incoming wave, either progressively changing the conductivity or shaping inner and outer conductor; in particular the latter aims at transforming the plane wave coming from the guided structure (e.g. cable) into a spherical wave that will expand in the loading section with quite a uniform power dissipation while moving along the line also at high frequency, reducing at minimum high-order modes. Such effect is obtained using a tractrix curve profile with a correct parabolic tapering to the incoming line.

Moreover, in many situations of non-coaxial standards, Loads are quite difficult to implement, e.g. much less easier than Shorts. Opens and Shorts are much easier especially at very high frequency and for non coaxial structures, such as planar transmission lines (see Chapter 6); the formers have some issue in their characterization at high frequency, due to the relevance of the stray capacitance terms. Open standards are more complex than a simple port left open, because i) the reference plane would be poorly defined and ii) there would be radiation in the portion of space surrounding the port (and thus more susceptibility and coupling to external elements). In order to reduce these inconveniences i) the inner conductor is terminated onto a dielectric plug

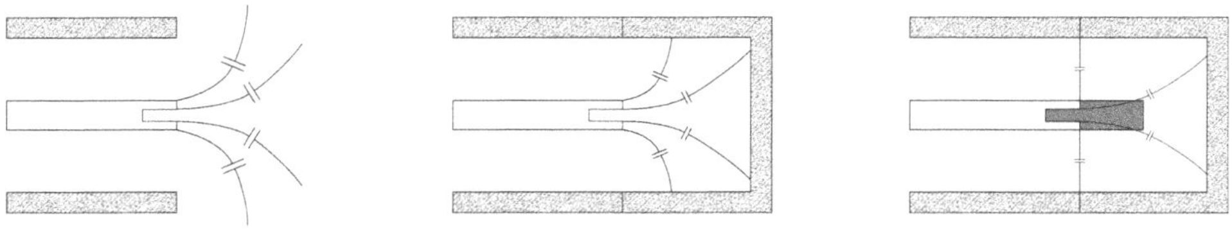

(a) (b) (c)

Figure 11.5.5 – Examples of Open terminations: (a) completely open port, (b) shielded open port, (c) shielded and plugged port.

or is prolonged in a suitably shaped short line section, and ii) open standards are usually built shielded extending the outer conductor, to reduce the evanescent radiating field (see Figure 11.5.5).

More than one Open or Short standard may be thus used, provided that the reflection coefficients (that are complex numbers and are not only consisting of magnitude) are different enough, easing the solution of the equation system: this is achieved using offset standards, that is Opens and Shorts with different electrical length with respect to their reference plane. We have just said that Shorts are preferred especially because of more control on stray parameters, so that we will hear of "Offset Shorts" more often than "Offset Opens". A calibration kit may then be composed of two Shorts and one Open and is called OSS, or three Short terminations with different phase delays may be used (SSS), solving effectively the problem of parasitic capacitance, as it is done for microwave waveguide systems.

Bianco et al. [38, 271] show that calibration using three Shorts is superior to the SOL method in terms of sensitivity and overall uncertainty, including intrinsic inaccuracy of standards and noise. Two configurations were tested, one for which at some frequency the three reflection coefficients are 120° apart, and another in which the phase relationship is 90° (two of them 180° apart). Whereas sensitivities in the sample variance formula (squared uncertainty) for short and open in the two procedures are quite similar (with variations that depend on the phase relationship and the frequency), the matched load in the SOL procedure, with a sensitivity that is about eight times larger, is responsible for the lower overall quality of the SOL calibration. Looking at eq. (11.5.31), any inaccuracy of the matched Load has a direct impact on directivity, that is probably the most relevant error term.

There are situations in which the accuracy of the matched Load and the limitation on directivity are not so relevant and conversely the matched Load is easier to build than a Short: it is the case of setups for the determination of dielectric properties of materials, e.g. solids or liquids, where a transmission line or resonant structure is build using the unknown material as dielectric and reflection coefficient is measured. In this case the Open might correspond to air as dielectric (i.e. no material), but the Short is quite hard to realize. Conversely, there are several reference dielectrics that have known and stable dielectric properties (e.g. methanol, pure water, PTFE, etc.).

A broad class of setups, such as Printed Circuit Boards that host semiconductor devices and connectors, may be defined "inaccessible test fixtures", for which there is no possibility of connecting the VNA calibration standards, but the standards shall be embedded at design time. In this case it is important that parasitics and design uncertainty is kept to a minimum; moreover, highly reflective loads are difficult to realize on PCBs, especially Opens for the same reason, as before: poor control of parasitics, worsened for inhomogeneous lines like microstrips by the varying electric field distribution between dielectric and air regions with frequency (dispersion). In this case it is advisable using as most as possible transmission lines (moving however to a two-port calibration), as in the extreme case of using only transmission lines, in the so-called Multiline method [100, 235]. In 2002 [100] the pioneering work by Bianco et al. [37] is acknowledged: while NIST Multiline and MultiCal® were explicitly hosting various calibration methods based on the use of line standards, in [37] the extreme three-line configuration was directly addressed, especially noting that lines were, and are, the most easy device to build and characterize in microstrip and planar transmission line applications. Quite accurate closed-form expressions exist for the characterization of planar transmission lines as reviewed in sec. 6.3; additionally, confirmation even for unusual geometries may be sought by finite element methods and in general full-wave solvers.

Line-based calibration methods are reviewed in the next section for two-port calibration: see sec. 11.5.3.6.

11.5.2.4 Correction of calibration standards

It is quite difficult to preserve an ideal behavior of calibration standards (Short, Open and Load) over a wide frequency range: parasitics, variability of material properties, geometry tolerance, etc. exert their influence, leading to non-ideal terminations. Their characteristics may be compensated for by supplying a suitable description: the VNA will mathematically remove these characteristics from the measurements. The characteristics are described and quantified in terms of a convenient representation, e.g. based on offset delay, offset loss and offset impedance [209]:

- *offset delay* describes the time delay between the calibration plane at the fixture input and the DUT position inside the fixture;
- *offset loss* takes into account the expected loss term due to skin effect in a coaxial structure; because the fixture is non-coaxial, losses as a function of frequency may differ from those of a coaxial transmission line, so that the entered value is only an approximation;
- *offset impedance* is necessary to account for any change of the fixture characteristic impedance, that is quite uncommon and may occur e.g. for an undesirable behavior of dielectric material, in particular for low-quality ones.

Manufacturers may have (and have) slightly different representations of correction parameters; consequently their instruments may not straightforwardly accept some parameters of a calibration kit that are not defined following the same approach. Tools or programs to manage calibration and uncertainty are normally able to load various standard definitions (see sec. 11.5.3.9).

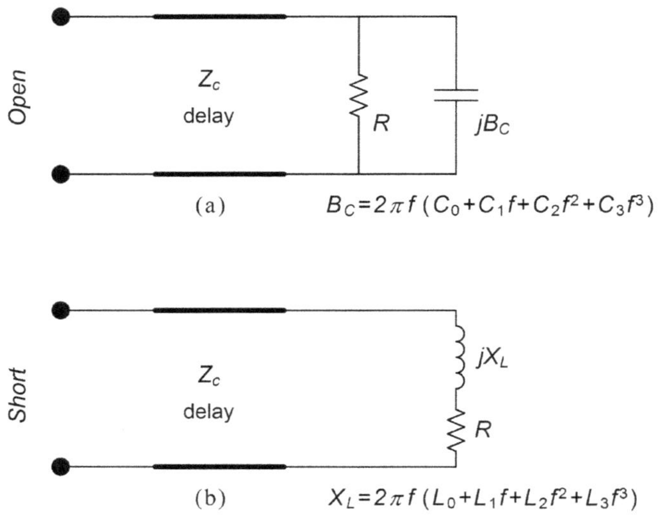

Figure 11.5.6 – Models for calibration standards: (a) Open, (b) Short.

Regarding impedance correction for parasitics and their change with frequency, a suitable frequency dependence scheme is used, as shown in Figure 11.5.6. For the Open, featuring an ideally infinite impedance, the relevant parasitics are the leakage conductance and the stray capacitance, shunting the otherwise infinite impedance; the capacitance terms simulate the more or less pronounced fringe field radiation and its dependency on frequency. Conversely, for the Short the most relevant parasitics are series elements, i.e. the additional series resistance (e.g. due to skin effect or losses) and the stray inductance, also applicable to most Load standards. For the Load standard, however, depending on the VNA model, it might be impossible to specify at the same time resistance and inductance value, so that the workaround is to use a delay value that gives the desired amount of phase rotation, once an arbitrary value of the characteristic impedance is assigned.

How to determine the coefficients of the models of the calibration standards, if the measurements that we can do are all functions of the same calibration standards? In the past, scaled models were approached, but today there is a wide use of simulation and Finite Element Methods. This in turn poses the question of how accurate such models are and which impact they have on the absolute accuracy of such a calibration. In general the cross-check of calibrations by standards of different origin and using different calibration algorithms ensures that uncertainty is under control and that flaws or mistakes are quickly identified.

In [272] the discontinuity that builds up when mating some "like" connectors[10] is compensated for assigning a capacitive behavior to it and determining the four C_0, C_1, C_2 and C_3 coefficients, by fitting the results of a set of finite element method simulations. The work reports several estimates of the capacitive model coefficients for most com-

[10] The 3.5 mm connector is an air filled high-performance connector that accepts 2.92 mm and SMA (dielectric filled) with which it is mechanically compatible; yet, a discontinuity is created that shall be accounted for by introducing compensation, by which the error is reduced.

Combination	C_0 [fF]	C_1 [aF/GHz]	C_2 [aF/GHz^2]	C_3 [aF/GHz^3]
3.5 mm/2.92 mm	6.956	−1.026	−0.014	0.0028
3.5 mm/SMA	5.959	−11.195	0.508	−0.0024
2.92 mm/SMA	13.420	−1.945	0.546	0.0159
2.4 mm/1.85 mm	8.984	−13.992	0.324	−0.0011

Table 11.5.1 – Coefficients of the capacitive model of adapters [272].

mon adapters modeled with nominal data appearing in the corresponding IEEE/ANSI specifications (see Table 11.5.1). However, for connectors such as SMA (in this case the lowest performance connector of the examined set), that may deviate from item to item with respect to nominal specifications, such a compensation with nominal values might produce worse than original results. Pollard reports some considerations on the reference mating plane and its relevance for microwave measurements; it is anyway underlined that such corrections and the related issues are significant above several GHz, as testified also by the very small value of C_0.

In general VNA manufacturers sell their own calibration kits together with their machines, that thus host pre-loaded values for correction under each calibration standard code and are accessible by menus. There is always some empty menu slot for "user defined kits", where custom values may be loaded or input by hand, once a new calibration kit is purchased, whatever its brand. In [115], section 9.1, it is described in detail how to characterize and determine the equivalent coefficients for a home-made microstrip calibration kit.

11.5.3 Two-port calibration

Analogous to the one-port calibration, the two-port calibration starts from the general model and equations of a complete scheme of two-port reflection/transmission measurement and calibration; not many details of equations are given, that can be found in sec. 11.5.1 and in some very good references that report extensively and completely the mathematical details: Dunsmore [115], Marks [236], Rytting [292], Silvonen [303], Speciale [307].

11.5.3.1 Scheme of two-port VNAs

The relevant elements of the system, as for one-port calibration, are reported in a block diagram, on which the error model is built: the block diagram includes a sinusoidal signal source (switched alternatively between port 1 and port 2) and the DUT, connected between ports 1 and 2 of the VNA; the resulting forward and backward measured terms are indicated by $m1$ and $m2$, as derived from the internal couplers that tap the forward and backward terms at the two ports, that for compactness may be indicated with numbers 3 and 4 (see Figure 11.5.7). The more complete two-

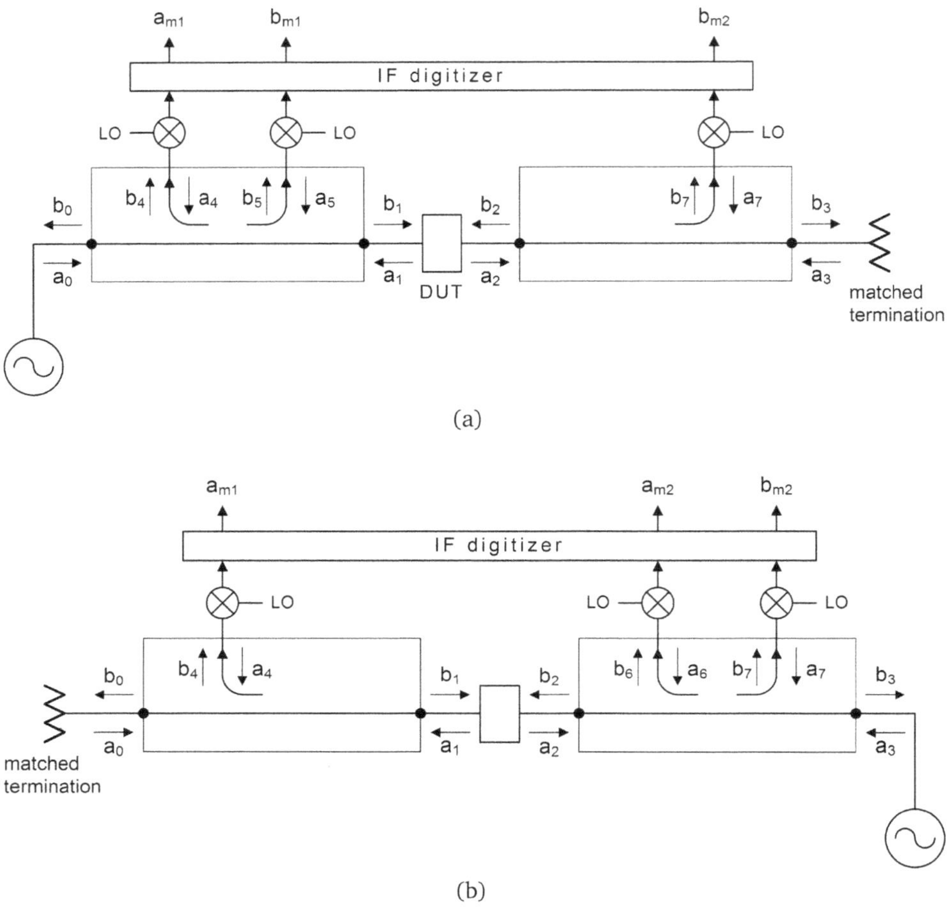

Figure 11.5.7 – VNA block diagram for two-port calibration: (a) forward, (b) backward.

reflectometer VNA architecture is considered, as it is the most common nowadays for two-port VNAs; it is derived readily from the previous one-port calibration model by mirroring it and accounting for the connection between the two ports (transmission).

11.5.3.2 Error model

Let's consider again the error box model that in sec. 11.5.1 was introduced with a general approach and then solved for one-port measurements in sec. 11.5.2.2, i.e. without considering any connection between ports. When performing two-port measurements, the two ports may still be considered separately achieving two one-port calibrations, one for each port, and then introducing the effect of the Through (or Line, or any other interconnecting element or device) to complete the solution of the equation system; this is quite intuitive, based on what has worked for one-port calibration.

The two diagrams appearing in Figure 11.5.2(a) and (b) for forward (indicated by "F") and backward (indicated by "R", for reverse) measurements contain the error terms for the most complete twelve-error term model. Calibration for this model is possible using one of the methods based on fully characterized sets of calibration standards.

The most known calibration method is the SOLT (it is available in all VNAs), but a relaxation of requirements regarding characterization of the Through was already proposed in 1992 [124]: the method was named RSOL, where the "R" stands for "reciprocal". The method is particularly useful to minimize cable movement during calibration and to cover non-insertable devices: its name is popularly "unknown thru", indicating that the Through standard could be only partially known, with reciprocity ($S_{12} = S_{21}$) required and a phase response known within a quarter of wavelength.

Other methods are possible, using less standards (three instead of four) and allowing for partially known standards, thus accommodating also for some poor characteristics of a standard that may be thus neglected and not included in the calculation, as they were unknown; examples of these poor characteristics might be dispersion effects of microstrip standards, non-ideal adapters, etc.. These methods are well described and classified in [121, 122, 303], where they are reviewed with respect to a general theoretical framework (in [122] the expression "12-term procedure" indicates the SOLT calibration method).

The possibility of using less standards, or a reduced number of known calibration standard features, resides in the need of determining less unknowns with respect to the full model (see sec. 11.5.1). The reduction of the full model to simpler and more usable error models is summarized here for clarity:

- in principle the number of the $\mathbf{S}_e$ matrix parameters in eq. (11.5.3) (or $\mathbf{T}_e$ matrix parameters in eq. (11.5.7)) is 16 and the complete solution would require 16 independent features of calibration standards; using four two-port calibration standards, or more, reapplying one-port standards to each VNA port separately, we reach the situation of 16 equations for 16 unknowns; however, the most complete model is reduced to twelve terms as explained in the comments in sec. 11.5.1.1 below eq. (11.5.7), not by invoking symmetry of the error matrix (it is not a physical matrix, as already commented at the end of sec. 11.5.2.1), but recognizing that some error terms are not relevant, especially with new VNA architectures and their performance;

- the most usable and complete VNA error model is made of 12 unknowns, having removed the information of the isolation between reflectometers, that is (almost) always negligible; the exact solution by classical methods requires 12 independent equations;

- when some symmetry is assumed, being able to merge tracking error terms for reflection and transmission in both directions, the error model is further reduced to 8 terms; in the next section 11.5.3.3 it is discussed how the relevant unknowns are in reality seven, because two terms appear in model equations always joined in a product (this is of course an ambiguity in the determination of the individual quantities, but once accepted, allows the simplification to seven unknowns without introducing any error).

Ending up with seven unknowns (error box parameters) and the possibility of collecting up to twelve equations, when the three two-port standards are fully used, gives rise to some redundancy and the chance to exploit it. This redundancy may result in a reduction of the standard parameters that are needed known, or in a different approach to the solution, e.g. using optimization techniques with respect to some goodness criterion, that e.g. minimizes some kind of error (mean square error, maximum error, inverse likelihood, etc.), accounting for and mitigating the unavoidable inaccuracies of measurements. We might call the former "reduced methods" (see sec. 11.5.3.8) and the latter "redundant methods" (see sec. 11.5.3.9).

11.5.3.3 Unknown thru, or RSOL

As underlined by Wong [342], the "unknown thru" method is well suited for "immovable" test ports, that is when non-insertables are to be measured, such as setups with hardly reachable parts or critically positioned, and "odd shape and multiport devices." This method thus is quite relevant when handling own setups and circuits that are non-standard, especially when the possibility of terminating the parts with standard connectors is limited, as it is the availability of a complete calibration set. The requisite for the unknown Through is that it is reciprocal, as almost all transmission lines, including planar transmission lines on PCB and wafer. It is possible thus to identify as "unknown" any transmission element that is not possible to fully characterize accurately, as would be required instead by a SOLT calibration method.

Considering the eight-term error box model of the VNA (see Figure 11.5.2), the matrices of the error term networks at the left and at the right of the DUT (or calibration standard) are indicated by $\boldsymbol{A}$ and $\boldsymbol{B}$, with the external device indicated by $\mathbf{T}$; the cascaded product is indicated as matrix $\mathbf{T}_m$, that is the resulting measured device:

$$\mathbf{T}_m = \mathbf{T}_{\mathrm{A}}\mathbf{T}\mathbf{T}_{\mathrm{B}} = \frac{\beta}{\alpha}\frac{1}{E_{RR}}\begin{bmatrix}\Delta_{\mathrm{A}} & E_{DF}\\ -E_{SF} & 1\end{bmatrix}[\mathbf{T}]\begin{bmatrix}\Delta_{\mathrm{B}} & E_{SR}\\ -E_{DR} & 1\end{bmatrix} \tag{11.5.37}$$

with $\Delta_{\mathrm{A}} = E_{RF} - E_{DF}E_{SF}$ and $\Delta_{\mathrm{B}} = E_{RR} - E_{DR}E_{SR}$ and the usual meaning of subscripts[11]: "D" for directivity, "S" for source match, "F" for forward and "R" for reverse; α and β are the known additional terms, one for each error box, that may be determined as ratio but represent the one degree of freedom, and include the remaining non-ideality of the VNA, in particular tracking error. The determinants of $\mathbf{T}_{\mathrm{A}}$ and $\mathbf{T}_{\mathrm{B}}$ are immediately calculated as

$$\det(\mathbf{T}_{\mathrm{A}}) = E_{RF} \qquad \det(\mathbf{T}_{\mathrm{B}}) = E_{RR} \tag{11.5.38}$$

Six of the eight terms can be determined by making separate one-port calibrations on each port 1 and 2 (using e.g. the SOL method of sec. 11.5.2.2) and then the other two are fixed by a Through measurement, observing that we need to fix only one, being possible a normalization on one of the parameters, what Eul and Schiek [122] call "one-dimensional ambiguity".

---

[11] This notation is preferred to the other with small e.

The relevance of the requirement of reciprocity of the Through reveals when it is invoked for the determinant of $\mathbf{T}$ that is unity ($\det(\mathbf{T})=1$); the ratio β/α may be determined observing that

$$\det(\mathbf{T}_m)=\left(\frac{1}{E_{RR}}\right)^2\left(\frac{\beta}{\alpha}\right)^2 E_{RF}E_{RR} \tag{11.5.39}$$

and that $\det(\mathbf{T}_m)$ with $\mathbf{T}_m=\begin{bmatrix} -\dfrac{\det(\mathbf{S}_m)}{S_{m21}} & \dfrac{S_{m11}}{S_{m21}} \\ -\dfrac{S_{m22}}{S_{m21}} & \dfrac{1}{S_{m21}} \end{bmatrix}$ calculated from the raw scattering parameters is $\dfrac{S_{m12}}{S_{m21}}$. Thus, the other requirement of "knowing the S_{21} phase shift or electrical length within a quarter wavelength" comes into play when the correct root of k shall be chosen, when solving the ratio $\beta/\alpha=kE_{RR}$, with k equal to one of the roots $\pm\sqrt{\dfrac{S_{12m}}{S_{21m}E_{RF}E_{RR}}}$.

Ferrero and Pisani, while formulating the two requirements, add an important consideration: if the DUT itself is reciprocal, well, the DUT may work as Through and serve as its own calibration standard!

Moreover, as pointed out by Wong and Hoffman [341], the "unknown thru" calibration is great at checking the quality of imperfect port connectors if the calibration standards are accurately modeled, since the assumption of a perfect Through is not necessary nor assumed in the calibration. Imperfections may be of various kinds, such as using low-quality adapters, or facing pin gap affecting the accurate position of the reference plane and causing various aberrations, etc., as detailed in sec. 7.1.1.7. Measuring the S parameters of the "thru" after performing an Unknown Thru calibration indicates the actual quality of the test port connectors; conversely if the port connectors are high precision parts, with no appreciable pin gap and other connector defects, the measurement of the thru is a good indication of the calibration accuracy. The combined effect of imperfections of the male and female pair may be treated in different ways, as for what is to be included in the calibration standard definitions and what instead ends up being assigned as test port inaccuracy.

As said, the "unknown through" method is particularly useful when working with setup elements that do not have good through connections, such as when including wafer probes or long cables subject to bending and flexure. The latter is quite common when measuring devices with many ports (such as splitters, directional couplers, etc.) with a two-port VNA, thus having the necessity of many connections and disconnections, with problems of connector, adapter and cable repeatability. Inaccuracy caused by cable movements is reduced with respect to the SOLT, or "known through" calibration method: a test reported by [342] shows a smoother curve for RSOL compared to SOLT, with an amplitude difference that is perceivable, eve if not so significant (about 0.05 dB at 50 GHz). Also when adapter removal is necessary, RSOL performs better, again with a smoother trace and much less dispersion, and an amplitude difference of about 0.02 dB at 50 GHz (15% relative error), when measuring a female-female 1.85 mm adapter.

11.5.3.4 TRL (Through-Reflect-Line) calibration

The TRL line method was introduced at the end of '70s in various contexts, e.g. as an alternative calibration method, but also as a preferred method for the new dual six-port VNA [119].

The first standard (the Through) determines the reference system impedance. For coaxial calibrations the transmission standard is intended as a zero-length (or "flush") device, made commonly using a female-female short barrel connecting together the male-ending cables; its center determines the reference plane with negligible error. When the Through has some length and is in reality a line, the TRL "degenerates" into a LRL calibration.

When calibrating for PCB and wafer applications, the use of the least number of reflect and matched standards is advisable for already commented problems of construction and characterization, taking into account possible stray parameters and non-ideality. A widespread procedure is thus the TRL, or LRL, using two transmission line standards and one reflect (TRL). The Through as a zero-length line directly connecting the ports (e.g. launchers or probes) is often impossible and this is why two lines of different length are realized going for the LRL method. The reference impedance of the calibration is that of the Line standard and this may be a source of error if its characteristic impedance $Z_{c,L}$ is not determined accurately: the Line standard needs to be symmetric (and a transmission line is), but if it is "imperfect" the TRL procedure becomes the TRM one, related by an impedance transformation [237], sec. 4.1. The definition or determination of the Line is the crucial point and Marks and Williams stigmatize that it shall be in the end determined by measuring it against some other standard: practically speaking, when working with board and wafer applications, often $Z_{c,L}$ is determined by careful design and/or by independent measurement, that is e.g. using time domain reflectometry or indirect measurements of line characteristics, using then known planar transmission line expressions and characteristics[12] (such as those described in Chapter 6 for microstrips and striplines, in sec. 6.3.1 and 6.3.4, respectively). To improve methods that assume frequency-independence or use low-frequency or even DC values of line parameters, there are methods using e.g. measurements on two lines of different length[13]. Additionally, the two lines might have a slightly different characteristic impedance, because of variability of design, fabrication and material characteristics (see sec. 6.3): this reflects in a deterioration of residual errors of directivity, port match and tracking. The same may be said for dispersion, that is variability of characteristic impedance with frequency for all kinds of planar

---

[12] Usually the total line capacitance is measured and under convenient assumptions of frequency independence of either resistance and inductance or conductance, the characteristic impedance may be estimated. See for example:
J.S. Kasten, M.B. Steer and R. Pomerlau, "Enhanced through-reflect-line characterization of two-port measuring systems using free-space capacitance calculation," *IEEE Transactions on Microwave Theory and Techniques*, Vol. 38, no. 2, Feb. 1990, pp. 215-217.
R.B. Marks and D.F. Williams, "Characteristic impedance determination using propagation constant measurements," *IEEE Microwave and Guided Wave Letters*, Vol. 1, June 1991, pp. 141-143.

[13] M.B. Steer, S.B. Goldberg, G. Rinnet, P.D. Franzon and I. Turlik, "Introducing the Through-Line deembedding procedure," IEEE MTT-S International Microwave Symposium Digest, Albuquerque, NM, USA, June 1-5, 1992, Vol. 3, pp. 1455-1458. doi: 10.1109/MWSYM.1992.188284

transmission lines, but in particular microstrips and coplanar waveguides (see sec. 6.3.1 and 6.3.3).

From a general standpoint the major drawback of the method is the lack of accuracy and periodic glitches in the resulting calibration curves when the instantaneous frequency is such that the Line standard is resonating (or, in other words, approaching integer multiples of half periods). The correct choice of the Line length was considered thoroughly [161], where short and long lines are considered, and in particular the frequency intervals of validity for using one long line and how to cover a wide frequency range without discontinuities with two lines. These rules have then been reported in manuals and papers, for the optimal selection of the length of Line standards, and are also mentioned by Marks, while evaluating the characteristics of the calibration set of the Multiline method.

The Reflect standard will be either a short or open, depending on the ease of implementation, on its stability and characterization, and so on: we already said that shorts are much better defined than opens (in particular for coaxial lines, but also for planar transmission lines, avoiding fringe field and evanescent radiating field, that might result even in unwanted coupling to nearby elements). On the other hand, when PCB or wafer probes are used, an open standard may be realized by lifting the probe tips. It shall be understood that the so realized reflect standard in all cases shall be "the same" at the two ports, relying mostly on uniformity of design and fabrication. We may observe that the Reflect standard is zero-length and may be used to set the reference plane rather than the Through, that might be unavailable: the reflect standard might include the launching structure until the planar transmission line structure is reached and then the short circuit (if it is a short) applied.

Finally, whenever for construction constraints straight nearly ideal planar transmission lines cannot be built and they shall contain e.g. bends or zig zag patterns, then the user should revert to an Unknown Through method, relying on the symmetry alone of such transmission lines, but necessitating then of a complete one-port calibration at each port, e.g. using SOL method. The problem is immediately evident that we are again in troubles with various calibration standards that might turn hard to design and characterize, but in reality any one-port calibration will work such as a SSL (two shorts, one offset, and a load) or SSS (three offset shorts), that may be effectively implemented using a limited board or wafer area.

11.5.3.5 LRL (Line-Reflect-Line) and LRM (Line-Reflect-Match) calibration

The LRL methods may be see as a variant of the TRL, in that the Through and the Line of TRL are two lines of different electrical length, such that at the largest frequency (shortest wavelength) the difference between the two is less than a quarter wavelength. It's a simple calibration scheme suitable for on-waver calibration, using a short as reflect standard (better defined and performing than an open) and lines, that, as already said, are quite convenient to build on wafer.

As for similar methods, increasing bandwidth towards lower frequency requires to add lines of increasing length.

The LRM method, then, may be seen as a degeneration of LRL where one of the lines has become of infinite length, absorbing the incident wave without a transmitted wave at the other end, that is a Matched load. This solves the problem of extension to lower frequency, where a matched load work satisfactorily and replaces bulky lines.

Of course the main problem of using a Matched load at high frequency remains: building a load in wafer or PCB applications often involves the use of lumped elements, such as chip resistors for the latter; in this case package inductance and physical length may bring in a significant stray inductance. A well known practice for PCBs is that of using two chip resistors connected in parallel and located symmetrically at the two sides of the planar line to terminate (usually it is said "two $100\,\Omega$ chip resistors" for the implicitly assumed $50\,\Omega$ reference impedance). It is however not forbidden to use LRM for the lower frequency interval and then switching to LRL when wavelength and necessary line length are conveniently small.

11.5.3.6 Multiline method

Making another step forward from TRL or LRL methods, the reflect standard may be replaced by another line standard, obtaining a method based solely on transmission lines (the TRL method uses different line standards to accommodate for the wide frequency range, but only one at a time). As anticipated when talking of the one-port calibration in sec. 11.5.2.3, the "multiline" denomination was introduced in 1991 [235], encompassing a set of methods that rely upon transmission lines as calibration standards, and was then implemented in the MultiCal software at NIST. The original idea may be led back to [37] and stemmed from the observation that line objects are much better characterized and easily implemented than opens, shorts and loads in a planar transmission line application (microstrips were taken as reference). We might say intuitively that we need only to build three line sections of different length, exactly with the same cross-section, that will undergo the same fabrication process, with same tolerances and material properties, thus characterized by a minimum variability among themselves and in general by negligible non-ideality (e.g. propagation modes and fringe field) and well matching the results of electromagnetic field solvers.

The mathematics of the method are well explained in [235], where there are also a few remarks regarding the significance of the used standards and of their reflection and transmission properties. Using lossy lines for example is beneficial and reduces the errors at the phase extremes close to the $0°$ and $180°$ conditions; the extreme case of infinite loss, that is the use of a matched load instead of a line (like in the LRM calibration method) is in this direction, but has the drawback that "it fails to provide the propagation constant". This is underlined in the same paper a few paragraphs before, observing that only reflection measurements of standards influence the first-order linear approximations of errors, and standards without transmission (e.g. perfect match or attenuator) might be used, but again without obtaining the determination of the propagation constant.

The effect of the additional line standard is to reduce the error standard deviation and to make it more uniform with respect to frequency (even if this does not demonstrate that this is the minimum achievable value). The reason is explained in [100]: the Multiline algorithm implemented in MultiCal selects always the best line to be used as common line at each frequency; on it the length differences of the other lines are formed and then the propagation constant determined and the eigenvalues assigned using a test of significance based on minimization of difference with respect to a common average value. A good example is given with the measurement of a short circuit in coplanar waveguide built on gallium arsenide: the resulting curve is very close to unity and smooth (deviations of less than 0.005), largely reducing the otherwise present peaks that swing $+0.03/-0.06$.

As a result of determination of the propagation constant, also attenuation constant α, phase constant β and an estimate of the effective relative dielectric permittivity $\varepsilon_{r,\mathrm{eff}}$ are available, the latter using a simplified model (we saw in sec. 6.3 that many terms and corrections may be necessary for planar transmission lines to reasonably fit experimental data, so such estimation might be not really accurate over the entire frequency interval).

MultiCal then determines the correction matrices A and B for the 8-term error box model as indicated in [100], eq. (35) and (36). As a result of the availability of the propagation constant, the calibration planes may be moved along the lines as desired; analogously, being the reference impedance arbitrary, the user may select a new reference impedance onto which the error coefficients of the error box model can be expressed, so to correct raw measurements in the desired reference impedance.

11.5.3.7 Electronic Calibration Units (ECUs)

Electronic calibrators are a temperature-controlled set of calibration standards that can be switched internally, avoiding problems of connector repeatability, cable flexure and in general wearing of connectors especially when subject to limitation of the number of mating/de-mating cycles. In general electronic calibration units (ECUs) have the least intrinsic uncertainty and behave also well for setup uncertainty, minimizing connections and movements. They have an albeit small uncertainty and recent research is considering the most relevant elements[14]:

- the VNA port and the ECU are at different temperatures and when connected heat flow occurs before reaching thermal equilibrium; the use of a connecting cable possibly reduces, or at least slows down, this effect, but its impact shall be accurately evaluated; additionally, the internal temperature controlling circuits of the VNA and ECU will have different dynamics, so that their responses may temporarily diverge during transients (slow thermal transients to some final environment temperatures should be monitored by means of repeated measurements to understand and evaluate their behavior);
- the low ECU uncertainty will highlight the residual connector repeatability, that removing the connecting cables, is the only relevant remaining factor;
- similarly, long term drift and aging may become quite relevant, especially if the presence of active devices (e.g. switches) inside the ECU is considered, with respect to a full mechanical calibration kit, recognized in principle as more stable.

[14] M. Zeier, "ECU Test Methods", presentation at HF-Circuits Workshop: Electronic Calibration Units, Boras, Sweden, Dec. 2013.

11.5.3.8 Reduced methods

In this section the possibility of derogate to a fully characterized set of calibration standards and a complete calibration is considered: depending on application, its construction details, frequency range and required accuracy we might be interested in reducing requirements related to characterization and diversification of standards and number of calibration steps. An example was the "unknown thru" method, that relax requirements on the Through standard.

Response-only calibration Another typical example is when a response-only calibration is performed for a two-port measurement: such method resembles a normalization or compensation, rather than a calibration, because straightforwardly the response of the through path between the two ports is subtracted from the measured response (see Figure 11.7.2 for a comparison of this method and a full two-port calibration applied to the characterization of 6 dB attenuators). It is evident that the results are subject to error due to unfavorable phase relationship of incident and reflected waves: ripple is caused by reflections due to port match that is here thus not taken into account, but compensated only for the particular phase relationship occurred during the response calibration; assuming a generic $-30\,\mathrm{dB}$ port mismatch the worst-case ripple will be for a $\pm 180°$ phase displacement of reflection terms that for an approximately $0\,\mathrm{dB}$ S_{21} response (e.g. for through, cable, adapter) amounts to $\pm 20\log_{10}(1+10^{-\mathrm{PM}/20}) = \pm 0.27\,\mathrm{dB}$.

Framework for calibration requisites A structured approach to the definition of requisites for various methods has appeared in the literature. As anticipated, only seven equations are needed to solve for the seven unknowns of the reduced two-port error box model, so that partly unknown calibration standards are allowed. Eul and Schiek [122] present the theoretical framework that justifies a series of conclusions:

- calibration standards identified as $\mathbf{N}_1$, $\mathbf{N}_2$ and $\mathbf{N}_3$ need to meet the following: $\mathbf{N}_1$ shall be fully known, $\mathbf{N}_2$ may have two unknowns (or degrees of freedom) and $\mathbf{N}_3$ three unknowns, for a total of 12 equations - 5 degrees of freedom = 7 useful equations for 7 error model unknowns;
- for $\mathbf{N}_1$ to be fully known, the choice is addressed normally to 1) a Through and these methods will begin with a "T"; ideally, $\mathbf{N}_1 = \begin{bmatrix} 1 & 0 \\ 0 & 1 \end{bmatrix}$; 2) a well characterized matched line is another possibility and is indicated with "L";
- for $\mathbf{N}_2$ 1) a transmission line is a possible choice (indicated by "L") that is however subject to phase shift constraints, requiring too a long line at low frequency and thus limiting the useful bandwidth, and featuring "forbidden" frequency intervals periodically occurring at each phase turn when sweeping frequency; the characteristic impedance of such line sets the reference impedance for the calibration; as an alternative, instead of pushing for an ideal low-loss line, 2) one may go for significant losses and the standard will be denoted by "A", as attenuation (with the "L" standard becoming a special case of very low or zero at-

tenuation); as another possibility, 3) we may sacrifice transmission property and opt for a reflect "R" standard or a matched "M" standard, that shall be known, because the two degrees of freedom have gone when transmission properties are left unspecified;

- for $\mathbf{N}_3$ a reflective standard shall be chosen, since the first two are both well matched; this standard may or may not have transmission, i.e. the two ports may or may not be connected; a typical choice is the use of a reflect Short or Open, that should be identical for the two ports for best performance, and thus the best solution is to measure subsequently the same reflect standard at the two ports; the use of the reflect standard lends itself to be implemented with unknown type of reflect indicated by "R", that however makes little sense because we should know if we have a short or an open; when this is known, the method will show a "S" or an "O", as usual.

Silvonen [303] moves only one remark about Eul and Schiek's work, that is the use of transmission matrices and some difficulty in handling homogeneously one-port and two-port standards; in his work he proposes the use of scattering parameters, always featuring finite values for the various kinds of standards, including those with zero transmission. Silvonen underlines some general facts about calibration, that complement those set forth by Eul and Schiek:

- at least two calibration standards are needed, if the fixture is symmetric; normally the procedure relies on three or more standards, with no upper limit for their number;
- one of the standards should be a known two-port or a through connection (the $\mathbf{N}_1$ in Eul and Schiek's work); the second standard can either be a known two-port or a known reflective or non-reflective termination with one of the S parameters allowed to be unknown (similar to $\mathbf{N}_2$ above);
- with redundant calibration standards, the approach is suitable for a least mean square solution or any other method for the solution of linear equation systems, with minimization of some kind of error.

Of all calibration standard combinations covered by the two approaches [122, 303], not all are really useful and effective for VNA calibration. For example Shoaib [302] demonstrates theoretically, supported by experimental results, that inconsistency occurs for 2-port VNA calibration when combinations with a Reciprocal standard are used.

It's with statistical methods (and in general with overdetermined methods) that a comprehensive description of degrees of freedom and of uncertainty of both calibration standards and calibration procedures may be suitably handled to derive optimal VNA calibration, also from an uncertainty and probabilistic viewpoint.

11.5.3.9 Redundant methods

In Eul and Schiek [122] the possibility of collecting additional pieces of information was formulated, in order to reduce the overall uncertainty of calibration using overdetermined calibration methods, as a means of correcting for error terms due to calibration standards and possibly instrumentation errors.

It might occur at one frequency only, as for the self-calibration method, or it may be a more global approach that leads to optimal calibration in the sense of some objective function, either deterministic (e.g. least mean square and non-linear constrained optimization) or statistic (e.g. maximum likelihood or Bayesian approach). These methods shouldn't be intended as optimal *per se*, because much depends on how cleverly the objective function was defined and how completely addresses calibration issues.

These methods began to appear nearly ten years ago [163, 222] and they are recently undergoing an intense study and development: their implementation and use is favored by the large computational capability available nowadays on computers and on-board VNAs themselves; manufacturers are making considerable efforts to embed similar tools with real-time response, able to include all aspects of uncertainty (calibration, repeatability, drift, etc.). In addition, optimized uncertainty management and calibrations are available also in complete programs used and proposed by manufacturers and most of all metrological institutes, reviewed in sec. 11.5.3.10 below.

Multi-frequency maximum likelihood The problem of optimal calibration formalizing it at a single frequency for a set of calibration standard using an approach that may be easily extended to other frequencies, thus finding a multi-frequency optimal solution [223]. By identifying a likelihood function to be maximized expressed in the unknown residual vector $\boldsymbol{\beta}$ and residual variance σ^2, the solution is found by solving a non-linear least square problem where the obtained residual variance value $\hat{\sigma}^2$ is the figure of merit quantifying the quality of the calibration. The approach may be readily extended to multiple frequencies, once the unknown residuals are decomposed in frequency dependent and frequency independent, and the former are grown up dimensionally to cover the assigned frequency range. An example of frequency-independent unknowns are the corrections for length and for pin gap of airlines. To reduce problem dimensionality and obtain a unique physically feasible solution, suitable constraints are added.

An example provided in the cited paper shows how the proposed method is able to improve the standard deviation of residuals by 25-50% with respect to a TRL calibration by applying only correction for length, while adding pin gap correction the improvement is slightly larger and more uniform, around 60%. This reflects into an improvement of source match by $8-12\,\mathrm{dB}$ and of tracking by nearly 60%.

Bayesian method Starting from overdetermined versions of some calibration algorithms and reviewing briefly MultiCal and StatisticCal programs, in [163] calibration standard and instrumentation errors are minimized by assuming no pre-assigned distribution. In reality, invoking the large number of internal VNA parameters determining its uncertainty, a multivariate Gaussian distribution is assumed. A minimum variance unbiased estimator of calibration correction is determined by a Bayesian approach: the measurement process is modeled as a statistical process, the calibration standards s and the raw measured S parameters r are used as input to the calibration algorithm, that returns an estimate of instrument parameters $\bar{v}$, by inversion of which any other S parameter of a DUT can be determined starting from the new raw measured S parameters. Uncertainties of standards and instrument are treated as random variables. The conditional probabilities are build assuming linearity and including additional constraints for the physical properties of standards, such as passivity and reciprocity; such constraints almost always lead to truncated distributions. A theoretical solution is calculated and reported in the paper, whose complexity grows quickly with the number of standards and that is numerically challenging. For these reasons Monte Carlo integration is followed.

In the reported synthetic example calibration standards are tested for two reference impedances ($50\,\Omega$ and $10\,\mathrm{k}\Omega$) and correspondingly quite different values of directivity and source match. The results report the mean square error of $\bar{v}$ compared to plain LRL calibration algorithm and cloned MultiCal and StatistiCal at two different frequencies: for $50\,\Omega$ the proposed Bayesian approach is slightly better or comparable to the two NIST algorithms, while for $10\,\mathrm{k}\Omega$ the method is even better, with other algorithms failing to converge in some conditions. When using real data the proposed method behaves visibly like StatistiCal (cloned) and only slightly better than MultiCal (cloned).

11.5.3.10 Calibration and uncertainty management software

As anticipated, recent projects have delivered very complete and powerful programs able to perform calibrations and additional functions selected by the user, together with management of instrument uncertainty and the possibility of interfacing directly to some existing VNAs. Three of them were selected and are briefly discussed in the following: StatistiCal at NIST [223], VNA Tools II at METAS [340] and MMS4 by Garelli and Ferrero [132]. A first comparison of VNA Tools II and MMS4 appeared in [298].

StatistiCal This method [223] is adopted by NIST in its StatistiCal software (see also sec. 5.7.1). The focus is on the geometrical uncertainty (line length uncertainty and conductor eccentricity, already reviewed in sec. 5.7.1 and 7.4.2) and its reduction by supplying additional information regarding calibration standards at all frequencies; the method is supported also by the formulation of additional constraints that ensure the feasibility of corrected geometrical parameters. The optimization is reached by maximum likelihood method applied to a likelihood function that contains frequency-dependent quantities (i.e. VNA calibration errors and calibration standard characterization), frequency-independent quantities (additional parameters of calibration standards that are invariant to frequency, such as the cited length and eccentricity) and residual variances (that result when the optimized solution is reached and characterize the residual uncertainty of the solution).

The method gives slight but significant corrections of calibration standard geometry, and for the reported example improves VNA source match by 4 to 9 dB and effective VNA tracking by as much as 50%. The authors comment also a few peaks in the residual standard deviation of the solution at about 9.5 and 11.5 GHz, attributing them to the "sampler-bounce effect" in the VNA reflectometer: the peaks are clearly visible and this gives an idea of the accuracy of the method; on the other hand, the robustness of the maximum likelihood approach is demonstrated by singling out outliers and weird data, rather than adapting the solution as in a straightforward least mean square approach. The improvement with respect to the existing NIST MultiCal is a factor of two of reduction of solution uncertainty in terms of residual standard deviation.

VNA Tools II It is a tool that performs evaluation of VNA calibration and related uncertainty for a N-port system using conventional error models, reviewed in sec. 11.5, including parameters drift: the tool takes into account calibration error terms as resulting from the adopted error model and calibration method (including switch error terms) and cable and connector repeatability. The solution is achieved by minimization of a weighted cost function, where weights are assigned on the basis of covariance matrix. The uncertainty contributions (noise, drift, cable and connector repeatability) are assumed uncorrelated, except linearity that depends among others on the used power level [340]; the tool follows a Type B approach, as pointed out in [298], and as coming out from authors' statement "uncertainties are assigned to all of them based on knowledge, specification or characterization of measurement setup and standards." They are propagated through the model using METAS UncLib (Uncertainty Library)[15] that calculates the resulting output given a generic model $\mathbf{Y} = \mathbf{f}(\mathbf{X})$, including the calculation of the Jacobian, so to determine the covariance of $\mathbf{Y}$ as $\mathbf{V}_Y = \mathbf{J}\mathbf{V}_X\mathbf{J}'$.

Drift of error model parameters, VNA linearity, cable and connector repeatability are included as definitions for the specific items in use, requiring the user to fill in a "measurement journal", where details such as VNA settings, changes to configuration, connections and disconnections and cable movements are annotated. They are then propagated through the measurement model, that is through calibration and error correction, using the UncLib library.

The output is both in textual and graphical form and consists of S-parameter plots resulting from calibration correction together with estimation of uncertainty, for which a tabular textual output with indication of the percentage of the overall uncertainty budget is also available.

MMS4 The theory behind the tool (MMS4 stands for Multiport Measurement Software ver. 4) is presented in [132]: the listed uncertainty sources are measurement noise (occurring during both calibration and measurements), standards definitions (as known affecting the estimate of the error model parameters and the correction of raw measurements), cable and connector repeatability. The adopted error model is a 10-term error model extended to N ports.

The uncertainty is evaluated and propagated differentiating the calibration equation and separating the contributions due to error-model coefficient uncertainty and those due to measurement noise. The proposed model takes into account various forms of measurement noise, distinguishing additive channel noise and phase noise (or jit-

15 M. Wollensack, METAS UncLib, 2009, available online: http://www.metas.ch/unclib.

ter noise) of local oscillator. Connector repeatability is considered assuming that the errors for each VNA port are uncorrelated: a very accurate model distinguishes the influence of repeatability on calibration and measurement phases. In all cases, for the effect of noise and repeatability, the evaluation of the covariance matrix is performed experimentally by repeated measurements, that is based on a Type A approach.

Throughout the paper there is constant attention to the numerical complexity and necessary computational capability, indicating as objective the reliable and efficient implementation on-board VNAs for real-time calculations: well, seeing the graphical representation of the uncertainty estimate updated in real-time while you are moving the cable has a pleasant effect.

Quite effectively [132], Fig. 6 shows the experimental repeatability resulting from connector repeatability alone (without movement of connected cables) and including the effect of cable movement (that largely hides, as expected, connector repeatability). When cable movement is considered, however, the assumption that the repeatability at ports and connectors is uncorrelated is not completely valid any longer, although the practical consequence is probably a second-order effect.

11.6 Correction for adapters

When measurements need adapters to mate different connector types and sexes, the user shall be aware that adapters worsen in general measurement performance. Different possibilities and relevant aspects may be considered:

- the VNA is calibrated on its own ports with the standard calibration kit and then the adapters are added to mate to the DUT, so that they are not included in the calibration and may or may not corrected separately; "post-correction" may be done in various ways, simply subtracting adapter attenuation for a specific transmission measurement (it's a response calibration, see sec. 11.5.3.8, and not so advisable) or including the complete characterization of the adapter for de-embedding (see sec. 11.7), or using time gating, excluding adapter response in the anti-transformed time-domain response (see sec. 11.9);

- adapters are included at the calibration time and VNA with adapters (and any other connecting element) is calibrated, thus moving the reference planes from the VNA ports to the DUT ports (DUT ports are an abstraction, and may correspond to the physical DUT ports or to any point in the setup where the user has decided to perform the calibration); a calibration kit is necessary for the adapter connector type (remember in-fixture calibration and other non-standard calibrations); in general, the operations during the calibration depend on connector types for VNA and DUT, and the availability of phase equal insertables: for simple configurations where the non-mateability problem consists of male/female mismatch, two "identical" adapters with e.g. male-male and male-female combination are used in the two distinct phases of calibration, e.g. during the initial one-port part and when connecting the two ports together;

- even including adapters in the calibrated setup, since the raw directivity is worse due to added adapter reflection (see sec. 11.5.2.2), the corrected parameters after calibration will be of lower metrological quality, e.g. slightly more susceptible to drift;
- some calibration methods are more robust and bear smaller uncertainty with respect to non-ideality of adapters, such as the Unknown Thru (see sec. 11.5.3.3), avoiding the need for mating/de-mating setup connectors and thus with great benefits in terms of repeatability and connector wearing.

Regarding correction for errors introduced by adapters, it is always possible to compensate for their effect by de-embedding (see sec. 11.7) or remove their effect from the measured response of the DUT by means of time gating (see sec. 11.9). In this section we consider how to estimate, and then compensate, the parameters that characterize an adapter, evaluating its performance for different types of measurements and applications.

Before proceeding, a preliminary example is shown of the kind of effect an adapter has on measurement results (see Example 11.1).

Example 11.1: Effect of adapters on VNA directivity

A DUT is measured, such as e.g. a filter, but it needs adapters to match connector type, N for the VNA and SMA for the DUT. We do not first include the adapter in the calibration, so that what we measure is affected by a significant error. The VNA measures correctly the return loss from port 1, where the N-SMA adapter is located, but the effect of the return loss from the adapter is to cause a fluctuating signal that depends on the mutual phase, resulting in a worst-case directivity given by the sum of the original VNA directivity and the return loss of the adapter, both measured in linear quantities. 1.05 e 1.25

If there is a significant difference between the original VNA directivity and the adapter return loss, the resulting directivity will be substantially the return loss of the adapter; for very good adapters, the result obtained with the complete calculation including the effect of the VNA directivity will be somewhat worse: for the poor N-SMA adapter the 19 dB return loss will be the final combined directivity of the setup; for the good adapter the 32.2 dB return loss gives an approximate estimate of the final directivity, that if combined with the original 40 dB VNA directivity (sum of the two reflection coefficients) gives 29.2 dB.

Including adapters in VNA calibration is always advisable, and in the case of the poor adapter we see a dramatic improvement. In general, however, there will be a slightly larger susceptibility to temperature effect and drift.

11.6.1 Non-insertable devices and adapters

Many devices are built with the same sex connector at each end. Cables are usually built with a terminating male connector; filters and isolators, conversely, have female connectors so that they may be inserted as through devices. An essential step in most calibrations, either scalar or vector, is to connect two measurement ports together, which for most connector types, dictates male and female test port connectors[16]. Several solutions are common:

- high quality adapters can be "swapped"; there exist special adapters called Phase Equal Insertables (PEIs), that are combinations of male/female pairs of some connector type all with the same electrical length and phase response: when using a female-female through connection to calibrate a test setup ending in a DUT with two male connectors, a PEI female-male through of the same type is then used during measurement in order to keep unaltered the effect of the extra length of the female-female through also during normal measurements. Phase match between PEIs is thus to be ensured over the most extended possible frequency range; using phase equal adapters ensures that errors associated with adapter exchange are very small. Attention shall be given to optimize PEI location so to keep low mating/de-mating wearing of original precision expensive connectors;
- a second calibration can be performed and adapter removal software will provide excellent result; this is included in the so-called Automatic Port Extension that is reviewed in sec. 11.7.1.2;
- a calibration using LRL or LRM will also provide an excellent result; however, this requires components not usually included in commercial calibration kits.

11.6.2 Adapter characterization

Uhlir [324] proposed in the '70s a simple procedure to estimate the S parameters of a "good enough" adapter — that, by the way, would not be of use, if it were not good enough! —. The basic procedure for the calibration of port 1 is repeated after the adapter is inserted into the port at the other end (port 2), not to be confused with the other VNA port (that is not considered in this example). What is required is a set of standard terminations for both connector types, at port 1 (the primary connector) and port 2 (the secondary connector). To visualize the problem, think of port 1 (VNA) of type N and of port 2 of SMA type at the other end of a male-N-to-female-SMA adapter.

Let S be the scattering matrix of the adapter and M the measurement at port 1, where Γ is the reflection coefficient of the device connected at port 2.

$$M = S_{11} + \frac{S_{21}^2 \Gamma}{1 - S_{22}\Gamma} \tag{11.6.1}$$

For a perfect matching termination at port 2, the reflection coefficient Γ is 0 and the resulting measured value, called M_L ("L" indicates that a matched load was connected), is $M_L = S_{11}$. From the conjugate matching theorem the maximum power transfer oc-

[16] Sexless connectors such as APC7 and the smaller 3.5 avoid this type of problem and in general better define the mating plane to be used as calibration plane, but their use has remained limited to a some applications and primary metrology, with the performance of other connector types equipping VNAs and in general instrument progressing quite fast.

curs if $\Gamma = S_{22}^*$. Assuming that the adapter is lossless (from here the need to assume a "good enough" adapter), the maximum power transfer occurs if we observe a zero reflection at the input port 1.

$$0 = S_{11} + \frac{S_{21}^2 S_{22}^*}{1 - |S_{22}|^2} \tag{11.6.2}$$

By invoking the conservation of power and reciprocity,

$$1 = |S_{11}|^2 + |S_{21}|^2 = |S_{22}|^2 + |S_{12}|^2 \tag{11.6.3}$$

and eq. (11.6.2) may be written as

$$S_{11} = -S_{22}^* \frac{S_{21}^2}{|S_{21}|^2} = M_L \tag{11.6.4}$$

or

$$S_{22} = -M_L e^{j2\theta} \tag{11.6.5}$$

where θ is the angle (or argument) of S_{21}, that in eq. (11.6.4) has normalized amplitude.

Combining all equations above we obtain an expression for Γ, for the difference of the measurement M at port 1 with respect to the value for maximum power transfer (that is zero reflection or $\Gamma = 0$).

$$\Gamma = \frac{M - M_L}{1 - M_L^* M} e^{j2\theta} \tag{11.6.6}$$

Then Uhlir's procedure gets too complicated, whereas by measuring any other reflective standard we have additional equations like eq. (11.6.1) that may be directly solved by numeric techniques; e.g. for Short and Open standards

$$M_S = S_{11} - \frac{S_{21}^2}{1 + S_{22}} \qquad M_O = S_{11} + \frac{S_{21}^2}{1 - S_{22}} \tag{11.6.7}$$

After substitution of $S_{11} = M_L$, the remaining two equations may be solved by isolating the terms S_{21}^2 on the left and then equating the right-hand sides, obtaining

$$S_{22} = \frac{M_O + M_S - 2M_L}{M_O - M_S} \qquad S_{21} = \sqrt{2 \frac{(M_L - M_S)(M_O - M_L)}{(M_O - M_S)}} \tag{11.6.8}$$

Relying on symbolic tools and packages, the system of equations is solvable also for unexpressed calibration standards, that may be assigned any value, taking into account suitable corrections and the possibility of expressing uncertainty and its propagation; the resulting expressions are however quite cumbersome and do not fit on two pages in readable form!

A more complete approach was proposed by Engen [118], requiring n measurements with different sliding short positions, m measurements with different sliding load positions and a matched load. The method is heavily based on sliding terminations typically used for waveguides and less common for coaxial lines; they are connected to one side of the adapter, the other side connected to the VNA port. Quite straightforwardly a set of equations is formulated, one for each termination and position of the slider, to determine the reflection coefficient at the adapter output:

$$\Gamma_i = r_i e^{j\phi_i} e^{-2(\alpha+j\beta)l} \tag{11.6.9}$$

where r_i and ϕ_i are the magnitude and phase of the reflection coefficient of the i-th standard and α and β are the attenuation and phase constant of the connecting line. The method was originally conceived for power measurements, but Γ_i may be easily related to S_{11} measurement. The system of equations may be solved numerically and returns the complete adapter characterization.

11.7 Embedding / De-embedding

Going a little farther, what was said for correction of adapters may be extended to correction of any fixture, probe or network, that is part of the test setup and that might be impossible to physically remove, or that it is desirable to remove mathematically rather than physically. Practical examples are chip carriers and sockets hosting integrated circuits and transistors, the connecting leads of transistors themselves (for which manufacturers sometimes give values of the main parasitics), or more complex devices, such as a matching network or a filter, without which the DUT cannot operate as prescribed or desired, or cannot function at all, or even simply known adapters and connecting cables. To better illustrate and justify the exigency of removing parts of a complex test fixture, it might stem from the need to determine the behavior of the DUT without these additional parts, e.g. to verify DUT performance alone against contractual or normative limits.

De-embedding is a mathematical operation of "subtraction" of the effect of these elements from measurements, when a satisfactory equivalent representation is available; such equivalent representation may be supplied in any of the various forms that describe two-port and three-port circuits (see sec. 1.3), and in general N-port networks (see sec. 10.3).

Symmetrically, there might be the need to include additional parts that have been separately characterized, for example to observe what the behavior of the DUT is, when it is in the final installation. This is the *embedding* operation. In this case additional elements are added and the overall response modifies accordingly. A typical example is the calculation of response by adding a cable or transmission line.

The initial matrix representation of the elements may be the most suited, e.g. admittance or impedance matrix representation or hybrid matrix representation for most of circuit elements and devices, being able to mathematically transform them in equivalent scattering or transmission matrix representations (see sec. 1.3.1.6). Embedding and de-embedding are performed mathematically by applying the matrices of the added or subtracted parts in a form suitable for a cascaded multiplication. Transmission T matrices are cascadable and the multiplication order is the one they are inserted into the system (see sec. 1.3 and 1.3.1.6). Embedding and de-embedding distinguish the parts to add or remove by multiplying by the direct or inverse T matrix.

Alternatively, with parasitics of components and devices, for which we distinguish between series and shunt parasitics, embedding and de-embedding may be done with a different matrix representation: parasitics are described by equivalent capacitance and inductance terms and the most straightforward approach is to add (or subtract) capacitance terms directly with a Y-matrix representation and inductance terms with a Z-matrix representation (whichever comes first).

Multiplication for T matrix or its inverse proceeds from the DUT ports to the VNA ports. In general, custom matrices can be input in the many simulators and VNAs by using the Touchstone .s2p, .s3p, .s4p formats; they contain S parameters versus frequency and their values may be the result of calculations, simulations, or measurements. These formats refer to a two-port, three-port and four-port configuration, respectively, and contain a header that specifies how the numeric values of S parameters are written, e.g. absolute value and angle. For mixed mode S parameters (i.e. for single-ended and differential ports altogether) there are e.g. .m3p and .m4p formats.

Several high-end VNAs allow the implementation of embedding and de-embedding in real time. However, there are no restrictions for an off-line embedding/de-embedding, since all the network representations are interchangeable. The heavy use of mathematical manipulation, as in the case of embedding/de-embedding and the possibility of doing it also on-board VNAs, clarifies the importance once more of mathematical processing of raw measurement data. Matlab®, for example, offers a couple of functions in the RF Toolbox: `deembedparams()` and `cascadeparams()`. With functions that add passive components to simulate stray parameters and create various kinds of circuits (e.g. attenuators) or can load file descriptions in Touchstone formats, de-embedding by cascading is possible.

Time gating (see sec. 11.9) is another form of de-embedding, working in time rather than frequency domain. Stinehelfer[17] presents a useful overview of time domain de-embedding using real examples.

11.7.1 De-embedding techniques

De-embedding is generally used to move the reference planes inside the test fixture, avoiding the need to fully calibrate inside the test fixture itself, for which we might lack calibration standards (think of a PCB test fixture with DUT soldered inside it). Because of the movement of reference planes, de-embedding is also called "port extension", when VNA ports are moved closer, next to the DUT. There are various possibilities to accomplish port extension:

---

[17] H.E. Stinehelfer Jr., "Discussion of de-embedding techniques using time-domain analysis," *Proceedings of the IEEE*, Vol. 74, no. 1, 1986, pp. 90 - 94. doi: 10.1109/PROC.1986.13409

- the simplest one is compensating for a line section of which nominal values are known (e.g. from datasheets or previous measurements), usually assuming it lossless and only responsible for phase shift, or delay;
- otherwise, the line section might be evaluated by measurements and the information to accomplish port extension be retrieved from experimental data in terms of loss and delay, that is the assumption of lossless line is removed, but the same line is always assumed homogeneous;
- last, more complex arrangements of distributed and lumped elements require a complete de-embedding, performed again by a mix of modeling and measurements, using the most appropriate matrix representation to apply corrections.

The first two methods rely on the assumption that the line to be de-embedded is seamlessly connected to the ports on which the original VNA calibration was made, or in other words that the adapters and connectors used to couple to the second line sections have no significant impact. From this the need to use always high-quality connectors, with low losses and good match.

11.7.1.1 Uniform ideal line, lossless adapters and same characteristic impedance

The most intuitive and simple method is to move the reference planes by assuming a uniform transmission line section whose length may be determined by physical measurement or asking the VNA: a TDR-like measurement will spot out discontinuities and their time distance (to get back to the physical distance an accurate estimate of the propagation velocity is needed); discontinuities may be pre-existing in our test fixture (such as the pad where the DUT is soldered), but for consistent and reliable results it is purposely inserted. For example a short circuit may be connected at the end of the line to de-embed, allowing a precise accurate location of line end: if simple insertion/removal of the short circuit is not possible, an identical transmission line shall be built next to the fixture, in order to perform the required measurements.

Assuming a uniform line, compensation may be limited to "moving the reference planes", that is compensating for delay, or phase displacement, but leaving attenuation unchanged: this approach is possible with lossless, or nearly lossless, elements, such as adapters, when measuring high-loss circuits. This is the extreme case of the problem of adapter characterization considered in sec. 11.6.2.

Let's assume that we have two adapters or line sections of length and phase displacement l_1 and φ_1, and l_2 and φ_2, connected to port 1 and port 2 of the VNA, respectively. In this case the return loss S_{11} shall be corrected for phase displacement of the adapter and for the reflected term on its way back, that is twice the phase displacement $e^{-j2\varphi_1}$; for the transmission term the wave incident on port 1 passes through the network and then reaches port 2 of the VNA, so that it will be compensated by $e^{-j(\varphi_1+\varphi_2)}$. In other words the S matrix of the DUT and the measured S_m matrix at the original VNA ports are related by

$$S_m = \Phi S_m \Phi \qquad S = \Phi^{-1} S_m \Phi^{-1} \qquad \Phi = \begin{bmatrix} e^{-j\varphi_1} & 0 \\ 0 & e^{-j\varphi_2} \end{bmatrix} \qquad (11.7.1)$$

11.7.1.2 Port extension by measurement

Many VNAs have the possibility of automatic port extension, calculating an estimate of the loss and delay terms by means of measurements of one or two reflect termination (an Open and a Short) to be connected where the ports shall be moved. It is possible in principle to determine both loss L and delay τ with one single measurement for each line section to de-embed, modifying the matrix Φ as

$$\Phi = \begin{bmatrix} L_1 e^{-j\varphi_1} & 0 \\ 0 & L_2 e^{-j\varphi_2} \end{bmatrix} \tag{11.7.2}$$

If the line section to de-embed is long enough to have more than one measurable ripple period and if the ripple period is such to have a good frequency resolution, then one measurement for each line section is usually enough. Conversely, for small frequency intervals or very short line sections, both reflect standards are needed to recover two ripple curves in phase opposition, as pointed out already by Pucic and Daywitt for the characterization of connector defects (see sec. 7.1.1.7).

11.7.1.3 Two-port network de-embedding for return loss measurement

For a two-port network that connects a simple one-port device to the VNA, it is possible to write explicitly relationships that rule de-embedding operation and experimental determination of parameters of the two-port network itself. Let's assume an impedance representation as in [30] and an unknown DUT Z_L:

$$Z_{in} = Z_{11} - \frac{Z_{12}Z_{21}}{Z_{22}+Z_L} \qquad Z_L = \frac{Z_{12}Z_{21}}{Z_{11}-Z_{in}} - Z_{22} \tag{11.7.3}$$

Z_L may be determined as shown once the impedance parameters of the two-port network are known. It is observed that Z_{21} and Z_{12} are always multiplied together and they cannot be separately determined, but this does not represent any relevant issue, accepting that the identified two-port network may be different, but perfectly equivalent for de-embedding. The identification of the two-port network is called unterminating and is based on the same equation, this time solved for Z_{ij} providing pairs of measured Z_{in} with known values of Z_L. To ease the solution, as done for the one-port error model, the expression above may be rewritten by making explicit the determinant of the impedance matrix:

$$Z_{in} = \frac{\Delta_Z + Z_{11}Z_L}{Z_{22}+Z_L} \tag{11.7.4}$$

With a large number of measurements, greater than three (that is the minimum number of equations to solve the system), least mean square and similar techniques may be used, to reduce the effect of noise and uncertainty.

11.7.1.4 Multi-stage de-embedding and DUT parasitics

Parasitics are a lumped-component interpretation of physical limitations of real components, often related to component leads, pads, vias: in Chapter 6 we saw that we can estimate stray inductance and capacitance. This approach holds until the wavelength gets so short that these elements shall be evaluated as distributed, rather than lumped elements: based on what was observed for vias behaving as stubs and the general knowledge of electromagnetic propagation, it is possible to assume that lumped parasitics may work well up to 10 or 20 GHz, with progressively coarser approximation above it. At wafer level [321] parasitics values experimentally determined work well up to about 50 GHz.

It is possible to include as well the correction for balanced-unbalanced transformation when using baluns and other types of coupling transformers.

Two-stage de-embedding Two-stage models for de-embedding are easily related to the equivalent circuit for a single device. More than two stages are needed for complex structures and especially for very high frequency, when even small parts of the fixture are relevant. As said, Y and Z matrices are best suited for treating shunt and series parasitics. For correspondence to physical elements, we shall distinguish between large devices at PCB level/package and small devices on wafer: for the latter usually stray capacitance terms are external, related to pads, and series stray inductance terms bring to the chip "internal" terminals; when considering the former, connecting wires are longer and internal stray capacitance due to soldering to land pads may be more relevant than, or of the same order of magnitude of, external pads/vias. To visualize this, we may consider a three-terminal devices such as a transistor, with common-emitter circuit (see sec. 1.3.3.4 at the end of Chapter 1).

Let's consider first an on-wafer configuration, with shunt capacitive terms preceding series inductive terms. Starting from measured results arranged in admittance matrix representation Y_1, we move past the shunt parasitics (capacitive elements) by subtracting their equivalent admittance representation Y_p: $Y_2 = Y_1 - Y_p$. To account for the series inductive parasitics we need an impedance matrix representation, so that we transform Y_2 into Z_2 and we move past the series inductive elements by subtracting their equivalent impedance representation Z_s: $Z_3 = Z_2 - Z_s$. Then Z_3 may be transformed into the desired de-embedded DUT representation, e.g. in S parameters.

It is necessary now to understand how the parasitics may be estimated: they may appear in the device datasheet, as typical for some simple configurations and fairly low frequency ranges (let's say up to some GHz); they may be also determined by simulation, using e.g. finite element methods and full-wave solvers; or they may be experimentally determined on the fixture itself, e.g. replacing the DUT with a short and an open. It is customary to build identical elements (e.g. pads, leads, lines) on the same substrate (e.g. PCB or wafer) to allow experimental characterization without interfering with the DUT. While measuring parasitics, rarely the various parts can be isolated: while measuring inductive parasitics in the short-circuit configuration, capacitive parasitics are still there, so that the equivalent impedance representation Z_s of the series inductive parasitics assumed above is available only if the the result of the short-circuit measurement Z'_s is compensated for the equivalent admittance representation Y_p obtained with the first open-circuit measurement; thus Y_p is transformed into an equivalent Z_p representation and $Z_s = Z'_s - Z_p$.

If a reversed order of shunt and series elements is assumed e.g. for larger devices, then the procedure to follow is the other way round.

The authors in [321] call the first configuration (first capacitance measured with open circuit, then inductance measured with short circuit) "open-short", and the other one "short-open": they consider that the impossibility of deciding which configuration is the best for a specific setup represents the uncertainty of the method (we may call it setup or model uncertainty); at low frequency it is negligible and the results of the two models coincide, starting to differ for increasing frequency. The expression of the ratio of the two de-embedding matrices of a hypothetical one-port DUT is

$$\frac{Y_{\mathrm{DUT},os}}{Y_{\mathrm{DUT},so}} = (1 - Y_{\mathrm{open}} Z_{\mathrm{short}})^2 \tag{11.7.5}$$

that begins to depart from unity when the parasitics approach the resonance: two octaves from the resonance they amount to about 1/16 and the error of the ratio is 12%. In [321] two test cases are considered, one transistor and one inductor, concluding that the de-embedding configuration is not critical up to about 120 GHz and 30 GHz, respectively.

If a π-cell model is more suitable, then the procedure is more complex and requires one more iteration (three-stage de-embedding). Moreover, the experimental determination of single elements also requires more steps, such as e.g. applying the short circuit at two different points, along the DUT leads and "inside" the DUT; since this is not possible, often a hybrid method is used, where the internal capacitance terms are derived from the datasheet, that is useful also to verify that the experimentally determined stray inductance of connecting leads matches that of the datasheet.

Three-stage and more de-embedding There are many situations in which more complex models are necessary, normally identified as three-stage and four-stage models, but nothing prohibits in principle to use even more stages. The method is always based on successive reductions, alternating Z and Y matrix representations, reducing series and shunt parasitics, usually identified in terms of inductance and capacitance, respectively (of course including a resistive part). The method is based on a work in 1991 (H. Cho and D.E. Burk, "A three-step method for the de-embedding of high-frequency S-parameter measurements," *IEEE Transactions on Electron Devices*, vol. 38, June 1991, pp. 1371–1375. doi: 10.1109/16.81628), later corrected and improved by Kolding (T.E. Kolding, "On-wafer calibration techniques for Giga-Hertz CMOS measurements," IEEE Int. Conf. Microelectronic Test Structures (ICMTS), Gothenburg, Sweden, Mar. 1999, pp. 105–110). An example of a four-stage de-embedding taken from a work by Kolding is shown in Figure 11.7.1. Improvements have been proposed in E.P. Vandamme, D.M.M.-P. Schreurs and C. van Dinther, "Improved three-step de-embedding method to accurately account for the influence of pad parasitics in silicon on-wafer RF test-structures," *IEEE Transactions on Electron Devices*, Vol. 48, no. 4, Apr. 2001, pp. 737-742. doi: 10.1109/16.915712.

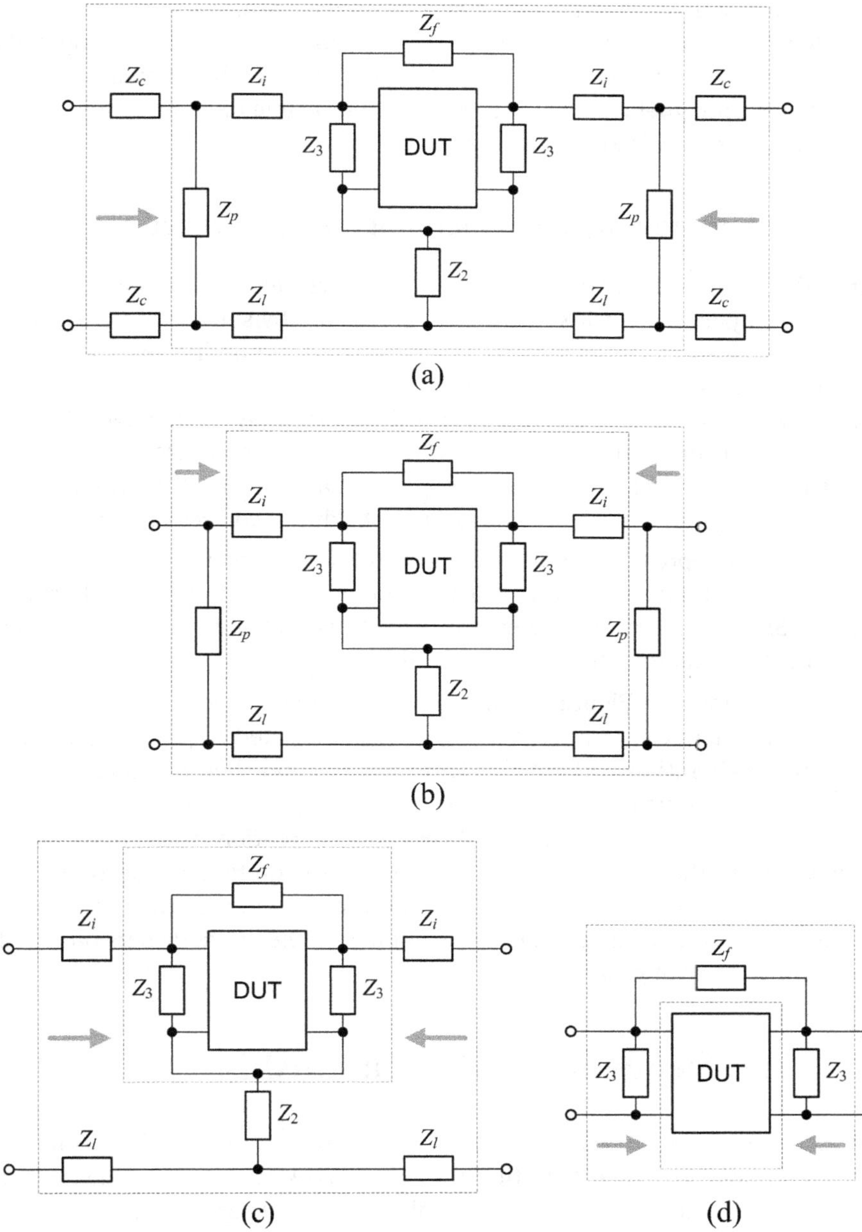

Figure 11.7.1 – Example of four-stage de-embedding as described by Kolding (see text).

11.7.2 Embedding techniques

Embedding is not so different from de-embedding, but is probably less commonly applied: while de-embedding aims at removing undesirable effects of parasitics, connections, even parts of the DUT, embedding may be used to include the effect of additional parts, with an approach that is however well covered by off-line processing and analysis, e.g. by simulation tools. Once the DUT is characterized by its own S-parameters, circuit and electromagnetic simulators are able in principle to embed "any" additional feature, provided that it is known enough to be modeled.

11.7.3 Amplitude-only de-embedding, or extraction

As part of the de-embedding operations performed frequently automatically (or semi-automatically) by modern VNAs, the simple correction by subtraction of the attenuation of one of the elements of the setup is considered: while in principle a complete calibration is necessary, when the errors introduced by the test fixture elements are very small with respect to DUT response, the former may be removed by a scalar measurement and some post-processing with an acceptable accuracy. This is what is in other words called "response calibration" and that lends itself to several variants, depending on the degree of improvement and introduced corrective actions.

A straightforward approach is considered, just to evaluate the impact of such approximation: a set of 6 dB attenuators is our DUT, connected to the VNA by a short low-loss cable and N-SMA adapters, altogether featuring a moderate insertion loss smaller than the 6 dB of the attenuators, up to a considerable frequency.

The results are shown in Figure 11.7.2, for two different approaches: an amplitude-only approach, where cables and adapters are compensated in post-processing, making a two-port calibration without the Through step and then subtracting the cable S_{21} from the measurements, and the correct approach, where the cable is used as the Through standard and included in a full two-port calibration, thus compensated automatically during the measurement. The accuracy is quite good up to 1 GHz, the difference between the two set of curves is not dramatic, but the attenuation curve obtained with the correct calibration is more stable and the curve dispersion is reduced by more than a factor of two.

11.8 Time-frequency basic relationships

This and the next sections focus on different aspects of relationships between time and frequency domains. We have seen that the native VNA measurement consists of frequency domain data and that these include all the elements of the measurement setup and associated unwanted reflections. By its nature a frequency-domain measurement made in steady-state conditions is the result of all the reflections and modes overlapped (see sec. 2.5.1) and is unable to discriminate from which point along the setup from port 1 to port 2 a specific time response is coming. However, there is a great deal of information in associating a specific response behavior to physical elements

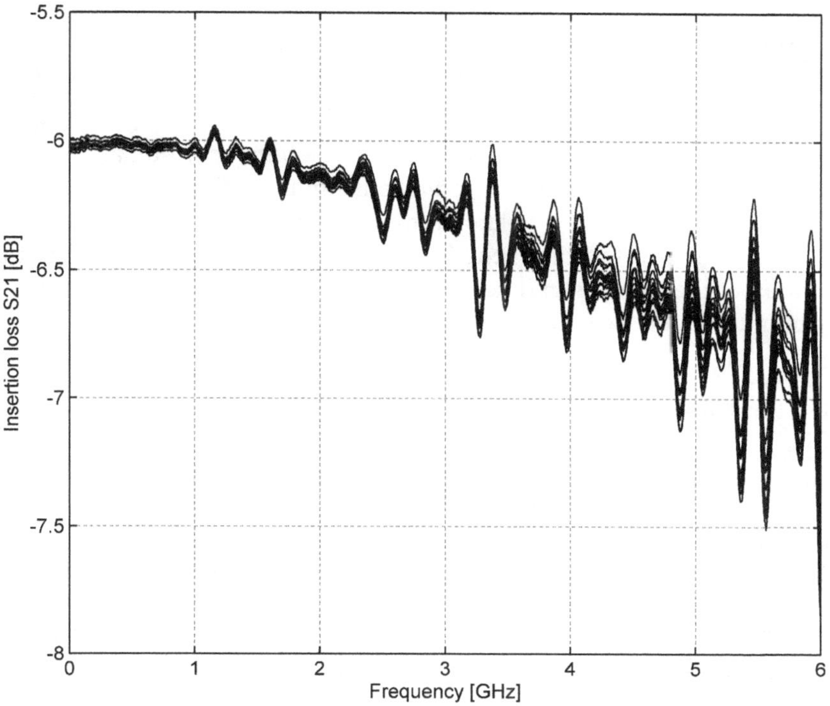

(a)

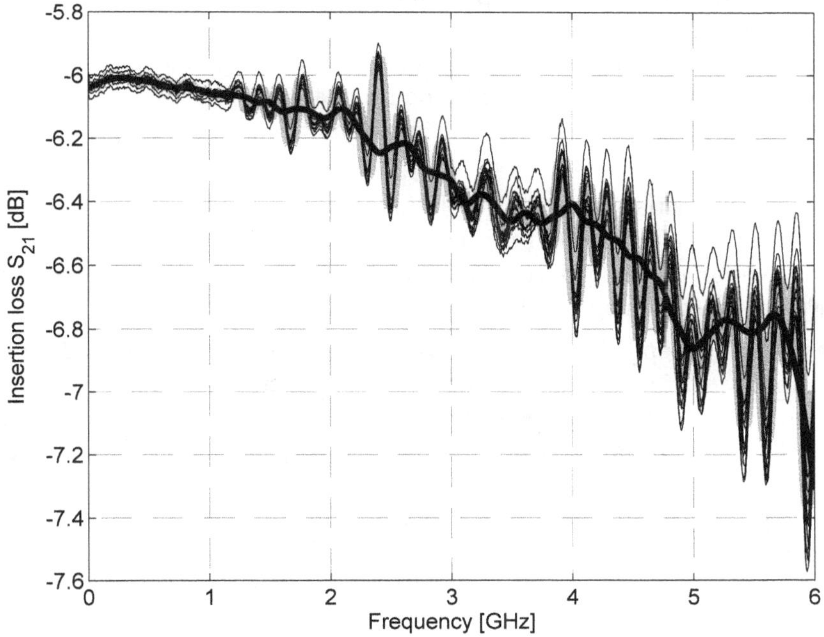

(b)

Figure 11.7.2 – Measurement of the insertion loss S_{21} of thirteen 6 dB attenuators: compensation of the feeding cable by (a) separate measurement and subtraction, (b) including it in the full two-port calibration (the thick gray curve is the mean of all the traces, the thick black curve is a second order robust interpolation).

and this is normally done by considering each element as a discontinuity, estimating the amplitude and arrival time of the related reflection term. Moreover, the shape of the reflected step waveform brings additional information on the nature of the discontinuity, that the impulse rarely allows (see sec. 2.6 and 11.10.3).

Being the time-domain response calculated from the frequency-domain response (that remains the original measurement), some constraints originate from the chosen frequency span and resolution: it is known from theory of DFT that the frequency resolution df is related to the time length of the transformation window and that the discretization of the samples in time domain (i.e. sampling frequency) sets the maximum span of the frequency interval (see sec. 3.1.1).

Example 11.2: Time and spatial resolution for a cable reflection measurement

Let's take a $L = 10\,\mathrm{m}$ RG213 cable with a velocity factor of 0.66. The traveling time is $T_{trv} = L/v = 10/0.66\,c = 45.5\,\mathrm{ns}$, with $c = 2.997925\,10^8\,\mathrm{m/s}$. For a transmission measurement (S_{21}) the best frequency resolution is $df = 1/T_{trv} = 22\,\mathrm{MHz}$; now, with 401 points available the maximum frequency span is 8.8 GHz. Conversely, for a reflection measurement (S_{11}) the traveled path is double (for the reflection at the end of the cable to get back to the source/measuring port) and the frequency resolution and the maximum frequency span, as a consequence, are half the previous ones. Increasing the number of points increases correspondingly the frequency span.

To recover the time-domain representation of signals from the frequency-domain response, the most straightforward approach is to apply the Inverse DFT (DFT and time-frequency transforms are considered in sec. 3.1). The Chirp Z-transform is normally used (see sec. 3.1.2): no clear explanation appears in the technical literature and instruments manuals, but looking into the original CZT paper [42, 277], it appears that it is not absolute speed, but frequency zooming and data interpolation to be the workhorses of this technique.

The exigency of spatial (or time) resolution comes from the necessity of resolving discontinuities along the setup (e.g. connectors, pads, and so on); the basic spatial or time resolution of the measurement establishes if separation of reflection terms is possible (see time gating in sec. 11.9).

Time resolution depends first on the chosen lowest frequency value and then on the type of window that is used for the frequency-to-time transformation (see sec. 11.10) and on the type of processing, i.e. low-pass or band-pass mode (see sec. 11.10). Here we state simply the facts about measurement settings and resulting time resolution, as reported in the literature: time resolution is shown in Table 11.8.1 and 11.8.2 for usual settings of the vast majority of VNAs. As already noted in the example above, time resolution is measured in multiples of the inverse of the frequency span. The time resolution value for comparison between the various settings of the table is determined as the one corresponding to the 10-90% rise time of a step signal or the 50% time width of an impulse signal.

The manufacturers specify time resolution as coming from a "cutoff" specification, that in turn is measured in multiples of the inverse of the frequency span. Where does all this come from?

Window type	Low-pass step response 10-90%	Low-pass impulse response	Band-pass impulse response
Minimum	0.45	0.60	1.20
Normal	0.99	0.98	1.95
Maximum	1.48	1.39	2.77

Table 11.8.1 – Smoothing windows applied to frequency domain data: relationship between frequency span, window type and time domain resolution (expressed in multiples of the inverse of the frequency span) for Agilent [7].

Window type	Low-pass step response	Low-pass impulse response	Bandpass impulse response
Rectangular	0.5	—	1.0
Nominal	1.0	—	2.0
Low sidelobe	1.5	—	3.0
Min sidelobe	2.0	—	4.0

Table 11.8.2 – Relationship between frequency span, window type and time domain resolution (expressed in multiples of the inverse of the frequency span) for Anritsu [17]; the reported values are evidently the theoretical ones with "round" multiples of the base bin.

First of all, the cutoff time is simply a quantification of the selectivity of gate function, that is its roll-off if we think in terms of frequency response of a filter. Here we are working ideally in time domain[18]: thus the roll off or selectivity of the gate function establishes the minimum resolution attainable.

Second, we observe that with the original data located in frequency domain, the roles of time and frequency domains are exchanged, so that the shape that we are used to "see" for frequency domain (i.e. the Fourier transform of a smoothing window, see sec. 3.1.3.2) is in reality to be assigned to time domain. This means that sidelobes will be present in the gate function for time domain and that the concept of selectivity and roll-off are to be evaluated by means of concepts of sidelobe suppression and equivalent bandwidth: manufacturers adopt the definition of 6 dB bandwidth (see Table 3.1.1) to quantify selectivity; to verify this, it is sufficient to observe the declared band-pass impulse response of the "rectangular" (also named "minimum" or "standard") window, that is 1.2 bin, where the "bin" is interpreted as time resolution, that is the inverse of frequency span.

[18] It is possible to apply gating of the time-domain response in time domain by multiplication, as it should naturally be, or in the frequency domain, by convolution. The latter is believed to be a faster approach because it does not require a transform/anti-transform operation, but it's exposed to more approximations (see sec. 11.10.7.4).

Example 11.3: Time and spatial resolution using time domain option

We want to identify the best setting for the smallest time resolution, given that our VNA has a maximum frequency span $\text{FS} = 6\,\text{GHz}$. From this information we have immediately that the basic time resolution is $dt_{base} = 1/\text{FS} = 167\,\text{ps}$. Now we must find the best settings for the effective time resolution and how this translates in spatial resolution. By inspection of Table 11.8.1 low-pass mode with its settings is the preferred one, together with the "minimum" window type (more on windows in sec. 11.10.2); the resulting time resolution is about 0.45 or 0.6 the basic time resolution: thus $dt = 0.45 \div 0.60\, dt_{base} = 75 \div 100\,\text{ps}$, depending on impulse or step-response mode. If we now consider the equivalent spatial resolution, after determining the velocity factor in the system, that we assume 0.66, for a transmission measurement (one-way) it is equivalent to 33.8 and 45.0 mm respectively. For a reflection measurement the spatial resolution is half, that is 16.9 and 22.5 mm. This is an estimate of the best spatial resolution provided by VNA frequency range. Doubling the frequency range, and thus the maximum frequency span, gives halved distances, so for the reflection measurement it will be 8.5-11.3 mm, still too large to discriminate reflections along a connector, but enough to remove the said connector from measurement by time gating. It is thus evident that, when designing the test fixture, there will be the need of physically separating elements that must be then discernible and discriminated by time gating, starting from VNA performance and frequency span as design input.

The "minimum" window appears to have the best resolution: "minimum" stands for minimum width, ensuring the best time resolution suitable when responses of equal magnitude shall be separated. However, the selection of the "minimum" window has its drawbacks: it causes the largest sidelobes of the three/four window types available, so that there is a limitation in the appreciable dynamic range of time response, before artifacts caused by the sidelobes become relevant. The minimum window is almost always the rectangular window, i.e. the window implicit in the cut of the sequence before transforming it, that thus corresponds to no smoothing window applied.

Third, data in Table 11.8.1 and 11.8.2 may be interpreted as at the origin of a limited rise time of the VNA that anyway may be compensated in the final measurement by assuming linearity of the system and invoking eq. (3.2.16) given in sec. 3.2.4.

Last, we should keep in mind the potential non-uniformity of propagation velocity in a typical transmission path, going through cables and connectors; this will limit our ability to precisely locate discontinuities. In dispersive media, such as waveguide, the non-linear phase response limits the ability to locate the actual peak of the response.

11.9 Time gating

We have seen that the native VNA measurements are in frequency domain and that measurement results include the unwanted reflections from all the discontinuities of test setup. By its nature a frequency domain measurement is unable to discriminate from which point along the setup from port 1 to port 2 unwanted artifacts are coming. If the obtained frequency-domain data are anti-transformed to time domain, then there will be the possibility of tracking the propagation of various signal components back and forth along the measurement chain and to define the time interval which we are interested in. In other words, cutting a time interval between t_{start} and t_{stop} instants is equivalent to remove reflections originating in time instants outside that interval, that corresponds to the spatial positions along the measurement chain before and after the two discontinuities. This is equivalent to set reference planes at the two positions where discontinuities are, or better slightly beyond, to account for frequency leakage.

Time-domain gating is analyzed now by considering a straightforward application of Inverse DFT and DFT. Multiplication by the gate function $g(t)$ in time domain is equivalent to convolution with its Fourier transform $G(f)$ in frequency domain. The frequency domain gating function $G(f)$ is centered on each frequency data point and the sum of the product term by term is computed (convolution).

$$X_g[k] = \sum_{j=-N/2}^{N/2} X[k-j]G[j] \tag{11.9.1}$$

When the index j approaches the boundaries with the argument of $X[\cdot]$ becoming negative or greater than N, circular (or cyclic) convolution wraps around and considers the other side of the vector X, while linear convolution put the values to 0: linear convolution is used here. When the $G(f)$ function moves towards the extreme of the measured data frequency interval more and more points are excluded from the sum of products and the resulting value is lower than it should be. At the last data point half of the data points beyond the frequency interval boundaries are excluded from the computation and the result is 6 dB lower; as a consequence any gating will distort the last points (and first points in bandpass mode) of gated frequency response. VNAs compensate for this by *post-gating normalization*. This phenomenon does not occur (and normalization is not necessary) if gating is operated as a straightforward multiplication in time domain, at the cost of a double transformation to time domain and back to frequency domain.

The $g(t)$ may be chosen arbitrarily sharp, if needed, in order to exclude elements that are very close to others, but correspondingly $G(f)$ will have ringing in frequency domain associated to these sharp transitions; conversely, for the rectangular $G(f)$ associated to direct truncation in frequency domain, ringing in time domain has $\sin(x)/x$ behavior [7]. A smoothing window alleviates this problem with a compromise between mainlobe width (i.e. effective resolution) and suppression of leakage and ringing, commonly addressed as *sidelobe suppression*. Several windows are available, the most common being the group of cosine functions (von Hann, Hamming, Hanning) and then those for improved performance and flexibility, Kaiser-Bessel and Dolph-Chebyshev (see sec. 3.1.3.2 and 11.10).

Applying time gating introduces errors in response due to the limited performance of the used window [228]:

- insufficient attenuation or sidelobe suppression is at the origin of the "out-of-gate attenuation error", proportional to the sidelobe attenuation level, so under 1% already for the Hamming window, usually corresponding to the second window choice ("nominal window"), the rectangular window having only -13 dB (or 22.4%); the aim of the out-of-gate attenuation is to reject reflection terms other than the presently gated one, so that the total error of this first kind for processing of a given discontinuity is the sum of all other reflection terms multiplied by the pertinent sidelobe attenuation;

- conversely, a particularly long and persistent reflection response when gated may undergo a truncation of its tail thus causing an underestimation of the desired response (see sec. 3.1.4); if the time gating is applied to embrace more than one reflection and some discontinuities are located at the edges of the time-gating function, then for them the truncation error will be larger;
- an additional error term due to truncation is identified [7], when gating is applied directly in frequency domain (convolution with the response of the filter equivalent to the time gating function); the reason is that when convolution is performed at the extremes of the frequency interval, the gating frequency function that multiplies the original frequency-domain response is overflowing, embracing non-existing points outside the frequency interval that give zero as partial result (the would-be value is halved, or $-6\,\mathrm{dB}$, in the worst case); the compensation is achieved by post-gate normalization, but at the expense of creating the well-known sensitivity to changes of the gate-center times (that is gate misalignment with respect to the target discontinuity ideally lying in the center); Agilent indicates that this error may affect the last 10% of the gated response in low-pass mode and the first and last 10% intervals in band-pass mode, but does not quantify its magnitude;
- when multiple discontinuities are present, a source of error is represented by the multiple reflection terms, i.e. those terms traveling in opposite direction with respect to the main terms, each hitting again back the previous discontinuity, where they undergo again further reflection; these terms are all second order terms being all coefficients multiplicative and smaller than one, especially when some of the discontinuities are associated to connectors and junctions (with a return loss of let's say $-20\,\mathrm{dB}$ or smaller); they might be relevant only with significant discontinuities and when very low side-lobe attenuation is used, reducing correspondingly the first type of error; these multiple reflections are in reality independent of the time gating and used window, but simply are captured by the time gating as extraneous terms, yet located in the right portion of the time axis.

The Agilent example of two capacitors connected by short line sections [115] was conceived to show the effects of both masking (one capacitor masking the other) and misalignment of time gate, right on top of the first capacitor and time shifted to the right. Definitely results are reliable when the discontinuity is aligned at the center of the time gate. This is in agreement with what was observed for the transform of step-like at the end of sec. 3.1.4.

11.10 Time domain analysis and windowing

Even though VNA provides a TDR-like display, there are differences between traditional TDR and VNA time domain techniques. Traditional TDR measurements are made by launching an impulse or step into the DUT and observing the time response with a wideband receiver, such as an oscilloscope. The transform used by the VNA resembles time domain reflectometry, but the VNA makes originally swept-frequency response measurements and mathematically transforms the data into a time-domain TDR-like trace. In *low-pass mode*, the VNA measures discrete positive frequency points, extrapolates dc, and assumes that the negative frequency response is the conjugate of the positive, i.e., that the response is hermitian. In *bandpass mode*, the VNA measures discrete frequency points centered between start and stop frequencies, uses CZT and work over any frequency range, giving an estimate of system impulse response (step response is not available if dc and low frequency are missing). Using narrowband receivers, the VNA allows for largely reduced system noise levels, and thus enhanced signal-to-noise ratio and therefore superior dynamic range with respect to TDR and oscilloscopes. This is significant in applications where low-level signals are measured.

It is quite known that when analyzing a time sequence for its spectral properties, phenomena such as frequency leakage (see sec. 3.1.3) shall be considered, especially when signal truncation occurs and a transformation time window T is used that is not a multiple of the period of signal components. Truncation is equivalent to multiplication of input sequence for a rectangular window and this creates the well-known pattern of sidelobes with *sinc* ($\sin(x)/x$ function) shape. This topic is treated extensively in the literature [153, 229, 256, 275] and is quite common in signal and spectral analysis. Using a VNA and aiming at resolving in the temporal (or spatial) dimension exposes us to a similar problem, but related to the IDFT, or ICZT, operation: it is known that the DFT and IDFT are a perfectly reversible transform pair, so the findings and criteria adopted in one domain (i.e. frequency domain) are perfectly reversible in the other domain (i.e. time domain), and the same may be said for the CZT/ICZT pair.

A brief comment follows on the correct identification of transformation operator: the difference between DFT and CZT is that the latter operates along a spiral of points described by a turning vector identical to that of the DFT, but with the possibility of specifying a varying amplitude, i.e. the radius of the spiral. In reality if the CZT is really used, this is done not to have a locus in the form of a spiral, but to have the already mentioned zooming capability and more freedom in calculating and retrieving only the points that are needed (for the latter also the Goertzel's recursive DFT implementation would be of help, without recurring to CZT). In the following the use of DFT is assumed for simplicity of expression.

11.10.1 Introduction and procedure

Time-domain signals are in all cases causal, except when doing the so called "zero phase digital filtering" [256], that consists of running a FIR filter in forward and then in reverse direction of time; the same of course applies to any other domain, but figuring out a time-domain signal and a convolution moving in the positive and negative direction of time is easier. This kind of filtering is not strictly possible if real-time operation is required, but it is always possible when data length spans over a short time interval, shorter than operator's reaction and user interface dynamics. It is thus possible to enforce zero delay whichever the filter phase response; the output is thus characterized by "zero-phase distortion, a filter transfer function, which equals the squared magnitude of the original filter transfer function, a filter order that is double the original filter order.[19]" This is particularly important when comparing time-domain waveforms resulting after different filtering options, that might be otherwise subject to different time delay.

When transforming from time to frequency domain, the resulting DFT has hermitian symmetry (even symmetry around dc with complex conjugate values) and this shall be always ensured before taking the inverse transform IDFT. This is a quite common necessity when taking real data, defined over a positively valued frequency axis, as for all our S-parameter measurements: such a data vector (with N data samples) shall undergo a two-step procedure, so that it can be correctly inverse-transformed to time domain using IDFT. A dc value is constructed by interpolation, because in no way a VNA can provide its user with a measured dc value (some kHz or tens of kHz, yes, but not dc), and the data vector grows to $N+1$ data samples; then, the hermitian symmetry is arranged by tilting the original data vector around the dc value and taking the complex conjugate of it, resulting in a $2N+1$ samples data vector.

Regarding the estimation of dc value, it is possible to use different orders of interpolation: for coarse frequency resolution system properties may change significantly over the first data samples, e.g. moving over tens or even a hundred MHz, so that it is advisable, first, to start from the lowest possible frequency, and, second, to apply higher order interpolation, such as spline or piecewise cubic interpolation[20]. This operation is however not required if using band-pass processing mode, that purposely ignores dc value and assumes a pulse-shaped, rather than a step-shaped, excitation. ICZT ensures that the dc value is not even required from an operational point of view: differently from the IDFT, the transform/inverse-transform operation is taken on a restricted arbitrary interval, that does not need to include dc or any low frequency point.

Moreover, CZT using no hermitian symmetry does not require to discard half of the output data points, and all time-domain data are good and can be used. The use of CZT has the advantage of making low-pass and the band-pass mode procedures look quite similar and the implementation is thus simplified.

Time resolution dt (i.e. the separation of time-axis points) is set by the maximum frequency F_2 of measured data: assuming a perfect matching of Nyquist criterion, a sampling frequency at twice F_2 is postulated and time resolution is the reciprocal of it: $dt = 1/(2F_2)$. The maximum span on the time axis T_2 cannot be larger than the inverse of the frequency resolution df, as it is for Fourier transform: $T_2 = 1/df$.

The peculiar characteristics that an observer spots out in the recorded TDR/TDT response are pulses and edges, both with a relatively extended frequency occupation. But what is the required frequency extension? or, in other words, what is the relation-

[19] From Matlab help of `filtfilt()` function.

[20] Both available in Matlab, as options of the interpolating function `interp1()`.

ship between the measured frequency range and the resulting effective time resolution? Basically, we have just seen that the basic time resolution is the reciprocal of the maximum frequency: $dt_{base} = 1/F_{max}$. Adequate reconstruction of a pulse or an edge for an accurate enough measurement requires something more than this: the correct measurement and reconstruction of harmonic components is necessary to adequately describe steep fronts; missing harmonics will cause the known Gibbs phenomenon. For this reason the advised frequency range is a multiple, usually three to ten times higher, depending on the available hardware; reducing the frequency range affects progressively the accuracy of amplitude estimation.

Before anti-transforming frequency-domain data, a taper window is applied to reduce leakage and increase the peak-to-floor distance (i.e. signal-to-noise ratio) of the so obtained time-domain curve and its intelligibility. It is remembered that time-domain transformation is mostly useful in return loss measurements, spotting out discontinuities and mismatches in the test setup by identifying the corresponding peaks in the time trace. The degrees of freedom are thus the number of points, the extension of the time interval (zooming parts of the test setup) and the possibility of controlling leakage and noise floor.

So, what are the available processing options for the original measured S parameters in frequency domain? they are two and correspond to the main branches of the flowchart shown in Figure 11.10.1:

1. frequency-domain S parameters are inverse-transformed to get a TDR/TDT-like curve of S_{11} and S_{21}, respectively, as shown on the left; the inverse transform may be performed by IDFT or ICZT, but a taper window is also applied to reduce leakage and increase the peak-to-floor distance of the time-domain curve;
2. or, it is desirable to remove the effect of unwanted responses (e.g. reflections caused by some element of the test fixture) by the so called "gated response" and the results are needed back in the frequency domain:
   (a) frequency-domain data may be filtered at the origin without any explicit representation in time domain and no inverse transform is needed; or,
   (b) frequency-domain data are brought to time domain, gated with a smoothing window that replaces the implicit rectangular gating, and then transformed back to frequency domain.

11.10.2 Windowing of frequency-domain data

Smoothing windows that may be found on most VNAs are the implicit rectangular, Hann (or von Hann), Hanning, Hamming, Gaussian, Kaiser-Bessel and Dolph-Chebyshev windows (see sec. 3.1.3 and 3.1.3.2). The different windows may have quite similar performance for some choices of their parameters, so that in many cases there are no appreciable differences if applying one or the other, following manufacturer's choice. Different manufacturers make some of them available in their VNAs, depending on their own judgment and philosophy: Agilent seems to prefer Kaiser-Bessel, while Rohde & Schwarz offers Hann, Bohman and Dolph-Chebyshev; a test made on Anritsu VNA indicates the use of Dolph-Chebyshev windows.

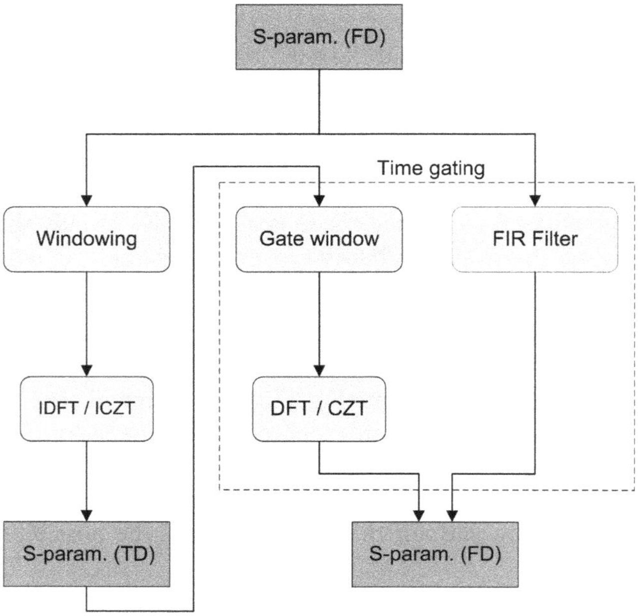

Figure 11.10.1 – Options for time-domain processing of original frequency domain (FD) S-parameter data.

In sec. 3.1.3 the question of which criteria and which windows are the best choice was already asked; the answer was simply that it depends on the application. VNAs use windowing for two tasks, windowing frequency-domain data when simulated time-domain response is required and windowing when applying time gating:

- for windowing of frequency-domain data before transforming them into time domain, VNAs offer three or four window choices, identified with various names, the first one being the rectangular and then other two or three with increasing sidelobe suppression, but coarser resolution; as said, depending on manufacturer, some families of high-performance windows may be preferred to others, but in the end differences are not dramatic;

- when time gating is applied, further windowing is necessary to avoid leakage phenomena due to the sharp cut of the gate, that is implicitly rectangular; normally three windows are offered, "minimum", "normal", "maximum", referring to the dynamic range given by the mainlobe-to-sidelobe amplitude ratio.

The characteristics of various windows are discussed altogether, because there are no substantial differences for windows applied to get a time-domain response and windows for cutting it and calculating back a gated frequency-domain response. We have understood that windowing is applied to reduce leakage, whichever the domain: our ears are well suited to the expression "frequency leakage", much less if our target is time domain.

We call "epoch" the original data set, whether it is in the time or frequency domain. If it is a time domain epoch that is going to be transformed into frequency domain, the epoch is a collection of time samples, with sampling time dt and the epoch duration is

$T = N\,dt$, N being the number of samples, assumed evenly spaced. If it is a frequency domain epoch, then the frequency extension is quantified by the maximum frequency (that is often called the "span", neglecting the fact that the sweep does not start exactly from dc, and we indicate it with F_{max}) and the resolution frequency df is given by $df = (F_{max} - F_{min})/M$, M being the number of frequency samples, possibly different from N, as it occurs when using CZT instead of DFT. F_{min} indicates a frequency value that may be more or less close to dc and may be prolonged to dc if the processing mode is "low-pass", otherwise, in band-pass processing mode its value may be any other value (see sec. 11.10.3).

The window features that deserve attention for selection or implementation are (see sec. 3.1.3.2):

- main lobe width, measured with respect to rectangular window, that sets the basic bin width, inversely proportional to the extension of the epoch that is going to be used; the main-lobe width may be made correspond to either ENBW, $-3\,\mathrm{dB}$ or $-6\,\mathrm{dB}$ bandwidths; as Harris pointed out, when considering coherent leakage terms, the $-6\,\mathrm{dB}$ criterion is relevant;

- the highest side-lobe combined with the side-lobe fall-off are important when aiming at separating two close features of similar amplitude, each one affecting the accuracy with which the other one is estimated in the convolution; not all windows have the first side-lobe that is also the highest one, so that the roll-off aiming downwards is the most relevant feature, but it is not an absolute indicator in all cases;

- the highest side-lobe, or side-lobe suppression alone, is indicative of the offered dynamic range; the "minimum" window is best when higher resolution is needed to resolve close responses with equivalent magnitudes, but has limitations due to leakage terms when components of different amplitude shall be compared (even if visually the improved resolution gives a false impression of better performance);

- the pass-band ripple may affect the accuracy with which the amplitude of features is estimated; normally the steepest roll-off is also associated with the largest ripple and a tradeoff is necessary.

For example, in Agilent's manuals the K-B window is adopted ([7], Table 1-2) and the indicated sidelobe attenuation for the "minimum" window confirms that it corresponds to the rectangular window (so, no smoothing window), "normal" is a K-B with $\beta = 3.0$ and "maximum" corresponds to $\beta = 6.0$. For Rhode & Schwarz the three window modes are implemented using Hann and Dolph-Chebyshev windows for normal and maximum window types. The "minimum" window corresponds always to the rectangular window.

11.10.3 Low-pass and band-pass processing modes

Time-domain low-pass mode simulates a traditional TDR measurement and supports both step and impulse excitation. There are some specific limitations for the measurement frequency range. It is required that the measured, positive data points be linearly spaced so that they are harmonically related from DC to the stop frequency F_{max}.

Additionally, since the Fourier Transform includes effects of dc value in the frequency response and VNAs do not measure dc response, the dc value must be extrapolated: the lower the first frequency points, the better the extrapolation to dc, so don't use low-pass mode if you cannot afford a frequency range sufficiently extended towards low frequency. For step-response results a conveniently low minimum frequency shall be selected for the extrapolated dc value to be consistent. The TDR response is calculated on a hermitian symmetrical spectrum (i.e. the negative portion of the spectrum, obtained mirroring the positive portion around dc, is the complex conjugate of the positive portion) and thus the time domain response is a real-valued function (as it should be). The frequency points are doubled and so the time-domain resolution: the resulting time step dt is half that of the band-pass mode, so that low-pass mode is to be preferred if maximum time resolution is required. The dc component is necessary for the complete representation of the spectrum before it is inverse-transformed: since a VNA cannot measure at dc, such component shall be reconstructed from the other data samples. This operation has been called "harmonic calibration" or "spectrum prolongation" and is nothing more than reconstructing dc by extrapolation with as suitable function (e.g. linear, cubic spline, etc.) using the first frequency samples: for correct reconstruction interpolation shall be able to follow the characteristics of the underlying spectrum profile and data samples shall be close enough to dc to minimize the frequency offset that is covered with extrapolation. It is suggested to include the minimum VNA frequency (e.g. 10 kHz or 100 kHz) and then space the data samples at the minimum, reducing the span and keeping the maximum number of samples available.

The advantage of including dc in low-pass mode is that it thus contains information that is useful in determining the type of impedance at the discontinuity.

However, low-pass mode must be abandoned if the nature of test setup and DUT precludes the use of an extended frequency range with harmonically related frequencies down to dc: this is the case of filters, narrow-band amplifiers, etc. But, not only: using band-pass mode is quite common also for fault location: if the user is not interested in determining accurately the nature of the discontinuity along a transmission line or RF channel, then band-pass mode may be used, compatible with the normal mode of transmission over the channel.

11.10.3.1 Low-pass mode (reflection response)

In low-pass mode both step and impulse stimulus are available, the former, as known, being able to give a complete information on where and which impedance discontinuity is. The real horizontal axis for low-pass measurement is the 2-way travel time. The used spectrum is hermitian and includes dc, so that the resulting time-domain response from IDFT is real; information is best represented in terms of the reflection coefficient (or impedance) versus time.

Since the dc component is available, a step excitation may be used (called "LP step" mode), that however requires the system under test to be able to propagate dc and low-frequency components. It is always possible to revert to the use of an impulse excitation ("LP impulse" mode), obtaining a representation similar to the one described for the band-pass mode (see sec. 11.10.3.2), but with twice the number of data points and so twice the resolution. Extension to dc is needed, however, to allow the correct processing of the signal and this is achieved by interpolation, that may be troublesome if using too few low-frequency samples. When operating off-line on an external PC there are many choices as for the interpolation function, fitting criteria and the possibility of treating differently amplitude and phase. Due to the many options and degrees of freedom of interpolation, the reconstructed time domain response is deemed by uncertainty.

The impulse mode reconstructs the time-domain impulse response from raw measured values $X(f)$, once the dc component $X(0)$ is added and the hermitian symmetry of the spectrum is enforced.

$$x_{lp,imp}(t) = \Im^{-1}\{X^*(-f)X(0)X(f)\} = \Im^{-1}\left\{\hat{X}(f)\right\} \qquad (11.10.1)$$

With a notation that makes the data samples and their indexes explicit

$$x_{lp,imp}[n] = \text{IDFT}\{[X^*[-k]X[0]X[k]]\} = \text{IDFT}\left\{\hat{X}[k]\right\} \quad n,\, k = -(N-1)\dots N-1 \qquad (11.10.2)$$

The step mode introduces the step function $u(t)$ transform

$$U(f) = \pi\delta(f) - j\frac{1}{2\pi f} \qquad (11.10.3)$$

written also as

$$U[k] = \begin{cases} \pi & k = 0 \\ -j\dfrac{1}{k} & k > 0 \end{cases} \qquad (11.10.4)$$

To obtain the time domain step response, thus, it is possible to multiply by $U[k]$ the spectrum $\hat{X}[k]$ before taking the anti-transform.

$$x_{lp,step}[n] = \text{IDFT}\{[Y^*[-k]Y[0]Y[k]]\} \qquad (11.10.5)$$

where $Y[k] = X[k]U[k]$.

Of course, the step response is the integral of the impulse response and thus it is always possible, in principle, to calculate the former from the latter directly, imposing a suitable constraint to determine the unknown integration constant.

The problem of correctly handling step-like excitation and the transformation process was already considered in sec. 3.1.4, where the various techniques proposed in the past are briefly reviewed and the attention is focused then on Cormack et al. work [90] and Waldmeyer's correction; it is worth noting, however, that the error between DFT and corrected DFT is smaller than about 0.5 dB (in Nicolson [251] the direct comparison between time and frequency domain results and theoretical solution is well within 0.1 dB at all points!). By the way errors due to frequency leakage when applying different tapering windows (see sec. 3.1.3.2) are of the same order of magnitude, and this is often considered a "permissible" degree of freedom.

Numeric implementation is discussed in sec. 11.10.7.

Step and impulse responses in low-pass mode are further considered in sec. 11.10.5 later on with the help of a reflective setup, with impedance change from Z_0 to about $2Z_0$ and back to Z_0 (an approximation of a Beatty standard, see sec. 11.12.3.2). The real circuit used for the test is built using coaxial cables of different characteristic impedance: RG58 is used for the 50 Ω base characteristic impedance and RG 62 increases the characteristic impedance to 93 Ω, nearly $2Z_0$. The step and impulse response are shown as black and gray traces. There is enough information to determine where (in time) the discontinuity is located and what type of discontinuity it is: the discontinuity at the first connection is a change in line impedance, where $Z_{L1} < Z_0$; the second discontinuity shows that $Z_{L1} = Z_0$. The impulse mode is best suited to clearly locate discontinuities spotting them out of the noise better than when the step is used. However it is in the step mode that full information of the transmission line and its discontinuities is collected (please see sec. 2.6 for a classification of line discontinuities and the associated TDR and TDT response).

Looking closely at the step trace shown in Figure 11.10.2, the response does not quite return to zero even though the impedance has changed back to Z_0; analogously, analyzing the impulse trace, the second discontinuity does not have quite the same absolute magnitude as the first discontinuity. Both of these illustrate the effects of *masking*, considered in sec. 11.10.4.

11.10.3.2 Band-pass mode (reflection response)

The band-pass mode is a general-purpose mode of operation. It gives the impulse response of the device, works on any device over any frequency range, and is relatively simple to use. Even with a coarse frequency span with fewer data points, its response is reliable, for what concerns the position on the time axis, while there is always some inaccuracy in the determination of the peak height. It is compulsory when the DUT is bandpass or if a large minimum frequency prohibits the use of low-pass mode, because the estimation by extrapolation of the dc component becomes highly inaccurate.

It is especially helpful for measuring band-limited devices and for making fault location measurements. Since the band-pass mode is the only mode that can be used over an arbitrary combination of start and stop frequencies, it is useful for devices that have a limited range of operating frequencies, such as filters, some couplers, amplifiers and in general communication channels. Not having any restrictions on the frequency range of the measurement is a distinct advantage over traditional TDR measurements

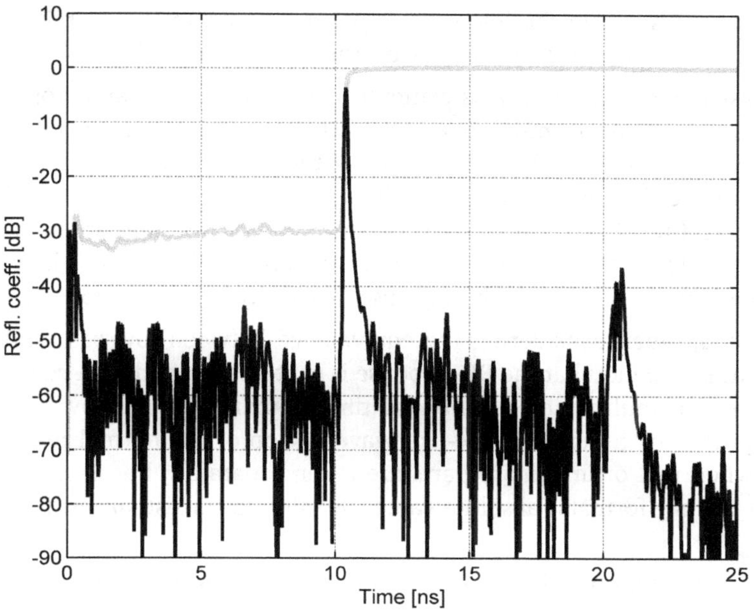

Figure 11.10.2 – Example of a characteristic impedance change (discontinuity) measured with step (gray curve) and impulse (black curve) response in low-pass mode.

(which require the DUT to be able to operate at dc!)[21]. Since the band-pass mode does not include a dc value, only the impulse excitation is supported (the so called "BP impulse" mode); the bandpass mode simulates a narrow-band TDR, that is quite common in finite element simulators solving electromagnetic problems in time domain (despite the shape of the impulses for time domain simulation is quite difficult to be reproduced in reality). It allows the identification of the locations of mismatches and the magnitude of the response (reliably, but deemed by some inaccuracy), but does not indicate whether the mismatches are capacitive, inductive, or resistive, or any combination.

The IDFT is calculated only on the measured data points, rather than taking the negative frequency response to be the conjugate of measured data, as in low-pass mode. This calculation gives a complex (real and imaginary parts) time-domain response and the magnitude (linear or log mag) of the response is to be displayed. Considering the frequency response of a low-pass system $X_{lp}(f)$ and its frequency translated band-pass version $X_{bp}(f) = X_{lp}(f)e^{-j2\pi f_0}$, it is easy to see that time-domain responses obtained for the two systems are equal in magnitude ($|x_{lp}(t)| = |x_{bp}(t)|$) and this establishes the equivalence criterion for the band-pass response: taking the absolute value leads to

[21] This characteristic is particularly useful not so much when transforming reflection responses, as when the transmission response is of interest: a band-pass device such as a filter or an amplifier may have a measurable and acceptable return loss over the entire frequency range, while relevant transmission may be limited to a narrow interval of frequencies.

meaningful results, while the low-pass response, thanks to the hermitian symmetry enforced in the frequency spectrum before transformation, is already real[22].

In band-pass mode, the window is centered between the start and stop frequencies; the IDFT is applied from minus one-half of the frequency span to plus one-half of the span, centered around the ideal center of the so-formed window. This windows both sides of the data which increases the impulse width and reduces the effective bandwidth; as a result, compared to low-pass mode for the same frequency span and number of points, band-pass mode has twice the impulse width and the resolution of the band-pass mode is half, which may obscure closely spaced responses.

As anticipated, when considering the equivalence between low-pass and band-pass time responses, the fact that the BP response is centered around VNA center frequency has the effect of multiplying the normal time domain response by a "modulation" function, producing a periodic sine-like wave on top of the normal response, that is apparent when real or imaginary parts are taken separately, but is eliminated in the log mag or lin mag formats (absolute value), leading to consistent results.

11.10.3.3 Low-pass and band-pass mode (transmission response)

When considering transmission response (i.e. insertion loss measurement using both VNA ports), a physical property of the measured devices or system is in our favor: almost always we will see frequency responses decreasing as the frequency increases (whether monotonically or oscillating is not relevant) and this assures that such responses meet the criteria for the existence of the inverse transform. Conversely, being return-loss response increasing towards unity for increasing frequency, they are not strictly Fourier transformable (as observed by Dunsmore in his thesis [114], sec. 3.3.3); yet, they can be transformed once truncation (with or without windowing) and implicit "periodization" are applied.

If the time-domain transform of reflection response (return loss measurement) with LP or BP mode has addressed the problem of performing TDR-like measurements with a VNA using frequency-domain data, now applying the same process to transmission response (insertion loss measurement) addresses TDT-like measurements. However, this part is not as thoroughly and extensively treated in the literature as the previous one, to the extent that in some cases the impossibility of getting time-domain transmission response from frequency-domain data is even suggested, or better "insinuated".

Note that in transmission measurements made in time domain, the value displayed on the x axis is always the actual electrical length, not the two-way travel time, as it might be in reflection measurements if no correction is applied. On the vertical axis we see the transmission coefficient, either in linear scale ("lin mag") or in dB ("log mag").

[22] In reality, also for the low-pass response a small residual of imaginary part remains due to numeric rounding and when applying windowing.

11.10.3.4 Alias in time domain response

The alias-free range is important to consider in all time-domain measurements and particularly for devices that are electrically long. It consists of the combined wrong setting of maximum frequency (or frequency span) and number of points, resulting in a too large frequency resolution and in the wrong time span, thus impeding that reflected terms to come back to the VNA port, for the complete picture of the overall response. As we see below the phenomenon of aliasing does not result in any artifacts or changes that appear evidently wrong, so there is no warning that it is occurring, unless we have another measurement at a reduced span or finer frequency resolution to use as reference: any shift in the main peaks indicates time aliasing, and of course also the amplitude of peaks is wrong, not only their position on the time axis. Similarly, when starting from time-domain waveforms fed to the DFT transform using an insufficient number of points for the used time span (the length of what we called "observation window"), aliasing may similarly occur as demonstrated in a LeCroy technical note[23]. Truncation of the otherwise useless tail of the original time-domain waveform is proposed (correspondingly reducing the frequency resolution of the frequency-domain representation, however without leaving out the relevant part of the signal), in order to fit back in the used number of samples. It is underlined that while operating in either domain, knowledge of the underlying waveforms, spectra and phenomena is highly necessary to operate correctly without leaving out relevant parts of signals.

A good example of frequency aliasing is found in cable fault location. Consider the measurement of a cable which is physically 10 meters long (so with a travel time of about 50 ns). Because this is a reflection measurement, the available time domain range must be greater than twice the actual electrical length of the cable, so about 100 ns. This is to allow the stimulus to travel to the discontinuity and return to the test port. Using a setup with 400 points and a frequency span of 2.5 GHz, the resolution Δf is 6.25 MHz and the available time range is about 160 nanoseconds. If the frequency span is increased to 6 GHz then the available time range becomes insufficient (about 67 ns), because Δf is more than twice larger. In this case, the time-domain actual response and aliased response will probably overlap.

When overlapping occurs, it is not generally possible to visually distinguish a real response from an aliased response without a test. If there is doubt whether a response is real or aliased, the first test is to increase the time span and see if there are obvious repetitions with observable spaces between them.

If one is still not sure that the response is real and aims at spotting out small amounts of time aliasing, then center the response on the display and store the trace in memory for comparison. Now reduce the frequency span and measure the device again. If the time domain response is valid, then it will remain in the center of the display with the same height; if the response is aliased, then it will move along the time axis when the frequency span is changed.

---

[23] A. Blankman, "SPARQ S-Parameter Measurements with Impulse Response Time Limiting," Teledyne LeCroy, Technical Note, rev. 2, June 2011.

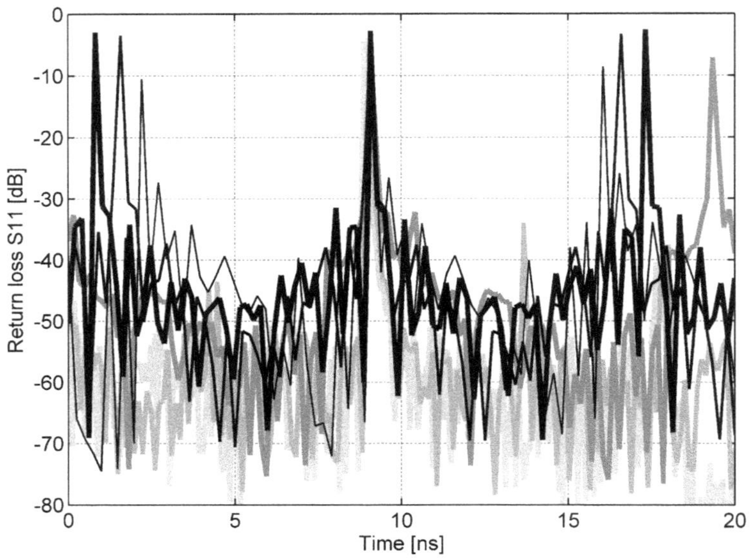

Figure 11.10.3 – Aliasing in time-domain response for the 1.86 m open-circuited RG 400 cable setup: basic curve for 323 points (thick light gray); additional curves for reduced number of points (shown from medium gray to dark gray for 162 and 122 points, respectively, and then black curves of reducing thickness for aliased situations with 98, 89 and 82 points). A low-sidelobe window was used to heavily remove leakage.

A practical example is shown in Figure 11.10.3, where the number of points is progressively reduced using a span of 6 GHz and a second peak appears: a simple setup is used where the cable has a length of 1.86 m (sufficient to see aliasing to appear) and is open-circuited at its far end; the number of data points used in the displayed traces is 323 (as an alias-free case), 162 (showing a second alias peak and beginning to shift the main peak, while still meeting the requirement on frequency resolution), 122, 98 and 89 (where progressively the alias becomes more and more relevant). The RG 400 cable has a velocity factor of 0.7 and the observed peak at 8.85 ns corresponds correctly to a propagation of 47.58 ps/cm for the 1.86 m length. With 162 points and 6 GHz span, the frequency resolution is 37 MHz and the alias free range is 27 ns, close to the total travel time of 17.7 ns, but theoretically still on the safe side. Going to 122, 98, 89 and 82 the resolution is even worse, 49.2 MHz, 61.2 MHz, 67.4 MHz and 73.2 MHz (corresponding to 20.3 ns, 16.3 ns, 14.8 ns and 13.7 ns), causing evident aliasing. It is important to note, thus, that aliasing may occur also when the frequency-resolution limit has not been violated, but when the margin is not sufficient.

11.10.3.5 Sweep time and long cables

When measuring long cables, from which long time delays are expected, besides the problem of time and frequency representation in terms of resolution and number of points, there is also an operative issue with the possibility that the adopted sweep time is too fast and that the measured insertion loss curve thus contains errors. When setting up a swept frequency-domain measurement over an extended frequency range, the frequency of the stimulus (i.e. the source frequency) is stepped by Δf in frequency every Δt in time. Knowing the number of points N, frequency and time steps are readily given by the frequency span divided by N and by the sweep time divided by N. If cable propagation delay τ is too long, the sourced frequency at port 1 will be moved by one (or more than one) frequency step, while the signal is still propagating along the cable and hasn't reached yet the other VNA port (port 2). For a correct measurement the minimum time-step Δt shall be longer than the expected propagation time, that is L/v, where L is the cable length and v is the propagation velocity. Of course simplified expressions involving cable velocity factor are possible, such as length of cable multiplied by $3.33\,\mathrm{ns/m}$ for the speed of light divided by the velocity factor k: $\tau \cong 3.33\,\mathrm{ns}\,L/k = 5.05\,\mathrm{ns/m}$ for a $k = 0.66$ cable, such as RG 58, RG 316 or RG 213. When violating the rule for the minimum time-step, the resulting error depends heavily on the characteristics of VNA and its settings (the used IF bandwidth, the more or less constant frequency step and sweep rate over different bands, etc.).

11.10.4 Mutual effect of discontinuities (masking)

"Masking occurs when the response of one discontinuity affects or obscures the response of subsequent discontinuities in the circuit." [7] So masking is defined as "obscuring" or "hiding", and this is what really happens; but why it happens? and under what circumstances?

Before going into the details, it shall be underlined that various phenomena are recollected under the term "masking". Let's consider a setup with a line bearing some discontinuities, such as junctions, shunt elements (e.g. branches and components), changes of characteristic impedance, etc. The signal transmitted from VNA port into the line and exciting successive discontinuities, after the partial reflection at the first one never reaches in its entirety the other discontinuities; the same may be said for the reflected signal terms going back to the VNA port from the various discontinuities, that are subject to reduction due to the same phenomenon. All the reflection and transmission coefficients are lower than unity and thus both the excitation signal and the reflected portion of it at each subsequent discontinuity are smaller than they should be if that discontinuity would be the first one (thing that the VNA assumes, knowing about only its own signal sent into the line). The same happens if the transmission medium has significant losses (e.g. cable with a large attenuation); this time the excitation signal and its reflection are gradually attenuated rather than cut at a lumped discontinuity. In all cases the reduction of the measured reflected signal leads the VNA to evaluate a discontinuity worse than it is in reality. Finally, if the discontinuities are quite close with respect to the available spatial resolution for the actual VNA settings, there is a mutual effect because of the processing methods (aliasing, leakage, loss of resolution with tapering windows). All these phenomena might contribute to the masking effect, more pronounced the farther we are from the VNA port and the higher the number of discontinuities we have encountered so far.

The physical explanation given above in qualitative terms may be better understood by taking the example of a transmission line with two discontinuities, each reflecting 50% of the incident voltage; the time domain response shows the correct reflection coefficient for the first discontinuity, $\rho_1 = 0.5$; correspondingly the transmission coefficient is also 0.5. However, the second discontinuity appears smaller because i) only a fraction of the applied voltage has reached it in the forward direction, and ii) some of the reflected voltage that should be measured by the VNA never reaches it, because it is reflected off by the first discontinuity when going in the reverse direction.

The exact change in the apparent reflection depends in part on the type of discontinuity that precedes it: changes in line impedance will appear somewhat differently than discrete reflections on the line. In general, the larger the reflection coefficient of the first discontinuity, the larger the masking effect on the second one [7]: the error approaches

$$\varepsilon_{\rho 2} = \frac{1}{1-\rho_1^2(f)} \tag{11.10.6}$$

having included the possibility that the reflection coefficient varies with frequency, e.g. when the discontinuity is caused by some reactive element.

In general, when a first discontinuity is responsible for the masking of the second one, the reflection coefficient of the latter results in

$$\rho_2' = (1-\rho_1^2)\rho_2 \tag{11.10.7}$$

where ρ_2' is the apparent reflection of the second discontinuity, including the effect of the first one; ρ_1 is the reflection of the first discontinuity and ρ_2 is the reflection of the second discontinuity alone. This expression is valid both for discontinuities represented by changes of line characteristic impedance or lumped elements.

Let's consider an example of characteristic impedance discontinuity: a $50-25-50\,\Omega$ Beatty standard. We may calculate ρ_1 and ρ_2: $\rho_1 = -0.333$ (due to the step change from $50\,\Omega$ to $25\,\Omega$) and the second transition has similar reflection coefficient $\rho_2 = -\rho_1 = +0.333$ when going back from $25\,\Omega$ to $50\,\Omega$. The application of eq. (11.10.7) predicts $\rho_2' = 0.889 \times 0.333 = 0.296$, as shown in [7], Fig. 19. This means that eq. (11.10.7) is correct to characterize characteristic impedance discontinuities, as well as discrete (lumped) discontinuities.

Regarding the error caused by neglecting the masking effect, it is possible to estimate it for a discontinuity occurring to the characteristic impedance, assumed constant with frequency: with a step change from Z_0 to Z_1 and then back to Z_0, as in the case of the Beatty standard, the error in the measurement of the reflection coefficient of the second discontinuity is given by

$$\varepsilon_{\rho 2} = \frac{4Z_0Z_1}{(Z_0+Z_1)^2} \tag{11.10.8}$$

If the discontinuity is caused by some discrete element, so with impedance possibly varying with frequency, the behavior may be different:

- if the discontinuity is due to shunt resistors (as for the loading effect of connected device or for pull-up/pull-down resistors) the effect is similar to the one discussed for constant line impedance;

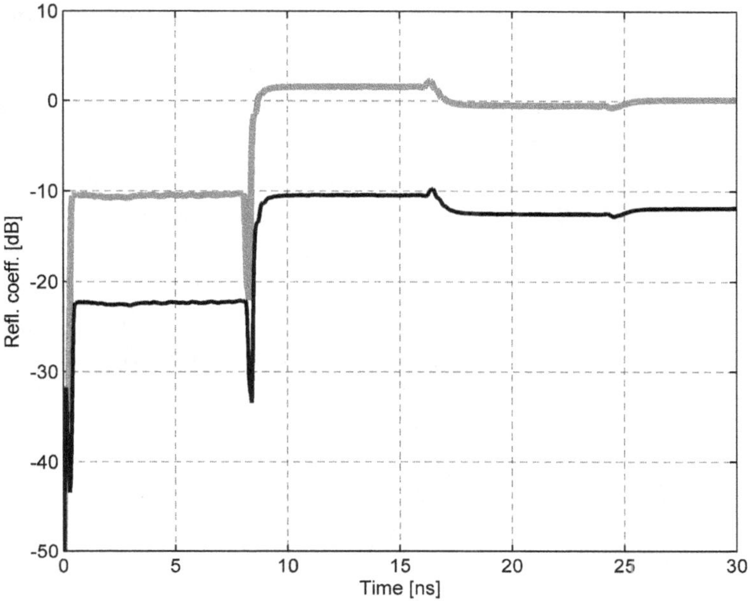

Figure 11.10.4 – Effect of line losses (simulated by a 6 dB attenuator) on return loss estimation (same as Figure 10.2.5 in Chapter 10).

- considering conversely two capacitors located along a multi-section line, the capacitors will increase the energy loss with frequency with their reactance getting smaller, and mismatch and resulting reflection getting worse (the typical Agilent example of [7], sec. 6, is exactly reported in [115]).

11.10.4.1 Effect of line losses

Let's consider now discontinuities occurring on a transmission line with significant losses, as it happens when the signal band is extended and we are using a low-quality coax cable or a dispersive planar transmission line. In a coax setup this may be simulated including an attenuator, that represents the assumed losses, yet lumped at the point of its insertion (see sec. 11.10.5 for an example); then, to evaluate the hiding effect of losses, we put a known discontinuity, like a short circuit, of which we know exactly the reflection characteristics. The impulse response of the short circuit alone has a return loss of 0 dB, as expected. However, when a 6 dB attenuator is included, the return loss is displayed as -12 dB, that actually represents the forward and return path loss through the attenuator (see Figure 11.10.4).

It is evident that losses affect the amplitude of the received signal on its two-way path and, if ignored, they affect also the accuracy when getting an estimate of reflection coefficient simply as the ratio of transmitted and received signals. This has nothing to do with the proximity of elements or line length, but depends only on the amount of losses. So, this is applicable to measurements of terminations at the other end of very long cables, that in all cases will exhibit significant losses.

11.10.4.2 Compensation of masking caused by reflections and attenuation

It is possible by knowing the characteristics of the transmission line (e.g. a coax cable of known p.u.l. losses), to apply a compensating corrective factor for masking. It is not known of any VNA that is offering the possibility of doing this automatically.

In order to draft a procedure for the necessary compensation, we shall first classify the phenomena for the effect they have on the measured reflection coefficients: we might have discontinuities and attenuation, the former lumped at a precise point along the setup, the latter distributed along the lines and/or lumped at each device.

Discontinuities are spotted out by a first S_{11} measurement, with which their longitudinal position is identified. Line losses can be measured with a S_{21} two-port measurement or reading values from datasheets. When two-port devices are inserted in the line, their insertion loss shall be also included.

What we have is thus a chain of objects along the line, each one with its own reflection and attenuation, having identified as "objects" both lumped elements and line sections: it is easy to recognize a graph, with branches (or segments) and nodes. We number the segments starting from 1 and preceded by "s", the nodes similarly preceded by "n".

Next, we know that the first discontinuity $n1$ bears a correct measurement of its own reflection coefficient ρ_{n1}, except for the attenuation of the first segment a_{s1} (if relevant, but assume that is quite short). It may thus be corrected, but in general at a first approximation we may say that the measured $\hat{\rho}_{n1}$ corresponds to the real ρ_{n1}. Going through the second segment to the second discontinuity, we shall correct the measurement by compensating for the effect of ρ_{n1} and of a_{s1} and a_{s2}. If using directly ρ_{n1}, and not $\hat{\rho}_{n1}$, the effect of a_{s1} is already compensated, and we may focus on the last two elements of the chain before the one we are considering: ρ_{n1} and a_{s2}. Compensation for ρ_{n2} is done by using eq. (11.10.7) and by multiplying for the reciprocal of twice the segment attenuation a_{s2}. Proceeding further along the setup we may apply recursively the procedure, provided that at each step the corrected reflection coefficients until the last previous discontinuity are used, that bear already the compensation of the attenuation of all segments behind. This is similar to the equivalent expression for reflection coefficient in frequency domain written by Lu and Brazil [228]. The compensation of masking may proceed first taking into account only reflections, that are easily identified both in LP and BP mode, and whether step or impulse response is used. Then attenuation of line segments may be added.

In order to derive an accurate, yet flexible, procedure and its algorithm, it is necessary to use the least amount of information that is not contained already in the measured S_{11} and S_{21}. The description above is intuitive for a human reader, but necessitates the identification of the lumped discontinuities along the setup (that we have called "nodes") and the assignment to segments of the correct parameters in terms of reflection and attenuation. In impulse response mode for S_{11}, the location of discontinuities is facilitated, using a routine for the identification of peaks (`findpeaks()` in Matlab is a good example). Impulse response mode is more flexible than step response mode, as commented in sec. 11.10.3.1 and 11.10.3.2 above. Yet, the information of the attenuation is recovered only by S_{21} measurements performed in LP step mode, thus limiting a lot the configurations where attenuation compensation is applicable. By the

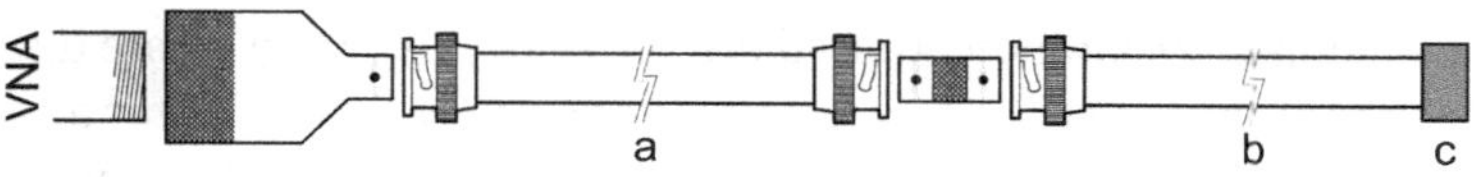

Figure 11.10.5 – Test setup for the cable reflection experiment: two types of cables are used, a RG 58 with 50 Ω characteristic impedance and RG 62 with 93 Ω characteristic impedance (an additional short semi-rigid 50 Ω cable is used for some tests); main cable sections "a" and "b" are 1 m length, the additional semi-rigid cable sample is 0.348 m; the termination "c" may be a 50 Ω load (L), a short circuit (S) or an open circuit (O).

Config.	"a" element	"b" element	"c" element
1A	RG 58	—	O
1B	RG 58	RG 62	O
2A	RG 62	—	O,S,L
2B	6 dB pad + RG 62	—	O,S
3	RG 62	semi-rigid	O,S,L

Table 11.10.1 – Configurations for cable reflection measurements.

way, except when using attenuators or high insertion loss devices, line attenuation is normally a second-order phenomenon with respect to reflections at discontinuities.

11.10.5 Low-pass mode with step and impulse response

A setup with multiple cable samples is considered, evaluating how VNA settings affect data representation and intelligibility of reflection measurements: the experiment requires a 1-port S_{11} measurement, performed on transmission lines created combining different cable samples and terminated on short (S), open (O) or matched load (L). The test setup is shown in Figure 11.10.5.

The tested configurations aim at verifying: i) low-pass (LP) mode with step response and impulse response for line discontinuities, ii) the typical voltage staircase/zig-zag at input for a short and open termination at the other side, iii) the reflections originating from mismatched characteristic impedance of the two cable sections, iv) the effect of various discontinuities along the test setup, taking into account the different propagation velocities, v) the effect of line attenuation by artificially increasing cable attenuation with an attenuator placed at the VNA port. The various configurations are summarized in Table 11.10.1 referring to relevant elements (sections and terminations indicated with "a", "b" and "c" as shown in Figure 11.10.5).

The simplest configuration 1A demonstrates the behavior of the LP mode using step and impulse response: two tests each are performed using a rectangular window with no leakage suppression and the "minimum" window with average leakage suppression (see Figure 11.10.6). The two windows behave similarly as expected, because there are no situations of close line discontinuities causing significant leakage and exerting mutual effect. Anyway, due to some leakage, the rectangular window (gray curve) produces slightly worse profiles of the reflection coefficient, with slightly larger values.

The impulse response, as already pointed out, is mostly useful to locate discontinuities: a slight reflection is visible at the beginning, due to the N-BNC adapter and amounts to about -27 dB (visible in both modes, step and impulse, with a nearly identical value); the third reflection is again caused by the N-BNC adapter at the VNA port, but is visible only in the impulse response curves, because it is not covered by the flat top of the step response. Comparing this third peak with the first one, again caused by the N-BNC adapter one may wonder why they do not have the same height, being the remaining elements all more or less matched to $50\,\Omega$: the difference in the height is caused not only by the slightly different impedance of the VNA port and the RG 58 cable (as said, almost identical and equal to $50\,\Omega$), but by cable attenuation and masking, being the third peak measured after the waves have propagated along the cable to the far end and back to the VNA port (this is due to a form of masking, caused by line attenuation; see sec. 11.10.4).

Comparing step and impulse mode traces, it is apparent that in step mode the noise floor is not visible and the trace is quite smooth and clear; in impulse mode the noise floor is around -50 dB and slightly higher around reflection peaks (due to leakage).

In configuration 1B a characteristic-impedance step change is caused by using two cascaded cable sections, with RG 58 section "a" followed by RG 62 section "b" of the same length (1 m). The results are shown in Figure 11.10.7. The interesting detail is the reflection caused by entering section "b" with the impedance mismatch between RG 62 (with $93\,\Omega$ characteristic impedance) and RG 58 (featuring $50\,\Omega$): a reflection of 0.3 appears, that corresponds to the observed -10.4 dB. When the wave reaches the far end of section "b", however, in case of a Short or Open termination the reflection is maximum whatever the characteristic impedance of its feeding section ("b" in our case). The impedance mismatch between "a" and "b" partly reflects the backward reflection and partly transmits it towards the VNA port; this time, coming from a line with a larger characteristic impedance ($Z_0 = 93\,\Omega$) into the RG 58 section, the influence of masking is that of giving an increased reflection coefficient, of about 1.5 dB.

The three terminations Open, Short and matched Load are tested in configuration 2, as shown in Figure 11.10.8. The curves are all characterized by an initial value around -35 dB and then a jump to about -10 dB: the former value indicates the good matching of the N-BNC adapter leaving the VNA port; the latter is due to the already observed impedance mismatch of RG 62, causing the already commented reflection of -10.4 dB. The exact value read on the curves of the upper half figure in the first ns is between -10.34 dB and -10.47 dB, that confirms the correct reflection estimation.

The three curves in the upper half of the figure refer to the termination with Short (light gray), Open (dark gray) and matched Load (black); the sudden and momentary jump to $-30/40$ dB indicates the well-matched BNC-SMA adapter used to apply S, O and L terminations, all with the same electrical length.

The matched load is not matched indeed when it is applied to the RG 62 cable and thus causes some reflection, visible after 7.9 ns: the amount of reflection is calculated to be 0.3, but what we see is not the expected twice -10.4 dB (that is -20.8 dB), but rather -24.5 dB. Why? the factor of two is due to the waves that cross twice the interface between the RG 62 and the $50\,\Omega$ load, but what is the reason for about 3.7 dB of difference, that becomes nearly 6.7 dB when we reach 15.8 ns? In reality

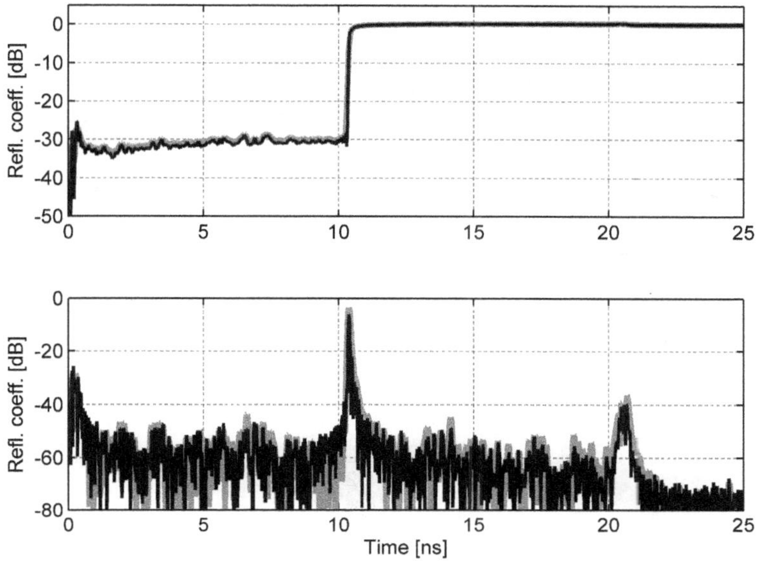

Figure 11.10.6 – Cable reflections configuration 1A: single RG 58 cable section matched to the VNA port impedance. Distinction of (above) step and (below) impulse response in low-pass (LP) mode: rectangular window (gray curve), minimum window (black curve).

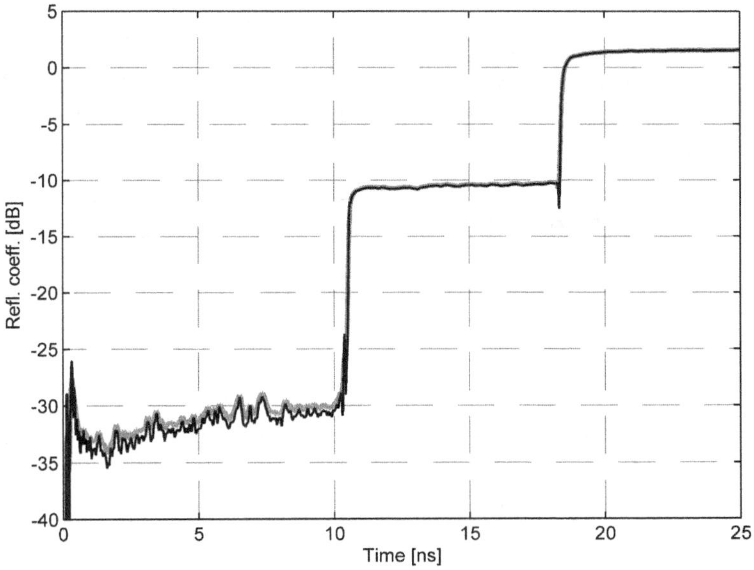

Figure 11.10.7 – Cable reflections configuration 1B: two cascaded cable sections "a" with RG 58 and "b" with RG 62. Distinction of (above) step and (below) impulse response in low-pass (LP) mode; the rectangular window is the black curve, the minimum window is the gray curve, and in the presence of distinct discontinuities they perfectly overlap.

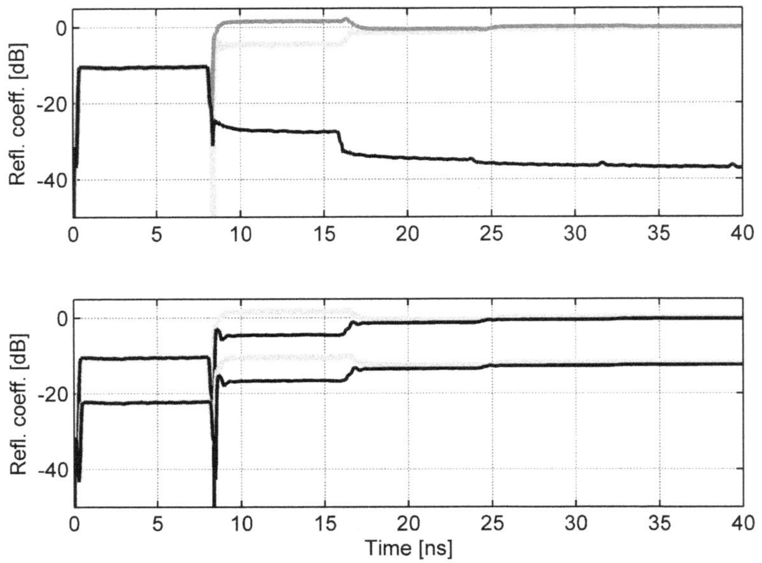

Figure 11.10.8 – Cable reflections configuration 2: (above) step response mode for Short (light gray), Open (dark gray) and matched Load (black) termination; (below) same two curves of short and open with an added 6 dB attenuator (config. 2B).

the reflection from the $50\,\Omega$ load going back to the VNA travels along the RG 62 with an amplitude of 0.3 (reflection coefficient), but is transmitted to the $50\,\Omega$ port of the VNA subject to another $1-0.3=0.7$ transmission coefficient, that is 0.21 in total: this corresponds to $-13.56\,\text{dB}$. Now we consider cable losses by inspecting the nominal values of the cables used for each section: the datasheet reads 12.5 to 28.5 dB of attenuation every 100 m of length between 200 and 1000 MHz for the RG 62 cable, that is about 0.2 dB for 1 m of cable, 0.4 dB for the wave traveling back and forth. This said, $-10.4-13.56-0.4 \cong -24.5\,\text{dB}$.

Going to the lower half of Figure 11.10.8, we see the effect of having added a 6 dB pad (attenuator) at the VNA port before feeding the RG 62 cable section: the two curves for O and S termination are shifted downward by 12 dB as expected, but they are quite similar, not to say identical; it is possible to see that the first edge has moved slightly to the right for the increased electrical length (attenuator contribution), but always starting from a very low value, indicating the good matching of the launcher while getting out of the VNA port.

By observing the time instants at which the two edges around the $-10.4\,\text{dB}$ pedestal are located, it is possible to estimate the RG 62 cable velocity: the time difference is $7.9\,\text{ns}$ and, compared to the speed of light for 1 m section length traveled back and forth (i.e. interpreted in the "two-way mode"), it gives a velocity factor of 0.84, that precisely corresponds to the nominal value of 0.83 given in datasheets[24].

Figure 11.10.9 shows with much more detail the phenomenon of multiple reflections: a short (0.35 m) cable section of $50\,\Omega$ semi-rigid cable was added at the end of section

[24] Pasternack, *RG 62A/U flexible cable datasheet*, rev. 1.1, 2013

"a" made of RG 62 and then terminated on the Open (light gray curve), Short (dark gray curve) and matched Load (black curve). Variations of the curve profile alternate more frequently: section "a" with RG 62 has the already estimated 7.9 ns of propagation time, while the added shorter section features a slightly slower cable (typical 0.7 velocity factor), that gives 1.5 ns of two-way propagation time (that is 0.75 ns of "true" one-way propagation time), as confirmed by measuring the time duration of each small "tooth" of the curve. It is observed that the small dip at 7.9 ns, when the wave enters for the first time the short added section of matched cable, indicates good matching, better than the one occurring when the wave travels within the added section: this is due to the BNC adapter that features quite a low return loss; the time duration of the dip corresponds in fact to about 100 ps.

By observing the light gray curve for the Open termination, we see the main curve oscillating every 7.9 ns, already seen in Figure 11.10.8, has a superimposed faster oscillation with the already commented 1.5 ns period. For the first 7.9 ns the curve shows the expected mismatch due to the RG 62 cable characteristic impedance, then when the wave enters the added section of matched cable, the mismatch drops to about −27 dB. This situation remains until the reflected wave reaches again the RG 62 cable section; soon after 10 ns, when the line ends on a reflective termination (either a short or open) the curve rises to, first −1.5 dB, and then after 1.5 ns to +0.7 dB (for the Open termination): in this case, as for the maximum that occurs between 17.5 ns and 19.0 ns, the "main" reflection back from the RG 62 cable section has completed and a reflection of the added short section superimposes with the "right" in phase relationship, to cause concurrent superposition of the two wave terms. For the black curve, when the added matched section is terminated onto a matched load, the resulting S_{11} is always quite low, around $-26/27$ dB.

The two-way representation was selected to interpret the curves and the reflection terms combination, but to locate discontinuities the one-way representation is more useful, because it gives readily a representation of the transmission line as it is seen by the traveling wave. This one-way representation of course is obtained by simply dividing by two the time axis of the received wave terms at the VNA port and it is done automatically by the VNA, once selected; as already pointed out amplitudes are not corrected and suffer attenuation and reflections as they occur twice during the propagation in the two directions.

11.10.6 Combination of low-pass and band-pass mode, windowing and time gating

The combined effects of the processing mode (low-pass or band-pass), of windowing and time-gating, and of setup discontinuities were tested on a purposely built setup, featuring several sections and interconnections, with the possibility of replacing terminations and including junctions and stubs. The diagram of the test setup is reported in Figure 11.10.10, where the various sections are identified with a letter: "a" is the N-SMA adapter fitting into the VNA port and protruding for 18 mm up to the female SMA end, "b" is a male-male SMA extension of 15 mm length, "c" is a female-female SMA extension of 12 mm length and "d" is either a short circuit or matched load

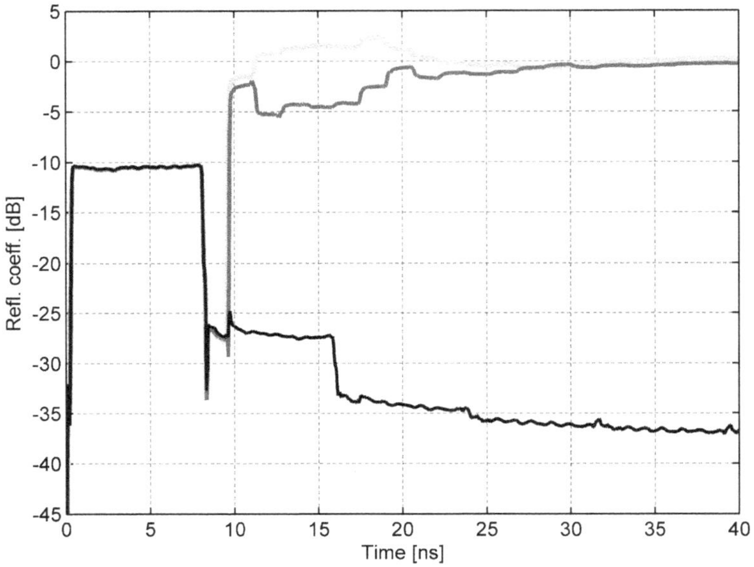

Figure 11.10.9 – Cable reflections configuration 3: step response mode for Open (light gray), Short (dark gray) and matched Load (black) termination.

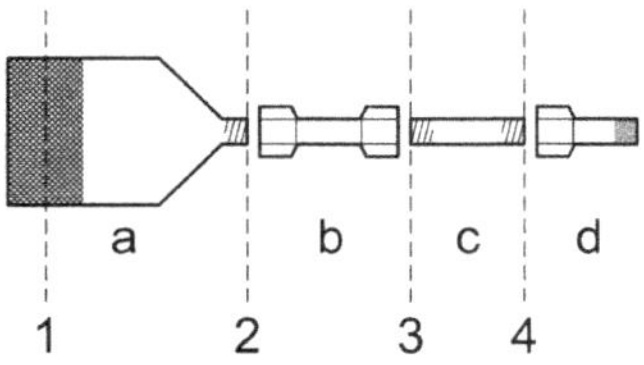

Figure 11.10.10 – Test setup for the experimental verification of processing mode, windowing and mutual effect of discontinuities: letters "a" through "d" indicate the setup elements, numbers "1" through "4" indicate the mating planes of setup elements.

termination; the short circuit terminations exist of 10 and 15 mm length, while the matched load has a negligible depth of 2 mm. In a case the extension "c" is replaced by a T junction of 18 mm, that branches onto a short circuit termination through the removed male-male SMA extension.

Assuming a 0.7 velocity factor (so $47\,\mathrm{ps/cm}$), the interfaces between the various sections are displaced in time with respect to the reference plane at the beginning of section "a" as: 84.8 ps, 70.7 ps, 56.5 ps and 47/70.7 ps. We will verify if the assumption of 0.7 velocity factor is correct by matching the measured peak positions.

11.10.6.1 Time domain S_{11}: electrical length and effect of line attenuation

The simple test setup shown in Figure 11.10.10 is used for the verification of time resolution, effect of line length on reflection peak position and of line attenuation on the peak height: it is composed of a N-SMA adapter connected to the VNA port on the N side and then terminated on various combinations of male-male and female-female

Config.	"a" element	"b" element	"c" element
1	M-M	F-F 10 mm F-F 15 mm	S
2	M-M	F-F 10 mm F-F 37 mm	L
3	T junction	F-F 37 mm	S

Table 11.10.2 – Configurations for time domain S_{11} measurements.

adapters ending in various types of terminations. As before the various sections are identified by a letter, as reported in Table 11.10.2: configuration 1 includes after the N-SMA adapter, a M-M adapter (section "a"), two F-F adapters of different lengths (10 and 15 mm) in section "b" and a male short termination (section "c"); configuration 2 replaces the short with a matched load in section "c" and uses two F-F adapters in section "b" of different length.

A simple case of short circuit termination is shown in Figure 11.10.11(a), where two SMA short circuits of different physical (and electrical) length are tested using LP and BP mode: in both modes, using the longest short circuit, the peak is moved to the right by the same amount, the peaks always corresponding between the two modes. The two thick curves refer to the highest resolution LP mode: counting the number of oscillations between the origin and one of the peaks and comparing the thin and the thick curve of the same color (corresponding to the LP and BP mode, respectively, for the same type of short circuit termination), we observe the effect of the 2:1 resolution ratio, i.e. the number of oscillations is double for the LP mode. By quantifying the shift in the peaks we have a first verification of the assumed velocity factor: the two peaks are at 239 ps and 261 ps, so with a time difference of 22 ps, that for the length difference of 5 mm gives a line velocity of $44\,\mathrm{ps/cm}$.

Going to Figure 11.10.11(b), where a $50\,\Omega$ matched termination is measured, in one case moving it farther from the VNA port by using a female-female extension, we observe two things: the peak at $-22\,\mathrm{dB}$ moves correctly to the right by the electrical length of the extension and a second minor peak appears, quite distinct from the main one, due to the transition from the line to the extension, whose matching is $5\,\mathrm{dB}$ better than the matched load termination. Finally, it may be worth underlying that the estimate of the peak position is inexact in both modes due to the unavoidable oscillation of leakage terms and there is no way to state that the LP mode is better than the BP mode in locating the real discontinuity (rather applying interpolation for the envelope of peaks is beneficial).

Regarding the height of the two peaks with $50\,\Omega$ termination, in the zoomed portion of the graph we may observe that they differ by nearly $0.5 - 0.8\,\mathrm{dB}$, a better estimate made impractical by the underlying ripple, lower in BP mode than in LP mode. The matched load is exactly the same and the peaks height shall be equal, except for the effect of the attenuation (quite small, honestly) of the female-female extension, about $0.3\,\mathrm{dB}$, and counting twice in a reflection measurement. This is a demonstration of

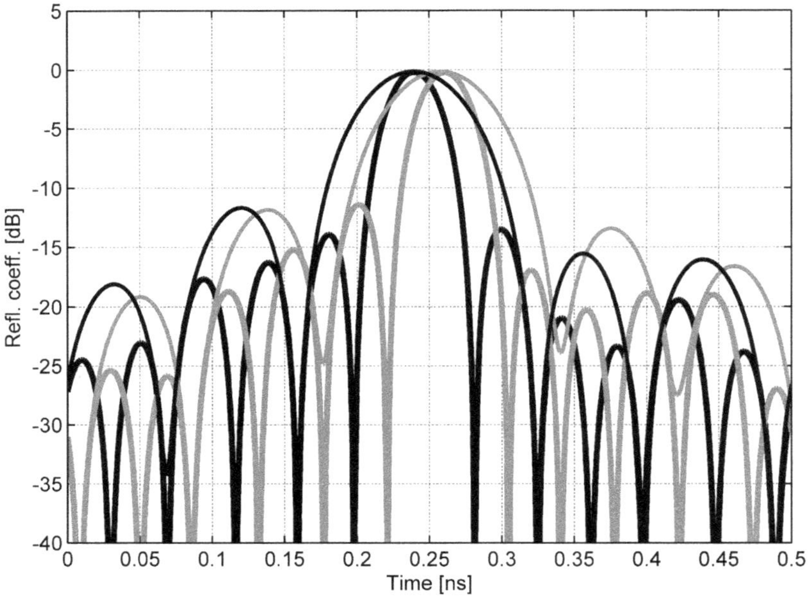

(a)

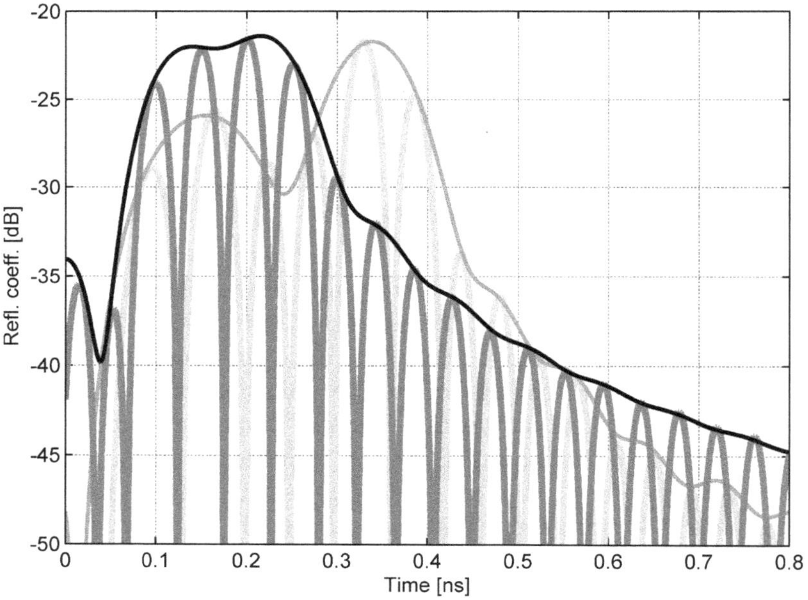

(b)

Figure 11.10.11 – Evaluation of reflection peaks: (a) config. 1, short circuit with extensions of 10 and 15 mm (black and gray, respectively) measured in LP (thick curves) and BP (thin curves) mode; (b) config. 2, matched 50 Ω load at plane 2 (black) and plane 4 (gray) in LP (thick curves) and BP (thin curves) impulse mode.

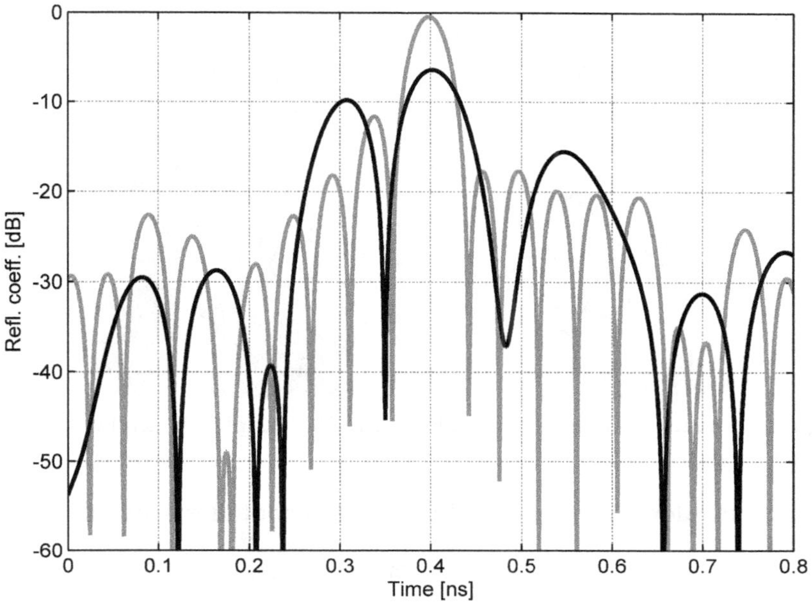

Figure 11.10.12 – Verification of masking effect in the measurement of S_{11}: the gray curve is obtained with the setup above with config. 1 (terminated on a short circuit and using a longer M-M adapter as element "b" with the same length of the T junction); the black curve is obtained replacing the "b" element with a T junction terminated on matched load, but halving the impedance seen at the interface between sections "a" and "b".

the influence of the extension attenuation on the accuracy of the reflection coefficient determination (see sec. 11.10.4).

Using the result of Figure 11.10.11(b) to estimate also the velocity factor, the distance between the two peaks in BP mode is 121 ps, related to a difference in physical length of extensions of about 27 mm: in this case the estimated propagation velocity is about $45\,\mathrm{ps/cm}$. We can thus establish that a good estimate of the velocity factor is 0.73-0.75.

In order to verify the masking effect due to a larger discontinuity preceding the one under measurement, the setup above was modified, replacing the "b" section with a T junction (the third branch terminated on a matched load) and trying to measure the short circuit termination connected at section "d". The result is shown in Figure 11.10.12 overlapping the configuration with T junction to the previous M-M adapter configuration (config. 1): the T junction lowers the impedance seen at the port between section "a" and "b" to $25\,\Omega$, creating a reflection coefficient of 1/3 (i.e.$-9.5\,\mathrm{dB}$), that is visible as a lower peak at 0.3 ns; the former peak at 0.4 ns is now no longer a neat $0\,\mathrm{dB}$ reflection (in "d" we have still the short circuit termination, that actually has a reflection of $-0.46\,\mathrm{dB}$), but is seen as a $-6.41\,\mathrm{dB}$ reflection. The reason is that the transmitted signal at port "a" is 2/3 and, after reaching the end of the line and being reflected back by the short circuit in "c", it is again cut by another 2/3 on the way back; at a first approximation we may neglect the higher order terms due to multiple reflections (see sec. 2.5.1 and Figure 2.5.1) obtaining rapidly nearly the result we look for, that is an attenuation of nearly 0.5 (4/9).

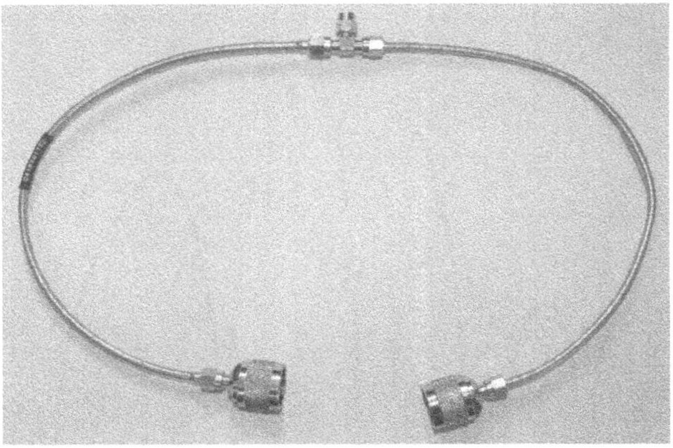

Figure 11.10.13 – Two semi-rigid cables and N-SMA adapters joined by T junction.

11.10.6.2 Time domain S_{11}: LP and BP mode, smoothing windows

Another setup is built using two cascaded semi-rigid cables of 360 and 375 mm length connected to the VNA ports with two N-SMA adapters and joined in the middle alternatively with the female-female SMA extension or with the T junction (see Figure 11.10.13). The test results appearing in Figure 11.10.14 compare how LP and BP modes using the four windows available detect the discontinuity represented by the T junction and the third branch left open; successively, this open branch termination is lengthened by inserting a male-male SMA extension of 12 mm (accounting for 54 ps propagation time).

The effects of using Low-Pass (LP) or Band-Pass (BP) mode are shown in Figure 11.10.14(a): above the LP mode is used with the four smoothing windows available on the VNA, from rectangular (black curve) to minimal (light gray); the same is repeated for BP mode. What is evident — as pointed out in the previous section — is the loss of resolution for BP mode, that suppresses the ripple by remarkably smoothing the curve and collapses the three peaks visible in the central part of the LP response into a single larger hump.

The question arises if the three peaks are really present in the measured response or if they are an artifact of the frequency-to-time domain transformation, that appears only in the LP mode thanks to the better resolution. This is clearly visible in Figure 11.10.14(b), where one LP and one BP curve are overlapped, both obtained using "minimum" window. An accurate measurement identifies three peaks due to the signal entering the T junction (at time 1.52 ns), hitting the end of the stub in open circuit (1.62 ns) and then coming back through the T (1.72 ns).

As a confirmation, and for better understanding of the phenomenon, a second experiment was done (see Figure 11.10.14(c)): the stub is lengthened by adding the male-male adapter, always ending in open circuit (this time on a male ending side). It is easy to see that, first, the central peak keeping the same height moves to the right from 1.62 ns to 1.70 ns; additionally, the curves overlap perfectly until the T is reached, because changing the stub has no effect on reflections coming from the first part of the line (except for some minor terms below -50 dB, that are absolutely negligible). It

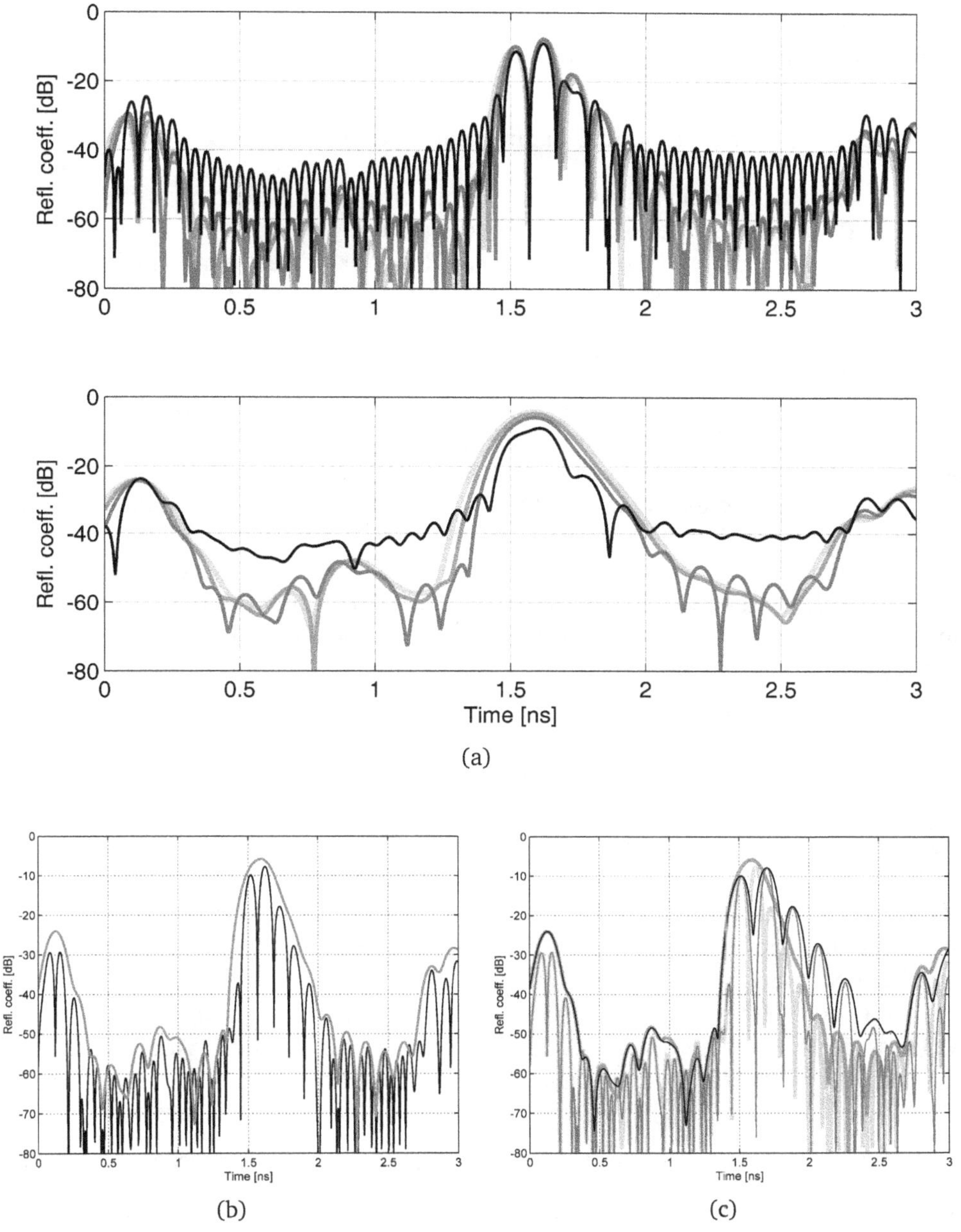

Figure 11.10.14 – Effectiveness of time gating to measure the reflection coefficient S_{11} in time domain for semi-rigid coax with intermediate T junction with stub: (a) LP (above) and BP (below) modes using the four windows ("minimum" to rectangular from lightest gray to black); (b) direct comparison of one LP and one BP curve with "minimum" window; (c) changing the position of the reflection peak by inserting a male-male extension with 54 ps propagation delay (the shape of the curve changes because two-way path is now longer than the basic time resolution).

may also be noted that the better resolution of the black curve in LP mode does not help in identifying the peak position, rather, the BP mode curves look more useful; it is however possible to create an artificial envelope by mathematically manipulating data during post-processing or using the smoothing option of many VNAs.

11.10.6.3 Frequency domain S_{11}: time gating and smoothing windows

While time gating in principle allows to spot out or reject any element of the line under test, a tradeoff between leakage suppression, resulting resolution and the remaining number of points that fall inside the time-gated interval is necessary. The last aspect is relevant in particular for extended setups, when aiming to zoom only specific details.

The curves shown in Figure 11.10.15(a) refer to the effect of applying time gating and recalculating the frequency domain S_{11} curve using a line section that is shrinking and closing around the discontinuity to analyze. The original curve (directly measured in frequency domain) is the thick light gray trace in the background, exhibiting the largest oscillations; the slightly darker curve with again evident oscillations is the frequency domain response calculated using time gating over the entire line section, so accounting only for the effects of frequency-to-time and time-to-frequency processing required by time gating. The other curves refer to line length reduction to 50%, 32%, 21%, 14%, the latter shown as black curve. It is evident that the last curve witnesses an inadequate situation (with about 140 points), underestimating the maxima and missing one of the minima.

When reconstructing the S_{11} curve, having removing by time gating (the so called 'notch gating") the effect of the T and the stub (see Figure 11.10.15(b)), the same effect occurs: the thick light gray curve is the S_{11} measurement of the transmission line without the T replaced by a male-male adapter of the same length and represents the reference S_{11} curve; the dark gray curve is calculated rejecting the central portion for 32% (the second case above) and the black curve rejects only 14% of the points around the T. The dark gray curve follows accurately the reference S_{11} curve, while the black one shows a significant overestimation of 5 to 10 dB. The dotted curves are obtained using the "minimum" window, thus ruling out as much as possible any leakage effect, demonstrating that the problem is not in the leakage. Even windows with less sidelobe suppression, but better resolution in terms of main lobe width, might have been used, providing acceptable rejection as well.

11.10.6.4 Application to a real case of a high frequency connector

To practically verify the usefulness of time gating, but also the intrinsic difficulty in the interpretation of results, we consider an example consisting of two low-loss semi-rigid cables connecting a test fixture for the measurement of a coax interposer connector. The test fixture is a small PCB with a controlled impedance stripline matching the characteristic impedance of the connector under test (CUT), that is $50\,\Omega$. The PCB is connected through two launchers (edge SMA connectors), to which two semi-rigid cables are connected.

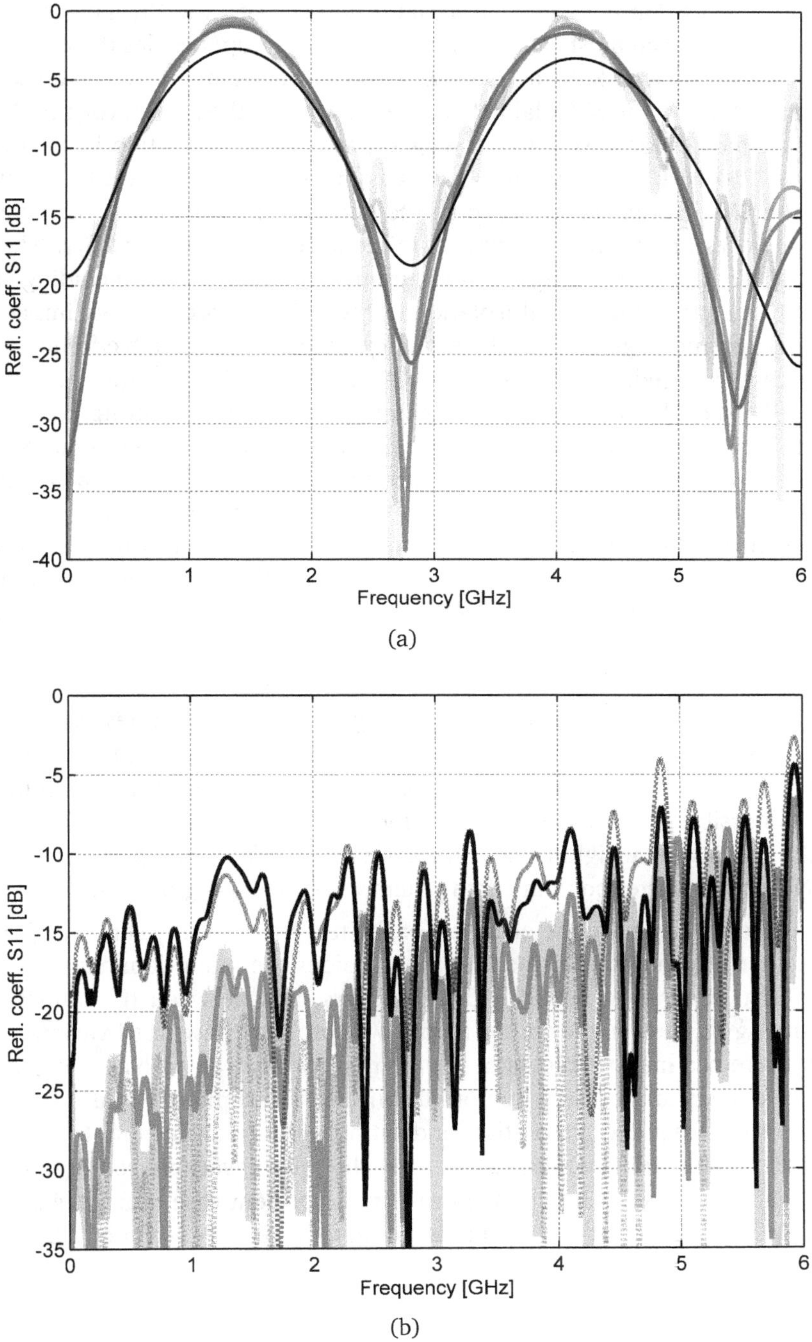

Figure 11.10.15 – Measurement of S_{11} for semi-rigid coax with intermediate T with stub: effectiveness of time gating to (a) spot out and (b) reject the effect of the discontinuity.

The total length between VNA ports is 0.79 m, but the distance between the two edge-connector reference planes is only 12.2 cm. Time gating will be operated first by taking two measurements that include the SMA launchers, biasing the S_{11} measurement with their discontinuity (the most significant of the whole setup, besides the one created in front of VNA ports by N-SMA adapters). Two different lengths are taken to demonstrate that as long as the SMA launchers are well inside the spatial (or time) interval, the recalculated response is not changing appreciably (see the two light gray curves in Figure 11.10.16(b)). Then, time gating is moved closer to the CUT by first, cutting out the SMA launchers and keeping the entire PCB stripline and then cutting the stripline almost in half, thus reducing its influence on the measured S_{11}. The last cut in terms of length is restricting the measurement to only 6.3 cm including the CUT; further refinement in the spatial isolation of the CUT has too a heavy impact on frequency leakage and in general on the accuracy, because the available frequency points are drastically reduced. When cutting at AA' (see Figure 11.10.16(a)) the number of points is 19.6% of the total (about 196 points), but when the cutting point is moved to DD' the samples are only 11% of the total (110 points) and the last curve for EE' is calculated with 96 points. Less points create problems of frequency leakage and edge effects, until now kept under control using a "nominal" smoothing window, being the results obtained with the rectangular window too confused and variable. To improve the S_{11} representation given by the black curve, smoothing and interpolation may be used (normally available in modern VNAs, but post-processing on an external laptop might be preferable) to fill the sag created around 4 GHz by the response zero, right where the return loss begins to increase significantly[25].

Two things are thus important when designing and using such a test fixture: leaving enough space between the SMA launchers and the CUT to ease time gating with respect to the time/space resolution available on our VNA, and using reference planes as close as possible to the SMA launchers for the calibration (the best choice is at the end of the feeding semi-rigid cables and not as in this first example at the VNA ports of type N), not to waste too many data points when operating the cut.

Another reason to separate the elements of the test fixture is explained in [3]: higher order modes are generated at the launcher, besides the main dominant mode; without "sufficient separation between the launchers, and between the launch and the DUT, coupling of these higher order modes will produce unwanted variations during the error-corrected measurements; a minimum of two wavelengths is recommended." Such 2λ rule leads to a prescription on distance that is similar to that related to time gating and time-frequency resolution, indicating a 0.5 to 1 bin time (or spatial) resolution for the step and impulse response in the most favorable conditions (see sec. 11.8 and Table 11.8.1, 11.8.2). The 2λ rule in the present case translates into $5\,\mathrm{cm}$ of wavelength in free space at $6\,\mathrm{GHz}$, that is $2.5\,\mathrm{cm}$ for a velocity factor of 2.5, and a total length of $5\,\mathrm{cm}$.

[25] The Matlab function `smooth()` was used, with `'rloess'` method and using a significant fraction of the total recalculated frequency points, between 25 and 30%.

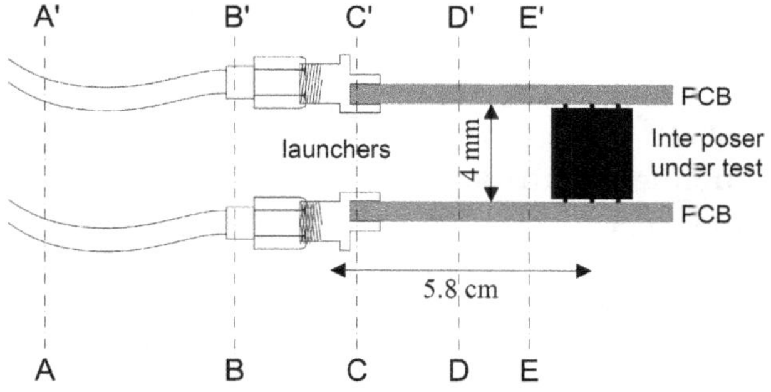

(a)

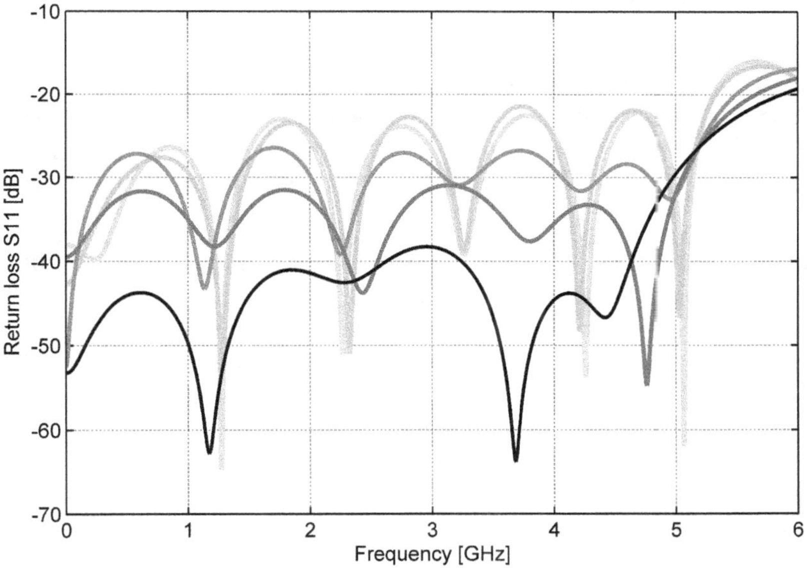

(b)

Figure 11.10.16 – Example of time gating and recalculation of the frequency domain response for a return loss measurement of a complete test fixture holding a coax interposer connector: (a) sketch of the test fixture and relevant points (similar to Figure 7.3.2(a)), (b) recalculated S_{11} curves for different time gating positions along the test fixture (AA' and BB': light gray curves; CC': medium gray curve; DD': dark gray curve; EE': black curve).

11.10.7 Details of time and frequency domain algorithms

Now we consider how time-frequency transformations may be implemented in VNAs and related software. The topic is introduced giving a historical overview of the methods appearing in the literature with enough details and information to be a starting point for implementation. Then, the attention is focused on writing clear and handy code to perform satisfactorily the required task, giving not too much importance to performance and compactness. To author's knowledge there is no real reference reporting clearly examples of code or detailed flowcharts that the user can use to perform his/her own post-processing. For this reason, various processing options were considered and the most relevant translated into code; the code was then cross-checked by comparing the results with those of a VNA for the same options. It is a long process because of the many settings whose effects overlap: tapering windows, frequency interpolation, zero padding, curve smoothing, etc. are all options that might more or less explicit and under user's control. Of course, there are situations in which the correspondence to the VNA output is only approximate, while in some cases the overlapping is perfect.

11.10.7.1 Approaches proposed in the literature

Interest for time- and frequency-domain relationships and their application to RF and microwave measurements dates back to the '60s, when time-domain equipment was much ahead in development and performance with respect to frequency-domain counterpart[26]. Later during the '70s the attention was turned to frequency-domain equipment and how to recover a time-domain representation of signals from frequency responses.

Different implementations were proposed in the past, since the pioneering work in 1974 by Hines and Stinehelfer [158]. There, the Fourier series terms (rather than those of the Fourier transform) are calculated by hand for the cases of "base-band impulse" (low-pass impulse mode), "square wave" (low-pass step mode) and "RF pulse" (band-pass impulse). Windowing is applied by choosing initially a Gaussian window and calculating the weighting coefficient sequence K_n for each mode. The authors insist on the band-limited assumption for input signals and remark the effect of leakage outside the transformation time interval. Going to the possibility of reconstructing the frequency domain response of a single "network segment", whose time domain response can be recognized and separated, they simply integrate the time domain coefficients for what they call the "integration interval $T_1 \rightarrow T_2$", stressing some assumptions about signal characteristics: there shouldn't be "significant interaction between these parts" (that is a combination of our considerations on time or spatial resolution and leakage effect), the response of a network segment shall be contained in the integration interval assuming that "the response is zero outside of this range", and signal

[26] A.M. Nicolson, "Wideband system function analyzer employing time-to-frequency domain translation," Proc. of WESCON Convention, Session n. 22, Aug. 1969.
B.J. Elliott. "System for precise observation of repetitive picosecond pulse waveforms," *IEEE Transactions on Instrumentation and Measurement*, Vol. 19, Nov. 1970, pp. 391-395.
H. M. Cronson and G.F. Ross, "Current status of time-domain metrology in material and distributed network research," *IEEE Transactions on Instrumentation and Measurement*, Vol. 21, Nov. 1972, pp. 495-500.

distortion due to transmission through other previous network segments should be kept to a minimum. What is quite relevant is that the authors state clearly that "The method is straightforward for reflection analysis. In transmission studies, the method is not usually applicable." This is an explicit reference for the applicability of time-to-frequency transformation to return loss measurements only, while it is simply alluded without further details or explanation in other cases.

Dunsmore [114] takes inspiration from this work 30 years later, giving some more details on impulse and step response expressions, but substantially confirming it as the method to adopt for frequency-to-time and time-to-frequency processing of VNA S-parameter curves; it is not explicitly said that it shall be applied to return loss only, but only examples based on return loss measurements are shown.

More than twenty years ago a different implementation was proposed, based on the Matrix Pencil Method, showing quite a remarkable accuracy[27]. The problem was solved by transforming VNA measured vectors into a pole expansion in the Laplace domain (where the poles z_i and the multiplying coefficients a_i, called residues, are the unknown quantities), which a series of exponential terms in time domain corresponds to. The authors state clearly that there is no difference between reflection and transmission VNA measurements. The poles in the z plane are combined with the residues in a set of matrices forming an eigenvalue problem: the solution for the poles is obtained by using a Singular Value Decomposition algorithm. The accuracy for the reported examples is remarkable, with errors of less than 1% in amplitude and position of reflection peaks (verification with return loss measurement of a Beatty[28] standard terminated in a short); even with a reduced frequency span (from 18 GHz to 2.5 GHz), the time-domain solution is quite accurate. However, no transmission measurement examples are given.

11.10.7.2 Time-frequency transform, low-pass and band-pass mode, step and impulse response

The implementation of the CZT itself is straightforward using Matlab, but the determination of the correct parameters to feed is a different matter: interpretation of resolution and span in either domain (frequency or time), adaptation of the number of points, normalization of results, to cite some. The function `td_response()` shown in Example 11.4 calculates the time-domain response starting from frequency-domain data. A few options may be selected by setting the corresponding parameter: step response or impulse response, CZT or DFT (for the latter the extrapolation to dc when step response is also selected), low-pass or band-pass mode, use of tapering, interpolation of data to increase resolution in frequency domain before transform. Step response is corrected as considered in sec. 3.1.4 is included. The results are shown in Figure 11.10.17 testing altogether with tapering window implementation that mimics an available VNA and the curves overlap almost perfectly.

---

[27] Z.A. Maricevic, T.K. Sarkar, Yingbo Hua and A.R. Djordjevic, "Time-Domain Measurements with the Hewlett-Packard Network Analyzer HP 8510 Using the Matrix Pencil Method," *IEEE Transactions on Microwave Theory and Techniques*, Vol. 39 n. 3, March 1991, pp. 538-547.

[28] For Beatty standard please see sec. 11.12.3.2.

Interpolation and zero padding may be included for completeness: they allow to artificially increase the resolution of original data and transformed data, respectively, and are more useful for DFT than CZT.

Example 11.4: Time domain response implementation using CZT

Processing of VNA frequency domain data is briefly shown for CZT (FFT version is not shown; the difference is the need to enforce hermitian symmetry of the spectrum before anti-transforming).

```
%% Excerpts of td_response() function
pm='LP'; %processing mode:  LP, BP
tr='FFT'; %inverse transform/transform pair:  FFT or CZT
wnd1='hann'; %type of window
wnd2=6; %optional window parameter
%
X=data(:,1)+1i*data(:,2);
Fmax=6e+9; Fmin=10e+3; N=length(X); df=(Fmax-Fmin)/(N-1); f=Fmin:df:Fmax;
Tmax=1/df; %Maximum time for alias-free response
dt_base=1/Fmax; %Base time resolution
dt=dt_base/2; %Time assuming fictitious sampling fs = 2*Fmax (Nyquist crit.)
%
% Desired time axis [T1 T2] and number of points M for CZT
T1=0.2e-9; T2=2.8e-9; M=600;
%
if strcmp(pm,'LP') %Estimation of dc value
    P=40;
    Xdc_r=interp1(f(1:P),real(X(1:P)),0,'pchip','extrap');
    Xdc_i=interp1(f(1:P),imag(X(1:P)),0,'pchip','extrap');
    Xdc=abs(Xdc_r+1i*Xdc_i);
end
%
if strcmp(pm,'LP') || strcmp(pm,'BP')
    if strcmp(tr,'CZT')
        Wt=W0*exp(-1i*2*pi*(T2-T1)/(M*Tmax)); At=exp(1i*2*pi*T1/Tmax);
        XX=conj(X); NW=length(X);
        switch wnd1
        case {'rectwin','bartlett','blackman','flattopwin','hamming','hann'}
            ws=window(wnd1,NW);
        case {'blackmanharris','bohmanwin','gausswin','kaiser'}
            ws=window(wnd1,NW,wnd2);
        end
        ws=ws/sum(ws)*(NW);
        x_t=conj(czt(XX.*ws,M,Wt,At));
        NW_max=length(x_t); %CZT has no symmetrical parts to remove
        t=T1:dt_CZT:T2-dt_CZT;
    end
%x_t=x_t(1:NW_max);
end
```

11.10.7.3 Tapering

Tapering as described in sec. 3.1.3.2 may be activated with one of the parameters of `td_response()` function; the variable `ws` contains window coefficients.

When trying to establish a relationship between the VNA available windows and the large number of combinations of windows and settings, the following problem arises:

a brute force technique would be successful by trying each window and each setting and then visually comparing the results, if the VNA curve were not showing a distance between its zeros not compatible with the expected main lobe width and spectral behavior of the considered windows. When a VNA implements direct frequency-domain processing using a digital filter (as it does), then the result will not be perfectly comparable. If a direct frequency domain implementation is to be considered, the number of possible settings increases dramatically, adding the number of the filter coefficients (the "taps" of the FIR filter implementation) and the filter specifications for the design, such as roll-off, attenuation in the stop-band, etc. Direct comparison and brute force thus become impractical.

To verify the implementation of time-domain and tapering functions some candidates for window type and parameters were selected and compared by trial-and-error with VNA output: the comparison with an Anritsu VNA in shown in Figure 11.10.17. It is evident that the overlap is nearly perfect for the rectangular window (Figure 11.10.17(a)); for the other windows (Figure 11.10.17(b,c,d)) there is a good correspondence with substantial overlap, especially at higher levels, the black curve indicating always the best window choice. It looks like Anritsu preferred to implement Dolph-Chebyshev windows, confirming the comment at the end of sec. 3.1.3.2, regarding the optimality of Dolph-Chebyshev windows for the main lobe width and sidelobe suppression.

11.10.7.4 Implementation of time gating

The `time_gating()` function may distinguish between notching and gating by one of the input parameters. Time gating may be implemented intuitively by applying a cut in time domain, using a tapering window when transforming back into the frequency domain, or by using a digital filter directly in the frequency domain, designed to realize the required cut in the fictitious time domain and implementing the necessary tapering. There are various techniques for the design of digital filters of the FIR type: the problem is enforcing not only the typical characteristics of a filter (roll-off and attenuation), but also leakage control. One of the design techniques upheld by Agilent uses the family of Kaiser-Bessel windows: FIR filter design is quite difficult if demanding roll-off and leakage suppression are considered together with a reduced number of data points; taking as example the K-B FIR filter design in Matlab, often the implementation is not feasible blocked by design constraints, one of which is the minimum data length with respect to filter length.

The exact determination of implementation details (filter order and/or type of window, parameter values and settings) corresponding to a specific menu option of the VNA is quite difficult, not to say impossible, and of no practical use in the end, once the overall method has been validated and the results are considered reliable. The user ends up with a time-gating function to perform the desired processing of VNA data, even data originally measured with no time-domain option. Because of the FIR filter implementation in VNAs, finding an exact correspondence with the result that can be obtained by using a tapering window in time domain is not possible.

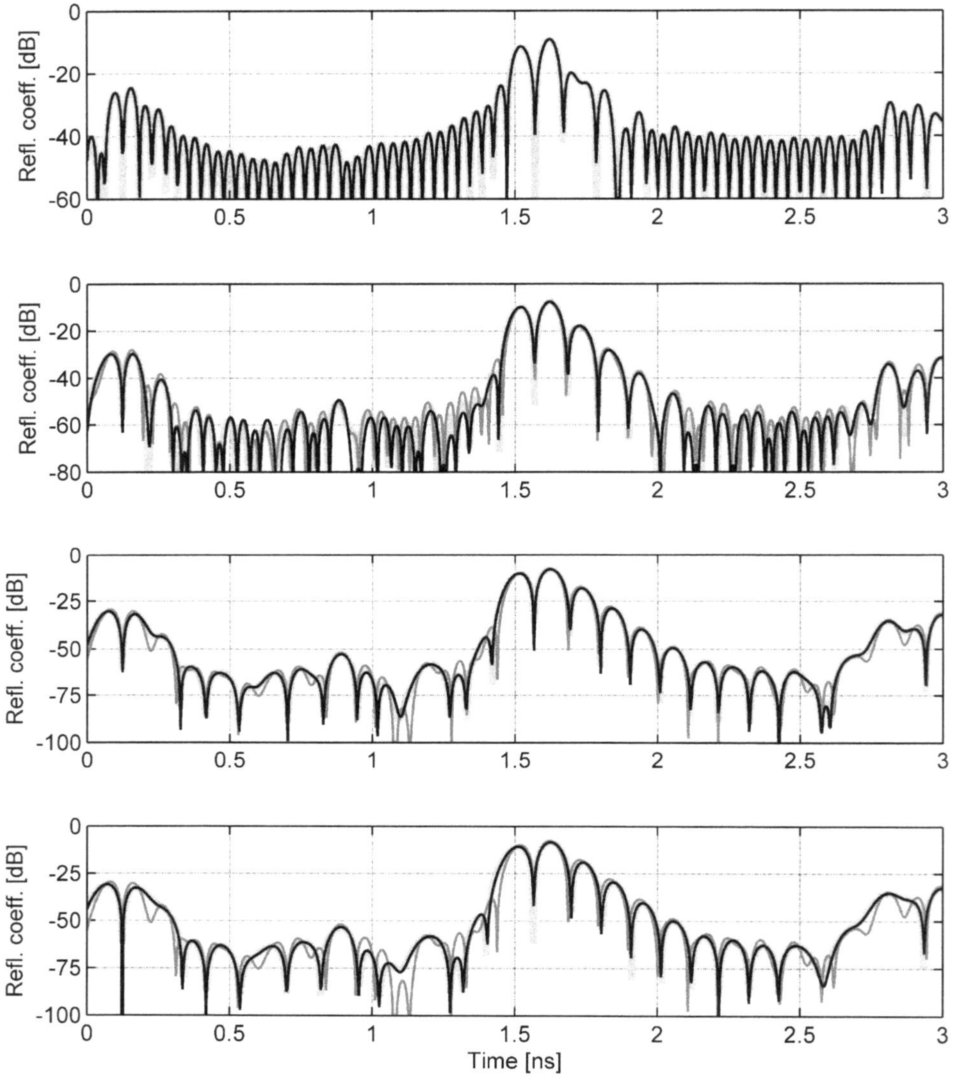

Figure 11.10.17 – Verification of tapering (identification of smoothing windows used in Anritsu VNA, the thick light gray trace in the background is always the original instrument trace): (a) rectangular window, aiming at verifying the operation of the inverse transform (here CZT); (b) nominal sidelobe window, Kaiser-Bessel $\beta = 3.5$ (dark gray), Dolph-Chebyshev $\alpha = 60\,\mathrm{dB}$ (black); (c) low sidelobe window, Kaiser-Bessel $\beta = 6.0$ (dark gray), Dolph-Chebyshev $\alpha = 80\,\mathrm{dB}$ (black); (d) minimum sidelobe window, Kaiser-Bessel $\beta = 6.0$ (dark gray), Dolph-Chebyshev $\alpha = 100\,\mathrm{dB}$ (black).

11.11 Uncertainty

Uncertainty is a broad subject treated at different levels of detail in some literature references [16, 39, 123, 309]. In modern measurements, raw measurement data are always corrected, compensated, post-processed and evaluated by a large amount of mathematics, typical of signal processing and statistics, and thus the rigorous definition and evaluation of uncertainty in all its aspects reveal their complexity.

Focusing on the use and operation of VNA, we may distinguish:

- *calibration uncertainty*, that is propagation of uncertainty during calibration from calibration standards to the mathematically obtained error terms and related correction;
- *raw measurement uncertainty*, characterizing the raw signals at VNA ports, just like for any other RF measurement (linearity, thermal and phase noise, etc.);
- *measurement uncertainty*, that is the effect of the first two on VNA output, consisting of one and two-port measured S parameters.

The problem of uncertainty management may be simplified, keeping under control the uncertainty of instrument, setup and adopted settings and procedures. It may be said that often VNA measurements are performed with much more attention to results than to characterization of metrological performance. In other words, whenever electromagnetic compatibility, signal integrity, material or component characterization, and not metrology or instrument characterization or calibration, are the fields of application, uncertainty, as well as repeatability and reproducibility[29], tend to be treated with some approximation. Probably one of the reasons is the complexity of VNA operation and its calibration. In a recent paper [140] the authors describe well — maybe exaggerating a little — the general attitude and the most common errors and metrology issues related to setup and measurement:

- uncertainty is not considered as a frequency- or DUT-dependent quantity;
- statistical analysis is approximate, calculating statistics on log-expressed quantities, on angles (see sec. 10.5) and not identifying truncated distributions;
- assuming independence in quantities and ignoring correlation between scattering parameters and repeated readings, as well as external biasing by common factors, may lead to erroneous quantification of the total uncertainty and its propagation; examples of common factors are environmental factors, external disturbance, setup resonances, etc.;

[29] It is worth recalling briefly the meaning of the two terms: repeatability addresses the question if under the "same", nominally unchanging, conditions repeated measurements give similar results; reproducibility conversely addresses if under typically varying conditions repeated measurements give similar results. The amount of variability in the two cases is the repeatability or reproducibility; both contribute to the overall system uncertainty. Furthermore, whilst environmental factors may be considered included in reproducibility, aging is not: stability, or the specular term "drift", addresses the problem of similarity of measurements taken at different moments in time; it is quantified by evaluating the variability over convenient time periods, such as hours, days, months or years. Also stability contributes to system uncertainty, and it is taken in due account by repeating calibrations and verifications.

- regarding the execution of the single measurement, the authors identify instrument noise, calibration uncertainty and connector repeatability, as the most relevant "metrology issues".

With some experience and memory of our own measurements, maybe we might add some other factors that have played a role in determining the "quality", not to say the uncertainty, of our measurements: loose connectors, defective cable-connector joints, cables too susceptible to bending and movement (maybe aged, worn, oxidized, or terminated on weak connectors), resonances, thermal instability. Cable and connector repeatability has been reviewed separately in their own chapters, in sec. 5.7.1 and 7.4.2. It is thus extremely important to establish our own verification procedure to identify and remove systematic defects and errors, in the widest meaning, that are responsible for a low-quality measurement: thus excessive noise due to oxidized surfaces, as well as distortion, are seen as systematic errors, especially because they have a clear origin and a known countermeasure, besides their spectral or statistical behavior; knowledge of thermal dynamics and time to equilibrium is extremely important, especially if new elements are introduced in the setup. What remains is a set of error terms and phenomena that we may loosely call "random errors", relaxing the rigorous mathematical interpretation proposed in GUM [39].

When approaching the measurement of a device or equipment (DUT) for its characterization, the uncertainty of the measurement system shall always be evaluated and clearly stated, in order not only to characterize what the stated results will be in terms of the associated confidence interval (see sec. 1.1.5), but also to preliminarily assess the capability of the measurement system to evaluate the desired DUT characteristic (Measurement Capability Assessment, or MCA). This is important in many cases, e.g. when VNA performance are barely sufficient for our task and we cannot afford to let them worsen due to additional error terms and uncertainty; or when performance of a new product is verified and quantified on the first samples and the results will have a strong impact on later production and corrective actions. Also when the measurement system is used to assess and monitor production routinely, uncertainty is relevant when sampling items out of the production chain; in this case the operator becomes a relevant parameter of the problem and shall be included in any thorough and complete statistical assessment of measurement system uncertainty and relevance of factors and parameters (by means e.g. of Analysis of Variance, ANOVA).

11.11.1 Calibration uncertainty

The aim is understanding and evaluating the contribution to the overall uncertainty budget of test fixture and calibration setup and procedure. Error models have been reviewed and the exact quantification of the uncertainty of the error correcting terms requires the evaluation of the propagation of uncertainty through the solution of the error model, starting from the uncertainty related to the input quantities, that is reviewed in the next section. Another source of uncertainty is that related to the calibration standards and their deviation with respect to the ideal Short, Open, Matched Load, etc. over the frequency range and the conditions of use (e.g. temperature) of calibration.

In [163] the authors list very clearly the types of errors that characterize calibration:

- *error in standards*, when calibration standards differ from their definition, or in other words there are deviations in their behavior, even after introducing the needed correction (see sec. 11.5.2.4);
- *instrumentation error*, due to errors introduced by variability of instrumentation between measurements (e.g. the drift that occurs between measurements during calibration or when measuring afterward);
- *scale reading error*, whenever an approximation is introduced due to limits of representation, the use or transcription of a limited number of digits, etc..

Some calibration procedures use a minimum number of standards, or a minimum number of known characteristics of the calibration standards, and do not try to correct such errors; other calibration methods exploit the intrinsic redundancy of the overdetermined equations of a full standard set to increase measurement accuracy (or, conversely, to reduce measurement uncertainty), as seen in sec. 11.5.3.9.

For example, LRL calibration method alone uses a minimum number of standard properties, but the overdetermined LRL method[30] employed by Multical software at NIST for example is able to include errors in standards through assumed statistical distribution (Gaussian); similarly VNA Tools II at METAS treats the characterization of errors in standards with assigned statistical distributions.

We might say that the first type of error is certainly peculiar of calibration, while for the other two types of error much depends on factors that are external to the calibration itself, such as environmental conditions, how the instrument is used, time that elapses between operations. Of all errors, it is easy to recognize that scale reading errors are mostly negligible, because modern instruments use high-precision digital hardware and a large number of digits when transferring or saving data (data truncation may happen, but it's quite uncommon). The method described in [163] addresses mathematically the first two types of errors in quite a general way, relaxing assumptions on distributions.

The propagation of uncertainty through calibration methods is reviewed in sec. 11.11.4.

11.11.2 Raw measurement uncertainty

VNA measures raw signals in frequency domain by means of its reflectometers and IF digitizing stage. Noise and linearity may be tackled in the same way as for Spectrum Analyzer in Chapter 9:

- noise is prevalently of the incoherent type, such as thermal, shot or flicker noise, due to internal devices, and in particular input attenuator, any amplifying stage and mixer;

[30] R.B. Marks, "A multiline method of network analyzer calibration," *IEEE Transactions on Microwave Theory and Techniques*, vol. 39, no. 7, pp. 1205–1215, Jul. 1991.

- phase noise is also present for the same reasons and with the same mechanisms considered in sec. 3.5 and then in sec. 9.5;
- deviations from linearity, such as signal compression, may occur for the same reasons and similarly to what was reported in sec. 3.4 and then in sec. 9.4.

Sharing the same external conditions of setup, the amount of noise entering the measurement is proportional to the IF bandwidth. As for the Spectrum Analyzer, a narrow bandwidth reduces the noise floor increasing the dynamic range at the expense of a slower measurement; this is true in general for incoherent or coherent noise, provided that the power spectral density is sufficiently flat. On the contrary, trace averaging is effective only if the noise is incoherent; moreover, trace averaging is extremely important during calibration to reduce the trace noise passed to the algorithm that determines the error-model parameters and correction of subsequent raw measurements. If speed is a relevant factor, trace averaging may be reduced during measurements, but taking full advantage of it during calibration is highly advisable.

For single-tone VNA measurements the influence of phase noise is less relevant than for the Spectrum Analyzer, but occurs in the same conditions, that is when a large dynamic range is necessary and the incoming signal is weak: during the measurement of a DUT featuring a large insertion loss or a well matched DUT, the signal arriving onto the far port, or the same port, respectively, may be quite small and the noise floor components become extremely important; one may be tempted to increase the power level of the launching port, but already above $0\,\mathrm{dBm}$ phase noise may begin to worsen, thus making nearly useless the adopted solution.

At the other end of the dynamic range, large signals applied to low attenuation devices or highly reflective loads may lead to linearity problems; for a single-tone measurement this mainly causes problems of signal compression, even if the appearing harmonics might slightly interfere with the demodulation and digitization process. Thus, large port power levels and large IF bandwidth are not advisable in such cases.

When the VNA is operated in time domain, or in time-gated frequency domain, another form of noise should be considered: leakage when transforming is a coherent interference caused by the combination of discontinuities and reflected terms, overlapping along the setup and modified and filtered by the transformation, including tapering. The whole subject was already extensively considered in sec. 3.1.3, and in sec. 11.10 specifically for VNA. The byproducts caused by leakage are much larger than the noise terms reviewed so far and thus are the first cause of inaccuracy when using time-domain operations. VNA accuracy is never declared for time-domain functions, since the result depends heavily on settings and test setup characteristics.

Anritsu quite concisely reports the uncertainty of one of its instrument defining the settings and conditions under which the reported performance apply. They use the most complete calibration method, using 12-term or ECU method, a low enough port power level of $-10\,\mathrm{dBm}$ (as a tradeoff between signal-to-noise ratio and linearity), a narrow IF bandwidth of $10\,\mathrm{Hz}$ (for many models it is the minimum bandwidth), the DUT return loss is minimum with $S_{11} = S_{22} = 0$ (matched load); they do not rely on trace averaging that is set to 1. Performances may be summarized as follows:

- the dynamic range is in excess of 100 dB starting at 300 kHz; above 2 MHz 110 – 115 dB are attainable;
- residual directivity is best between 2.5 and 40 GHz (50 dB), between 40 and 45 dB elsewhere;
- residual source match is similar, maybe slightly better, because more uniformly distributed around 47 dB;
- residual load match is limited by residual directivity in its quantification and is thus declared same as residual directivity; additionally, its determination is influenced by the uncertainty contribution of the cable connecting the two ports, so that they declare a derating of about 8 dB using the suggested test port cable;
- reflection and transmission tracking are quite small, typically ± 0.03 dB up to 40 GHz, ± 0.09 dB up to 67 GHz; for the low frequency range up to a few MHz, performance is slightly worse and declared as ± 0.08 dB.

In general performance for low-frequency range are not fully characterized and often estimated values are reported; by the way it is not expected that a VNA is heavily used in the tens or hundreds kHz range and calibration and traceability in this frequency range have received attention from time to time[31].

From these figures and assumptions on connector and cable repeatability, as well as on environmental conditions, the manufacturers (adopting a Type B approach [39]) derive curves of uncertainty for a specific measurement (e.g. reflection measurement or transmission measurement), usually parametrized with frequency and with the nominal value of the measured DUT parameter. We already pointed out that directivity and source port match errors are much more relevant in particular DUT conditions, that is when a well matched and a highly reflective DUT are measured, respectively (this is considered in detail in the two following sections 11.11.3.1 and 11.11.3.2).

Last, it is underlined that, being the VNA an instrument that heavily relies on calibration and error correction, its performance are largely influenced by the calibration algorithm, by the calibration standards and in general the quality of the setup elements: the curves reported in instruments manuals and datasheets should be considered as the best performance attainable. For this reason it is extremely important that everyone ought to introduce verification methods (see sec. 11.12) to monitor the

---

[31] U. Stumper, "Tracing the complex RF reflection in the MHz range back to dc resistance standards by utilizing planar NiCr thin-film resistors," *IEEE Transactions on Instrumentation and Measurement*, Vol. 42, no. 2, pp. 351–355, Apr. 1993.
M. G. Cox, M. P. Dainton, and N.M. Ridler, "An interpolation scheme for precision reflection coefficient measurements at intermediate frequencies. Part 1: Theoretical development," 18th IEEE IMTC, Budapest, Hungary, May 21–23, 2001, pp. 1720–1725.
N. M. Ridler, M. J. Salter, and P. R. Young, "An interpolation scheme for precision reflection coefficient measurements at intermediate frequencies. Part 2: Practical implementation," 18th IEEE IMTC, Budapest, Hungary, May 21–23, 2001, pp. 1731–1735.
F. Ziadé, A. Poletaeff and D. Allal, "Primary Standard for S-Parameter Measurements at Intermediate Frequencies (IFs)," *IEEE Transactions on Instrumentation and Measurement*, Vol. 62, no. 3, Mar. 2013, pp. 659–666.

quality of calibrations and the behavior of their own equipment, rather than blindly rely on official figures or just send periodically the equipment for external calibration and certification.

11.11.3 Measurement uncertainty

Focusing on VNA frequency domain measurements and related settings errors and uncertainty may be broadly classified as noise and linearity, uncertainty from calibration and imperfect error-model parameters correction, and setup uncertainty.

The amount of broadband noise getting into the single reading is directly proportional to the used IF bandwidth and with respect to the attainable signal-to-noise ratio it is possible to operate also on the source power level. However, source power level increases phase noise (sec. 11.2.4 and 11.11.2) and affects in general linearity [340]; higher source power levels should be used only when strictly necessary, while a cleaner trace may be surely obtained by some averaging. Phase noise is a form of coherent noise and does not reduce with averaging.

Regarding the effect of VNA calibration it is quite customary to address VNA accuracy in terms of residuals of error-model parameters, namely directivity, port match and tracking to cite the most relevant.

Setup uncertainty embraces broadly cable and connector repeatability, including cable flexure, temperature, aging, drift and external noise. This is quite a complex problem: environmental factors influence all VNA measurements, including calibration, and may be quite variable (e.g. randomly or cyclically). To the aim of controlling and weighting them appropriately, a considerable amount of experience is necessary, as well as a framework for the propagation of uncertainty and an experimental approach to their characterization. Recently, programs for automated calibration and uncertainty management have begun to appear and they were reviewed in sec. 11.5.3.10: in addition to all the calibration algorithms, these programs propose a method for uncertainty management, that may be purely Type B, or based on a Type A processing of continuously acquired measurements, or hybrid.

The most relevant error-model parameters are now reviewed: directivity and port match are considered, to be continued in sec. 11.12 for the verification of VNA accuracy. It is only underlined that these two VNA performance indexes, as well as others, are not guaranteed to be independent quantities; some degree of correlation is unavoidably established by the adopted calibration method, that propagates environmental and calibration standard uncertainties.

Propagation of uncertainty is considered in a dedicated section (sec. 11.11.4) for the most diffused one- and two-port measurements.

11.11.3.1 Directivity error

The directivity of VNA reflectometer (see sec. 11.5) is probably the most relevant parameter: it is indicated by E_{DF} in the error box model and by δE_{DF} for the residual error after calibration. Roughly speaking, directivity is the remaining signal when the reflectometer is terminated on a perfectly matched load and there is no reflected signal (it was defined in sec. 11.3.3 and considered as for VNA calibration in sec. 11.5): the reason for such reading being not 0 are e.g. geometry errors, connector mismatches and imperfections, besides any error related to calibration standards, because the measurement of directivity cannot be better than the matching error of the matched Load standard (see sec. 11.5.2.2). The relevance of the directivity term is that, when performing a measurement in normal use, the directivity term sums to the truly reflected signal with an arbitrary phase relationship, that causes a ripple to appear while the frequency is swept: the amplitude of the ripple is proportional to the ratio between the directivity and the expected reflection from the measured DUT, so that directivity-caused limitations are mostly relevant when measuring low-reflection DUTs [14].

Let's consider a DUT with a given return loss $RL\,[\mathrm{dB}]$: measuring it with a residual directivity $D\,[\mathrm{dB}]$ of the VNA port brings to a fluctuation of the received signal at the digitizing internal port due to the varying phase relationship as the frequency sweeps; the summation of the true RL term and the superimposed directivity error is to be made taking into account the variable phase, so as a vector summation of the linear quantities (that are the wave amplitudes).

$$V_{DUT} = \Gamma_{DUT} V = V\,10^{-RL/20} \tag{11.11.1}$$

$$V_{dir} = d\,V = V\,10^{-D/20} \tag{11.11.2}$$

$$\Delta V = V^{+}_{DUT} - V^{-}_{DUT} = (V_{DUT} + V_{dir}) - (V_{DUT} - V_{dir}) = 2V_{dir} \tag{11.11.3}$$

$$\frac{\Delta V}{V_{DUT}} = \frac{2V_{dir}}{V_{DUT}} = 2\frac{10^{-D/20}}{10^{-RL/20}} = 2\,10^{-(D-RL)/20} \tag{11.11.4}$$

$$\begin{aligned} \frac{V^{+}_{DUT}}{V_{DUT}} &= 1 + \frac{V_{dir}}{V_{DUT}} = 1 + 10^{-(D-RL)/20} \\ \frac{V^{-}_{DUT}}{V_{DUT}} &= 1 - \frac{V_{dir}}{V_{DUT}} = 1 - 10^{-(D-RL)/20} \end{aligned} \tag{11.11.5}$$

In this way the values appearing in the chart shown in [14] are obtained; the directivity error for the most common values of RL and D is shown in Figure 11.11.1.

Example 11.5: Numerical example for the directivity error

Considering a 20 dB return loss for the DUT (that corresponds to a 0.1 reflection coefficient or VSWR=1.22), what is the effect of the 35 dB directivity value, only 15 dB smaller than that? the error calculated for the two opposite phase relationships would be -1.42 dB and $+1.70$ dB, for a total peak-to-peak ripple of 3.12 dB within which the error will swing for varying frequency. The reading will thus range from 18.58 dB and 21.70 dB, that corresponds to a nearly 18% error!

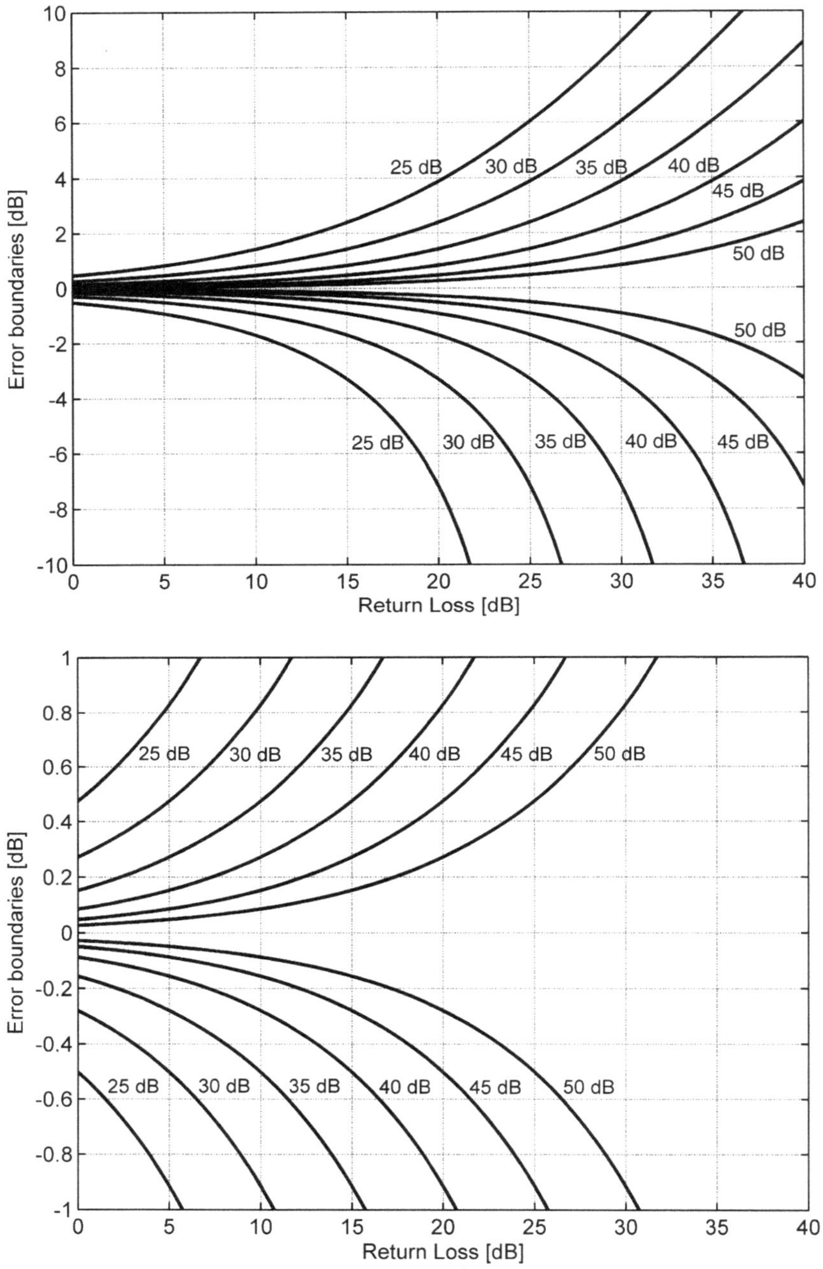

Figure 11.11.1 – Curves of maximum positive and negative error terms for Directivity influence on Return Loss measurement; positive and negative error terms correspond to V^+_{DUT} and V^-_{DUT}, respectively. Curves are plotted for Directivity values starting from 25 dB every 5 dB.

11.11.3.2 Port match error

The reflectometer non-ideality will cause also a port match error (Source Port Match was indicated as E_{SF} when considering the error box model and resulting in the residual port match error δE_{SF} after calibration). The port match error may be seen as a deviation from the full reflection reading when applying a Short or Open calibration standard; the reason is that the reflected signal from the calibration standard will be re-reflected due to port mismatch and the resulting readout will be smaller, thus underestimating the standard reflection coefficient (port match was also considered in relationship to VNA and its calibration in sec. 11.5). The portion of signal reflected back at the VNA port reaches the Short or Open calibration standard, where it is fully reflected back to the VNA test port, that it reaches undergoing a delay and arriving with an arbitrary phase relationship while the frequency is sweeping. The error due to the port match becomes less significant when measuring DUTs with good matching, as often occurs.

As for the Directivity error, calculation may be done by writing down the reflected term due to Source Port Match as a function of the expected signal, and then summing or subtracting it from that signal, simulating the two extreme phase relationships. Assuming no losses in the feeding lines, the reflected term is thus directly summed or subtracted from the measured signal and the error is thus inversely proportional to Source Port Match. When measuring DUT with better matching than a full reflection (as for the Short or Open), also Port Match error improves. In the following Source Port Match is assumed however good enough to in turn assume that the transmission coefficient at the port is unity and that the reflected wave from DUT to VNA port is no further reflected.

$$V_{PM} = \Gamma_{\text{DUT}}\, PM\, \Gamma_{\text{DUT}}\,(1-PM)\, V \cong V\, 10^{-(PM+2RL)/20} \tag{11.11.6}$$

$$\Delta V = V^{+}_{\text{DUT}} - V^{-}_{\text{DUT}} = (V_{\text{DUT}} + V_{PM}) - (V_{\text{DUT}} - V_{PM}) = 2V_{PM} \tag{11.11.7}$$

$$\frac{\Delta V}{V_{\text{DUT}}} = \frac{2V_{PM}}{V_{\text{DUT}}} = 2\frac{10^{-(PM+2RL)/20}}{10^{-RL/20}} = 2\,10^{-(PM+RL)/20} \tag{11.11.8}$$

$$\begin{aligned} \frac{V^{+}_{\text{DUT}}}{V_{\text{DUT}}} &= 1 + \frac{V_{PM}}{V_{\text{DUT}}} = 1 + 10^{-(PM+RL)/20} \\ \frac{V^{-}_{\text{DUT}}}{V_{\text{DUT}}} &= 1 - \frac{V_{PM}}{V_{\text{DUT}}} = 1 - 10^{-(PM+RL)/20} \end{aligned} \tag{11.11.9}$$

As seen, Source Port Match has less influence than Directivity on overall measurement uncertainty and is relevant only for high reflection DUT, such as a Short or Open, or maybe a filter outside its operating bandwidth. In all other cases, when the DUT has a moderate match and the reflected signal is not maximum as assumed above, the uncertainty due to the source port match will be proportionally lower.

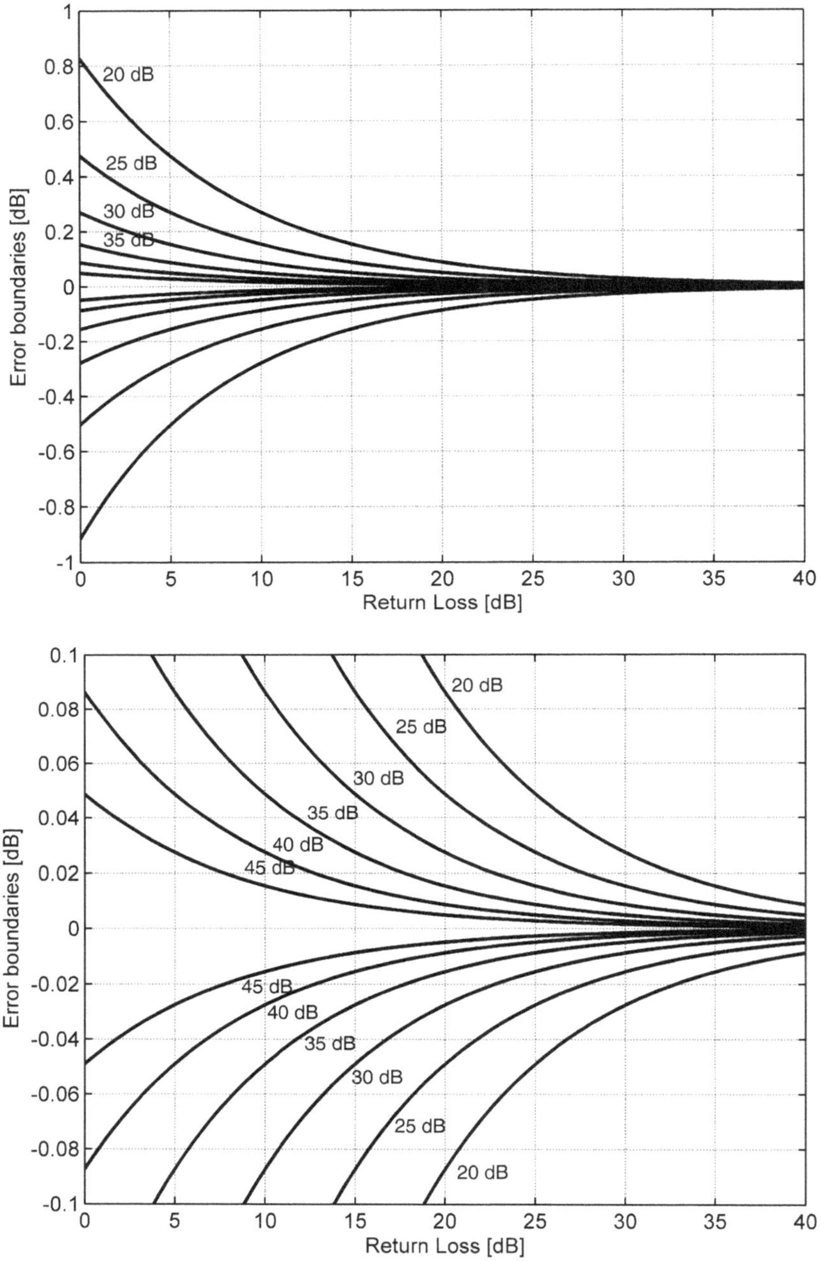

Figure 11.11.2 – Curves of maximum positive and negative error terms for Source Port Match influence on Return Loss measurement; positive and negative error terms correspond to V^+_{DUT} and V^-_{DUT}, respectively. Curves are plotted for Source Port Match values starting from 20 dB every 5 dB.

Having assumed a straight connection between VNA port and DUT, we have ruled out attenuation: line attenuation is influential because terms resulting by more than one reflection will undergo attenuation a proportionally larger number of times (even if line attenuation is in general limited to a fraction of dB); conversely DUT attenuation will not influence results, whereas it will be beneficial, reducing the effect of the far port (Load Port Match).

11.11.3.3 Effect of added adapters on Directivity and Port Match

When adding adapters, both Port Match and Directivity worsen (see Example 11.1). As already pointed out, when VNA and test fixture have different connector types, an adapter is needed and whatever its impedance match and reflection coefficient, it worsens the original VNA port match. The resulting reflection coefficient is the sum of the two single reflection coefficients and the overall effect is a significant increase of the overall uncertainty, even if the added adapter is a high-quality one [14].

The resulting directivity is computed by summing in the worst-case combination the reflection coefficients of VNA port, Γ_{DIR}, and of the adapter, Γ_{ADPT}: when taking a 40 dB directivity port (with $\Gamma_{\mathrm{DIR}} = 0.01$) and a good adapter with VSWR of 1.05 (that is $\Gamma_{\mathrm{ADPT}} = 0.025$), the resulting equivalent directivity is that of a reflection coefficient of 0.035, that is only 30 dB!

Similarly, with a Port Match of 30 dB, the resulting reflection coefficient is $\Gamma_{\mathrm{PM}} = 0.025$ and adding the previous $\Gamma_{\mathrm{ADPT}} = 0.025$ of the adapter gives a resulting equivalent port match with a reflection coefficient of 0.05, i.e. 26 dB.

So, the most affected, as we might expect, is Directivity, that represents the most relevant performance parameter during return loss measurements. Even good adapters worsen performance and they shall be included in the calibration, that in turn requires that calibration standards are available in the new connector type.

11.11.3.4 Two-port measurements

If transmission measurements are considered, thus including insertion loss, the analysis is more complex: two ports are involved, so there are more combinations of mismatch error signals, some dependent on DUT characteristics; moreover, the range of measured levels is large, practically from 0 down to a minus hundred of dB and the signal-to-noise ratio becomes a significant factor. The measurement of active devices featuring a positive gain widens the range of levels and brings in the power level as an additional factor; however, this will not be considered further.

When measuring very small or very large insertion loss values, the largest errors are observed. For example, when measuring low values of insertion loss of about a tenth of dB or less, mismatch errors and drifts after calibration cause positive values to be read as if the passive device has gain at those frequencies. Mid-range values (6 to 30 dB) effectively reduce mismatch errors and results in the best measurement behavior.

When considering small attenuation such as for high quality short cables or PCB transmission lines, or when verifying a Through, even a Port Match of 30 dB at each VNA port (that is Source and Load Port Match), can cause fluctuations of some hundreds of dB, thus crossing the 0 dB line at some frequencies.

For larger attenuation the effect of noise (and resulting limited signal-to-noise ratio) comes into play: the adopted countermeasures reduce the IF bandwidth (e.g. 100 Hz or lower) and/or averaging of VNA trace (that we know reduces the variance of the measured noise trace, see sec. 9.2.10.1 and 9.3.1). When manufacturers declare equipment uncertainty or performance, quite often the IF bandwidth is the narrowest attainable and averaging is heavily used, (see sec. 11.11.2).

The main terms that cause error of a two-port insertion loss measurement are graphically represented in Figure 11.11.3 [14]. The first term on top is the desired insertion loss reading, while the successive terms next to the arrows entering the "VNA port 2" block represent the effect of further reflections due to non-ideal port match at the VNA or the DUT, having indicated with a simple subscript "1" or "2" the terms for the VNA ports 1 and 2 and "D1" and "D2" the terms for the DUT seen from the port towards VNA port 1 and towards VNA port 2. The resulting error terms sum with arbitrary phase relationships, together with the external noise term V_n. For each term a linear quantity may be calculated even if original quantities are expressed in dB and summed together to obtain an all-comprehensive expression that collects transmitted and reflected terms (through one or multiple bounces) and noise terms: DUT return and insertion loss are thus transformed into an equivalent reflection coefficient ρ_D or transmission coefficient T, whereas for VNA ports only reflection coefficients ρ_1 and ρ_2 are defined; the noise term may be quantified separately or calculated from the signal-to-noise ratio value and the applied signal amplitude V.

The "wanted" received amplitude for a transmission measurement is $\mathrm{V}\,T$; all other terms due to reflections contribute to an overall error term V_{err}:

$$V_m = \mathrm{V}\,T + V_{err} \tag{11.11.10}$$

Looking at Figure 11.11.3, the sum of all terms received at port 2 (represented by arrows entering the right-hand block of VNA port 2) gives:

$$V_{err} = \mathrm{V}\left[T\,\Gamma_2\Gamma_{D2} + T^3\Gamma_1\Gamma_2 + T\,\Gamma_{D1}\Gamma_1 + T\,\Gamma_{D1}\Gamma_1\Gamma_{D2}\Gamma_2\right] + V_n \tag{11.11.11}$$

where quantities are complex and sum with arbitrary phase relationship.

It may be easily seen that for low-loss DUTs transmission is nearly unity and reflection coefficients are small, maybe of the same order of magnitude of VNA port reflection, so that no terms may be neglected and simplified; conversely, for a non-negligible loss (e.g. the DUT is an attenuator) the second term with T^3 might be less influential and in general also DUT reflection coefficients are smaller and prevail on the VNA ones.

The worst-case composition of terms (all summed in phase with positive or negative sign) gives the expected peak-to-peak ripple of the trace while sweeping frequency.

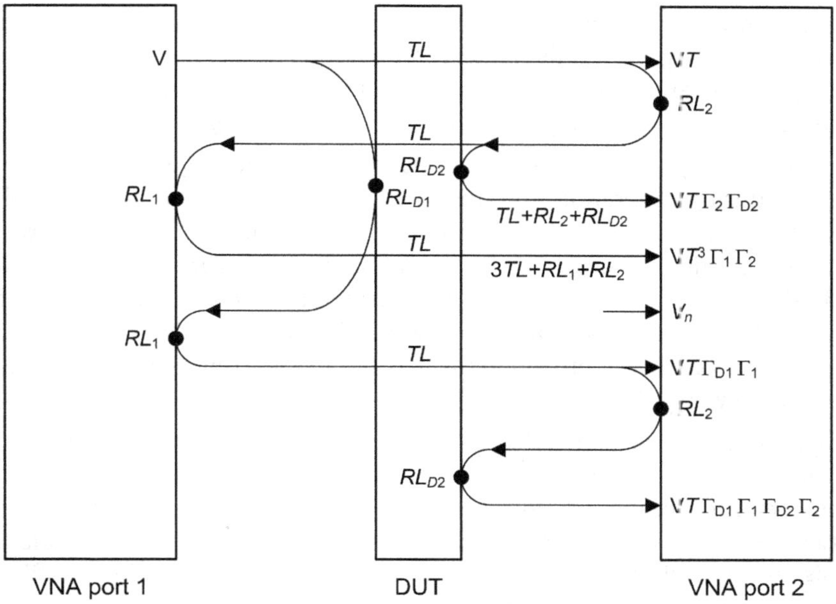

Figure 11.11.3 – Diagram of reflected and transmitted terms causing error in two-port insertion loss measurement.

11.11.4 Propagation of uncertainty

Propagation of uncertainty as indicated by GUM requires the formulation of an uncertainty model and the calculation of sensitivities of output quantities with respect to basic quantities. The fact that the involved quantities are complex quantities subject to constraints regarding their magnitude and underlying correlation between components makes the analysis more difficult; the problem is being addressed considering the best procedures to follow, suitable graphical representations and statistical properties of derived quantities (please see sec. 10.5). The uncertainty VNA model is usually drawn based on the already discussed calibration error model (see sec. 11.5), mostly the 8-term model, but also the 10- or 12-term ones, and even the full 16-term; the specific calibration method determines how the calibration standards are "used" to quantify the error terms and thus how the uncertainty of calibration standards diffuses to the said error terms, that in turn lead to the corrected measured quantities, starting from the raw measured quantities.

Sensitivity coefficients are the usual way of formally approaching the propagation of uncertainty (in jargon, LPU, Law of Propagation of Uncertainty), as indicated in the Guide to the Measurement of Uncertainty [39]. With a general first-order approximation (usually named "Bode sensitivity") the sensitivity coefficients are based on first-order derivatives. Their calculation doesn't straightforwardly allow their use in the propagation of uncertainty, unless some conditions are met. Depending on the used methods and tools, it might be necessary to assume that the quantities are independent, including the real and imaginary part of the same complex variable, usually receiving an equal uncertainty (see e.g. VNA Tools II in sec. 11.5.3.9). These conditions are rather strict and hard to fulfill completely: the real and imaginary part of a variable are only a representation and are inextricably linked by the underlying physical process and by e.g. the mathematical conservation of the intensity while the phase is rotating; moreover, assuming equal uncertainty may work in general, but not when there are extreme variations e.g. at resonances.

We saw that for one-port and two-port measurements (and in general for N-port measurements), during calibration, the measured wave quantities a and b are related to the calibration standard quantities by multiplying for a left and a right matrix called $\mathbf{T}_{\mathrm{A}}$ and $\mathbf{T}_{\mathrm{B}}$, containing e.g. the error terms of the 8-term error model. The calibration method calculates the error terms so that measured and original quantities of calibration standards "coincide"; we said that the residual errors, indicated by a leading δ and due to unavoidable approximations and inaccuracy, represent the "errors" from the uncertainty viewpoint.

Since calibration standards are affected by uncertainty regarding the "true" S-parameter values, one approach may prefer to express error box uncertainties starting from the S parameters of calibration standards; conversely one may argue that the accessible quantities are the measured ones at VNA reflectometers and that they should be used as a basis for uncertainty expressions. In general, being the uncertainty small, the two formulations may be considered equivalent.

11.11.4.1 Calibration uncertainty of one-port reflection measurements

The simpler case of a one-port return loss measurement lends itself to be treated analytically to evaluate uncertainty related to calibration standards [345]. The approach starts from assumed known calibration standards as for their reflection coefficient and uncertainty. When performing one-port measurements of DUT impedance Z, the error in the determination of the resulting reflection coefficient Γ with respect to the reference impedance Z_0 may be put in relationship with the errors in the calibration standards used for VNA calibration. The measured reflection coefficient is indicated as Γ_m (that is the measured S_{11}, or S_{11m}) and the error terms of the one-port calibration model are directivity error D $(=E_{DF})$, source match error M $(=E_{SF})$ and frequency response error R $(=E_{TF})$. We have seen that when applying three known calibration standards (their reflection coefficients being Γ_1, Γ_2 and Γ_3) errors are determined based on the three corresponding readings Γ_{m1}, Γ_{m2} and Γ_{m3}. Therefore, when measuring the DUT impedance, its reflection coefficient Γ is determined from these six basic quantities used in the calibration plus the measured reflection coefficient Γ_m.

The so determined DUT Γ may be put in relationship with the error quantities above and the measured reflection coefficient Γ_m as

$$\Gamma = \frac{\Gamma_m - D}{M\,(\Gamma_m - D) + R} \tag{11.11.12}$$

rewritten more explicitly as

$$\Gamma\Gamma_m M - \Gamma M D + \Gamma R - \Gamma_m + D = 0 \tag{11.11.13}$$

Using this expression for the three calibration standards one at a time, as they are connected during the calibration, it is possible to express D, M and R as a function of the three standards themselves:

$$D=\frac{1}{F}\left[\Gamma_{m1}\Gamma_{m2}\Gamma_3(\Gamma_1-\Gamma_2)+\Gamma_{m2}\Gamma_{m3}\Gamma_1(\Gamma_2-\Gamma_3)+\Gamma_{m3}\Gamma_{m1}\Gamma_2(\Gamma_3-\Gamma_1)\right] \quad (11.11.14)$$

$$M=\frac{1}{F}\left[\Gamma_{m3}(\Gamma_2-\Gamma_1)+\Gamma_{m1}(\Gamma_3-\Gamma_2)+\Gamma_{m2}(\Gamma_1-\Gamma_3)\right] \quad (11.11.15)$$

$$R=\frac{1}{F^2}\left[(\Gamma_1-\Gamma_2)(\Gamma_{m1}-\Gamma_{m2})(\Gamma_2-\Gamma_3)(\Gamma_{m2}-\Gamma_{m3})(\Gamma_3-\Gamma_1)(\Gamma_{m3}-\Gamma_{m1})\right] \quad (11.11.16)$$

with

$$F=\Gamma_{m3}\Gamma_3(\Gamma_2-\Gamma_1)+\Gamma_{m1}\Gamma_1(\Gamma_3-\Gamma_2)+\Gamma_{m2}\Gamma_2(\Gamma_1-\Gamma_3) \quad (11.11.17)$$

Evidently, the expression is written in order to make explicit index rotation.

Then applying the differential operator eq. (11.11.13) in order to have explicitly the variation (or uncertainty) of each quantity, the error in the correction of the three error quantities D, M and R may be determined. The expression "error in the correction of the three error quantities" looks circular and not clear, but indicates the error made in determining these three quantities (as we know, named "errors") during the calibration, quantities that are compensated for during successive VNA measurements; if their determination would be error-free (in the sense of the metrological error), then the corrected VNA measurements would be exact as far as the uncertainty of calibration standards and their measurement during calibration itself.

$$(1-\Gamma M)\,\delta D+\Gamma(\Gamma_m-D)\,\delta M+\Gamma\,\delta R+\left[R+M(\Gamma_m-D)\,\delta\Gamma+(\Gamma M-1)\,\delta\Gamma_m\right]=0 \quad (11.11.18)$$

Thus solving for each differential of D, M and R as depending on the three calibration standards

$$\begin{aligned}\delta D\ =&\frac{1}{F^2}\{(\Gamma_{m1}-\Gamma_{m2})(\Gamma_{m2}-\Gamma_{m3})(\Gamma_{m3}-\Gamma_{m1})\\ &[(\Gamma_2-\Gamma_3)\Gamma_2\Gamma_3\,\delta\Gamma_1+(\Gamma_3-\Gamma_1)\Gamma_3\Gamma_1\,\delta\Gamma_2+(\Gamma_1-\Gamma_2)\Gamma_1\Gamma_2\,\delta\Gamma_3]+\\ &+(\Gamma_{m2}-\Gamma_{m3})^2(\Gamma_2-\Gamma_1)(\Gamma_3-\Gamma_1)\Gamma_2\Gamma_3\,\delta\Gamma_{m1}+\\ &+(\Gamma_{m3}-\Gamma_{m1})^2(\Gamma_3-\Gamma_2)(\Gamma_1-\Gamma_2)\Gamma_3\Gamma_1\,\delta\Gamma_{m2}+\\ &+(\Gamma_{m1}-\Gamma_{m2})^2(\Gamma_1-\Gamma_3)(\Gamma_2-\Gamma_3)\Gamma_1\Gamma_2\,\delta\Gamma_{m3}\}\end{aligned} \quad (11.11.19)$$

$$
\begin{aligned}
\delta M = & \frac{1}{F^2}\{(\Gamma_{m1}-\Gamma_{m2})(\Gamma_{m3}-\Gamma_{m1})(\Gamma_2-\Gamma_3)^2\delta\Gamma_1+ \\
& +(\Gamma_{m2}-\Gamma_{m3})(\Gamma_{m1}-\Gamma_{m2})(\Gamma_3-\Gamma_1)^2\delta\Gamma_2+ \\
& +(\Gamma_{m3}-\Gamma_{m1})(\Gamma_{m2}-\Gamma_{m3})(\Gamma_1-\Gamma_3)^2\delta\Gamma_3+ \\
& -(\Gamma_1-\Gamma_2)(\Gamma_2-\Gamma_3)(\Gamma_3-\Gamma_1) \\
& [(\Gamma_{m2}-\Gamma_{m3})\delta\Gamma_{m1}+(\Gamma_{m3}-\Gamma_{m1})\delta\Gamma_{m2}+(\Gamma_{m1}-\Gamma_{m2})\delta\Gamma_{m3}]\}
\end{aligned}
\tag{11.11.20}
$$

$$
\begin{aligned}
\delta R = & \frac{1}{F^3}\{[F+2(\Gamma_{m1}-\Gamma_{m2})\Gamma_2(\Gamma_1-\Gamma_3)] \\
& \big[(\Gamma_{m1}-\Gamma_{m2})(\Gamma_{m2}-\Gamma_{m3})(\Gamma_{m3}-\Gamma_{m1})(\Gamma_2-\Gamma_3)^2\delta\Gamma_1+ \\
& -(\Gamma_1-\Gamma_2)(\Gamma_2-\Gamma_3)(\Gamma_3-\Gamma_1)(\Gamma_{m2}-\Gamma_{m3})^2\delta\Gamma_{m1}\big]+ \\
& +[F+2(\Gamma_{m2}-\Gamma_{m3})\Gamma_3(\Gamma_2-\Gamma_1)] \\
& \big[(\Gamma_{m1}-\Gamma_{m2})(\Gamma_{m2}-\Gamma_{m3})(\Gamma_{m3}-\Gamma_{m1})(\Gamma_3-\Gamma_1)^2\delta\Gamma_2+ \\
& -(\Gamma_1-\Gamma_2)(\Gamma_2-\Gamma_3)(\Gamma_3-\Gamma_1)(\Gamma_{m3}-\Gamma_{m1})^2\delta\Gamma_{m2}\big]+ \\
& +[F+2(\Gamma_{m3}-\Gamma_{m1})\Gamma_1(\Gamma_3-\Gamma_2)] \\
& \big[(\Gamma_{m1}-\Gamma_{m2})(\Gamma_{m2}-\Gamma_{m3})(\Gamma_{m3}-\Gamma_{m1})(\Gamma_1-\Gamma_2)^2\delta\Gamma_3+ \\
& -(\Gamma_1-\Gamma_2)(\Gamma_2-\Gamma_3)(\Gamma_3-\Gamma_1)(\Gamma_{m1}-\Gamma_{m2})^2\delta\Gamma_{m3}\big]\}
\end{aligned}
\tag{11.11.21}
$$

The total error in the determination of Γ is obtained by differentiating eq. (11.11.12)

$$
\delta\Gamma = \frac{-R\,\delta D-(\Gamma_m-D)^2\delta M-(\Gamma_m-D)\delta R+R\,\delta\Gamma_m}{[M(\Gamma_m-D)+R]^2}
\tag{11.11.22}
$$

When calibrating the VNA with a SOL method (where Short, Open and matched Load are used in the sequence 1, 2 and 3), using the relationships between D, M and R and the Γ coefficients of standards reported before in sec. 11.5.2.2 (shown here below for reader's ease), the expressions above simplify to:

$$
D=\Gamma_{m3} \qquad M=\frac{\Gamma_{m1}+\Gamma_{m2}-2\Gamma_{m3}}{\Gamma_{m2}-\Gamma_{m1}} \qquad R=2\,\frac{(\Gamma_{m1}-\Gamma_{m3})(\Gamma_{m3}-\Gamma_{m2})}{\Gamma_{m2}-\Gamma_{m1}}
\tag{11.11.23}
$$

$$
\delta D = -2\,\frac{(\Gamma_{m1}-\Gamma_{m3})(\Gamma_{m3}-\Gamma_{m2})}{(\Gamma_{m2}-\Gamma_{m1})}\,\delta\Gamma_3+\delta\Gamma_{m3}
\tag{11.11.24}
$$

$$\begin{aligned}\delta M =& \frac{\Gamma_{m1}-\Gamma_{m3}}{\Gamma_{m2}-\Gamma_{m1}}\delta\Gamma_1+\frac{\Gamma_{m3}-\Gamma_{m2}}{\Gamma_{m2}-\Gamma_{m1}}\delta\Gamma_2+4\frac{(\Gamma_{m3}-\Gamma_{m2})(\Gamma_{m1}-\Gamma_{m3})}{(\Gamma_{m2}-\Gamma_{m1})^2}\delta\Gamma_3+\\ &-2\frac{\Gamma_{m3}-\Gamma_{m2}}{(\Gamma_{m2}-\Gamma_{m1})^2}\delta\Gamma_{m1}-2\frac{\Gamma_{m1}-\Gamma_{m3}}{(\Gamma_{m2}-\Gamma_{m1})^2}\delta\Gamma_{m2}-2\frac{1}{\Gamma_{m2}-\Gamma_{m1}}\delta\Gamma_{m3}\end{aligned} \quad (11.11.25)$$

$$\begin{aligned}\delta R =& \frac{(\Gamma_{m1}-\Gamma_{m3})(\Gamma_{m3}-\Gamma_{m2})}{(\Gamma_{m2}-\Gamma_{m1})}\delta\Gamma_1-2\frac{(\Gamma_{m3}-\Gamma_{m2})^2}{(\Gamma_{m2}-\Gamma_{m1})^2}\delta\Gamma_{m1}+\\ &-\frac{(\Gamma_{m1}-\Gamma_{m3})(\Gamma_{m3}-\Gamma_{m2})}{(\Gamma_{m2}-\Gamma_{m1})}\delta\Gamma_2+2\frac{(\Gamma_{m1}-\Gamma_{m3})^2}{(\Gamma_{m2}-\Gamma_{m1})^2}\delta\Gamma_{m2}\\ &+4\frac{[\Gamma_{m2}-\Gamma_{m1}+2(\Gamma_{m3}-\Gamma_{m2})]\,(\Gamma_{m3}-\Gamma_{m2})(\Gamma_{m1}-\Gamma_{m3})}{(\Gamma_{m2}-\Gamma_{m1})^2}\delta\Gamma_3+\\ &-2\frac{[\Gamma_{m2}-\Gamma_{m1}+2(\Gamma_{m3}-\Gamma_{m2})]}{(\Gamma_{m2}-\Gamma_{m1})}\delta\Gamma_{m3}\end{aligned} \quad (11.11.26)$$

11.11.4.2 Calibration uncertainty of two-port measurements

For the case of a two-port measurement it is worth considering briefly two issues: the use of internal error-model coefficients as catch-all quantities to express uncertainty, and the distinction between theoretical values of calibration standards or measured accessible values. Sensitivity for propagation of uncertainty may be calculated with respect to either the S parameters of the calibration standards (thus accounting for uncertainty of their performance, of their model correction or of the connecting elements, e.g. cables and connectors) or more directly to error-model quantities (as considered in sec. 11.11.3.1 and 11.11.3.2 above).

Sensitivity to error-model quantities Under the assumption that error terms E are small or unity[32] (as appropriate for a nearly ideal VNA operation), a first-order simplification may be used: the complete expressions[33] of derivatives of eq. (11.5.20) to (11.5.23) in sec. 11.5.1.2 may be simplified significantly, obtaining the following for reflection and transmission measurement (the other coefficients being nearly zero):

$$\begin{aligned}
&\frac{\partial S_{11}}{\partial E_{DF}} \simeq -1 \quad \text{(I)} && \frac{\partial S_{21}}{\partial E_{SF}} \simeq -S_{11}^m S_{21}^m \quad \text{(V)}\\
&\frac{\partial S_{11}}{\partial E_{SF}} \simeq -(S_{11}^m)^2 \quad \text{(II)} && \frac{\partial S_{21}}{\partial E_{LF}} \simeq -S_{22}^m S_{21}^m \quad \text{(VI)}\\
&\frac{\partial S_{11}}{\partial E_{RF}} \simeq -S_{11}^m \quad \text{(III)} && \frac{\partial S_{21}}{\partial E_{TF}} \simeq -S_{21}^m \quad \text{(VII)}\\
&\frac{\partial S_{11}}{\partial E_{LF}} \simeq -S_{21}^m S_{12}^m \quad \text{(IV)} && \frac{\partial S_{21}}{\partial E_{XF}} \simeq -1 \quad \text{(VIII)}
\end{aligned} \quad (11.11.27)$$

[32] For clarity this means assuming $E_{DF}\simeq 0$, $E_{SF}\simeq 0$, $E_{LF}\simeq 0$, $E_{RF}\simeq 1$, $E_{TF}\simeq 1$ and analogously $E_{DR}\simeq 0$, $E_{SR}\simeq 0$, $E_{LR}\simeq 0$, $E_{RR}\simeq 1$, $E_{TR}\simeq 1$.

[33] The complete expressions are available in [143] as the result of calculation with a symbolic math tool.

The overall uncertainty of reflection S_{11} or transmission S_{21} quantities is obtained by separating the differentials, or in other words by multiplying each sensitivity for the corresponding error of the independent quantity. For transmission measurement Hall suggests to use immediately relative components of uncertainty, dividing the expressions (V) to (VIII) in eq. (11.11.27) by S_{21m}.

Sensitivity with respect to calibration standards Stumper [312, 313] wrote ten years ago two papers on the uncertainty of VNA full two-port measurements depending on the calibration procedure (SOLT or TRL), identifying the uncertainty that characterizes the measurement of DUT S parameters. The papers are quite compact and report both theoretical approach and some sample measurements for experimental confirmation. Not all expressions are reported and their implementation requires some hand work to calculate the last derivatives and to put plain on paper all expressions for all configurations. By the way Stumper's expressions are the same as those appearing in EURAMET Guide [123]. As pointed out by Hall [143], Stumper's equations were expressed in terms of DUT S parameters, rather than measured S parameters; the reason is that Stumper was linking the imperfections of the calibration standards to the resulting uncertainty in the corrected measurements returned by the VNA. As said, assuming a general satisfactory accuracy, DUT S parameters may be approximated with measured S parameters, and thus the two formulations are inter-related with only a change of sign of expressions. The reason, of course, of using measured S parameters rather than the calibration ones, is that they are accessible, even when measuring a generic DUT whose characteristics are not known until measured.

S_{jk} indicate the S parameters that characterize the DUT, while m_{jk} are the raw measurements of the same parameters before correction from VNA calibration is applied. Raw measurements of S_{jk} parameters are in reality called "normalized raw values" in [312], and are related to the raw values read by the digitizing IF module in a four-sampler full two-port measurement: $m_{11} = m_1/m_2$, $m_{21} = m_4/m_2$, $m_{12} = m_1'/m_3'$ and $m_{22} = m_4'/m_3'$, where Stumper uses the prime to indicate measurements referred to port 2 as the active port (reverse), while without the prime the active port is port 1 (forward).

SOLT calibration The three SOL standards have three reflection coefficients M_i, K_i and L_i respectively, with $i = 1, 2$ indicating the VNA port; their deviations (or errors, for the determination of the uncertainty) are indicated by a δ in front of the respective symbols: δM_i, δK_i and δL_i. In reality distinguishing for the VNA port by means of the subscript i is of no practical use, being the configuration symmetrical.

$$m_{11} = \frac{[S_{11}(e_R - e_D e_S) + e_D]\,(1 - S_{22} e_L) + S_{21} S_{12} e_L (e_R - e_D e_S)}{F} \qquad (11.11.28)$$

$$m_{21} = \frac{S_{21} e_T}{F} \qquad (11.11.29)$$

where

$$F = (1 - S_{11}e_S)(1 - S_{22}e_L) - S_{21}S_{12}e_Se_L \tag{11.11.30}$$

What is affected by deviations in the calibration standard coefficients M_i, K_i and L_i ($i = 1, 2$ is the VNA port index) is the correction applied to the raw measurement m giving the final values of the S parameters S_{jk}. Raw measurements remain the same and thus their deviation is assumed to be zero (of course, non linearity or noise may affect raw measurements, but they are neglected). Thus it is possible to write two expressions for the total differentials equating the deviation of m to zero:

$$\begin{aligned} 0 = \delta m_{11} &= \left(\frac{\partial m_{11}}{\partial S_{11}}\right)\delta S_{11} + \left(\frac{\partial m_{11}}{\partial S_{22}}\right)\delta S_{22} + \cdots \\ &+ \left(\frac{\partial m_{11}}{\partial e_D}\right)\delta e_D + \left(\frac{\partial m_{11}}{\partial e_S}\right)\delta e_S + \cdots \end{aligned} \tag{11.11.31}$$

$$\begin{aligned} 0 = \delta m_{21} &= \left(\frac{\partial m_{21}}{\partial S_{11}}\right)\delta S_{11} + \left(\frac{\partial m_{21}}{\partial S_{22}}\right)\delta S_{22} + \cdots \\ &+ \left(\frac{\partial m_{21}}{\partial e_D}\right)\delta e_D + \left(\frac{\partial m_{21}}{\partial e_S}\right)\delta e_S + \cdots \end{aligned} \tag{11.11.32}$$

extended to the four S parameters (S_{11}, S_{22}, S_{21} and S_{12}) and to the five error terms (e_D, e_S, e_R, e_L and e_T). The five error terms are those of the VNA calibration model already examined in sec. 11.5.1: e_D is the directivity error, e_S is the source match error, e_R is the reflection tracking error, e_L is the load match error and e_T is the transmission tracking error. These five quantities are those referred to port 1 and are complemented by another five quantities with prime, referred to port 2, for a total of ten terms of the calibration error model.

Analogously, thus, for port 2 there may be defined two "m" equations and two total differentials equations, having exchanged subscript "1" with "2" and terms with prime with those without, and viceversa.

$$m_{22} = \frac{\left[S_{22}(e'_R - e'_De'_S) + e'_D\right](1 - S_{11}e'_L) + S_{21}S_{12}e'_L(e'_R - e'_De'_S)}{F'} \tag{11.11.33}$$

$$m_{12} = \frac{S_{12}e'_T}{F'} \tag{11.11.34}$$

where

$$F' = (1 - S_{22}e'_S)(1 - S_{11}e'_L) - S_{21}S_{12}e'_Se'_L \tag{11.11.35}$$

and the total differentials become

$$
\begin{aligned}
0=\delta m_{22} &= \left(\frac{\partial m_{22}}{\partial S_{11}}\right)\delta S_{11}+\left(\frac{\partial m_{22}}{\partial S_{22}}\right)\delta S_{22}+\cdots \\
&+\left(\frac{\partial m_{22}}{\partial e'_D}\right)\delta e'_D+\left(\frac{\partial m_{11}}{\partial e'_S}\right)\delta e'_S+\cdots
\end{aligned}
\tag{11.11.36}
$$

$$
\begin{aligned}
0=\delta m_{12} &= \left(\frac{\partial m_{12}}{\partial S_{11}}\right)\delta S_{11}+\left(\frac{\partial m_{12}}{\partial S_{22}}\right)\delta S_{22}+\cdots \\
&+\left(\frac{\partial m_{12}}{\partial e'_D}\right)\delta e'_D+\left(\frac{\partial m_{12}}{\partial e'_S}\right)\delta e'_S+\cdots
\end{aligned}
\tag{11.11.37}
$$

Deviations of error terms are related back to deviations of calibration standards for a total of four SOLT standards: there will be six deviations for the reflection terms (SOL) referred to the two VNA ports, namely δM_1, δM_2, δK_1, δK_2, δL_1 and δL_2, and four deviations of the S parameters of the Through standard, namely δS^T_{11}, δS^T_{21}, δS^T_{12} and δS^T_{22} (where the capital T stands for Through and not for tracking, that is a subscript).

$$m^M_{11}=\frac{M_1(e_R-e_Se_D)+e_D}{1-M_1e_S} \tag{11.11.38}$$

$$m^K_{11}=\frac{K_1(e_R-e_Se_D)+e_D}{1-K_1e_S} \tag{11.11.39}$$

$$m^L_{11}=\frac{L_1(e_R-e_Se_D)+e_D}{1-L_1e_S} \tag{11.11.40}$$

$$m^T_{11}=\frac{\left[S^T_{11}(e_R-e_De_S)+e_D\right](1-S^T_{22}e_L)+S^T_{21}S^T_{12}e_L(e_R-e_De_S)}{G} \tag{11.11.41}$$

$$m^T_{21}=\frac{S^T_{21}e_T}{G} \tag{11.11.42}$$

where

$$G=(1-S^T_{11}e_S)(1-S^T_{22}e_L)-S^T_{21}S^T_{12}e_Se_L \tag{11.11.43}$$

Finally the sensitivity coefficients between DUT S parameters and deviations of the used calibration standards are calculated by combining the above expressions. Stumper reports the sensitivity of the four S_{jk} parameters (S_{11}, S_{21}, S_{12} and S_{22}) with respect to the deviation of the Matched load δM_1 and δM_2, indicating how to calculate the other sensitivities for the Short and Open standards, that is for δK_1 and δK_2, and for δL_1 and δL_2, respectively.

For the influence of Through there are more expressions to consider, given by the combination of the four S_{jk} DUT parameters with the four Through S^T_{jk} parameters and only the error term e_L.

$$\delta S_{11} = \frac{S_{12}S_{21}(1-S_{11}e'_L - e_L e'_L)}{N}\,\delta S^T_{11} \tag{11.11.44}$$

$$\delta S_{11} = \frac{-S_{11}S_{12}S_{21}e_L e'_L}{N}\,\delta S^T_{12} \tag{11.11.45}$$

$$\delta S_{11} = \frac{-S_{11}S_{12}S_{21}e_L e'_L}{N}\,\delta S^T_{21} \tag{11.11.46}$$

$$\delta S_{11} = \frac{-S_{11}S_{12}S_{21}e_L}{N}\,\delta S^T_{22} \tag{11.11.47}$$

$$\delta S_{12} = \frac{S_{12}e'_L(1-S_{22}e_L - S_{21}S_{12})}{N}\,\delta S^T_{11} \tag{11.11.48}$$

$$\delta S_{12} = \frac{S_{12}(1-S_{22}e_L - S_{21}S_{12}e_L e'_L)}{N}\,\delta S^T_{12} \tag{11.11.49}$$

$$\delta S_{12} = \frac{S_{11}S_{12}e'_L(1-S_{22}e_L)}{N}\,\delta S^T_{21} \tag{11.11.50}$$

$$\delta S_{12} = \frac{S_{11}S_{12}(1-S_{22}e_L)}{N}\,\delta S^T_{22} \tag{11.11.51}$$

The remaining relationships for the VNA port 2 are obtained, as indicated by Stumper, by exchanging subscripts 1 and 2 and primed with non-primed e quantities.

$$\delta S_{22} = \frac{S_{12}S_{21}(1-S_{22}e_L - e_L e'_L)}{N}\,\delta S^T_{22} \tag{11.11.52}$$

$$\delta S_{22} = \frac{-S_{22}S_{12}S_{21}e_L e'_L}{N}\,\delta S^T_{21} \tag{11.11.53}$$

$$\delta S_{22} = \frac{-S_{22}S_{12}S_{21}e_L e'_L}{N}\,\delta S^T_{12} \tag{11.11.54}$$

$$\delta S_{22} = \frac{-S_{22}S_{12}S_{21}e'_L}{N}\,\delta S^T_{11} \tag{11.11.55}$$

$$\delta S_{21} = \frac{S_{21}e_L(1-S_{11}e'_L - S_{21}S_{12})}{N}\,\delta S^T_{22} \tag{11.11.56}$$

$$\delta S_{21} = \frac{S_{21}(1-S_{11}e'_L - S_{21}S_{12}e_L e'_L)}{N}\,\delta S^T_{21} \tag{11.11.57}$$

$$\delta S_{21} = \frac{S_{22}S_{21}e_L(1-S_{11}e'_L)}{N}\,\delta S^T_{21} \tag{11.11.58}$$

$$\delta S_{21} = \frac{S_{22}S_{21}(1-S_{11}e'_L)}{N}\delta S^T_{11} \qquad (11.11.59)$$

The paper concludes reporting experimental results for a range of calibration standards that may be assumed typical, yes, but of a metrology laboratory level of performance.

11.11.4.3 Time-domain extraction of one-port calibration errors

Wubbeler et al. [343] report a method for correction of one-port calibrations, that determines explicitly complex directivity and source match residuals of the error model using a second-order function and may be used to verify VNA calibration and to correct measurements.

The publication is first of all very useful because clearly and succinctly reviews the frequency and time domain relationships for a one-port reflection measurement, where the error terms of the one-port error model find their place in the interpretation of spectra and waveforms. The authors explain the post-processing of a reflection measurement of a short-circuited airline with reflection coefficient Γ_a, performed with a SOL-calibrated VNA; the frequency-domain ripple, already commented in sec. 7.1.1.7 for test of connector defects, is analyzed for its constitutive terms[34]:

$$\Gamma_m = \delta + (1+\tau)\Gamma_a + \mu(1+\tau)\Gamma_a^2 \qquad (11.11.60)$$

where the three parameters are related to the residual VNA error-model parameters as $\delta = \delta E_{DF}$, $\mu = \delta E_{SF}$, $\tau = \delta E_{RF}$, i.e. directivity, port match and reflection tracking error.

Remembering that for generality airline losses shall be accounted for, the propagation constant is written as $\gamma = \alpha + j\beta = \alpha + j\omega/v$ and $\Gamma_a = -S_{12}S_{21} = -e^{-2\alpha l}e^{-2j\omega l/v} = -\lambda e^{-2j\omega l/v}$, having thus assumed that the short circuit is perfect with a unity reflection coefficient and that any deviation is due to the airline insertion loss, assumed symmetrical.

$$\Gamma_m = \delta - (1+\tau)\lambda e^{-2j\omega l/v} + \mu(1+\tau)\lambda^2 e^{-4j\omega l/v} \qquad (11.11.61)$$

The three terms are then identified as A, B and C and it is observed that for a sufficiently long airline the two frequency-dependent phase shifts are significantly different to distinguish the three components: when transformed into time domain, they become three distinct time-delayed peaks. Limited time resolution, some leakage in time-frequency transformations and transmission line dispersion (minimal when using an airline) smear and widen such peaks, but they remain visible when inspecting the time-domain waveform. The authors, however, propose a method for the auto-

---

34 This expression is obtained by simplification of the more complete equation $\Gamma_m = \delta + (1+\tau)\frac{\Gamma_a}{1-\mu\Gamma_a}$, where the $(1+\tau)$ term is preserved in front of all terms containing Γ_a. EURAMET [123] on the contrary reports its "voltage reflection coefficient" expression dropping the $(1+\tau)$ term for the Γ_a^2 term and similarly [115], eq. [3.68].

matic determination of the three terms, based on successive down-conversions and low-pass filtering; to ease calculations and improve resolution, extrapolation by fitting an autoregressive model and enforcing data symmetry before anti-transforming are proposed. Following the steps described in section II.B of the paper, one is able to setup an automatic method for the extraction of peak amplitudes.

Besides the availability of an automatic method, what is important is how to determine the peak height with minimum error, even when visually inspecting the time-domain TDR waveform: without stressing the issue, the authors perform sample averaging by low-pass filtering the down-converted terms, thus reducing the error that may be done when isolating a single value in the waveform peaks. Second, quite cleverly they underline that avoiding to use directly the λ value (that might be only approximately known) increases the accuracy, and they obtain this by up-converting and then subtracting directly the estimated intermediate term of eq. (11.11.61). However, an additional measurement is needed for an accurate estimate of τ, that should be otherwise assumed zero.

When the results are evaluated by comparing $\tilde{\Gamma}_a$ calculated from eq. (11.11.60), once the error-model parameters are estimated, with the "true" value of the short circuit terminated airline[35], the authors underline that the most relevant algorithm parameter is the width of the low-pass filter. The order of the autoregressive model and the amount of data samples extrapolated and added to improve the resolution are two other less important parameters, that however were set to nearly optimal values with a trial and error approach (order 3 and 30% of data added to each side of the original vector, respectively). The rms error of $\left|\tilde{\Gamma}_a - \Gamma_a\right|$ confirms that the method is quite helpful in correcting low-grade (or low-quality) calibrations: 10 to 20 dB of improvements are visible, with the rms error going from $-40\,\mathrm{dB}$ to nearly $-60\,\mathrm{dB}$; when the VNA is effectively and accurately calibrated the improvement is minimal, but still perceivable (let's say up to 8 dB).

The same method is explored by Savin et al.[36], who propose a Kalman filter implementation based on the unscented transform and analyze the effect on time resolution of a reduced number of samples and of different smoothing windows (for frequency leakage reduction).

11.12 Verification of VNA calibration and accuracy

A few tests to use as routine checks for verification of correct calibration of VNA, of its accuracy and in general of its raw performance are extremely useful. For a matter of clarity distinction is made between return loss (reflection) verifications and insertion loss (transmission) verifications.

---

35 Determined using a traceable cross-ratio (or quarter-wave) method.

36 A.A. Savin, V.G. Guba and B.D. Maxson, "Residual Errors Determination for Vector Network Analyzer at a Low Resolution in the Time Domain," 82nd ARFTG Microwave Measurement Conference, Nov. 18-21, 2013, Columbus, OH, USA, pp. 1-5. doi: 10.1109/ARFTG-2.2013.6737335

11.12.1 Return loss (reflection) verifications

11.12.1.1 VNA port left open

The first experiment is to perform a measurement with VNA port left open; it shall result into a 0 dB S_{11} value within a few dB. Why a few dB and not something closer to zero? All depends on the calibration that the VNA is storing and using: if the N port of the VNA was calibrated with a standard N-type SOLT calibration set, it is quite likely that, leaving the VNA port open, the result will be quite close. On the contrary, if offset calibration standards were used, or adapters were included changing e.g. to a 3.5 mm type calibration set, then measuring at the VNA port plane will give quite a different result for the offset and for the inclusion of the adapter response. If for each case a calibration standard is used with the correct connector type, then S_{11} will always be quite close to 0 dB. The S_{11} curve will show an oscillation that is a combination of physical length and reference plane position; because of directivity error the S_{11} curve may be not entirely negative as it might be expected and physically reasonable.

11.12.1.2 Termination on a matched load

For a perfect matched load the correct S_{11} value is toward $-\infty$. In practice we will read the effective directivity of the port, about $-30/-40$ dB. The problem of the availability of a good matched load for a broad frequency range reflects into the impossibility of assessing directivity values better than about $-30/35$ dB. Depending on the IF bandwidth, also the amount of incoming noise will worsen this figure, resulting in a slight increase of the measured value.

11.12.1.3 Termination on a known mismatched load

This is the case when a calibration matched load for a different reference impedance is available, such as $75\,\Omega$. The resulting reflection coefficient is easy to calculate (in this case it is 0.2, or -13.98 dB). If needed, a resistive load may be built by taking a connector of the required type and soldering SMT resistors with circular symmetry around the central hot conductor. Of course, such a Do It Yourself practice may result in poor performance as the frequency increases: a terminated microstrip, for example, might be working satisfactorily up to a few or several GHz.

11.12.1.4 Port match measurement using offset shorts

The VNA Port Match might be in general more difficult to measure and the return-loss ripple technique (see Figure 11.11.2) may be exploited [263]. What is needed is an offset short and for correct interpretation of the peak-to-peak ripple over the set frequency interval, the offset length shall be such to have not too few nor too many complete phase rotations due to port match of reflection term vector, so to have a clearly visible ripple. A length of 1 to 3 cm works well over the 1 to 18 GHz frequency interval (similarly to what was already observed for the characterization of connectors and adapters in sec. 7.1.1.7).

Why an offset short? If we are going to use the same Short standard used in the calibration, any systematic error due to incorrect calibration (or any other error) will compensate again and cancel out; any other well defined termination (such as an offset Short from the same or another calibration kit) will on the contrary be effective.

The resulting port match is obtained by reading the observed peak-to-peak ripple and finding the corresponding port match in Figure 11.11.2; the readability of the curves may be improved if a pair of offset Short and Open is measured, with unavoidable frequency shifts, however, due to slight offset differences. Again the noise might affect the accuracy of the estimate, so that narrow IF bandwidth and trace averaging help in reducing the uncertainty.

In reality, what is measured is the combination of the residual directivity and port match, the latter involved for the use of a reflect standard.

11.12.2 Insertion loss (transmission) verifications

11.12.2.1 Through cable or barrel

If a good quality through is available (e.g. a low-, or well known, attenuation cable, or a barrel to connect two port extensions), the resulting S_{21} will be very small and nearly 0 dB. What a short barrel does is to offset the reference plane and it is thus often characterized by its phase delay: when such through is part of a calibration kit (e.g. for non-insertable devices, as part of a set of phase equal insertables), then its delay is quite accurately declared. Of course the quality of such Through once inserted in the measurement setup is measured also by its repeatability (as said, sensitivity of the overall setup to flexure, bending and shaking).

High-performance oscilloscopes are often equipped with a calibration kit for channel gain calibration and deskew when used for TDR applications: one of the supplied parts is a flexible cable long enough to connect the farthest connectors (e.g. 50 cm) always coming with a certified attenuation.

11.12.2.2 Cable and attenuator

If a good attenuator is available (e.g. calibrated, with known frequency response, as those included in verification kits), a combined cable + attenuator test may be performed, looking substantially for the attenuator insertion loss (in the presence of at least 6 dB of attenuation, there is no need for an accurate determination of the cable insertion loss). This is probably the most popular test that is made to verify VNA calibration and performance: the attenuator offers a well-matched impedance and is quite stable to be used as reference for a long time, provided that it is stored adequately and the environment temperature is not so different from that of its calibration/certification.

Moreover, the adoption of an attenuator as an impedance matching device is also favored by the fact that some attenuation between ports removes the effects of port match at the remote port (i.e. load port match) when performing a verification of return loss.

11.12.2.3 Isolation test removing through connection

If both VNA ports are correctly terminated onto matched loads and not connected by any through connection, the measurement of S_{21} gives an indication of leakage and effective isolation between VNA ports. If matched loads are not used, reflections might worsen the results, even if some degree of correction may be applied; in general, well matched attenuators may replace matched loads.

Well-matched attenuators may be verified first with a standard SOLT calibration, measuring return and insertion loss (see for example Figure 11.7.2). Then, once the input impedance is known to a given accuracy, VNA isolation can be checked and measured. Indeed, this is a check rather than a measurement, because the uncertainty in the original determination of the attenuator return loss is influenced by the VNA port isolation itself! (port isolation and insertion losses, as part of the port directivity error term, were considered in sec. 11.5.2.2)

The signal leaking through onto the reflectometers is in any case quite small and its measurement might be hindered by noise, so that the use of the lowest IF bandwidth and trace averaging are absolutely advisable: by the way this is simply a verification of abnormal situations and a general check of worst-case upper limits rather than an accurate measurement of port isolation.

11.12.3 Combined return and insertion loss verifications

11.12.3.1 T-check verification

T-check method was promoted by Rohde&Schwarz [157, 257] and a specific software for VNA verification now accompanies their equipment[37]. T-check is based on the particular form of the equations of a three-port device terminated onto a load, possibly matched, at one of the ports[38]. The use of C_T and the T-checker is a rapid verification of VNA calibration: ideally equal to unity, it is allowed to deviate by as much as $\pm 10\%$ while keeping a good accuracy, a further $\pm 5\%$ is marked in yellow to indicate acceptable, but critical accuracy, and outside this $\pm 15\%$ band the accuracy of the VNA calibration is compromised.

A straightforward implementation is that of separate devices for the T junction and the terminating load on port 3, e.g. a $50\,\Omega$ matched load; a more compact form is preferable for improved performance and to extend the frequency range. The T-checker may consist thus of a conductive-resistive path between hot and return conductors of a transmission line: it may be made with lumped resistors or conductive film deposition. When dealing with planar transmission lines, such as microstrips, the usual arrangement is that of two twin symmetrical resistors on each side going to the return

---

[37] It can be downloaded from the Rohde & Schwarz website: the name has evolved to VNAMUC®, that is VNA Measurement Uncertainty Calculator and among the various attempts a link that works is `http://www.rohde-schwarz.la/es/service_and_support/Downloads/Software/`.

[38] About the termination of port 3 Rohde & Schwarz gives different versions: they suggest $50\,\Omega$ as a possibility, but underline that it may be any impedance, even including reactive component [257]; in a presentation titled "Calibration techniques and measurement accuracy" they on the contrary indicate $30\,\Omega$ as a preferred value.

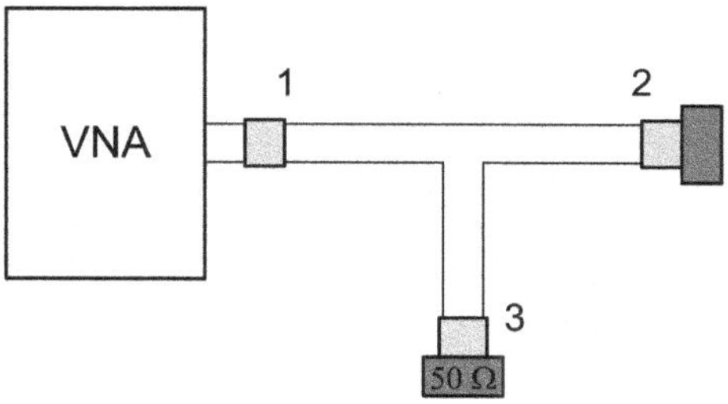

Figure 11.12.1 – T-checker connection and port numbering.

plane beneath through vias. If considering a coaxial line, a resistive disk may be created by film deposition on a supporting dielectric bead, that is then placed between the inner and outer conductors, with care to ensure a uniform and effective electric contact; this is what is offered by Rohde & Schwarz as coaxial type-N T-check device.

A generic T junction with 3x3 S-parameter matrix needs only to be lossless, without strictly requiring symmetry or reciprocity. A generally applicable unitary condition was reported in Chapter 10 as eq. (10.1.31). Applied to the three-port T junction, it gives two sets of relationships between its S parameters: the sum of squared magnitude of S parameters in a row (resulting from the product of a S parameter by its complex conjugate) gives one of the 1s on the unitary diagonal matrix, and the sum of the product of the S parameter in a row by those of a column with different number is zero, because leading to an off-diagonal term of the unitary matrix. In particular the following expressions are useful for the demonstration found in [257]; the selection of subscripts and port number is such that port 3 is the branched one, where the matched load is located, port 2 is the one at which the load (any load) is connected, and port 1 is the one connected to the VNA (see Figure 11.12.1).

$$|S_{11}|^2+|S_{12}|^2+|S_{13}|^2=1 \tag{11.12.1}$$

$$|S_{21}|^2+|S_{22}|^2+|S_{23}|^2=1 \tag{11.12.2}$$

$$S_{11}S_{21}^*+S_{12}S_{22}^*+S_{13}S_{23}^*=0 \tag{11.12.3}$$

After isolating the two terms $|S_{13}|^2$ and $|S_{23}|^2$ in the first two equations to be used later with the product $S_{13}S_{23}^*$ transformed in modulus in the third equation, it is obtained:

$$C_T=\frac{|S_{13}|\,|S_{23}|}{\sqrt{|S_{13}|^2\,|S_{23}|^2}}=\frac{|S_{11}S_{21}^*+S_{12}S_{22}^*|}{\sqrt{\left(1-|S_{11}|^2-|S_{12}|^2\right)\left(1-|S_{21}|^2-|S_{22}|^2\right)}}=\frac{N}{\sqrt{A\,B}} \tag{11.12.4}$$

and this ratio is by definition equal to unity, whatever the load connected to port 3. It is preferable anyway to avoid the use of highly reflective loads, because in that case $|S_{33}|$ would be close to unity and the remaining S_{13} and S_{23} too small.

		Uncertainty S_{11} %			
		0.01	0.02	0.05	0.1
Uncertainty S_{21} %	0.01	1.001 (0.0336) [0.100]	1.0015 (0.0425) [0.123]	1.004 (0.081) [0.170]	1.016 (0.156) [0.261]
	0.02	1.004 (0.0624) [0.177]	1.004 (0.0678) [0.191]	1.008 (0.097) [0.271]	1.02 (0.166) [0.350]
	0.05	1.026 (0.159) [0.424]	1.027 (0.161) [0.430]	1.03 (0.177) [0.515]	1.04 (0.229) [0.665]
	0.1	1.125 (0.382) [0.970]	1.122 (0.383) [1.000]	1.127 (0.395) [1.06]	1.15 (0.445) [1.27]

Table 11.12.1 – Statistical parameters (mean, (standard deviation), [skewness]) of C_T, resulting for the matched condition $Z = Z_0 = 50\,\Omega$, applying uniform distributions for the uncertainty of S_{11}, S_{21} S_{12} and S_{22}; only one reflection and one transmission S parameter is shown. Distributions were obtained with 100000 random generations.

Any deviation from unity of C_T is interpreted as a quantification of the VNA accuracy, ideally perfect for $C_T = 1$. What is the variation of C_T as a function of port 1 and 2 S parameters and in which conditions it is useful to verify VNA accuracy?

The maximum of C_T corresponds to maximum numerator and minimum denominator. It is evident that both terms in multiplication under square root A and B may be zero for plausible values of S_{11}, S_{12} and S_{21}, S_{22}. The T-checker is symmetric, so when measuring at its port 1 and 2

$$S_{11} = S_{22} = -\frac{Z_0}{2Z_3 + Z_0} \qquad S_{21} = S_{12} = \frac{2Z_3}{2Z_3 + Z_0} \tag{11.12.5}$$

where Z_3 is the impedance connected to port 3 and Z_0 the reference impedance, i.e. $50\,\Omega$.

Since the T-checker is a three-port device, port 1 and 2 S parameters alone do not fulfill the normality condition. For the matched condition $Z_3 = Z_0 = 50\,\Omega$ that gives $S_{11} = S_{22} = -1/3$ and $S_{21} = S_{12} = +2/3$, the value of C_T is studied in terms of distribution as a function of the S-parameter uncertainty; some results are shown in Table 11.12.1.

It is evident that for an accurate measurement characterized by a low uncertainty (up to 0.02), the resulting C_T distribution is sufficiently symmetric; for larger data dispersion all the distributions are heavily skewed and the mean value tends slightly to increase. The values of 0.02 and 0.05 correspond fairly well to the values of uncertainty of 0.2 and 0.4 dB assumed by Rohde & Schwarz.

11.12.3.2 Beatty standard

A Beatty standard may be built on a PCB support (e.g. with a microstrip or a stripline) or using cables of different characteristic impedance. Even if a Beatty standard is normally used to verify TDR measurements, its known return and insertion loss values may be used for a one-shot verification of VNA settings and calibration, provided that it is insertable in the current setup. A 50-25-50 Ω Beatty standard has a total return loss of -6 dB given by the multiple reflections, provided that a convenient multiple of the propagation time is taken as the dwell time for each frequency point (see sec. 11.10.3.5); the insertion loss, of course, is exactly the same, that is -6 dB.

A satisfactory Beatty standard may be built with affordable PCB techniques, determining its characteristics by construction: high performance materials are available operating up to several GHz and line geometry may be quite accurately defined with modern PCB CAD programs and PCB fabrication techniques (see microstrips and striplines at sec. 6.3.1 and sec. 6.3.4, respectively).

One example of a 25 ΩBeatty standard appears in [301], where a 1-inch of 25 Ω inner section is shown with two feeding 50 Ω microstrip sections long enough to support two reference planes cut at a quarter-inch distance from the 25 Ω discontinuities. One of the most relevant things observed is that the dielectric permittivity is not constant and needs some adjustment; the errors in determining the resonance frequency of the 1-inch segment are not as large as expected because there is some compensation between microstrip dispersion and permittivity variation (if a stripline were used, the consequences of permittivity changes with frequency would have been more evident). The observed permittivity variation is 10% between 2 and 20 GHz; the material is a standard FR4 that matches a general $\varepsilon_r = 4.2$ specification at 1 GHz.

11.12.3.3 Directivity and port match with "assurance airline"

Directivity and Port Match of VNA port may be tested with an assurance airline adopting a time or a frequency domain technique: the height of reflection terms in the former and the amount of ripple in the latter give indication of the two parameters, or better of the residual error of the two parameters δE_{DF} and δE_{SF} after calibration (EURAMET calls them "effective" and clearly states that they correspond to the residual terms after calibration).

Time-domain measurement Anritsu [17] shows a very intuitive procedure to display directivity and port match by using a time-domain reflection measurement of an "assurance airline" terminated on a Short (or it may be an open circuit, leaving the airline simply open, but it may have in general poorer results, for two reasons: leaving the airline open is not like terminating it on an Open and beadless airlines need a termination to keep in place the center conductor). The mathematical base is that already shown in sec. 11.11.4.3 and that may be found in [343]. The signal from the VNA port (in impulse response measurement mode, and possibly low-pass mode to increase resolution) enters the airline and shows a first peak (corresponding to directivity), travels along the airline and is reflected at its end (this is the largest peak due to the short/open termination, as close to unity as possible) and then reaches back the VNA port where a fraction of it might be reflected back (port match).

In Figure 11.12.2 a simplified TDR waveform is drawn where the three main reflections can be seen: the reflection of the Short at the end of the airline has been exaggeratedly lowered, to indicate masking effect due to slight line losses; artifacts due to reflections

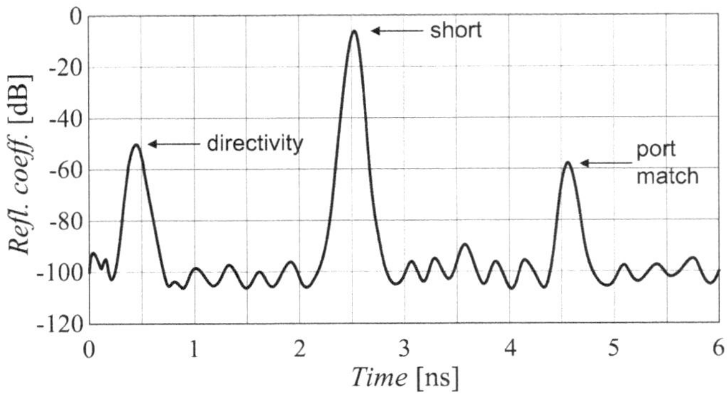

Figure 11.12.2 – Hand-drawn example of TDR for VNA verification with airline terminated on a short: directivity is assumed slightly worse than port match; noise floor and leakage terms have been conveniently kept low.

from the nearest connector entering the airline are always present, increasing the noise floor and reducing the visibility of the directivity peak (also because of the proximity to the beginning of time axis).

A brief comment is necessary on the representativeness of such a measurement: being a time domain curve, it is quite effective in spotting out the point of the test setup where each reflection term belongs to (by the way we know where reflections are coming from, except in case of mistakes), but the displayed value is the average across the swept frequency range in the original frequency-domain measurement. Cutting the whole frequency range into smaller intervals increases the confidence in assigning a frequency domain meaning to a specific time-domain value, but may have a significant impact on time (and spatial) resolution (see sec. 11.8).

While testing of directivity and port match is in principle easy, the accurate and reliable estimation of these two characteristics of VNA and its calibration cannot be better than the available standards. The airline and a calibration short are probably the best combination in terms of frequency extension and uncertainty. A cheaper alternative might be a carefully designed PCB transmission line, such as a stripline (see sec. 6.3.4), using high performance substrate materials and thus ensuring good performance and a stable characteristic impedance over an extended frequency range. Variability with temperature and other environmental factors is however an issue, much more than if using airlines. All depends on the use of the resulting information (confirmation, internal traceability, other calibrations), on the required performance (i.e. the intrinsic uncertainty of the method) and on the available budget.

An identical frequency-domain procedure is proposed by EURAMET, where the airline is doubly terminated on a matched Load and a Short.

Frequency-domain measurement As proposed by EURAMET [123], directivity and port match are tested in frequency domain using an assurance airline terminated on a matched Load and on a Short, respectively. The method is based on the quantification of the ripple resulting from the said residual errors, after having applied a correction, if necessary, for interacting factors (when more than one VNA parameter contributes to the error in a specific configuration). The ripple will show a periodicity that is given by the speed of light in the airline divided by twice the length (as already occurred for the evaluation of connector defects in sec. 7.1.1.7). EURAMET advices to have at least ten data points per ripple period: longer lines will cause many ripple periods to appear, tracking better any frequency dependency of the VNA parameter, but requiring a larger number of swept points. The ripple is always measured as peak-to-peak ripple between a pair of adjacent peak and trough (provided that the ripple period is short enough) and will be indicated as X_{pp}. Of course, numeric techniques of interpolation, as well as upsampling to increase the number of points are possible and advisable to increase the accuracy of results.

When terminating on Load for the Directivity measurement, the ripple X_{pp} appearing in the S_{11} shown with vertical linear scale (i.e. the reflection coefficient Γ_1) gives an estimate of the directivity error indicated by EURAMET with the letter D (δE_{DF}): $D = X_{pp}/2$. The directivity error (or residual directivity) is given by the reflection coefficient of the Load used in the calibration (if a SOL type calibration was undertaken) plus any additional approximation in the determination of error model parameters and drift. The airline has the function to offset the Load: even using the same Load to do this test the reason for the ripple is that the airline and the load might not have the same impedance; if they are the same the calibration would be perfect with zero residual, otherwise we trust the airline as the correct impedance value and we are going to measure the mismatch between the VNA assumed reference impedance (that of the calibration Load) and that of the airline, resulting in the observed ripple. The airline is an impedance transformer as described by eq. (2.3.12) and the input impedance seen at the VNA port will swing between Z_t and Z_0^2/Z_t every quarter period, with $Z_t = Z_{\text{Load}}$ used to terminate the airline. Assuming for example that the VNA calibration at $50.5\,\Omega$ and that the airline is truly $50\,\Omega$, the two extreme values are right $50.5\,\Omega$ and $49.5\,\Omega$, with an almost linear relationship. Going to dB this example would have displayed a $-40\,\text{dB}$ peak-to-peak ripple for a $-46\,\text{dB}$ mismatch and residual directivity.

Using a different matched Load will lead to the problem of verifying that it has a better reflection coefficient otherwise confusing the result (see [123], sec. 6.2.1.4); by the way with a better matched Load available in the correct connector type why we didn't use it to calibrate the VNA?

As known, the directivity error tends to increase with frequency; a good definition of its frequency dependency does not require more than a few values, in favor of not too a frequent ripple and thus of too a long assurance airline: EURAMET suggests to have a value for the range $1-8\,\text{GHz}$, one for $8-12\,\text{GHz}$ and one for $12-18\,\text{GHz}$. Of course a few more values may be advisable for better representation and as a further verification of consistency. Thus the required ripple periods are between approximately four and seven, two or three in the first 10 GHz.

Daywitt[39] considers in detail the problem of graphical interpolation with too a few points and too a short airline: the conclusion is that the length of the airline should be equal to the shortest wavelength to check, so that there is a sufficient number of oscillations to track the envelope fluctuations (that carry the information we are inter-

[39] W.C. Daywitt, "Determining Adapter Efficiency by Envelope Averaging Swept Frequency Reflection Data," *IEEE Transactions on Microwave Theory and Techniques*, vol. 38 no. 11, Nov. 1990, pp. 1748-1752.

ested in). In the EURAMET examples the used airline is 30 cm long for a maximum frequency of 18 GHz, so tuned with more than a half-wavelength criterion (but not a full wavelength) to the maximum frequency. The paper considers also the errors in general related to estimate of connector efficiency by the ripple method: while the error terms are related to the assumptions made to derive the efficiency, the envelope averaging error has a broader application.

As already pointed out, what is important is that the reflection coefficient of the matched load be lower than the directivity we aim to quantify, otherwise the resulting ripple will be mainly influenced by the poor matching of the load itself.

When, conversely, the airline is terminated on a Short, the ripple gives indication of the sought port match value δE_{SF} indicated as $M = X_{pp}/2$. There is the concomitant influence of directivity D and tracking error T, or δE_{RF}, (the latter usually small and negligible), but maximizing the reflection coefficient Γ (as when using a good short), we will carry out a good estimate of M.

It is also evident that the two estimates regard only magnitude: the extension to phase quantification is still under development. However, Persson[40] proposed an extension to estimate also the phase of directivity error. Wübbeler et al. have proposed five years later an extension to treat complex numbers rather than magnitudes, operating in both time and frequency and artificially moving the quantities at the beginning and the end of the airline by multiplying for the estimated phase shift. The method was developed between 2007 and 2009[41]; the reader is invited to have a look at the references in footnote for further details.

Characteristics of airline An airline is a calculable coaxial transmission line with minimum loss and characteristic impedance set by careful design and trimming of its geometry: expressions for the line characteristic impedance and admittance (see sec. 5.3.1.4) and that quantify the sensitivity of coaxial lines to variability of geometrical and electrical properties (see sec. 5.3.2) are put to work to minimize and quantify airline variability and thus its uncertainty. The calculation of accuracy starting from known and reliable formulas and the measurement of geometrical properties is called by Juroschek and Free "dimensional approach" [203]. There exist accurate expressions [97, 98, 203] for the determination of the characteristic impedance and other line characteristics, as well as the sensitivity to tolerances of geometry and basic electrical properties (see sec. 5.3.1.4); with some slight changes they are the same included in the METAS VNA Tools II program described in sec. 11.5.3.10. Alternatively, the direct measurement of line characteristics such as the propagation constant and the

[40] P. Persson, "An Algorithm for the Evaluation of the Residual Directivity "ripple" trace", ANAMET Report 034, May 2002.

[41] G. Wübbeler, C. Elster, T. Reichel, and R. Judaschke, "Determination of complex residual error parameters of a calibrated vector network analyser," 69th ARTFG Conference, Honololu, HI, USA, June 8, 2007.
G. Wübbeler, R. Judaschke and C. Elster, "Estimation of residual error parameters of vector network analyzers," XIX IMEKO World Congress Fundamental and Applied Metrology Sept. 6-11, 2009, Lisbon, Portugal, pp. 950-952.
G. Wübbeler, C. Elster, T. Reichel, and R. Judaschke, "Determination of the Complex Residual Error Parameters of a Calibrated One-Port Vector Network Analyzer," *IEEE Transactions on Instrumentation and Measurement*, Vol. 58, no. 9, Sept. 2009, pp. 3238-3244.

capacitance (or admittance) by independent methods (e.g. VNA and bridge measurement) allows similarly the quantification of the airline accuracy. Whereas the length of the line is ruled out by the ratio of the propagation constant and capacitance to form the characteristic impedance ($Z_c = \gamma/(j\omega C + G)$), the accuracy of the method is improved for long lines. In general dimensional and gamma/capacitance methods give comparable results [203] to better than 0.1%, if some rare outliers are excluded.

A thorough description of the verification program based on dimensional measurement is presented by Horibe *et. al* [164] for the National Metrology Institute of Japan. More than one method is used (air gauge measurement system and laser gauge measurement system among others, with expanded uncertainty in the range of $1\,\mu\mathrm{m}$. Dimensional stability is also considered when the airline is connected and connectors tightened properly at the prescribed torque: variability of $0.8\,\mu\mathrm{m}$ and $2.5\,\mu\mathrm{m}$ for the male and female ends were observed.

Also repeatability due to connection and disconnection is an issue and influences all the measurement results that are not obtained with dimensional measurements.

A source of error that is particularly evident in beadless airlines (that is airlines where the inner conductor is held in place only by the inserted connectors) is the necessity of building the inner conductors slightly shorter than the outer cylinder to avoid axial forces during tightening of terminal connectors; the location of the center pin with respect to the mating plane is thus variable and a little amount of recess is unavoidable. Adjusting the center pin is not an operation that all users can make in the necessary controlled conditions (to give the order of magnitude $10\,\mu\mathrm{m}$ makes the difference!). Depending on how the airline is assembled to the test port, the airlines center conductor would most likely either be butted up to the test port adapters center conductor (resulting in a protruding pin gap at the test port) or the airlines center conductor would be butted up to the offset terminations center conductor (resulting in a largely recessed pin gap at the test port). Beaded airlines have one side with the inner conductor centered by a beaded section that ensures also nearly zero pin gap; the other side of the coin is of course the distortion of the characteristic impedance caused by the dielectric material of which the beaded support is made (it was discussed when considering partially-filled coax cables in sec. 5.3.1.2).

In [285] it is shown that the beaded airline suffers of slight systematic errors (called "linear" quantities in the paper) as for directivity is concerned, but that repeatability is much better than that of a beadless airline, the two resulting in almost equal performance if the two error terms are summed together (overall error). When considering the source match error, the systematic error terms are nearly equal, but repeatability is again better for the beaded airline by about a factor of two. In this work measurements were made using the ripple method suggested by EURAMET, using a low-reflection matched termination to estimate directivity and then a reflecting short termination to estimate source match error.

Bibliography

[1] H.A. Aebischer, B. Aebischer, "Improved Formulae for the Inductance of Straight Wires," *Advanced Electromagnetics*, Vol. 3, No. 1, Sept. 2014, pp. 31-43.

[2] Agilent, "Applying Error Correction to Network Analyzer Measurements", *Application Note 1287*-3, code 5965-7709E, 2002.

[3] Agilent, "Agilent Network Analysis Applying the 8510 TRL Calibration for Non-Coaxial Measurements," *Product Note 8510-8A*, 2000.

[4] Agilent, "Spectrum Analyzer – Measurements and Noise", *Application Note 1303*, code 5966-4008E, Apr. 2, 2008.

[5] Agilent, "8 Hints for Better Spectrum Analysis", *Application Note 1286-1*, code 5965-7009E, May 13, 2008.

[6] Agilent, "Spectrum Analysis Basics," *Application Note 150*, 1971 (revised in 2014).

[7] Agilent, "Time Domain Analysis Using a Network Analyzer", *Application Note 1287-12*, code 5989-5723EN, May 2, 2012.

[8] Agilent, "In-Fixture Measurements Using Vector Network Analyzers", *Application Note 1287-9*, code 5968-5329E, 2000.

[9] Agilent, "De-embedding and Embedding S-Parameter Networks Using a Vector Network Analyzer", *Application Note 1364-1*, code 5980-2784EN, 2001.

[10] D. Agrez, "Weighted Multipoint Interpolated DFT to Improve Amplitude Estimation of Multifrequency Signal," *IEEE Transactions on Instrumentation and Measurements*, vol. 51 n. 2, Apr. 2002, pp. 287-292. doi: 10.1109/19.997826.

[11] D.W. Allan, "Statistics of atomic frequency standards," *Proceedings of IEEE*, vol. 54, pp. 221-230, Feb. 1966. doi: 10.1109/PROC.1966.4634

[12] D.W. Allan and J.A. Barnes, "A modified "Allan variance" with increased oscillator characterization ability," 35th Annual Frequency Control Symposium, Philadelphia, PA, USA, May 27-29, 1981, pp. 470-475. doi: 10.1109/FREQ.1981.200514

[13] M.B. Amin and I.A. Benson, "Nonlinear Effects in Coaxial Cables at Microwave Frequencies," *Electronics Letters*, Vol. 13, No. 25, Dec. 1977, pp. 768-770. doi: 10.1049/el:19770543

[14] Anritsu, "Reflectometer Measurements – Revisited," *Application Note 11410-00214*, rev. C, April 2000.

[15] Anritsu, "Three and Four Port S-parameter Measurements", rev. B, 2002.

[16] Anritsu, "Calculating VNA Measurement Accuracy," *Application Note 11410-00464*, Rev. A, Dec. 2008.

[17] Anritsu, "Time Domain Measurements Using Vector Network Analyzers," *Application Note 11410-00206*, rev. D, March 2009.

[18] ANSI/IEEE Std. 376, "IEEE Standard for the Measurement of Impulse Strength and Impulse Bandwidth", 1975.

[19] ANSI/IEEE Std. 1004, *IEEE Standard Definitions of Planar Transmission Lines*, 1987.

[20] ANSI Std. C63.2, "Electromagnetic Noise and Field Strength, 10 kHz to 40 GHz - Specifications", 1987.

[21] Arlon Guide, "Everything you ever wanted to know about laminates ... but were afraid to ask", Nov. 2008, [on-line] https://www.circuitboards.com/pdfs/EverythingYouWanted.pdf (last access Nov. 2014)

[22] I.J. Bahl and R. Garg, "Simple and Accurate Formulas for a Microstrip with Finite Strip Thickness," *Proceedings of IEEE*, Vol. 65, n. 11, Nov. 1977, pp. 1611-1612. doi: 10.1109/PROC.1977.10783

[23] I. Bahl and P. Bhartia, *Microwave Solid State Circuit Design*, 2nd ed., John Wiley & Sons, New Jersey, 2003.

[24] M. Bailey, Guidance on Selecting and Handling Coaxial RF Connectors, 1MA99, v1.0, Rohde & Schwarz, 2012-12-12.

[25] N. Balabanian and T.A. Bickart, *Electrical Network Theory*, John Wiley & Sons, New York, 1968.

[26] D. Ballo, "Network Analyzer Basics", *1998 Back to Basics Seminar*, Hewlett Packard, code 5965-7917E.

[27] J.A. Barnes, A.R. Chi, L.S. Cutler, D.J. Healy, D.B. Leeson, T.E. McGunigal, J.A. Mullen, W.L. Smith, R.L. Sydnor, R.F.C. Vessot and G.M.R. Winkler, "Characterization of frequency stability," *IEEE Transactions on Instrumentation and Measurement*, vol. 20, Apr. 1971, pp. 105–120. doi: 10.1109/TIM.1971.5570702

[28] J.A. Barnes, "The Generation and Recognition of Flicker Noise," National Bureau of Standards, Boulder, CO, Rep. 9284, 1967.

[29] J.A. Barnes and S. Jarvis Jr., "Efficient Numerical and Analog Modeling of Flicker Noise Processes," National Bureau of Standards, Boulder, CO, NBS Tech. Note 604, 1971.

[30] R.F. Bauer Jr. and P. Penfield, "De-Embedding and Unterminating," *IEEE Transactions on Microwave Theory and Techniques*, Vol. 22, no. 3, pp. 282-288, March 1974. doi: 10.1109/TMTT.1974.1128212

[31] J.S. Bendat and A.G. Piersol, *Engineering Applications of Correlation and Spectral Analysis*, John Wiley & Sons, Canada, 1980.

[32] J.S. Bendat and A.G. Piersol, *Random Data - Analysis and Measurement procedures*, 2nd ed., John Wiley & Sons, Canada, 1986.

[33] F.A. Benson, P.A. Cudd and J.M. Tealby, "Leakage from coaxial cables," *IEE Proceedings A*, Vol. 139, No. 6, Nov. 1992, pp. 285-303.

[34] D. Bergrfried and H. Fischer, "Insertion-Loss Repeatability Versus Life of Some Coaxial Connectors," *IEEE Transactions on Instrumentation and Measurement*, Vol. 19, no. 4, Nov. 1970, pp. 349-353. doi: 10.1109/TIM.1970.4313926.

[35] R.M. Bethea, B.S Duran, T.L. Boullion, *Statistical Methods for Engineers and Scientists*, Marcel Dekker, Inc., New York, 1985.

[36] P. Bhartia and P. Pramanick, "A New Microstrip Dispersion Model," *IEEE Transactions on Microwave Theory and Techniques*, Vol. 32, no 10, Oct. 1984, pp. 1379-1384. doi: 10.1109/TMTT.1984.1132855

[37] B. Bianco, M. Parodi, S. Ridella and F. Selvaggi, "Launcher and Microstrip Characterization," *IEEE Transactions on Instrumentation and Measurement*, Vol. 25, no. 4, Dec. 1976, pp. 320-323. doi: 10.1109/TIM.1976.6312235

[38] B. Bianco. A. Corana, S. Ridella and C. Simicich, "Evaluation of Errors in Calibration Procedures for Measurements of Reflection Coefficient," *IEEE Transactions on Instrumentation and Measurement*, Vol. 27, No. 4, pp 354-358, Dec. 1978. doi: 10.1109/TIM.1978.4314711

[39] BIPM, *Evaluation of measurement data — Guide to the expression of uncertainty in measurement*, JCGM 100:2008.

[40] G. Blando, J. Miller, I. Novak, J. DeLap and Cheryl Preston, "Attenuation in PCB Traces due to Periodic Discontinuities," DesignCon 2006, Feb. 6-9, 2006, Santa Clara, CA, USA.

[41] W.R. Blood Jr., *Motorola MECL System Design Handbook*, 4th ed, 1980.

[42] L.I. Bluestein, "A linear filtering approach to the computation of the discrete Fourier transform," Northeast Electronics Research and Engineering Meeting Record, Vol. 10, 1968, pp. 218-219. (later published as "A linear filtering approach to the computation of the discrete Fourier transform,", *IEEE Transactions on Audio Electroacoustics*, Vol. 18, no. 4, pp. 451-455, 1970. doi: 10.1109/TAU.1970.1162132)

[43] D.E. Bockelman and W.R. Eisenstadt, "Combined Differential and Common-Mode Scattering Parameters: Theory and Simulation," *IEEE Transactions on Microwave Theory and Techniques*, Vol. 43, No. 7, July 1995, pp. 1530-1539. doi: 10.1109/22.392911

[44] E. Bogatin, "Design Rules for Microstrip Capacitance," *IEEE Transactions on Components, Hybrids and Manufacturing Technology*, Vol. 11, n. 3, Sept. 1988, pp. 253-259. doi: 10.1109/33.16649.

[45] E. Bogatin, *Signal and Power Integrity - Simplified*, 2nd Ed., Prentice Hall, Englewood Cliffs, New Jersey, 2010.

[46] J. Bongiorno and A. Mariscotti, "The variability of microstrips formulas as a source of uncertainty in microwave setups," *Measurement*, Vol. 73, Sept. 2015, pp. 33–54. doi:10.1016/j.measurement.2015.05.014

[47] G. Boyd, "How to convert 2 port S-parameters to 4 port S-parameters", PMC Sierra white paper, 05/05/2004.

[48] Bracewell, *The Fourier Transform and Its Applications*, 2nd Ed., McGraw-Hill, New York, 1986.

[49] O. Breitenbach, T. Hähner and B. Mund, "Screening of Cables in the MHz to GHz Frequency Range Extended Application of a Simple Measuring Method", IEE Colloquium on Screening Effectiveness Measurements (Ref. No. 1998/452), 1998, pp. 7/1-7/15. doi: 10.1049/ic:19980730

[50] G. Brist, S. Hall, S. Clouser and T. Liang, "Non-classical conductor losses due to copper foil roughness and treatment," Proc. of Electronic Circuits World Convention, Vol. 10, 2005, pp. 22-24.

[51] T.G. Bryant and J.A. Weiss, "MSTRIP (Parameters of Microstrip)," *IEEE Transactions on Microwave Theory and Techniques*, (Computer Program Description), Vol. 19, no. 4, Apr. 1971, pp. 418-419, doi: 10.1109/TMTT.1971.1127535

[52] J.H. Broxon and D.K. Linkhart, "Twisted-Wire Transmission Lines," *RF Design*, June 1990, pp. 73-76.

[53] H. Carlin, "The scattering matrix in network theory," *IRE Transactions on Circuit Theory*, vol. 3, no. 2, pp. 88–97, Jun 1956. doi: 10.1109/TCT.1956.1086297 (correction, doi: 10.1109/TCT.1957.1086351)

[54] B. Carlson, *Communication Systems – An Introduction to Signals and Noise in Electrical Communication*, 3rd ed., New York, McGraw Hill, 1988.

[55] CENELEC EN 50117-4-1, *Coaxial cables Part 4-1: Sectional specification for cables for BCT cabling in accordance with EN 50173 - Indoor drop cables for systems operating at 5 MHz - 3 000 MHz*, 2008.

[56] CENELEC EN 50121-2 (IEC 62236-2), *Railway applications - Electromagnetic compatibility - Part 2: Emission of the whole railway system to the outside world*, 2006-07.

[57] CENELEC EN 50121-3-1 (IEC 62236-3-1), *Railway applications - Electromagnetic compatibility - Rolling stock - Part 3.1: Train and complete vehicle*, 2006-07.

[58] CENELEC EN 50173-1, *Information Technology - Generic Cabling Systems - Part 1: General requirements*, 2007.

[59] CENELEC EN 50174-1, *Information technology - Cabling installation Part 1: Installation specification and quality assurance*, 2009-05 + Amd. A1, 2012-03.

[60] CENELEC EN 50288-2-1, *Multi-element metallic cables used in analog and digital communication and control - Part 2-1: Sectional specifications for screened cables characterized up to 100 MHz - Horizontal and building backbone cables*, 2003-12.

[61] CENELEC EN 50288-9-1, *Multi-element metallic cables used in analog and digital communication and control - Part 9-1: Sectional specifications for screened cables characterized up to 1000 MHz - Horizontal and building backbone cables*, 2010.

[62] CENELEC EN 50288-10-1, *Multi-element metallic cables used in analog and digital communication and control - Part 10-1: Sectional specifications for screened cables characterized up to 500 MHz - Horizontal and building backbone cables*, 2010.

[63] CENELEC EN 50289-1-5, *Communication cables - Specifications for test methods - Part 1-5: Electrical test methods - Capacitance*, 2001-06.

[64] CENELEC EN 50289-1-8, *Communication cables - Specifications for test methods - Part 1-8: Electrical test methods - Attenuation*, 2001-06.

[65] CENELEC EN 50289-1-10, *Communication cables - Specifications for test methods - Part 1-8: Electrical test methods - Crosstalk*, 2001-06.

[66] CENELEC EN 50290-2-1, *Communication cables Part 2-1: Common design rules and construction*, 2005-02.

[67] CENELEC EN 50346, *Information technology - Cabling installation - Testing of installed cabling*, 2002.

[68] CENELEC EN 55024, *Information technology equipment - Immunity characteristics - Limits and methods of measurement*, 2010-11.

[69] CENELEC EN 60512-25-1, *Connectors for electronic equipment - Tests and measurements – Part 25-1: Test 25a - Crosstalk ratio*, 2001-10.

[70] CENELEC EN 60512-25-2, "*Connectors for electronic equipment - Tests and measurements – Part 25-2: Test 25b - Attenuation (Insertion loss)*, 2002-06.

[71] CENELEC EN 60512-25-3, *Connectors for electronic equipment - Tests and measurements – Part 25-3: Test 25c - Rise time degradation*, 2001-10.

[72] CENELEC EN 60512-25-4, *Connectors for electronic equipment - Tests and measurements – Part 25-4: Test 25d - Propagation delay*, 2001-10.

[73] CENELEC EN 60512-25-5, *Connectors for electronic equipment - Tests and measurements – Part 25-5: Test 25e - Return loss*, 2004-09.

[74] CENELEC EN 60512-25-6, *Connectors for electronic equipment - Tests and measurements – Part 25-6: Test 25f - Eye pattern and jitter*, 2004-07.

[75] CENELEC EN 60512-25-7, *Connectors for electronic equipment - Tests and measurements – Part 25-7: Test 25g - Impedance, reflection coefficient and standing voltage wave ratio (VSWR)*, 2005-03.

[76] CENELEC EN 61188-1-2, *Printed boards and printed board assemblies - Design and use. Part 1-2: Generic requirements – Controlled impedance*, 1998.

[77] CENELEC EN 61000-6-1, *Electromagnetic compatibility (EMC) Part 6-1: Generic standards - Immunity for residential, commercial and light-industrial environments*, 2007-01.

[78] CENELEC EN 61000-6-2, *Electromagnetic compatibility (EMC) Part 6-2: Generic standards - Immunity for industrial environments*, 2005-08.

[79] CENELEC EN 61000-6-3, *Electromagnetic compatibility (EMC) Part 6-3: Generic standards – Emission for residential, commercial and light-industrial environments*, 2007-01.

[80] CENELEC EN 61000-6-4, *Electromagnetic compatibility (EMC) Part 6-4: Generic standards – Emission for industrial environments*, 2007-01.

[81] B. Chia, R. Kollipara, D. Oh, C. Yuan and L.S. Boluna, "Study of PCB Trace Crosstalk in Backplane Connector Pin Field," IEEE Electrical Performance of Electronic Packaging, Scottsdale, AZ, USA, Oct. 2006, pp. 281-284. doi: 10.1109/EPEP.2006.321155

[82] R.A. Chipman, *Schaum's outline of Theory and Problems of Transmission Lines*, McGraw-Hill, 1968.

[83] CISPR 16-2-1, *Specification for radio disturbance and immunity measuring apparatus and methods – Part 2-1: Methods of measurement of disturbances and immunity - Conducted disturbance measurements*, 2008-10.

[84] CISPR 16-4-2, *Specification for radio disturbance and immunity measuring apparatus and methods – Part 4-2: Uncertainties, statistics and limit modelling - Uncertainty in EMC measurements*, 2003-11.

[85] S.B. Cohn, "Characteristic Impedance of the Shielded-Strip Transmission Line," *IRE Transactions on Microwave Theory and Techniques*, Vol. 2, July 1954, pp. 52-57. doi: 10.1109/TMTT.1954.1124875

[86] S.B. Cohn, "Shielded coupled-strip transmission line," *IRE Transactions on Microwave Theory and Techniques*, Vol. 3, Oct. 1955, pp. 29-38. doi: 10.1109/TMTT.1955.1124973

[87] R. Collier and D. Skinner eds., *Microwave Measurements*, 3rd ed., IET Electrical Measurement Series Vol. 12, 2007.

[88] R.E. Collin, *Foundations for Microwave Engineering*, 2nd ed., IEEE Press, New York, 2001.

[89] J. Coonrod, "Ambiguous Influences Affecting Insertion Loss of Microwave Printed Circuit Boards," *IEEE Microwave Magazine*, July/Aug. 2012, pp. 66-75. doi: 10.1109/MMM.2012.2197145

[90] G.D. Cormack and J.O. Binder, "The extended function fast Fourier transform (EF-FFT)," *IEEE Transactions on Instrumentation and Measurement*, Vol. 38, no. 3, March 1989, pp. 730-735. doi: 10.1109/19.32182

[91] G.D. Cormack, D.A. Blair and J.N. McMullin, "Enhanced spectral resolution FFT for step-like signals," *IEEE Transactions on Instrumentation and Measurement*, Vol. 40, no. 1, Feb. 1991, pp. 34-36. doi: 10.1109/19.69945

[92] G.D. Cormack and J.N. McMullin, "Frequency de-aliased FFT analysis of step-like functions," *IEEE Transactions on Instrumentation and Measurement*, Vol. 40, no. 4, Aug. 1991, pp. 773-775. doi: 10.1109/19.85351

[93] P.A. Cudd, F.A. Benson and J.E. Sitch, "Prediction of leakage from single braid screened cables," IEE Proceedings, Vol. 133, Part A, No. 3, May 1986, pp. 144-151. doi: 10.1049/ip-a-1.1986.0026

[94] W. C. Daywitt, "First-order symmetric modes for a slightly lossy coaxial transmission line", IEEE Trans. Microwave Theory & Tech., vol. 38, no. 11, pp. 1644-1650, November 1990" doi: 10.1109/22.60011

[95] W.C. Daywitt, "A Simple Technique for Investigating Defects in Coaxial Connectors," *IEEE Transactions on Microwave Theory and Techniques*, vol. 35 no. 4, Apr. 1987, pp. 460-464. doi: 10.1109/TMTT.1987.1133673.

[96] W.C. Daywitt, "A Simple Technique for Determining Joint Losses on a Coaxial Line from Swept-Frequency Reflection," *IEEE Transactions on Instrumentation and Measurement*, Vol. 36, no. 2, June 1987, pp. 468-473. doi: 10.1109/TIM.1987.6312721.

[97] W.C. Daywitt, "The Propagation Constant of a Lossy Coaxial Line with a Thick Outer Conductor," *IEEE Transactions on Microwave Theory and Techniques*, vol. 43 no. 4, Apr. 1995, pp. 907-911. doi: 10.1109/22.375243

[98] W.C. Daywitt, "Complex admittance of a lossy coaxial open circuit with hollow center conductor," *Metrologia*, Vol. 24, no. 1, 1987, pp. 1-10. doi: 10.1088/0026-1394/24/1/003

[99] J. Deane, "Relations for conversion amongst two-port network parameters", technical brief, Sept. 1998, [online] http://personal.ee.surrey.ac.uk/Personal/J.Deane/mat/mat.html (last access March 2014).

[100] D.C. DeGroot, J.A. Jargon and R.B. Marks, "Multiline TRL revealed," 60th ARFTG Conference Digest, Washington D.C., USA, Dec. 2002, pp. 131-155. doi: 10.1109/ARFTGF.2002.1218696

[101] S. Deibele and J.B. Beyer, "Measurements of Microstrip Effective Relative Permittivities," *IEEE Transactions on Microwave Theory and Techniques*, Vol. 35, no. 5, pp. 535-538, May 1987. doi: 10.1109/TMTT.1987.1133696

[102] A. Demir, A. Mehrotra and J. Roychowdhury, "Phase Noise in Oscillators: A Unifying Theory and Numerical Methods for Characterization," *IEEE Transactions on Circuits and Systems — I: Fundamental Theory and Applications*, Vol. 47, no. 5, May 2000, pp. 655-674. doi: 10.1109/81.847872

[103] B. Démoulin and L. Koné, "Shielded Cables Transfer Impedance Measurement High frequency range 100 MHz-1 GHz", IEEE-EMC Newsletter, 2011, pp. 42-50.

[104] E.J. Denlinger, "A frequency dependent solution for microstrip transmission lines," *IEEE Transactions on Microwave Theory and Techniques*, vol. 19, no. 1, pp. 30-39, Jan. 1971. doi: 10.1109/TMTT.1971.1127442

[105] A. Deutsch, R.S. Krabbenhoft, K.L. Melde, C.W. Surovic, G.A. Katopis, G.V. Kopcsay, Z. Zhou, Z. Chen, Y.H. Kwark, T.-M. Winkel, X. Gu and T.E. Standaert, "Application of the Short-Pulse Propagation Technique for Broadband Characterization of PCB and Other Interconnect Technologies," *IEEE Transactions on Electromagnetic Compatibility*, vol. 52, no. 2, May 2010, pp. 266-287. doi: 10.1109/TEMC.2009.2037971

[106] A. Deutsch, T.-M. Winkel, G.V. Kopcsay, C.W. Surovic, B.J. Rubin, G.A. Katopis, B.J. Chamberlin and R.S. Krabbenhoft, "Extraction of $\varepsilon_r(f)$ and $\tan\delta(f)$ for Printed Circuit Board Insulators up to 30 GHz Using the Short-Pulse Propagation Technique," *IEEE Transactions on Advanced Packaging*, Vol. 28, no. 1, Feb. 2005, pp. 4-12. doi: 10.1109/TADVP.2004.841679

[107] A. Deutsch, "Electrical Characteristics of Interconnections for High-Performance Systems," *Proceedings of IEEE*, Vol. 86, no. 2, Feb. 1998, pp. 315-355. doi: 10.1109/5.659489

[108] F. Di Paolo, *Networks and Devices Using Planar Transmission Lines*, CRC Press, Boca Raton, FL, USA, 2000.

[109] Y. Ding, Y.H. Kwark, Lei Shan, C. Baks and Ke Wu, "Techniques for De-embedding a High Port Count Connector to PCB Via Interposer," IEEE 61th Electronic Components and Technology Conference, Lake Buena Vista, FL, USA, May 31-June 3, 2011, pp. 467-472. doi: 10.1109/ECTC.2011.5898552

[110] A.R. Djordjevic, R.M. Biljie, V.D. Likar-Smiljanic and T.K. Sarkar, "Wideband frequency-domain characterization of FR-4 and time-domain causality," *IEEE Transactions on Electromagnetic Compatibility*, Vol. 43, no. 4, Nov. 2001, pp. 662-667. doi: 10.1109/15.974647

[111] A.W. Drake, *Fundamentals of Applied Probability Theory*, McGraw-Hill, New York, 1967.

[112] K. Duda, "DFT Interpolation Algorithm for Kaiser–Bessel and Dolph–Chebyshev Windows," *IEEE Transactions on Instrumentation and Measurements*, vol. 60 n. 3, Mar. 2011, pp. 784-790. doi: 10.1109/TIM.2010.2046594.

[113] D. Dunham, J. Lee, S. McMorrow and Y. Shlepnev, "Design and Optimization of a Novel 2.4 mm Coaxial Field Replaceable Connector Suitable for 25 Gbps System and Material Characterization up to 50 GHz," DesignCon 2011, Santa Clara, CA, USA, Jan. 31 - Feb. 3, 2011.

[114] J.P. Dunsmore, "The Time-Domain Response of Coupled-Resonator Filters with Applications to Tuning," Ph.D. thesis, University of Leeds, Jan. 2004.

[115] J.P. Dunsmore, *Handbook of microwave component measurements: with advanced VNA techniques*, John Wiley & Sons, Hoboken, N.J, USA, 2012.

[116] T.C. Edwards and R.P. Owens, "2-18 GHz Dispersion Measurements on 10-100 Ω Microstrip Lines on Sapphire," *IEEE Transactions on Microwave Theory and Techniques*, Vol. 24, no. 8, pp. 506-513, Aug.1976. doi: 10.1109/TMTT.1976.1128888

[117] B. Eicher and L. Boillot, "Very Low Frequency to 40 GHz Screening Measurements on Cables and Connectors: Line Injection Method and Mode Stirred Chamber," IEEE International Symposium on Electromagnetic Compatibility, Anaheim, CA, Aug. 1992, pp. 302-307. doi: 10.1109/ISEMC.1992.626099

[118] G.F. Engen, "Calibration technique for automated network analyzers with application to adapter evaluation," IEEE Transactions on Microwave Theory and Techniques, Vol. 22, no. 12, Dec. 1974, pp. 1255–1260. doi: 10.1109/TMTT.1974.1128472

[119] G.F. Engen and C.A. Hoer, "Thru-Reflect-Line: An Improved technique for calibrating the Dual Six-Port ANA," *IEEE Transactions on Microwave Theory and Techniques*, Vol. 27, No. 12, Dec. 1979, pp. 987-993. doi: 10.1109/TMTT.1979.1129778

[120] A.J. Estin, "Scattering Parameters of SMA Coaxial Connector Pairs," *IEEE Transactions on Instrumentation and Measurement*, Vol. 25 no. 4, Dec. 1976, pp. 329-334. doi: 10.1109/TIM.1976.6312237

[121] A.J. Estin, J.R. Juroshek, R.B. Marks, F.R. Clague and J. Wayde Allen, "Basic RF and Microwave Measurements: a Review of Selected Programs," *Metrologia*, Vol. 29, 1992, pp. 135-151. doi: 10.1088/0026-1394/52/1/121

[122] H.-J. Eul and B. Schiek, "A Generalized Theory and New Calibration Procedures for Network Analyzer Self-Calibration," *IEEE Transactions on Microwave Theory and Techniques*, vol. 39 no. 4, Apr. 1991, pp. 724-731. doi: 10.1109/22.76439

[123] EURAMET, *Guidelines on the Evaluation of Vector Network Analyzers (VNA)*, European Co-operation for Accreditation, cg-12, ver. 2.0, 03-2011.

[124] A. Ferrero, U. Pisani "Two-port Network Analyzer Calibration Using an Unknown 'Thru'," *IEEE Microwave and Guided Wave Letters*, Vol. 2, No. 12, pp. 505-507, Dec. 1992. doi: 10.1109/75.173410

[125] A. Ferrero and M. Pirola, "Generalized mixed-mode S-parameters," *IEEE Transactions on Microwave Theory and Techniques*, Vol. 54, no. 1, Jan. 2006, pp. 458-463. doi: 10.1109/TMTT.2005.860497

[126] F. Filippone, A. Mariscotti and P. Pozzobon, "The Internal Impedance of Traction Rails for DC Railways in the 1-100 kHz Frequency Range," *IEEE Trans. on Instrumentation and Measurement*, vol. 55 n. 5, Oct. 2006, pp. 1616-1619. doi: 10.1109/TIM.2006.880912

[127] W.H. Fonger, *Transistors I*, Radio Corporation of America (RCA) Laboratories, Princeton, NJ, 1956 (pp. 239-297). (p. 354).

[128] E.P. Fowler, "Screening measurements in the time domain and their conversion into the frequency domain," *Journal of the Institution of Electronic and Radio Engineers*, Vol. 55, No. 4, pp. 127-132, April 1985. doi: 10.1049/jiere.1985.0043

[129] E.P. Fowler, "Measurements of Screening Effectiveness of Multi-Pin and Coaxial Connectors to 300 and 1000 MHz," IEE Colloquium on Screening Effectiveness Measurements (Ref. No. 1998/452), 6 May 1998, London, UK. doi: 10.1049/ic:19980734.

[130] D.A. Frickey, "Using the Inverse Chirp-Z Transform for Time-Domain Analysis of Simulated Radar Signals," International Conference on Signal Processing Applications and Technology (ICSPAT), Oct. 18-21, 1994.

[131] D.A. Frickey, "Conversions between S, Z, Y, H, ABCD, and T parameters which are valid for complex source and load impedances," *IEEE Transactions on Microwave Theory and Techniques*, Vol. 42, no. 2, Feb. 1994, pp. 205-211. doi: 10.1109/22.275248

[132] M. Garelli and A. Ferrero, "A Unified Theory for S-Parameter Uncertainty Evaluation," *IEEE Transactions on Microwave Theory and Techniques*, Vol. 60, no. 12, Dec. 2012, pp. 3844-3855. doi: 10.1109/TMTT.2012.2221733

[133] G.G. Gentili, and M. Salazar-Palma, "The definition and computation of modal characteristic impedance in quasi-TEM coupled transmission lines," *IEEE Transactions on Microwave Theory and Techniques*, Vol. 43, no. 2, March 1995, pp. 338-343. doi: 10.1109/22.348093

[134] W.J. Getsinger, "Microstrip Dispersion Model," *IEEE Transactions on Microwave Theory and Techniques*, Vol. 21, no.1, Jan. 1973, pp. 34-39. doi: 10.1109/TMTT.1973.1127911

[135] G. Ghione and C. Naldi, "Analytical Formulas for Coplanar Lines in Hybrid and Monolithic MICs," Electronics Letters, Vol. 20, 1984, pp. 179-181. doi: 10.1049/el:19840120

[136] M.E. Goldfarb and R.A. Pucel, "Modeling Via Hole Grounds in Microstrips," *IEEE Microwave and Guided Wave Letters*, Vol. 1, no. 6, June 1991, pp. 135-137. doi: 10.1109/75.91090

[137] G. Gonzalez, *Microwave Transistor Amplifiers - Analysis and Design*, Prentice Hall, Englewood Cliffs, NJ, 1984.

[138] T. Grandke, "Interpolation Algorithms for Discrete Fourier Transforms of Weighted Signals," *IEEE Transactions on Instrumentation and Measurements*, vol. 32 n. 2, June 1983, pp. 350-355. doi: 10.1109/TIM.1983.4315077.

[139] C.A. Greenhall, "Spectral ambiguity of Allan variance," *IEEE Transactions on Instrumentation and Measurement*, vol. 47 n. 3, June 1998, pp. 623-627. doi: 10.1109/19.744312

[140] B. Grossman, M. Peterson and J. Torres, "Multiport VNA Measurement Uncertainty – Signal Integrity Applications," IEEE International Symposium on Electromagnetic Compatibility, Long Beach, CA, USA, Aug. 14-19, 2011. doi: 10.1109/ISEMC.2011.6038385

[141] F.W. Grover, *Inductance calculations*, Dover, Mineola, NY, 2004.

[142] K.C. Gupta, R. Garg and I.J. Bahl, *Microstrip Lines and Slotlines*, Artech House, Norwood, MA, USA, 1979.

[143] B.D. Hall, "VNA error models: Comments on EURAMET/cg-12/v.01," Industrial Research Limited Report 2444, Measurement Standards Laboratory of New Zealand, June 2010.

[144] B.D. Hall, "On the propagation of uncertainty in complex-valued quantities," *Metrologia*, Vol. 41, no. 3, pp. 173–177, 2004. doi:10.1088/0026-1394/41/3/010

[145] B.D. Hall, "Calculations of measurement uncertainty in complex-valued quantities involving "uncertainty in the uncertainty"," 64th ARFTG Microwave Measurements Conference, 2004, 15–22. doi: 10.1109/ARFTGF.2004.1427565

[146] B.D. Hall, "Some considerations related to the evaluation of measurement uncertainty for complex-valued quantities in radio frequency measurements," *Metrologia*, Vol. 44, no. 6, pp. 62–67. doi:10.1088/0026-1394/44/6/N04

[147] L. Halme, "Development of IEC Cable Shielding Effectiveness Standards," IEEE International Symposium on Electromagnetic Compatibility, Anaheim, CA, Aug. 1992, pp. 321-328. doi: 10.1109/ISEMC.1992.626102

[148] L. Halme and R. Kytonen, "Background and Introduction to EM Screening (Shielding) Behaviours and Measurements of Coaxial and Symmetrical Cables, Cable Assemblies and Connectors," IEE Colloquium on Screening Effectiveness Measurements (Ref. No. 1998/452), May 6, 1998, pp. 4/1-4/28. doi: 10.1049/ic:19980727

[149] L. Halme and B. Mund, "EMC of Cables, Connectors and Components with Triaxial Test set-up," 62nd IWCS International Wire & Cable Symposium, Charlotte, NC, USA, Nov. 10-13, 2013, pp. 83-90.

[150] E.O. Hammerstad, "Equations for microstrip circuit design," Proc. European Microwave Conference, 1975, pp. 268-272.

[151] E.O. Hammerstad and F. Bekkadal, Microstrip Handbook, University of Trondheim, ELAB report STF44 A74169, Feb. 1975.

[152] E. Hammerstad and O. Jensen, "Accurate Models for Microstrip Computer-Aided Design," IEEE International Symposium on Microwave Theory and Techniques, Washington, DC, May 28-30, 1980, pp. 407-409. doi: 10.1109/MWSYM.1980.1124303

[153] F.J. Harris, "On the Use of Windows for Harmonic Analysis with the Discrete Fourier Transform," *Proceedings of IEEE*, Vol. 66 n. 1, Jan. 1978, pp. 51-83. doi: 10.1109/PROC.1978.10837

[154] J. Henrie, A. Christianson and W.J. Chappell, "Prediction of Passive Intermodulation From Coaxial Connectors in Microwave Networks," *IEEE Transactions on Microwave Theory and Techniques*, Vol. 56, no. 1, Jan. 2008, pp. 209-216. doi: 10.1109/TMTT.2007.912166

[155] Hewlett Packard, "S-Parameters Design", *Application Note 154*, 1990.

[156] Hewlett Packard, "S-Parameters Techniques for Faster, More Accurate Network Design", *Application Note 95-1*, 1997.

[157] M. Hiebel, *Fundamentals of Vector Network Analysis*, Rohde & Schwarz, 2007.

[158] M.E. Hines and H.E. Stinehelfer, "Time-domain Oscillographic Microwave Network Analysis Using Frequency-Domain Data," *IEEE Transactions on Microwave Theory and Techniques*, Vol. 22, No. 3, March 1974, pp. 276-282. doi: 10.1109/TMTT.1974.1128211

[159] L.O. Hoeft, T.M. Salas and W.D. Prather, "Comments on the Line Injection Method for Measuring Surface Transfer Impedance of Cables," 12th Intern. Zurich Symposium on Electromagnetic Compatibility, Zurich, Switzerland, Feb. 1997, pp. 263-268.

[160] L.O. Hoeft and J.S. Hofstra, "Measured Electromagnetic Shielding Performance of Commonly Used Cables and Connectors," *IEEE Transactions on Electromagnetic Compatibility*, vol. 30, No. 3, Aug. 1988, pp. 260-275. doi: 10.1109/15.3304

[161] C.A. Hoer, "Choosing Line Lengths for Calibrating Network Analyzers," *IEEE Transactions on Microwave Theory and Techniques*, Vol. 31, No. 1, pp 76-78, Jan. 1983. doi: 10.1109/TMTT.1983.1131433

[162] C. Hoer and C. Love, "Exact Inductance Equations for Rectangular Conductors With Applications to More Complicated Geometries," Journal of Research of the National Bureau of Standards – C, Engineering and Instrumentation, Vol. 69C, no. 2, April-June 1965, pp. 127-137.

[163] J. Hoffmann, P. Leuchtmann, J. Ruefenacht and R. Vahldieck, "A Stable Bayesian Vector Network Analyzer Calibration Algorithm," *IEEE Transactions on Microwave Theory and Techniques*, Vol. 57, no. 4, Apr. 2009, pp. 869-880. doi: 10.1109/TMTT.2009.2015096

[164] M. Horibe, M. Shida, and K. Komiyama, "Development of Evaluation Techniques for Air Lines in 3.5- and 1.0-mm Line Sizes," *IEEE Transactions on Instrumentation and Measurement*, vol. 58, no. 4, April 2009, pp. 1078–1083, April 2007. doi: 10.1109/TIM.2008.2008084

[165] P. Horowitz and W. Hill, *The art of electronics*, 2nd ed., Cambridge University Press, 1989.

[166] R.M. Howard, *Principles of Random Signal Analysis and Low Noise Design*, John Wiley & Sons, New York, 2002.

[167] D.A. Howe, "The Total Deviation Approach to Long-Term Characterization of Frequency Stability," *IEEE Transactions on Instrumentation and Measurement*, Vol. 47 no. 5, Sept. 2000, pp. 1102-1110. doi: 10.1109/58.869040.

[168] A. Huynh, P. Håkansson and S. Gong, "Mixed-mode S-parameter conversion for networks with coupled differential signals," European Microwave Conference, July 2007, pp. 238-241.

[169] IEC 60096-1, Radio-frequency cables – Part1: General requirements and measuring methods, 1986-01.

[170] IEC 96-1 Amd. 2, Radio-frequency cables – Part1: General requirements and measuring methods, Amendment 2, 1993-06.

[171] IEC 61935-1, Specification for the testing of balanced and coaxial information technology cabling Part 1: Installed balanced cabling as specified in the standards series EN 50173, 2009-07.

[172] IEEE Std. 287, *IEEE Standard for Precision Coaxial Connectors (DC to 110 GHz)*, 2007.

[173] IEEE Std. 748, *IEEE Standard for Spectrum Analyzers*, 1979.

[174] IEEE Std. 1139, *Standard Definition of Physical Quantities for Fundamental and Time Metrology*, 2003.

[175] IEEE Std. 1139, *Standard Definition of Physical Quantities for Fundamental and Time Metrology - Random Instabilities*, 2008.

[176] IPC-D-317A, *Design Guidelines for Electronic Packaging Utilizing High-Speed Techniques*, 1995-01.

[177] IPC-SM-782A, *Surface Mount Design and Land Pattern Standard*, Aug. 1993 (+ Amd. 1, Oct. 1996, + Amd. 2, Apr. 1999).

[178] IPC-TM-650-2.5.5.1, *Permittivity (Dielectric Constant) and Loss Tangent (Dissipation Factor) of Insulating Material at 1 MHz (Contacting Electrode Systems)*, rev. B, 1986-05.

[179] IPC-TM-650-2.5.5.2, *Dielectric Constant and Dissipation Factor of Printed Wiring Board Material – Clip Method*, rev. A, 1987-12.

[180] IPC-TM-650-2.5.5.3, *Permittivity (Dielectric Constant) and Loss Tangent (Dissipation Factor) of Materials (Two Fluid Cell Method)*, rev. C, 1987-12.

[181] IPC-TM-650-2.5.5.4, *Dielectric Constant and Dissipation Factor of Printed Wiring Board Material-Micrometer Method*, rev. -, 1985-10.

[182] IPC-TM-650-2.5.5.5, *Stripline Test for Permittivity and Loss Tangent (Dielectric Constant and Dissipation Factor) at X-Band*, rev. C, 1998-03.

[183] IPC-TM-650-2.5.5.6, *Non-Destructive Full Sheet Resonance Test for Permittivity of Clad Laminates*, rev. C, 1989-05.

[184] IPC-TM-650-2.5.5.7, *Characteristic Impedance of Lines on Printed Boards by TDR*, rev. A, 2004-03.

[185] IPC-TM-650-2.5.5.9, *Permittivity and Loss Tangent, Parallel Plate, 1 MHz to 1.5 GHz*, rev. -, 1998-11.

[186] IPC-TM-650-2.5.5.10, *High Frequency Testing to Determine Permittivity and Loss Tangent of Embedded Passive Materials*, rev. -, 2005-07.

[187] IPC-TM-650-2.5.5.11, *Propagation Delay of Lines on Printed Boards by TDR*, 2009-04.

[188] IPC-TM-650-2.5.5.12, *Test Methods to Determine the Amount of Signal Loss on Printed Boards*, rev. A, 2012-07.

[189] IPC-TM-650-2.5.5.13, *Relative Permittivity and Loss Tangent Using a Split-Cylinder Resonator*, rev. -, 2007-01.

[190] IPC-2141, *Controlled Impedance Circuit Boards and High Speed Logic Design*, 1996-04.

[191] IPC-2141, *Controlled Impedance Circuit Boards and High Speed Logic Design*, 2004-03.

[192] IPC-2221A, *Generic Standard on Printed Board Design*, 2003-05.

[193] ITU-T G.810, SERIES G: *Transmission Systems and Media: Digital transmission systems – Digital networks – Design objectives for digital networks*, 08-1996.

[194] Recommendation ITU-R P.1057-1, *Probability distributions relevant to radiowave propagation modelling*, 2001.

[195] V.K. Jain, W.L. Collins and D.C. Davis, "High-Accuracy Analog Measurements via Interpolated FFT," *IEEE Transactions on Instrumentation and Measurements*, vol. 28 n. 2, June 1979, pp. 113-122. doi: 10.1109/TIM.1979.4314779.

[196] R.H. Jansen and N.H.L. Koster, "New aspects concerning the definition of microstrip characteristic impedance as a function of frequency," IEEE International Microwave Symposium Digest, Dallas, TX, USA, June 15-17, 1982, pp. 305-307. doi: 10.1109/MWSYM.1982.1130700

[197] R.L. Jesch, "Repeatability of SMA Coaxial Connectors," *IEEE Transactions on Instrumentation and Measurement*, Vol. 25 no. 4, Dec. 1976, pp. 314-320. doi: 10.1109/TIM.1976.6312234

[198] R. John, "Selecting Microwave/RF Cable Assemblies for Reliable Performance Over Time," Gore, March 2014.

[199] H.W. Johnson and M. Graham, *High-Speed Digital Design - A Handbook of Black Magic*, Prentice Hall, Englewood Cliffs, New Jersey, 1988.

[200] Johnson Components, Cambridge Products, *RF connector application guide*, JCI 161, March 1999.

[201] D. Jorgesen and C. Marki, "Directivity and VSWR Measurements," Marki Microwave white paper, [online]: http://www.markimicrowave.com/Assets/appnotes/ directivity_and_vswr_measurements.pdf (last access March 2015).

[202] J. Juroshek, "A study of measurements of connector repeatability using highly reflecting loads (short paper)," *IEEE Transactions on Microwave Theory and Techniques*, vol. 35, no. 4, Apr. 1987, pp. 457-460. doi: 10.1109/TMTT.1987.1133672

[203] J.R. Juroshek and G.M. Free, "Measurements of the Characteristic Impedance of Coaxial Air Line Standards," *IEEE Transactions on Microwave Theory and Techniques*, Vol. 42, no. 2, Feb. 1994, pp. 186-191. doi: 10.1109/22.275245

[204] J.F. Kaiser, "Nonrecursive Digital Filter Design Using the I0-sinh Window Function," Proc. IEEE Symp. Circuits and Systems, Apr. 1974, pp. 20-23.

[205] D.G. Kam, M.B. Ritter, T.J. Beukema, J.F. Bulzacchelli, P.K. Pepeljugoski, Y.H. Kwark, Lei Shan, Xiaoxiong Gu, C.W. Baks, R.A. John, G. Hougham, C. Schuster, R. Rimolo-Donadio and Boping Wu, "Is 25 Gb/s On-Board Signaling Viable?" *IEEE Transactions on Advanced Packaging*, Vol. 32, no. 2, May 2009, pp. 328-344. doi: 10.1109/TADVP.2008.2011138

[206] I. Kasa, "A circle fitting procedure and its error analysis," *IEEE Transactions on Instrumentation and Measurement*, vol. 25 no. 1, p. 8-14, Mar. 1976. doi: 10.1109/TIM.1976.6312298

[207] H.R. Kaupp, "Characteristics of Microstrip Transmission Lines", *IEEE Trans. on Electronics and Computers*, Vol. 16, No. 2, April 1967, pp. 185-193. doi: 10.1109/PGEC.1967.264815

[208] S.M. Kay and S.L. Marple, "Spectrum Analysis – A Modern Perspective," Proceedings of the IEEE, Vol. 69, No. 11, Nov. 1981, pp. 1380-1419. doi: 10.1109/PROC.1981.12184.

[209] Keysight Technologies, "Specifying Calibration Standards and Kits for Keysight Vector Network Analyzers," code 5989-4804EN, Aug. 2014.

[210] M. Kirschning and R.H. Jansen, "Accurate Model for Effective Dielectric Constant of Microstrip with Validity up to Millimiter-Wave Frequencies," *Electronics Letters*, (18 March 1982) Vol 18, No. 6, pp 272-273. doi: 10.1049/el:19820186

[211] T. Kley, "Measuring the Coupling Parameters of Shielded Cables," *IEEE Transactions on Electromagnetic Compatibility*, vol. 35, No. 1, 1993, pp. 10-20. doi: 10.1109/15.249391

[212] T. Kley, "Optimised Single-Braided Cable Shields," *IEEE Transactions on Electromagnetic Compatibility*, vol. 35, No 1, 1993, pp. 1-9. doi: 10.1109/15.249390

[213] E.D. Knowles and L.W. Olson, "Cable Shielding Effectiveness Testing," *IEEE Transactions on Electromagnetic Compatibility*, vol. 16, No. 1, Feb. 1974, pp. 16-23. doi: 10.1109/TEMC.1974.303318

[214] M. Kobayashi, "A Dispersion Formula Satisfying Recent Requirements in Microstrip CAD," *IEEE Transactions on Microwave Theory and Techniques*, Vol. 36, Aug. 1988, pp. 1246-1250. doi: 10.1109/22.3665

[215] P.A. Kok and D. De Zutter, "Prediction of the excess capacitance of a via-hole through a multilayered board including the effect of connecting microstrips or striplines," *IEEE Transactions on Microwave Theory and Techniques*, Vol. 42, no. 12, Dec. 1994, pp. 2270-2276. doi: 10.1109/22.339752

[216] K. Kurokawa, "Power waves and the scattering matrix," *IEEE Transactions on Microwave Theory and Techniques*, vol. 13, no. 2, pp. 194-202, Mar 1965. doi: 10.1109/TMTT.1965.1125964

[217] Y.S. Lee, W.J. Getsinger and L.R. Sparrow, "Barium Tetratitanate MIC Technology," *IEEE Transactions on Microwave Theory and Techniques*, vol. 27, no. 7, pp. 655-660, Jul. 1979. doi: 10.1109/TMTT.1979.1129696

[218] P. Lesage and C. Audoin, "Characterization of frequency stability: Uncertainty due to the autocorrelation function of the frequency fluctuations," *IEEE Transactions on Instrumentation and Measurement*, Vol. 22, no. 6, pp. 157-161, June 1973. doi: 10.1109/TIM.1973.4314128 (see also corrections published Mar. 1974 and Sept. 1976).

[219] P. Leuchtmann and J. Rüfenacht, "On the Calculation of the Electrical Properties of Precision Coaxial Lines," *IEEE Transactions on Instrumentation and Measurement*, Vol. 52 no. 2, Apr. 2004, pp. 392-397. doi: 10.1109/TIM.2003.822719.

[220] P. Leuchtmann and J. Rüfenacht, "Remarks on the Accurate Calculation of Air Lines," METAS Tech. Rep. 2002-250-483, Jun 21, 2002.

[221] B.M. Levin, "Calculation of Electrical Parameters of Two-Wire Lines in Multiconductor Cables," *IEEE Transactions on Electromagnetic Compatibility*, Vol. 50, no. 3, Aug. 2008, pp. 697-703. doi: 10.1109/TEMC.2008.927924

[222] A. Lewandowski, "Multi-frequency approach to vector-network-analyzer scattering-parameter measurements," Ph.D. thesis, Warsaw University of Technology – Faculty of Electronics and Information Systems, 2010. [online]: https://repo.pw.edu.pl/docstore/download.seam?fileId=WEiTI-fef766f9-1f26-4d16-bd25-18c48b1846ba (last access March 2015).

[223] A. Lewandowski and W. Wiatr, "Correction for line-length errors and center conductor gap variation in the coaxial multiline through-reflect-line calibration," 74th ARTFG Conf. Digest, 2009. doi: 10.1109/ARFTG74.2009.5439110

[224] Y. Liu, L. Tong, W.-X. Zhu, Y. Tian, and B. Gao, "Impedance Measurements of Nonuniform Transmission Lines in Time Domain using an Improved Recursive Multiple Reflection Computation Method," Progress In Electromagnetics Research (PIER), Vol. 117, pp. 149-164, 2011.

[225] W. Lowdermilk and F. Harris, "Finite Arithmetic Considerations for the FFT implemented in FPGA-based Embedded Processors in Synthetic Instruments," *IEEE Instrumentation & Measurement Magazine*, Vol. 10, no. 4, Aug. 2007, pp. 44-49. doi: 10.1109/MIM.2007.4291222

[226] J. Loyer and R. Kunze, "SET2DIL: Method to Derive Differential Insertion Loss from Single-Ended TDR/TDT Measurements," DesignCon 2010, Santa Clara, CA, USA, Feb. 1-4, 2010.

[227] J. Loyer, "Bidirectional SET2DIL", Intel, May 29, 2013. [online] http://www.qtechinstrument.com/eWebEditor/uploadfile/20130610161218661.pdf (last access July 2015).

[228] Ke Lu and T.J. Brazil, "A Systematic Error Analysis of HP 8510 Time-domain Gating Techniques with Experimental Verification," IEEE MTT-S International Microwave Symposium Digest, June 14-18, 1993, pp. 1259-1262. doi: 10.1109/MWSYM.1993.277102.

[229] R.G. Lyons, *Understanding Digital Signal Processing*, Prentice Hall, 2001.

[230] T.E. MacKenzie and A.E. Sanderson, "Some fundamental design principles for the development of precision coaxial standards and components," *IEEE Transactions on Microwave Theory and Techniques*, vol. 14, no. 1, pp. 29-39, Jan. 1966. doi: 10.1109/TMTT.1966.1126148

[231] N. Marcuvitz, *Waveguide Handbook*, Peter Peregrinus, 1986.

[232] A. Mariscotti, "Measuring the stray capacitance of solenoids with a transmitting and a receiving coil", *Metrology and Measurement Systems*, vol. XVIII n. 1, 2011, pp. 47-56 (ISSN 0860-8229).

[233] A. Mariscotti, "A Magnetic Field Probe with MHz Bandwidth and 7 decades Dynamic Range", *IEEE Transactions on Instrumentation and Measurements*, vol. 58 n. 8, Aug. 2009, pp. 2643-2652. doi: 10.1109/TIM.2009.2015693.

[234] A. Mariscotti, "A low cost capacitive bridge based on voltage drop balance," *Measurement*, Elsevier, vol. 43 n. 9, Nov. 2010, pp. 1094-1098. doi: 10.1016/j.measurement.2010.04.007

[235] R.B. Marks, "A multiline method of network analyzer calibration," *IEEE Transactions on Microwave Theory and Techniques*, Vol. 39, no. 7, July 1991, pp. 1205-1215. doi: 10.1109/22.85388

[236] R.B. Marks, "Formulations of the basic vector network analyzer error model including switch-terms," 50th ARFTG Conference, vol. 32, 1997, pp. 115-126. doi: 10.1109/ARFTG.1997.327265

[237] R.B. Marks and D.F. Williams, "A General Waveguide Circuit Theory," Journal of Research of the National Institute of Standards and Technology, Vol. 97 no. 5, Sept.-Oct. 1992, pp. 533-562.

[238] S.L. Marple, *Digital Spectral Analysis*, Prentice Hall, Englewood Cliffs, NJ, 1987.

[239] W. Marshall Leach, Jr., *Fundamentals of Low-Noise Electronic Analysis and Design*, Kendall Hunt publishing, 2000.

[240] L. Martens, *High-Frequency Characterization of Electronic Packaging*, Springer, New York, 1998.

[241] L. Martens, An Madou, L. Koné, B. Demoulin, Per Sjöberg, A. Anton, Jan Van Koetsem, H. Hoffmann and U. Schricker, "Comparison of Test Methods for the Characterization of Shielding of Board-to-Backplane and Board-to-Cable Connectors," *IEEE Transactions on Electromagnetic Compatibility*, vol. 42, No. 4, Nov. 2000, pp. 427-440. doi: 10.1109/15.249391

[242] I. Martinez-Garcia, "In-situ calibration and direct de-embedding of RF integrated circuits and microwave structures using self-compensating techniques," Ph.D. dissertation, Stanford University – Department of Electrical Engineering, 2010. [online]: http://purl.stanford.edu/vd527pg7763 (last access April 2015).

[243] R. Mavaddat, *Network Scattering Parameters*, World Scientific, Singapore, 1996.

[244] Y.S. Meng and Y. Shan, "Measurement uncertainty of complex-valued microwave quantities," Journal of Progress in Electromagnetic Research (JPIER), Vol. 136, 2013, pp. 421–433.

[245] E. Milotti, "1/f noise: a pedagogical review," [online] http://arxiv.org/abs/physics/0204033 or http://courses.washington.edu/phys431 (last access Oct. 2014).

[246] MIL-STD-461-F, Requirements for the Control of Electromagnetic Interference Characteristics of Subsystems and Equipment, 2007-10.

[247] C.G. Montgomery, R.H. Dicke, and E.M. Purcell, *Principles of Microwave Circuits*, MIT Rad. Lab. Ser., vol. 8, pp. 176-179 (McGraw-Hill, New York, 1948).

[248] D.C. Montgomery and G.C. Runger, *Applied statistics and probability for engineers*, John Wiley & Sons. New York, 2003.

[249] J. Nadolny, "Correlation Between Measured and Simulated Parameters of a Proposed Transfer Standard," *AMP Journal of Technology*, Vol. 5, June 1996, pp. 60-64.

[250] N.S. Nahman and M.E. Guillaume, "Deconvolution of time domain waveforms in the presence of noise," National Bureau of Standards Technical Note 1047, US Department of Commerce, Oct. 1981. [online]: https://archive.org/details/deconvolutionoft1047nahm (last access January 2015).

[251] A.M. Nicolson, "Broad-band microwave transmission characteristics from a single measurement of the transient response," *IEEE Transactions on Instrumentation and Measurement*, Vol. 17 no. 4, Dec. 1968, pp. 395-402. doi: 10.1109/TIM.1968.4313741

[252] A.M. Nicolson, "Forming the Fast Fourier Transform of a Step Response in Time-Domain Metrology," *Electronics Letters* 9, pp. 317-318, 1973. doi: 10.1049/el:19730228

[253] A.M. Niknejad, "Two-Port Networks and Ampliers," Handouts of EECS 142 course, Berkeley Wireless Research Center, University of California, Berkeley, Sept. 22, 2008.

[254] NIST/SEMATECH, *e-Handbook of Statistical Methods*, [online] http://www.itl.nist.gov/div898/handbook/ (last access Aug. 2014).

[255] A. Ogunsola and A. Mariscotti, *Electromagnetic Compatibility in Railways – Analysis and Management*, Springer, 2012.

[256] A.V. Oppenheim and R.W. Schafer, *Discrete-Time Signal Processing*, Prentice Hall, Englewood Cliffs, NJ, 1989.

[257] O. Ostwald, "T-Check Accuracy Test for Vector Network Analyzers utilizing a Tee-junction," Rohde & Schwarz, code 1EZ43_0E, 1998.

[258] M. Pajovic, Jinghan Yu and D. Milojkovic, "Analysis of Via Capacitance in Arbitrary Multilayer PCBs," *IEEE Transactions on Electromagnetic Compatibility*, Vol. 49, no. 3, Aug. 2007, pp. 722-726. doi: 10.1109/TEMC.2007.902382

[259] M.M. Pajovic, "A Closed-Form Equation for Estimating Capacitance of Signal Vias in Arbitrarily Multilayered PCBs, *IEEE Transactions on Electromagnetic Compatibility*, Vol. 50, no. 4, Aug. 2008, pp. 966-973. doi: 10.1109/TEMC.2008.2004606

[260] P. Pakay and A. Torok, "Analysis of insertion loss repeatability of coaxial connectors," *The Radio and Electronic Engineer*, Vol. 47, No. 7, pp. 315-319, July 1977.

[261] J. Palecek, M. Vestenický, P. Vestenický and J. Spalek, "Examination of SMA Connector Parameters," IEEE 16th Intern. Conf. on Intelligent Engineering Systems, June 13-15, 2012, Lisbon, Portugal, pp. 259-263. doi: 10.1109/INES.2012.6249841.

[262] A. Papoulis, Probability, *Random variables and Stochastic processes*, 2nd ed., McGraw-Hill, 1987.

[263] B. Pastori, "Verifying VNA source match using coaxial offset shorts," Maury Microwave, 5C-027, Feb. 27, 2006.

[264] C.R. Paul, *Multiconductor Transmission Lines*, 2nd. ed., John Wiley & Sons, New Jersey, 2008.

[265] C.R. Paul, *Introduction to Electromagnetic Compatibility*, 2nd ed., John Wiley & Sons, New Jersey, 2006.

[266] D.B. Percival, "Characterization of Frequency Stability: Frequency-Domain Estimation of Stability Measures," *Proceedings of IEEE*, Vol. 79, no. 6, June 1991, pp. 961-972. doi: 10.1109/5.84973

[267] J.M. Peterson and B. Grossman, "A Process to Reduce Reproducibility Error in VNA Measurements," Microwave Measurement Symposium (ARFTG), 2010 76th ARFTG, Nov. 30-Dec. 3, 2010, Clearwater Beach, Florida, pp. 1-5. doi: 10.1109/ARFTG76.2010.5700052

[268] Picosecond Pulse Labs, "Time Domain Reflectometry (TDR) and Time Domain Transmission (TDT) Measurement Fundamentals," App. Note AN-15, Nov. 2004.

[269] P. Pino, "Distortion Inherent to VNA Test Port Cable Assemblies," Gore, July 2005.

[270] P. Pino, "Intermateability of SMA, 3.5-mm, and 2.92-mm Connectors," Gore, Jan. 2007 (appearing also in the *Microwave Journal*, March 2007).

[271] R. Pollard, "Verification of System Specifications of a High Performance Network Analyzer," 23rd ARFTG Conference Digest-Spring, June 1984, Vol. 5, pp. 38-50. doi: 10.1109/ARFTG.1984.323576

[272] R.D. Pollard, "Compensation Technique Improves Measurements for a Range of Mechanically Compatible Connectors," *Microwave Journal*, Oct. 1994.

[273] D.M. Pozar, *Microwave Engineering*, John Wiley & Sons, 2nd ed., 1998.

[274] J.G. Proakis, *Digital Communications*, 3rd ed., New York, McGraw Hill, 1995.

[275] J.G. Proakis and D.G. Manolakis, *Digital Signal Processing - Principles, Algorithms, and Applications*, 3rd ed., Prentice Hall, 1996.

[276] P.J. Pupalaikis, "Validation Methods for S-parameter Measurement Based Models of Differential Transmission Lines," DesignCon, 2008, Santa Clara, CA, USA, Feb. 4-7, 2008.

[277] L.R. Rabiner, R.W. Shafer and C.M. Rader, "The chirp z-transform algorithm," *IEEE Transactions on Audio Electroacoustics*, Vol. 17 no. 2, pp. 86-92, 1969. doi: 10.1109/TAU.1969.1162034

[278] G.L. Ragan, *Microwave Transmission Circuits*, MIT Radiation Lab., 1st ed., 1948. [online]: https://archive.org/details/MicrowaveTransmissionCircuits [ark:/13960/t6xw5j422] (last access August 2014).

[279] S. Ramo, J.R. Whinnery and T. Van Duzer, *Fields and waves in communication electronics*, New York: J. Wiley & Sons, 1965, pp. 291-297.

[280] C. Rauscher, V. Janssen and R. Minihold, Fundamentals of Spectrum Analysis, Rohde & Schwarz, 2001.

[281] J.C. Rautio, "A New Definition of Characteristic Impedance," IEEE International Microwave Symposium Digest, Vol. 2, pp. 761-764, Jul. 1991. doi: 10.1109/MWSYM.1991.147116

[282] N.M. Ridler and M.J. Salter, "An approach to the treatment of uncertainty in complex S-parameter measurements," *Metrologia*, vol. 39, no. 3, pp. 295–302, 2002. doi: 10.1088/0026-1394/39/3/6

[283] N.M. Ridler and M.J. Salter, "A generalised approach to the propagation of uncertainty in complex S-parameter measurements," 64th ARFTG Microwave Measurement Symposium, Fall 2004, Dec. 2-3, 2004, Orlando, Florida, pp. 1-15. doi: 10.1109/ARFTGF.2004.1427564

[284] W.J. Riley, *Handbook of Frequency Stability Analysis*, NIST Special Publication 1065, July 2008.

[285] T.H. Roberts and Yeou-Song (Brian) Lee, "A Superior Solution to Control the Connection Repeatability of Coaxial Airline Impedance Standards used for Network Analyzer Verification," 77th ARFTG Microwave Measurement Conference (ARFTG), Baltimore, MD, USA, June 10, 2011, pp. 1-6. doi: 10.1109/ARFTG77.2011.6034575

[286] Rogers leaflet TM 3.3.3, "Width and Effective Dielectric Constant Data for Design of Microstrip Transmission Lines on Various Thicknesses, Types and Claddings of TMM® Microwave Laminates," Nov. 1999.

[287] Rohde & Schwarz, "Calculate the mixed-mode S-parameter from an S4P file", [online] http://www3.rohde-schwarz.com/www/rs_sc.nsf/faq/ZVBCalculatet.html (last access Aug. 2014).

[288] Rohde & Schwarz, "Time Domain Measurements using Vector Network Analyzer ZVR", code 1EZ44_0E, 1998.

[289] Rohde & Schwarz, "Frequently Asked Questions about Vector Network Analyzer ZVR", code 1EZ38_3E, 1998.

[290] E. Rubiola and F. Vernotte, "The cross-spectrum experimental method," arXiv:1003.0113v1 [physics.ins-det], Feb. 27, 2010, pp. 1-39.

[291] J. Rutman and F.L. Walls, "Characterization of Frequency Stability in Precision Frequency Sources," *Proceedings of the IEEE*, Vol. 79, No. 6, June 1991, pp. 952-960. doi: 10.1109/5.84972

[292] D. Rytting, "Network Analyzer Error Models and Calibration Methods", Agilent, 1998.

[293] D. Rytting, "Let Time Domain Provide Additional Insight in Network Behavior," Hewlett-Packard RF & Microwave Measurement Symposium and Exhibition, Apr. 1984.

[294] S. Sali, "An Improved Model for the Transfer Impedance Calculations of Braided Coaxial Cables," *IEEE Transactions on Electromagnetic Compatibility*, vol. 33, No 2, 1991, pp. 139-143. doi: 10.1109/15.78351

[295] S.A. Schelkunoff, "The Electromagnetic Theory of Coaxial Transmission Lines and Cylindrical Shields," *The Bell System Technical Journal*, Vol. 13, n. 4, Oct. 1934, pp. 532-579. doi: 10.1002/j.1538-7305.1934.tb00679.x

[296] M.V. Schneider, B. Glance and W.F. Bodtmann, "Microwave and millimeter wave hybrid integrated circuits for radio systems," *Bell System Technical Journal*, Vol. 48, n. 6, July-Aug. 1969, pp. 1703-1726. doi: 10.1002/j.1538-7305.1969.tb01147.x

[297] L. Schnell ed., *Technology of Electrical Measurements*, Wiley, Chichester, UK, 1993.

[298] M. Sellone, N. Shoaib, L. Callegaro and L. Brunetti, "Two different ways in evaluating the uncertainty of S-parameter measurements," 20th Imeko TC4 International Symposium, Benevento, Italy, Sept. 15-17, 2014.

[299] Lei Shan, Young Kwark, Christian Baks, and Mark Ritter, Boping Wu, "Layer Misregistration in PCB and Its Effects on Signal Propagation," IEEE 60th Electronic Components and Technology Conference, Las Vegas, NV, USA, June 1-4, 2010, pp. 605-611. doi: 10.1109/ECTC.2010.5490909

[300] Y. Shlepnev and C. Nwachukwu, "Practical methodology for analyzing the effect of conductor roughness on signal losses and dispersion in interconnects," DesignCon 2012, Santa Clara, CA, USA, Jan. 30 - Feb. 2, 2012.

[301] Y. Shlepnev, A. Neves, T. Dagostino and S. McMorrow, "Measurement-Assisted Electromagnetic Extraction of Interconnect Parameters on Low-Cost FR-4 boards for 6-20 Gb/sec Applications," DesignCon 2009, Santa Clara, CA, USA, Feb. 2-5, 2011.

[302] N. Shoaib, "A Novel Inconsistency Condition for 2-port Vector Network Analyzer Calibration," *Microwave and Optical Technology Letters*, Vol. 54, no. 10, Oct. 2012, pp. 2372-2375. doi: 10.1002/mop

[303] K.J. Silvonen, "A general approach to network analyzer calibration," *IEEE Transactions on Microwave Theory and Techniques*, vol. 40, no. 4, April 1992, pp. 754-759. doi: 10.1109/22.127526

[304] D. Skinner, "Guidance on using Precision Coaxial Connectors in Measurement," NPL guide, 3rd ed., Aug. 2007. [online]: www.npl.co.uk/content/ConMediaFile/8750 (last access February 2015).

[305] N. Sladek, "Fundamental considerations in the design and application of high precision coaxial connectors," Proc. IRE International Convention Record, vol. 13, Mar 1965, pp. 182–189. doi: 10.1109/IRECON.1965.1147493

[306] P.I. Somlo, "The Computation of Coaxial Line Step Capacitances," *IEEE Transactions on Microwave Theory and Techniques*, Vol. 15, no. 1, pp. 48-53, Jan. 1967. doi: 10.1109/TMTT.1967.1126368

[307] R.A. Speciale, "A Generalization of the TSD Network-Analyzer Calibration Procedure, Covering n-port Scattering-Parameter Measurements, Affected by Leakage Errors," *IEEE Transactions on Microwave Theory and Techniques*, vol. 25 No. 12, pp. 1100-1115, Dec. 1977. doi: 10.1109/TMTT.1977.1129282

[308] R. Stassen, "Einsatz eines Mikrowellen-SPitzenmeßplatzes zur Charakterisierung von Transistore direkt auf dem Wafer im Frequenzbereich von 0,045 bis 26,5GHz", Diplomarbeit am Institut für Schicht- und Ionentechnik am Forschungszentrum Jülich GmbH, available from Technische Informationsbibliothek, Hannover.

[309] J. Stenarson and K. Yhland, "Uncertainty Propagation Through Network Parameter Conversions," *IEEE Transactions on Instrumentation and Measurement*, Vol. 58, no. 4, Apr. 2009, pp. 1152-1157. doi: 10.1109/TIM.2008.2008578.

[310] J. Stimple, *Clock Synthesis, Phase Locked Loops, and Clock Recovery, Digital Communications Test and Measurement*, Prentice Hall, 2008.

[311] P. Stoica, and R.L. Moses, *Introduction to Spectral Analysis*, Prentice Hall, 1997.

[312] U. Stumper, "Influence of TMSO Calibration Standards Uncertainties on VNA S-Parameter Measurements," *IEEE Transactions on Instrumentation and Measurement*, Vol. 52 no. 2, Apr. 2003, pp. 311-315. doi: 10.1109/TIM.2003.810041

[313] U. Stumper, "Uncertainty of VNA S-Parameter Measurement Due to Nonideal TRL Calibration Items," *IEEE Transactions on Instrumentation and Measurement*, Vol. 54 no. 2, Apr. 2005, pp. 676-679. doi: 10.1109/TIM.2005.843521

[314] D. B. Sullivan, D. W. Allan, D. A. Howe, and F. L. Walls, Eds., "Characterization of clocks and oscillators," *NIST Tech Note 1337*, Mar. 1990.

[315] B.T. Sztenkuti, "Shielding Quality of Cables and Connectors: some Basics for Better Understanding of Test Methods," IEEE International Symposium on Electromagnetic Compatibility, Anaheim, CA, Aug. 1992, pp. 294-301. doi: 10.1109/ISEMC.1992.626098

[316] D.G. Swanson Jr., "Grounding Microstrip Lines with Via Holes," *IEEE Transactions on Microwave Theory and Techniques*, Vol. 40, no. 8, Aug. 1992, pp. 1719-1721. doi: 10.1109/22.149532

[317] V. Teppati, A. Ferrero and M. Sayed, *Modern RF and Microwave Measurement Techniques*, Cambridge University Press, 2013.

[318] M. Thorburn, A. Agoston and V.K. Tripathi, "Computation of frequency-dependent propagation characteristics of microstrip-like propagation structures with discontinuous layers," *IEEE Transactions on Microwave Theory and Techniques*, Vol. 38, n. 2, Feb. 1990, pp. 148-153. doi: 10.1109/22.46424

[319] X. Tian, Y.-J. Zhang, J. Lim, K. Qiu, R. Brooks, Ji Zhang and Jun Fan, "Numerical Investigation of Glass-Weave Effects on High-Speed Interconnects in Printed Circuit Board," IEEE International Symposium on Electromagnetic Compatibility, Raleigh, NC, USA, Aug. 4-8, 2014, pp. 475-479. doi: 10.1109/ISEMC.2014.6899019

[320] R. Tiedemann, "Current Flow in Coaxial Braided Cable Shields," *IEEE Transactions on Electromagnetic Compatibility*, Vol. 45, no. 3, Aug. 2003, pp. 531-537. doi: 10.1109/TEMC.2003.815562

[321] L.F. Tiemeijer and R.J. Havens, "A Calibrated Lumped-Element De-Embedding Technique for On-Wafer RF Characterization of High-Quality Inductors and High-Speed Transistors," *IEEE Transactions on Electron Devices*, Vol. 50, no. 3, March 2003, pp. 822–829. doi: 10.1109/TED.2003.811396

[322] Chih-Chun Tsai, Yung-Shou Cheng, Ting-Yi Huang, Yungping Alvin Hsu and Ruey-Beei Wu, "Design of Microstrip-to-Microstrip Via Transition in Multilayered LTCC for Frequencies up to 67 GHz," *IEEE Transactions on Components, Packaging and Manufacturing Technology*, Vol. 1, no. 4, Apr. 2011, pp. 595-601. doi: 10.1109/TCPMT.2011.2104416

[323] M. Tyni, "Transfer impedance of coaxial cables with braided outer conductors," Research Proceedings of the Wroclaw Polytechnic Institute of Telecommunication and Acoustics, vol. 27, pp. 410-419, 1975.

[324] A. Uhlir, "Correction for Adapters in Microwave Measurements," *IEEE Transactions on Microwave Theory and Techniques*, Vol. 22 no. 3, March 1974, pp. 330-332. doi: 10.1109/TMTT.1974.1128219

[325] G.E. Valley and H. Wallman, "Vacuum Tube Amplifiers", MIT Radiation Laboratory Series 18, McGraw-Hill, 1948.

[326] E.F. Vance "Shielding Effectiveness of Braided-Wire Shields," *IEEE Transactions on Electromagnetic Compatibility*, Vol. 17, N. 2, May 1975, pp 71–77. doi: 10.1109/TEMC.1975.303389

[327] Hugo Van hamme and Marc Vanden Bossche, "Flexible Vector Network Analyzer Calibration with Accuracy Bounds Using an 8-Term or a 16-Term Error Correction Model," *IEEE Transactions on Microwave Theory and Techniques*, Vol. 42, no. 6, June 1994, pp. 976-987. doi: 10.1109/22.293566

[328] M.E. van Valkenburg, *Reference Data for Engineers: Radio, Electronics, Computer and Communication*, 8th ed., Newnes, 1998.

[329] K. Vaz and M. Caggiano, "Measurement Technique for the Extraction of Differential S-Parameters from Single-Ended S-Parameters," IEEE 27th Intern. Spring Seminar on Electronics Technology, Vol. 2, May 13-16, 2004, Bankya, Bulgaria, pp. 313-317. doi: 10.1109/ISSE.2004.1490442

[330] K. Vaz, K.M. Ho and M. Caggiano, "Error Reducing Techniques for the Scattering Parameter Characterization of Differential Networks Using a Two-Port Network Analyzer," IEEE 28th Intern. Spring Seminar on Electronics Technology, May 19-22, 2005, Wiener Neustadt, Austria, pp. 342-347. doi: 10.1109/ISSE.2005.1491052

[331] VDE 0472, part 507, *Testing of cables, wires and flexible cords – Reduction factor*, Apr. 1983.

[332] B.C. Wadell, *Transmission Line Design Handbook*, Artech House, Norwood, MA, USA, 1991.

[333] H.A. Wheeler, "Transmission-Line properties of a Round Wire in a Polygon Shield," *IEEE Transactions on Microwave Theory and Techniques*, Vol. 27, No. 8, Aug. 1979, pp. 717-721. doi: 10.1109/TMTT.1979.1129712

[334] H.A. Wheeler, "Transmission-Line properties of a Strip Line between Parallel Planes," *IEEE Transactions on Microwave Theory and Techniques*, Vol. 26, No. 11, Nov. 1978, pp. 866-876. doi: 10.1109/TMTT.1978.1129505

[335] H.A. Wheeler, "Transmission-Line properties of a Strip on a Dielectric Sheet on a Plane," *IEEE Transactions on Microwave Theory and Techniques*, Vol. 25, No. 8, Aug. 1977, pp. 631-647. doi: 10.1109/TMTT.1977.1129179

[336] H.A. Wheeler, "Transmission Line Properties of Parallel Strips Separated by a Dielectric Sheet," *IEEE Transactions on Microwave Theory and Techniques*, Vol. 13, no. 2, Apr. 1965, pp 172-185. doi: 10.1109/TMTT.1965.1125962

[337] J.R. Whinnery and H.W. Jamieson, "Equivalent Circuits for Discontinuities in Transmission Lines," Proceedings of the IRE, Vol. 32, no. 2, Feb. 1944, pp. 98-114. doi: 10.1109/JRPROC.1944.229737

[338] J.R. Whinnery, H.W. Jamieson and T.E. Robbins, "Coaxial Line Discontinuities," Proceedings of the IRE, Vol. 32, no. 11, Nov. 1944, pp. 695-709. doi: 10.1109/JRPROC.1944.234027

[339] M. Wollensack and J. Hoffmann, "METAS VNA Tools II - Math Reference V1.4," Oct. 2014, [online]: http://www.metas.ch/dam/data/metas/ Fachbereiche/Hochfrequenz/VNA_Tools/VnaToolsMath V1.4.pdf. (last access July 2015)

[340] M. Wollensack, J. Hoffmann, J. Ruefenacht and M. Zeier, "VNA Tools II: S-parameter uncertainty calculation," 79th ARFTG Microwave Measurement Conference, Montreal, Quebec, Canada, June 22, 2012, pp. 1-5. doi: 10.1109/ARFTG79.2012.6291183

[341] K. Wong and J. Hoffman, "Improving VNA Measurement Accuracy by Including Connector Effects in the Models of Calibration Standards," IEEE 82nd ARFTG Microwave Measurement Conference, Columbus, OH, USA, Nov. 18-21, 2013, pp. 1-7. doi: 10.1109/ARFTG-2.2013.6737334

[342] K. Wong, "The "Unknown Thru" Calibration Advantage," 63rd ARFTG Conference, Fort Worth, TX, USA, June 11, 2004, pp. 73-81. doi: 10.1109/ARFTG.2004.1387858

[343] G. Wübbeler, C. Elster, T. Reichel, and R. Judaschke, "Determination of complex residual error parameters of a calibrated vector network analyser," *IEEE Transactions on Microwave Theory and Techniques*, vol. 58, no. 9, pp. 3238-3244, Sept. 2009. doi: 10.1109/TIM.2009.2017170

[344] E. Yamashita, K. Atsuki, and T. Ueda, "Microstrip dispersion in a wide-frequency range," *IEEE Transactions on Microwave Theory and Techniques*, vol. 29, no. 6, pp. 610-611, June 1981. doi: 10.1109/TMTT.1981.1130403

[345] N. Yannopoulou and P. Zimourtopoulos, "Total Differential Errors in One-Port Network Analyzer Measurements with Application to Antenna Impedance," arXiv: physics/0703204 [physics.ins-det], Dec. 9, 2007.

[346] K. Yhland and J. Stenarson, "A simplified treatment of uncertainties in complex quantities," Conference on Precision Electromagnetic Measurements, London, UK, June 27-July 2, 2004, pp. 652–653. doi: 10.1109/CPEM.2004.305464

[347] Yaojiang Zhang, Erping Li, Zawzaw Oo, Wenzu Zhang, Enxiao Liu, Xinchang Wei and Jun Fan, "Analytical Formulas for the Barrel-plate and Pad-plate Capacitance in the Physics-based Via Circuit Model for Signal Integrity analysis of PCBs," Asia Pacific Microwave Conference APMC 2009, Dec. 7-10, pp. 2432-2435. doi: 10.1109/APMC.2009.5385476

[348] L. Zhu, K. Wu, "Revisiting characteristic impedance and its definition of microstrip line with a self-calibrated 3-D MoM scheme," *IEEE Microwave and Guided Wave Letters*, Vol. 8, no.2, pp. 87-89, Feb. 1998. doi: 10.1109/75.658650

Index

The Answer to the Great Question of . . . Life, the Universe and Everything
. . . Forty-two.

Deep Thought, *The Hitchhiker's Guide to the Galaxy*

www.ingramcontent.com/pod-product-compliance
Lightning Source LLC
Chambersburg PA
CBHW081208220326
41598CB00037B/6709
9788894109108